Robotics

Robotics

Science and Systems V

edited by Jeff Trinkle, Yoky Matsuoka, and Jose A. Castellanos

The MIT Press
Cambridge, Massachusetts
London, England

Library of Congress Cataloging-in-Publication Data

Robotics: Science and Systems Conference (5th : 2009 : Seattle, Wash.)
Robotics : science and systems V / edited by Jeff Trinkle, Yoky Matsuoka, and Jose A. Castellanos.
    p. cm.
Includes bibliographical references.
ISBN 978-0-262-51463-7 (pbk. : alk. paper)
1. Robotics—Congresses. I. Trinkle, Jeffrey C. II. Matsuoka, Yoky. III. Castellanos, José A., 1969– IV. Title.
TJ210.3.R6435 2010
629.8'92—dc22

                                                                                                    2009052777

The MIT Press is pleased to keep this title available in print by manufacturing single copies, on demand, via digital printing technology.

Contents

Preface

We are delighted to showcase papers presented at *Robotics: Science and Systems (RSS) 2009*, held at the University of Washington in Seattle, USA, from June 28 to July 1, 2009. This year's meeting brought together more than 300 researchers to Seattle from Europe, Asia, North America, and Australia. Within the last five years, RSS has established its position as a single-track main-stream robotics conference with papers of the highest quality.

The paper review process was rigorous. All papers received a minimum of four high-quality reviews (that is over 600 reviews from over 170 program committee members). After the reviews were completed, the program committee members and area chairs discussed reviews for each paper in an on-line forum within the conference management system. Then the authors were invited to rebut the reviews. Following the rebuttals, the program committee members and area chairs discussed further and finalized their reviews. Final acceptance decisions and presentation categories were made at the area chair meeting in Seattle with the program chair. Of the 154 submissions, 19 were selected for poster presentation and 20 were selected for podium presentation. The papers cover a wide range of topics in robotics spanning manipulation, locomotion, machine learning, localization, visual SLAM, haptics, biologically inspired design and control, etc.

The conference spanned two and a half days. Following the RSS tradition, there were four invited talks by leaders in fields that inspire robotics.

- Prof. Michael Dickinson from the California Institute of Technology, USA presented "Visually-Mediated Behaviors of the Fruit Fly."
- Prof. John Delaney from the University of Washington presented "Next Generation Ocean Sciences: The Leading Edge of an Environmental Renaissance."
- Prof. Marc Ernst from the Max Planck Institute for Biological Cybernetics, Germany presented "The Puzzle of Human Multisensory Perception: Optimal Integration for Action."
- Prof. John Doyle from the California Institute of Technology, USA presented "Rules of Engagement: The Architecture of Robust Evolvable Networks."

The Early Career Spotlight Talk highlighted a rising star in the robotics community.

- Katherine Kuchenbecker from the University of Pennsylvania, USA presented "Haptography: Creating Authentic Haptic Feedback from Recordings of Real Interactions."

Eleven well-attended workshops spanned one and a half days preceding the conference:

- "Bridging the Gap Between High-Level Discrete Representations and Low-Level Continuous Behaviors" organized by Dana Kulic, Jan Peters, and Pieter Abbeel
- "Good Experimental Methodology in Robotics" organized by John Hallam, Angel P. del Pobil, and Fabio Bonsignorio
- "Understanding the Human Hand for Advancing Robotic Manipulation" organized by Ravi Balasubramanian and Yoky Matsuoka
- "Regression in Robotics—Approaches, and Applications" organized by Christian Plagemann, Jo-Anne Ting, and Sethu Vijayakumar
- "Algorithmic Automation" organized by Ken Goldberg, Vijay Kumar, Todd Murphey, and Frank van der Stappen
- "Integrating Mobility and Manipulation" organized by Brian Gerkey, Kurt Konolige, Odest Chadwicke, Robert Platt, and Neo Ee Sian
- "Creative Manipulation: Examples Using the WAM" organized by William Townsend and Yoky Matsuoka
- "Aquatic Robots and Ocean Sampling" organized by Gaurav Sukhatme and Kanna Rajan

- "Autonomous Flying Vehicles: Fundamentals and Applications" organized by Srikanth Saripalli and Peiter Abbeel
- "Protein Structure, Kinematics, and Motion Planning" organized by Lydia Tapia, Nancy Amato, and Mark Moll
- "Introduction to Microsoft Robotics Developer Studio" organized by Trevor Taylor and Stewart Tansley

RSS 2009 was a success thanks to the efforts of many people. We gratefully acknowledge the enormous time spent by the program committee and 16 area chairs. The area chairs were: Srinivas Akella (University of North Carolina at Charlotte), Hajime Asama (University of Tokyo), Devin Balkcom (Dartmouth College), Ayanna Howard (Georgia Institute of Technology), Tetsunari Inamura (National Institute of Informatics), Eric Klavins (University of Washington), Jana Kosecka (George Mason University), Danica Kragic (Royal Institute of Technology), Jun Morimoto (Advanced Telecommunications Research Institute International), Jose Neira (Universidad de Zaragoza), Allison Okamura (Johns Hopkins University), Toru Omata (Tokyo Institute of Technology), Lynne Parker (University of Tennessee at Knoxville), Stergios Roumeliotis (University of Minnesota), Nicola Simeon (LAAS-CNRS), and Russ Tedrake (Massachusetts Institute of Technology). Together, their expertise covered an extraordinarily broad swath of the robotics landscape.

The workshops chair was Lydia Kavraki (Rice University), and the publicity chair was Devin Balkcom (Dartmouth University). The local arrangement chairs were Dieter Fox and Rajesh Rao (University of Washington). They and their support staff handled the local arrangement details beautifully, including a stunning banquet on an island shore, accessible only by boats. We particularly want to thank Mary Jane Shirakawa and Jan Kvamme from the University of Washington Education Outreach for handling local details with grace and Scott Rose for managing the conference webpage with light-speed efficiency.

RSS 2009 was only possible with industrial sponsors:
- Gold sponsors ($15,000 or more): The Boeing Company
- Bronze sponsors ($5,000–$10,000): Willow Garage, Naval Research Laboratory
- Other sponsors: Google, Intel, Microsoft Research
- Best paper award ($2,500): iRobot
- Best student paper award ($2,500): Springer

We also would like to thank our technical sponsors: IEEE Robotics and Automation Society, Association for the Advancement of Artificial Intelligence, International Foundation of Robotics Research, and the Robotics Society of Japan.

Finally, we would like to express our gratitude to the members of the robotics community who have adopted RSS and its philosophy, and who have made RSS 2009 an outstanding meeting by submitting manuscripts, participating in the review process, and by attending RSS 2009. We look forward to many more exciting meetings of RSS in the years to come.

Jeff Trinkle, Rensselaer Polytechnic Institute, General Chair
Yoky Matsuoka, University of Washington, Program Chair
Jose A. Castellanos, Universidad de Zaragoza, Publications Chair

July 2009

Cooperative Manipulation and Transportation with Aerial Robots

Nathan Michael, Jonathan Fink, and Vijay Kumar
University of Pennsylvania
Philadelphia, Pennsylvania
Email: {nmichael, jonfink, kumar}@grasp.upenn.edu

Abstract—In this paper we consider the problem of controlling multiple robots manipulating and transporting a payload in three dimensions via cables. We develop robot configurations that ensure static equilibrium of the payload at a desired pose while respecting constraints on the tension and provide analysis of payload stability for these configurations. We demonstrate our methods on a team of aerial robots via simulation and experimentation.

I. INTRODUCTION

Aerial transport of payloads by towed cables is common in emergency response, industrial, and military applications for object transport to environments inaccessible by other means. Examples of aerial towing range from emergency rescue missions where individuals are lifted from dangerous situations to the delivery of heavy equipment to the top of a tall building. Typically, aerial towing is accomplished via a single cable attached to a payload. However, only limited controllability of the payload is achievable with a single attachment point [1].

In this work we address the limitations of aerial towing by designing cooperative control laws for multiple aerial robots that enable manipulation of a payload in three dimensions. While we formulate the general conditions for system equilibrium at the desired pose for an arbitrary number of robots, we focus on a system of three aerial robots for discussions of workspace and payload stability. We show that despite the fact that such a system is underactuated and limited by unilateral tension constraints, we are able to manipulate a payload to a desired pose (position and orientation) in simulation and experimentation. We extend the analysis to motion planning by exploring the set of possible robot configurations for a given payload pose.

While examples of cooperative multi-robot planar towing exist in the literature, there are significant differences between interacting with an object on the ground and in the air. In [2], the authors consider the problem of cooperative towing with a team of ground robots, where under quasi-static assumptions there is a unique solution to the motion of the object given the robot motions. This approach is not directly extensible to three-dimensional manipulation where frictional ground forces are absent and gravity introduces dynamics into the problem. The cooperative aerial towing problem is similar to the problem of controlling cable-actuated parallel manipulators in three dimensions, where in the former the payload pose

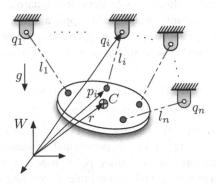

Fig. 1. A rigid body suspended by n cables with world-frame pivot points q_i. Analysis techniques for cable-actuated parallel manipulators assume that q_i is fixed while l_i varies in magnitude, while for cooperative aerial manipulation we fix l_i and vary q_i by changing the positions of the aerial robots.

is affected by robot positions and in the latter pose control is accomplished by varying the lengths of multiple cable attachments (see Fig. 1). Thus the work on workspace analysis [3, 4], control [5], and static analysis [6] of such parallel manipulators is directly relevant to this paper.

More generally, we are interested in the mechanics of payloads suspended by n cables in three dimensions. The $n = 6$ case is addressed in the literature on cable-actuated platforms. When $n = 5$, if the line vectors are linearly independent and the cables are taut, the line vectors and the gravity wrench axis must belong to the same linear complex [7]. The payload is free to instantaneously twist about the reciprocal screw axis. When $n = 4$, under similar assumptions on linear independence and positive tension, the line vectors and the gravity wrench must belong to the same linear congruence. The unconstrained freedoms correspond (instantaneously) to a set of twists whose axes lie on a cylindroid. In the $n = 3$ case, all three cables and the gravity wrench axis must lie on the same regulus - the generators of a hyperboloid which is a ruled surface [8]. Of course, in all of these cases there are special configurations in which the screw systems assume special forms [7] which are not discussed in this paper. The arguments for the $n = 1$ and the $n = 2$ cases are similar, but in these cases, the cables and the center of mass must lie on the same vertical plane for equilibrium.

We approach the development of controllers for cooperative aerial manipulation and transportation as follows: in Sect. II we formulate conditions for general static equilibrium of an

1

object in three dimensions, after which we focus our discussion on systems with three robots. By deriving conditions of static equilibrium and analyzing these conditions based on bounded tension models and stability, in Sect. III we are able to identify valid system configurations for aerial manipulation that ensure that the payload achieves a desired pose. Through simplifying assumptions guided by our robot model, we are able to arrive at closed-form analytic solutions for the tensions in the cables and payload stability for a given system configuration. For the derivation of valid configurations we assume point-model robots. In Sect. IV, we develop and experimentally validate controllers for dynamic robots that respect this point-model and enable application of our methods to a team of quadrotors. We review and analyze simulation and experimental results in Sects. V and VI and conclude in Sect. VII.

II. PROBLEM FORMULATION

A. Mechanics of a cable-suspended payload

We begin by considering the general problem with n robots (quadrotors in our experimental implementation) in three dimensions. We consider point robots for the mathematical formulation and algorithmic development although the experimental implementation requires us to consider the full twelve-dimensional state-space of each quadrotor and a formal approach to realizing these point abstractions, which we provide in Sect. IV. Thus our configuration space is given by $\mathcal{Q} = \mathbb{R}^3 \times \ldots \times \mathbb{R}^3$. Each robot is modeled by $q_i \in \mathbb{R}^3$ with coordinates $q_i = [x_i, y_i, z_i]^\mathrm{T}$ in an inertial frame, W (Fig. 2). The ith robot cable with length l_i is connected to the payload at the point P_i with coordinates $p_i = [x_i^p, y_i^p, z_i^p]^\mathrm{T}$ in W. We require P_1, P_2, and P_3 to be non-collinear and span the center of mass. The payload has mass m with the center of mass at C with position vector $r = [x_C, y_C, z_C]^\mathrm{T}$. The payload's pose $\mathbf{A} \in SE(3)$ can be locally parameterized using the components of the vector r and the Euler angles with six coordinates: $[x_C, y_C, z_C, \alpha, \beta, \gamma]^\mathrm{T}$. The homogeneous transformation matrix describing the pose of the payload is given by:

$$\mathbf{A} = \begin{bmatrix} \mathbf{R}(\alpha, \beta, \gamma) & \begin{pmatrix} x_C \\ y_C \\ z_C \end{pmatrix} \\ 0 & 1 \end{bmatrix}. \qquad (1)$$

Note that $\mathbf{R}$ is the rotation matrix going from the object frame B to the world frame W (as depicted in Fig. 2). Additionally, for this work we follow the *Tait-Bryan* Euler angle parameterization for $\{\alpha, \beta, \gamma\}$.

The equations of static equilibrium can be written as follows. The cables exert zero-pitch wrenches on the payload which take the following form after normalization:

$$\mathbf{w}_i = \frac{1}{l_i} \begin{bmatrix} q_i - p_i \\ p_i \times q_i \end{bmatrix}.$$

The gravity wrench takes the form:

$$\mathbf{g} = -mg \begin{bmatrix} e_3 \\ r \times e_3 \end{bmatrix},$$

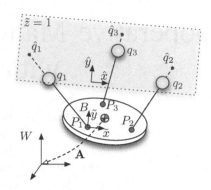

Fig. 2. A team of three point-model robots manipulate a payload in three dimensions. The coordinates of the robots in the inertial frame W are $q_i = [x_i, y_i, z_i]$ and in the body-fixed frame (attached to the payload) B are $\tilde{q}_i = [\tilde{x}_i, \tilde{y}_i, \tilde{z}_i]$. The rigid body transformation from B to W is $\mathbf{A} \in SE(3)$. Additionally, we denote the projection of the robot position $\tilde{q}_i$ along $q_i - p_i$ to the plane $\tilde{z} = 1$ as $\hat{q}_i = [\hat{x}_i, \hat{y}_i, 1]$.

where g is the acceleration due to gravity and $e_3 = [0, 0, 1]^\mathrm{T}$. For static equilibrium:

$$\begin{bmatrix} \mathbf{w}_1 & \mathbf{w}_2 & \cdots & \mathbf{w}_n \end{bmatrix} \begin{bmatrix} \lambda_1 \\ \lambda_2 \\ \vdots \\ \lambda_n \end{bmatrix} = -\mathbf{g}, \qquad (2)$$

where $\lambda_i \geq 0$ is the tension in the ith cable.

When $n = 3$, in order for (2) to be satisfied (with or without non-zero tensions), the four line vectors or zero pitch wrenches, $\mathbf{w}_1$, $\mathbf{w}_2$, $\mathbf{w}_3$, and $\mathbf{g}$ must belong to the same *regulus*. The lines of a regulus are points on a 2-plane in $\mathbb{PR}^5$ [9], which implies that the body is underconstrained and has three degrees of freedom. Instantaneously, these degrees of freedom correspond to twists in the *reciprocal screw system* that are reciprocal to $\mathbf{w}_1$, $\mathbf{w}_2$, and $\mathbf{w}_3$. They include zero pitch twists (pure rotations) that lie along the axes of the *complementary regulus* (the set of lines each intersecting all of the lines in the original regulus). Geometrically, (2) simply requires the gravity wrench to be reciprocal to the reciprocal screw system, a fact that will be exploited in our calculations in the next section.

B. Cooperative manipulation with three aerial robots

From this point on, we will discuss the special case of a payload transported by three robots. The analysis for $n \neq 3$ is not very different. However the $n = 3$ case is the smallest n for which we can achieve equilibrium for a large set of specified three-dimensional poses of the payload.[1]

We will make the following simplifying assumptions for the $n = 3$ case:

1) The payload is a homogeneous, planar object and the center of mass lies in the plane of the three pivot points.

---

[1] The $n = 1$ case is a simple pendulum with only one stable equilibrium with a single orientation. The $n = 2$ case limits our ability to achieve a desired angle of rotation about the line joining P_1 and P_2.

2) The mass of the object is sufficiently small that three robots are able to lift the object.
3) The payload does not flip during manipulation, restricting the orientation to $|\alpha| < \frac{\pi}{2}$ and $|\beta| < \frac{\pi}{2}$.
4) We require the robots to assume positions that are on one side of the plane of the three pivot points.

For the analysis, we will use a local frame, B, attached to the payload that is defined with the origin at P_1, the x axis pointing toward P_2 and the $x - y$ plane coincident with the plane formed by P_1, P_2, and P_3. In this local coordinate system, the components are denoted by $(\tilde{\cdot})$ and given as:

$$P_1 = [0, 0, 0]^{\mathrm{T}}, \quad P_2 = [\tilde{x}_2^p, 0, 0]^{\mathrm{T}}, \quad P_3 = [\tilde{x}_3^p, \tilde{y}_3^p, 0]^{\mathrm{T}};$$
$$\tilde{q}_1 = [\tilde{x}_1, \tilde{y}_1, \tilde{z}_1]^{\mathrm{T}}, \quad \tilde{q}_2 = [\tilde{x}_2, \tilde{y}_2, \tilde{z}_2]^{\mathrm{T}}, \quad \tilde{q}_3 = [\tilde{x}_3, \tilde{y}_3, \tilde{z}_3]^{\mathrm{T}}.$$

Note that without loss of generality, we assume that $\tilde{x}_2^P > \tilde{x}_3^P$, restricting the possible permutations of P_i. Equation (2) takes the form:

$$\tilde{\mathbf{W}}\lambda = -\tilde{\mathbf{g}}, \tag{3}$$

with

$$\tilde{\mathbf{W}} = \begin{bmatrix} \tilde{x}_1 & \tilde{x}_2 - \tilde{x}_2^p & \tilde{x}_3 - \tilde{x}_3^p \\ \tilde{y}_1 & \tilde{y}_2 & \tilde{y}_3 - \tilde{y}_3^p \\ \tilde{z}_1 & \tilde{z}_2 & \tilde{z}_3 \\ 0 & 0 & \tilde{y}_3^p \tilde{z}_3 \\ 0 & -\tilde{x}_2^p \tilde{z}_2 & -\tilde{x}_3^p \tilde{z}_3 \\ 0 & \tilde{x}_2^p \tilde{y}_2 & \tilde{x}_3^p \tilde{y}_3 - \tilde{y}_3^p \tilde{x}_3 \end{bmatrix}$$

$$\tilde{\mathbf{g}} = -mg \begin{bmatrix} \mathbf{R}^{\mathrm{T}} e_3 \\ \begin{bmatrix} \frac{1}{3}(\tilde{x}_2^p + \tilde{x}_3^p) \\ \frac{1}{3}(\tilde{y}_3^p) \\ 0 \end{bmatrix} \times \mathbf{R}^{\mathrm{T}} e_3 \end{bmatrix}.$$

III. MECHANICS OF 3-D MANIPULATION WITH CABLES

A. Robot positions for desired payload pose

The first problem that must be solved is the analog to the inverse kinematics problem in parallel manipulators:

Problem 1 (Inverse Problem). *Given the desired payload position and orientation (1), find positions of the robots, q_i, that satisfy the kinematics of the robots-cables-payload system and the equations of equilibrium (3).*

The inverse kinematics problem is underconstrained. If the cables are in tension, we know that the following constraints must be true:

$$(\tilde{x}_i - \tilde{x}_i^p)^2 + (\tilde{y}_i - \tilde{y}_i^p)^2 + (\tilde{z}_i - \tilde{z}_i^p)^2 = l_i^2, \tag{4}$$

for $i = \{1, 2, 3\}$. We impose the three equations of static equilibrium (3) to further constrain the solutions to the inverse problem. This further reduces the degrees of freedom to three. Finally, we require that there exist a *positive*, 3×1 vector of multipliers, λ, that satisfies (3).

We solve this problem by finding the three screws (twists) that are reciprocal to the three zero pitch wrenches. Define the 6×3 matrix $\mathbf{S}$ of twists with three linearly independent twists such that the vectors belong to the null space of $\tilde{\mathbf{W}}^{\mathrm{T}}$:

$$\tilde{\mathbf{W}}^{\mathrm{T}} \mathbf{S} = 0.$$

$\mathbf{S}(\tilde{x}_i, \tilde{y}_i, \tilde{z}_i)$ is an algebraic function of the positions of the quadrotors. In order to satisfy (3), q_i must satisfy the three algebraic conditions:

$$\mathbf{S}(\tilde{x}_i, \tilde{y}_i, \tilde{z}_i)^{\mathrm{T}} \tilde{\mathbf{g}} = 0. \tag{5}$$

The inverse problem reduces to the problem of solving for the three-dimensional set $Q_c \subset Q$ by solving for the nine variables $\{(\tilde{x}_i, \tilde{y}_i, \tilde{z}_i), i = 1, 2, 3\}$ subject to (4, 5).

We now restrict our attention to a reduced space of possible configurations based upon the assumptions (3, 4). We introduce the notion of normalized components, denoted by $(\hat{\cdot})$, with $\hat{q}_i = [\hat{x}_i, \hat{y}_i, 1]$ to define the position of the ith robot projected to a constant height $\tilde{z}_i = 1$ above the payload. Based on this simplification (we now only need to solve for three planar positions), we redefine the static equilibrium condition as

$$\mathbf{S}(\hat{x}_i, \hat{y}_i, 1)^{\mathrm{T}} \tilde{\mathbf{g}} = 0. \tag{6}$$

Solving the system of equations in (6) yields algebraic solutions for $\{\hat{x}_2, \hat{x}_3, \hat{y}_3\}$ as functions of $\{\hat{x}_1, \hat{y}_1, \hat{y}_2\}$. Note that any solution to (6) is indeed a solution to (5) after a scaling of cable tension values. Further, we may compute the position of the robots $\tilde{q}_i$ given the normalized coordinates $\hat{q}_i$ and the kinematic constraint (4) as

$$\tilde{q}_i = l_i \frac{\hat{q}_i}{\|\hat{q}_i\|} + P_i.$$

By using $\hat{q}_i$ as coordinates, we can obtain closed-form analytic solutions for the positions of all robots that respect the kinematic constraints and the condition for static equilibrium.

If $\tilde{q}_i$ are chosen to satisfy the equations of equilibrium (3), the multipliers can be obtained by:

$$\lambda = -\tilde{\mathbf{W}}^{\dagger} \tilde{\mathbf{g}},$$

where $\tilde{\mathbf{W}}^{\dagger}$ is the Moore-Penrose inverse of $\tilde{\mathbf{W}}$. This allows us to check to see if the equilibrium pose yields tensions that satisfy (non-negative) lower and upper bounds. Figure 3 depicts a representation of the bounded and positive tension workspace for various configurations.

B. The pose of the payload for hovering robots

Problem 2 (Direct Problem). *Given the actual positions of the robots, $\{q_1, q_2, q_3\}$, find the payload position(s) and orientation(s) satisfying the kinematics of the robots-cables-payload system (4) and the equations of equilibrium (3).*

With hovering robots and cables in tension, we can treat the object as being attached to three stationary points through rigid rods and ball joints to arrive at the constraints in (4). Accordingly the object has three degrees of freedom. Imposing the three equilibrium conditions in (5) we can, in principle, determine a finite number of solutions for this analog of the direct kinematics problem. While this theoretical setting allows the analysis of the number of solutions using basic tools in algebraic geometry, very little can be established when the idealization of rigid rods is relaxed to include limits on cable tensions. Therefore we pursue a numerical approach to the

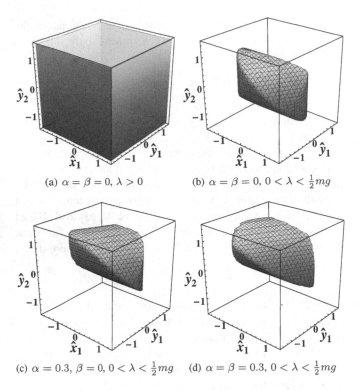

(a) $\alpha = \beta = 0, \lambda > 0$ (b) $\alpha = \beta = 0, 0 < \lambda < \frac{1}{2}mg$

(c) $\alpha = 0.3, \beta = 0, 0 < \lambda < \frac{1}{2}mg$ (d) $\alpha = \beta = 0.3, 0 < \lambda < \frac{1}{2}mg$

Fig. 3.   For a payload with mass $m = 0.25\,\mathrm{kg}$ and $\tilde{x}_2^p = 1\,\mathrm{m}$, $\tilde{x}_3^p = 0.5\,\mathrm{m}$, and $\tilde{y}_3^p = 0.87\,\mathrm{m}$, the numerically determined workspace of valid tensions in the normalized coordinates space $\{\hat{x}_1, \hat{y}_1, \hat{y}_2\}$. Any point selected in these valid regions meets the conditions of static equilibrium while ensuring positive and bounded tensions for all robots. Note that in these plots we do not consider inter-robot collisions (as compared to Fig. 9(a)).

	λ_1	λ_2	λ_3
1	0.06	0.80	5.48
2	−2.72	0.37	1.57
3	−3.11	0.03	2.56
4	−2.64	0.27	7.50
5	−5.54	−1.19	−0.61

TABLE I
EIGENVALUES, λ_i, OF (9) DEFINED BY THE EQUILIBRIUM
CONFIGURATIONS FOR THE REPRESENTATIVE EXAMPLE (FIG. 4(B)).

problem by considering an alternative formulation. To numerically solve the direct problem, we formulate an optimization problem which seeks to minimize potential energy of the payload for a given robot configuration and payload geometry:

$$\underset{p_i}{\arg\min} \quad mgz_C$$

$$\text{s.t.} \quad \|p_i - q_i\| \leq l_i, \ i = \{1,2,3\}$$
$$\|p_i - p_j\| = L_{ij}, \ i,j = \{1,2,3\}, i \neq j$$

where the objective function is linear in terms of p_i (assuming a planar payload), and L_{ij} is the metric distance between two anchor points p_i and p_j.

C. Stability analysis

Assuming that each robot, q_i, remains stationary, we wish to find the change in the potential energy of the payload. Define the height of the payload z_C as the center of mass, which we assume corresponds to the geometric center of the payload.

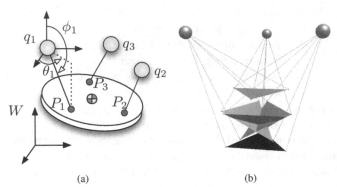

(a)   (b)

Fig. 4.   Determining the pose of the payload for hovering robots via spherical coordinates (Fig. 4(a)). Given a representative robot configuration, five equilibrium solutions for the pose of the payload are found via the methods in Sect. III-B (Fig. 4(b)). The bottommost payload pose is the only stable configuration (as noted in Table I).

Therefore,

$$z_C = \frac{1}{3} \sum_{i=1}^{3} (z_i + l_i \cos \phi_i), \qquad (7)$$

where ϕ_i is defined in terms of spherical coordinates as depicted in Fig. 4(a). From (7), we may derive the potential energy of the payload as

$$V = mgz_C = \frac{mg}{3} \sum_{i=1}^{3} (z_i + l_i \cos \phi_i), \qquad (8)$$

where m is the mass of the payload and g is acceleration due to gravity. To study the stability of the payload, we consider the second-order changes of the potential energy. We first note that the robot-cables-payload system has three degrees of freedom when requiring that cables be in tension. Therefore, we choose to parameterize the solution of the potential energy with respect to $\{\phi_1, \theta_1, \phi_2\}$ and define the Hessian of (8) as

$$\mathcal{H}(\phi_1, \theta_1, \phi_2) = \begin{bmatrix} \frac{\partial^2 V}{\partial \phi_1^2} & \frac{\partial^2 V}{\partial \phi_1 \partial \theta_1} & \frac{\partial^2 V}{\partial \phi_1 \partial \phi_2} \\ \frac{\partial^2 V}{\partial \theta_1 \partial \phi_1} & \frac{\partial^2 V}{\partial \theta_1^2} & \frac{\partial^2 V}{\partial \theta_1 \partial \phi_2} \\ \frac{\partial^2 V}{\partial \phi_2 \partial \phi_1} & \frac{\partial^2 V}{\partial \phi_2 \partial \theta_1} & \frac{\partial^2 V}{\partial \phi_2^2} \end{bmatrix}. \qquad (9)$$

Due to space constraints, we defer the presentation of the analytic derivation of the entries of (9) to [10] where we show that a closed-form analytic representation of (9) is possible, enabling stability analysis of robots-cables-payload configurations. In Fig. 4(b), we provide the resulting set of equilibrium solutions to the direct problem for a representative example robot configuration and the corresponding numerical stability analysis in Table I.

IV. EXPERIMENTATION

A. The quadrotor robots

In Sect. II, we assume point-model aerial robots but in Sect. V provide results on a team of commercially available quadrotors (see Fig. 5). In this section we develop a transformation from desired applied forces to control inputs required by the hardware platform. Some of this discussion is specific to the quadrotor used for our experiments.

Fig. 5. The aerial robots and manipulation payload for experimentation. The payload is defined by $m = 0.25$ kg and $\tilde{x}_2^p = 1$ m, $\tilde{x}_3^p = 0.5$ m, and $\tilde{y}_3^p = 0.87$ m, with $l_i = 1$ m.

1) Model: We begin by considering an aerial robot in $\mathbb{R}^3$ with mass m and position and orientation, $q = [x, y, z] \in \mathbb{R}^3$, and $\mathbf{R}(\alpha, \beta, \gamma) \in SO(3)$, respectively. We do not have direct access to the six control inputs for linear and angular force control, but instead have access to four control inputs $\nu = [\nu_1, \ldots, \nu_4]$ defined over the intervals $[\nu_1, \nu_2, \nu_3] \in [-1, 1]$ for the orientations and $\nu_4 \in [0, 1]$ for the vertical thrust.

Based on system identification and hardware documentation, we arrive at the following dynamic model relating the control inputs and the forces and moments produced by the rotors in the *body* frame:

$$\tau_x = -K_{v,x}^q \dot{\alpha} - K_{p,x}^q (\alpha - \nu_1 \alpha_{max})$$
$$\tau_y = -K_{v,y}^q \dot{\beta} - K_{p,y}^q (\beta - \nu_2 \beta_{max})$$
$$\tau_z = -K_{v,z}^q \dot{\gamma} - K_{p,z}^q \nu_3 \gamma_{inc}$$
$$f_z = \nu_4 f_{max}$$

where τ_x, τ_y, and τ_z are the torques along the body-fixed x, y, and z axes, respectively, and f_z is the force (thrust) along the body-fixed z axis.

The relevant parameters above are defined as:

- K_p^q, K_v^q — feedback gains applied by the aerial robot;
- f_{max} — maximum achievable thrust as a non-linear function of battery voltage;
- $\alpha_{max}, \beta_{max}$ — maximum achievable roll and pitch angles;
- γ_{inc} — conversion parameter between the units of ν_3 and radians.

All of these parameters were identified through system identification. Therefore, in the discussion that follows, we consider the following desired inputs:

$$\begin{bmatrix} \alpha^d \\ \beta^d \\ \gamma^d \\ f_z^d \end{bmatrix} = \begin{bmatrix} \nu_1 & \nu_2 & \nu_3 & \nu_4 \end{bmatrix} \begin{bmatrix} \alpha_{max} \\ \beta_{max} \\ \gamma_{inc} \\ f_{max} \end{bmatrix}, \quad (10)$$

noting that we can compute the control inputs ν given the desired values assuming the parameters are known.

B. Controllers

We now derive a transformation from the control inputs for our point-model abstraction in the world frame to the control inputs ν in the body-frame described in Sect. IV-A.

The dynamics of the point robot in the world frame can be written as:

$$m I_3 \ddot{q} = F_g + \begin{bmatrix} F_x \\ F_y \\ F_z \end{bmatrix}. \quad (11)$$

where F_x, F_y, and F_z are the control input forces in the world frame and F_g is the force due to gravity. These can be related to the forces in the robot body frame:

$$\begin{bmatrix} 0 \\ 0 \\ f_z \end{bmatrix} = \mathbf{R}(\alpha, \beta, \gamma)^{-1} \begin{bmatrix} F_x \\ F_y \\ F_z \end{bmatrix}. \quad (12)$$

Thus, for a given (current) γ, there is a direct relationship between the triples (α, β, f_z) and (F_x, F_y, F_z). Solving (12) results in sixteen solutions for (α, β, f_z), eight of which require $f_z \leq 0$. Requiring that the robot always apply thrust with $F_z > 0$ reduces the solution set to the following:

$$f_z = \sqrt{F_x^2 + F_y^2 + F_z^2}$$
$$\alpha = \pm \cos^{-1}\left(\frac{\sqrt{\alpha_n}}{\sqrt{2} f_z} \right)$$
$$\alpha_n = F_x^2 + 2F_x F_y \sin(2\gamma) + F_y^2 + 2F_z^2 + \left(F_x^2 - F_y^2\right) \cos(2\gamma)$$
$$\beta = \pm \cos^{-1}\left(\frac{F_z}{\sqrt{F_z^2 + (F_x \cos(\gamma) + F_y \sin(\gamma))^2}} \right). \quad (13)$$

From (13), it is clear that two solution sets are valid, but only for specific intervals. For example, consider pitch (β) when applying controls in the yaw directions $[-\pi/2, 0, \pi/2, \pi]$ with $\{F_x, F_y, F_z\} > 0$. For each of the four cases respectively, we expect $\beta < 0$, $\beta > 0$, $\beta > 0$, and $\beta < 0$. A similar argument may be made for α. Therefore, we must construct a piecewise-smooth curve as a function of the external forces $\{F_x, F_y\}$ and γ over the intervals $[-\pi, \gamma_1]$, $[\gamma_1, \gamma_2]$, $[\gamma_2, \pi]$, defined by the zero points of the conditions in (13). For α, we find that

$$[\gamma_{\alpha_1}, \gamma_{\alpha_2}] = -\tan^{-1}\left(\frac{F_x}{F_y} \right)$$
$$+ \begin{cases} [-\frac{3\pi}{2}, -\frac{\pi}{2}] & \text{if } -\tan^{-1}\left(\frac{F_x}{F_y}\right) > \frac{\pi}{2} \\ [-\frac{\pi}{2}, \frac{\pi}{2}] & \text{otherwise} \end{cases}$$

In a similar manner, we find for β

$$[\gamma_{\beta_1}, \gamma_{\beta_2}] = 2 \tan^{-1}\left(\sqrt{1 + \frac{F_x^2}{F_y^2}} - \frac{F_x}{F_y} \right)$$
$$+ \begin{cases} [-\frac{3\pi}{2}, -\frac{\pi}{2}] & \text{if } \tan^{-1}\left(\sqrt{1 + \frac{F_x^2}{F_y^2}} - \frac{F_x}{F_y}\right) > \frac{\pi}{4} \\ [-\frac{\pi}{2}, \frac{\pi}{2}] & \text{otherwise} \end{cases}$$

We now construct piecewise-smooth solutions that reflect the expected inputs for all values of γ. From (13), we denote the positive α solution as α_+ and the negative solution as α_-. The piecewise-smooth input for α is

$$\alpha = \begin{cases} \{\alpha_+, \alpha_-, \alpha_+\} & \text{if } F_y > 0 \\ \{\alpha_-, \alpha_+, \alpha_-\} & \text{otherwise} \end{cases}$$

In a similar manner, for β we find

$$\beta = \begin{cases} \{\beta_-,\ \beta_+,\ \beta_-\} & \text{if } F_x > 0 \\ \{\beta_+,\ \beta_-,\ \beta_+\} & \text{otherwise} \end{cases}$$

Both α and β become singular when $F_y = 0$. However, it is clear from the definitions in (13) that at this singularity for the intervals $[-\pi,\ 0]$ and $[0,\ \pi]$,

$$\alpha = \begin{cases} \{\alpha_-,\ \alpha_+\} & \text{if } F_x > 0 \\ \{\alpha_+,\ \alpha_-\} & \text{otherwise} \end{cases}$$

and for the intervals $[-\pi,\ -\pi/2]$, $[-\pi/2,\ \pi/2]$, and $[\pi/2,\ \pi]$

$$\beta = \begin{cases} \{\beta_-,\ \beta_+,\ \beta_-\} & \text{if } F_x > 0 \\ \{\beta_+,\ \beta_-,\ \beta_+\} & \text{otherwise} \end{cases}$$

Of course, there remains the trivial solution of $\alpha = \beta = 0$ when $F_x = F_y = 0$.

Therefore, we may define a force $F = [F_x,\ F_y,\ F_z]^{\mathrm{T}}$ in the world frame that is transformed into appropriate control inputs via (10, 12). To this end, we compute F in implementation based on proportional-integral-derivative (PID) feedback control laws determined by the desired robot configurations with feedforward compensation based on (11). For the purposes of this work, we additionally control ν_3 to drive γ to zero.

As a limitation of this approach, we note that the controller above results in thrust f_z for any desired input. Therefore, when the robot is hovering at a level pose ($\alpha = \beta = 0$), a desired horizontal motion for which F_x must be nonzero and $F_z = mg$, results in an unintended vertical thrust $f_z > mg$ until the robot rotates from the level pose.

C. Payload model

The payload is a rigid frame with cable attachment points given by $\tilde{x}_2^p = 1\,\mathrm{m}$, $\tilde{x}_3^p = 0.5\,\mathrm{m}$, and $\tilde{y}_3^p = 0.87\,\mathrm{m}$ with mass $m = 0.25\,\mathrm{kg}$. The cable lengths are equal with $l_i = 1\,\mathrm{m}$.

D. Simulation and experiment design

Our software and algorithms are developed in C/C++ using Player/Gazebo [11] and interfaced with MATLAB for high-level configuration specification for both simulation and experiments.

Experiments are conducted with the AscTec Hummingbird quadrotor [12] from Ascending Technologies GmbH with localization information being provided by a Vicon motion capture system [13] running at 100 Hz with millimeter accuracy. Control commands are sent to each robot via Zigbee at 20 Hz. The quadrotor is specified to have a payload capacity of 0.2 kg with a physical dimension conservatively bound by a sphere of radius $R = 0.3\,\mathrm{m}$.

Robots can be positioned inside the $6.7\,\mathrm{m} \times 4.4\,\mathrm{m} \times 2.75\,\mathrm{m}$ workspace with an error bound of approximately $\pm 0.05\,\mathrm{m}$ in each direction. However, as the robots experience the effects of unmodelled external loads and interactions during manipulation, errors increase to approximately $\pm 0.15\,\mathrm{m}$.

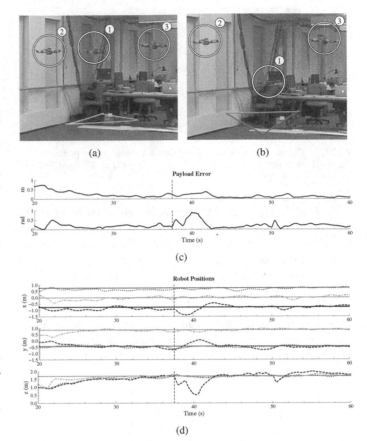

Fig. 6. In Fig. 6(a) the robots assume the desired configuration while in Fig. 6(b) one robot has suffered a large actuation failure. Figure 6(c) depicts the mean squared position and orientation error of the payload while Fig. 6(d) provides data on the individual robot control error. Vertical dashed lines in Figs. 6(c) and 6(d) show the exact time of the robot actuator failure.

V. RESULTS

Using the solution to the inverse problem presented in Sect. III and the abstraction of a fully dynamic aerial vehicle to a point robot in Sect. IV, we are able to experimentally verify our ability to lift, transport, and manipulate a six degree of freedom payload by controlling multiple quadrotors.

A. Cooperative lifting

We start with a symmetric configuration similar to that in Fig. 4(b) for benchmarking the performance of the system. Figure 6(a) depicts the team of robots in this configuration raising the payload to a specified pose in the workspace. During the course of experimentation in this configuration, in one of the trials, a robot suffers a momentary actuator failure causing it to lose power and drop (see Fig. 6(b)). We use the data from this trial to demonstrate the team's ability to quickly recover and return to the desired configuration. Analysis of the data in Figs. 6(c) and 6(d) suggests that the time constant for the closed loop response of the system is around $1 - 1.5\,\mathrm{s}$.

B. Cooperative manipulation and transport

In this experiment, the system is tasked with manipulating the payload through a sequence of poses. For each pose, we

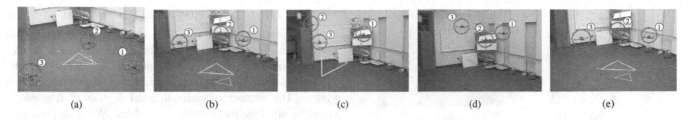

Fig. 7. Snapshots demonstrating cooperative manipulation and transportation. The team starts from initial conditions before takeoff (Fig. 7(a)), stabilizes the platform at each desired pose (Figs.7(b) – 7(d)), and returns the payload to the first pose before landing (Fig. 7(e)). Circles highlight individual robot positions during the evolution of the experiment. Videos of the experiments are available at http://kumar.cis.upenn.edu/movies/RSS2009.flv

compute the desired configuration (q_1, q_2, q_3) for the robots and drive each robot to the desired goal. The underlying control system avoids inter-robot collisions via a simple potential field controller derived from the point-robot model in (11) with each robot modeled as a sphere of radius R, guaranteeing $\|q_i - q_j\| > 2R$ for all pairs of robots. This results in a smooth, though unplanned, trajectory for the payload. Figures 7 and 8 depict several snapshots during this experiment and the resulting system performance, respectively.

This experiment demonstrates that even though the manipulation system is underactuated for $n = 3$, it is able to position and orient the payload as commanded. The small oscillations around each equilibrium configuration are inevitable because of the low damping and perturbations in the robot positions.

VI. PLANNING MULTI-ROBOT MANIPULATION AND TRANSPORTATION TASKS

In this section, we address motion planning for the aerial manipulation tasks and the generation of trajectories for the robots that respect (a) the kinematic workspace constraints; (b) the conditions of *stable* equilibrium for the payload; (c) constraints on the cable tensions ($\lambda_{max} \geq \lambda_i > 0$); and (d) the geometric constraints necessary to avoid collisions ($\|q_i - q_j\| > 2R$). In Sect. III, we computed Q_c to to be the set of robot positions satisfying (a, b) with positive tension in each cable. We now derive the effective workspace for the robots $Q_M \subset Q_c$ consisting of robot positions that satisfy (a-d) above. The planned trajectories of the robots must stay within Q_M which has a complex shape owing to the nonsmooth constraints in (a-d).

Figure 9 illustrates the effective workspace Q_M parameterized by $(\hat{x}_1, \hat{y}_1, \hat{y}_2)$. Indeed for a given payload pose, there are multiple points in the workspace that satisfy the conditions of stable equilibrium. Three representative conditions are shown in the figure. Thus during motion planning it is possible to optimize any number of design goals, including equal sharing of loads by the cooperating robots, robustness to disturbances to the payload, and maximizing the stability of the robots-payload system. To demonstrate this, we consider the response of the payload to an external disturbance in experimentation given two distinct robot configurations selected from the space of valid solutions Q_M depicted in Fig. 9(a). The first mirrors the simulation configuration shown in Fig. 9(b) while the second configuration is selected to maximize the smallest natural frequency of the payload given by the smallest eigenvalue of

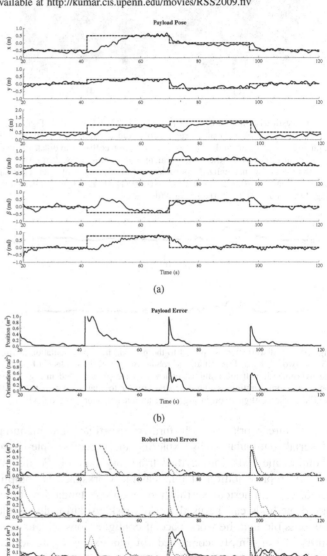

Fig. 8. Data from the manipulation and transportation experiment, including: pose data for the payload overlaid with a dashed line representing desired values (Fig. 8(a)), aggregate position and orientation error of the payload (Fig. 8(b)), and individual robot control mean squared errors (Fig. 8(c)).

the Hessian in (9). Figure 10 shows that the robot configuration determined by the natural frequency-based measure attenuates the payload error more quickly than the other configuration in which the lower natural frequency results in oscillations that take longer to damp out.

7

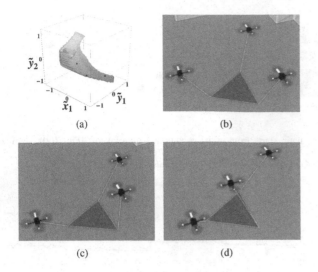

(a)   (b)

(c)   (d)

Fig. 9.   Various points in Q_M for $\alpha = \beta = 0$ (Figs. 9(b)–9(d)). Figure 9(a) depicts numerically determined regions of valid tensions in the space of $\hat{q}$ requiring $\lambda_i < \frac{1}{2}mg$ and $\|q_i - q_j\| > 1\,\text{m}$ (for collision avoidance), with black points indicating the configurations selected in Figs. 9(b)-9(d). For completeness, the normalized coordinates, $\{\hat{x}_1, \hat{y}_1, \hat{y}_2\}$, of each configuration follows: $\{-0.2724, -0.3054, -0.3054\}$ (Fig. 9(b)), $\{0, 0, -0.9\}$ (Fig. 9(c)), $\{0.6, 0.45, -0.9\}$ (Fig. 9(d)).

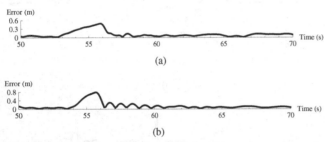

(a)

(b)

Fig. 10.   A disturbance is applied to the payload in experimentation at time 55 s in two trials. In Fig. 10(a), the robot configuration is selected based on the maximization of the natural frequency of the payload and in Fig. 10(b), the robot configuration is chosen to be similar to that shown in Fig. 9(b). Note that the configuration in Fig. 10(a) attenuates error more quickly.

In future work we will further investigate the planning of aerial manipulation by exploring the use of sample-based planning methods for payload trajectory control. While the analytic representation of our feasible workspace is complicated, fast numerical verification suggests standard sample-based methods can be used for this task. Additionally, while it is possible that the workspace of configurations with tension limits is not simply connected, in our extensive numerical studies, we have not found this to happen with realistic values of geometric parameters and tension bounds. This suggests that it should be possible to transition smoothly from one configuration to another without loss of cable tension. Finally, the fact that the direct problem has multiple stable solutions is a potential source of concern in real experimentation since positioning the robots at a desired set of positions does not guarantee that the payload is at the desired position and orientation. In a forthcoming paper [14], we show that by further constraining Q_M, we can reduce the direct problem to a Second Order Cone Program (SOCP). We plan to incorporate these constraints into our motion planning algorithm.

VII. CONCLUSION AND FUTURE WORK

We presented a novel approach to aerial manipulation and transport using multiple aerial robots. We derived a mathematical model that captures the kinematic constraints and the mechanics underlying stable equilibria of the underactuated system. The number of unconstrained degrees of freedom is equal to six less the number of robots. We also presented an experimental implementation and results that suggest that cooperative manipulation can be used as an effective way of manipulating and transporting payloads that are beyond the capability of individual micro UAVs.

The main limitation of our approach lies in our inability to damp out oscillations in the underdamped system. Because of this, our trajectory following capabilities are limited to slow motions with harmonics well under the fundamental frequencies of around $3 - 5\,\text{Hz}$. One possibility is to use the robots to actively damp out oscillations using methods analogous to controlling flexible manipulators. We are currently engaged in a more thorough study of the underlying joint configuration space and the effects of cable constraints with a view to developing motion planning algorithms. We are also considering the application of control and estimation methods that relax our current reliance on globally available state information and enable a better understanding of the effects of sensing and actuation uncertainty on control performance.

REFERENCES

[1] R. M. Murray, "Trajectory generation for a towed cable system using differential flatness," in *IFAC World Congress*, San Francisco, CA, July 1996.
[2] P. Cheng, J. Fink, S. Kim, and V. Kumar, "Cooperative towing with multiple robots," in *Proc. of the Int. Workshop on the Algorithmic Foundations of Robotics*, Guanajuato, Mexico, Dec. 2008.
[3] E. Stump and V. Kumar, "Workspaces of cable-actuated parallel manipulators," *ASME Journal of Mechanical Design*, vol. 128, no. 1, pp. 159–167, Jan. 2006.
[4] R. Verhoeven, "Analysis of the workspace of tendon-based stewart platforms," Ph.D. dissertation, University Duisburg-Essen, Essen, Germany, July 2004.
[5] S. R. Oh and S. K. Agrawal, "A control lyapunov approach for feedback control of cable-suspended robots," in *Proc. of the IEEE Int. Conf. on Robotics and Automation*, Rome, Italy, Apr. 2007, pp. 4544–4549.
[6] P. Bosscher and I. Ebert-Uphoff, "Wrench-based analysis of cable-driven robots," in *Proc. of the IEEE Int. Conf. on Robotics and Automation*, vol. 5, New Orleans, LA, Apr. 2004, pp. 4950–4955.
[7] K. H. Hunt, *Kinematic Geometry of Mechanisms*. Oxford University Press, 1978.
[8] J. Phillips, *Freedom in Machinery*. Cambridge University Press, 1990, vol. 1.
[9] J. M. Selig, *Geometric Fundamentals of Robotics*. Springer, 2005.
[10] N. Michael, S. Kim, J. Fink, and V. Kumar, "Kinematics and statics of cooperative multi-robot aerial manipulation with cables," in *ASME Int. Design Engineering Technical Conf. & Computers and Information in Engineering Conf.*, San Diego, CA, Aug. 2009.
[11] B. P. Gerkey, R. T. Vaughan, and A. Howard, "The Player/Stage Project: Tools for multi-robot and distributed sensor systems," in *Proc. of the Int. Conf. on Advanced Robotics*, Coimbra, Portugal, June 2003, pp. 317–323.
[12] "Ascending Technologies, GmbH," http://www.asctec.de.
[13] "Vicon Motion Systems, Inc." http://www.vicon.com.
[14] J. Fink, N. Michael, S. Kim, and V. Kumar, "Planning and control for cooperative manipulation and transportation with aerial robots," in *Int. Symposium of Robotics Research*, Luzern, Switzerland, Aug. 2009.

Learning of 2D Grasping Strategies
from Box-based 3D Object Approximations

Sebastian Geidenstam, Kai Huebner, Daniel Banksell and Danica Kragic

Computer Vision & Active Perception Lab

KTH – Royal Institute of Technology, Stockholm, Sweden

Email: {sebbeg,khubner,banksell,danik}@kth.se

Abstract—In this paper, we bridge and extend the approaches of 3D shape approximation and 2D grasping strategies. We begin by applying a shape decomposition to an object, i.e. its extracted 3D point data, using a flexible hierarchy of minimum volume bounding boxes. From this representation, we use the projections of points onto each of the valid faces as a basis for finding planar grasps. These grasp hypotheses are evaluated using a set of 2D and 3D heuristic quality measures. Finally on this set of quality measures, we use a neural network to learn good grasps and the relevance of each quality measure for a good grasp. We test and evaluate the algorithm in the GraspIt! simulator.

I. INTRODUCTION

In the field of intelligent grasping and manipulation, a robot may recognize an object first and then reference an internal object model. For unknown objects, however, it needs to evaluate from data it can collect on the spot. How to grasp a novel object is an ongoing field of research. Difficulties in this area include (i) the high dimensionality of the problem, (ii) incomplete information about the environment and the objects to be grasped, and also (iii) generalizable measures of quality for a planned grasp.

Since contacts and forces of the fingers on an object's surface make up a grasp, it is very important to have good information both about the hand and the object to be grasped. Both hand and object constraints together with the constraints for the task to be performed need to be considered [1]. Though there is interesting work on producing grasp hypotheses from 2D image features only, e.g. [2, 3], most techniques rely on 3D data. Due to the complexity of the task, much work has been done for simplifications of 3D shape, such as planar [4] or 3D-contour-based [5] representations. Other approaches involve modelling an object perfectly, i.e. known a-priori, or with high-level shape primitives, such as the use of grasp pre-shapes or Eigengrasps [6, 7, 8]. One work that uses high-level shape primitives, and is similar to ours in terms of learning, but by using an SVM approach, is [9]. Another approach to learning from 2D grasp qualities, using neural networks and genetic algorithms, is presented in [10].

This paper builds on the work of Huebner *et al.* [11, 12], which uses a hierarchy of minimum volume bounding boxes to approximate an object from a set of 3D points delivered by an arbitrary 3D sensor, e.g. laser scanners or stereo camera setups. Grasping is then done by approaching each face of a box until contact, backing up, and then grasping the object. What this work lacked however, was a way to explicitly place the fingers

of the hand and to choose the best configuration of the hand. Learning to predict successful grasps was done only with raw data from the projections of points inside a box onto the face to be grasped. Secondly, our work makes use of an algorithm for finding and predicting the success of a grasp, but for planar objects, as proposed by Morales *et al.* [4, 13]. The approach uses 2D image analysis to find contact point configurations that are valid given specific kinematic hand constraints. From the geometrical properties of an object, it then calculates a set of quality measures that can later be used for learning to predict the success of found grasp hypotheses. The limitations of this work lie mainly in the fact that really 'planar' objects and representations are discussed in which information about 3D shape is discarded.

In this paper, we bridge and extend these two methods to enable 2D grasping strategies for 3D object representations.

II. 3D BOX APPROXIMATION

We will shortly revisit the pre-computation of approximating a 3D point cloud by a constellation of minimum volume bounding boxes (MVBBs). The fit-and-split approach starts with fitting a root bounding box and estimating a best split by using the 2D projections of the enclosed points to each of the box surfaces. Depending on a volume gain parameter t, two child boxes might be produced and then be tested for splitting. To provide an insight to this algorithm as a base for the experiments in this paper, the two core algorithms have been sketched in Fig. 1. For more details and examples, we refer to Huebner *et al.* [11]. However, it is important to note that in that work (i) 2D projections have been used to estimate a split and (ii) only edge-parallel planar splits have been tested.

From these constraints, three main problems were evident relating to the original split estimation. These problems are outlined as follows.

1) Splitting of non-convex regions, e.g. u-shapes: As shown in [11], the presented algorithm will not do any splitting in case of u-shaped 2D projections. This is due to the fact that it uses upper bounds and area minimization, which are constant in such cases. This means that a split does not result in a substantial change in the area of a region. A solution for this problem remains a challenge [11], especially when sparse and noisy data is provided. For 3D data from real vision systems or laser scanners, such distortions are unavoidable, in part because of occlusion or sensor inaccuracies. Thus,

Algorithm II.1: BOXAPPROXIMATE($points^{3D}$)

$box \leftarrow findBoundingBox(points^{3D})$
$faces \leftarrow nonOppositeFaces(box)$
$(p,q) \leftarrow split(\text{FINDBESTSPLIT}(faces, points^{3D}))$
if $(percentualVolume(p+q, box) < t)$
 then $\begin{cases} \text{BOXAPPROXIMATE}(p) \\ \text{BOXAPPROXIMATE}(q) \end{cases}$
 else return (box)

Algorithm II.2: FINDBESTSPLIT($faces, points^{3D}$)

for $i \leftarrow 1$ **to** 3
$\mathbf{do}\begin{cases} p^{2D} \leftarrow project(points^{3D}, faces[i]) \\ \mathbf{for}\ x \leftarrow 1\ \mathbf{to}\ width(faces[i]) \\ \quad \mathbf{do}\begin{cases} (p1,p2) \leftarrow verticalSplit(p^{2D}, x) \\ a1 \leftarrow boundArea^{2D}(p1) \\ a2 \leftarrow boundArea^{2D}(p2) \\ \mathbf{if}\ (a1+a2 < minArea) \\ \quad \mathbf{then}\begin{cases} minArea \leftarrow (a1+a2) \\ bestSplit \leftarrow (i,x) \end{cases} \end{cases} \\ \mathbf{for}\ y \leftarrow 1\ \mathbf{to}\ height(faces[i]) \\ \quad \mathbf{do}\begin{cases} (p1,p2) \leftarrow horizontalSplit(p^{2D}, y) \\ a1 \leftarrow boundArea^{2D}(p1) \\ a2 \leftarrow boundArea^{2D}(p2) \\ \mathbf{if}\ (a1+a2 < minArea) \\ \quad \mathbf{then}\begin{cases} minArea \leftarrow (a1+a2) \\ bestSplit \leftarrow (i,y) \end{cases} \end{cases} \end{cases}$
return $(bestSplit)$

Fig. 1. Pseudocode (original algorithm): a point set and its bounding box, respectively, are recursively split (II.1). A good split was estimated through analysis of 2D splits of the projected points onto each of the box faces (II.2).

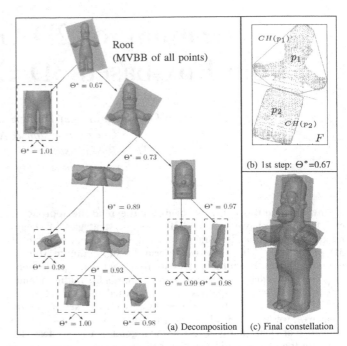

Fig. 2. (a) Example of a decomposition hierarchy, using a gain parameter of t=0.98. With $\Theta^* < t$, a valid cut is detected, as presented for the first step in (b). Otherwise, the box is a leaf box (*dashed*), i.e. a part of the final constellation which is plotted in (c).

how to distinguish between a real non-convex object region and just incompleteness of the data becomes a critical issue. The models used in [11] were ideal models, extracted from simulated 3D mesh data. As it is our aim to evaluate our algorithm also on real sensory data, we can not generally assume such ideal conditions.

2) Splitting along non-edge-parallel directions: The minimum volume box fitting approach naturally fits extensions of the shape into corners of a box, as this keeps the box smaller. The handle of a cup, for example, will fit best diagonally into one of the box corners. However, such diagonal structures in particular can rarely be cut parallel to one of the box edges as proposed in the previous algorithm.

3) Sensitivity to noise: The box decomposition's robustness showed the splitting to be very sensitive to noise. This is not a main issue in terms of single box or face grasping in general, since any constellation of boxes will produce grasp hypotheses. However, if one would like to take into account and learn from a whole constellation of boxes, then robustness and repeatability are necessary.

A. Improved Split Algorithm Using 2D Convex Hulls

For the experiments presented in this paper, we have therefore implemented a new algorithm based on convex hulls. The new algorithm replaces II.2, solving the above mentioned issues, and in addition producing much more confident splitting

results. For efficiently computing convex hulls on a set of 2D points p, like our projections (see Fig. 2), we use a monotone chain algorithm [14]. Starting from the convex hull $CH(p)$ of the whole projection set p, we select those segments of the hull that exceed a given threshold in length. We thereby assume that those segments either span a non-convex region of the outer contour of the data, or that they represent a very straight edge. On these segments, we interpolate a number of sample points. Between each pair of points on each pair of segments, we simulate a cut that splits the point set p into two subsets p_1 and p_2. The two segment points that minimize,

$$\Theta = [A(CH(p_1)) + A(CH(p_2))]/F, \tag{1}$$

where A is the area function for a convex hull and F the overall rectangular area of the face (see Fig. 2b), define our best split. An example of such a decomposition tree produced with the new hull algorithm is presented in Fig. 2.

B. Evaluation

To be able to make a large scale test of the box decompositions stability, an algorithm was developed that estimates if two box decompositions are similar or not. First, the algorithm summarizes the total volume V_i of all boxes which a decomposition i is composed of. Second, it calculates the Euclidean distances between the centers of all pairs of leaf boxes and summarizes them as D_i. In order to determine if two compositions i, j are similar, the differences in overall volume and distance measures between the decompositions are simply compared with empirically found thresholds:

if $|D_i - D_j| \leq 0.1 \wedge |V_i - V_j| \leq 0.9$ **then** $similar(i,j)$. (2)

TABLE I

PERCENTAGE OF DECOMPOSITIONS WITH SIMILAR COUNTERPARTS. MAXIMALLY, 33 COMPARISONS ARE PERFORMED PER OBJECT.

3D	Bunny	Car	Cup	Duck	Goblet	Goose	Heart	Homer	Horse	Human	Mug	PaperCup	Pen	Pillow	Radio	Squirrel	ToyDog
Old	78,62	100,00	77,78	61,59	21,21	24,28	100,00	55,07	53,99	19,93	80,43	100,00	91,67	91,67	84,62	59,42	27,54
New	94,00	100,00	97,78	91,67	22,77	46,00	100,00	66,67	30,33	58,00	62,67	74,24	100,00	100,00	100,00	75,00	28,33

To test the stability of the box decomposition algorithms, we simulated 17 different object models and added various levels of noise in terms of close proximity noise. Using 19 different noise levels and 14 different levels of point removal, 33 modified point clouds emerge from each original object point cloud. Both algorithms were then executed (t=0.9) on each of those point clouds before comparing the resulting box decompositions of each unmodified with its modified models. The results presented in Table I show that the previous algorithm is quite sensitive to noise. However, simpler objects like Car or Pen gave very good results. This is mainly because they all produced only one box due to their compact shape. On the other hand, more complex models like the toydog or the human model gave quite poor results. We note that the bound-based algorithm tends to produce a single large box enveloping the whole object also in such cases. This raises the similarity rate significantly, but is not preferred in our application.

The new hull-based algorithm produces much better approximation for the objects, very few single-box decompositions, and a significantly better similarity rate. The models that produced single-box decompositions with the bound-based algorithm produce worse values in some cases. This is caused by better approximations with multiple boxes that are more sensitive to the comparison than a single-box-to-single-box comparison. Since we prefer multi-box decompositions which give better object approximation, this is a good improvement, while the new algorithm is considerably less affected by noise.

The old and the new techniques are also compared to each other in Fig. 3 according to robustness to the change of the gain parameter t (for t, see II.1 and Fig. 2), e.g. the duck model decomposition repeatedly shows the same constellation. Another visible effect is that the decompositions seem more intuitive, e.g. in case of the cup handle.

III. 2D GRASP HYPOTHESES

In this paper, we are concerned with finding 2D grasps for 3D objects. Thus, we need to find a suitable grasping strategy based on the above mentioned box decomposition. We base our grasping hypotheses on the faces of the final box decomposition. The set of hypotheses is further reduced by including geometrical heuristics on which faces are valid in terms of visibility, reachability, and more [12]. For each leaf box in the hierarchy, the points enveloped by it are projected onto the valid faces of the box and stored in a grayscale image. The distance of the closest point to each pixel cell onto which it is projected is stored as a grayscale value between 0 and 255, where 1 is the depth of the box and 255 means zero depth (see Fig. 4a). This provides us with 2.5D representations of the object parts. The decomposition captures symmetries of objects quite well, resulting in faces and thus projections that are often perpendicular to the axes of most variance. This yields suitable information about approach directions of planar grasps and a good dissection of the object. In short, for each of the projections attained, grasps will be planned similarly to a top-view on a planar object. Thus grasp points on the contour of the projection images need to be found.

For grasp hypotheses from 2D contours, we will use an algorithm that is closely related to the work of Morales [4]. This algorithm involves a four step procedure for finding a number of grasp hypotheses, followed by a fifth to disqualify unfeasible grasps and selecting the best of the hypotheses.

A. Finding Good Regions to Grasp

We use the notion of grasp region, as defined in [4] and assume that a good region for grasping is a region that is as straight as possible. The fact that studies have shown that slightly concave curvature may be better suited [15] is left as a possible extension to the work. For this task a combination of the Canny edge detector and the k-angular bending algorithm [16] was used. First, the projection images described above are preprocessed by erosion and dilation steps. By removing pixels with fewer than 2 neighbours the number of outliers in the image is reduced. Expanding each remaining pixel (a projected point from 3D) to its neigbouring 8 pixels, gaps caused by sparse 3D point information are filled. Without these steps internal contours will be found that do not actually exist in the object and many grasp regions will be invalid. By using

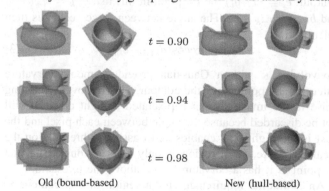

Old (bound-based)   New (hull-based)

Fig. 3. Results of the box approximation for two models. Compared to the results produced with the bound-based algorithm [11] to the left, new hull-based constellations (right) stay more robust despite of different decomposition granularities (described by gain thresholds t in each row).

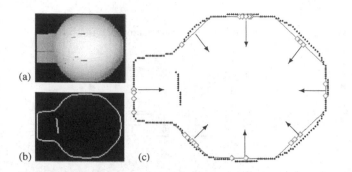

Fig. 4. (a) Projection image of the Duck model's head box (from above). (b) Canny edges image. Size of the Gaussian, lower and higher thresholds are automatically chosen by the Matlab edge algorithm. (c) A set of grasp regions (contour and 2D normal vectors) and grasp points (diamonds). The regions are found with $\sigma = 2.5$, curvature threshold $t_\kappa = 0.4$ (max. angle in radians), accumulated curvature threshold $T_\kappa = 4$, minimum length $l_{min} = 20$mm (similar to Barrett finger width), and maximum length $l_{max} = 50$mm.

the edge detector with comparably high smoothing, see Fig. 4 for an example, we extract edges in the image.[1] These edges correspond to inner and outer contours of the projected object part as well as places where depth is rapidly changing. We assume that for these edges, grasps can be executed similarly to planar objects.

With the discrete edge points detected, these are ordered in a list following the contour. From the contour we will extract regions that satisfy four main conditions:

1) The curvature in any point of the region should be low,
2) the total accumulated curvature of a region should not be too high,
3) a minimum length of a region should be achieved, in order to reduce the number of hypotheses and reduce the effect of positioning errors,
4) a maximum length of a region should not be exceeded, in order to break long straight parts of the contour into several regions such that two fingers can be placed at the same side of an object such as a cube.

The curvature part is handled by the k-angular bending algorithm [16], that considers k neighbours in each direction of a point to determine its curvature by calculating the angles to these neighbours. Let C be the ordered list of points on the contour, $c_i = (x_i, y_i)$ the ith point in this list, $\vec{a}_{i,k} = c_{i+k} - c_i$ and $\vec{b}_{i,k} = c_{i-k} - c_i$. The angle between these vectors is then calculated as,

$$\kappa_i = \arccos(\vec{a}_{ki} \cdot -\vec{b}_{ki}). \tag{3}$$

Convolving κ with a Gaussian provides smooth curvature values at any point along the contour. This removes remaining noise in the image so that a pixelated straight diagonal will not be discarded because the angle between each pixel and the next is too high. This enables us to assign a threshold on the local curvature, i.e. a region is only chosen considering that no point in it has a curvature value above the threshold.

An additional requirement for a region on the contour to be accepted is that the accumulated curvature of a region is

not higher than a chosen threshold. This condition is checked by summing up all κ-values for the region and comparing to the threshold. This takes care of problems with low constant curvature such as for a circle. Without the use of accumulated curvature, the circle would be regarded to have either no feasible regions to grasp or one region going straight through the center. With accumulated curvature, the circle will be broken into several regions. This also applies to other shapes with regions of low curvature with the same sign. Each region is approximated with the line connecting the endpoints of that region. Thus, each region is only represented by these two points and the inwards pointing normal, see Fig. 4.

The two remaining conditions for a region to be considered are the minimum and maximum length of a region. A minimum length is needed in order to account for positioning errors and to have a value close to the finger width. A maximum length is needed so that the representation does not become too simplified. If for a simple object like a cube no maximum length of regions was set, its projection images would only be represented by four regions that would be very hard to combine into a working grasp. By dividing the regions such that none are larger than the assigned maximum length produces more regions and therefore enables the possibility to place two fingers on one single side of the square, for example. Lower maximum length gives more regions and thus higher number of hypotheses, which means more possibilities. One should be cautious, however, since computation time increases rapidly with the number of regions.

B. Determining Finger Positions on the Regions

For each possible triplet (in the case of a 3-finger hand such as the Barrett hand [17] that is used) of regions, two criteria must be met. The normals must positively span the plane and finger placement must be such that all the friction cones of these fingers intersect. In this paper we will assume Coulomb friction and point contacts. By considering the union of all friction cones of one region and looking at the intersection of such 'combined friction cones', one can determine if the intersections of all three regions are empty and the hypothesis discarded, or non-empty and considered. This becomes a geometrical problem for each triplet and can be solved with standard linear programming methods. In the case of non-empty intersections, the centroid of the intersection area is calculated and projected back to the regions. These points will be used for finger positions, as discussed in [4].

C. Determining Hand Configuration

From the finger positions and the Barrett hand kinematics one can test if there is a configuration that can reach the selected points. By varying the angle of the thumb[2] to the surface and searching for those angles that correspond to configuration of the hand that can reach all three grasp points, one can find hypotheses for grasps. The angle of the thumb is varied on the interval $(-\arctan \mu, \arctan \mu)$ in 100 steps,

---

[1]We use the Matlab standard function *edge* with parameter 'Canny' and a value for $\sigma = 2.5$ and automatically calculated threshold.

[2]Note that the thumb of the Barrett hand does not allow rotation. Thus, the angle of the thumb to the object is closely connected to the hand orientation.

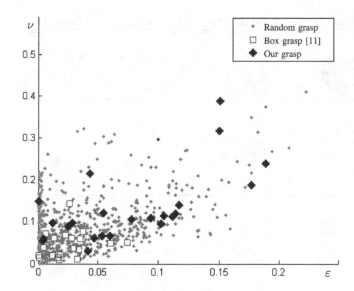

Fig. 5. Plot of 2000 random grasps, hypotheses found with the method from [11], and hypotheses found with our new approach, on the Bunny model. On the axes are the two built-in quality measures from GraspIt!. Only one grasp per finger positioning is chosen randomly and plotted. As can be seen, the best hypotheses are close to the best of the 2000 random grasps, suggesting that with only these 23 hypotheses one could get one or more good grasp.

where μ is the same friction coefficient used for calculating the friction cones in the previous section. Those configurations that satisfy the conditions are stored as grasp hypotheses. For a more detailed description, see [13]. A plot of the quality for grasp hypotheses found for the Bunny compared to 2000 random grasps is shown in Fig. 5.

D. Determining the Quality of a Grasp

From the algorithm presented one can generate a number of grasp hypotheses. However, we still need to determine which grasps are more likely to be successful. This is done by different measures of quality. Firstly however one needs to discard grasp hypotheses that are not reachable. This means that all grasp hypotheses outside of the physical reach of the hand will be discarded. This includes those grasps where one part of the object is in the way for grasping another part of the object. One example for the duck appears when a top grasp is attempted, but with finger positionings on the body of the duck: the head would be occluding the body, thus this grasp is discarded even before attempting it.

Many of the quantitative quality measures are the same as the ones developed by Morales *et al.* [4, 13], and will thus only be mentioned by name. There is one important difference, however: the empirical normalization constants used by Morales *et al.* will not be used here, as an artificial neural network will be used to determine the weights of each measure instead.

The measures derived from Morales are the following:

q_1: Grasp Triangle Size,	q_5: Finger Spread,
q_2: Point Arrangement,	q_6: Focus Deviation,
q_3: Force Line,	q_7: 2D Force Focus.
q_4: Finger Extension,	

These measures however are developed for planar objects. To adapt to non-planar objects to be grasped in our case we add two extra quality measures. These are:

1) Finger Depth Difference: The projection image contains information about the depth of the shape. Thus, it is possible to compare the selected grasp point depths d_i for each finger i with the linear approximations of the real finger extensions $g(e_i)$ by

$$q_8 = (g(e_1) - d_1)^2 + (g(e_2) - d_2)^2 + (g(e_3) - d_3)^2, \quad (4)$$

where $g(\cdot)$ is the linear depth approximation function.[3] This measure depicts how close to the desired grasp points the grasp is likely to be. Note that this measure is the one that explicitly takes into account the 2.5D information provided by the box approximation and projection steps from Section II.

2) 3D Force Focus: Ideally, one would like to measure the distance from the force focus in three dimensions to the actual center of gravity. This is, however, not possible since the information about the object is incomplete and the representations of grasps are only in two dimensions. We provide a rough approximation of this quality by using the center of the root box in the decomposition (containing all points in the point cloud) and the mean of the calculated finger positions in three dimensions,

$$q_9 = \|\bar{p}_{finger} - p_{rootCenter}\|. \quad (5)$$

IV. EVALUATION

The evaluation of the algorithm has mainly been made with data from simulation. This object data consists of 42 different 3D models, consisting of 14 different objects in 3 different scales to provide more data to train on. First, each model was decomposed with the box decomposition algorithm, using a gain threshold t=0.90. Over all models, this resulted in 570 projections from leaf boxes. 5951 grasp triplets were finally found from those projections and used as the data set for evaluation of grasps. Different types of results for the used models and decompositions are presented in Fig. 6. In a next step, the presented quality measures were computed, and grasp success measures extracted by simulating the grasps in GraspIt! [18]. The correlation between these quality measures and success measures is going to be learned by a neural network. We also explore how different network architectures affect the overall result.

A. Measure for Success

We want to produce a set of grasp hypotheses where the outcome is known in order to supervise the training. Since to do this with a real robot would be both time-consuming and costly in order to get enough data to train on, a more time and cost-efficient simulation option was used. By simulating the grasp hypotheses found for different objects, and by measuring the success for these in the simulator a set of input / output

---

[3]For the Barrett hand: $g(e) = 0.953 * e + 128.8$, empirically found. Using this linear approximation causes little loss in precision compared to calculating the actual inverse kinematics for the hand.

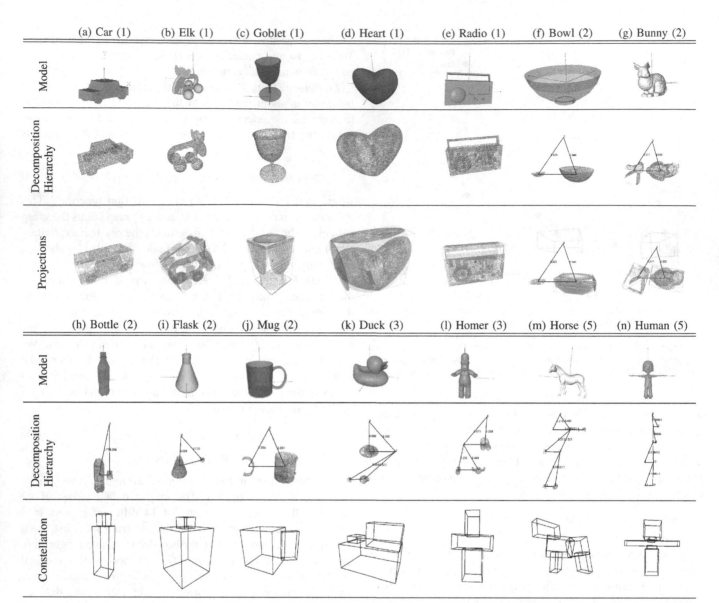

Fig. 6. The 14 models used in the experiment in order of complexity. The number of boxes resulting from the final decomposition hierarchies (2nd and 5th row) is assigned to each model in brackets. Since (a)-(e) are very compact and only the root box is included, the 3rd row visualizes examples of box face projections that will be used for the experiments. For the more complex models (h)-(n), the 6th row depicts the final box constellations. We refrained from showing constellations for (a)-(g), since they would only show 1-2 boxes, as also from showing projections for (h)-(n), since they would be hard to recognize.

pairs was found. The GraspIt! simulator [18] provides two different measures of success, introduced in [19].

The first measure, denoted ν here, measures the volume of the intersection of friction cones from the finger contacts. The second measure, called ε here, is a measure of the radius of the largest sphere that can fit in this space. For the purpose of learning it is more suitable to use only one measure since it is easier to learn a one-dimensional function than a two-dimensional. However, in order to utilize all of the information provided and since there is no firm consensus on which measure is best [20], we use a combination of the two:

$$s_i = (\nu_i/\nu_{max})^2 + (\varepsilon_i/\varepsilon_{max})^2 \; , \qquad (6)$$

where s_i is the success of grasp i, ν_{max} and ε_{max} are the maximum ν and ε values found for the current object and all hypotheses, respectively.

Other measures that could be used would be the division of grasps into two classes, namely successful and unsuccessful, or to train the system using only one of the above measures. Since potentially this could be a waste of useful grasp quality information, the above combination measure was chosen.

Given that the net is used to grasp an unknown object in the final application, the system will continue learning from newly observed hypotheses. However, since we have the possibility to initially gather vast amounts of simulation data to use for training, one additional example will not impact the prediction results noticeably. Combining this with the possibility to retrain an eager learner when the system is offline makes the advantages of a lazy learner, like kNN, diminish. Therefore, we used a feed-forward neural network to implement a supervised eager learning approach. The training algorithm used was the Levenberg-Marquardt backpropagation algorithm included in the Matlab Neural Networks Toolbox. It is a fast training algorithm with good generalization properties, which is needed for predicting unknown objects.

B. Leave-One-Object-Out Validation

For evaluation, we apply a leave-one-object-out validation. This method validates by picking out one of the 14 objects that is not the test object, while all grasp hypotheses that belong to this object will be used as part of the validation set. There are both advantages and disadvantages to this approach. Considering that the prediction error for an unknown validation object is at a minimum, it would be intuitive to assume that the prediction error for an unknown test object would also be minimal. However, these two objects can be very different, both in complexity and suitable grasps, more than the ones in the training set and the test set. Another drawback with this technique is that the size of the validation set is different for each object used for validation. An advantage to using this approach is that it seldom overfits and thus stops the training when generalization still performs well.

C. Network Architecture

We used a network architecture with 9 input nodes, from the 9 quality measures, and 1 output node corresponding to the success measure s. From here, we still must decide how many hidden layers and how many hidden nodes in each layer to use. To be able to decide what is a good architecture and what is not we need to measure the overall success of the network. This measure should incorporate the s-measure for a grasp, but being as independent of an object as possible, e.g. an easy object to grasp will give better s-measures, but should (with the same prediction success) give the same success of the net, S_{net}. We want to rank the grasp hypotheses and only use the ones highest ranked. This is reflected in the network success measure by using only the top-ranked 10% hypotheses and comparing them with the lowest-ranked 10%:

$$S_{net} = \frac{(s_{high} - s_{low})}{0.1n} = \frac{\left(\sum_{i=1}^{0.1n} r_i - \sum_{i=0.9n}^{n} r_i\right)}{0.1n}, \quad (7)$$

where r_i is the ranked list of hypotheses, the hypothesis with highest predicted success being at index 1, and n is the total number of hypotheses.

To test the effect of adding hidden layers, three different setups were used: the first had only one hidden layer with 10 hidden nodes, the second had two hidden layers each with 10 hidden nodes, and the third had three hidden layers each with 10 hidden nodes. There was no improvement of prediction success for more than one hidden layer. The time for training, however, increased dramatically.

Extensive testing was performed in order to choose the number of nodes in the one hidden layer. This testing was done with the leave-one-object-out validation described above. Tests were made with 1 to 30 hidden nodes. After studying the performance results depicted in Fig. 7, the optimal number of hidden nodes was chosen to be 8. Using this architecture, no training phase in our experiment required more than 100

epochs. As can be seen in Fig. 7, the time an exemplary training time is around 3 seconds.

D. Learning to Grasp Unknown Objects

In order to learn to grasp unknown objects, the leave-one-object-out validation method was applied again. Each time the network is trained one object is left out of the training data to be used for testing. This unknown object will be used to determine how well the algorithm has performed. In order to get a reliable result these tests were run 10 times. As such, one can conclude that the method for grasp synthesis, the quality measures and the learning approach used can indeed find and rank a set of grasps for an unknown object. Fig. 8 shows distribution of predicted success measures for each model. Some of the high-ranked grasps are presented in Fig. 9. These were encountered after separately performing a training on all other models in the set (except for one validation object). Note that the input for the overall approach is only a 3D point cloud representation that could also be delivered from real sensor input. The approach therefore does neither need training on every possible object model, nor does it rely on connected surface structure, like triangle meshes.

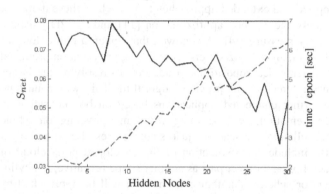

Fig. 7. Success (solid) and training time (dashed) for different number of hidden nodes with the leave-one-object-out validation, averaged over 10 runs. The maximum success value of $S_{net} = 0.078$ was detected at 8 nodes.

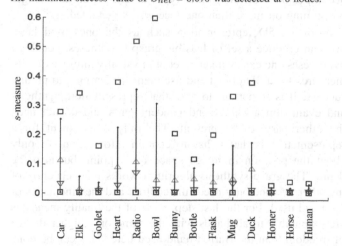

Fig. 8. Prediction results for the 14 models used (see Fig. 6). Upwards pointing triangles represent mean of the best 10% grasps, downwards the worst. Lines correspond to the possible span of predictions, with a perfect prediction of the best in the top and a perfect prediction of the worst in the bottom. The squares represent the best and the worst grasp for each object.

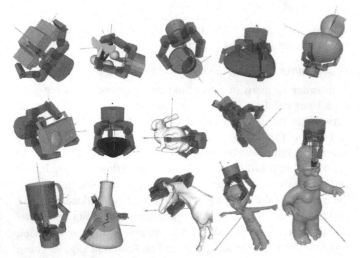

Fig. 9. Visualization of some high-ranked predicted grasps for all models.

V. Conclusions and Future Work

In this paper, we presented an approach for grasping 3D objects by using 2D grasping strategies and heuristics. We applied and extended approaches from each of these domains, namely the 3D box approximation [11] and the 2D grasping quality measures [4]. We showed that, given a point cloud of 3D data, an optimized version of box approximation produced repeatable decompositions in addition to resolving the issues encountered in our previous algorithm. This will contribute to further connected applications based on box constellation representations, e.g. learning from and grasping on whole constellations instead of just single boxes. Learning might also include a classification of the enveloped point cloud of each box and object part as another shape primitive, i.e. cylinders or spheres. Another classification will be approached by learning from the box constellation itself. Not only similarities between constellations could be used, e.g. all 'duck'-like box decompositions afford similar types of grasps, but also finger positioning on more than one face will be enabled.

From a 2.5D representation such as the ones used here, one can produce a set of feasible grasp hypotheses. For these hypotheses one can evaluate a set of physically intuitive quality measures for a 3D object and use them for learning to predict success. It is important to note that representation, synthesis and evaluation are three independent parts and do not need the other parts to be present. The only requirement for a representation is that it has to contain information not only about the position in image space for a point, but also the depth. The grasp synthesis algorithm works independently of the other two and only needs the contour and the kinematics of the hand used. For the last step, most of the quality measures are extendible to all hands with the same or a higher degree of freedom than the Barrett hand used here. This can be done either by the use of virtual fingers, or by an extension of the measures themselves to include sums and differences for more than three fingers. A continuation of the work could include an extension of the quality measures to better take into account

3D shape. With the use of more flexible hands the complete inverse kinematics could be used for finding reachable points in 3D space. A natural extension to the learning part is to include not only data from simulation, but to continue learning from real-world objects. By retraining the network with the increased data set, the evaluation would get more precise and be a useful learning system.

Acknowledgment

This work was supported by EU through PACO-PLUS, IST-FP6-IP-027657.

References

[1] B.-H. Kim, B.-J. Yi, S.-R. Oh, and I. H. Suh, "Non-Dimensionalized Performance Indices based Optimal Grasping for Multi-Fingered Hands," *Mechatronics*, vol. 14, pp. 255–280, 2004.
[2] A. Saxena, J. Driemeyer, and A. Y. Ng, "Robotic Grasping of Novel Objects using Vision," *International Journal of Robotics Research*, vol. 27, no. 2, pp. 157–173, 2008.
[3] G. M. Bone and E. Y. Du, "Multi-Metric Comparison of Optimal 2D Grasp Planning Algorithms," in *IEEE International Conference on Robotics & Automation*, 2001, pp. 3061–3066.
[4] A. Morales, P. J. Sanz, A. P. del Pobil, and A. Fagg, "Vision-based three-finger grasp synthesis constrained by hand geometry," *Robotics and Autonomous Systems*, vol. 54, pp. 496–512, 2006.
[5] D. Aarno et al., "Early Reactive Grasping with Second Order 3D Feature Relations," in *From Features to Actions*, 2007, pp. 319–325.
[6] C. Goldfeder, P. K. Allen, C. Lackner, and R. Pelossof, "Grasp Planning Via Decomposition Trees," in *IEEE International Conference on Robotics and Automation*, 2007, pp. 4679–4684.
[7] M. Ciocarlie, C. Goldfeder, and P. Allen, "Dexterous Grasping via Eigengrasps: A Low-Dimensional Approach to a High-Complexity Problem," in *Sensing and Adapting to the Real World*, 2007.
[8] A. T. Miller, S. Knoop, H. I. Christensen, and P. K. Allen, "Automatic Grasp Planning Using Shape Primitives," in *IEEE International Conference on Robotics and Automation*, 2003, pp. 1824–1829.
[9] R. Pelossof, A. Miller, P. Allen, and T. Jebara, "An SVM Learning Approach to Robotic Grasping," in *IEEE International Conference on Robotics and Automation*, 2004, pp. 3512–3518.
[10] A. Chella, H. Dindo, F. Matraxia, and R. Pirrone, "Real-Time Visual Grasp Synthesis Using Genetic Algorithms and Neural Networks," in *AI*IA 2007: Artificial Intelligence and Human-Oriented Computing*, 2007, pp. 567–578.
[11] K. Huebner, S. Ruthotto, and D. Kragic, "Minimum Volume Bounding Box Decomposition for Shape Approximation in Robot Grasping," in *IEEE Int. Conf. on Robotics and Automation*, 2008, pp. 1628–1633.
[12] K. Huebner and D. Kragic, "Selection of Robot Pre-Grasps using Box-Based Shape Approximation," in *IEEE International Conference on Intelligent Robots and Systems*, 2008, pp. 1765–1770.
[13] A. Morales, "Learning to Predict Grasp Reliability with a Multifinger Robot Hand by using Visual Features," Ph.D. dissertation, Department of Computer and Engineering Science, Universitat Jaume I, 2004.
[14] A. M. Andrew, "Another Efficient Algorithm for Convex Hulls in Two Dimensions," *Information Processing Letters*, vol. 9, pp. 216–219, 1979.
[15] D. Montana, "The Condition for Contact Grasp Stability," in *IEEE Int. Conference on Robotics and Automation*, 1991, pp. 412–417.
[16] A. Rosenfeld and E. Johnston, "Angle Detection on Digital Curves," in *IEEE Transactions on Computers*, vol. C-22, 1973, pp. 875–878.
[17] W. T. Townsend, "The BarrettHand Grasper – Programmably Flexible Part Handling and Assembly," *Industrial Robot: An Int. Journal*, vol. 27, no. 3, pp. 181–188, 2000.
[18] A. T. Miller and P. K. Allen, "Graspit! A Versatile Simulator for Robotic Grasping," *IEEE Robotics & Automation Magazine*, vol. 11, no. 4, pp. 110–122, 2004.
[19] C. Ferrari and J. Canny, "Planning Optimal Grasps," in *IEEE International Conference on Robotics and Automation*, 1992, pp. 2290–2295.
[20] C. Goldfeder, M. Ciocarlie, H. Dang, and P. K. Allen, "The Columbia Grasp Database," in *IEEE International Conference on Robotics and Automation*, 2009.

LQR-Trees: Feedback Motion Planning on Sparse Randomized Trees

Russ Tedrake
Computer Science and Artificial Intelligence Lab
Massachusetts Institute of Technology
Cambridge, MA 02139
Email: russt@mit.edu

Abstract— Recent advances in the direct computation of Lyapunov functions using convex optimization make it possible to efficiently evaluate regions of stability for smooth nonlinear systems. Here we present a feedback motion planning algorithm which uses these results to efficiently combine locally valid linear quadratic regulator (LQR) controllers into a nonlinear feedback policy which probabilistically covers the reachable area of a (bounded) state space with a region of stability, certifying that all initial conditions that are capable of reaching the goal will stabilize to the goal. We investigate the properties of this systematic nonlinear feedback control design algorithm on simple underactuated systems and discuss the potential for control of more complicated control problems like bipedal walking.

I. INTRODUCTION

Consider the problem of stabilizing a periodic (limit cycle) trajectory for a bipedal walking robot. Although many well-developed tools exist for local stabilization[25, 15], dynamic constraints due to actuator saturation and/or underactuation limit the validity of these solutions to a small neighborhood around the nominal trajectory. Dynamic programming approaches based on discretizing the state and action spaces require potentially very fine resolution to deal with the discontinuous dynamics of impact, and require many simplifications for application to even the simplest walking models[5].

This paper aims to build on recent advances from control theory and from randomized motion planning to design efficient and general algorithms for nonlinear feedback control synthesis in nonlinear underactuated systems like bipedal walking. Specifically, the controls community has recently developed a number of efficient algorithms for direct computation of Lyapunov functions for smooth nonlinear systems, using convex optimization [9, 17]. These tools can plug into motion planning algorithms to automatically compute planning "funnels" for even very complicated dynamical systems, and open a number of interesting possibilities for algorithm development. In particular, we present the LQR-Tree algorithm, which uses locally optimal linear feedback control policies to stabilize planned trajectories computed by local trajectory optimizers, and computational Lyapunov verification based on a sum-of-squares method to create the funnels.

The aim of this work is to generate a class of algorithms capable of computing verified feedback policies for under-actuated systems with dimensionality beyond what might be

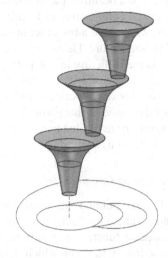

Fig. 1: Cartoon of motion planning with funnels in the spirit of [4].

accessible to grid-based algorithms like dynamic programming. The use of local trajectory optimizers and local feedback stabilization scales well to higher-dimensions, and reasoning about the feedback "funnels" allows the algorithm to cover a bounded, reachable subset of state space with a relatively sparse set of trajectories. In addition, the algorithms operate directly on the continuous state and action spaces, and thus are not subject to the pitfalls of discretization. By considering feedback during the planning process, the resulting plans are certifiably robust to disturbances and quite suitable for implementation on real robots. Although scaling is the driving motivation of this approach, this paper focuses on the coverage properties of the LQR-Tree algorithm by carefully studying a simple 2D example (the torque-limited simple pendulum), which reveals the essential properties of the algorithm on a problem where the control synthesis procedure can be easily visualized.

II. BACKGROUND

A. Feedback motion planning

For implementation on real robots, open-loop trajectories generated by a motion planning system are commonly stabilized by a feedback control system.[1] While this decoupled approach works for most problems, it is possible that a

---

[1]Note that an increasingly plausible alternative is real-time, dynamic replanning.

planned trajectory is not stabilizable, or very costly to stabilize compared to other, more desirable trajectories. Algorithms which explicitly consider the feedback stabilization during the planning process can avoid this pitfall, and as we will see, can potentially use a local understanding of the capabilities of the feedback system to guide and optimize the search in a continuous state space.

Mason popularized the metaphor of a funnel for a feedback policy which collapses a large set of initial conditions into a smaller set of final conditions[16]. Burridge, Rizzi, and Koditschek then painted a beautiful picture of feedback motion planning as a sequential composition of locally valid feedback policies, or funnels, which take a broad set of initial conditions to a goal region[4] (see Figure 1). At the time, the weakness of this approach was the difficulty in computing, or estimating by trial-and-error, the region of applicability - the mouth of the funnel, or preimage - for each local controller in a nonlinear system. Consequently, besides the particular solution in [4], these ideas have mostly been limited to reasoning about vector-fields on systems without dynamics[12].

B. Direct computation of Lyapunov functions

Burridge et al. also pointed out the strong connection between Lyapunov functions and these motion planning funnels[4]. A Lyapunov function is a differentiable positive-definite output function, $V(\mathbf{x})$, for which $\dot{V}(\mathbf{x}) < 0$ as the closed-loop dynamics of the system evolve. If these conditions are met over some ball in state space, B_r, containing the origin, then the origin is asymptotically stable. The ball, B_r, can then be interpreted as the preimage of the funnel. Lyapunov functions have played an incredibly important role in nonlinear control theory, but can be difficult to discover analytically for complicated systems.

The last few years has seen the emergence of a number of computational approaches to discovering Lyapunov functions for nonlinear systems, often based on convex optimization(e.g., [9, 17]). One of these techniques, which forms the basis of the results reported here, is based on the realization that one can check the uniform positive-definiteness of a polynomial expression (even with constant coefficients as free parameters) using a *sums of squares* (SOS) optimization program[17]. Sums of squares programs can be recast into semidefinite programs and solved using convex optimization solvers (such as interior point methods); the freely available SOSTOOLS library makes it quite accessible to perform these computations in MATLAB[18]. As we will see, the ability to check uniform positive (or negative) definiteness will offer the ability to verify candidate Lyapunov functions over a region of state space for smooth (nonlinear) polynomial systems.

These tools make it possible to automate the search for Lyapunov functions. Many researchers have used this capability to find stability proofs that didn't previously exist for nonlinear systems[17]. In this paper, we begin to explore the implications for planning of being able to efficiently compute planning funnels.

C. Other related work

The ideas presented here are very much inspired by the randomized motion planning literature, especially rapidly-exploring randomized trees (RRTs)[11] and probabilistic roadmaps (PRMs)[10]. This work was also inspired by [14] and [19] who point out a number of computational advantages to using sample-paths as a fundamental representation for learning policies which cover the relevant portions of state space.

In other related work, [1] used local trajectory optimizers and LQR stabilizers with randomized starting points to try to cover the space, with the hope of verifying global optimality (in the infinite resolution case) by having consistent locally quadratic estimates of the value function on neighboring trajectories. The conditions for adding nodes in that work were based on the magnitude of the value function (not the region of guaranteed stability). In the work described here, we sacrifice direct attempts at obtaining optimal feedback policies in favor of computing good-enough policies which probabilistically cover the reachable state space with the basin of attraction. As a result, we have stronger guarantees of getting to the goal and considerably sparser collections of sample paths.

III. THE LQR-TREE ALGORITHM

Like many other randomized planning algorithms, the proposed algorithm creates a tree of feasible trajectories by sampling randomly over some bounded region of state space, and growing the existing tree towards this random sample point. Here, when each new trajectory "branch" is added to the tree, we do some additional work by creating a trajectory stabilizing controller and by immediately estimating the basin of attraction of this controller using semi-definite programming. Because both the feedback design and the stability analysis work backwards in time, we perform these computations on only a backwards tree, starting from the goal. The result is that the backwards tree becomes a large web of local controllers which grab initial conditions and pull them towards the goal (with formal certificates of stability for the nonlinear, continuous state and action system). We terminate the algorithm when we determine (probabilistically) that all initial conditions which are capable of reaching the goal are contained in the basin of attraction of the tree.

Although many trajectory stabilizing feedback controller designs are possible (and potentially compatible with this approach), we have selected to use a time-varying linear quadratic regulator (LQR) design. LQR, iterative LQR (iLQR)[21, 23], and the closely related differential dynamic programming (DDP)[8] are common tools for roboticists, and have demonstrated success in a number of applications. LQR control synthesis has the additional benefit that it returns the quadratic cost-to-go function for the linear system, which is also a valid Lyapunov function for the nonlinear system over some region in the vicinity of the trajectory. We design a conservative approximation of this region using sums-of-squares optimization. Finally, we use the computed basin of attraction to influence the way that our tree grows, with the goal of filling the reachable state space with the basin of attraction of a sparse set of trajectories.

The details of each of these steps are described in the remainder of this section.

A. Essential components

1) Time-varying LQR feedback stabilization: Let us first consider the subproblem of designing a time-varying LQR feedback based on a time-varying linearization along a nominal trajectory. Consider a controllable, smoothly differentiable, nonlinear system:

$$\dot{\mathbf{x}} = \mathbf{f}(\mathbf{x}, \mathbf{u}), \qquad (1)$$

with a stabilizable goal state, $\mathbf{x}_G$. Define a nominal trajectory (a solution of equation 1) which reaches the goal in a finite time: $\mathbf{x}_0(t), \mathbf{u}_0(t)$, with $\forall t \geq t_G, \mathbf{x}_0(t) = \mathbf{x}_G$ and $\mathbf{u}_0(t) = \mathbf{u}_G$. Define

$$\bar{\mathbf{x}}(t) = \mathbf{x}(t) - \mathbf{x}_0(t), \quad \bar{\mathbf{u}}(t) = \mathbf{u}(t) - \mathbf{u}_0(t).$$

Now linearize the system around the trajectory, so that we have

$$\dot{\bar{\mathbf{x}}}(t) \approx \mathbf{A}(t)\bar{\mathbf{x}}(t) + \mathbf{B}(t)\bar{\mathbf{u}}(t).$$

Define a quadratic regulator (tracking) cost function as

$$J(\mathbf{x}', t') = \int_{t'}^{\infty} \left[\bar{\mathbf{x}}^T(t)\mathbf{Q}\bar{\mathbf{x}}(t) + \bar{\mathbf{u}}^T(t)\mathbf{R}\bar{\mathbf{u}}(t) \right] dt,$$
$$\mathbf{Q} = \mathbf{Q}^T \geq 0, \mathbf{R} = \mathbf{R}^T > 0, \mathbf{x}(t) = \mathbf{x}'.$$

In general, $\mathbf{Q}$ and $\mathbf{R}$ could easily be made a function of time as well. With time-varying dynamics, the resulting cost-to-go is time-varying. It can be shown that the optimal cost-to-go, J^*, is given by

$$J^*(\bar{\mathbf{x}}, t) = \bar{\mathbf{x}}^T \mathbf{S}(t)\bar{\mathbf{x}}, \quad \mathbf{S}(t) = \mathbf{S}^T(t) > \mathbf{0}.$$

where $\mathbf{S}(t)$ is the solution to

$$-\dot{\mathbf{S}} = \mathbf{Q} - \mathbf{S}\mathbf{B}\mathbf{R}^{-1}\mathbf{B}^T\mathbf{S} + \mathbf{S}\mathbf{A} + \mathbf{A}^T\mathbf{S}, \qquad (2)$$

and the boundary condition $\mathbf{S}(t_G)$ is the positive-definite solution to the equation:

$$0 = \mathbf{Q} - \mathbf{S}\mathbf{B}\mathbf{R}^{-1}\mathbf{B}^T\mathbf{S} + \mathbf{S}\mathbf{A} + \mathbf{A}^T\mathbf{S},$$

(given by the MATLAB `lqr` function). The optimal feedback policy is given by

$$\bar{\mathbf{u}}^*(t) = -\mathbf{R}^{-1}\mathbf{B}^T(t)\mathbf{S}(t)\bar{\mathbf{x}}(t) = -\mathbf{K}(t)\bar{\mathbf{x}}(t).$$

2) LTI verification: We first estimate the basin of attraction of the linear time-invariant (LTI) feedback controller, $\mathbf{K}(t_G)$, executed for $t \geq t_G$. We verify that this controller stabilizes the fixed point given by $(\mathbf{x}_G, \mathbf{u}_G)$ by demonstrating that a function, $V(\mathbf{x})$, is a valid Lyapunov function for the nonlinear system over a bounded region of state-space, $\mathcal{B}$, defined by

$$\mathcal{B}(\rho) : \{\mathbf{x} | 0 \leq V(\mathbf{x}) \leq \rho\}$$

where ρ is a positive scalar. The origin is asymptotically stable if

- $V(\mathbf{x})$ is positive definite in $\mathcal{B}(\rho)$,

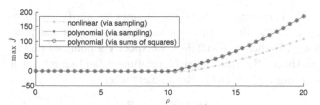

Fig. 2: Polynomial verification of LTI feedback on the damped simple pendulum ($m = 1kg, l = .5m, b = .1m^2kg/s, g = 9.8m/s^2, \mathbf{Q} = diag([10, 1]), \mathbf{R} = 15, N_f = 3, N_m = 2$).

- $\dot{V}(\mathbf{x}) < 0$ in $\mathcal{B}(\rho)$.

Furthermore, all initial conditions in $\mathcal{B}(\rho)$ will converge to 0[22].

Here we use $V(\mathbf{x}) = J^*(\mathbf{x})$; the linear optimal cost-to-go function is (locally) a Lyapunov function for the nonlinear system. The first condition is satisfied by the LQR design. For the second condition, first observe that

$$\dot{J}(\bar{\mathbf{x}}) = 2\bar{\mathbf{x}}^T \mathbf{S} \mathbf{f}(\mathbf{x}_G + \bar{\mathbf{x}}, \mathbf{u}_G - \mathbf{K}\bar{\mathbf{x}}).$$

In the case where $\mathbf{f}$ is polynomial in $\mathbf{x}$ and $\mathbf{u}$, we can verify this condition exactly by specifying a sums-of-squares (SOS) feasibility program[17]:

$$\dot{J}^*(\bar{\mathbf{x}}) + h(\bar{\mathbf{x}})(\rho - J^*(\bar{\mathbf{x}})) < 0$$
$$h(\bar{\mathbf{x}}) = \mathbf{m}^T(\bar{\mathbf{x}})\mathbf{H}\mathbf{m}(\bar{\mathbf{x}}), \quad \mathbf{H} > 0,$$

where $\mathbf{m}$ is a vector of monomials of order N_m. Note that some care must be taken because $J^*(0) = 0$; we use a slack variable approach and search for solutions were $\dot{J}^*$ is uniformly less than some numerical tolerance above zero.

In many cases (including the manipulator dynamics considered in this paper), even if $\mathbf{f}$ is not polynomial it is still possible to perform the verification algebraically through a change of coordinates. However, for simplicity and generality, in the algorithm presented here we simply approximate the stability condition using a Taylor expansion of $\mathbf{f}$, with order N_f greater than one. We use $\hat{\mathbf{f}}$ to denote the Taylor expansion of $\mathbf{f}$ and $\hat{J}^*$ for the resulting approximation of $\dot{J}^*$.

Finally, we estimate the basin of attraction by formulating a convex optimization to find find the largest region $\mathcal{B}(\rho)$ over which the second condition is also satisfied:

$$\max \rho \quad \text{subject to}$$
$$\hat{\dot{J}}^*(\bar{\mathbf{x}}) + \mathbf{m}^T(\bar{\mathbf{x}})\mathbf{H}\mathbf{m}(\bar{\mathbf{x}})(\rho - J^*(\bar{\mathbf{x}})) < 0$$
$$\rho > 0, \quad \mathbf{H} > 0.$$

The estimated basin of attraction is a conservative approximation of the true basin of attraction in every way, except that the nonlinear dynamics are approximated by the polynomial expansion. This limits our analysis to smooth nonlinear systems, and restricts our strict claims of verification in this paper to truly polynomial systems. In practice, the algorithm acquires conservative, but impressively tight approximations of the basin of attraction for the system in detailed tests with the pendulum, as illustrated in Figure 2, and the cart-pole.

3) *LTV verification:* Next we attempt to verify the performance of the linear time-varying feedback over the time $t \in [0, t_G]$. Rather than stability, we specify a bounded region of state space, $\mathcal{B}_f$, (the outlet of the funnel) and search for a time-varying region, $\mathcal{B}(t)$, (the funnel) where

$$\mathcal{B}(t) : \{\mathbf{x} | \mathbf{F}(\mathbf{x}, t) \in \mathcal{B}_f\}, \tag{3}$$

and $\mathbf{F}(\mathbf{x}, t)$ is defined as the simulation function which integrates the closed-loop dynamics from t to t_f. When $\mathcal{B}_f$ is chosen as the LTI basin of attraction from the previous section, this funnel becomes the basin of attraction of the infinite-horizon trajectory. As before, we will use the cost-to-go as a (now time-varying) storage function, $V(\mathbf{x}, t)$, and search for the largest positive time-varying level-set, $\rho(t)$, over the interval $[t_0, t_f]$, which defines a region,

$$\mathcal{B}(\rho(\cdot), t) : \{\mathbf{x} | 0 \le V(\mathbf{x}, t) \le \rho(t)\},$$

satisfying condition 3. Similarly, we use

$$\mathcal{B}_f : \{\mathbf{x} | 0 \le V(\mathbf{x}, t_f) \le \rho_f\},$$

where ρ_f is a positive constant representing the constraint on final values (specified by the task). Note that this naturally implies that $\rho(t_f) \le \rho_f$.

A sufficient, but conservative, verification of our bounded final value condition can be accomplished by verifying that $B(\rho(\cdot), t)$ is a closed set over $t \in [t_0, t_f]$. The set is closed if $\forall t \in [t_0, t_f]$ we have

- $V(\mathbf{x}, t) \ge 0$ in $\mathcal{B}(\rho(\cdot), t)$,
- $\dot{V}(\mathbf{x}, t) \le \dot{\rho}(t)$ in $\mathcal{B}^\sharp(\rho(\cdot), t)$,

where $\mathcal{B}^\sharp$ is the boundary of the region $\mathcal{B}$,

$$\mathcal{B}^\sharp(\rho(\cdot), t) : \{\mathbf{x} | V(\mathbf{x}, t) = \rho(t)\}.$$

Again, we choose here to use $V(\mathbf{x}, t) = J^*(\mathbf{x}, t)$; the first condition is again satisfied by the LQR derivation which ensures $\mathbf{S}(t)$ is uniformly positive definite. Now we have

$$\dot{J}^*(\bar{\mathbf{x}}, t) = 2\bar{\mathbf{x}}^T \mathbf{S}(t) \mathbf{f}\left(\mathbf{x}_0(t) + \bar{\mathbf{x}}, \mathbf{u}_0(t) - \mathbf{K}(t)\bar{\mathbf{x}}\right) + \bar{\mathbf{x}}^T \dot{\mathbf{S}}(t)\bar{\mathbf{x}}. \tag{4}$$

Here, even if $\mathbf{f}$ is polynomial in $\mathbf{x}$ and $\mathbf{u}$ and the input tape $\mathbf{u}_0(t)$ was polynomial, our analysis must make use of $\mathbf{x}_0(t)$, $\mathbf{S}(t)$, and $\mathbf{K}(t)$ which are the result of numerical integration (e.g., with `ode45` in Matlab). We will approximate this temporal dependence with (elementwise) piecewise polynomials using splines of order N_t, where N_t is often chosen to be 3 (cubic splines), with the knot points at the timesteps output by the variable step integration, which we denote $t_0, t_1, ..., t_N$, with $t_N = t_f$, e.g.:

$$\forall t \in [t_k, t_k + 1], \quad S_{ij}(t) \approx \sum_{m=0}^{N_t} \alpha_{ijm}(t - t_k)^m = \hat{S}_{ij}(t),$$

$$\hat{J}^*(\bar{\mathbf{x}}, t) = \bar{\mathbf{x}}^T \hat{\mathbf{S}} \bar{\mathbf{x}}.$$

Once again, we substitute a Taylor expansion of the dynamics to obtain the estimate $\hat{J}^*$.

Now we approximately verify the second condition by formulating a series of sums-of-squares feasibility programs

$$\hat{J}^*(\bar{\mathbf{x}}, t) - \dot{\rho}(t) + h_1(\bar{\mathbf{x}}, t)\left(\rho(t) - \hat{J}^*(\bar{\mathbf{x}}, t)\right)$$
$$+ h_2(\bar{\mathbf{x}}, t)(t - t_k) + h_3(\bar{\mathbf{x}}, t)(t_{k+1} - t) \le 0, \tag{5}$$

$$h_1(\bar{\mathbf{x}}, t) = \mathbf{h}_1^T \mathbf{m}(\bar{\mathbf{x}}, t), \tag{6}$$

$$h_2(\bar{\mathbf{x}}, t) = \mathbf{m}^T(\bar{\mathbf{x}}, t)\mathbf{H}_2\mathbf{m}(\bar{\mathbf{x}}, t), \quad \mathbf{H}_2 = \mathbf{H}_2^T > 0, \tag{7}$$

$$h_3(\bar{\mathbf{x}}, t) = \mathbf{m}^T(\bar{\mathbf{x}}, t)\mathbf{H}_3\mathbf{m}(\bar{\mathbf{x}}, t), \quad \mathbf{H}_3 = \mathbf{H}_3^T > 0, \tag{8}$$

for $k = N - 1, ..., 1$.

We attempt to find the largest $\rho(t)$ satisfying the verification test above by defining a piecewise-polynomial of order N_ρ given by

$$\rho_k(t) = \sum_{m=0}^{N_\rho} \beta_{km}(t - t_k)^m,$$

$$\rho(t) = \begin{cases} \rho_k(t), & \forall t \in [t_k, t_{k+1}) \\ \rho_f, & t = t_f, \end{cases}$$

and we formulate the optimization:

$$\max_{\beta} \int_{t_k}^{t_{k+1}} \rho_k(t) dt, \quad \text{subject to}$$

$$\rho_k(t_{k+1}) \le \rho_{k+1}(t_{k+1}), \quad \text{equations (5) - (8),}$$

for all $k = N - 1, ..., 1$.

4) *Growing the tree:* Another essential component of the LQR-tree algorithm is the method by which the backwards tree is extended. Following the RRT approach, we select a sample at random from some distribution over the state space, and attempt to grow the tree towards that sample. Unfortunately, RRTs typically do not grow very efficiently in differentially constrained (e.g., underactuated) systems, because simple distance metrics like the Euclidean distance are inefficient in determining which node in the tree to extend from. Further embracing LQR as a tool for motion planning, in this section we develop an affine quadratic regulator around the sample point, then use the resulting cost-to-go function to determine which node to extend from, and use the open-loop optimal policy to extend the tree.

Choose a random sample (not necessarily a fixed point) in state space, $\mathbf{x}_s$ and a default $\mathbf{u}_0$, and use $\bar{\mathbf{x}} = \mathbf{x} - \mathbf{x}_s$, $\bar{\mathbf{u}} = \mathbf{u} - \mathbf{u}_0$.

$$\dot{\bar{\mathbf{x}}} = \frac{d}{dt}(\mathbf{x}(t) - \mathbf{x}_s) = \dot{\mathbf{x}}(t)$$
$$\approx \mathbf{f}(\mathbf{x}_s, \mathbf{u}_0) + \frac{\partial \mathbf{f}}{\partial \mathbf{x}}(\mathbf{x}(t) - \mathbf{x}_s) + \frac{\partial \mathbf{f}}{\partial \mathbf{u}}(\mathbf{u} - \mathbf{u}_0)$$
$$= \mathbf{A}\bar{\mathbf{x}} + \mathbf{B}\bar{\mathbf{u}} + \mathbf{c}.$$

Now define an affine quadratic regulator problem with a hard constraint on the final state, but with the final time, t_f, left as a free variable[13]:

$$J(\bar{\mathbf{x}}_0, t_0, t_f) = \int_{t_0}^{t_f} \left[1 + \frac{1}{2}\bar{\mathbf{u}}^T(t)\mathbf{R}\bar{\mathbf{u}}(t)\right] dt,$$

$$\text{s.t.} \quad \bar{\mathbf{x}}(t_f) = 0, \quad \bar{\mathbf{x}}(t_0) = \bar{\mathbf{x}}_0, \quad \dot{\bar{\mathbf{x}}} = \mathbf{A}\bar{\mathbf{x}} + \mathbf{B}\bar{\mathbf{u}} + \mathbf{c}.$$

Without loss of generality (since the dynamics are autonomous), we will use $J(\bar{\mathbf{x}}_0, t_f - t_0)$ as a shorthand for $J(\bar{\mathbf{x}}_0, t_0, t_f)$. It can be shown that the optimal (open-loop) control is

$$\bar{\mathbf{u}}^*(t) = -\mathbf{R}^{-1}\mathbf{B}^T e^{\mathbf{A}^T(t_f - t)}\mathbf{P}^{-1}(t_f)\mathbf{d}(\bar{\mathbf{x}}(t_0), t_f),$$

where

$$\dot{\mathbf{P}}(t) = \mathbf{A}\mathbf{P}(t) + \mathbf{P}(t)\mathbf{A}^T + \mathbf{B}\mathbf{R}^{-1}\mathbf{B}^T, \quad \mathbf{P}(t_0) = \mathbf{0}$$

$$\mathbf{d}(\bar{\mathbf{x}}, t) = \mathbf{r}(t) + e^{\mathbf{A}t}\bar{\mathbf{x}}, \quad \dot{\mathbf{r}}(t) = \mathbf{A}\mathbf{r}(t) + \mathbf{c}, \quad \mathbf{r}(\bar{\mathbf{x}}, t_0) = \mathbf{0}$$

and the resulting cost-to-go is

$$J^*(\bar{\mathbf{x}}, t_f) = t_f + \frac{1}{2}\mathbf{d}^T(\bar{\mathbf{x}}, t_f)\mathbf{P}^{-1}(t_f)\mathbf{d}(\bar{\mathbf{x}}, t_f).$$

Thanks to the structure of this equation, it is surprisingly efficient to compute the cost-to-go from many initial conditions (here the existing vertices in the tree) simultaneously. For each $\bar{\mathbf{x}}$ the horizon time, $t_f^* = \operatorname{argmin}_{t_f} J^*(\bar{\mathbf{x}}, t_f)$, is found by selecting the minimum after integrating $\mathbf{P}(t)$ and $\mathbf{r}(t)$ over a fixed horizon. This cost-to-go function provides a relatively efficient *dynamic* distance metric[2] for the RRT expansion which performs much better than Euclidean metrics for underactuated systems[6].

Once the "closest" node in the existing tree is identified, by this LQR distance metric, the tree is extended by applying a series of actions backwards in time from the closest node. The initial guess for this series of actions is given by $\bar{\mathbf{u}}^*(t)$ from the LQR distance metric, but this estimate (which is only accurate in the neighborhood of the sample point) can be further refined by a fast, local, nonlinear trajectory optimization routine. In the current results, we use a direct collocation[24, 2] implementation using the formulation from equation III-A.4, but with the nonlinear dynamics. If the direct collocation method cannot satisfy the final value constraint, then the point is considered (temporarily) unreachable, and is discarded. Interestingly, using the LQR open-loop control to initialize the nonlinear optimization appears to help overcome many of the local minima in the nonlinear optimization process.

5) A sampling heuristic: Finally, we take advantage of the Lyapunov verification by changing the sampling distribution. Adding branches of the tree that will be contained by the existing basin of attraction has little value. The sampling heuristic used here is implemented by sampling uniformly over the desired subset of state space, then rejecting any sample which are already in the basin of attraction of any of the tree branches. This "collision checking" is very inexpensive; it is far more expensive to add a useless node into the tree. Other sampling distributions are possible, too. One interesting alternative is sampling from states that are just at the edges of the basin of attraction, e.g., $\forall_i J^*(\mathbf{x} - \mathbf{x}_0^i, t) > \rho_i(t), \exists_j J^*(\mathbf{x} - \mathbf{x}_0^i, t) \leq 1.5\rho_j(t)$.

[2]Note that it is not technically a distance metric, since it is not symmetric, but the RRT does not require symmetry.

B. The algorithm

The algorithm proceeds by producing a tree, T, with nodes containing the tuples, $\{\mathbf{x}, \mathbf{u}, \mathbf{S}, \mathbf{K}, \rho_c, i\}$, where $J^*(\bar{\mathbf{x}}, t) = \bar{\mathbf{x}}^T\mathbf{S}\bar{\mathbf{x}}$ is the local quadratic approximation of the value function, $\bar{\mathbf{u}}^* = -\mathbf{K}\bar{\mathbf{x}}$ is the feedback controller, $J^*(\bar{\mathbf{x}}, t) \leq \rho(t)$ is the funnel, $\rho(t)$ is described by the vector of polynomial coefficients ρ_c, and i is a pointer to the parent node.

Algorithm 1 LQR-Tree ($\mathbf{x}_G, \mathbf{u}_G, \mathbf{Q}, \mathbf{R}$)

1: $[\mathbf{A}, \mathbf{B}] \Leftarrow$ linearization of $\mathbf{f}(\mathbf{x}, \mathbf{u})$ around $\mathbf{x}_G, \mathbf{u}_G$
2: $[\mathbf{K}, \mathbf{S}] \Leftarrow$ LQR($\mathbf{A}, \mathbf{B}, \mathbf{Q}, \mathbf{R}$)
3: $\rho_c \Leftarrow$ level-set computed as described in section III-A.2
4: T.init($\{\mathbf{x}_g, \mathbf{u}_g, \mathbf{S}, \mathbf{K}, \rho_c,$ NULL$\}$)
5: **for** $k = 1$ to K **do**
6:     $\mathbf{x}_{rand} \Leftarrow$ random sample as described in section III-A.5; if no samples are found, then FINISH
7:     $\mathbf{x}_{near}$ from cost-to-go distance metric described in section III-A.4
8:     $\mathbf{u}_{tape}$ from extend operation described in section III-A.4
9:     **for each** $\mathbf{u}$ in $\mathbf{u}_{tape}$ **do**
10:         $\mathbf{x} \Leftarrow$ Integrate backwards from $\mathbf{x}_{near}$ with action $\mathbf{u}$
11:         $[\mathbf{K}, \mathbf{S}]$ from LQR derivation in section III-A.1
12:         $\rho_c \Leftarrow$ level-set computed as in section III-A.3
13:         $i \Leftarrow$ pointer to node containing $\mathbf{x}_{near}$
14:         T.add-node($\mathbf{x}, \mathbf{u}, \mathbf{S}, \mathbf{K}, \rho_c, i$)
15:         $\mathbf{x}_{near} \Leftarrow \mathbf{x}$
16:     **end for**
17: **end for**

Execution of the LQR-tree policy is accomplished by selecting any node in the tree with a basin of attraction which contains the initial conditions, $\mathbf{x}(0)$, and following the time-varying feedback policy along that branch all of the way to the goal.

IV. SIMULATIONS

Simulation experiments on a two-dimensional toy problem have proven very useful for understanding the dynamics of the algorithm. Figure 3 tells the story fairly succinctly. The algorithm was tested on a simple pendulum, $I\ddot{\theta} + b\dot{\theta} + mgl\sin\theta = \tau$, with $m = 1, l = .5, b = .1, I = ml^2, g = 9.8$. Here $\mathbf{x} = [\theta, \dot{\theta}]^T$ and $\mathbf{u} = \tau$. The parameters of the LQR-tree algorithm were $\mathbf{x}_G = [\pi, 0]^T$, $\mathbf{u}_G = 0$, $\mathbf{Q} = diag([10, 1])$, $\mathbf{R} = 15$, $N_f = 3, N_m = 2, N_x = 3, N_S = 3$.

Figure 3(a) shows the basin of attraction (blue oval) after computing the linear time-invariant (LTI) LQR solution around the unstable equilibrium. Figure 3(b) shows the entire trajectory to the first random sample point (red dot), and the funnels that have been computed so far for the second-half of the trajectory. Note that the state-space of the pendulum lives on a cylinder, and that the trajectory (and basin of attraction) wraps around from the left to the right. Plots (c-d) show the basin of attraction as it grows to fill the state space. The final tree in Figure 3(d) also reveals three instances where

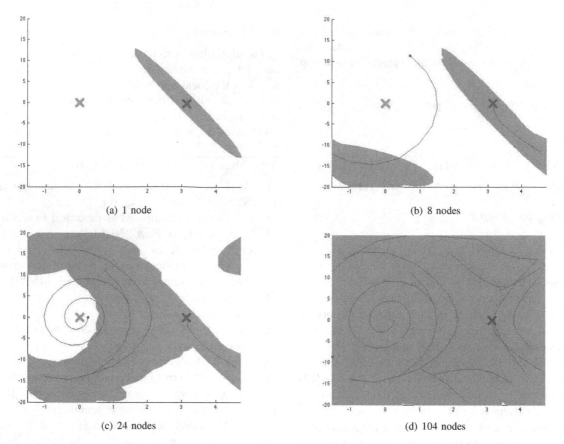

(a) 1 node

(b) 8 nodes

(c) 24 nodes

(d) 104 nodes

Fig. 3: An LQR-tree for the simple pendulum. The x-axis is $\theta \in [-\pi/2, 3\pi/2]$ (note that the state wraps around this axis), and the y-axis is $\dot{\theta} \in [-20, 20]$. The X on the left represents the stable fixed point; the X on the right represents the unstable (upright) fixed point. The ovals represent the "funnels," sampled at every node.

the trajectories on the tree cross - this is a result of having an imperfect distance metric.

Note that state $\mathbf{x} = [0, 0]^T$, corresponding to the stable fixed-point of the unactuated pendulum, is covered by the basin of attraction after 32 nodes have been added. The algorithm was not biased in any way towards this state, but this bias can be added easily. The entire space is probabilistically covered (1000 random points chosen sequentially were all in the basin of attraction) after the tree contained just 104 nodes. On average, the algorithm terminates after 146 nodes for the simple pendulum with these parameters. For contrast, [3] shows a well-tuned single-directional RRT for the simple pendulum which has 5600 nodes. However the cost of adding each node is considerably greater here than in the traditional RRT, dominated by the line search used to maximize the estimated region of stability. The entire algorithm runs in about two minutes on a laptop, without any attempt to optimize the code.

V. DISCUSSION

A. Properties of the algorithm

Recall that for nonlinear systems described by a polynomial of degree $\leq N_f$, the verification procedures used here are conservative; the true basin of attraction completely contains the estimated stability region. In practice, this is often (but not provably) the case for more general smooth nonlinear systems.

Proposition 1: For nonlinear systems described by a polynomial of degree $\leq N_f$, the LQR-tree algorithm probabilistically covers the sampled portion of the reachable state space with a stabilizing controller and a Lyapunov function, thereby guaranteeing that all initial conditions which are capable of reaching the goal will stabilize to the goal.

Proving proposition 1 carefully requires a proof that the local trajectory optimizer is always capable of solving a trajectory to a reachable point in the state space that is in an ϵ-region outside the existing basin of attraction. This is likely the case, seeing as the nonlinear optimizer is seeded by a linear optimal control result which will be accurate over some region of similar size to the basin of attraction ellipse. However, the full proof is left for future work.

Perhaps even more exciting is the fact that, in the model explored, this coverage appears to happen rapidly and allow for fast termination of the algorithm. The pendulum is a surprisingly rich test system - for example, as key parameters such as $\mathbf{R}$ or b change, the size of the funnels can change dramatically, resulting in quite different feedback policy coverings of the state space, and always resulting in rapid coverage.

It is also worth noting that the trajectories out of a more standard RRT are typically smoothed. Trajectories of the closed-loop system which result from the LQR algorithm are (qualitatively) quite smooth, despite coming from a randomized algorithm. The LQR stabilizing controller effectively smoothes the trajectory throughout state space.

B. Straight-forward variations in the algorithm

- **Compatible with optimal trajectories**. The LQR-tree algorithm provides a relatively efficient way to fill the reachable state space with funnels, but does not stake any claim on the optimality of the resulting trajectories. If tracking particular trajectories, or optimal trajectories, is important for a given problem, then it is quite natural to seed the LQR-tree with one or more locally optimal trajectories (e.g., using [1]), then use the random exploration to fill in any missing regions.
- **Early termination**. For higher dimensional problems, covering the reachable state space may be unnecessary or impractical. Based on the RRTs, the LQR-trees can easily be steered towards a region of state space (e.g., by sampling from that region with slightly higher probability) containing important initial conditions. Termination could then occur when some important subspace is covered by the tree.
- **Bidirectional trees**. Although LQR-trees only grow backwards from the goal, a partial covering tree (from an early termination) could also serve as a powerful tool for real-time planning. Given a new initial condition, a forward RRT simply has to grow until it intersects with the *volume* defined by the basin of attraction of the backwards tree.
- **Finite-horizon trajectories**. The LQR stabilization derived in section III-A.1 was based on infinite horizon trajectories. This point was necessary in order to use the language of basins of attraction and asymptotic stabilization. Finite-horizon problems can use all of the same tools (though perhaps not the same language), but must define success as being inside some finite volume around the goal state at t_G. Funnels connecting to this volume are then computed using the same Riccati backup.

C. Controlling walking robots

A feedback motion planning algorithm like the LQR-tree algorithm could be a very natural control solution for walking robots, or other periodic control systems. In this case, rather than the goal of the tree being specified as a point, the goal would be a periodic (limit cycle) trajectory. This could be implemented in the tree as a set of goal states, which happen to be connected, and the basin of attraction of this goal would emerge from the periodic steady-state solution of the Riccati equation and verification process on the limit cycle. Limit cycles for walking systems in particular are often described as a hybrid dynamics punctuated by discrete impacts. These discrete jump events must be handled with care in the feedback design and verification, but are not fundamentally incompatible with the approach[20].

Figure 4 cartoons the vision of how the algorithm would play out for the well-known compass gait biped[7]. On the left is a plot of the (passively stable) limit cycle generated by the compass gait model walking down a small incline. This trajectory can be stabilized using a (periodic) time-varying linearization and LQR feedback, and the resulting basin of attraction might look something like the shaded region in Figure 4(a). The goal of the LQR-tree algorithm would then be to fill the remaining portion of state space with transient "maneuvers" to return the system to the nominal limit cycle. A potential solution after a few iterations of the algorithm is cartooned in Figure 4(b). This work would naturally build on previous work on planning in hybrid systems (e.g.,[3]).

D. Multi-query algorithms

Another very interesting question is the question of reusing the previous computational work when the goal state is changed. In the pendulum example, consider having a new goal state, $\mathbf{x}_G = [\pi + 0.1, 0]^T$ - this would of course require a non-zero torque to stabilize. To what extent could the tree generated for stabilizing $\mathbf{x}_G = [\pi, 0]^T$ be used to stabilize this new fixed point? If one can find a trajectory to connect up the new goal state near the root of the tree, then the geometry of the tree can be preserved, but naively, one would think that all of the stabilizing controllers and the verification would have to be re-calculated. Interestingly, there is also a middle-road, in which the existing feedback policy is kept for the original tree, and the estimated funnels are not recomputed, but simply scaled down to make sure that the funnels from the old tree transition completely into the funnel for the new tree. This could be accomplished very efficiently, by just propagating a new ρ_{max} through the tree, but might come at the cost of losing coverage. One reason why this multi-query question is so exciting is that the problem of controlling a robot to walk on rough terrain could be nicely formulated as a multi-query stabilization of the limit cycle dynamics from Figure 4.

E. A software distribution

A MATLAB toolbox implementing the LQR-Tree algorithm is available at http://groups.csail.mit.edu/locomotion/software.html.

VI. SUMMARY AND CONCLUSIONS

Recent advances in direct computation of Lyapunov functions have enabled a new class of feedback motion planning algorithms for complicated dynamical systems. This paper presented the LQR-Tree algorithm which uses Lyapunov computations to evaluate the basins of attraction of randomized trees stabilized with LQR feedback. Careful investigations on a torque-limited simple pendulum revealed that, by modifying the sampling distribution to only accept samples outside of the computed basin of attraction of the existing tree, the result was a very sparse tree which covered the state space with a basin of attraction.

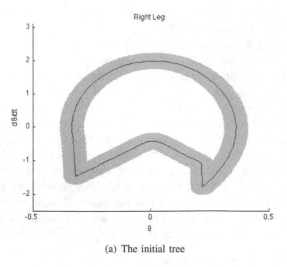

(a) The initial tree

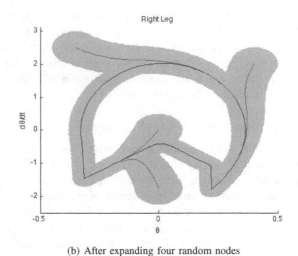

(b) After expanding four random nodes

Fig. 4: Sketch of LQR-trees on the compass gait biped.

Further investigation of this algorithm will likely result in a covering motion planning strategy for underactuated systems with dimensionality greater than what is accessible by discretization algorithms like dynamic programming, and early termination strategies which provide targeted coverage of state space in much higher dimensional systems. The resulting policies will have certificates guaranteeing their performance on the system model.

ACKNOWLEDGMENT

The author gratefully acknowledges Alexandre Megretski for introducing me to sums of squares optimization and for many helpful discussions.

REFERENCES

[1] Christopher G. Atkeson and Benjamin Stephens. Random sampling of states in dynamic programming. In *Advances in Neural Information Processing Systems*, 2008.

[2] John T. Betts. *Practical Methods for Optimal Control Using Nonlinear Programming*. SIAM Advances in Design and Control. Society for Industrial and Applied Mathematics, 2001.

[3] Michael Branicky and Michael Curtiss. Nonlinear and hybrid control via RRTs. *Proc. Intl. Symp. on Mathematical Theory of Networks and Systems*, 2002.

[4] R. R. Burridge, A. A. Rizzi, and D. E. Koditschek. Sequential composition of dynamically dexterous robot behaviors. *International Journal of Robotics Research*, 18(6):534–555, June 1999.

[5] Katie Byl and Russ Tedrake. Approximate optimal control of the compass gait on rough terrain. In *Proc. IEEE International Conference on Robotics and Automation (ICRA)*, 2008.

[6] Elena Glassman and Russ Tedrake. Rapidly exploring state space. *In Progress*, 2009.

[7] A. Goswami, B. Espiau, and A. Keramane. Limit cycles and their stability in a passive bipedal gait. pages 246–251. IEEE International Conference on Robotics and Automation (ICRA), 1996.

[8] David H. Jacobson and David Q. Mayne. *Differential Dynamic Programming*. American Elsevier Publishing Company, Inc., 1970.

[9] Tor A. Johansen. Computation of lyapunov functions for smooth nonlinear systems using convex optimization. *Automatica*, 36(11):1617 – 1626, 2000.

[10] L.E. Kavraki, P. Svestka, JC Latombe, and M.H. Overmars. Probabilistic roadmaps for path planning in high-dimensional configuration spaces. *IEEE Transactions on Robotics and Automation*, 12(4):566–580, August 1996.

[11] S. LaValle and J. Kuffner. Rapidly-exploring random trees: Progress and prospects. In *Proceedings of the Workshop on the Algorithmic Foundations of Robotics*, 2000.

[12] Steven M. LaValle. *Planning Algorithms*. Cambridge University Press, 2006.

[13] Frank L. Lewis. *Applied Optimal Control and Estimation*. Digital Signal Processing Series. Prentice Hall and Texas Instruments, 1992.

[14] Sridhar Mahadevan and Mauro Maggioni. Proto-value functions: A laplacian framework for learning representation and control in markov decision processes. Technical Report TR-2006-35, University of Massachusetts, Department of Computer Science, July 2006.

[15] Ian R. Manchester, Uwe Mettin, Fumiya Iida, and Russ Tedrake. Stable dynamic walking over rough terrain: Theory and experiment. In *Proceedings of the International Symposium on Robotics Research (ISRR)*, 2009.

[16] M.T. Mason. The mechanics of manipulation. In *Proceedings of the IEEE International Conference on Robotics and Automation*, pages 544–548. IEEE, 1985.

[17] Pablo A. Parrilo. *Structured Semidefinite Programs and Semialgebraic Geometry Methods in Robustness and Optimization*. PhD thesis, California Institute of Technology, May 18 2000.

[18] Stephen Prajna, Antonis Papachristodoulou, Peter Seiler, and Pablo A. Parrilo. *SOSTOOLS: Sum of Squares Optimization Toolbox for MATLAB Users guide*, 2.00 edition, June 1 2004.

[19] Khashayar Rohanimanesh, Nicholas Roy, and Russ Tedrake. Towards feature selection in actor-critic algorithms. Technical report, Massachusetts Institute of Technology Computer Science and Artificial Intelligence Laboratory, 2007.

[20] A.S. Shiriaev, L.B. Freidovich, and I.R. Manchester. Can we make a robot ballerina perform a pirouette? orbital stabilization of periodic motions of underactuated mechanical systems. *Annual Reviews in Control*, 2008.

[21] Athanasios Sideris and James E. Bobrow. A fast sequential linear quadratic algorithm for solving unconstrained nonlinear optimal control problems, February 2005.

[22] Jean-Jacques E. Slotine and Weiping Li. *Applied Nonlinear Control*. Prentice Hall, October 1990.

[23] Emanuel Todorov and Weiwei Li. Iterative linear-quadratic regulator design for nonlinear biological movement systems. volume 1, pages 222–229. International Conference on Informatics in Control, Automation and Robotics, 2004.

[24] Oskar von Stryk. Numerical solution of optimal control problems by direct collocation. In *Optimal Control, (International Series in Numerical Mathematics 111)*, pages 129–143, 1993.

[25] E. R. Westervelt, B. Morris, and K. D. Farrell. Analysis results and tools for the control of planar bipedal gaits using hybrid zero dynamics. *Autonomous Robots*, 23:131–145, Jul 2007.

Human Motion Database with a Binary Tree and Node Transition Graphs

Katsu Yamane
Disney Research, Pittsburgh
kyamane@cs.cmu.edu

Yoshifumi Yamaguchi
Dept. of Mechano-Informatics
University of Tokyo
yamaguti@ynl.t.u-tokyo.ac.jp

Yoshihiko Nakamura
Dept. of Mechano-Informatics
University of Tokyo
nakamura@ynl.t.u-tokyo.ac.jp

Abstract— Database of human motion has been widely used for recognizing human motion and synthesizing humanoid motions. In this paper, we propose a data structure for storing and extracting human motion data and demonstrate that the database can be applied to the recognition and motion synthesis problems in robotics. We develop an efficient method for building a binary tree data structure from a set of continuous, multi-dimensional motion clips. Each node of the tree represents a statistical distribution of a set of human figure states extracted from the motion clips. We also identify the valid transitions among the nodes and construct node transition graphs. Similar states in different clips may be grouped into a single node, thereby allowing transitions between different behaviors. Using databases constructed from real human motion data, we demonstrate that the proposed data structure can be used for human motion recognition, state estimation and prediction, and robot motion planning.

I. INTRODUCTION

Using a collection of human motion data has been a popular approach for both analyzing and synthesizing human figure motions, especially thanks to recent improvements of motion capture systems. In the graphics field, motion capture data have been widely used for producing realistic animations for films and games. A body of research efforts have been directed to techniques that allow reuse and editing of existing motion capture to new characters and/or scenarios. In the robotics field, there are two major applications of such databases: building a human behavior model for robots to recognize human motions, and synthesizing humanoid robot motions.

Databases for robotics applications are required to perform at least the following two functions: First, they have to be able to categorize human motion into distinct behaviors and recognize the behavior of a newly observed motion. This is commonly achieved by constructing a mathematical model that bridges the continuous motion space and the discrete behavior space. The problem is that it is often difficult to come up with a robust learning algorithm for building the models because raw human motion data normally contain noise and error. The efficiency of search also becomes a problem as the database size increases.

Another function is to synthesize a robot motion that adapts to current situation, which is computationally demanding because of the large configuration space of humanoid robots. Motion database is a promising approach because they can reduce the search space by providing the knowledge on how

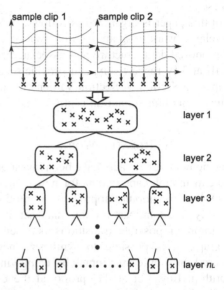

Fig. 1. A binary tree with human motion data. Each x mark represents a frame in one of the sample motion clips. The frames are iteratively split into two descendant nodes. Each frame therefore appears once in each layer. Database size can be reduced by making leaf nodes from multiple frames and keep only the statistical information of the frames included in each node.

human-like robots should move. For this purpose, however, motion capture data must be organized so that the planner can effectively extract candidates of motions and/or configurations. A database should also be able to generate high-quality motion data, which is also difficult because sample motion data are usually compressed to reduce the data size.

In this paper, we propose a highly hierarchical data structure for human motion database. We employ the binary-tree structure as shown in Fig. 1, which is a classical database structure widely used in various computer science applications because of search efficiency. However, constructing a binary-tree structure from human motion data is not a trivial problem because there is no straightforward way to split the multi-dimensional, continuous motion data into two descendant nodes. Our first contribution is a simple, efficient clustering algorithm for splitting a set of sample frames into two descendant nodes to construct a binary tree from human motion data.

We also develop algorithms for basic statistical computations based on the binary tree structure. Using these algorithms, we can recognize a newly observed motion sequence, estimate the current state and predict future motions, and plan new sequences that satisfy given constraints.

Another minor but practically important aspect of the proposed database is the ability to incorporate motion data from different sources. For example, we may want to include motion data captured with different marker sets, or include animation data from a 3D CG film. It is therefore important to choose a good motion representation to allow different marker sets and/or kinematic models. In this paper, we propose a scheme called *virtual marker set* so that motion data from different sources can be represented in a uniform way and stored in a single database.

The rest of this paper is organized as follows. In Section II, we review related work in graphics and robotics. We then present the proposed data structure and associated algorithms in Sections III and IV respectively, and provide several application examples in Section V. Section VI demonstrates experimental results using human motion capture data, followed by concluding remarks.

II. RELATED WORK

In the graphic field, researchers have been investigating how to reuse and edit motion capture data for new scenes and new characters. One of the popular approaches is motion graphs, where relatively large human motion data set is analyzed to build a graph of possible transitions between postures. Using the graph, it is possible to synthesize new motion sequences based on simple user inputs by employing a graph search algorithm. Kovar et al. [1] proposed the concept of Motion Graphs where similar postures in a database are automatically detected and connected to synthesize new motion sequences. They presented an application of synthesizing new locomotion sequence that follows a user-specified path. Lee et al. [2] employed a two-layered statistical model to represent a database, where the higher-level (coarse) layer is used for interacting with user inputs and the lower-level (detail) layer is used for synthesizing whole-body motions. Arikan et al. [3] also proposed a planning algorithm based on a concept similar to motion graphs. Related work in the graphics field mostly focuses on synthesizing new motions from simple user inputs using, for example, interpolation and numerical optimization [4].

In robotics, learning from human demonstration, or imitation, has been a long-term research issue [5]–[7] and a number of algorithms have been developed for storing human motion data and extracting appropriate behaviors. Because human motion varies at every instance even if the subject attempts to perform the same motion, it is necessary to model human behaviors by either statistical models [7], [8], nonlinear dynamical systems [9], [10], or a set of high-level primitives [11]. Related work relevant to this paper includes the Hidden Markov Model (HMM) representation of human behaviors [8] and the hierarchical motion database based on HMM [12]. Another hierarchical motion categorization method is also proposed using neural network models [10].

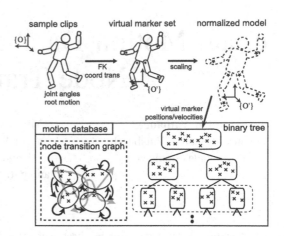

Fig. 2. Constructing the database.

However, most work in the robotics field is still focused on robust learning of human behaviors. Scalability of the motion database or synthesizing transitions between different behaviors have not been investigated well.

In the vision community, human motion database has been used to construct human motion models for human tracking in videos. Sidenbladh et al. [13] proposed a binary-tree data structure for building a probabilistic human motion model. They first performed PCA on the entire motion sequence and then projected each frame data to the principal axes to construct a binary tree.

III. BUILDING THE DATABASE

The process for building the proposed database is summarized in Fig. 2. The user provides one or more sample motion clips represented as a pair of root motion and joint angle sequences, typically obtained by motion capture or hand animation. The joint angle data are then converted to virtual marker data through forward kinematics (FK) computation to obtain the marker positions and velocities, coordinate transformation to remove the trunk motion in the horizontal plane, and scaling to normalize the subject size. The positions and velocities of the virtual markers are used to represent the state of the human figure in each sample frame.

To construct a binary tree, we first create the root node that contains all frames in the sample motion clips. We then iteratively split the frames into two descendant nodes using the method described in Section III-B. After the tree is obtained, we count the number of transitions among the nodes in each layer to construct the node transition graphs as described in Section III-C. The binary tree and node transition graphs are the main elements of the proposed motion database.

A. Motion Representation

There are several choices for representing the state of a frame in sample motion clips. A reasonable choice would be to use joint angles [2], [8] because they uniquely define a configuration by minimum number of parameters. However, joint angle representation strictly depends on the skeleton model and it is difficult to map the states between different models. In addition, joint angle representation may not be consistent with visual appearance of the human figure because the contribution of each joint angle to the Cartesian positions of the links can vary.

Another possibility is to use point clouds [1]. This method is independent of underlying skeleton models and also more intuitive because it directly represent the overall shape of the figure. The problem is that it is difficult compute the distance between two poses because registration between two point clouds is required.

In our implementation, we use a much simpler approach called *virtual marker set*, where all motions are represented by the trajectory (position and velocity) of N_v virtual markers. The virtual marker set is defined by the database designer so that it well represents the motions contained in the database, and can be different from any physical marker sets. The virtual marker set approach would become similar to the point cloud method as the number of virtual markers increases, although each marker in the virtual marker set should be labeled.

If a motion is represented by joint angle trajectories of a skeleton model, it can be easily converted to the virtual marker set representation by giving the relationship between each marker in the virtual marker set and the skeleton used to represent the original motions. The relationship between a virtual marker and a skeleton is defined by specifying the link to which the marker is attached, and giving the relative position in its local frame. Although this approach requires some work on the user's side, it allows the use of multiple skeleton models with simple distance computation.

After converting the motion data to the virtual marker set representation, we perform a coordinate transformation to remove the horizontal movement in the motion and scaling to normalize the subject size. Each marker position is represented in a new reference coordinate whose origin is located on the floor below the root joint, z axis is vertical, and x axis is the projection of the front direction of the subject model onto the floor. The marker velocities are also converted to the reference coordinate.

Formally, a sample motion data matrix X with N_S frames is defined by

$$X = (x_1 \ x_2 \ \dots x_{N_S}) \tag{1}$$

where x_i is the state vector of the i-th sample frame defined by

$$x_i = \left(p_{i1}^T \ v_{i1}^T \ p_{i2}^T \ v_{i2}^T \ \dots p_{iN_v}^T \ v_{iN_v}^T\right)^T \tag{2}$$

and p_{il} and v_{il} are the position and velocity of marker l in sample frame i. If multiple motion clips are to be stored in a database, we simply concatenate all state vectors horizontally and form a single sample motion data matrix.

B. Constructing a Binary Tree

A problem of constructing a binary tree for motion data is how to cluster the sample frames into groups with similar states. Most clustering algorithms require a large amount of computation because they check the distances between all

pairs of frames. This process can take extremely long time as the database size increases.

Here we propose an efficient clustering algorithm based on principal component analysis (PCA) and minimum-error thresholding technique. The motivation for using PCA is that it determines the axes that best characterize the sample data. In particular, projecting all samples onto the first principal axis gives a one-dimensional data set with the maximum variance, which can then be used for separating distinct samples using adaptive thresholding techniques developed for binarizing images.

The process to split node k into two descendant nodes is as follows. Assume that node k contains n_k frames whose mean state vector is $\bar{x}_k$. Also denote the sample motion data matrix of node k by X_k. We compute the zero-mean singular value decomposition of X_k as

$$X_k^{'T} = U\Sigma V^T \tag{3}$$

where each column of X_k' is obtained by subtracting $\bar{x}_k$ from the original state vectors, Σ is a diagonal matrix whose elements are the singular values of X_k sorted in the descending order, and U and V are orthogonal matrices. The columns of V represents the principal axes of X_k. We obtain the one-dimensional data set s_k by projecting X_k' onto the first principal axis by

$$s_k = X_k^{'T} v_1 \tag{4}$$

where v_1 denotes the first column of V.

Once the one-dimensional data is obtained, the clustering problem is equivalent to determining the threshold that minimizes classification error. We shall use a minimum-error thresholding technique [14] to determine the optimal threshold. After sorting the elements of s_k and obtaining the sorted vector s_k', this method determines the index m such that the data should be divided between samples m and $m + 1$ using the following equation:

$$m = \underset{i}{\mathrm{argmax}} \left\{ i \log \frac{\sigma_1}{i} + (n_k - i) \log \frac{\sigma_2}{n_k - i} \right\} \tag{5}$$

where σ_1 and σ_2 denote the variance of the first i and last $n_k - i$ elements of s_k', respectively.

We obtain a binary tree by repeating this process until the division creates a node containing fewer frames than a predefined threshold. To ensure the statistical meaning of each node, we also avoid nodes with small number of sample frames by setting a threshold for minimum frame number. If Eq.(5) results in a node with fewer number of frames than the latter threshold, we do not perform the division. Therefore, a node may not be divided if it contains many similar frames.

Some branches may be shorter than others because we extend each branch as much as possible and some of them may hit the thresholds earlier. In such cases, we simply extend shorter branches by attaching a copy of the leaf node so that the length of all branches become the same. Each node therefore can have 0 (leaf nodes), 1 or 2 descendant nodes.

C. Node Transition Graphs

After constructing the binary tree, we then build the node transition graphs based on the transitions observed between nodes in each layer. Because we know the set of frames included in each node, we can easily determine the transition probabilities by dividing the number of transitions to a specific node by the total number of frames in the node.

We build two kinds of node transition graphs at each layer. The *global transition graph* describes the average node transition probabilities observed in all sample clips. The transition probability from node m to node n in the same layer is computed by

$$p_{m,n} = \frac{t_{m,n}}{\Sigma_i t_{m,i}} \qquad (6)$$

where $t_{k,l}$ denotes the total number of transitions from nodes k to l observed in all sample clips. A *clip transition graph* describes the node transition probabilities observed in a specific sample clip. We can use the same equation (6) to compute the transition probabilities, except that $t_{k,l}$ only considers transitions within the same sample clip.

The global transition graph at each layer is similar to motion graphs [1] in the sense that all possible transitions between nodes are included. However, the way we construct the graph is different from existing motion graph techniques, resulting in a more efficient database construction. Our method generally requires $O(N \log N)$ for a database with N sample frames because the depth of the tree is typically $O(\log N)$ and splitting the frames at each layer requires $O(N)$ computations, while most motion graph and other clustering techniques require $O(N^2)$ computations because they usually compute the distance between each pair of frames in the database.

The clip transition graphs are similar to human behavior models using HMMs [8]. In most HMM-based approaches, a simple left-to-right model or single-chain cyclic model with fixed number of nodes is assumed because it is difficult to train an HMM with arbitrary length or arbitrarily interconnected nodes. In our method, we do not assume the structure of the node transition or the number of nodes used to represent a sample clip. If a sample clip includes a cyclic pattern, for example, our method automatically models the cycle by producing a single transition loop, while a left-to-right model tries to encode the whole sequence within a given number of nodes.

IV. Algorithms

For a given tree and node transition graph, we should be able to perform the following two basic operations:

- find the optimal node transition to generate a newly observed motion clip, and
- compute the probability that a newly observed motion clip is generated by a node transition graph.

In the rest of the section, we shall denote the newly observed motion comprising M frames by $\hat{X} = (\hat{x}_1 \ \hat{x}_2 \ \ldots \ \hat{x}_M)$ where $\hat{x}_i$ is the state vector at frame i. Here we assume that both positions and velocities of virtual markers are given.

A. Optimal Node Transition

The probability that the observed motion $\hat{X}$ was produced by a node transition $\mathcal{N} = \{n_1, n_2, \ldots, n_M\}$ is given by

$$P(\mathcal{N}|\hat{X}) = \prod_{i=1}^{M} P_t(n_{i-1}, n_i) P_s(n_i | \hat{x}_i) \qquad (7)$$

where $P_t(k, l)$ is the transition probability from node k to l ($P(n_0, n_1) = 1$) and $P_s(k|x)$ is the probability that the state was at node k when the observed state was x. $P_s(k|x)$ is obtained by the Bayesian inference:

$$P_s(k|x) = \frac{P_o(x|k)P(k)}{P(x)} = \frac{P_o(x|k)P(k)}{\sum_i P_o(x|i)} \qquad (8)$$

where $P_o(x|k)$ is the likelihood that state vector x is output from node k and $P(k)$ is the *a priori* probability that the state is at node k.

The actual form of $P_o(x|k)$ depends on the probability distribution used for each node. In this paper, we assume a simple Gaussian distribution with mean $\bar{x}_k$ and covariance Σ_k, in which case $P_o(x|k)$ can be computed by

$$P_o(x|k) = \frac{1}{(\sqrt{2\pi})^N |\Sigma_k|} \exp\left(-\frac{1}{2}(x - \bar{x}_k)^T \Sigma_k^{-1}(x - \bar{x}_k)\right).$$

$P(k)$ can be either a uniform distribution among the nodes, or weighted according to the number of sample frames included in the nodes. In our implementation, we use a uniform distribution for the global transition graph. We also use a uniform distribution for each clip transition graph, excluding the nodes that did not appear in the sample clip.

Obtaining the optimal node transition $\mathcal{N}^*$ is the problem of finding the node sequence that maximizes Eq.(7). A common method for this purpose is to perform forward-backward algorithm or dynamic programming. However, such algorithms can be computationally expensive for long sequences or densely connected graphs. We could omit nodes far enough from each observed frame using a threshold, but it is difficult to determine the threshold so that enough number of candidates are left for the search.

Instead of searching the entire node sequence at a single layer, we utilize the binary tree data structure by starting from the top layer. Because the top layer only contains the single root node r, the trivial optimal sequence at the top layer, $\mathcal{N}_1^*$, is to visit the root node M times, i.e., $\mathcal{N}_1^* = \{r, r, \ldots, r\}$. Starting from this initial sequence, we perform a dynamic programming to find the best way to trace the descendants of the nodes in the sequence all the way down to the bottom layer. We could also terminate at an intermediate layer if we do not need precise results, in which case the result would be obtained faster.

B. Motion Generation Probability

Motion generation probability is defined as the probability that a node transition graph generates the observed motion. This probability can be used for identifying the type of behavior. We can compute the motion generation probability by summing up the probability of generating the motion by all possible node transitions. However, there may be huge number of possible node transitions for long motions or large transition graphs.

An alternative used in this paper is to use the dynamic programming described in the previous subsection to find multiple node sequences. Because the algorithm returns node sequences in the descending order of probability, we can approximate the total motion generation probability by using the top few node sequences.

V. APPLICATIONS

A. State Estimation and Prediction

Estimating the current state is accomplished by taking the last node in the most likely node transition in the global node transition graph. Once the node transition is estimated with high probability, we can then predict the next action by tracing the node transition graph. By combining the probability of the node transition and the probability of the future transition candidate, we can also obtain the confidence of the prediction.

B. Motion Recognition

Motion recognition is the process to find a motion clip in the database that best matches a newly observed motion sequence. This is accomplished by comparing the motion generation probability from the clip transition graphs.

C. Motion Planning

The global transition graph can also be used for planning a new motion sequence subject to kinematic constraints. In addition, because the tree has multiple layers that model the sample clips at different granularities, motion planning can also be performed at different precision levels. The only issue is how to compute the motion of the root in the horizontal plane because those information are removed from the database.

Our solution is to use the marker velocity data to obtain the root velocity, and then integrate the velocity to obtain root trajectory. The velocity can be obtained by employing a numerical inverse kinematics algorithm, which essentially solves the following linear equation:

$$\dot{\theta} = Jv \qquad (9)$$

where $\dot{\theta}$ is the joint velocities including the root linear and angular velocities, v is the vector of all marker velocities extracted from the mean vector of a node, and J is the Jacobian matrix of the marker positions with respect to joint angles. This is usually an overconstrained IK problem because the virtual marker set contains more markers than necessary to determine the joint motion. We therefore apply the singularity-robust (SR) inverse [15] of J to obtain $\dot{\theta}$.

VI. RESULTS

A. Sample Data Set

We use two different sets of data to demonstrate the generality of our approach. The minimum number of sample

TABLE I
PROPERTIES OF THE TWO DATABASES.

	Database 1	Database 2
# of frames	5539	11456
# of layers	17	16
# of nodes	1372	1527
# of nodes in bottom layer	167	205

frames in a node is set to 16 in both databases. The properties of the databases are summarized in Table I.

The first set (*Database 1*) consists of 19 motion clips including a variety of jog, kick, jump, and walk motions, all captured separately in the motion capture studio at University of Tokyo. The motions were captured using our original marker set consisting of 34 markers. This marker set is also used as the virtual marker set to construct all the databases in the experiments. *Database 1* is used to demonstrate the analysis experiments.

The second set (*Database 2*) is generated from selected clips in a publicly available motion library [16] and will be used for the planning experiment. The database includes two long locomotion sequences with forward, backward and sideward walk of the same subject. Although *Database 2* contains twice as many sample frames as *Database 1*, it resulted in relatively small number of layers and nodes probably because they consist of similar motions and hit the minimum frame number threshold earlier.

Because the virtual marker set consists of 34 markers, the dimension of the state vector is 204.

B. Properties of the Database

We first visualize *Database 1* to investigate its properties. In Fig. 3, the spheres (light-blue for readers with color figures) represent all nodes in the bottom layer and the red lines represent the transitions among the nodes. The white mesh represents the plane of first and second principal axes. The projection onto the first-second principal axes plane is also shown in Fig. 4. Figure 5 shows the nodes and global transition graphs at layers 4 and 7, with the top layer numbered layer 1.

The location of each node is computed by projecting its mean vector onto the first three principal components of the root node of the tree. The size of the node is proportional to the maximum singular value of the sample motion data matrix of the node.

Figure 6 shows the mean marker positions and velocities of the nodes marked in Fig. 4. These images clearly show the geometrical meaning of the first and second principal axis: the first axis represents the vertical velocity and the second axis represents which leg is raised. Apparently there is no verbally explainable meaning in the third axis.

Figure 7 shows the node transition graphs of two representative clips. As expected from the above observation, motions such as jump that involves mostly the vertical motion stay in the first and third axes plane, while jog, kick and walk motions stay in the second and third axes plane.

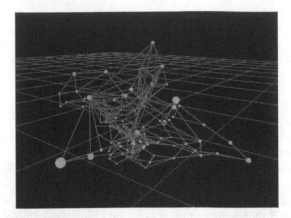

Fig. 3. Visualization of the database in the three principal axes space.

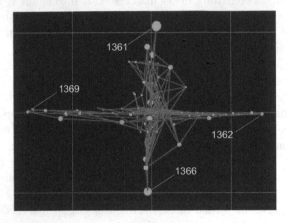

Fig. 4. Visualization of the database in the first (horizontal) and second (vertical) principal axes space.

C. State Estimation and Prediction

We experimented the state estimation ability by providing the first 0.2 s of a novel kick motion by the same subject and a jump motion by a different subject. The results are shown in Figures 8 and 9 respectively. Note that the global node transition is used to compute the best node sequence, although only the node transition in relevant sample clip is drawn in each figure for clarity.

In both cases, the database could predict the subjects' actions before they actually occurred. In Fig. 8, all nodes on the identified node sequence (marked as "estimated states")

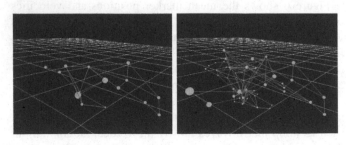

Fig. 5. Nodes and global transition graphs at layers 4 and 7.

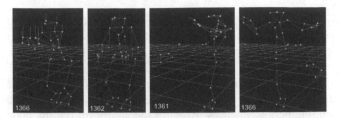

Fig. 6. Mean marker positions and velocities of selected nodes. The lines rooted at markers denote the velocities. From left to right: nodes 1369, 1362, 1361, and 1366 as marked in Fig. 4.

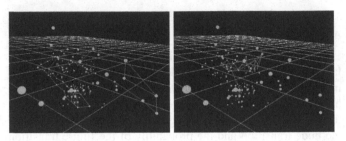

Fig. 7. Node transition graphs of selected clips. Left: jump, right: kick.

are included in the transition graph of the kick samples. The last node in the sequence is likely to transit to the dark blue node in the left figure with the probability of 0.36, which corresponds to the marker positions and velocities in the right figure. Similarly, the result of Fig. 9 indicates that the database can correctly identify that the subject is in preparation for a jump.

D. Motion Recognition

Figures 10–11 show the results of motion recognition experiments. The three graphs in Fig. 10 show differences between layers for the same observation. We computed the motion generation probability of new motions with respect to the node transition graph of each sample motion clip. Because the new motion is 2.5 s long and computation of the probabilities takes long time, we computed the probability of generating the

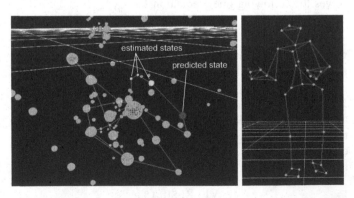

Fig. 8. Result of state estimation and prediction for a kick motion of the same subject. Left: the nodes on the identified node transition are marked as "estimated states." Right: the marker positions and velocities corresponding to the node marked as "predicted state" in the left figure.

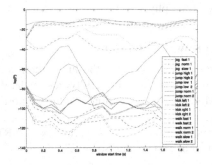 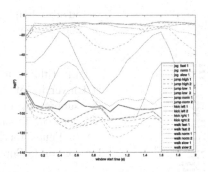 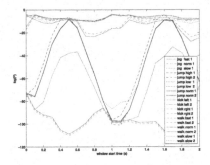

Fig. 10. Time profile of generation probability of a new walk motion of the same subject. From left to right: bottom layer, layer 10, layer 7.

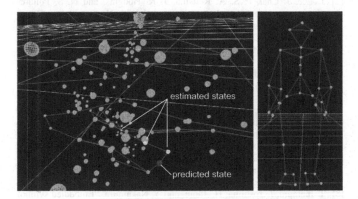

Fig. 9. Result of state estimation and prediction for a jump motion of a different subject. Left: the nodes on the identified node transition are marked as "estimated states." Right: the marker positions and velocities corresponding to node marked as "predicted state" in the left figure.

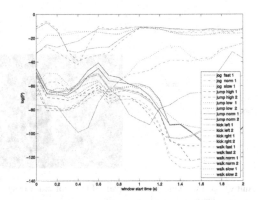

Fig. 11. Time profile of generation probability of a walk motion of a different subject (bottom layer).

sequences within a sliding window of width 0.4 s. The graphs depict the time evolution of probabilities when we moved the window from $[0.0s, 0.4s]$ to $[2.0s, 2.4s]$. The lines are colored according to the type of motions. We use same line color and type for very similar sample clips for clarity of presentation.

The first three graphs show the probability of a new walk motion by the same subject as the database, while a walk motion from a different subject was used for Fig. 11. These results show that the model can successfully recognize the type of observed motion even for different subjects. It is also suggested that the statistical models at different layers have different granularity levels because the variation of probabilities is smaller for upper layers. In particular, similar motions such as walk and jog begin to merge at layer 7.

Figure 12 show the result at the bottom layer for a completely unknown motion (lifting an object from the floor). It is clear that the database cannot tell whether the motion is jump or kick, which is intuitively reasonable because these three motions start from small bending.

E. Motion Planning

We performed a simple motion planning test with only start and goal position constraints. Figure 13 shows the planned motion when the goal position is given as 2.0 m front and

1.0 m left of the current position. The planner outputs a reasonable motion under the available samples, which is to walk forward for 2 m and do a side walk to the left for 1 m. We can observe that the feet occasionally penetrate the floor. We would need the contact status information to fix this problem.

VII. CONCLUSION

In this paper, we proposed a new data structure for storing multi-dimensional, continuous human motion data, and demonstrated its basic functions through experiments. The main contributions of the paper are summarized as follows:

1) We proposed a method for constructing a binary tree data structure from human motion data. We applied PCA and a minimum-error thresholding technique to efficiently find the optimal clustering of sample frames.
2) We proposed to build a global node transition graph representing the node transitions in all sample clips, as well as a clip transition graph representing the node transitions in each sample motion clip.
3) We developed two algorithms for computing the most probable node transition and generation probability for a given new motion sequence, based on the binary tree data structure.

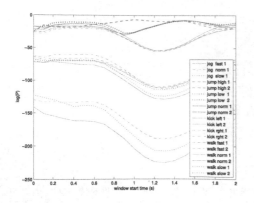

Fig. 12. Time profile of generation probability of a lifting motion (bottom layer).

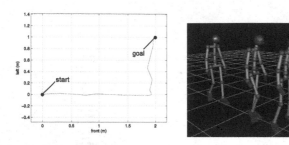

Fig. 13. Result of motion planning. Left: path of the root in horizontal plane. Right: postures at selected frames.

4) We demonstrated three applications of the proposed database through experiments using real human motion data.

There are several functions yet to be addressed by our database. We currently do not support incremental learning of new motions because all sample frames must be available to perform the PCA. If the new sample clips do not drastically change the distribution of whole samples, we could apply one of the extensions of PCA for online learning techniques [17]. Some techniques for balancing binary trees for multi-dimensional data could also be employed to reorganize the tree [18].

It should be relatively easy to add segmentation and clustering functions because the sample motion clips are abstracted by node transition graphs. We can easily detect same or similar node transitions in different motion clips, which could be used for segmentation. Clustering of sample clips can be achieved by evaluating the distance between motion clips based on their node transition graphs and applying standard clustering algorithms [19].

Our current database only contains the marker position and velocity data. It would be interesting to add other modalities such as contact status, contact forces, and muscle tensions. In particular, we would be able to solve the contact problem in our planning experiment if we have access to the contact status information. Contact force and muscle tension information

would also help generating physically feasible motions for humanoid robots.

ACKNOWLEDGEMENTS

Part of this work was conducted while the first author was at University of Tokyo. The authors gratefully acknowledge the support by the Ministry of Education, Culture, Sports, Science and Technology, Japan through the Special Coordination Funds for Promoting Science and Technology, "IRT Foundation to Support Man and Aging Society."

REFERENCES

[1] L. Kovar, M. Gleicher, and F. Pighin, "Motion graphs," *ACM Transactions on Graphics*, vol. 21, no. 3, pp. 473–482, 2002.
[2] J. Lee, J. Chai, P. S. A. Reitsma, J. K. Hodgins, and N. S. Pollard, "Interactive Control of Avatars Animated With Human Motion Data," *ACM Transactions on Graphics*, vol. 21, no. 3, pp. 491–500, July 2002.
[3] O. Arikan and D. A. Forsyth, "Synthesizing Constrained Motions from Examples," *ACM Transactions on Graphics*, vol. 21, no. 3, pp. 483–490, July 2002.
[4] A. Safonova and J. Hodgins, "Interpolated motion graphs with optimal search," *ACM Transactions on Graphics*, vol. 26, no. 3, p. 106, 2007.
[5] C. Breazeal and B. Scassellati, "Robots that imitate humans," *Trends in Cognitive Science*, vol. 6, no. 11, pp. 481–487, 2002.
[6] S. Schaal, A. Ijspeert, and A. Billard, "Computational approaches to motor learning by imitation," *Phylosophical Transactions of the Royal Society of London B: Biological Sciences*, vol. 358, pp. 537–547, 2003.
[7] A. Billard, S. Calinon, and F. Guenter, "Discriminative and adaptive imitation in uni-manual and bi-manual tasks," *Robotics and Autonomous Systems*, vol. 54, pp. 370–384, 2006.
[8] T. Inamura, I. Toshima, H. Tanie, and Y. Nakamura, "Embodied symbol emergence based on mimesis theory," *International Journal of Robotics Research*, vol. 24, no. 4/5, pp. 363–378, 2004.
[9] A. Ijspeert, J. Nakanishi, and S. Schaal, "Movement imitation with nonlinear dynamical systems in humanoid robots," in *Proceedings of International Conference on Robotics and Automtation*, 2002, pp. 1398–1403.
[10] H. Kadone and Y. Nakamura, "Symbolic memory for humanoid robots using hierarchical bifurcations of attractors in nonmonotonic neural networks," in *Proceedings of International Conference on Intelligent Robots and Systems*, 2005, pp. 2900–2905.
[11] D. Bentivegna, C. Atkeson, and G. Cheng, "Learning tasks from observation and practice," *Robotics and Autonomous Systems*, vol. 47, no. 2–3, pp. 163–169, 2004.
[12] D. Kulić, W. Takano, and Y. Nakamura, "Incremental learning, clustering and hierarchy formation of whole body motion patterns using adaptive hidden markov chains," *International Journal of Robotics Research*, vol. 27, no. 7, pp. 761–784, 2008.
[13] H. Sidenbladh, M. Black, and L. Sigal, "Implicit probabilistic models of human motion for synthesis and tracking," in *European Conference on Computer Vision*, 2002, pp. 784–800.
[14] J. Kittler and J. Illingworth, "Minimum error thresholding," *Pattern Recognition*, vol. 19, no. 1, pp. 41–47, 1986.
[15] Y. Nakamura and H. Hanafusa, "Inverse Kinematics Solutions with Singularity Robustness for Robot Manipulator Control," *Journal of Dynamic Systems, Measurement, and Control*, vol. 108, pp. 163–171, 1986.
[16] "CMU graphics lab motion capture database," http://mocap.cs.cmu.edu/.
[17] M. Artač, M. Jogan, and A. Leonardis, "Incremental pca or on-line visual learning and recognition," in *Proceedings of the 16 th International Conference on Pattern Recognition*, 2002, pp. 30 781–30 784.
[18] V. K. Vaishnavi, "Multidimensional balanced binary trees," *IEEE Transactions on Computers*, vol. 38, no. 7, pp. 968–985, 1989.
[19] J. Ward, "Hierarchical grouping to optimize an objective function," *Journal of the American Statistical Association*, vol. 58, pp. 236–244, 1963.

Explicit Parametrizations of the Configuration Spaces of Anthropomorphic Multi-linkage Systems

Li Han and Lee Rudolph
Department of Mathematics and Computer Science
Clark University
Worcester, MA 01610
Email: [lhan, lrudolph]@clarku.edu

Abstract— **Multi-fingered manipulation systems are important in the study of robotics. These are also challenging systems, in part because of the loop closure constraints required of several (virtual) loops each formed by two fingers and the grasped object. Most existing work describes system configurations using joint parameters, in which loop closure constraints are expressed by highly nonlinear equations. Such a formulation amounts to an implicit parametrization of the configuration space (*CSpace*) as a lower-dimensional semi-algebraic subset embedded in a higher-dimensional ambient joint parameter space. The non-zero difference between the two dimensions is the codimension of *CSpace* as seen in the given parametrization.**

In this paper, we point out that, quite generally, parametrizations leading to lower codimensional configuration spaces provide multi-faceted advantages over those producing higher codimensions. For two example manipulation system—a 3-fingered hand and a planar star-manipulator with any number of fingers— we present explicit parameterizations, which are effectively of codimension 0. We base these parametrizations on our recently developed construction trees of simplices (such as triangles and tetrahedra) for multi-object systems; such a tree gives simplex-based parameters on *CSpace*, in which loop closure constraints become simplex formation constraints (such as triangle inequalities and Cayley–Menger determinant constraints). Both example systems are very difficult to deal with using joint angle parameters. Our results here further demonstrate the generality and effectiveness of the simplex-based approach.

I. OVERVIEW

Multi-linkage robotic systems, such as robotic hands and legs, have great potential in diverse areas; they have been intensively researched for several decades(see *e.g.* books [4], [5], [18]–[20], [22], the recent white paper [7], and references therein). These challenging systems are subject to complicated constraints, among them loop closure constraints arising from (virtual) loops formed by multiple limbs all touching one objects. Conventional parameters for these systems are joint parameters, in which loop closure constraints are formulated as nonlinear equations. In such a formulation, the configuration space (*CSpace*) of a system is a semi-algebraic or semi-analytic set, generically a smooth manifold, embedded in its ambient joint space (for an example, see the study of single-loop systems with spherical-type joints [21], [28]). Such an embedding gives an **implicit** parametrization of *CSpace*: by the Implicit Function Theorem, any given configuration point has a *CSpace* neighborhood on which some subset of the ambient space parameters are coordinates for a local chart,

but to effectively find an atlas of such charts covering *CSpace*, and to compute properties of these charts like their geometries and mutual intersections, is not easy. The lack of methods to access such local information becomes a huge hurdle in the study of the global structure of *CSpace*.

The **codimension** of a subset in an ambient space is the difference between its dimension and that of the ambient space; thus, in 3-dimensional Euclidean space, a curve has codimension 2, a surface has codimension 1, and a non-empty open subset (such as the set of points enclosed by a sphere) has codimension 0. By an **explicit** parametrization of an open subset U of *CSpace*, we mean one whose chart has codimension 0: that is, U is (explicitly) identified with an open set in Euclidean space of the same dimension as *CSpace*. An **explicit atlas** for *CSpace* is one in which all charts have codimension 0; with an explicit atlas in hand, we do not have to study *CSpace* as a lower-dimensional subspace of a higher-dimensional space. As an example, consider the unit circle C in the (x, y)-plane. With x and y coordinates as implicit parameters for C, the smallest atlas comprises four charts (semicircles). In contrast, the polar angle θ (on any interval of length less than 2π) is an explicit parameter, and yields an explicit atlas comprising just 2 charts. More generally, any simple closed curve Γ has a 2-chart explicit atlas coordinatized by arc length, but for any $n > 2$ there exists Γ for which any atlas with implicit parameters x and y needs n charts.

As briefly discussed in Section II, there are many advantages to the development and use, when possible, of parameters for *CSpace* that reduce codimension (explicit parameters being optimal). There are also advantages—which can sometimes conflict with codimension reduction—to parameters in which constraint formulations are particularly simple.

In this paper, we present explicit parametrizations for two example anthropomorphic linkage systems. (1) The 3-fingered manipulation system in Fig. 1 has 14 degrees of freedom, dropping to 11 when it grasps an object with all 3 fingers at fixed contacts. In joint angle parameters, the 11-dimensional configuration space of the grasping hand has codimension 3 in a 14-dimensional torus, and the 3 loop closure constraint equations are complicated trigonometric expressions. In joint Cartesian coordinates, the same 11-dimensional *CSpace* is realized as a codimension-22 submanifold; although the defining equations are quadratic polynomials, the loop closure

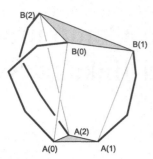

Fig. 1. A model grasp system with 3 fingers; each has a spherical base joint, and intermediate revolute joints with parallel axes.

constraints make the implicit parameters highly coupled and difficult to solve. By contrast, our simplex-based parameters for this *CSpace* are explicit, and the loop closure constraints are straightforward simplex-formation constraints (triangle and Cayley–Menger determinant inequalities). This parametrization provides a solid foundation for the study of system configurations, and can be used to solve the global structure of *CSpace* and develop efficient manipulation planning algorithms. We discuss it in section III. (2) Recent papers [17], [25] have studied motion planning for *star-manipulators*, a class of planar closed-chain manipulators. In section IV we present explicit parametrizations for all star-manipulators.

II. EXPLICIT PARAMETERS: MOTIVATIONS, PRIOR WORK

In view of complex closure constraints in the joint parameters, recent results on the **explicit parametrization and solved structures** of one type of systems with loops become even more remarkable. In short, for a planar single-loop linkage with n revolute joints (equivalently, a planar n-sided polygon), fixed link lengths, and having 3 "long links" (a technical condition defined in [21], [28]), it is proved [14], [21], [28] that one set of explicit configuration parameters consists of joint angles of $n - 3$ "short links", and the configuration space under such a parametrization comprises two disjoint $(n - 3)$-dimensional tori (the two components correspond to two possible orientations of the 3 long links—"elbow up" and "elbow down"). The existence of two components in *CSpace* reflects the fact that, while a loop without 3 long links can be reoriented (if link crossings are allowed), one with 3 long links cannot be. We will call this property **un-reorientability**.

The explicit parametrization and solved structures just mentioned greatly facilitate computations for un-reorientable planar loops. For example, any values for the $n - 3$ parameter angles will correspond to valid loop configurations. Also, a torus can be cut open into a cube, which, being convex, allows trivial path planning (any two points can be connected by a line segment); it follows that given two identically oriented configurations of an un-reorientable planar loop, it is trivial to find a path between them (ignoring collision or any other constraints), by computing and "linearly" interpolating parameter angle values. (Of course, whereas two points in a truly convex set are joined by a unique line segment, in a torus there are many such segments; all but one cross one or more "cuts", and all are equally trivial to compute.)

A linear path of a un-reorientable planar loop satisfies the loop closure constraints and stays on one of the two *CSpace* tori. But it may involve link penetrations. Extensive research by the motion planning community has made it clear that collision-avoidance is very difficult to deal with, and that it is very hard to find analytic descriptions of the subset *CFree* of **valid** configurations, and its complement *CObstacle*, for general obstacles. Recent successes of randomized path planners (see books [3], [15], [16] and references therein) suggest that **sampling-based approaches** may be an important framework in which to integrate efficient sampling methods while also dealing with such difficult factors in configuration space and path planning problems as high dimensionality and complicated linkage constraints. Sampling-based approaches can handle complicated constraints (by checking a sample point against the constraints, rather than trying to find all solutions to the constraints) and can be used with implicit surfaces in ambient spaces, but sampling efficiency can be improved by explicit parametrization (or, more generally, low codimension). This is due to two facts: for a purely random sampling scheme, the statistical success rate for generating valid samples in a subset equals the volume of the valid subset divided by the volume of the entire sample space; and, given an acceptable error $\varepsilon > 0$, the volume of an ε-neighborhood of a submanifold of codimension $k > 0$ tends to 0 like ε^k.

Thus, for example, consider the problem of generating one collision-free loop configuration, that is, a point in *CFree*. In general *CFree* is open in *CSpace*, and thus has the same dimension as *CSpace*. Consequently, if we have an explicit parametrization of *CSpace*, we can take any explicit chart as a sample space, and our success rate for generating a collision-free loop configuration in that chart will be the volume of the part *CFree* in that chart divided by the volume of the chart. If this rate is 0 for a particular chart, it means that the chart actually misses *CFree*, and there may even be an understandable structural reason why (see [13] for examples); if this ratio is positive, valid configurations should be found with positive probability. In contrast, if we only have parameters for the ambient space (*e.g.*, the joint parameter space) and an implicit parametrization for the configurations (as defined by the closure constraints in joint parameters), then the sampling space is the ambient space, the configuration space has codimension $d \geq 1$, and as stated above this means that the probability of finding any loop configuration— let alone a collision-free configuration—by random sampling in the ambient space approaches 0 like the dth power of the acceptable error. By analogy with the well-known "curse of dimensionality", that the number of samples needed to adequately cover a sample space increases exponentially with the dimension of the space, we propose the name **"curse of codimensionality"** for the situation just described: that, other things being equal, the higher the codimension of a target space in its ambient space, the more challenging it is to sample it efficiently (the challenge increasing with the codimension).

Given that parameters corresponding to lower-codimension parametrizations tend to provide multifaceted advantages for configuration related problems (aiding analytical study of global *CSpace* structure, facilitating sampling, etc.), finding such parameter systems is a problem of considerable interest. Systems without loops, like rigid bodies and open chains, have well-known and widely used explicit parametrizations, *e.g.*, various 6-DOF parameters for a $3D$ object or joint parameters for an open chain. However, to the best of our knowledge, rather few systems involving loops have been given explicit parametrizations in prior work: (a) single-loop planar linkages with revolute joints and 3 long links [14], [21], [28]; (b) trilaterable manipulators [24]; (c) flag manipulators [1], [27].

Loosely speaking, a system is "trilaterable" in the sense of [24] if it can be decomposed into tetrahedra in such a way that all unknown edge-lengths of the tetrahedra can be systematically computed from known edge-lengths using distance constraints (triangle inequalities and Cayley–Menger constraints; see below); flag manipulators generalize trilaterable systems. Note that the solutions in [1], [24], [27] assume systems already given in trilaterated form, with kinematic structures explicitly including all distance parameters needed to determine system configurations (*e.g.*, lengths of the legs between base and platform of a parallel manipulator).

Our Work. We have recently developed simplex-based approaches for multibody systems that allow construction trees of simplices (such as triangles and tetrahedra). The main representative systems used in our published papers, which describe their explicit paramatrizations and some other issues, are these: (a) single-loop planar linkages with revolute joints [8], [11]–[13]; (b) single-loop spatial linkages with spherical joints [8], [9]; (c) single-vertex rigid origami folds [10]. Along with some simple examples, we have pointed out in these papers that the described approaches are directly applicable to multi-loop systems having simplicial construction trees. (Our approaches can also easily accomodate link length constraints, which can be used to model prismatic joints along the links, link length tolerance and uncertainty, and so on.)

In this paper we present explicit parametrizations for examples of multi-loop linkage systems with anthropomorphic structures, which would have been very hard to study using the conventional joint parameters. Our first example system comprises 3 fingers with spherical base joints, revolute intermediate joints, and tips that we model (when the hand is grasping an object) as spherical joints; one motivational problem for this system is gaited 3-fingered manipulation (see Fig. 2). Our second example system is a general star-manipulator (as in Fig. 6). The example systems, and their

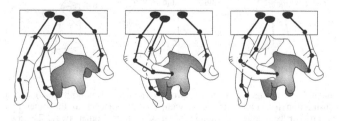

Fig. 2. Manipulation: gaiting between two 2-finger grasps via a 3-finger grasp.

simplicial construction trees, are substantially different from those in our earlier. Here we will focus on their explicit simplex-based parameters. Our results further demonstrate the versatility and effectiveness of the simplex-based approach.

Remarks. (1) We use the term "grasp" purely kinematically, and thus ignore such important practical issues as force-closure (see [23]). (2) Multi-limb systems using finger gaiting and legged locomotion have been studied before, as kinematic stratified systems [6], [29]; the main focus of that work was control issues, and those papers describe configuration spaces as coarse-grained stratified spaces, with each grasp mode (having the object grasped by one particular subset of fingers) corresponding to one conceptual, implicit stratum, embedded (as usual) in the ambient parameter space. (3) The "reachable distances" approach to motion planning for closed chains given in [26] is closely related to our work on closed chains, and it would be interesting to see how such an approach might apply to the example systems in this paper.

III. EXAMPLE SYSTEM I: AN ANTHROPOMORPHIC HAND

Our example 3-fingered hand system as drawn in Fig. 1 is an idealized model of the human hand shown in Fig. 2, where we model each finger link by the line segment between its two end points and do not consider the rotational movement about the links. Further, we consider the fixed base joints B_i ($i = 0, 1, 2$, modulo 3) to be spherical, and all other link joints to be revolute; further, on each finger, the axes of all the revolute joints are parallel to each other and perpendicular to the finger plane. Of the three fingers, one (thumb) has 3 links, and the other two have 4 links each, so in total the system has 14 degrees of freedom. As shown in Fig. 3, we use the following 14 inter-joint diagonals to define a tree of simplices (triangles and tetrahedra in this case) for such a system.

- 3 diagonals between the fingertips, $[A(0) A(1)]$, $[A(1) A(2)]$, $[A(2) A(0)]$
- 3 diagonals between the finger tips and bases, $[A(i) B(i)]$, $i = 0, \ldots, 2$
- 3 diagonals for decomposing the (virtual) parallel 3-platform formed by $\{A(i), B(i), i = 0, \ldots, 2\}$ into three tetrahedra (as shown in the top row of Fig. 3, $[A(1) B(0)]$, $[A(2) B(0)]$, $[A(20 B(1)]$
- 5 diagonals for decomposing each virtual finger loop (closed by the base-to-tip diagonal) into a tree of triangle, such as those shown at the bottom row of Fig. 3 for trees of triangles anchored at the tip joints—a planar n-gon has $n - 3$ diagonals for a triangle tree (for more details, see [8], [11]); in total, the three fingers have 5 such diagonals, one for the thumb, and two each for the other two fingers.

A. Simplex-based Parameters

The simplices in Fig. 3 form the major part (with a subtlety and some additional simplices to be discussed in subsection III-C) of a **simplicial construction tree** of this system, that is, a tree of simplices satisfying the following three conditions. (1) Each link in the linkage system is an edge of at least one simplex in the tree. (2) The set of joints of the linkage system equals the set of all vertices of all simplices in the tree. (3) The configurations of the linkage system can be constructed from the shapes of the simplices and relative configurations of simplices adjacent in the tree.

These three properties enable us to construct configurations using the **simplex assembly process**. To see this, note that placing a simplex in an ambient space is equivalent to determining the coordinates of its vertices. Now, given the shapes and relative configurations of the simplices, we construct the corresponding configuration of the system recursively as follows. (I) Fix the initial tetrahedron (along with the fixed base joints) in place. (II) While there is an edge in the tree such that one simplex has already been placed in space but the other simplex has not yet been placed, then use the relative configuration between the two to place the other one. When this simplex assembly process terminates, the configuration has been constructed.

Our **simplex based parameters** for the system comprise

- shape parameters (for the shapes of the simplices): lengths of the diagonals
- orientation parameters (for the relative configurations of adjacent simplices).

There are generally two types of orientation parameters—either a continuous angle (a dihedral angle), *e.g.*, for two spatial simplices sharing one edge, or a discrete (essentially binary-valued) orientation sign, *e.g.*, for two planar triangles sharing an edge or two spatial tetrahedra sharing a face.

These binary orientations of these simplices are not extrinsic orientations relating the simplices to external objects (*e.g.*, fixed reference frames). Rather, the orientation of a simplex as used here is an intrinsic property of the relative distributions of the simplex vertices among themselves. For a planar triangle, its orientations (denoted by "+" and "−" here as in our earlier papers) indicate whether the vertices of the triangle form a counterclockwise or clockwise cycle when traversed in some specified order. The tetrahedron orientation signs are likewise

defined. As an example, Fig. 4 shows two configurations and one tetrahedral tree of a triangular prism, *i.e.* a three-legged parallel platform with spherical joints at both ends of the legs. In the figure, the fixed links and diagonals are drawn in black and in gray respectively. The pictured configurations differ only in the opposite orientations of $Tet(3)$. For more about orientation parameters, see [8], [12].

These simplex-based parameters can be used to parametrize the example hand system, either open or grasping, in a **unified** way. When the hand grasps an object with all three fingers at fixed contact points on the objects, we model the grasp by fixed link lengths of the diagonal among the tips. Then the system loses 3 DOF but can still be explicitly parametrized by the simplex-based parameters, by eliminating from the parameter set the now fixed lengths of the diagonals between the fingertips, $[A(0)\,A(1)]$, $[A(0)\,A(2)]$, $[A(1)\,A(2)]$. (When only two fingers grasp an object, the explicit parametrization set is obtained by eliminating just one diagonal length.)

Also note that the simplex assembly process as described earlier can be considered as an algorithm for the **forward kinematics** for determining the joint positions and thus system configurations from the simplex-based parameters. The **inverse kinematics** problem, when considered for some fixed fingertip positions or desired inter-fingertip distances (such as those arising from grasping an object at fixed contacts), is roughly equivalent to finding the valid parameter values of the system *CSpace*, in other words, the *CSpace* as parametrized by the simplex-based parameters, which is the main topic of this paper and will be discussed next.

B. Simplex Formation Constraints

In our new simplex-based parameters, the formulations of loop closure constraints are **fundamentally different** from the conventional ones for joint parameters or Cartesian coordinates. The key point is that

each loop can be formed if and only if all the simplices in its part of the construction tree can be successfully formed.

Therefore, we formulate the loop closure constraints as simplex formation constraints, such as triangle inequalities for triangles and Cayley-Menger determinant constraints [2] for general simplices in Euclidean space. Further, these constraints are in shape parameters only, since only the shape parameters, along with fixed link lengths, determine the existences and shapes of simplices. Orientation parameters are independent of the loop closure constraints.

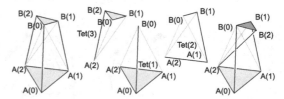

Fig. 4. A triangular prismoid: two configurations (at both ends) and a construction tree of 3 tetrahedra, with $Tet(2)$ adjacent to the other two (because it has common triangular faces with the other two). The two configurations differ only in the orientation of $Tet(3)$.

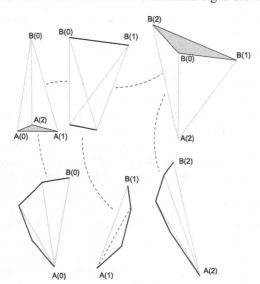

Fig. 3. Major part of a simplicial construction tree for the system in Fig. 1.

Below we recite the well-known simplex formation constraints on the side lengths of triangles and tetrahedra in Euclidean space, which are the only simplices used in the example systems in this paper. Denote the vertices of a triangle by $P(i)$, $i = 1, 2, 3$. Their interpoint distances must satisfy the following triangle inequalities.

$$
\begin{aligned}
d(1,2) - d(2,3) - d(3,1) &\leq 0 \\
d(2,3) - d(3,1) - d(1,2) &\leq 0 \quad (1) \\
d(3,1) - d(1,2) - d(2,3) &\leq 0
\end{aligned}
$$

Equivalently, we can require non-negativity of the following Cayley-Menger determinant, which is proportional to the area of the triangle.

$$
2\left(\frac{-1}{2}\right)^3
\begin{vmatrix}
0 & 1 & 1 & 1 \\
1 & 0 & D(1,2) & D(1,3) \\
1 & D(2,1) & 0 & D(2,3) \\
1 & D(3,1) & D(3,2) & 0
\end{vmatrix} \geq 0 \quad (2)
$$

(Here $D(i,k) = d^2(i,k)$ is the squared distance between ex points $P(i)$ and $P(k)$.) Similarly, for a tetrahedron with vertices $P(i)$, $i = 1, \ldots, 4$ and squared side lengths $D(i,j)$ ($0 \leq i, j \leq 3$), existence is equivalent to the CMD constraint

$$
\begin{vmatrix}
0 & 1 & 1 & 1 & 1 \\
1 & 0 & D(0,1) & D(0,2) & D(0,3) \\
1 & D(1,0) & 0 & D(1,2) & D(2,3) \\
1 & D(2,0) & D(2,1) & 0 & D(2,3) \\
1 & D(3,0) & D(3,1) & D(3,2) & 0
\end{vmatrix} \geq 0. \quad (3)
$$

In both (2) and (3), equality is equivalent to singularity of the corresponding simplex (collinearity of the vertices of a triangle, coplanarity of the vertices of a tetrahedron).

C. An Example Simplicial Construction Tree

For the simplices drawn in Fig. 3 for the three-fingered grasping system in Fig. 1, the main idea is to use the base-to-tip diagonals $[A(i)\,B(i)]$ and inter-tip diagonals $[A(i)\,A(i+1)]$, together with the links $[B(i)\,B(i+1)]$ between the fixed bases, to form a virtual triangular prism. Then define three more diagonals, such as $[A(1)\,B(0)]$, $[A(2)\,B(0)]$, and $[A(2)\,B(1)]$ (see Fig. 3) to form a construction tree of three tetrahedra for the prism, in a way similar to Fig. 4. (Clearly other choices of three diagonals could be made, resulting in different trees.)

Now for each finger, its base-to-tip diagonal forms a virtual loop with its links. Because all the internal joints are modeled to be revolute, with the rotational axes parallel to each other and perpendicular to the finger plane, this finger loop can be considered to be a loop with revolute joints contained in a plane (that rotates with the base spherical joint); and we have studied construction trees for planar loops with revolute joints in our previous papers, to which we refer the reader for further details ([11], [12] deal only with the loop closure constraint, [13] extends our study to self-collision avoidance). Briefly, a planar loop with n revolute joints has a triangle construction tree with $n - 3$ diagonals and $n - 2$ triangles (for $n > 3$ this tree is not unique; in Fig. 3 we have used the construction tree

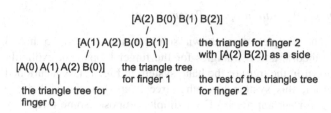

Fig. 5. A construction tree for the example 3-fingered manipulation system (the triangle trees for finger loops are not given in detail).

of triangles anchored at the tip joints, mostly for simplicity of the figures—other trees certainly could be used instead).

To keep the figure simple and use it to reflect main ideas in our approach, we do not reflect in Fig. 3 one subtlety existing in the simplicial construction for the hand system drawn in Fig. 1, namely, how to determine the finger planes. Note that each finger shares its base-to-tip diagonal with an appropriate tetrahedron and is adjacent to the tetrahedron in the construction tree. Suppose that in the simplex assembly process, we place the tetrahedra first. Then to put down the triangles for the finger loops, we need to be able to determine the orientations of the finger planes. These orientations will largely depend on the particular realizations (actuations) of the spherical joints. In our preliminary study, we have found that we can build upon the diagonals and the construction trees in Fig. 1 to accommodate various possible realizations of the spherical joints, sometimes by introducing additional diagonals. These new diagonals increase the codimension of the corresponding configuration spaces in their new ambient spaces; but their codimensions are still generally smaller than they would be in joint angles space. We will leave these issues to a future paper. Here we'd like to introduce an extra degree of freedom for each finger, which is to allow each finger plane to rotate freely about its end-to-tip diagonal.

Note that, with our human fingers, when we keep the base and tip of a finger fixed in space, we can still bend (rotate) the finger plane a little bit around the base-to-tip diagonal, probably allowed by the tendons in the hand. While the models of these behaviors and their engineering implications need more careful examination, here we will use this model to demonstrate the applicability of the simplex-based approach to highly complex manipulation type systems. We also note that such a model can still be useful for applications in areas such as computer animation, where actuations and engineering feasibilities are not as critical as for physical systems.

For this extra degree of freedom for each finger, we use a dihedral angle between the plane of the finger with one triangular face of its adjacent tetrahedron. The overall system now has 17 degrees of freedom; its simplicial construction tree is indicated in Fig. 5. The simplex-based parameters are

- shape parameters (for the shapes of the simplices): lengths of the 14 diagonals, and
- orientation parameters (for the relative configurations of adjacent simplices): the 3 dihedral angles and 11 essentially binary orientations of the simplices.

D. Configuration Space Parametrizations

The simplex tree shown schematically in Fig. 5 has 3 tetrahedra, and 8 triangles for the finger loops (not explicitly shown in the tree) of the hand system in Fig. 1. (As mentioned earlier, this system has other trees, but this one will do to explain our approach.) Each simplex imposes some inequality constraints on the shape parameters involved in it. Taken together, all these simplex formulation constraints define the set of **feasible shape parameters**, which we call *DStretch*. For this system, squared diagonal lengths can serve well as shape parameters, partly because both triangle and tetrahedron constraints can be formulated in them as the Cayley-Menger determinant constraints (2)(3). But it is also fine to use a combination of squared diagonal lengths (for those involved in tetrahedra) and direct diagonal lengths(for those only involved in triangles). More generally, it is possible to use non-degenerate (diffeomorphic) functions to map the diagonal-length-based shape parameters to other parameters.

Of the 14 diagonals listed at the beginning of section III, some will have fixed lengths for some grasp configurations. For example, when the hand grasps an object with all three fingers at fixed contact points on the objects, the three diagonals $[A(0) A(1)]$, $[A(0) A(2)]$, $[A(1) A(2)]$ between the fingertips will have fixed lengths and should be eliminated from the shape parameter set for the system configurations in this grasp mode. Similarly, for all configurations with two fingers grasping an object at fixed contacts, the diagonal between the corresponding two fingers has a fixed length and is no longer needed for the shape parameterization. Similarly, if we are interested in fixing the base-to-end distance of some finger, we can use that distance for the relevant diagonal and exclude the diagonal from the shape parameter set.

In summary, the set of 14 diagonals and its appropriate subsets provide explicit parametrizations of the shapes of simplices in the construction tree. These shape parameters are subject to simplex formation constraints, as well as fixed length constraints for suitable formulation of grasping modes.

As mentioned above, the loop closure constraints impose no constraints on the set of **feasible orientation parameters**, which we call *DFlip*. For the hand system and its simplicial tree in Fig. 5, $DFlip = (S^1)^3 \times \{+, -\}^{11}$, where the three-dimensional torus $(S^1)^3$ reckons dihedral angles and $\{+, -\}^{11}$ intrinsic simplex orientations. Note that if we want to fix the value of one dihedral angle or the orientation sign of one simplex, we can take out its corresponding orientation parameter from the set of system orientation parameters. Also note that the joint limits existing in human hands (for most of us) generally allow each finger to bend only toward the palm, with its fully extended configuration as one extreme feasible configuration. If we impose similar joint angle constraints on a robotic hands, and use anchored triangles in the construction tree (as in our example), then we need consider just one orientation for each triangle, so that the triangle-related orientation signs drop out of *DFlip*. As will become clear soon, a smaller set of orientation parameters will substantially simplify the overall structure of *CSpace* (by reducing the number of copies of *DStretch* in *CSpace*).

Singular configurations play an important role in constructing the overall configuration spaces. Their shape and orientation parametrization need careful treatments. Refer to papers [9], [11], [12] for our approaches to similar issues for loops with spherical type joints. Here, by labeling singular simplices—like a tetrahedron (resp., triangle) with 4 (resp., 3) coplanar (resp., collinear) vertices—with both + and −, we obtain (roughly) an identification of the **configuration space** with $DStretch[\mathfrak{J}] \times DFlip$. In other words, using the simplicial construction tree in Fig. 5, the configuration space of the hand system in Fig. 1 consists of 2^{11} copies of $DStretch \times (S^1)^3$. The overall parametrization and stratified structures of the *CSpace* here is very similar to that for a planar loop with n revolute joints described in [11], [12], but with some important technical differences, such as more complicated constraints on *DStretch* and more complex gluing of different strata. Further, the overall structures of the configuration spaces will depend on the identification (gluing) of different strata along proper singular subsets; and the identification of the hand system will be more complicated than that of a planar polygon. We will address these topics in future papers. Here we summarize the main parametrization results whose proofs we have sketched in this section.

Theorem 1. Consider the 3-fingered hand system in Fig. 1 and its simplicial construction tree in Fig. 5. Then:

(A) The configurations of the system are described by simplex-based parameters.

(B) The system configuration space *CSpace* is essentially the product of *DStretch* and *DFlip*, where (1) *DStretch* comprises shape parameters satisfying explicit, simply evaluated constraints (triangle or Cayley–Menger determinant inequalities, and range inequalities) required for successful simplex formation, and is a convex body in squared diagonal lengths, while (2) *DFlip* comprises relative orientation parameters, and is independent of loop closure constraints.   □

IV. EXAMPLE SYSTEM II: STAR MANIPULATOR

Our second example system is the general star-manipulator, as defined in [17], [25]. Such a manipulator is formed by joining k planar chains with revolute joints to a common point (like the thorax of an insect) and then fixing the base of each chain to the ground (an example is shown, in heavy black lines, in the leftmost subfigure of Fig. 6). The result is a planar linkage system with multiple loops (for $k \geq 3$). The designers of these manipulators call the chains "legs". We will call them "fingers" for the sole purpose of using the same hand terminology for the example systems in this paper.

We assume that each finger has at least 2 joints so as to give a nontrivial reachable workspace for the thorax A. Consider the thorax A to be the tip of all the fingers. Its position is determined by the finger link lengths and joint angles. For a manipulator with k fingers, the condition that all k fingertips meet at one common point in the plane imposes $2(k-1)$ constraints on the joint angles. So if we use joint angles as the parameters for the study of a star-shaped manipulator with k fingers, the configuration space is an semi-analytic set of codimension $2(k-1)$ in the ambient joint angle space.

Alternatively, if we know the position of the thorax A, we can define an auxiliary link between A and each base $B(i)$ (drawn as dashed lines in Fig. 6). Each such auxiliary link forms a (virtual) loop with the original links of the corresponding finger, and the loop is a planar loop with revolute joints, whose configurations can be studied via its construction trees of triangles (for instance, anchored triangles as shown in Fig. 3). See [8], [11]–[13] for more details.

The position of A is parametrized by its Cartesian coordinates (A_x, A_y), and its range (the workspace of A) is easily found. Specifically, the reachable workspace of the fingertip of the open chain with revolute joints and fixed base $B(i)$ is generally an annulus (exceptionally, a disk) with center at $B(i)$, having inner and outer radii that are easily computed from the finger link lengths. Such an annulus workspace corresponds to the minimum and maximum bounds on the distance between the finger tip and the base.

$$\text{(min-radius-}i)^2 \le iD(A, B(i)) \le \text{(max-radius-}i)^2 \quad (4)$$

When the fingers are assembled to form the star-manipulator, the workspace of A is the intersection of these annuli given by inequalities (4), $i = 0, \ldots, k-1$. The explicitness (and comparative simplicity) of this workspace, along with some critical insights and novel ideas, was used in [17], [25] to develop a smart polynomial time algorithm for a complete path planner (ignoring all constraints other than loop closure) for these manipulators; the algorithm uses Cartesian coordinates on the workspace, and joint angle parameters for the fingers.

Here we focus on parametrization rather than motion planning, using generalized simplex-based parameters for a star-manipulator, as follows:

- Cartesian coordinates for the thorax A: (A_x, A_y);
- shape parameters for the triangles in the construction trees of the finger loops: lengths of the diagonals; and
- orientation parameters for the triangles in the construction trees of the finger loops: essentially binary variables for the intrinsic triangle orientations.

In these parameters, the system constraints are the radius constraints (4) (on (A_x, A_y)), and triangle formation constraints (parametrized by (A_x, A_y)) on triangle shape parameters. There are no constraints on the orientation parameters.

Our general methods can be used to show that this parametrization has the following properties.

Theorem 2 Consider a star-shaped manipulator and its generalized simplex-based parameters. Then:

(A) The configurations of the system are described by the generalized simplex-based parameters.

(B) The system configuration space *CSpace* essentially the product of *DStretch* and *DFlip*, where (1) *DStretch*, the set of feasible length parameters, comprises the Cartesian coordinates of the thorax and triangle shape parameters satisfying explicit, simply evaluated constraints (the radii, and triangle or

Cayley–Menger determinant inequalities) required for reachable thorax positions and successful simplex formation, while (2) *DFlip* comprises triangle orientation parameters, and is independent of loop closure constraints. □

Again *CSpace*, as parametrized by the generalized simplex-based parameters, consists of copies of *DStretch*. Fig. 6 shows a schematic figure of *DStretch* for a 3-DOF star-manipulator. The 3-jointed finger has one diagonal, which has an interval of feasible values for each thorax position; the 4 jointed finger has 2 diagonals, with a convex polygonal region of feasible values. So *DStretch* is a family of 3-dimensional prisms parametrized by the points of the reachable workspace $WS(A)$ of A.

The similarity in the parametrization and structures of the configuration space of these two example systems are common for **linkage systems allowing simplicial construction trees**. The example systems and their simplicial construction trees presented here are substantially different from those in our earlier papers [8]–[13], and the encouraging results to these challenging multi-loop systems here further demonstrate the versatility of the simplex-based approach.

Many important linkage systems (*e.g.*, general $6R$ manipulators and many multivertex origamis) can be shown not to allow simplicial construction trees. Nonetheless, using methods related to simplicial construction trees, we have derived some interesting results for several such systems; for example, we can recover the number of inverse kinematics solutions for $6R$ manipulators in a natural way. Here we have no space for further details, which will appear in later papers.

V. SUMMARY

Linkage systems such as multi-fingered hands and multi-legged mobile systems are important in the study of robotics. They are also very challenging – in part because that the limbs form loops when their tips contact some common objects such as an grasped object for a hand or a common ground for legs. Conventionally the configurations of these systems are described in joint parameters, and the corresponding configuration spaces are lower-dimensional semi-algebraic subsets (as defined by loop closure constraints) embedded in joint parameter space. Such an **implicit parametrization**, coupled with complex nonlinear formulation of the loop closure constraints, poses challenges for diverse problems such as computing *CSpace* structures and using sampling for motion planning.

In contrast, **explicit parametrization**, where the cardinality of the set of parameters is equal to the dimension of the configuration space under study, offers advantages over implicit parametrization, especially when the constraints on the explicit parameters have nice properties. In this paper, we present explicit parametrization for two example anthropomorphic manipulation systems. Our parametrization are based on our recently developed simplicial construction trees for multi-object systems; and the constraints on the simplex-based parameters are mainly simplex formation constraints (such as triangle inequalities and Cayley-Menger determinant constraints). The model systems here are difficult to deal with using the joint angle parameters; and our results here, along

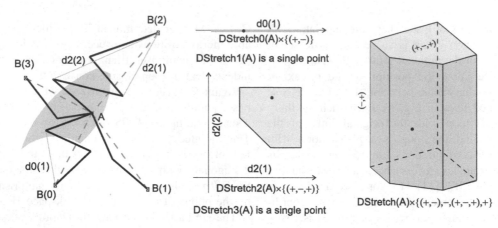

Fig. 6. Schematic view of *DStretch* of a star-manipulator.

with those in our earlier papers [8]–[13], demonstrate the generality and effectiveness of the simplex-based approach.

Not every linkage system has a simplicial construction tree. But in our preliminary study of other linkages, joint types and system constraints, we have found that many kinematic structures can accommodate triangle or tetrahedral decompositions, sometimes by first ignoring some constraints in the systems. Such structures often lead to parametrizations of configuration spaces with **lower codimensions** than those parametrized by joint parameters. In ongoing research, we are generalizing our simplex-based approach, and developing algorithms to identify and create systems with simplicial construction trees.

ACKNOWLEDGMENT

We are partially supported by NSF grant IIS0713335. We thank the anonymous referees for their helpful remarks.

REFERENCES

[1] M. Alberich, F. Thomas, and C. Torras, *Flagged parallel manipulators*, IEEE Trans. Robotics (2007), 1013–1023.

[2] L. M. Blumenthal, *Theory and applications of distance geometry, 2nd edition*, American Mathematical Society, Providence, 1970.

[3] H. Choset, W. Burgard, S. Hutchinson, G. Kantor, L. E. Kavraki, K. Lynch, and S. Thrun, *Principles of robot motion: Theory, algorithms, and implementation*, MIT Press, Cambridge, MA, 2005.

[4] John J. Craig, *Introduction to robotics: Mechanics and control, 2nd edition*, Addison-Wesley Publishing Company, Reading, MA, 1989.

[5] M. Cutkosky, *Grasping and fine manipulation for automated manufacturing*, Kluwer Academic Publishers, Boston, 1986.

[6] B. Goodwine and J. Burdick, *Motion planning for kinematic stratified systems with application to quasi-static legged locomotion and finger gaiting*, Algorithmic and Computational Robotics – New Directions (WAFR 2000), 2000, pp. 109–127.

[7] R. Grupen and O. Brock (signatories R. Ambrose, C. Atkeson, O. Brock, R. Brooks, C. Brown, J. Burdick, M. Cutkosky, A. Fagg, R. Grupen, J. Hoffman, R. Howe, M. Huber, O. Khatib, P. Khosla, V. Kumar, L. Leifer, M. Mataric, R. Nelson, A. Peters, K. Salisbury, S. Sastry, and R. Savely), *White paper: Integrating manual dexterity with mobility for human-scale service robotics – the case for concentrated research into science and technology supporting next-generation robotic assistants*, http://robotics.cs.umass.edu/~oli/publications/index.html, 2004.

[8] L. Han and L. Rudolph, *The inverse kinematics of a serial chain with joints under distance constraints*, Robotics: Science and Systems, 2006.

[9] _____, *A unified geometric approach for the inverse kinematics of a serial chain with spherical joints*, Proc. IEEE Int. Conf. Robot. Autom. (ICRA), 2007.

[10] _____, *Simplex-tree based kinematics of foldable objects as multi-body systems involving loops*, Robotics: Science and Systems, 2008.

[11] L. Han, L. Rudolph, J. Blumenthal, and I. Valodzin, *Stratified deformation space and path planning for a planar closed chain with revolute joints*, Proc. Seventh International Workshop on Algorithmic Foundation of Robotics (WAFR), 2006.

[12] _____, *Convexly stratified deformation space and efficient path planning for a planar closed chain with revolute joints*, Int. J. Robot. Res. **27** (2008), 1189–1212.

[13] L. Han, L. Rudolph, S. Dorsey-Gordon, D. Glotzer, D. Menard, J. Moran, and J. R. Wilson, *Bending and kissing: Computing self-contact configurations of planar loops with revolute joints*, Proc. IEEE Int. Conf. Robot. Autom. (ICRA), 2009.

[14] M. Kapovich and J. Millson, *On the moduli spaces of polygons in the euclidean plane*, J. Diff. Geom. **42** (1995), 133–164.

[15] J. C. Latombe, *Robot motion planning*, Kluwer Academic Publishers, Boston, 1991.

[16] S. M. LaValle, *Planning algorithms*, Cambridge University Press, Cambridge, U.K., 2006, http://planning.cs.uiuc.edu/.

[17] G.F. Liu, J.C. Trinkle, and N. Shvalb, *Motion planning for a class of planar closed-chain manipulators*, Proc. IEEE Int. Conf. Robot. Autom. (ICRA), 2006, pp. 133–138.

[18] M. Mason, *Mechanics of robotic manipulation*, MIT Press, Cambridge, 2001.

[19] M. Mason and K. Salisbury, *Robot hands and the mechanics of manipulation*, MIT Press, Cambridge, 1985.

[20] J.-P. Merlet, *Parallel robots*, Springer, New York, 2000.

[21] R.J. Milgram and J.C. Trinkle, *The geometry of configuration spaces for closed chains in two and three dimensions*, Homology Homotopy and Applications (2002).

[22] R. M. Murray, Z. Li, and S. S. Sastry, *A mathematical introduction to robotic manipulation*, CRC Press, Boca Raton, 1994.

[23] J. Ponce, S. Sullivan, A. Sudsang, J.-D. Boissonat, and J.-P. Merlet, *On computing four-finger equilibrium and force-closure grasps of polyhedral objects*, Int. J. Robot. Res. **16** (1996), 11–35.

[24] J. Porta, L. Ros, and F. Thomas, *On the trilaterable six-degree-of-freedom parallel and serial manipulators*, Proc. IEEE Int. Conf. Robot. Autom. (ICRA), 2005.

[25] N. Shvalb, L.G. Liu, M. Shoham, and J.C. Trinkle, *Motion planning for a class of planar closed-chain manipulators*, Int. J. Robot. Res. **26** (2007), no. 5, 457–474.

[26] X. Tang, S. Thomas, and N. Amato, *Planning with reachable distances*, Proc. Eighth International Workshop on Algorithmic Foundation of Robotics (WAFR), 2008.

[27] C. Torras, F. Thomas, and M. Alberich, *Stratifying the singularity loci of a class of parallel manipulators*, IEEE Trans. Robotics (2006), 23–32.

[28] J.C. Trinkle and R.J. Milgram, *Complete path planning for closed kinematic chains with spherical joints*, Int. J. Robot. Res. **21** (2002), no. 9, 773–789.

[29] Y. Wei and B. Goodwine, *Stratified manipulation on non-smooth domains*, IEEE Trans. Robotics and Automation **20** (Feb. 2004), no. 1, 128–132.

Approximating Displacement with the Body Velocity Integral

Ross L. Hatton and Howie Choset
Carnegie Mellon University
{rlhatton, choset}@cmu.edu

Abstract—In this paper, we present a technique for approximating the net displacement of a locomoting system over a gait without directly integrating its equations of motion. The approximation is based on a volume integral, which, among other benefits, is more open to optimization by algorithm or inspection than is the full displacement integral. Specifically, we develop the concept of a *body velocity integral* (BVI), which is computable over a gait as a volume integral via Stokes's theorem. We then demonstrate that, given an appropriate choice of coordinates, the BVI for a gait approximates the displacement of the system over that gait. This consideration of coordinate choice is a new approach to locomotion problems, and provides significantly improved results over past attempts to apply Stokes's theorem to gait analysis.

I. INTRODUCTION

Locomotion is everywhere. Snakes crawl, fish swim, birds fly, and all manner of creatures walk. The facility with which animals use internal joint motions to move through their environments far exceeds that which has been achieved in artificial systems; consequently there is much interest in raising the locomotion capabilities of such systems to match or surpass those of their biological counterparts. A fundamental aspect of animal locomotion is that it is primarily composed of gaits – cyclic shape motions which efficiently transport the animal. Examples of such gaits include a horse's walking, trotting, and galloping, a fish's translation and turning strokes, and a snake's slithering and sidewinding. The efficacy of these motions, along with the abstraction that they allow from shape to position changes, suggests that gaits will form an equally important part of artificial locomotion.

Here, we are specifically interested in producing tools for designing gaits for mechanical systems which result in desired net position changes. Much prior work in gait design has taken the approach of choosing parameterized basis functions for gaits and simulating the motion of the system while executing the gaits, optimizing the input parameters to find gaits which meet the design requirements. Such optimization with forward simulation is computationally expensive and vulnerable to local minima. Therefore, there is growing interest in using curvature analysis tools, such as Stokes's theorem, to replace the simulation step with a simple volume integration, which is more amenable to optimization. Unfortunately, these Stokes's theorem methods as previously developed are not completely applicable to most interesting systems; either they are restricted to designing small, inefficient motions, or they provide incomplete information about the actual displacement

of the system over the course of a gait.

In this paper we address these limitations by developing the concept of a *body velocity integral* (BVI), which provides an expanded and physically meaningful interpretation of previous locomotion work based on Stokes's theorem. We then identify conditions under which this body velocity integral is a good estimate of the true displacement resulting from a gait. We finish by introducing the notion that rather than being intrinsic to the system, the presence of these conditions is dependent on the choice of parameterization of the system, and demonstrating that this choice of parameterization can be manipulated to ensure their existence.

II. PRIOR WORK

Our work builds on the body of locomotion literature which uses geometric mechanics to separate internal shape changes from the external motions they produce. The application of geometric mechanics to locomotion, pioneered by Shapere and Wilczek [1] and further developed by Murray and Sastry [2] and Kelly and Murray [3], provides a powerful mathematical framework for analyzing locomotion. A key product of this work is the development of the *reconstruction equation* for nonholonomic systems, which relates body velocity to changes in internal shape for a broad class of locomoting systems. We will not rederive the reconstruction equation here; for a thorough treatment, see [4]–[6].

This reconstruction equation has been used in a variety of locomotion contexts. Ostrowski *et al.* [5], [7] combined the reconstruction equation with Lie bracket theory to generate sinusoidal gaits which translate and rotate a variety of snake-like systems. Bullo and Lynch used the reconstruction equation to decouple the locomotion of kinodynamic systems and design kinematic gaits [8]. More recently, there has been interest in applying these techniques to swimming robots, such as McIsaac and Ostrowski's work on anguilliform (eel-like) robots [9] and Morgansen *et al.*'s work on fish [10], both of which combine the geometric approach with biomimetic elements. In [11], we introduced the *connection vector field* as a tool for visualizing the reconstruction equation differentially.

It is not generally possible to integrate the reconstruction equation in closed form, raising difficulties for the inverse problem of finding shape changes which result in desired translations. In special cases, however, Stokes's theorem can be used to find the net motion resulting from gaits [3]. Mukherjee [12] used this principle to analyze the motion of rolling disks, and Walsh and Sastry [13] applied it to the case of an isolated three-link robot. Shammas *et al.* [6], [14] combined this approach with the reconstruction equation to define functions on the shape space of their three-link robots, which allowed the design of gaits resulting in specified rotations. A similar technique was used by Melli *et al.* [15] and later by Avron and Raz [16] to generate gaits for swimming robots.

III. Background

The present work makes use of several key techniques borrowed from geometric mechanics, the salient points of which we review here.

A. Three-link Kinematic Snake

While the principles we are investigating are relevant to a wide range of systems, including fish, differential drive cars, and satellites, in this paper we focus on a particular system, the *three-link kinematic snake* investigated by Shammas *et al.* [6], [14]. Illustrated in Fig. 1, this system has a passive wheelset on each link, preventing lateral translation while freely allowing rotation and longitudinal translation. The joint angles α_1 and α_2 are actively controlled, and the overall position and orientation of the system are taken as those of the middle link.

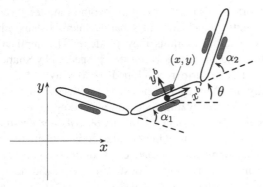

Fig. 1: Model for the three-link kinematic snake. The overall location of the system is the x, y position and orientation θ of the middle link with respect to the fixed coordinate system. The shape coordinates are α_1 and α_2, the angles of the two joints. The passive wheels on each link constrain lateral but not longitudinal or rotational motion.

B. The Reconstruction Equation and the Local Connection

When analyzing a multi-body locomoting system, it is convenient to separate its configuration space Q into a position space G and a shape space M, such that the position $g \in G$ locates the system in the world, and the shape $r \in M$ gives the relative arrangements of its bodies, and then consider how manipulating the shape affects the position. The geometric mechanics community [2]–[6] has addressed this question with the development of the *reconstruction equation* and the *local connection*, tools for relating the body velocity of the system, ξ, to its shape velocity $\dot{r}$, and accumulated momentum p.

The general reconstruction equation is of the form

$$\xi = -\mathbf{A}(r)\dot{r} + \mathbf{\Gamma}(r)p, \tag{1}$$

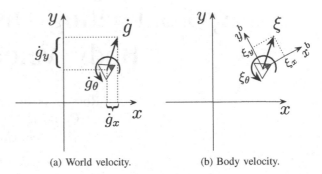

(a) World velocity.　　　(b) Body velocity.

Fig. 2: Two representations of the velocity of a robot. The robot, represented by the triangle, is translating up and to the right, while spinning counterclockwise. In (a), the *world velocity*, $\dot{g}$, is measured with respect to the global frame. The *body velocity*, ξ, in (b) is the world velocity represented in the robot's instantaneous local coordinate frame.

where ξ is the body velocity of the system, $\mathbf{A}(r)$ is the *local connection*, a matrix which relates joint to body velocity, $\mathbf{\Gamma}(r)$ is the *momentum distribution function*, and p is the *generalized nonholonomic momentum*, which captures how much the system is "coasting" at any given time [4]. The body velocity, illustrated for a planar system in Fig. 2(b), is the position velocity as expressed in the instantaneous local coordinate frame of the system. For systems which translate and rotate in the plane, *i.e.* which have an $SE(2)$ position space, the body and world velocities are related to each other as

$$\xi = \begin{bmatrix} \cos\theta & \sin\theta & 0 \\ -\sin\theta & \cos\theta & 0 \\ 0 & 0 & 1 \end{bmatrix} \dot{g}, \tag{2}$$

where θ is the system's orientation.

For systems which are sufficiently constrained, the generalized momentum drops out, and the system behavior is dictated by the *kinematic reconstruction equation*,

$$\xi = -\mathbf{A}(r)\dot{r}, \tag{3}$$

in which the local connection thus acts as a kind of Jacobian, mapping from velocities in the shape space to the corresponding body velocity. For the rest of this paper, we will limit our attention to these kinematic systems.

Kinematic Snake Example 1: For the kinematic snake, we take the position as $g = (x, y, \theta) \in SE(2)$ of the middle link and the shape as the joint angles $r = [\alpha_1\ \alpha_2]^T \in \mathbb{R}^2$, as in Fig 1. Outside of singularities, the passive wheelsets constitute three independent constraints, equal in number to the position coordinates, so the system has no direction in which to coast and behaves kinematically. As detailed in [6], the constraints define the local connection for the system, such that the reconstruction equation, normalized for link length, is

$$\xi = \begin{bmatrix} \xi_x \\ \xi_y \\ \xi_\theta \end{bmatrix} = -\frac{1}{D} \begin{bmatrix} 1 + \cos(\alpha_2) & 1 + \cos(\alpha_1) \\ 0 & 0 \\ -\sin(\alpha_2) & -\sin(\alpha_1) \end{bmatrix} \begin{bmatrix} \dot{\alpha}_1 \\ \dot{\alpha}_2 \end{bmatrix}, \tag{4}$$

where $D = \sin(\alpha_1) - \sin(\alpha_2) + \sin(\alpha_1 - \alpha_2)$.

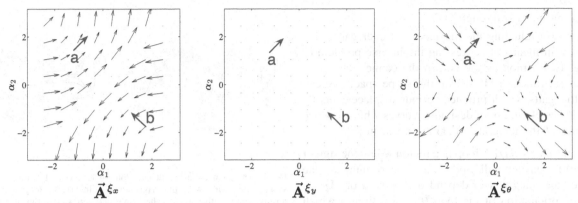

Fig. 3: The connection vector fields for the kinematic snake. Due to singularities in the vector fields at the lines $\alpha_1 = \pm\pi$, $\alpha_2 = \pm\pi$, and $\alpha_1 = \alpha_2$, the magnitudes of the vector fields have been scaled to their arctangents for readability. The $\dot{r}$ vectors at **a** and **b** produce pure forward translation and pure negative rotation, respectively.

C. Connection Vector Fields

Each row of the local connection $\mathbf{A}(r)$ can be considered as defining a vector field on the shape space whose dot product with the shape velocity produces the corresponding component of the body velocity,

$$\xi_i = \vec{\mathbf{A}}^{\xi_i}(r) \cdot \dot{r} \tag{5}$$

where, for convenience, we wrap the negative sign into the vector field definition.

Considering the local connection as a set of vector fields with the dot product operator provides strong geometric intuition for understanding the relationship between shape and position motions. The geometric interpretation of the dot product in (5) is

$$\xi_i = \vec{\mathbf{A}}^{\xi_i}(r) \cdot \dot{r} = \|\vec{\mathbf{A}}^{\xi_i}(r)\| \|\dot{r}\| \cos\Theta , \tag{6}$$

where Θ is the angle between the vectors. Taking the $\cos\Theta$ term as a measure of the alignment of $\vec{\mathbf{A}}^{\xi_i}(r)$ and $\dot{r}$, ξ_i is positive, negative, or zero when the two vectors have correspondingly positive, negative, or zero alignment, and is scaled by the magnitudes of $\vec{\mathbf{A}}^{\xi_i}$ and $\dot{r}$.

Kinematic Snake Example 2: The connection vector fields for the kinematic snake with reconstruction equation as in (4) are shown in Fig. 3 along with two example $\dot{r}$ vectors. At position **a**, $\dot{r}$ is aligned with $\vec{\mathbf{A}}^{\xi_x}$ and orthogonal to $\vec{\mathbf{A}}^{\xi_\theta}$, and will thus produce forward (positive longitudinal) motion with no rotation. At **b**, $\dot{r}$ is orthogonal to $\vec{\mathbf{A}}^{\xi_x}$ and anti-aligned with $\vec{\mathbf{A}}^{\xi_\theta}$, and will thus produce negative rotation with no translation.

D. Shape Changes

To describe operations in the shape space of a robot, we define shape changes, gaits, and image-families of gaits.

Definition 3.1 (Shape change): A *shape change* $\psi \in \Psi$ is a trajectory in the shape space M of the robot over an interval $[0, T]$, *i.e.*, the set of all shape changes is

$$\Psi = \{\psi \in C^1 \mid \psi : [0, T] \to M\} \tag{7}$$

where $\psi(0), \psi(T) \in M$ are respectively the start and end shapes.

Definition 3.2 (Gait): A *gait* $\phi \in \Phi$ is a cyclic shape change, *i.e.*

$$\Phi = \{\phi \in \Psi \mid \phi(0) = \phi(T)\}. \tag{8}$$

Note that a gait has a defined start shape $\phi(0)$; two gaits whose images in M are the same closed curve, but with different start points, are distinct.

Definition 3.3 (Image-family): The *image-family* of a gait is the set of all gaits which share its image (*i.e.*, trace out the same closed curve) in M.

E. Stokes's Theorem

In general, to calculate the displacement resulting from a gait, we must use the local connection to find the body velocity and then integrate this velocity over time. In some cases, however, we can use Stokes's theorem to replace this time integral with an area integral. By this theorem, the line integral along a closed curve on a vector field is equal to the integral of the curl of that vector field over a surface bounded by the curve. For example, for systems with two shape variables, the integral of a component of the body velocity over a gait is thus

$$\int_0^T \xi_i(\tau) \, d\tau = \int_0^T \vec{\mathbf{A}}^{\xi_i}(\phi(\tau)) \cdot \dot{\phi}(\tau) \, d\tau \tag{9}$$

$$= \int_\phi \vec{\mathbf{A}}^{\xi_i}(r) \, dr \tag{10}$$

$$= \iint_{\phi_a} \operatorname{curl} \vec{\mathbf{A}}^{\xi_i}(r) \, dr, \tag{11}$$

where ϕ_a is the area on M enclosed by the gait. If a component of the world velocity is always equal to a component of the body velocity, *i.e.*, $\dot{g}_j = \xi_j$ for a given j, then we can apply Stokes's theorem to find the net displacement in that direction over the course of a gait, by identifying

$$\Delta g_j = \int_0^T \dot{g}_j(\tau) \, d\tau = \int_0^T \xi_j(\tau) \, d\tau \tag{12}$$

and substituting into (9) through (11).

In addition to evaluating the displacement over gaits, (11) offers a powerful means of addressing the inverse problem of designing gaits. By plotting the curl of the connection vector field as a height function $\mathbf{H}^{\zeta_j}(r)$ on the shape space, we can easily identify gaits which produce various displacements by inspection. Rules of thumb for designing curves which produce desired values of the integral in (11) are given in [6]:

1) *Non-zero integral (I)* A loop in a region where the sign of $\mathbf{H}^{\zeta_j}(r)$ is constant will produce a non-zero integral. The sign of the integral will depend on the sign of $\mathbf{H}^{\zeta_j}(r)$ and the orientation of the loop (the direction in which it is followed).
2) *Non-zero integral (II)* A figure-eight across a line $\mathbf{H}^{\zeta_j}(r) = 0$, where each half is in a region where the sign of $\mathbf{H}^{\zeta_j}(r)$ is constant, will produce a non-zero integral.
3) *Zero integral* A loop that encloses equally positive and negative regions of $\mathbf{H}^{\zeta_j}(r)$, such that the integrals over the positive and negative regions cancel each other out, will produce a zero integral.

Kinematic Snake Example 3: From (2), we see that $\xi_\theta = \dot{g}_\theta$ for systems with an $SE(2)$ position space, as is the case for the kinematic snake. For these systems we can thus use Stokes's theorem to design gaits which produce desired net rotations of these systems. The rotational height function for the kinematic snake,

$$\mathbf{H}^{\zeta_\theta} = \frac{\partial \vec{\mathbf{A}}_2^{\xi_\theta}}{\partial \alpha_1} - \frac{\partial \vec{\mathbf{A}}_1^{\xi_\theta}}{\partial \alpha_2} \tag{13}$$

is plotted in Fig. 4 along with two gait image-families. In Fig. 4(a), the loop encircles an area over which the integral $\iint_{\phi_a} \mathbf{H}^{\zeta_\theta}\, d\alpha_1 d\alpha_2$ is zero, so the net rotation resulting from any gait in that image-family will be zero. In Fig. 4(b), each half of the figure-eight pattern is on a different side of the $\mathbf{H}^{\zeta_\theta} = 0$ line, and the two halves have opposite orientations. The integrals over the two areas will thus sum together, and the snake will gain a net rotation from any gait in the family. The sign of this net rotation will depend on the orientation of the curve (the direction in which the curve is followed) [6], [15].

IV. THE BODY VELOCITY INTEGRAL

The convenience and simplicity of using the $\mathbf{H}^{\zeta_\theta}$ height function to design gaits resulting in desired rotations makes it tempting to apply the same techniques to designing gaits which produce specified rotations. Unfortunately, this is not generally possible, as the integral of body velocity provided by the translational height functions does not correspond to the resulting displacement. To see why this is the case, consider that there are two senses in which "integrating the body velocity over time" can be interpreted. In the first, most natural sense, we integrate to find the resulting displacement. This

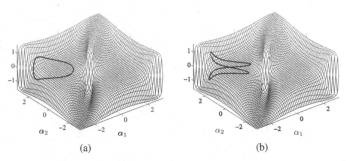

Fig. 4: The connection height function for rotation, $\mathbf{H}^{\zeta_\theta}$ for the kinematic snake, along with two gait image-families. In (a), the loop encircles equal positive and negative areas of the height function, so any gait from this family will result in zero net rotation. The two loops of the image-family in (b) have opposite orientations and encircle oppositely-signed regions of $\mathbf{H}^{\zeta_\theta}$, so gaits from this family will result in a non-zero net rotation. To accommodate the singularity along the $\alpha_1 = \alpha_2$ line, the height function is scaled to its arctangent for display.

quantity is found by the nonlinear iterative integral

$$g(t) = \begin{bmatrix} x(t) \\ y(t) \\ \theta(t) \end{bmatrix} = \int_0^t \begin{bmatrix} \cos\theta(\tau) & -\sin\theta(\tau) & 0 \\ \sin\theta(\tau) & \cos\theta(\tau) & 0 \\ 0 & 0 & 1 \end{bmatrix} \begin{bmatrix} \xi_x(\tau) \\ \xi_y(\tau) \\ \xi_\theta(\tau) \end{bmatrix} d\tau, \tag{14}$$

in which the x and y components of the body velocity are rotated into the world frame at each time.

The second sense of integrating ξ is to take its simple vector integral ζ,

$$\zeta(t) = \begin{bmatrix} \zeta_x(t) \\ \zeta_y(t) \\ \zeta_\theta(t) \end{bmatrix} = \int_0^t \begin{bmatrix} \xi_x(\tau) \\ \xi_y(\tau) \\ \xi_\theta(\tau) \end{bmatrix} d\tau, \tag{15}$$

where we term ζ the *body velocity integral*, or BVI. Physically, the BVI corresponds to the raw odometry for the robot, *i.e.*, the net forwards minus backwards motion in each body direction, and it does not account for changes in the alignment between body and world directions, *i.e.* the relative orientation of the body and world frames.

The area integrals under the height functions correspond to this second sense of integration, and as such cannot generally be used to determine the displacement resulting from a given gait. For instance, consider how the kinematic snake moves while executing gaits from the gait image family depicted in Fig. 5. On the height functions, the gait family encircles a positive area on $\mathbf{H}^{\zeta_x}$, a zero area on $\mathbf{H}^{\zeta_y}$, and a net-zero area on $\mathbf{H}^{\zeta_\theta}$. As in the example from Section III-E, we can conclude from the net zero integral on $\mathbf{H}^{\zeta_\theta}$ that the three-link robot will undergo no *net* rotation over the course of any gait from this family. However, it will undergo *intermediate* rotation, and, by comparing the translational BVI (ζ_x, ζ_y) against the actual (x, y) displacement resulting from gaits in this family, as in Fig. 6, we see that the BVI does not equal the displacement. Further, as the BVI has a single value for all gaits in an image family, but the displacement is dependent on the choice of start/end shape, it is apparent that we can have no expectation of finding a map from BVI to displacement.

There is a bright side, however. While our example shows that we cannot always take the BVI as an indication of the

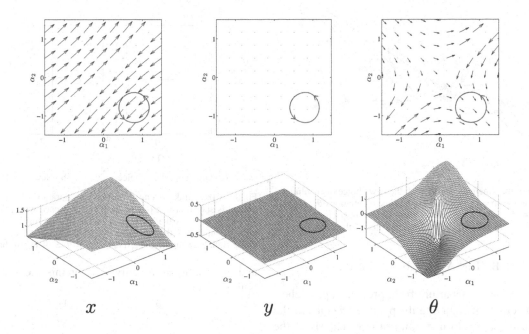

$$x \qquad\qquad y \qquad\qquad \theta$$

Fig. 5: Example gait image-family overlaid on the connection vector fields and height functions of the kinematic snake.

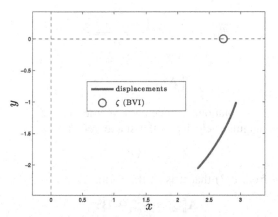

Fig. 6: The BVI and locus of displacements corresponding to the image family of gaits for the kinematic snake robot depicted in Fig. 5 are represented respectively by the circle on the x-axis and the arc-section. The BVI only depends on an area integral over the area enclosed by the gait and is thus the same for all gaits in the family. The displacement, however, also depends on the start/end shape of the gait, and there is thus a range of displacements, each corresponding to one starting point in the image family. The displacements are on the order of one link length of the robot.

net displacement, it does not show that we can *never* do so. To investigate the existence of gaits for which the BVI is a reliable measure of displacement, we consider the error ε_ζ between the BVI and displacement over an arbitrary gait. By subtracting (14) from (15) and combining the integrals, we can express this error as

$$\varepsilon_\zeta = \zeta - g = \int_0^t \begin{bmatrix} 1 - \cos\theta & \sin\theta & 0 \\ -\sin\theta & 1 - \cos\theta & 0 \\ 0 & 0 & 0 \end{bmatrix} \begin{bmatrix} \xi_x \\ \xi_y \\ \xi_\theta \end{bmatrix} \, d\tau, \quad (16)$$

with an implicit dependence on τ for all variables inside the integral. From (16), we can easily identify a condition which will guarantee that the error term is small: If $\theta(t)$ remains

small for all t, then the matrix in the integrand will remain close to a zero matrix for all t, and the error ε_ζ will thus be close to a zero vector. We will thus turn our attention to finding conditions for which this small angle condition holds true.

Trivially, small gaits offer θ no opportunity to grow large; these gaits have been studied extensively in the controls literature with the aid of Lie algebra analysis [2], [5], [10], [15]. These gaits, however, spend more energy in high-frequency oscillations than they do in producing net motion, and are thus inefficient; we are interested in designing larger, more efficient gaits. To this end, we observe that as the orientation of the system over a gait is

$$\theta(t) = \int_0^t \xi_\theta(\tau) \, d\tau = \int_{\phi(0)}^{\phi(t)} \vec{\mathbf{A}}^{\xi_\theta}(r) \, dr, \quad (17)$$

its value at any time is bounded to

$$|\theta(t)| \leq \int_{\phi(0)}^{\phi(t)} |\vec{\mathbf{A}}^{\xi_\theta}(r) \, dr|. \quad (18)$$

For gaits in regions for which $\|\vec{\mathbf{A}}^{\xi_\theta}(r)\|$ vanishes, $\theta(t)$ and $\varepsilon_\zeta(t)$ are thus guaranteed to be small, and the BVIs of those gaits are thus good approximations of their displacements.

The $\vec{\mathbf{A}}^{\xi_\theta}(r)$ vector field for the kinematic snake in Fig. 5 clearly does not become small at any point, so there are no regions over which we can use the BVI to design gaits. However, we will see below that the connection vector fields depend on the representation of the system, and that there are sets of connection vector fields for the kinematic snake in which $\vec{\mathbf{A}}^{\xi_\theta}(r)$ does vanish over a region of M.

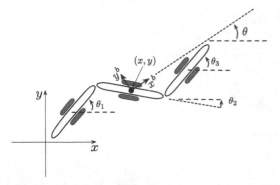

Fig. 7: Orientation of the kinematic snake with the new choice of coordinates. The orientation θ is the mean of the individual link orientations θ_1, θ_2, and θ_3. Note that the body frame directions x^b and y^b are respectively aligned with and perpendicular to the orientation line, and not to the central axis of the middle link.

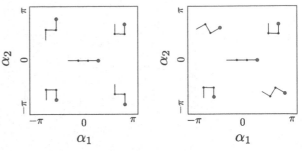

(a) Old coordinate choice   (b) New coordinate choice

Fig. 8: A three-link system in various configurations. In both (a) and (b), the system is shown at five points on the shape space with reference orientation $\theta = 0$. In (a), θ is the orientation of the middle link, while in (b), θ is the mean orientation of the links. The system is much more consistently "pointing to the right" (an intuitive interpretation of $\theta = 0$) in (b) than in (a).

V. THE BVI IN A NEW BODY FRAME

We now make a major departure from previous approaches, and consider the effect of changing the parameterization of the problem. For instance, up to this point in our analysis of the kinematic snake we have been considering g as the position and orientation of its middle link, as shown in Fig. 1. This choice of coordinates is customary in the treatment of three-link systems [3], [6], [11], [15], [16], and has the advantage of providing for the simple representation of the system constraints in the body frame. An equally valid choice of coordinates, however, would be to take θ not as the orientation of the middle link, but as the mean orientation of all three links, *i.e.*,

$$\theta = \frac{\theta_1 + \theta_2 + \theta_3}{3}, \tag{19}$$

where θ_i is the orientation of the ith link, as in Fig. 7.

This choice of coordinates has two key benefits over the previous choice. First, the orientation θ matches an intuitive understanding of the "orientation" of the system much more strongly with the new choice than for the old choice, as illustrated in Fig. 8. Second, and more importantly, the $\mathbf{A}^{\xi_\theta}$ connection vector field vanishes in key regions, allowing the use of the height functions to design gaits with specific displacements.

To gain insight into this second benefit, consider that the orientations θ_1 and θ_3 of the outer links of the kinematic snake in Fig. 7 can be expressed in terms of the orientation θ_2 of the middle link and the joint angles α_1 and α_2 from Fig. 1 as

$$\theta_1 = \theta_2 - \alpha_1 \tag{20}$$
$$\theta_3 = \theta_2 + \alpha_2. \tag{21}$$

Consequently, θ_{new}, the mean orientation of the three links in (19), is thus

$$\theta_{\mathrm{new}} = \frac{(\theta_2 - \alpha_1) + \theta_2 + (\theta_2 + \alpha_1)}{3} \tag{22}$$
$$= \theta_2 + \frac{(-\alpha_1) + \alpha_2}{3}, \tag{23}$$

and the rotational velocity $\dot{\theta}_{\mathrm{new}}$ of this mean orientation line is then

$$\dot{\theta}_{\mathrm{new}} = \dot{\theta}_2 + \frac{(-\dot{\alpha}_1) + \dot{\alpha}_2}{3}. \tag{24}$$

Given that θ_2 and $\dot{\theta}_2$ are respectively the orientation and angular velocity of the old body frame, θ_{old} and $\dot{\theta}_{\mathrm{old}}$, the new angular velocity is

$$\dot{\theta}_{\mathrm{new}} = \dot{\theta}_{\mathrm{old}} + \frac{(-\dot{\alpha}_1) + \dot{\alpha}_2}{3} \tag{25}$$
$$= \vec{\mathbf{A}}_{\mathrm{old}}^{\xi_\theta} \cdot \dot{r} + \begin{bmatrix} -\frac{1}{3} & \frac{1}{3} \end{bmatrix} \cdot \dot{r} \tag{26}$$
$$= \left(\vec{\mathbf{A}}_{\mathrm{old}}^{\xi_\theta} + \begin{bmatrix} -\frac{1}{3} & \frac{1}{3} \end{bmatrix} \right) \cdot \dot{r}. \tag{27}$$

As the new rotational connection vector field, $\vec{\mathbf{A}}_{\mathrm{new}}^{\xi_\theta}$, relates the new angular velocity to the shape velocity by

$$\dot{\theta}_{\mathrm{new}} = \vec{\mathbf{A}}_{\mathrm{new}}^{\xi_\theta} \cdot \dot{r}, \tag{28}$$

we see from (27) that it is of the form

$$\vec{\mathbf{A}}_{\mathrm{new}}^{\xi_\theta} = \vec{\mathbf{A}}_{\mathrm{old}}^{\xi_\theta} + \vec{\mathbf{B}}^{\xi_\theta}, \tag{29}$$

with $\vec{\mathbf{B}}^{\xi_\theta} = \begin{bmatrix} -\frac{1}{3} & \frac{1}{3} \end{bmatrix}$. Representing this sum graphically, as in Fig. 9, we see the effect of the change of coordinates on the local connection: The connection modifier $\vec{\mathbf{B}}^{\xi_\theta}$ is approximately equal to the negative of the old rotational connection vector field $\vec{\mathbf{A}}_{\mathrm{old}}^{\xi_\theta}$ in the circled regions, and thus nullifies it when they are summed together.

By a similar derivation, the (row vector) translational components of of the new local connection are related to those of the old local connection by the rotation

$$\begin{bmatrix} \vec{\mathbf{A}}_{\mathrm{new}}^{\xi_x} \\ \vec{\mathbf{A}}_{\mathrm{new}}^{\xi_y} \end{bmatrix} = \begin{bmatrix} \cos\beta & \sin\beta \\ -\sin\beta & \cos\beta \end{bmatrix} \begin{bmatrix} \vec{\mathbf{A}}_{\mathrm{old}}^{\xi_x} \\ \vec{\mathbf{A}}_{\mathrm{old}}^{\xi_y} \end{bmatrix}, \tag{30}$$

where

$$\beta = \theta_{\mathrm{new}} - \theta_{\mathrm{old}} = \frac{(-\dot{\alpha}_1) + \alpha_2}{3}. \tag{31}$$

The connection vector fields and height functions corresponding to the new coordinate choice are shown in Fig. 10, along with the same gait image-family as in Fig. 5. As the gait image-family shown is in the null region of $\vec{\mathbf{A}}_{\mathrm{new}}^{\xi_\theta}$, the

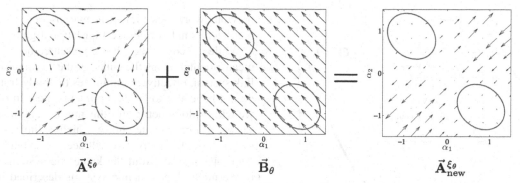

$$\vec{\mathbf{A}}^{\xi_\theta} \qquad \vec{\mathbf{B}}_\theta \qquad \vec{\mathbf{A}}^{\xi_\theta}_{new}$$

Fig. 9: The effect of adding $\vec{\mathbf{B}}^{\xi_\theta}$ to $\vec{\mathbf{A}}^{\xi_\theta}_{old}$. In the circled regions, $\vec{\mathbf{B}}^{\xi_\theta}(r) \approx -\vec{\mathbf{A}}^{\xi_\theta}_{old}(r)$, so $\|\vec{\mathbf{A}}^{\xi_\theta}_{new}\| \approx 0$ in these regions, and the BVI will be a good estimate of the displacement resulting from gaits located there. Note that for visual clarity within each plot, the vectors in the different fields are not to scale.

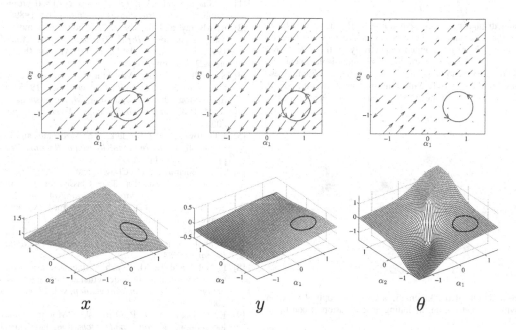

$$x \qquad\qquad y \qquad\qquad \theta$$

Fig. 10: Example gait image-family with connection vector fields and height functions for the kinematic snake with the new mean orientation coordinate choice.

orientation θ of the system remains close to zero for all time during any of the gaits in the image-family as per (18). From (16), the error term ε_ζ over the course of these gaits is thus small, and the BVI ζ is a good approximation of the displacement g resulting from any of the gaits in the image-family, as illustrated in Fig. 11.

VI. DEMONSTRATION

We applied the gaits from the image family in Figs. 5 and 10 to a physical instantiation of the kinematic snake shown in Fig. 12, and plotted the resulting displacements in Fig. 13. During the experiments, we observed some backlash in the joints and slip in the constraints, to which we attribute the differences between the experimental and calculated loci in Figs. 13 and 11. Even with this error, the BVI was a considerably more effective estimate of the displacement with the new coordinate choice than with the old choice. Under the old coordinate choice, the error in the the net direction of travel between the BVI estimate and the actual displacement ranged

from 0.72 to 0.25 radians. With the new choice, the error in the estimate ranged from 0.10 to −0.04 radians, reducing the maximum error magnitude by 86% and eliminating the minimum error.

VII. CONCLUSION

In this paper, we have shown that the choice of coordinates used to represent a mobile system directly affects the computational cost of evaluating its motion properties. By choosing a set of coordinates which minimizes the rotation of the body frame in response to changes in system shape, we have advanced a technique for approximating the net displacement resulting from a gait by means of Stokes's theorem, rather than by the considerably more expensive integration of the equations of motion. In turn, this approximation provides a means of designing gaits by encircling sign-definite regions of well-defined height functions. The technique constitutes a significant improvement over previous efforts in this area, as it is applicable to efficient macroscopic translational gaits, while

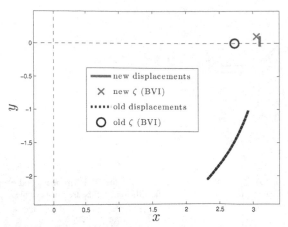

Fig. 11: The BVI and displacements corresponding to the image-family of gaits in Fig. 10 for the kinematic snake robot with the new measure of orientation depicted in Fig. 7 are represented respectively by the cross above the x-axis and the short arc-section, with the BVI and displacements as measured in the old coordinate choice presented for reference. With the new measure, the orientation of the system remains almost constant over the course of any of the gaits in the family, and the BVI ζ is thus a good approximation of the displacements.

Fig. 12: Physical instantiation of the kinematic snake. Two optical mice on the middle link provide odometry data, including an estimation of constraint slip.

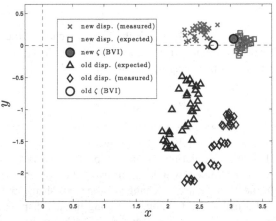

Fig. 13: BVI and displacements for the physical kinematic snake while executing the example image-family of gaits. The boxes show the expected displacements for the robot, calculated from the actual trajectories followed by the joints, while the ×es show the actual measured displacement, which differs from the expected displacement because of backlash and constraint slip. As in Fig. 11, the BVI is a much better estimate of the displacement with the new coordinate choice than with the old choice.

previous work was limited to macroscopic rotational gaits or inefficient small-amplitude translational gaits.

While the coordinate choice presented here, which takes the mean orientation of the links as the orientation of the three-link system, effectively minimizes the body frame rotation, we do not believe it to be the optimal choice. In our future work, we will explore metrics for this minimization, and methods for selecting a coordinate choice which optimizes it for arbitrary systems. Additionally, we will seek to extend the benefits of coordinate choice from the kinematic systems described here to the mixed kinodynamic systems described in [17].

REFERENCES

[1] A. Shapere and F. Wilczek, "Geometry of self-propulsion at low reynolds number," *Geometric Phases in Physics*, Jan 1989.

[2] R. Murray and S. Sastry, "Nonholonomic motion planning: steering using sinusoids," *IEEE Transactions on Automatic Control*, Jan 1993. [Online]. Available: http://eavr.u-strasbg.fr/~bernard/education/ensps_3a/tmp/murray.pdf

[3] S. Kelly and R. M. Murray, "Geometric phases and robotic locomotion," *J. Robotic Systems*, Jan 1995. [Online]. Available: ftp://avalon.caltech.edu/pub/murray/preprints/cds/cds94-014.ps.gz

[4] A. M. Bloch *et al.*, *Nonholonomic Mechanics and Control*. Springer, 2003.

[5] J. Ostrowski and J. Burdick, "The Mechanics and Control of Undulatory Locomotion," *International Journal of Robotics Research*, vol. 17, no. 7, pp. 683 – 701, July 1998.

[6] E. A. Shammas, H. Choset, and A. A. Rizzi, "Geometric Motion Planning Analysis for Two Classes of Underactuated Mechanical Systems," *The International Journal of Robotics Research*, vol. 26, no. 10, pp. 1043–1073, 2007. [Online]. Available: http://ijr.sagepub.com/cgi/content/abstract/26/10/1043

[7] J. Ostrowski, J. Desai, and V. Kumar, "Optimal Gait Selection for Nonholonomic Locomotion Systems," *International Journal of Robotics Research*, 2000.

[8] F. Bullo and K. M. Lynch, "Kinematic controllability for decoupled trajectory planning in underactuated mechanical systems," *IEEE Transactions on Robotics and Automation*, vol. 17, no. 4, pp. 402–412, August 2001.

[9] K. McIsaac and J. P. Ostrowski, "Motion planning for anguilliform locomotion," *Robotics and Automation*, Jan 2003. [Online]. Available: http://ieeexplore.ieee.org/xpls/abs_all.jsp?arnumber=1220714

[10] K. Morgansen, B. Triplett, and D. Klein, "Geometric methods for modeling and control of free-swimming fin-actuated underwater vehicles," *Robotics*, Jan 2007. [Online]. Available: http://ieeexplore.ieee.org/xpls/abs_all.jsp?arnumber=4399955

[11] R. L. Hatton and H. Choset, "Connection vector fields for underactuated systems," *IEEE BioRob*, October 2008.

[12] R. Mukherjee and D. Anderson, "Nonholonomic Motion Planning Using Stokes' Theorem," in *IEEE International Conference on Robotics and Automation*, 1993.

[13] G. Walsh and S. Sastry, "On reorienting linked rigid bodies using internal motions," *Robotics and Automation, IEEE Transactions on*, vol. 11, no. 1, pp. 139–146, January 1995.

[14] E. Shammas, K. Schmidt, and H. Choset, "Natural Gait Generation Techniques for Multi-bodied Isolated Mechanical Systems," in *IEEE International Conference on Robotics and Automation*, 2005.

[15] J. B. Melli, C. W. Rowley, and D. S. Rufat, "Motion Planning for an Articulated Body in a Perfect Planar Fluid," *SIAM Journal of Applied Dynamical Systems*, vol. 5, no. 4, pp. 650–669, November 2006.

[16] J. Avron and O. Raz, "A geometric theory of swimming: Purcell's swimmer and its symmetrized cousin," *New Journal of Physics*, 2008.

[17] E. A. Shammas, H. Choset, and A. A. Rizzi, "Towards a Unified Approach to Motion Planning for Dynamic Underactuated Mechanical Systems with Non-holonomic Constraints," *The International Journal of Robotics Research*, vol. 26, no. 10, pp. 1075–1124, 2007. [Online]. Available: http://ijr.sagepub.com/cgi/content/abstract/26/10/1075

Tactile Texture Recognition with a 3-axial Force MEMS Integrated Artificial Finger

Florian de Boissieu, Christelle Godin,
Bernard Guilhamat and Dominique David
CEA, LETI, MINATEC
17 rue des Martyrs
38054 Grenoble Cedex 9
France

Christine Serviere and Daniel Baudois
GIPSA-Lab
961 rue de la Houille Blanche
Domaine universitaire - BP 46
F - 38402 Saint Martin d'Heres cedex
France

Abstract—Recently, several three-axial MEMS-based force sensors have been developed. This kind of force micro sensor is also called tactile sensor in literature for its similarities in size and sensitivity with human mechanoreceptors. Therefore, we believe these three-axial force sensors being able to analyse textures properties while sliding on a surface, as would do a person with his finger. In this paper, we present one of these sensors packaged as an artificial finger, with a hard structure for the bone and a soft rubber for the skin. Preliminary experiments show a good sensitivity of the finger, as its ability to sense the periodic structure of fabrics or to differentiate papers from fabrics calculating a friction coefficient. Its performance for discrimination of different surfaces is then estimated on fine textures of 10 kinds of paper. Supervised classification methods are tested on the data. They lead to an automatic classifier of the 10 papers showing good performances.

I. INTRODUCTION

Research on tactile sense has stirred up a growing interest in the past few years. Giving a robot the perception of forms and textures would open to lots of applications, as object manipulation or objective texture recognition, in fields as different as paper and fabric manufactures, surgery [1] or cosmetics [2].

During the ten last years, several artificial finger prototypes were developed for texture recognition. To reproduce texture sensing, an approach is to develop sensors by mimicking the structural features of a human finger. Howe [3] and Tanaka [4] developed sensors composed of several layers of different elasticity to imitate the bone and the layers of skin (dermis, epidermis...). Mechanoreceptors were represented by strain gauges and piezoelectric elements as PVDF (polyvinylidene fluoride) integrated in the artificial skin. Howe manages to detect a 6.5 μm high knife-edge probe. Tanaka discriminates rubbers of different hardness and roughness (450-100 μm grain size). Mukaibo [5] went a little further adding skin ridges to its artificial finger, as well as a cantilever system measuring the overall normal and shear forces. Hosoda [6] integrated two kinds of receptor, strain gauges and PVDF films, in the two layers of an anthropomorphic fingertip, reproducing by this way the four kinds of mechanoreceptors of human fingertips. These two last prototypes integrate mono-axial strain sensors and the studies were limited to differentiate textures of quite

different materials (cork, aluminium, wood...). Moreover, these prototypes size is quite large compared to a human finger. This can be a problem for finer texture studies.

Recently, several three-axial MEMS (Micro Electro Mechanical Systems) measuring forces in the three dimensions of space have been developed [7], [8], [9], [10]. Except Beccai et al. [11], who report some slip detection results with a tactile device for an artificial hand, very little results have been published on the use of three-axial MEMS force sensor in artificial tactile sensing experiments. On the basis of Yao's model [12], CEA-LETI is producing one of these innovative sensors as individual elements and as arrays of 10×1 and 10×3 elements.

Among the four kinds of mechanoreceptors we own in the fingertip, Pacinian Corpuscules (PC) are responsible for the detection of vibrations occuring while rubbing a surface. The size of a PC is about 1 mm large and its sensitivity about 1 mN. As it is implanted quite deeply in the skin, its receptive field is large. But what makes the PC interesting is its high frequency response (15 Hz-400 Hz) and therefore its capability to detect small vibrations coming from fine texture exploration [13].

The sensor produced by CEA-LETI is similar in size and sensitivity to those of a PC. As it is a silicon MEMS it has a high frequency response. Our approach is to use such a sensor to study fine textures discrimination. We integrated this silicon MEMS to a finger shaped structure and covered it with a soft rubber skin to protect it. We expect this artificial finger to be able to discriminate fine textures exploring surfaces as would do a human person.

The first part of this paper describes the sensor and the tactile exploration system. The second part presents two preliminary experiments to state on the sensitivity of the sensor: the discrimination of coarse textures such as paper and fabric, and the reconstruction of forces and friction coefficient images exploring a printed paper. The last part describes an experiment of fine texture classification for 10 different kinds of paper. This document presents the results of two classification algorithm among several tested. First, a supervised classification method, using spectral features, gave us a classification rate of 60%, which is good compared

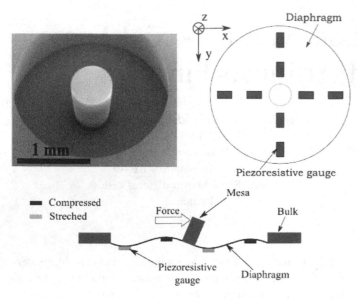

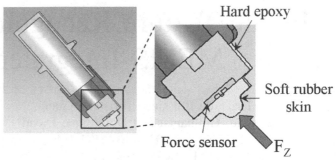

Fig. 2. Artificial finger composed of the 3-axial force sensor flush-mounted on a hard epoxy layer and covered by a soft rubber skin

Fig. 1. Three-axial sensor made in a silicon mono-crystal. Top,Left: Scanning Electrnical Microscopy (SEM) photo of the three-axial sensor. Top, Right: bottom view of the diaphragm. Bottom: Diaphragm deflection when a force is applied. Here, light gauges are stretched by the deflection, dark gauges are compressed.

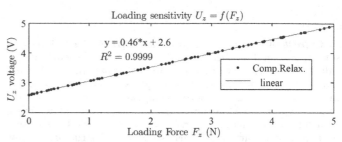

Fig. 3. Loading response of the artificial finger covered with 80 shore A rubber, for three loading-unloading cycles.

to the 10% random classification rate. The second applying neural networks shows similar results but using only 5 selected features.

II. ARTIFICIAL FINGER AND EXPLORATION SYSTEM

A. Artificial finger

The artificial finger has three components: a hard structure for the bone, a tactile element for the mechanoreceptor and a rubber for the artificial skin.

The silicon force sensor (see figure 1) is a sensor measuring stress in three dimensions of space. It consists of a mono-cristal silicon element, composed of a small mesa surrounded by a thin diaphragm under which 8 piezoresistors are diffused. The piezoresistors are implanted on orthogonal axis x and y. As a force is applied on the mesa, the diaphragm will deflect. The resulting strain of the diaphragm induces a compression or a stretch of the piezoresistors, changing their resistor values. Connecting these resistors to Wheatstone bridges, we can measure three voltages U_x, U_y, U_z respectively proportional to the three components F_x, F_y, F_z of an equivalent force on the mesa as demonstrated in [12]. Figure 1 (bottom) illustrates the strain of the diaphragm when a force is applied on the mesa. This silicon sensor is able to support a $1 MPa$ pressure and a $2N$ tangential force on top of the mesa.

The silicon tactile sensor is flush-mounted on a cylindrical piece of hard epoxy and wire-bonded to its electronic. The electronic is composed of three amplifiers, one for each output voltage.

To protect the silicon sensor and also to transmit friction forces when exploring a texture, the sensor is covered by a soft rubber skin made of polyurethane as shown in figure 2. The rubber shape is composed of two parts: a cylindrical

basis over the silicon sensor and a semi-spherical top part that will be in contact with the sample to be explored. The semi-spherical shape of the top part was designed to keep the contact surface quite constant even with the wear. The size of this top part was chosen to keep a contact surface small enough ($\phi \simeq 3\ mm$) to detect texture details such as fine roughness and small friction events of textures. The basis of the rubber, which is in contact with the silicon sensor, is chosen just a little larger than the size of the silicon diaphragm, so that it optimizes the measurement of the global force applied at on the top when exploring a texture. At first the rubber was made of polyurethane AXSON UR5801/UR5850 of hardness 50 shore A. This rubber was used for preliminary experiments presented in section III. But as it weared out fast, we decided to change it for a polyurethane LOCTITE 3957 of hardness 80 shore A. This rubber is hard enough to limit considerably the wear, but it is still soft enough to keep a contact surface sufficient to be sensitive to friction without deteriorating paper or fabrics samples while rubbing them. The 80 shore A rubber was used for the fine texture classification experiment presented in section IV. The artificial finger covered with poluyrethane was characterized in loading, showing a very good linearity with both kind of rubber. Figure 3 presents the loading characteristic of the sensor covered with the 80 shore A rubber.

B. Exploration and acquisition systems

The artificial finger is fixed on the dart of an HP plotter machine (sort of printer) used as the sample exploration system (see figure 4). The plotter has two step motors allowing to

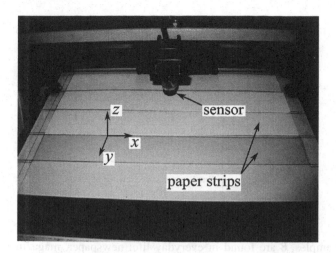

Fig. 4. Plotter machine used as the exploration system. The artificial finger is fixed on the dart usually dedicated to take a pencil. On this picture, the last experiment paper samples presented in section IV is installed on the plotter.

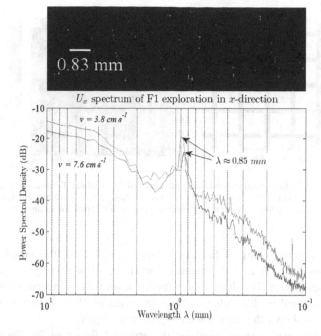

Fig. 5. Top: Jean texture (F1) photo. In x-direction the Jean weave is about 0.83 mm periodic. Bottom: Spectrum of U_x voltage for $v = 3.8 \ cm \cdot s^{-1}$ in light gray and $v = 7.6 \ cm \cdot s^{-1}$ in dark gray. The common line ($\lambda \approx 0.85 \ mm$) for both speed exploration spectra corresponds to the weave periodicity. The extensibility and the deformability of the material explain the difference between theoretical and measured wavelength.

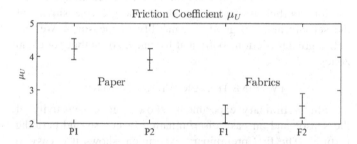

Fig. 6. Friction coefficient (μ_U) of paper samples (P1, P2) and fabrics samples (F1, F2).

explore surfaces in both x and y directions. One is dedicated to move the sensor along a bar representing the x-axis. The other one is used to move the samples along the y-axis thanks to rollers. An electromechanical system lifts up the sensor when not exploring. A 0.4N normal force is applied thanks to a string on the electromechanical system. The plotter is connected to a PC workstation by a National Instrument GPIB-USB cable in order to control the moving speed and the position of the artificial finger. The acquisition system is composed of an electronic circuit of amplification, an analogue Nyquist filter (SCXI 1000 Nat. Inst.) and a data acquisition card (DAQCard 6036E Nat. Inst.). The exploration and acquisition systems are synchronously controlled by a dedicated Labview software.

III. PRELIMINARY EXPERIMENTS

To state on the sensitivity of the artificial finger two experiments were built up as a preliminary to the classification experiment. The first experiment aims at testing the ability of the sensor to discriminate coarse textures easily discriminable by touch. The purpose of the second experiment is to reconstruct an image of the forces felt by the artificial finger when exploring a printed paper.

A. Coarse texture discrimination

In this experiment four samples were explored: two papers and two fabrics. The papers (P1 and 2) have a special texture called Soft Touch (ArjoWiggins) which is quite soft and grips a little. The samples of fabric are both made of cotton, but each one has its own periodic weave form: one with a jean weave (T1), the other with a plain weave (T2). The experiment consisted in exploring the four samples in x and y directions with two different speeds ($v = 3.8 \ cm \cdot s^{-1}, 7.6 \ cm \cdot s^{-1}$).

Firstly, observing the spectrum of U_x voltage for fabrics, we detect easily the lines corresponding to the periodic weave forms. Figure 5 presents a photo of the jean texture and the corresponding spectrum of an x-direction exploration. We can note the common line for both speed of exploration corresponding to the weave wavelength ($\lambda \approx 0.85 \ mm$).

Secondly, we calculated, for each sample, a friction coefficient defined by $\mu_U = \frac{\sqrt{U_x{}^2 + U_y{}^2}}{U_z}$, that we consider to be proportional to the usual friction coefficient defined by $\mu_F = \frac{F_T}{F_N}$, where F_T and F_N are the tangential and normal forces at the contact surface. We find a higher friction coefficient for the Soft Touch papers than for the fabrics. The difference in the friction coefficients seems quite representative of the difference in stickiness we feel rubbing Soft Touch paper and fabric samples. Figure 6 shows the mean friction coefficient with bars representing standard deviation.

B. Image of forces

The second preliminary experiment consists in reconstructing the images of forces during the exploration of a texture. With images of forces we could expect a better visualization of

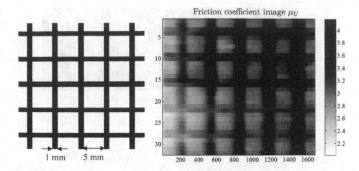

Fig. 7.   Left: Original printed grid explored horizontally (x-direction) at $v = 38 \; mm \cdot s^{-1}$. Right: 1680×32 pixels image of the friction coefficient μ_U. The friction coefficient is greater for inked zones than for clean zones.

TABLE I

10 PAPER SAMPLES TO BE CLASSIFIED. EACH SAMPLE IS ASSOCIATED WITH A CLASS NUMBER ARBITRARLY CHOSEN

Class n	Paper sample
1	Printer paper
2	Soft Touch
3	Skin Touch
4	Tracing paper
5	Large grain size drawing paper
6	Blotting paper
7	Newspaper
8	Coated paper
9	Small grain size drawing paper
10	Photo paper

what is occurring to the sensor during exploration. The sample explored was a grid printed on white paper with a classical laser printer. The thickness of the ink layer was estimated, with an optical profilometer, to be of the order of 10 μm. Figure 7 (left) shows the dimensions of the printed grid.

The grid was scanned by the artificial finger on 32 lines in x-direction finger with a speed $v = 38 \; mm \cdot s^{-1}$ and a sampling frequency of 2 kHz. Scanned lines are spaced of 1 mm. After slightly low filtering signal to avoid artefacts, we reconstructed a 1680×32 pixels image of friction coefficient μ_U. Figure 7 (right) presents the reconstructed image of the friction coefficient. The image shows the great sensitivity of the sensors and the good repeatability of measures. We can note a greater friction coefficient for inked zones than for clean zones.

IV. FINE TEXTURE CLASSIFICATION

Both preliminary experiments show a good sensitivity of the sensor and an easy discrimination of coarse and periodic textures. The first preliminary experiment shows it is easy to discriminate papers from fabrics with the friction coefficient. But it seems quite difficult on figure 6 to discriminate one paper from the other. The next step of this study was then to evaluate the ability of the artificial finger to discriminate quite similar, fine and random textures, like paper textures.

This new experiment consists in classifying 10 textures of paper, each texture of paper representing a class. We constructed an important database to evaluate the efficiency of several classification algorithms. One simple and fast algorithm using spectrum of U_x gives rise to good results. The second presented below reduces signal to 5 features and uses a neural network as classification algorithm.

A. Samples and database

The study of classification was based on paper samples because it is particularly adapted to our exploration system and it offers lots of possibilities in terms of texture. Paper is also a material we are used to manipulate. For this classification experiment we selected 10 samples that we could identify relatively easily blind-rubbing them (easy for experts but much more difficult for a common person). Among the 10 paper

samples, 8 are found in everyday life: newspaper, magazines (coated paper), printer paper, tracing paper, drawing paper with two grain size, photo paper and blotting paper. The last two papers are special Soft Touch and Skin Touch textures (ArjoWiggins), which are soft and grip a little. An arbitrarly chosen class number is associated to each paper sample, as shown in Table I.

The experiment consists in scanning the samples in different zones along one direction, here x-axis of the plotter. The wear of the artificial finger can be considered as insignificant when sliding on a few centimetres. But for the experiment, the artificial finger is sliding on a total distance of 20 m (2 m on each sample) making the wear more important. To make following classification independent of the wear of the artificial finger rubber, the paper samples must be explored randomly. The trick we found to achieve this goal easily and to avoid any manipulation of samples during the experiment consisted in placing 10 paper strips of $25 \times 185 \; mm$, one of each sample, on the same support (see figure 4). For the support, an A4 sheet of transparency is chosen for its non-compressibility and smoothness, so that it doesn't disturb the sample perception. The strips are fixed along the x-axis with a 3M spray adhesive repositionable that won't penetrate the samples. Thanks to its rollers, the plotter can move the transparency support along the y-axis and choose randomly the sample to be explored without any manipulation.

The sample exploration method consists in scanning 4 cm segments on the sample, along the x-direction, with the artificial finger. The samples were explored at a speed of $3.8 \; mm \cdot s^{-1}$ with a sampling frequency of 1 kHz. Compared to previous experiments, the speed of exploration was reduced in order to reduce rubber wear. The samples are randomly scanned 50 times each resulting in a database of 50 acquisitions of (U_x, U_y, U_z) per class.

B. Feature selection

Among the three voltage components, U_x (measuring the force F_x parallel to the movement) seemed to be the more relevant since U_y, measuring the force F_y perpendicular to the movement, is quite reduced, and U_z, measuring the loading force F_z controlled by the string, is almost constant. Hence, to simplify the analysis, we restricted ourselves to the use of

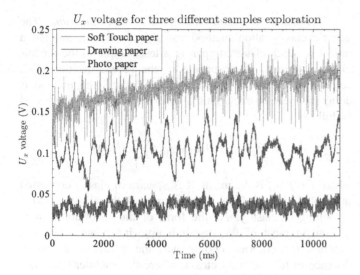

Fig. 8. U_x voltages for the exploration of Soft Touch paper, grained drawing paper and photo paper. These signals illustrate the differences found in friction coefficient, variance and kurtosis features which can be identified respectively as easiness in sliding, texture roughness and stick-slip.

U_x for this classification study.

To identify characteristic features of the sample explored, signals are analysed in the complementary temporal and spectral domains. Three features were found to be characteristic of the samples in the temporal domain:

- Friction coefficient μ_U: for a constant normal force μ_U is identified as the mean tangential force necessary to slide.
- U_x variance σ_x^2 representing the variations of forces around its mean. It can be identified as a representation of texture roughness.
- U_x kurtosis (4th statistical order) measuring the peackedness of a distribution. A gaussian distribution has a kurtosis value of 3. A flat distribution is giving a kurtosis value below 3 and a peack distribution a kurtosis value over 3. The kurtosis can be interpreted as a measurement of the amount of infrequent extreme deviations of the signal, as opposed to frequent modestly-sized deviations measured by variance. We identify the kurtosis as a measure for the stick-slip effect occurring when exploring a flat adherent surface, as photo paper.

Figure 8 presents U_x signals of Soft Touch paper (black), grained drawing paper (dark gray) and photo paper (light gray). This figure illustrates well the three features presented above. Figure 9 presents the mean temporal features of the 10 classes sorted in ascending order. Soft touch paper (class 2) is soft but grips a little. Its corresponding signal, bottom signal on figure 8, has a low U_x mean and a quite important density of small peaks. This is characterized by a low friction coefficient but a kurtosis value slightly over 3, as shown on figure 9. Grained drawing paper (class 5) has an important roughness which is characterized by an important variance (important U_x variations on middle signal). Photo paper (class 10) is very adherent which can be seen on top signal as a high density of big peaks. It shows the stick-slip characteristic of

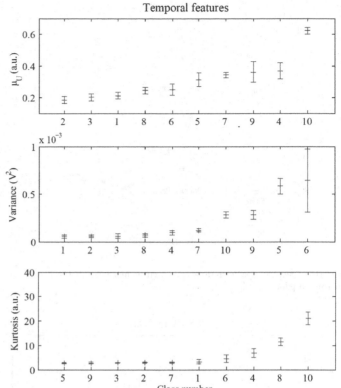

Fig. 9. Mean values of the three temporal features, friction coefficient, variance and kurtosis, versus class number. The mean values are calculated on the training data. Bars represent the intra-class standard deviation of the feature. Classes are sorted in ascending order. One can note that the friction coefficient is almost 10 times smaller than the one presented in figure 6. It could be attributed to the 80 shore A rubber that is about 10 times harder than the 50 shore A rubber used in previous experiments. The 80 shore A rubber slips much easier on the paper surfaces giving a lower tangential force for the same normal force applied.

the movement. Therefore, it has an important kurtosis value. On figure 8 the discrimination between the three samples seems easy. But as one can see on figure 9, some classes are overlapping in feature space making the classification much more difficult.

The spectrum of each acquisition was extracted from the U_x component. As the plotter exploration system was quite noisy, due to the step motors, we extracted the band $20 - 250\ Hz$ from the spectrum and filtered out motors vibrations. Figure 10 presents the preprocessed spectra of Soft Touch, drawing and photo paper samples converted in decibel. On this figure we can observe three features that seem to differ from a sample to another:

- the decibel spectrum mean
- the decibel spectrum slope
- the decibel spectrum form, that is to say the resonances

Figure 11 illustrates the mean spectral features of the 10 classes sorted in ascending order. Paper samples submited to a stick-slip movement are showing a higher spectrum mean and a lower spectrum slope than the other classes of samples.

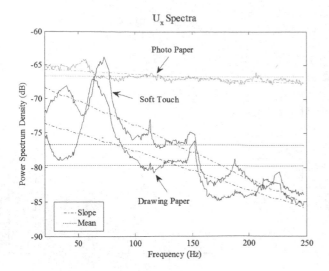

Fig. 10. Spectrum, spectrum slope and spectrum mean of Soft Touch, drawing and photo paper samples.

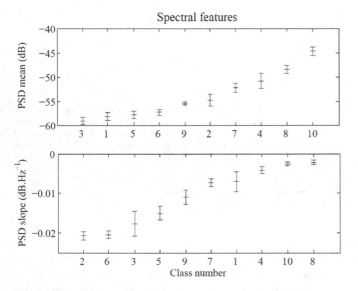

Fig. 11. Mean values of the two spectral features, spectrum mean and slope, versus class number. The mean values are calculated on the training data. Bars represent the intra-class standard deviation of the feature. Classes are sorted in ascending order.

C. Classification algorithms

The data features are of two types. On one side the 5 features that are the friction coefficient, variance, kurtosis, spectrum slope and mean. On the other side the spectrum form that is composed of 822 frequency coefficient. Therefore, two classification algorithms were tested to classify the ten paper samples. The first consists in classifying the spectrum forms with a minimum euclidean distance to the mean spectrum. The second uses the 5 features left as an input to an artificial neural network. This last method permits to reduce the space of classification inputs and to take into account non-linearities. The following is presenting these two methods.

1) Minimum euclidean distance to mean spectrum: The generic classification method usually consists in training a model for each class, comparing a test sample to the models via a criterion and, according to the result, deciding if it belongs to a class or another. Here, the training method is doing the mean characteristic of each class, which is expressed as follows:

$$P^j(f)_{dB} = 10 \cdot \log_{10} \left(\sum_{i=1}^{N} S_i^j(f) \right) \quad (1)$$

where $P^j(f)_{dB}$ is the mean of N spectra of class j converted in decibel and $S_i^j(f)$ is the spectrum of the ith acquisition belonging to class j. As the form of the spectrum seemed to contain most of the discriminative information between samples, the criterion chosen was the minimum euclidean distance to the mean spectrum, where the euclidean distance between two spectra S_i and S_j is defined as:

$$d(S_i, S_j) = \sqrt{\sum_f \left(S_i(f)_{dB} - S_j(f)_{dB} \right)^2} \quad (2)$$

Hence the decision is given by:

$$C = \min_j \left[d(S, P^j) \right] \quad (3)$$

where S is a spectrum to be classified and C the attributed class.

2) Neural Network algorithm: Multi-layer perceptron (MLP) neural networks are known to be well suited for non-linear classification problems. But its structure must be chosen carefully. For multi-class classification tasks several configuration can be tested. The configuration that gave us the best results was a one-against-rest classification method. That is to say 10 MLPs, one for each class, each MLP having a unique output neuron with target 1 for its associated class examples and 0 for others. Each MLP is composed of a 5 inputs layer, one input for each of the 5 features, which distributes the data to a hidden layer itself connected to an output neuron. A log sigmoid transfer function is chosen for both hidden layer and ouptut neurons, giving then an output in the interval [0;1]. Between the 10 outputs of the 10 MLPs, the class attributed is the class associated with the MLP having the greatest output. To avoid overfitting, we decided to restrict to a hidden layer with a maximum of 5 neurons.

The weights of the networks are trained using a Bayesian regularization backpropagation algorithm. This algorithm trains fast and produces networks that generalize well.

D. Classification evaluation

To test the classification algorithms, the 50 acquisitions/class database is first split in 40 acquisitions *training* dataset and 10 acquisitions *generalisation* dataset. To evaluate accurately the classifications rates we performed a 10-fold cross-validation on the *training* dataset [14]. The basic processing steps can be summed up as follows:

1) separate the feature dataset into 10 folds per class
2) use 9 folds for training

TABLE II
SPECTRUM CLASSIFICATION RATE OF 10 DIFFERENT PAPERS

	Training	Validation	Test
Original spectrum	$67.5 \pm 0.7\%$	$58 \pm 2\%$	$53 \pm 5\%$
Centered spectrum	$71.8 \pm 0.7\%$	$60 \pm 2\%$	$61 \pm 5\%$

3) use the remaining fold for testing
4) repeat the two previous steps until all folds are used
5) store the average classification score

The different datasets are equally distributed between the 10 classes. But each MLP gives a one-against-rest output. To avoid an over training of the rest-class, the training dataset is partially replicated before entering each MLP. This way the binary output targets get equally distributed between the one-class and the rest-class. This replication is only applied during the training of MLP. It does not present any interest for classification evaluation, neither for the whole Euclidean classification (training nor evaluation). The original datasets as described above are used in these cases.

E. Results

1) Spectrum classification: We noted a quite important variance in decibel spectrum means between acquisition of the same class. Therefore, the classification algorithm was tested with original spectrum but also with centered spectrum (meanless decibel spectrum). Table II shows the classification results. In both cases, original or centered spectrum, we obtain good results with a classification rate of about 60% to be compared to the 10% classification rate of a random classification. As expected, the classification works slightly better on centered spectrum with more equilibrated results between validation and test.

2) 5 features classification: The spectra extracted from the signal are of large dimension (822 frequency components). With the aim of enlarging the database with much more samples of all sorts of material, it can be useful to reduce the number of features. The euclidean distance algorithm applied on the 5 features left gave worse results than spectrum classification. Taking into account non-linearities thanks to the neural network classification algorithm, we obtain results similar to spectrum classification but with much less features.

An essential parameter to be chosen in neural networks is the number of neurons in the hidden layer. A too large hidden layer will overfit the training dataset without increasing validation and test classification rates. The amount of data limits us to a maximum of 5 neurons. Figure 12 shows classification results of training and validation datasets as a function of the number of neurons. More than 2 neurons increases only the training classification rate giving then the limit over which the MLP is overfitting the training dataset. Table III gives the training, validation and test classification rates for the 10-class MLPs with 2 hidden neurons each.

The two hidden neuron classifier reaches a 71% validation rate but limits to 58% test rate. This difference between validation and test enhance the limit of classifier generalisation.

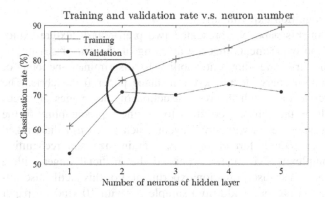

Fig. 12. Classification rates function of the number of neurons of the hidden layer. Note that the training classification rate increases continuously with number of hidden neurons while the validation rate stays constant over 2 hidden neurons. This is significant of overfitting over 2 hidden neurons.

TABLE III
10-CLASS WITH 2 HIDDEN NEURONS NETWORK CLASSIFICATION RATE

Training	Validation	Test
$74.1 \pm 0.7\%$	$71 \pm 2\%$	$58 \pm 5\%$

The training and validation datasets would need to be enlarged with the examples that are misclassified, and test dataset to be enlarge to better estimate the generalisation rate.

However, taking the 58% as a minimum classification rate of this last classifier, we can conclude to a similar and good performance of both spectrum-based and neural-network based algorithms. To compare the two algorithms, the differences between them are to be found in there performances for a particular application more than on these results. One major argument in favour of neural network is its ability to encircle classes by a non-linear limits. Therefore it should be more efficient as texture classes are getting more and more similar. One major drawback is a long training time due to the use of the whole database for MLPs weights training. Therefore, for a large and evolutive database of quite different textures, the euclidean distance algorithm would be more interesting.

We are now thinking of an algorithm merging the two precedent to increase the classification performances. At the moment, we are also working at increasing the sensitivity of our artificial finger prototype to enhance its texture perception and classification results.

V. CONCLUSION

During this study, we managed to package the three-axial force sensor as an artificial finger, with a quite biomimetic structure and size. Studying different shapes and hardness for the rubber-skin tested, we selected one that would be optimal for a dynamic texture recognition task. The characterisation of the artificial finger showed a great linearity in loading. For this study, we have dedicated the artificial finger to explore everyday life textures, like papers and fabrics. And the use of a plotter as an exploration system allowed us to automatize the exploration procedure and lead various kinds of texture recognition experiments.

In this paper we presented two preliminary texture exploration experiments that are attesting of the sensor sensitivity. The first one shows its ability to discriminate periodic or random coarse textures, for instance paper from fabric. The second one, with the reconstruction of the images of forces felt by the sensor, permitted to see the reproducibility of the measures, even with small events such as 10 μm thick inked zones. Going further in texture grain size and recognition complexity, we finally evaluated the artificial finger with a multi-class discrimination experiment. For this tactile discrimination task we build up a sample set with 10 kinds of paper ((photo, drawing, blotting papers ...), a material commonly manipulated in every day life and which texture is usually tactily controlled for various reasons such as applications or marketing purposes. We explored these samples several times and applied different classification algorithms to the signal database acquired. One algorithm was based on euclidean distance to the mean spectrum, and another one on a neural network with 5 features as input. Both showed good results and testified of the sensor ability to discriminate fine textures. The next step will be to go beyond classification and characterize textures tactily, i.e. measuring softness, roughness, stickiness with the artificial finger. This opens up perspectives for various interesting applications.

ACKNOWLEDGEMENT

This work was partly supported by the EU-NEST programme, MONAT project (contract 21 number 29000). We would like to thank Patrice Rey (CEA-LETI) for providing the MEMS sensors necessary to make all our artifical finger prototypes and ArjoWiggins for providing Soft and Skin Touch paper samples. We would also like to thank J. Scheibert, A. Prevost and G. Debregeas (ENS Paris) for their very helpful advices.

REFERENCES

[1] M.E.H. Eltaib and J.R. Hewit, "Tactile sensing technology for minimal access surgery a review", *Mechatronics*, vol. 13, pp. 1163-1177, Dec. 2003.

[2] M. Tanaka, J.L. Lvque, H. Tagami , K. Kikuchi and S. Chonan, "The Haptic Finger - a new device for monitoring skin condition", *Skin Res. Technol.*, vol. 9, pp. 131-136, May 2003.

[3] R. D. Howe and M. R.Curtosky, "Dynamic Tactile Sensing : Perception of Fine Surface Feature with stress rate Sensing", *IEEE Trans. Robotics and Automation*, Vol. 9, no. 2, pp. 140-151, Apr. 1993.

[4] M. Tanaka, H. Sugiura, J. L. Leveque, H. Tagami, K. Kikuchi and S. Chonan, "Active haptic sensation for monitoring skin conditions", *J. Materials Process. Techno.*, Vol. 161, pp. 199-203, Apr. 2005.

[5] Y. Mukaibo, H. Shirado, M. Konyo and T. Maeno, "Development of a Texture Sensor Emulating the Tissue Structure and Perceptual Mechanism of Human Fingers", in *Proc. 2005 IEEE Int. Conf. Robot. Autom.*, pp. 2565-2570.

[6] K. Hosoda, Y. Tada and M. Asada, "Anthropomorphic robotic soft fingertip with randomly distributed receptors", *Robot. Auton. Syst.*, vol. 54, no. 2, pp. 104-109, Feb. 2006.

[7] L. Beccai et al., "Design and fabrication of a hybrid silicon three-axial force sensor for biomechanical applications", *Sens. Actuators A*, vol. 120, no. 2, pp. 370382, May 2005.

[8] B. J. Kane, M. R. Cutkosky and G. T. A. Kovacs, "A Traction Stress Sensor Array for Use in High-Resolution Robotic Tactile Imaging", *J. Microelectromech. Syst.*, Vol. 9, no .4, Dec. 2000.

[9] K. Kim et al, "A silicon-based flexible tactile sensor for ubiquitous robot companion applications", *2006 J. Phys.: Conf. Ser. 34*, pp. 399-403.

[10] G. Vasarhelyi et al., "Characterisation of an integrable single-crystalline 3-D Tactile Sensor", *IEEE Sens. J.*, Vol. 6.4, pp. 928-934, Aug. 2006.

[11] L. Beccai et al., "Development and Experimental Analysis of a Soft Compliant Tactile Microsensor for Anthropomorphic Artificial Hand", *IEEE/ASME Trans. Mechatronics*, vol. 13, pp. 158-168, Apr. 2008.

[12] C.T. Yao, M.C. Peckerar, J.H. Wasilik, C. Amazeen and S. Bishop, "A novel three-dimensional microstructure fabrication technique for a triaxial tactile sensor array", in *Proc. IEEE Micro Robot. Teleop. Workshop*, 1987.

[13] M. Hollins and S. J. Bensmaia, "The coding of roughness", *Can. J. Exp. Psychol.*, vol. 61, pp. 184-95, Sep. 2007.

[14] R. O. Duda, P. E. Hart and D. G. Stork, "Cross-Validation", in *Pattern Recognition*, 2nd ed., Wiley Interscience, 2001, pp. 483-484.

On the Complexity of the Set of Three-finger Caging Grasps of Convex Polygons

Mostafa Vahedi
Department of Information and Computing Sciences
Utrecht University
Utrecht, The netherlands
Email: vahedi@cs.uu.nl

A. Frank van der Stappen
Department of Information and Computing Sciences
Utrecht University
Utrecht, The netherlands
Email: frankst@cs.uu.nl

Abstract—We study three-finger caging grasps of convex polygons. A part is caged with a number of fingers when it is impossible to rigidly move the part to an arbitrary placement far from its initial placement without penetrating any finger. A convex polygon with n vertices and a placement of two fingers —referred to as the base fingers— are given. The caging region is the set of all placements of the third finger that together with the base fingers cage the polygon. We derive a novel formulation of caging in terms of visibility in three-dimensional space. We use this formulation to prove that the worst-case combinatorial complexity of the caging region is close to $O(n^3)$, which is a significant improvement of the previously known upper bound of $O(n^6)$. Moreover we provide an algorithm with a running time close to $O(n^3 \log n)$ that considerably improves the current best known algorithm, which runs in $O(n^6)$ time.

I. INTRODUCTION

The caging problem (or: capturing problem) was posed by Kuperberg [7] as a problem of finding placements for a set of fingers that prevent a polygon from moving arbitrarily far from its given position. In other words, a polygon is caged with a number of fingers when it is impossible to continuously and rigidly move it to infinity without intersecting any finger. A set of placements of fingers is called a grasp.

Caging grasps are related to the notions of form (and force) closure grasps (see e.g. Mason's text book [8]), and immobilizing and equilibrium grasps [13]. A part is immobilized by a number of fixed fingers (forming an immobilizing grasp) when any motion of the part violates the rigidity of the part or the fingers. An equilibrium grasp is a grasp whose grasping fingers can exert wrenches (not all of them zero) through grasping points to balance the object.

Rimon and Blake [12] introduced the notion of the *caging set* (also known as *inescapable configuration space* [16, 20], and recently regularly referred to as the *capture region* [4, 10, 11]) of a hand as all hand configurations which maintain the object caged between the fingers. They proved that in a multi-finger one-parameter gripping system, the hand's configuration at which the cage is broken corresponds to a frictionless equilibrium grasp.

Caging has been applied to a number of problems in manipulation such as grasping and in-hand manipulation [16, 20], mobile robot motion planning [5, 6, 9, 18, 19], and error-tolerant grasps of planar objects [1, 2, 12]. Caging grasps are particularly useful in scenarios where objects just need to be transported (and not subjected to e.g. high-precision machining operations). The fact that the object cannot escape the fingers guarantees that —despite some freedom to move— the object travels along with the fingers as these travel to their destination. The set of caging grasps is significantly larger than the set of immobilizing grasps (as the latter forms a lower-dimensional subset of the former). The additional options for finger placements can be of great value when maneuvering the object amidst obstacles. Moreover, caging grasps are considerably less sensitive to finger misplacements.

In this paper we consider algorithms for caging grasps by robotic systems with two degrees of freedom. The first two papers by Rimon and Blake [12] and Davidson and Blake [1] consider systems with a single degree of freedom. Several other papers [11, 12, 14, 17, 23] study two-finger caging grasps as a special case of robotic systems with one degree of freedom. There are also papers [3, 16, 19, 21] that consider robotic systems with more degrees of freedom. All these papers present approximate algorithms for computing caging grasps. As such they differ from our work, as these algorithms compute a subset of the set of caging grasps. We consider computation of *all* caging grasps for a given placement of the base fingers.

There are two papers [4, 23] on three-finger caging grasps of polygons that propose algorithms for robotic systems with two degrees of freedom that report the entire solution set. In these papers, a polygon with n edges and a placement of two fingers—referred to as the *base fingers*—are given. It is required to compute the *caging region* for the third finger, which is the two-dimensional set of all placements of the third finger that together with the base fingers cage the polygon. (Consider Figure 1.) Erickson et al. [4] provided the first algorithm for the exact computation of the caging region of convex polygons, running in $O(n^6)$ time. In their paper the base fingers were assumed to be placed along the boundary of the polygon. They also established an upper bound of $O(n^6)$ on the worst-case complexity of the caging region of convex polygons, where the caging region was shown to be the visible scene of $O(n^3)$ constant-complexity surfaces in a three-dimensional space. Vahedi and van der Stappen [23] proposed another algorithm generalizing the previous results to compute the caging region of arbitrary polygons for *any* given placement of the base fingers, that runs in $O(n^6 \log^2 n)$ time. They established the same $O(n^6)$ upper bound on the worst-case complexity of the caging region of non-convex polygons, where the caging region was shown to be a subset of the arrangement of $O(n^3)$ constant-complexity curves defined by equilibrium grasps. (The *arrangement* of a set X of two-dimensional curves is the set of maximally-connected zero-, one-, and two-dimensional subsets induced by the curves of X not intersecting any of the subsets.) However, in both cases the mentioned upper bound on the worst-case complexity of the caging region was due to the proposed algorithms, and it remained an open problem to establish a better upper bound for convex or non-convex polygons. In this paper, we tackle the problem for convex polygons. We prove that the worst-case complexity of the caging region of convex polygons is $O(\lambda_{10}(n^3))$, which significantly improves the already known upper bound of $O(n^6)$, as $\lambda_{10}(n^3)$ is known [15] to be $O(n^3 \log^* n)^1$ and thus very close to $O(n^3)$. To establish the upper bound, firstly we have narrowed down the types of surfaces introduced by Erickson et al. [4] that play a role in the caging region complexity. Secondly, we have formulated a new way to compute the caging region using those surfaces. In addition, we develop an efficient algorithm to compute the caging region in $O(\lambda_s(n^3) \log n)$ time using a divide-and-conquer technique.

In Section II we introduce some definitions and assumptions used in the paper, define a three-dimensional space called *canonical grasp space*, and explain a formulation to compute the caging region in canonical grasp space. In Section III, we use this formulation to prove an upper bound on the complexity of the caging region and present an efficient algorithm to compute it. We conclude the paper with a discussion of future work.

II. Definitions and Assumptions

A *convex polygon* is an intersection of a number of half planes; Throughout the paper P is a bounded convex polygon without parallel edges that has a fixed reference frame and n edges.

We assume that the fingers are points. The distance between the base fingers is d. We assume that vertices and edges of P are in general position, i.e. no two vertices are at distance exactly d from each other, no vertex has (shortest) distance exactly d to an edge, and the angle between the altitude lines drawn from any vertex to any pair of edges is not equal to the supplement of the angle between the corresponding pair of edges (i.e. they do not add up to π).

Instead of considering rigid placements of the base fingers around the fixed polygon P, we can equivalently fix the base fingers at $b_1 = (0,0)$ and $b_2 = (d,0)$ and consider possible placements $q \in \mathbb{R}^2 \times [0, 2\pi)$ of P (with respect to these fingers). Let $P[q]$ denote the set of points covered by P when

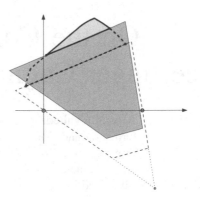

Fig. 1. $P[q]$ and $P[\alpha]$ are rigid translates. Caging region of $P[\alpha]$ is a superset of the caging region of $P[q]$.

placed at q. Let $P[\alpha]$ be the polygon P rotated by the angle α around its reference point for which both base fingers are in contact with it and the extensions of the edges touched by the base fingers intersect each other below the x-axis. We refer to $P[\alpha]$ as the *canonical placement* of any other placement q of P with orientation α. A canonical placement can be specified by a single parameter, which is orientation. In Figure 1, $P[\alpha]$ is the canonical placement of $P[q]$. The polygon $P[\alpha]$ may not be defined for all orientations α, which we explain about when we define the base-diameter. In this paper, we consider canonical placements of P because the caging region of any other placement $q = (x, y, \alpha)$ of P is a subset of the caging region of its canonical placement $P[\alpha]$ [22]. Figure 1 shows one example, in which the caging region of $P[q]$ is the area surrounded by bold solid curves and the caging region of $P[\alpha]$ is the union of both the area surrounded by bold solid curves and the area surrounded by bold dashed curves. Moreover, given the caging region of $P[\alpha]$, the caging region of $P[q]$ can be computed easily [22].

Consider P at a placement $q = (x, y, \alpha)$. Every horizontal line intersects the polygon in at most two points. The *base-diameter* of the polygon is the maximum length intersection among all horizontal lines. The *base-diameter* of any $P[q']$ with $q' = (x', y', \alpha)$ equals that of $P[q]$. When the base-diameter of $P[q]$ with $q = (x, y, \alpha)$ is less than d, $P[\alpha]$ is not defined, as it is not possible to place a translated copy of P with orientation α such that it touches both b_1 and b_2. A *critical angle* is an angle α for which d is equal to the base-diameter of $P[\alpha]$.

Every three-finger grasp in which the polygon is at $P[\alpha]$ can be specified by $(p, \alpha) \in \mathbb{R}^2 \times [0, 2\pi)$, to which we refer as a canonical grasp. (The canonical grasp of a grasp is uniquely defined by the definition of canonical placement.) The parameter α specifies the orientation of the polygon and $p = (x, y)$ specifies the location of the third finger. We refer to the space $\mathbb{R}^2 \times [0, 2\pi)$ of all such three-finger grasps as the *canonical grasp space*. Throughout the paper the y-axis is the vertical axis both in object plane and also in canonical grasp space. The canonical grasp (p, α) is the canonical grasp of every other grasp of P at a placement q with orientation α.

$^1\log^* n = \min\{i \geq 0 : \log^{(i)} n \leq 1\}$, where i is a nonnegative integer, and $\log^{(i)} n$ is the logarithm function applied i times in succession, starting with argument n.

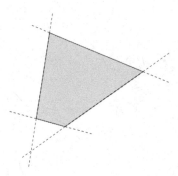

Fig. 2. The triple of edges (e_1, e_2, e_3) is non-triangular, while the triple of edges (e_3, e_4, e_1) is triangular.

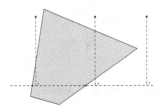

Fig. 3. (p_2, α) is an enclosing grasp while both (p_1, α) and (p_3, α) are non-enclosing grasps.

A triple of edges of P is called *triangular* if the supporting lines of the edges form a triangle that encloses P, and is called *non-triangular* otherwise. Consider Figure 2 to see examples of triangular and non-triangular triples of edges.

Consider a canonical grasp (p, α). The downward vertical ray emanating from the point $p = (x, y)$ in the plane $\theta = \alpha$ may intersect $P[\alpha]$ at an edge. This edge together with the edges of $P[\alpha]$ touched by the base fingers forms a triple of edges. The canonical grasp (p, α) is an *enclosing grasp* if the vertical ray intersects $P[\alpha]$ and if the triple of edges is triangular. Otherwise if the vertical ray does not intersect $P[\alpha]$, or if the triple of edges is non-triangular, then the canonical grasp is a *non-enclosing grasp*. See Figure 3 for examples of enclosing and non-enclosing grasps.

The following two lemmas explain two important facts about the canonical grasps of caging grasps.

Lemma 2.1: [22] The canonical grasp of every caging grasp exists and it is an enclosing grasp.

Lemma 2.2: [22] A caging grasp and its canonical grasp are reachable from each other by a sequence of translations.

If a canonical grasp is a non-enclosing grasp then it is a non-caging grasp by Lemma 2.1. The caging region of any placement q of P is a subset of the caging region of its canonical placement by Lemma 2.2. Moreover, if the canonical grasp of a three-finger grasp is non-caging then the grasp itself is non-caging too.

Let $C(\alpha)$ be the set of all placements of the third finger that together with the base fingers form a caging grasp of $P[\alpha]$. In this paper we are interested in the combinatorial complexity of $C(\alpha)$ and its computation in the worst case.

The boundary of $C(\alpha)$ consists of two x-monotone chain of curves of which the lower one is a subset of the boundary of $P[\alpha]$ [22]. Let $K(\alpha)$ be the upper part of the curves on the boundary of $C(\alpha)$. Clearly the complexity of $C(\alpha)$ is proportional to the complexity of $K(\alpha)$ plus $O(n)$ in the worst case.

Vahedi and van der Stappen [23] have proven that the placement of the third finger on $K(\alpha)$ corresponds to equilibrium grasps, which we mention here in form of a lemma.

Lemma 2.3: [23] Every placement of the third finger on $K(\alpha)$ corresponds to a two-finger equilibrium grasp or a three-finger equilibrium grasp.

Every three-finger equilibrium grasp is a canonical grasp and, thus, corresponds to a point in canonical grasp space. Two-finger equilibrium grasps, however, are not defined inside canonical grasp space because only one of the base fingers contacts the polygon. Instead, as we explain in Subsection II-C, we can represent them with their canonical grasp.

A. Visibility in Canonical Grasp Space

In this subsection we define $\mathcal{P}$ in canonical grasp space as a 3D object defined by sliding the polygon P on both base fingers (i.e. keeping the contact with both base fingers). The surface patches of $\mathcal{P}$ play an important role in the next sections in establishing a bound on the complexity of the caging region. Recall that every canonical grasp corresponds to a point in canonical grasp space. As we consider the surface patches of $\mathcal{P}$ as obstacles, we define visibility between two points in canonical grasp space (with the same x and y coordinates) as a sufficient condition that the corresponding canonical grasps have similar caging properties. Then we explain a number of properties of the surface patches of $\mathcal{P}$.

Consider an edge e of $P[\alpha]$. The edge e together with the edges e_1 and e_2 touched by the base fingers forms a triple of edges. Consider a motion of P at orientation α in which P is rotated and translated while keeping the contact with the base fingers. We refer to this motion as the sliding of P at orientation α. Clearly it is possible to slide P at orientation α in either clockwise or counterclockwise directions. As we slide P at orientation α in each direction one of the base fingers will eventually reach a vertex of the polygon, and the pair of edges touched by the base fingers change. Meanwhile, the trace of the edge e forms a surface patch, $s(e_1, e_2, e) \subset \mathbb{R}^2 \times [0, 2\pi)$, in canonical grasp space, that corresponds to the triple of edges. Clearly, $s(e_1, e_2, e)$ has a constant complexity. Let $\bar{s}(e_1, e_2, e)$ be part of $s(e_1, e_2, e)$ that are induced by all angles α for which the polygon $P[\alpha]$ is below the edge e along the y-axis. The surface patch $\bar{s}(e_1, e_2, e)$ has a constant complexity as well. If the triple (e_1, e_2, e) of edges is triangular then we call $\bar{s}(e_1, e_2, e)$ a triangular surface patch.

We define $\mathcal{P}$ in canonical grasp space as the set of patches $\bar{s}(e_1, e_2, e)$ for all edges e of the polygon P and all pairs e_1 and e_2 of edges touched by the base fingers. In other words, $\mathcal{P}$ is the set of all surface patches that are formed by considering the upper part of $P[\alpha]$ for all angles α for which d is less than the base diameter of $P[\alpha]$. The intersection of the plane $\theta = \alpha$ with $\mathcal{P}$ is the upper part of $P[\alpha]$ where α is an angle

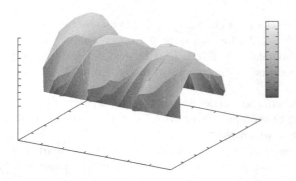

Fig. 4. The surface patches of $\mathcal{P}$.

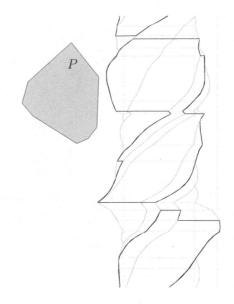

Fig. 5. Triangular borders of P projected to the space (x, θ).

for which d is less than the base diameter of $P[\alpha]$. Every surface patch in $\mathcal{P}$ corresponds to a set of three-finger grasps whose finger placements are on a unique triple of edges of P two of which are touched by the base fingers and the other one is touched by the third finger. Figure 4 displays $\mathcal{P}$ for the convex polygon shown in Figure 5. In Figure 5 the surface patches of $\mathcal{P}$ are projected to the space (x, θ). The horizontal gray lines display the orientations in which the pair of edges touched by the base fingers change. The surface patches of $\mathcal{P}$ form a number of connected terrains that are bounded by critical angles along the θ-axis. Since there are $O(n^3)$ triples of edges, there are $O(n^3)$ surface patches in $\mathcal{P}$.

Consider an angle α for which $P[\alpha]$ is defined. The point (p, α) in canonical grasp space corresponds to a valid canonical grasp provided that p is not inside $P[\alpha]$.

Consider two points (p, α) and (p, β) in canonical grasp space that correspond to valid canonical grasps. Here we define *visibility* between two such grasps. In this paper, the visibility is defined only along the θ-axis. In other words, we define visibility between two points in the canonical grasp space only when the line connecting the two points is parallel to the θ-axis. There are two different line segments that connect (p, α) and (p, β), which correspond to positive and negative directions along the θ-axis. We define (p, β) to be visible from (p, α) if and only if at least one of the two line segments that connect the two points intersect no surface patches of $\mathcal{P}$. According to the definition, when (p, β) and (p, α) are visible from each other, they are reachable from each other by rotation and translation; thus, they are either both caging or both non-caging. In fact, the visibility condition is a sufficient condition for the two grasps to be both caging or both non-caging but it is not a necessary condition.

B. Triangular Borders

The set $\mathcal{P}$ of surface patches consists of a number of triangular and non-triangular surface patches. The set of triangular surface patches forms a number of connected components. *Triangular borders* are the outer boundary of these connected components. In other words, triangular borders are the common boundary between the triangular surface patches and non-triangular surface patches, and also the boundary between triangular surface patches and critical angles. Every canonical

grasp that corresponds to a point on the triangular borders is a non-caging grasp.

In Figure 5 the surface patches of $\mathcal{P}$ are projected to the space (x, θ), in which the θ-axis is the vertical axis and the x-axis is the horizontal axis. The triangular borders are displayed in bold and the loci of vertices are displayed in gray; the distance between the two dotted vertical lines equals the distance between the base fingers. As it is displayed there are two connected components enclosed by triangular borders. (Recall that the θ-axis is circular.)

C. Two-finger Equilibrium Grasps

In this subsection we explain about two-finger equilibrium grasps of convex polygons. Due to a general position assumption, two-finger equilibrium grasps involve the third finger and one of the base fingers.

Consider a two-finger equilibrium grasp of P at placement q with orientation α. We can show that if $P[\alpha]$ is defined, the equilibrium grasp and its canonical grasp are reachable from each other by translation, and thus they are non-caging because two-finger equilibrium grasp is non-caging. We state this fact in form of a lemma.

Lemma 2.4: Every two-finger equilibrium grasp and its canonical grasp are reachable from each other, and thus they are non-caging.

According to Lemma 2.4, since every two-finger equilibrium grasp is reachable from its canonical grasp, we can represent the two-finger equilibrium grasps in canonical grasp space by their canonical grasps.

Since for every edge there is at most one vertex with which it can form a two-finger equilibrium grasp, there are at most n pairs consisting of a vertex and an edge that induce a two-finger equilibrium grasp. Each pair of an edge and a vertex that forms a two-finger equilibrium grasp induces four curves in canonical grasp space based on the finger that is at the vertex and the base finger that is involved. We refer to these curves as the two-finger equilibrium curves each of which has a constant complexity. Similarly there are $O(n)$ pairs consisting of two vertices that define a two-finger equilibrium grasp. Two vertices that define a two-finger equilibrium grasp are necessarily antipodal.

The two-finger equilibrium curves in canonical grasp space can be classified into two groups, such that in each group the curves are circular centered around one of the base fingers in the projection to the space (x, y) [23]. The two-finger equilibrium curves of each group involve the same base finger either b_1 or b_2.

D. Escaping by Translation

In this subsection we first present some definitions and a result by Erickson et al. [4] to identify all placements of the third finger that prevent $P[\alpha]$ from escaping by pure translation; these placements form a two-dimensional region. Then we investigate the relationship between this region and the caging region.

Let the convex polygon $Q[\alpha, b_1]$ be the union of the set of all translated copies of $P[\alpha]$ touching the base finger b_1. The polygon $Q[\alpha, b_1]$ has twice number of edges of $P[\alpha]$. Every edge of $P[\alpha]$ is parallel to exactly two edges of $Q[\alpha, b_1]$. Similarly let $Q[\alpha, b_2]$ be the union of the set of all translated copies of $P[\alpha]$ touching b_2.

Let $X[\alpha]$ be the points inside the intersection of $Q[\alpha, b_1]$ and $Q[\alpha, b_2]$ that are above $P[\alpha]$. Figure 6 shows $Q[\alpha, b_1]$ and $Q[\alpha, b_2]$; the polygon is displayed in dark-gray and $X[\alpha]$ is displayed in light-gray.

Erickson et al. [4] have proven that the set $X[\alpha]$ is the set of all placements of the third finger that prevents $P[\alpha]$ from escaping by pure translation.

Lemma 2.5: [4] The polygon $P[\alpha]$ can escape by pure translation if and only if the third finger placement is outside $X[\alpha]$.

By Lemma 2.5, each edge of $P[\alpha]$ on the boundary of $X[\alpha]$ forms an enclosing triple of edges together with the edges touched by the base fingers. Let $\partial X^u[\alpha]$ be the upper-boundary of $X[\alpha]$. The set of edges of $\partial X^u[\alpha]$ consists of two continuous sets of edges: one set of edges belonging to the boundary of $Q[\alpha, b_1]$ and one set of edges belonging to the boundary of $Q[\alpha, b_2]$.

Clearly, $X[\alpha]$ is a superset of $C(\alpha)$. The following lemma proves that if a point on $\partial X^u[\alpha]$ is on the caging boundary then either the corresponding grasp corresponds to a two-finger equilibrium grasp or the corresponding grasp is on triangular borders, from which we omit the proof.

Lemma 2.6: A point on $\partial X^u[\alpha]$ that belongs to $K(\alpha)$ either corresponds to a two-finger equilibrium grasp or it is on a triangular borders.

In Figure 6, the points on $\partial X^u[\alpha]$ that can possibly be on $K(\alpha)$ are marked with small gray circles.

E. Caging and Non-caging in Canonical Grasp Space

In this subsection we prove that it is possible to compute $C(\alpha)$ for a given orientation α by using the surface patches of

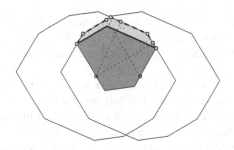

Fig. 6. Illustration of Lemma 2.6.

$\mathcal{P}$, and two types of non-caging grasps: grasps on triangular borders and canonical grasps of two-finger equilibrium grasps.

If $p \in C(\alpha)$, then the grasp (p, α) in canonical grasp space is only visible from other caging grasps. In other words, if (p, β) is a non-caging grasp and it is visible from (p, α) then $p \notin C(\alpha)$. In this subsection, we formulate a way to identify all points in the plane $\theta = \alpha$ that are visible from non-caging grasps. In previous sections, we have introduced two groups of non-caging grasps: grasps on triangular-borders and two-finger equilibrium grasps. In this subsection, we define vertical walls on grasps that are on triangular borders, and also on canonical grasps of two-finger equilibrium grasps (which we explain more in the next paragraphs). These walls represent a set of non-caging grasps, such that if a point on the plane $\theta = \alpha$ is visible from a point on one of these walls then that point represents a non-caging grasp. We prove the important fact that if an enclosing grasp is not visible from any of the mentioned vertical walls, then it is a caging grasp. Therefore, these walls and the surface patches of $\mathcal{P}$ provide enough information to compute $C(\alpha)$.

Consider a non-caging grasp $((x, y'), \beta)$ in canonical grasp space. Consider another grasp $((x, y), \beta)$ where $y > y'$; thus the point $((x, y), \beta)$ is vertically above $((x, y'), \beta)$ in canonical grasp space, and $((x, y), \beta)$ is a non-caging grasp as well. Consider another enclosing grasp $((x, y), \alpha)$ in canonical grasp space. If $((x, y), \alpha)$ is visible from $((x, y), \beta)$, then since $((x, y), \beta)$ is non-caging, $((x, y), \alpha)$ is non-caging as well.

The set of all grasps $((x, y), \beta)$ in canonical grasp space in which $y > y'$ and $((x, y'), \beta)$ is on triangular borders, defines a number of vertical walls, to which we refer as the *triangular-border walls*. No point in $C(\alpha)$ in the plane $\theta = \alpha$ can be visible from a point on a triangular-border wall. Therefore, the set of points in the plane $\theta = \alpha$ that are visible from no point on the triangular-border walls is a superset of $C(\alpha)$.

Recall that the canonical grasps of the two-finger equilibrium grasps form a number of curves in canonical grasp space. We define the upward vertical walls on all points on these curves, to which we refer as the *two-finger equilibrium walls*. Similar to the triangular-border walls, if (p, α) is visible from a point on a two-finger equilibrium wall then (p, α) is non-caging. The two-finger equilibrium walls intersect no surface patches of $\mathcal{P}$ by Lemma 2.4.

We define the *non-caging walls* as the union of the set of triangular-border walls and two-finger equilibrium walls. Let

$V(\alpha)$ be the set of points in the plane $\theta = \alpha$ that are visible from no point on the non-caging walls. We already know that $C(\alpha)$ is a subset of $V(\alpha)$. We prove that $C(\alpha)$ is equal to $V(\alpha)$.

First we provide a lemma which we use to prove the main result of this subsection. The following lemma states that the local minima of y-coordinates in the interior of the intersection of a triangular surface patch of $\mathcal{P}$ with a plane $x = x$ are immobilizing grasps, from which we omit the proof. In its proof we have used a number of results proven by Vahedi and van der Stappen [22]. (Every immobilizing grasp is a caging grasp.)

Lemma 2.7: The local minima of y-coordinates in the interior of the intersection of a triangular surface patch of $\mathcal{P}$ with a plane $x = x$ are immobilizing grasps.

The following lemma states that $C(\alpha)$ is equal to $V(\alpha)$. This result will be the foundation of our approach to establish the complexity bound. To prove the claim, we prove the equivalent lemma that every non-caging enclosing grasp $((x, y), \alpha)$ is visible from a non-caging wall. Note that if $((x, y), \alpha)$ is visible from a non-caging wall then $((x, y'), \alpha)$ with $y' > y$ is also visible from that wall. Therefore, it suffices to prove the claim for the points on $K(\alpha)$.

Lemma 2.8: For every point (x, y) on $K(\alpha)$, the point $((x, y), \alpha)$ is visible from a point on a non-caging wall.

Proof: If the point $((x, y), \alpha)$ corresponds to a two-finger equilibrium grasp then $((x, y), \alpha)$ is visible from a two-finger equilibrium wall by Lemma 2.4, and the claim follows. Assume that $((x, y), \alpha)$ does not correspond to a two-finger equilibrium grasp. As a result, the point $((x, y), \alpha)$ must correspond to a three-finger equilibrium grasp by Lemma 2.3. The number of three-finger equilibrium grasps in which the x-coordinate of the third finger is fixed and also the distance between the base fingers is d, is limited. Therefore, assume that $((x, y), \alpha)$ is a point in canonical grasp space that corresponds to a three-finger equilibrium grasp, $(x, y) \in K(\alpha)$, it is not visible from a two-finger equilibrium wall or a triangular-border wall, and has a y-coordinate that is minimal among all such three-finger equilibrium grasps.

Since $((x, y), \alpha)$ is non-caging, the canonical placement of P at orientation α can escape by first sliding along the base fingers and then by translating, according to Erickson et al. [4]. Let $((x, y), \beta)$ be the closest reachable canonical grasp at which it is possible for the canonical placement of P at orientation β to escape by translation. The set of walls visible from $((x, y), \beta)$ is the same as the walls visible from $((x, y), \alpha)$. Since the polygon can escape by translation through the grasp $((x, y), \beta)$, (x, y) is neither a point inside $C(\beta)$ nor a point inside $X[\beta]$. If (x, y) is on $K(\beta)$, then it corresponds to a two-finger equilibrium grasp or it is on a triangular border by Lemma 2.6. Therefore, assume that (x, y) is outside $C(\beta)$. Let Q be the set of all points in canonical grasp space whose corresponding canonical grasps are reachable from $((x, y), \beta)$ by sliding the polygon on the base fingers while allowing the third finger to monotonically be squeezed. Every non-caging wall visible from a point in Q is also visible from the points

$((x, y), \beta)$ and $((x, y), \alpha)$. The set Q contains a number of local minima along the θ-axis with respect to the y-coordinate. Since $((x, y), \beta)$ is not visible from a triangular-border wall the local minima of Q are immobilizing grasps by Lemma 2.7. Let $((x, y_m), \alpha_m)$ be one of those immobilizing grasps and consider $(x, y') \in K(\alpha_m)$. We have $((x, y'), \alpha_m) \in Q$ and thus all non-caging walls that are visible from $((x, y'), \alpha_m)$ are also visible from $((x, y), \alpha)$. If $((x, y'), \alpha_m)$ corresponds to a two-finger equilibrium grasp the claim follows. Otherwise, the grasp $((x, y'), \alpha_m)$ corresponds to a three-finger equilibrium grasp by Lemma 2.3 for which $y' < y$. The existence of the grasp $((x, y'), \alpha_m)$ contradicts the assumption. ∎

Corollary 2.9: $K(\alpha)$ is the lower boundary of the non-caging walls projected to the plane $\theta = \alpha$ not obstructed by the surface patches of $\mathcal{P}$.

III. COMPLEXITY AND COMPUTATION OF THE CAGING REGION

In this section we prove that the complexity of the caging region is close to $O(n^3)$ in the worst case. We also propose an algorithm that efficiently computes the boundary of the caging region in a time that is close to $O(n^3 \log(n))$ in the worst case. The main fact we prove is that the complexity of the visible part of a surface patch of $\mathcal{P}$ not hindered by the non-caging walls is constant. This fact gives us both a way to establish an upper bound on the complexity of the caging region and obtain a solution to compute the caging region efficiently.

According to Corollary 2.9, $K(\alpha)$ is the lower boundary of the non-caging walls not obstructed by the patches of $\mathcal{P}$ projected onto the plane $\theta = \alpha$. Here, however, we formulate a slightly different way to compute $K(\alpha)$. We consider the clockwise and counterclockwise viewing direction separately. We consider only the surface patches and non-caging walls within the same connected component of triangular surface patches of $\mathcal{P}$ intersected by the plane $\theta = \alpha$. For each direction, we project the visible part of each surface patch to the plane $\theta = \alpha$ by considering only the non-caging walls as obstacles, and compute the upper boundary of the projections. Without considering the surface patches of $\mathcal{P}$, we separately project the non-caging walls to the plane $\theta = \alpha$ and compute the lower boundary of the projections. Let $V^+(\alpha)$ be the maximum of the two resulting boundaries in the clockwise viewing direction. Define $V^-(\alpha)$ similarly for the counterclockwise direction. We have the following lemma, from which we omit the proof.

Lemma 3.1: $K(\alpha)$ is the minimum of $V^+(\alpha)$ and $V^-(\alpha)$.

As we consider the non-caging walls as obstacles, we prove in Lemmas 3.5 and 3.6 that the complexity of the visible part of each surface patch is constant. Before that, we mention three results that can be easily verified.

Lemma 3.2: The two-finger equilibrium walls involving the same base finger do not intersect each other.

Observation 3.3: The triangular-border walls do not intersect each other.

Lemma 3.4: The non-caging walls do not intersect the surface patches of $\mathcal{P}$.

First we prove that the visible part of a surface patch not obstructed by triangular-border walls has constant complexity. In Lemma 3.6 we consider the two-finger equilibrium walls as well.

Lemma 3.5: The visible part of a surface patch of $\mathcal{P}$ not obstructed by triangular-border walls has constant complexity.

Proof: Since the triangular-border walls are built on the surface patches of $\mathcal{P}$ we can regard them as unbounded in both upward and downward directions. To see, consider a point which is visible behind and below a triangular-border wall. Then that point is hindered by the surface patch upon which the triangular-border wall is built.

We divide the triangular-border walls into two groups being on the left or on the right with respect to the x-axis. The walls in each group have a complete order along θ-axis according to their distance from the plane $\theta = \alpha$. Consider the left group. (The right group can be treated similarly.) Since we cannot see the side that does not face the plane $\theta = \alpha$ we consider each local maximum with respect to the x-coordinate as a wall perpendicular to the θ-axis and ignore the walls in between two consecutive local maxima. We consider the set of local maxima of the triangular borders along x-axis and then we consider the sub-sequence of local maxima in increasing order. The reason is that a local maximum is completely invisible behind another local maximum with a larger x-coordinate.

Since the complexity of a surface patch is constant it has a constant number of local minima with respect to the x-axis. Consider a local minimum that is hindered by a triangular-border wall (from the increasing sub-list). The hindering wall is the wall that is the closest to the local minimum with respect to θ-axis and is between the plane $\theta = \alpha$ and the local minimum. Then every other point of the surface patch that its x-coordinate is larger than the local maximum of the triangular-border wall and it is connected by a x-monotone curve (on the surface patch) to the hindered point (i.e. the local minimum point), is not hindered by any other triangular-border wall of the same group. Every other point of the surface patch that its x-coordinate is smaller than the local maximum of the triangular-border wall and it is connected by a x-monotone curve to the hindered point (i.e. the local minimum point), is hindered. Every point on the surface patch is connected to at least one local minimum with a x-monotone curve. ∎

The following lemma is the main lemma we use to provide an upper bound on the complexity of caging region.

Lemma 3.6: The visible part of a surface patch of $\mathcal{P}$ not obstructed by non-caging walls has constant complexity.

Proof: Consider one side of the plane $\theta = \alpha$ and an arbitrary surface patch of $\mathcal{P}$. Consider the visible part of the surface patch not obstructed by the triangular-border walls. By Lemma 3.5 the complexity of this visible part is constant. Therefore, we compute the visible part of the surface patch not obstructed by the triangular-border walls and then remove (or ignore) the triangular-border walls. Consider the two-finger equilibrium walls that involve the base finger b_1. (We can show that there is no need to consider part of the surface patch on the right or left side of a two-finger equilibrium wall, but we

do not explain it here.)

There is a complete order between the two-finger equilibrium walls along θ-axis according to their distance from the plane $\theta = \alpha$. We traverse the two-finger equilibrium walls according to that order and we compute a sub-sequence with decreasing radii. The reason is that, a two-finger equilibrium wall with a larger radius is completely invisible behind a two-finger equilibrium wall with a smaller radius by Lemma 3.2.

Consider the local maxima of the surface patch with respect to the distance to the line $(0, 0, \theta)$. Since the complexity of a surface patch is constant the number of such local maxima is constant. If a local maximum is visible then all points that are monotonically connected to the local maximum and have have less distance with respect to the local maximum are visible too. If a local maximum is not visible, then it is hindered by a number of two-finger equilibrium walls from which consider the wall (from the decreasing sub-list) that is the furthest away from the plane $\theta = \alpha$. All points of the surface patch that are monotonically connected to the local maximum (by a path on the surface patch) and have distance less than the radius of this wall are not hindered by any other two-finger equilibrium wall of this group. All points of the surface patch that are monotonically connected (by a path on the surface patch) to the local maximum and have distance larger than the radius of this wall, are hindered.

We can similarly argue about the other group of two-finger equilibrium walls that involve the base finger b_2. ∎

In the following lemma and theorem we provide an upper bound on the worst-case complexity of $K(\alpha)$.

Lemma 3.7: The complexity of $K(\alpha)$ is at most the sum of complexities of $V^+(\alpha)$ and $V^-(\alpha)$.

Proof: The set $K(\alpha)$ is the minimum of $V^+(\alpha)$ and $V^-(\alpha)$ by Lemma 3.1. Let the complexity of $V^+(\alpha)$ be of order $O(f(n))$. Consider the sequence of the breaking points of both $V^+(\alpha)$ and $V^-(\alpha)$ in increasing order with respect to their x-coordinates. Between every two consecutive breaking point, exactly one sub-curve of $V^+(\alpha)$ and exactly one sub-curve of $V^-(\alpha)$ lie within the interval. These two sub-curves intersect each other a constant number of times. Since, the total number of breaking points of both $V^+(\alpha)$ and $V^-(\alpha)$ is of order $O(f(n))$, the total number of breaking points of $K(\alpha)$ is also of order $O(f(n))$. ∎

A Davenport-Schinzel sequence, $DS(m, s)$-sequence, is a sequence of m symbols in which no two symbols alternate more than s times. The lower boundary of m two-dimensional x-monotone curve segments in which no two curve segments intersect each other more than $s - 2$ times is a $DS(m, s)$-sequence. The maximum length of a $DS(m, s)$-sequence is $\lambda_s(m)$ [15].

Theorem 3.8: The complexities of both $V^+(\alpha)$ and $V^-(\alpha)$ are of order $O(\lambda_{10}(n^3))$.

Proof: The degree of the silhouette curves of the visible part of each surface patch is at most four [4]. Therefore each two curves intersect each other at most eight times. The complexity of the upper-boundary of the visible part of each surface patch projected to the plane $\theta = \alpha$ is $O(\lambda_{10}(n^3))$ by

Lemma 3.6. The complexity of the lower-boundary of the non-caging walls projected to the plane $\theta = \alpha$ is $O(\lambda_{10}(n^3))$ too. The complexity of the maximum of the two resulting chain of arcs is $O(\lambda_{10}(n^3))$ too. The proof for the last part is the same as the proof explained in Lemma 3.7. ∎

We briefly explain a way to compute $K(\alpha)$ in $O(\lambda_{10}(n^3) \log n)$ time. The visible parts of all surface patches can be computed in $O(n^3 \log n)$ time. To compute the upper-boundary of the projected visible parts we use a divide-and-conquer technique. We divide the projected visible parts into two groups and compute the upper-boundary for each group separately. Then we merge the results to compute the final upper-boundary. We use the same technique to compute the lower-boundary for the projected patches of non-caing walls.

Theorem 3.9: $K(\alpha)$ can be computed in $O(\lambda_{10}(n^3) \log n)$ time.

IV. CONCLUSION

We have provided a worst-case bound of almost $O(n^3)$ on the combinatorial complexity of the caging region of a convex polygon with n vertices, and an algorithm with a running time close to $O(n^3 \log n)$ to compute the caging region. Both results present a major improvement over previous results. Our results have been obtained by exploiting a novel formulation of caging in terms of visibility in canonical grasp space.

The first question that comes to mind is whether the bounds reported here are tight. To gain insight into this question we will aim to construct convex polygons that have a caging region with a complexity that approaches the upper bound of $O(n^3)$. Another challenge is to extend the bounds obtained in this paper to the caging region of non-convex polygons.

ACKNOWLEDGMENT

M. Vahedi is supported by the Netherlands Organisation for Scientific Research (NWO). A. F. van der Stappen is partially supported by the Dutch BSIK/BRICKS-project.

REFERENCES

[1] C. Davidson and A. Blake. Caging planar objects with a three-finger one-parameter gripper. In *ICRA*, pages 2722–2727. IEEE, 1998.

[2] C. Davidson and A. Blake. Error-tolerant visual planning of planar grasp. In *Sixth International Conference on Computer Vision (ICCV)*, pages 911–916, 1998.

[3] Rosen Diankov, Siddhartha Srinivasa, David Ferguson, and James Kuffner. Manipulation planning with caging grasps. In *IEEE International Conference on Humanoid Robots*, December 2008.

[4] J. Erickson, S. Thite, F. Rothganger, and J. Ponce. Capturing a convex object with three discs. *IEEE Tr. on Robotics*, 23(6):1133–1140, 2007.

[5] J. Fink, M. A. Hsieh, and V. Kumar. Multi-robot manipulation via caging in environments with obstacles. In *ICRA*, Pasedena, CA, May 2008.

[6] J. Fink, N. Michael, and V. Kumar. Composition of vector fields for multi-robot manipulation via caging. In

[7] W. Kuperberg. Problems on polytopes and convex sets. *DIMACS Workshop on Polytopes*, pages 584–589, 1990.

[8] M. Mason. *Mechanics of Robotic Manipulation*. MIT Press, August 2001. Intelligent Robotics and Autonomous Agents Series, ISBN 0-262-13396-2.

[9] G. A. S. Pereira, V. Kumar, and M. F. M. Campos. Decentralized algorithms for multirobot manipulation via caging. *Int. J. Robotics Res.*, 2004.

[10] P. Pipattanasomporn and A. Sudsang. Two-finger caging of concave polygon. In *ICRA*, pages 2137–2142. IEEE, 2006.

[11] P. Pipattanasomporn, P. Vongmasa, and A. Sudsang. Two-finger squeezing caging of polygonal and polyhedral object. In *ICRA*, pages 205–210. IEEE, 2007.

[12] E. Rimon and A. Blake. Caging 2d bodies by 1-parameter two-fingered gripping systems. In *ICRA*, volume 2, pages 1458–1464. IEEE, 1996.

[13] E. Rimon and J. W. Burdick. Mobility of bodies in contact—part i: A 2nd–order mobility index for multiple–finger grasps. *IEEE Tr. on Robotics and Automation*, 14(5):696–717, 1998.

[14] A. Rodriguez and M. Mason. Two finger caging: squeezing and stretching. In *Eighth Workshop on the Algorithmic Foundations of Robotics (WAFR)*, 2008.

[15] M. Sharir and P. K. Agarwal. *Davenport-Schinzel sequences and their geometric applications*. Cambridge University Press, New York, NY, USA, 1996.

[16] A. Sudsang. Grasping and in-hand manipulation: Geometry and algorithms. *Algorithmica*, 26(3-4):466–493, 2000.

[17] A. Sudsang and T. Luewirawong. Capturing a concave polygon with two disc-shaped fingers. In *ICRA*, pages 1121–1126. IEEE, 2003.

[18] A. Sudsang and J. Ponce. A new approach to motion planning for disc-shaped robots manipulating a polygonal object in the plane. In *ICRA*, pages 1068–1075. IEEE, 2000.

[19] A. Sudsang, J. Ponce, M. Hyman, and D. J. Kriegman. On manipulating polygonal objects with three 2-dof robots in the plane. In *ICRA*, pages 2227–2234, 1999.

[20] A. Sudsang, J. Ponce, and N. Srinivasa. Grasping and in-hand manipulation: Experiments with a reconfigurable gripper. *Advanced Robotics*, 12(5):509–533, 1998.

[21] A. Sudsang, F. Rothganger, and J. Ponce. Motion planning for disc-shaped robots pushing a polygonal object in the plane. *IEEE Tr. on Robotics and Automation*, 18(4):550–562, 2002.

[22] M. Vahedi and A. F. van der Stappen. Geometric properties and computation of three-finger caging grasps of convex polygons. In *IEEE International Conference on Automation Science and Engineering (CASE)*, 2007.

[23] M. Vahedi and A. F. van der Stappen. Caging polygons with two and three fingers. *Int. J. Robotics Res.*, 27(11/12):1308–1324, 2008.

Proc. of Robotics: Science and Systems, Atlanta, GA, June 2007.

On the Consistency of Multi-robot Cooperative Localization

Guoquan P. Huang*, Nikolas Trawny*, Anastasios I. Mourikis†, and Stergios I. Roumeliotis*

*Dept. of Computer Science and Engineering, University of Minnesota, Minneapolis, MN 55455
Email: {ghuang|trawny|stergios}@cs.umn.edu
†Dept. of Electrical Engineering, University of California, Riverside, CA 92521
Email: mourikis@ee.ucr.edu

Abstract—In this paper, we investigate the consistency of extended Kalman filter (EKF)-based cooperative localization (CL) from the perspective of observability. To the best of our knowledge, this is the first work that analytically shows that the error-state system model employed in the standard EKF-based CL always has an observable subspace of higher dimension than that of the actual nonlinear CL system. This results in unjustified reduction of the EKF covariance estimates in directions of the state space where no information is available, and thus leads to inconsistency. To address this problem, we adopt an observability-based methodology for designing consistent estimators and propose a novel Observability-Constrained (OC)-EKF. In contrast to the standard EKF-CL, the linearization points of the OC-EKF are selected so as to ensure that the dimension of the observable subspace remains the same as that of the original (nonlinear) system. The proposed OC-EKF has been tested in simulation and experimentally, and has been shown to significantly outperform the standard EKF in terms of both accuracy and consistency.

I. INTRODUCTION

In order for multi-robot teams to perform tasks such as exploration, surveillance, and search and rescue, their members need to precisely determine their positions and orientations (poses). In GPS-denied areas and in the absence of robust landmarks, teams of robots can still jointly estimate their poses by sharing relative position measurements (cf. [1], [2], [3]). Current approaches to solving the cooperative localization (CL) problem, in either centralized or distributed fashion, are based on the extended Kalman filter (EKF) [3], maximum likelihood estimation (MLE) [4], maximum a posteriori (MAP) estimation [5], or particle filtering (PF) [6]. Among these algorithms, the EKF arguably remains a popular choice due to its relatively low computational cost and easy implementation.

While recent research efforts have focused on reducing the computational complexity of EKF-CL [7], [8], [9], the fundamental issue of *consistency* has received little attention. As defined in [10], a state estimator is consistent if the estimation errors are zero-mean and have covariance smaller than or equal to the one calculated by the filter. Consistency is one of the primary criteria for evaluating an estimator's performance; if an estimator is inconsistent, then the accuracy of the produced state estimates is unknown, which renders them unreliable.

In this paper, we present a study of the consistency of EKF-based CL from the perspective of observability. Based on the

study, we introduce a novel EKF estimator that significantly improves consistency as well as accuracy. In particular, the major contributions of this work are the following:

• We investigate the observability properties of the error-state system model employed by the EKF, and show that its observable subspace has *higher dimension* than that of the underlying nonlinear CL system. As a result, the covariance estimates of the EKF undergo reduction in directions of the state space where no information is available, hence leading to *inconsistency*. To the best of our knowledge, this is the first work to identify and report the inconsistency of EKF-CL.

• Based on the observability analysis, we introduce the Observability-Constrained (OC)-EKF to improve consistency. Specifically, the linearization points of the OC-EKF's system model are judiciously selected to ensure that the linearized CL system has an observable subspace of the *same dimension* as the nonlinear CL system. We show that this requirement is satisfied by evaluating the state-propagation Jacobians at the state estimates *before* (instead of after) each update, while the measurement Jacobians are computed in the same way as for the standard EKF. Through extensive simulation and experimental tests, we verify that the OC-EKF outperforms the standard EKF in terms of both accuracy and consistency, even though it uses older (and thus less accurate) state estimates to compute the filter Jacobians. This result in turn indicates that the observability properties of the system model play a key role in determining the filter's accuracy and consistency.

II. RELATED WORK

To the best of our knowledge, no work has yet *analytically* addressed the consistency issue in CL. In contrast, recent research has focused on the consistency of EKF-SLAM (cf. [11], [12], [13], [14], [15]) showing that the computed state estimates tend to be inconsistent. Specifically, Julier and Uhlmann [11] first observed that when a stationary robot measures the relative position of a new landmark multiple times, the estimated variance of the robot's orientation becomes smaller. Since the observation of a previously unseen feature does not provide any information about the robot's state, this reduction is "artificial" and thus leads to inconsistency. Bailey *et al.* [13] examined several symptoms of the inconsistency of the standard EKF SLAM algorithm, and argued, based on simulation results, that the uncertainty in the robot orientation is the main cause of the inconsistency of EKF-SLAM. Huang and Dissanayake [14] extended the analysis of [11] to the case of a robot observing a landmark from two positions (i.e., the robot observes a landmark, moves and then re-observes the landmark), and proposed a constraint that the filter Jacobians must satisfy to allow for consistent estimation. They also showed that this condition is generally violated, due to the fact that the filter Jacobians at different time instants are computed using different estimates for the same state variables.

In our previous work [15], we conducted a theoretical analysis of EKF-SLAM inconsistency and identified as a fundamental cause the mismatch in the dimensions of the observable subspaces between the linearized system, employed by the EKF, and the underlying nonlinear system. Furthermore, we introduced the First Estimates Jacobian (FEJ)-EKF, which significantly outperforms the standard EKF and the robocentric mapping algorithm [12], in terms of both accuracy and consistency. The FEJ-EKF solution to the SLAM inconsistency problem was reached by imposing constraints inferred from the system observability analysis, and can serve as a model for a new methodology for designing consistent estimators for nonlinear systems. Such an approach is therefore employed in this work for addressing the inconsistency of EKF-CL.

We should note that a recent publication by Bahr *et al.* [16] addresses a related but different problem, namely the consistency of a distributed CL algorithm due to re-use of information. In the decentralized estimation scheme of [16], the cross-correlations between the state estimates of different robots are not estimated. However, it is well-known that if cross-correlations between robots are not properly taken into account during filter updates, inconsistency can arise [6], [17]. The algorithm in [16] avoids inconsistency by maintaining a careful record of past robot-to-robot measurement updates. In contrast to the above fully decentralized scenario, in our work the cross-correlation terms are maintained in the filter, and the EKF employed for estimation is optimal, except for the inaccuracies introduced by linearization. Our work focuses on identifying and addressing the cause of inconsistency of this EKF-based CL estimator.

III. STANDARD EKF-BASED CL

In this section, we present the equations of 2D EKF-CL with general system and measurement models.[1] In particular, in the standard formulation of CL, the state vector comprises the N robot poses expressed in the global frame of reference. Thus, at time-step k the state vector is given by:

$$\mathbf{x}_k = \begin{bmatrix} \mathbf{x}_{1_k}^T & \cdots & \mathbf{x}_{N_k}^T \end{bmatrix}^T \tag{1}$$

where $\mathbf{x}_{i_k} \triangleq [\mathbf{p}_{i_k}^T \quad \phi_{i_k}]^T$ denotes the ith robot pose (position and orientation). In general, EKF-CL recursively evolves in two steps: propagation and update, based on the discrete-time system and measurement models, respectively.

A. EKF Propagation

In the propagation step, each robot processes its odometry measurements to obtain an estimate of the pose change between two consecutive time steps, which is then employed in the EKF to propagate the robot state estimate. The EKF propagation equations are given by:[2]

$$\hat{\mathbf{p}}_{i_{k+1|k}} = \hat{\mathbf{p}}_{i_{k|k}} + \mathbf{C}(\hat{\phi}_{i_{k|k}}) \cdot {}^k\hat{\mathbf{p}}_{i_{k+1}} \tag{2}$$

$$\hat{\phi}_{i_{k+1|k}} = \hat{\phi}_{i_{k|k}} + {}^k\hat{\phi}_{i_{k+1}} \tag{3}$$

for all $i = 1, \ldots, N$. In the above expressions, $\mathbf{C}(\cdot)$ denotes the 2×2 rotation matrix, and ${}^k\hat{\mathbf{x}}_{i_{k+1}} \triangleq [{}^k\hat{\mathbf{p}}_{i_{k+1}}^T \quad {}^k\hat{\phi}_{i_{k+1}}]^T$ is the odometry-based estimate of the ith robot's motion between time-steps k and $k+1$, expressed with respect to the robot's frame of reference at time-step k. This estimate is corrupted by zero-mean, white Gaussian noise $\mathbf{w}_{i_k} = {}^k\mathbf{x}_{i_{k+1}} - {}^k\hat{\mathbf{x}}_{i_{k+1}}$, with covariance $\mathbf{Q}_k$. Clearly this system model is nonlinear, and can be described by the following generic nonlinear function:

$$\mathbf{x}_{i_{k+1}} = \mathbf{f}(\mathbf{x}_{i_k}, {}^k\hat{\mathbf{x}}_{i_{k+1}} + \mathbf{w}_{i_k}) \tag{4}$$

In addition to the state propagation equations, the linearized error-state propagation equation is necessary for the EKF:

$$\tilde{\mathbf{x}}_{i_{k+1|k}} = \mathbf{\Phi}_{i_k}\tilde{\mathbf{x}}_{i_{k|k}} + \mathbf{G}_{i_k}\mathbf{w}_{i_k} \tag{5}$$

In the above expression, $\mathbf{\Phi}_{i_k}$ and $\mathbf{G}_{i_k}$ are obtained by linearization of the state propagation equations (2)-(3):

$$\mathbf{\Phi}_{i_k} = \begin{bmatrix} \mathbf{I}_2 & \mathbf{J}\left(\hat{\mathbf{p}}_{i_{k+1|k}} - \hat{\mathbf{p}}_{i_{k|k}}\right) \\ \mathbf{0}_{1\times 2} & 1 \end{bmatrix} \tag{6}$$

$$\mathbf{G}_{i_k} = \begin{bmatrix} \mathbf{C}(\hat{\phi}_{i_{k|k}}) & \mathbf{0}_{2\times 1} \\ \mathbf{0}_{1\times 2} & 1 \end{bmatrix} \tag{7}$$

where we have employed the identity $\mathbf{C}(\hat{\phi}_{i_{k|k}}) \cdot {}^k\hat{\mathbf{p}}_{i_{k+1}} = \hat{\mathbf{p}}_{i_{k+1|k}} - \hat{\mathbf{p}}_{i_{k|k}}$, and $\mathbf{J} \triangleq \begin{bmatrix} 0 & -1 \\ 1 & 0 \end{bmatrix}$.

By stacking all N robots' states to create the state vector for the entire system, we have

$$\tilde{\mathbf{x}}_{k+1|k}$$
$$= \begin{bmatrix} \mathbf{\Phi}_{1_k} & \cdots & \mathbf{0} \\ \vdots & \ddots & \vdots \\ \mathbf{0} & \cdots & \mathbf{\Phi}_{N_k} \end{bmatrix} \begin{bmatrix} \tilde{\mathbf{x}}_{1_{k|k}} \\ \vdots \\ \tilde{\mathbf{x}}_{N_{k|k}} \end{bmatrix} + \begin{bmatrix} \mathbf{G}_{1_k} & \cdots & \mathbf{0} \\ \vdots & \ddots & \vdots \\ \mathbf{0} & \cdots & \mathbf{G}_{N_k} \end{bmatrix} \begin{bmatrix} \mathbf{w}_{1_k} \\ \vdots \\ \mathbf{w}_{N_k} \end{bmatrix}$$
$$\triangleq \mathbf{\Phi}_k\tilde{\mathbf{x}}_{k|k} + \mathbf{G}_k\mathbf{w}_k \tag{8}$$

It is important to point out that the form of the propagation equations presented above is general, and holds for any robot kinematic model (e.g., unicycle, bicycle, or Ackerman).

---

[1]For the purpose of the consistency study and in order to simplify the derivations, in this paper we focus on centralized EKF-CL. Note that a distributed implementation [3] does not alter the system properties.

[2]Throughout this paper the subscript $\ell|k$ refers to the estimate of a quantity at time-step ℓ, after all measurements up to time-step k have been processed, while the superscript (ij) refers to the relative measurement from robot i to robot j. $\hat{x}$ is used to denote the estimate of a random variable x, while $\tilde{x} = x - \hat{x}$ is the error in this estimate. $\mathbf{0}_{m\times n}$ and $\mathbf{1}_{m\times n}$ denote $m \times n$ matrices of zeros and ones, respectively, while $\mathbf{I}_n$ is the $n \times n$ identity matrix.

B. EKF Update

The measurements used for updates in CL are always a function of the relative pose of the observed robot j with respect to the observing robot i:

$$\mathbf{z}_k^{(ij)} = \mathbf{h}(\mathbf{x}_{i_k}, \mathbf{x}_{j_k}) + \mathbf{v}_k^{(ij)} = \mathbf{h}\left(^i\mathbf{x}_{j_k}\right) + \mathbf{v}_k^{(ij)} \qquad (9)$$

where $^i\mathbf{x}_{j_k} \triangleq \begin{bmatrix} ^i\mathbf{p}_{j_k} \\ ^i\phi_{j_k} \end{bmatrix} = \begin{bmatrix} \mathbf{C}^T(\phi_{i_k})(\mathbf{p}_{j_k} - \mathbf{p}_{i_k}) \\ \phi_{j_k} - \phi_{i_k} \end{bmatrix}$ is the relative pose of the observed robot j with respect to the observing robot i at time-step k, and $\mathbf{v}_k^{(ij)}$ is zero-mean Gaussian noise with covariance $\mathbf{R}_k^{(ij)}$ associated with the measurement $\mathbf{z}_k^{(ij)}$. In this work, we allow $\mathbf{h}$ to be *any* measurement function. For instance, $\mathbf{z}_k^{(ij)}$ can be a direct measurement of relative pose, a pair of distance and bearing measurements, bearing-only measurements from monocular cameras, etc. In general, the measurement function is nonlinear, and hence it is linearized for use in the EKF. The linearized measurement-error equation is given by:

$$\tilde{\mathbf{z}}_k^{(ij)} \simeq \begin{bmatrix} \mathbf{0} & \cdots & \mathbf{H}_{i_k}^{(ij)} & \cdots & \mathbf{H}_{j_k}^{(ij)} & \cdots & \mathbf{0} \end{bmatrix} \tilde{\mathbf{x}}_{k|k-1} + \mathbf{v}_k^{(ij)}$$
$$\triangleq \mathbf{H}_k^{(ij)} \tilde{\mathbf{x}}_{k|k-1} + \mathbf{v}_k^{(ij)} \qquad (10)$$

where $\mathbf{H}_{i_k}^{(ij)}$ and $\mathbf{H}_{j_k}^{(ij)}$ are the Jacobians of $\mathbf{h}$ with respect to the ith and jth robot poses, respectively, evaluated at the state estimate $\hat{\mathbf{x}}_{k|k-1}$. Using the chain rule of differentiation, these are computed as:

$$\mathbf{H}_{i_k}^{(ij)} = -(\nabla \mathbf{h}_k^{(ij)}) \mathbf{A}(\hat{\phi}_{i_{k|k-1}}) \begin{bmatrix} \mathbf{I}_2 & \mathbf{J}(\hat{\mathbf{p}}_{j_{k|k-1}} - \hat{\mathbf{p}}_{i_{k|k-1}}) \\ \mathbf{0}_{1\times2} & 1 \end{bmatrix}$$
$$(11)$$

$$\mathbf{H}_{j_k}^{(ij)} = (\nabla \mathbf{h}_k^{(ij)}) \mathbf{A}(\hat{\phi}_{i_{k|k-1}}) \qquad (12)$$

where $\mathbf{A}(\hat{\phi}_{i_{k|k-1}}) \triangleq \begin{bmatrix} \mathbf{C}^T(\hat{\phi}_{i_{k|k-1}}) & \mathbf{0}_{2\times1} \\ \mathbf{0}_{1\times2} & 1 \end{bmatrix}$, and $\nabla \mathbf{h}_k^{(ij)}$ denotes the Jacobian of $\mathbf{h}$ with respect to the vector $^i\mathbf{x}_{j_k}$ evaluated at the state estimate $\hat{\mathbf{x}}_{k|k-1}$.

IV. CL OBSERVABILITY ANALYSIS

In this section, we perform an observability analysis for the EKF-CL system derived in the previous section, and compare its observability properties with those of the underlying nonlinear system. Based on this analysis, we draw conclusions about the consistency of the filter.

Martinelli and Siegwart [18] have shown that the underlying nonlinear system of CL in general has three unobservable degrees of freedom, corresponding to the global position and orientation. Thus, when the EKF is used for state estimation in CL, we would expect that the system model employed by the EKF also shares this property. However, in this section we show that this is not the case, since the unobservable subspace of the linearized error-state model of the standard EKF is generally only of dimension two.

Note that, in general, the Jacobian matrices $\mathbf{\Phi}_k$, $\mathbf{G}_k$, and $\mathbf{H}_k$ used in the EKF-CL linearized error-state model (cf. (8) and (10)) are defined as:

$$\mathbf{\Phi}_k = \nabla_{\mathbf{x}_k} \mathbf{f}|_{\{\mathbf{x}_k^\star, \mathbf{0}\}}, \quad \mathbf{G}_k = \nabla_{\mathbf{w}_k} \mathbf{f}|_{\{\mathbf{x}_k^\star, \mathbf{0}\}}, \quad \mathbf{H}_k = \nabla_{\mathbf{x}_k} \mathbf{h}|_{\mathbf{x}_k^\star} \quad (13)$$

In these expressions, $\mathbf{x}_k^\star$ denotes the *linearization point* for the state $\mathbf{x}_k$, used for evaluating the Jacobians, while a linearization point equal to the zero vector is chosen for the noise. The EKF employs the above linearized system model for propagating and updating the estimates of the covariance matrix, and thus the observability properties of this model affect the performance of the estimator.

The observability properties of the linearized error-state model of EKF-CL can be studied by examining the observability matrix for the time interval between time-steps k_o and $k_o + m$, defined as [19]:

$$\mathbf{M} \triangleq \begin{bmatrix} \mathbf{H}_{k_o} \\ \mathbf{H}_{k_o+1}\mathbf{\Phi}_{k_o} \\ \vdots \\ \mathbf{H}_{k_o+m}\mathbf{\Phi}_{k_o+m-1}\cdots\mathbf{\Phi}_{k_o} \end{bmatrix} \qquad (14)$$
$$= \mathbf{M}(\mathbf{x}_{k_o}^\star, \mathbf{x}_{k_o+1}^\star, \ldots, \mathbf{x}_{k_o+m}^\star) \qquad (15)$$

The last expression makes explicit the fact that the observability matrix is a function of the linearization points used in computing all the Jacobians within the time interval $[k_o, k_o+m]$. In turn, this implies that *the choice of linearization points affects the observability properties* of the linearized error-state system of the EKF. This key fact is the basis of our analysis. In what follows, we discuss different possible choices for linearization, and the observability properties of the corresponding linearized systems.

A. Ideal EKF-CL

Before considering the rank of the matrix $\mathbf{M}$, which is constructed using the *estimated* values of the state in the filter Jacobians, it is interesting to study the observability properties of the "oracle," or "ideal" EKF (i.e., the filter whose Jacobians are evaluated using the *true* values of the state variables). In the following, all matrices evaluated using the true state values are denoted by the symbol " $\breve{}$ ".

For our derivations, it will be useful to define:

$$\delta\mathbf{p}_{ij}(k, \ell) \triangleq \mathbf{p}_{i_k} - \mathbf{p}_{j_\ell} \qquad (16)$$

which is the difference between two robot positions at time-steps k and ℓ. Using the above definition, we note that (cf. (6))

$$\breve{\mathbf{\Phi}}_{i_{k_o+1}}\breve{\mathbf{\Phi}}_{i_{k_o}} = \begin{bmatrix} \mathbf{I}_2 & \mathbf{J}\delta\mathbf{p}_{ii}(k_o+2, k_o) \\ \mathbf{0}_{1\times2} & 1 \end{bmatrix} \qquad (17)$$

Based on this identity, it is easy to show by induction that

$$\breve{\mathbf{\Phi}}_{i_{k_o+\ell-1}}\breve{\mathbf{\Phi}}_{i_{k_o+\ell-2}}\cdots\breve{\mathbf{\Phi}}_{i_{k_o}} = \begin{bmatrix} \mathbf{I}_2 & \mathbf{J}\delta\mathbf{p}_{ii}(k_o+\ell, k_o) \\ \mathbf{0}_{1\times2} & 1 \end{bmatrix} \qquad (18)$$

which holds for all $\ell > 0$.

In the ensuing analysis, we study the case of a two-robot team, and assume that both robots continuously observe each other during the time interval $[k_o, k_o + m]$. Note that this is not a necessary assumption and is made only to simplify the notation. The generalized analysis for the case where the team consists of an arbitrary number of robots is presented in [20].

The measurement Jacobian $\breve{\mathbf{H}}_{k_o+\ell}$ in this case can be written as (cf. (11) and (12)):

$$\breve{\mathbf{H}}_{k_o+\ell} = -\mathbf{Diag}\Big((\nabla\breve{\mathbf{h}}_{k_o+\ell}^{(12)})\mathbf{A}(\phi_{1_{k_o+\ell}}), (\nabla\breve{\mathbf{h}}_{k_o+\ell}^{(21)})\mathbf{A}(\phi_{2_{k_o+\ell}})\Big) \quad (19)$$

$$\times \begin{bmatrix} \mathbf{I}_2 & \mathbf{J}\delta\mathbf{p}_{21}(k_o+\ell, k_o+\ell) & -\mathbf{I}_2 & \mathbf{0}_{2\times 1} \\ \mathbf{0}_{1\times 2} & 1 & \mathbf{0}_{1\times 2} & -1 \\ -\mathbf{I}_2 & \mathbf{0}_{2\times 1} & \mathbf{I}_2 & \mathbf{J}\delta\mathbf{p}_{12}(k_o+\ell, k_o+\ell) \\ \mathbf{0}_{1\times 2} & -1 & \mathbf{0}_{1\times 2} & 1 \end{bmatrix}$$

while the following identity is immediate (cf. (8) and (18)):

$$\breve{\boldsymbol{\Phi}}_{k_o+\ell-1}\breve{\boldsymbol{\Phi}}_{k_o+\ell-2}\cdots\breve{\boldsymbol{\Phi}}_{k_o} = \quad (20)$$

$$\begin{bmatrix} \mathbf{I}_2 & \mathbf{J}\delta\mathbf{p}_{11}(k_o+\ell, k_o) & \mathbf{0}_{2\times 2} & \mathbf{0}_{2\times 1} \\ \mathbf{0}_{1\times 2} & 1 & \mathbf{0}_{1\times 2} & 0 \\ \mathbf{0}_{2\times 2} & \mathbf{0}_{2\times 1} & \mathbf{I}_2 & \mathbf{J}\delta\mathbf{p}_{22}(k_o+\ell, k_o) \\ \mathbf{0}_{1\times 2} & 0 & \mathbf{0}_{1\times 2} & 1 \end{bmatrix}$$

Multiplication of equations (19) and (20) yields:

$$\breve{\mathbf{H}}_{k_o+\ell}\breve{\boldsymbol{\Phi}}_{k_o+\ell-1}\breve{\boldsymbol{\Phi}}_{k_o+\ell-2}\cdots\breve{\boldsymbol{\Phi}}_{k_o} = \quad (21)$$

$$-\mathbf{Diag}\Big((\nabla\breve{\mathbf{h}}_{k_o+\ell}^{(12)})\mathbf{A}(\phi_{1_{k_o+\ell}}), (\nabla\breve{\mathbf{h}}_{k_o+\ell}^{(21)})\mathbf{A}(\phi_{2_{k_o+\ell}})\Big) \times$$

$$\begin{bmatrix} \mathbf{I}_2 & \mathbf{J}\delta\mathbf{p}_{21}(k_o+\ell, k_o) & -\mathbf{I}_2 & -\mathbf{J}\delta\mathbf{p}_{22}(k_o+\ell, k_o) \\ \mathbf{0}_{1\times 2} & 1 & \mathbf{0}_{1\times 2} & -1 \\ -\mathbf{I}_2 & -\mathbf{J}\delta\mathbf{p}_{11}(k_o+\ell, k_o) & \mathbf{I}_2 & \mathbf{J}\delta\mathbf{p}_{12}(k_o+\ell, k_o) \\ \mathbf{0}_{1\times 2} & -1 & \mathbf{0}_{1\times 2} & 1 \end{bmatrix}$$

Using this result, the observability matrix of the ideal EKF-CL, $\breve{\mathbf{M}}$, can be obtained as (cf. (14)):

$$\breve{\mathbf{M}} = \underbrace{-\mathbf{Diag}\Big((\nabla\breve{\mathbf{h}}_{k_o}^{(12)})\mathbf{A}(\phi_{1_{k_o}}), \cdots, (\nabla\breve{\mathbf{h}}_{k_o+m}^{(21)})\mathbf{A}(\phi_{2_{k_o+m}})\Big)}_{\breve{\mathbf{D}}} \times \quad (22)$$

$$\underbrace{\begin{bmatrix} \mathbf{I}_2 & \mathbf{J}\delta\mathbf{p}_{21}(k_o, k_o) & -\mathbf{I}_2 & \mathbf{0}_{2\times 1} \\ \mathbf{0}_{1\times 2} & 1 & \mathbf{0}_{1\times 2} & -1 \\ -\mathbf{I}_2 & \mathbf{0}_{2\times 1} & \mathbf{I}_2 & \mathbf{J}\delta\mathbf{p}_{12}(k_o, k_o) \\ \mathbf{0}_{1\times 2} & -1 & \mathbf{0}_{1\times 2} & 1 \\ \\ \mathbf{I}_2 & \mathbf{J}\delta\mathbf{p}_{21}(k_o+1, k_o) & -\mathbf{I}_2 & -\mathbf{J}\delta\mathbf{p}_{22}(k_o+1, k_o) \\ \mathbf{0}_{1\times 2} & 1 & \mathbf{0}_{1\times 2} & -1 \\ -\mathbf{I}_2 & -\mathbf{J}\delta\mathbf{p}_{11}(k_o+1, k_o) & \mathbf{I}_2 & \mathbf{J}\delta\mathbf{p}_{12}(k_o+1, k_o) \\ \mathbf{0}_{1\times 2} & -1 & \mathbf{0}_{1\times 2} & 1 \\ \vdots & \vdots & \vdots & \vdots \\ \mathbf{I}_2 & \mathbf{J}\delta\mathbf{p}_{21}(k_o+m, k_o) & -\mathbf{I}_2 & -\mathbf{J}\delta\mathbf{p}_{22}(k_o+m, k_o) \\ \mathbf{0}_{1\times 2} & 1 & \mathbf{0}_{1\times 2} & -1 \\ -\mathbf{I}_2 & -\mathbf{J}\delta\mathbf{p}_{11}(k_o+m, k_o) & \mathbf{I}_2 & \mathbf{J}\delta\mathbf{p}_{12}(k_o+m, k_o) \\ \mathbf{0}_{1\times 2} & -1 & \mathbf{0}_{1\times 2} & 1 \end{bmatrix}}_{\breve{\mathbf{U}}}$$

Lemma 4.1: The rank of the observability matrix, $\breve{\mathbf{M}}$, of the ideal EKF-CL is equal to 3.

Proof: The rank of the product of the matrices $\breve{\mathbf{D}}$ and $\breve{\mathbf{U}}$ is given by (cf. (4.5.1) in [21]), $\text{rank}(\breve{\mathbf{D}}\breve{\mathbf{U}}) = \text{rank}(\breve{\mathbf{U}}) - \dim(\mathcal{N}(\breve{\mathbf{D}})\bigcap\mathcal{R}(\breve{\mathbf{U}}))$. By denoting $\breve{\mathbf{U}} \triangleq \begin{bmatrix} \breve{\mathbf{u}}_1 & \cdots & \breve{\mathbf{u}}_6 \end{bmatrix}$, it is evident that $\breve{\mathbf{u}}_1 = -\breve{\mathbf{u}}_4$, $\breve{\mathbf{u}}_2 = -\breve{\mathbf{u}}_5$, while $\breve{\mathbf{u}}_3 + \breve{\mathbf{u}}_6 = \alpha_1\breve{\mathbf{u}}_4 + \alpha_2\breve{\mathbf{u}}_5$, where $\begin{bmatrix} \alpha_1 & \alpha_2 \end{bmatrix}^T \triangleq -\mathbf{J}\delta\mathbf{p}_{21}(k_o, k_o)$. We also note that $\{\breve{\mathbf{u}}_i\}_{i=4}^6$ are linearly independent. Therefore, $\mathcal{R}(\breve{\mathbf{U}}) = \text{span}\begin{bmatrix} \breve{\mathbf{u}}_4 & \breve{\mathbf{u}}_5 & \breve{\mathbf{u}}_6 \end{bmatrix}$. Thus, $\text{rank}(\breve{\mathbf{U}}) = 3$. We now observe that in general $\breve{\mathbf{D}}\mathbf{u}_i \neq \mathbf{0}$, for $i = 4, 5, 6$. Moreover any vector $\mathbf{y} \in \mathcal{R}(\breve{\mathbf{U}}) \setminus \mathbf{0}$ can be written as $\mathbf{y} = \beta_4\breve{\mathbf{u}}_4 + \beta_5\breve{\mathbf{u}}_5 + \beta_6\breve{\mathbf{u}}_6$ for some $\beta_i \in \mathbb{R}$, where β_i, $i = 4, 5, 6$, are not simultaneously equal to zero. Thus, in general, $\breve{\mathbf{D}}\mathbf{y} = \beta_4\breve{\mathbf{D}}\breve{\mathbf{u}}_4 + \beta_5\breve{\mathbf{D}}\breve{\mathbf{u}}_5 + \beta_6\breve{\mathbf{D}}\breve{\mathbf{u}}_6 \neq \mathbf{0}$, which

implies that $\mathbf{y}$ does not belong to the nullspace of $\breve{\mathbf{D}}$. Therefore, $\dim(\mathcal{N}(\breve{\mathbf{D}})\bigcap\mathcal{R}(\breve{\mathbf{U}})) = 0$, and, finally, $\text{rank}(\breve{\mathbf{M}}) = \text{rank}(\breve{\mathbf{U}}) - \dim(\mathcal{N}(\breve{\mathbf{D}})\bigcap\mathcal{R}(\breve{\mathbf{U}})) = \text{rank}(\breve{\mathbf{U}}) = 3$. ∎

Most importantly, by inspection, it can be verified that a basis for the right nullspace of $\breve{\mathbf{U}}$ (and thus for the right nullspace of $\breve{\mathbf{M}}$) is given by:

$$\mathcal{N}(\breve{\mathbf{M}}) = \underset{\text{col.}}{\text{span}} \begin{bmatrix} \mathbf{I}_2 & \mathbf{J}\mathbf{p}_{1_{k_o}} \\ \mathbf{0}_{1\times 2} & 1 \\ \mathbf{I}_2 & \mathbf{J}\mathbf{p}_{2_{k_o}} \\ \mathbf{0}_{1\times 2} & 1 \end{bmatrix} \triangleq \text{span}\begin{bmatrix} \mathbf{n}_1 & \mathbf{n}_2 & \mathbf{n}_3 \end{bmatrix} \quad (23)$$

From the structure of the vectors $\mathbf{n}_1$ and $\mathbf{n}_2$ we see that a change in the state by $\Delta\mathbf{x} = \alpha\mathbf{n}_1 + \beta\mathbf{n}_2$, $\alpha, \beta \in \mathbb{R}$ corresponds to shifting the $x-y$ plane by α units along x, and by β units along y. Thus, if the two robots are shifted equally, the states $\mathbf{x}$ and $\mathbf{x}' = \mathbf{x} + \Delta\mathbf{x}$ will be indistinguishable given the relative measurements. To understand the physical meaning of $\mathbf{n}_3$, we consider the case where the $x-y$ plane is rotated by a small angle $\delta\phi$. Rotating the coordinate system transforms any point $\mathbf{p} = [x \ y]^T$ to a point $\mathbf{p}' = [x' \ y']^T$, given by:

$$\begin{bmatrix} x' \\ y' \end{bmatrix} = \mathbf{C}(\delta\phi)\begin{bmatrix} x \\ y \end{bmatrix} \simeq \begin{bmatrix} 1 & -\delta\phi \\ \delta\phi & 1 \end{bmatrix}\begin{bmatrix} x \\ y \end{bmatrix} = \begin{bmatrix} x \\ y \end{bmatrix} + \delta\phi\begin{bmatrix} -y \\ x \end{bmatrix}$$

where we have employed the small angle approximations $\cos(\delta\phi) \simeq 1$ and $\sin(\delta\phi) \simeq \delta\phi$. Using this result, we see that if the plane containing the two robots is rotated by $\delta\phi$, the CL state vector will change to $\mathbf{x}' \simeq \mathbf{x} + \delta\phi\mathbf{n}_3$, which indicates that $\mathbf{n}_3$ corresponds to rotation of the $x-y$ plane. This result implies that any such global rotation is unobservable, and will cause no change to the measurements. The preceding analysis for the meaning of the basis vectors of the unobservable subspace agrees with [18] as well as with intuition, which dictates that the *global coordinates* of the robots (rotation and translation) are unobservable, since the measurements only depend on the relative robot configurations.

B. Standard EKF-CL

We now study the observability properties of the standard EKF-CL, in which the Jacobians are evaluated at the estimated state (i.e., $\mathbf{x}_k^\star$ is the latest state estimate). The following definitions will be useful for our derivations:

$$\mathbf{d}\hat{\mathbf{p}}_i(k) \triangleq \hat{\mathbf{p}}_{i_{k|k}} - \hat{\mathbf{p}}_{i_{k|k-1}} \quad (24)$$

$$\boldsymbol{\Delta}\hat{\mathbf{p}}_{ij}(k, \ell) \triangleq \hat{\mathbf{p}}_{i_{k|k-1}} - \hat{\mathbf{p}}_{j_{k_o|k_o-1}} - \sum_{\kappa=k_o}^{\ell}\mathbf{d}\hat{\mathbf{p}}_j(\kappa) \quad (25)$$

$$\delta\hat{\mathbf{p}}_{ij}(k, \ell) \triangleq \hat{\mathbf{p}}_{i_{k|k-1}} - \hat{\mathbf{p}}_{j_{\ell|\ell-1}} \quad (26)$$

where k_o is the first time instant of interest, and $k, \ell \geq k_o$. In the above expressions, $\mathbf{d}\hat{\mathbf{p}}_i$ is the correction in the ith robot position estimate due to the EKF update at time-step k, while $\delta\hat{\mathbf{p}}_{ij}$ is the estimated difference between two robot positions (cf. (16)) evaluated using the uncorrected estimates immediately after the respective propagation steps.

We start by deriving an expression analogous to that of (17):

$$\boldsymbol{\Phi}_{i_{k_o+1}}\boldsymbol{\Phi}_{i_{k_o}} = \begin{bmatrix} \mathbf{I}_2 & \mathbf{J}\boldsymbol{\Delta}\hat{\mathbf{p}}_{ii}(k_o+2, k_o+1) \\ \mathbf{0}_{1\times 2} & 1 \end{bmatrix} \quad (27)$$

Using induction, we can show that:

$$\Phi_{i_{k_o+\ell-1}}\Phi_{i_{k_o+\ell-2}}\cdots\Phi_{i_{k_o}} = \begin{bmatrix} \mathbf{I}_2 & \mathbf{J}\Delta\hat{\mathbf{p}}_{ii}(k_o+\ell,k_o+\ell-1) \\ \mathbf{0}_{1\times2} & 1 \end{bmatrix}$$

for $\ell > 0$. As a result, the following identity is immediate:

$$\Phi_{k_o+\ell-1}\Phi_{k_o+\ell-2}\cdots\Phi_{k_o} = \tag{28}$$
$$\begin{bmatrix} \mathbf{I}_2 & \mathbf{J}\Delta\hat{\mathbf{p}}_{11}(k_o+\ell,k_o+\ell-1) & \mathbf{0}_{2\times2} & \mathbf{0}_{2\times1} \\ \mathbf{0}_{1\times2} & 1 & \mathbf{0}_{1\times2} & 0 \\ \mathbf{0}_{2\times2} & \mathbf{0}_{2\times1} & \mathbf{I}_2 & \mathbf{J}\Delta\hat{\mathbf{p}}_{22}(k_o+\ell,k_o+\ell-1) \\ \mathbf{0}_{1\times2} & 0 & \mathbf{0}_{1\times2} & 1 \end{bmatrix}$$

The measurement Jacobian now is given by (cf. (19)):

$$\mathbf{H}_{k_o+\ell} = \tag{29}$$
$$-\text{Diag}\Big((\nabla\mathbf{h}_{k_o+\ell}^{(12)})\mathbf{A}(\hat{\phi}_{1_{k_o+\ell|k_o+\ell-1}}), \ (\nabla\mathbf{h}_{k_o+\ell}^{(21)})\mathbf{A}(\hat{\phi}_{2_{k_o+\ell|k_o+\ell-1}})\Big)$$
$$\times \begin{bmatrix} \mathbf{I}_2 & \mathbf{J}\delta\hat{\mathbf{p}}_{21}(k_o+\ell,k_o+\ell) & -\mathbf{I}_2 & \mathbf{0}_{2\times1} \\ \mathbf{0}_{1\times2} & 1 & \mathbf{0}_{1\times2} & -1 \\ -\mathbf{I}_2 & \mathbf{0}_{2\times1} & \mathbf{I}_2 & \mathbf{J}\delta\hat{\mathbf{p}}_{12}(k_o+\ell,k_o+\ell) \\ \mathbf{0}_{1\times2} & -1 & \mathbf{0}_{1\times2} & 1 \end{bmatrix}$$

Multiplication of (29) and (28) yields:

$$\mathbf{H}_{k_o+\ell}\Phi_{k_o+\ell-1}\cdots\Phi_{k_o} = \tag{30}$$
$$-\text{Diag}\Big((\nabla\mathbf{h}_{k_o+\ell}^{(12)})\mathbf{A}(\hat{\phi}_{1_{k_o+\ell|k_o+\ell-1}}), \ (\nabla\mathbf{h}_{k_o+\ell}^{(21)})\mathbf{A}(\hat{\phi}_{2_{k_o+\ell|k_o+\ell-1}})\Big) \times$$
$$\begin{bmatrix} \mathbf{I}_2 & \mathbf{J}\Delta\hat{\mathbf{p}}_{21}(k_o+\ell,k_o+\ell-1) & -\mathbf{I}_2 & -\mathbf{J}\Delta\hat{\mathbf{p}}_{22}(k_o+\ell,k_o+\ell-1) \\ \mathbf{0}_{1\times2} & 1 & \mathbf{0}_{1\times2} & -1 \\ -\mathbf{I}_2 & -\mathbf{J}\Delta\hat{\mathbf{p}}_{11}(k_o+\ell,k_o+\ell-1) & \mathbf{I}_2 & \mathbf{J}\Delta\hat{\mathbf{p}}_{12}(k_o+\ell,k_o+\ell-1) \\ \mathbf{0}_{1\times2} & -1 & \mathbf{0}_{1\times2} & 1 \end{bmatrix}$$

Thus, the observability matrix $\mathbf{M}$ (cf. (14)) can be written as:

$$\mathbf{M} = \tag{31}$$
$$\underbrace{-\text{Diag}\Big((\nabla\mathbf{h}_{k_o}^{(12)})\mathbf{A}(\hat{\phi}_{1_{k_o|k_o-1}}), \ \cdots, \ (\nabla\mathbf{h}_{k_o+m}^{(21)})\mathbf{A}(\hat{\phi}_{2_{k_o+m|k_o+m-1}})\Big)}_{\mathbf{D}} \times$$
$$\underbrace{\begin{bmatrix} \mathbf{I}_2 & \mathbf{J}\delta\hat{\mathbf{p}}_{21}(k_o,k_o) & -\mathbf{I}_2 & \mathbf{0}_{2\times1} \\ \mathbf{0}_{1\times2} & 1 & \mathbf{0}_{1\times2} & -1 \\ -\mathbf{I}_2 & \mathbf{0}_{2\times1} & \mathbf{I}_2 & \mathbf{J}\delta\hat{\mathbf{p}}_{12}(k_o,k_o) \\ \mathbf{0}_{1\times2} & -1 & \mathbf{0}_{1\times2} & 1 \\ \mathbf{I}_2 & \mathbf{J}\Delta\hat{\mathbf{p}}_{21}(k_o+1,k_o) & -\mathbf{I}_2 & -\mathbf{J}\Delta\hat{\mathbf{p}}_{22}(k_o+1,k_o) \\ \mathbf{0}_{1\times2} & 1 & \mathbf{0}_{1\times2} & -1 \\ -\mathbf{I}_2 & -\mathbf{J}\Delta\hat{\mathbf{p}}_{11}(k_o+1,k_o) & \mathbf{I}_2 & \mathbf{J}\Delta\hat{\mathbf{p}}_{12}(k_o+1,k_o) \\ \mathbf{0}_{1\times2} & -1 & \mathbf{0}_{1\times2} & 1 \\ \vdots & \vdots & \vdots & \vdots \\ \mathbf{I}_2 & \mathbf{J}\Delta\hat{\mathbf{p}}_{21}(k_o+m,k_o+m-1) & -\mathbf{I}_2 & -\mathbf{J}\Delta\hat{\mathbf{p}}_{22}(k_o+m,k_o+m-1) \\ \mathbf{0}_{1\times2} & 1 & \mathbf{0}_{1\times2} & -1 \\ -\mathbf{I}_2 & -\mathbf{J}\Delta\hat{\mathbf{p}}_{11}(k_o+m,k_o+m-1) & \mathbf{I}_2 & \mathbf{J}\Delta\hat{\mathbf{p}}_{12}(k_o+m,k_o+m-1) \\ \mathbf{0}_{1\times2} & -1 & \mathbf{0}_{1\times2} & 1 \end{bmatrix}}_{\mathbf{U}}$$

By proceeding similarly to the proof of Lemma 4.1, it can be verified (cf. [20]) that in general rank($\mathbf{M}$) = rank($\mathbf{U}$) = 4. We thus see that the linearized error-state system model of the standard EKF-CL has different observability properties from that of the ideal EKF-CL. In particular, by processing the measurements collected in the time interval $[k_o, k_o+m]$, the EKF acquires information along the 4 directions of the state space corresponding to the observable subspace of the linearized system. However, the measurements actually provide information in only 3 directions of the state space (i.e., the

robot-to-robot relative pose). This shows that the filter gains "spurious information" along unobservable directions of the underlying nonlinear CL system, which leads to inconsistency.

To probe further, we note that a basis of the right nullspace of $\mathbf{M}$ is given by (cf. (23)):

$$\mathcal{N}(\mathbf{M}) = \text{span}\begin{bmatrix} \mathbf{n}_1 & \mathbf{n}_2 \end{bmatrix} \tag{32}$$

Recall that these two vectors correspond to shifting the $x-y$ plane, while the direction corresponding to the rotation is "missing" from the unobservable subspace of the EKF system model (cf. (23)). Therefore, we see that the filter will gain "nonexistent" information about the robots' global orientation. This will lead to an unjustified reduction in the orientation uncertainty, which will, in turn, further reduce the uncertainty in all state variables. We point out that the *root cause* of the problem is that the linearization points used for computing the Jacobians in the standard EKF-CL are the latest state estimates (i.e., the linearization point corresponding to the same state variable changes after each propagation). This increases the dimension of the observable subspace, and thus fundamentally alters the properties of the estimation process.

V. OBSERVABILITY CONSTRAINED (OC)-EKF CL

In the preceding section, it was shown that when the state-propagation Jacobians are evaluated using the latest state estimates, the EKF linearized error-state system model has an observable subspace of dimension higher than that of the actual CL system. This will always lead to unjustified reduction of the covariance estimates, and thus inconsistency. This issue is related to the problem of inconsistency of EKF-SLAM, which was identified in [15]. In that work, to address the problem, we proposed an observability-based methodology for designing consistent estimators for EKF-SLAM. The key idea of this approach is to select linearization points to ensure that the EKF linearized error-state system model has appropriate observability properties, even at the cost of sacrificing linearization accuracy if necessary (i.e., choosing a linearization point which may increase the linearization error). By doing so, the influx of spurious information along the erroneously observable direction of the state space is avoided, thus improving the consistency of the estimates. It is important to note that this solution can be used for designing consistent estimators for any nonlinear system. Therefore, in this work, we adopt this observability-based methodology to design a consistent EKF estimator, termed *Observability-Constrained (OC)-EKF* for CL. In particular, to guarantee the appropriate observability properties of the EKF linearized system model, we propose selecting the linearization points based on the following lemma:

Lemma 5.1: If the linearization points, $\mathbf{x}_{i_{k+1}}^\star$, $\mathbf{x}_{i_k}^\star$ and $\mathbf{x}_{j_k}^\star$, at which the filter Jacobian matrices $\Phi_{i_k} = \Phi_{i_k}(\mathbf{x}_{i_{k+1}}^\star, \mathbf{x}_{i_k}^\star)$ and $\mathbf{H}_k^{(ij)} = \mathbf{H}_k(\mathbf{x}_{i_k}^\star, \mathbf{x}_{j_k}^\star)$, are evaluated, are selected as

$$\mathbf{x}_{i_{k+1}}^\star = \hat{\mathbf{x}}_{i_{k+1|k}} \quad \mathbf{x}_{i_k}^\star = \hat{\mathbf{x}}_{i_{k|k-1}} \quad \mathbf{x}_{j_k}^\star = \hat{\mathbf{x}}_{j_{k|k-1}} \tag{33}$$

then it is guaranteed that the EKF linearized error-state model has unobservable subspace of dimension 3.

Proof: Using the linearization points (33), the state-propagation Jacobian $\mathbf{\Phi}_{i_k}$ (cf. (6)) is now computed as:

$$\mathbf{\Phi}_{i_k} = \begin{bmatrix} \mathbf{I}_2 & \mathbf{J}\left(\hat{\mathbf{p}}_{i_{k+1|k}} - \hat{\mathbf{p}}_{i_{k|k-1}}\right) \\ \mathbf{0}_{1\times 2} & 1 \end{bmatrix} \quad (34)$$

The difference compared to (6) in the standard EKF is that the robot position estimate prior to updating, $\hat{\mathbf{p}}_{i_{k|k-1}}$, is used in place of the updated estimate, $\hat{\mathbf{p}}_{i_{k|k}}$. In contrast, the measurement Jacobian, $\mathbf{H}_k^{(ij)}$, is computed in the same way as for the standard EKF (cf. (10)). As a result, using the definition of $\delta\hat{\mathbf{p}}_{ij}$ (26), the observability matrix of the OC-EKF becomes:

$$\mathbf{M'} = \qquad (35)$$

$$\underbrace{-\mathbf{Diag}\left((\nabla\mathbf{h}_{k_o}^{(12)})\mathbf{A}(\hat{\phi}_{1_{k_o|k_o-1}}), \cdots, (\nabla\mathbf{h}_{k_o+m}^{(21)})\mathbf{A}(\hat{\phi}_{2_{k_o+m|k_o+m-1}})\right)}_{\mathbf{D}}$$

$$\times \underbrace{\begin{bmatrix} \mathbf{I}_2 & \mathbf{J}\delta\hat{\mathbf{p}}_{21}(k_o,k_o) & -\mathbf{I}_2 & \mathbf{0}_{2\times 1} \\ \mathbf{0}_{1\times 2} & 1 & \mathbf{0}_{1\times 2} & -1 \\ -\mathbf{I}_2 & \mathbf{0}_{2\times 1} & \mathbf{I}_2 & \mathbf{J}\delta\hat{\mathbf{p}}_{12}(k_o,k_o) \\ \mathbf{0}_{1\times 2} & -1 & \mathbf{0}_{1\times 2} & 1 \\ \mathbf{I}_2 & \mathbf{J}\delta\hat{\mathbf{p}}_{21}(k_o+1,k_o) & -\mathbf{I}_2 & -\mathbf{J}\delta\hat{\mathbf{p}}_{22}(k_o+1,k_o) \\ \mathbf{0}_{1\times 2} & 1 & \mathbf{0}_{1\times 2} & -1 \\ -\mathbf{I}_2 & -\mathbf{J}\delta\hat{\mathbf{p}}_{11}(k_o+1,k_o) & \mathbf{I}_2 & \mathbf{J}\delta\hat{\mathbf{p}}_{12}(k_o+1,k_o) \\ \mathbf{0}_{1\times 2} & -1 & \mathbf{0}_{1\times 2} & 1 \\ \vdots & \vdots & \vdots & \vdots \\ \mathbf{I}_2 & \mathbf{J}\delta\hat{\mathbf{p}}_{21}(k_o+m,k_o) & -\mathbf{I}_2 & -\mathbf{J}\delta\hat{\mathbf{p}}_{22}(k_o+m,k_o) \\ \mathbf{0}_{1\times 2} & 1 & \mathbf{0}_{1\times 2} & -1 \\ -\mathbf{I}_2 & -\mathbf{J}\delta\hat{\mathbf{p}}_{11}(k_o+m,k_o) & \mathbf{I}_2 & \mathbf{J}\delta\hat{\mathbf{p}}_{12}(k_o+m,k_o) \\ \mathbf{0}_{1\times 2} & -1 & \mathbf{0}_{1\times 2} & 1 \end{bmatrix}}_{\mathbf{U'}}$$

It becomes evident that, compared to the observability matrix of the ideal EKF-CL (cf. (22)), the only difference arising in $\mathbf{U'}$ is that $\delta\mathbf{p}_{ij}$ is replaced by its estimate $\delta\hat{\mathbf{p}}_{ij}$, for $i, j = 1, 2$. Thus, by a proof analogous to that of Lemma 4.1, we can show that rank$(\mathbf{M'}) = 3$. Moreover, the nullspace of $\mathbf{M'}$ becomes:

$$\mathcal{N}(\mathbf{M'}) = \underset{\text{col.}}{\text{span}} \begin{bmatrix} \mathbf{I}_2 & \mathbf{J}\hat{\mathbf{p}}_{1_{k_o|k_o-1}} \\ \mathbf{0}_{1\times 2} & 1 \\ \mathbf{I}_2 & \mathbf{J}\hat{\mathbf{p}}_{2_{k_o|k_o-1}} \\ \mathbf{0}_{1\times 2} & 1 \end{bmatrix} \quad (36)$$

This implies that the error-state system model used by the OC-EKF has unobservable subspace of dimension 3. The above proof is for the case of a two-robot team. In the more general case of an N-robot team ($N \geq 2$), it can be shown [20] that the corresponding observability matrix is of rank $3N - 3$, and thus its nullspace is also of dimension 3. ∎

It is important to point out that, compared to the standard EKF, the *only* change in the OC-EKF is the way in which the state-propagation Jacobians are computed (cf. (34) vs. (6)), while the state estimate and covariance in the OC-EKF are propagated and updated in the same way as in the standard EKF. We also stress that the OC-EKF estimator is realizable and causal, as it does not require knowledge of the true state.

VI. SIMULATION RESULTS

A series of Monte-Carlo comparison studies were conducted under various conditions to validate the preceding theoretical analysis and the proposed OC-EKF. The metrics used to evaluate filter performance are: (i) the average root mean square (RMS) error, and (ii) the average normalized (state) estimation error squared (NEES) [10]. We compute these error metrics by averaging over all Monte Carlo runs for each time step. It is known that the NEES of an M-dimensional Gaussian random variable follows a χ^2 distribution with M degrees of freedom. Therefore, if a filter is consistent, we expect that the average NEES for each robot pose will be close to 3 for all k. The larger the deviation of the NEES from this value, the larger the inconsistency of the filter.

In the simulation tests, we consider the CL scenario in which four robots randomly move in an area of size 20 m × 20 m. 100 Monte Carlo simulations were performed, and during each run, all filters process the same data, to ensure a fair comparison. The three estimators compared are: (i) the ideal EKF, (ii) the standard EKF, and (iii) the OC-EKF.

For the results presented in this section, four identical robots with a simple differential drive model move on a planar surface, at a constant linear velocity of $v = 0.25$ m/sec, while the rotational velocity is drawn from the uniform distribution over $[-0.5, 0.5]$ rad/sec. The two drive wheels are equipped with encoders, which measure their revolutions and provide measurements of velocity (i.e., right and left wheel velocities, v_r and v_l, respectively), with standard deviation equal to $\sigma = 5\%v$ for each wheel. These measurements are used to obtain linear and rotational velocity measurements for the robot, which are given by $v = \frac{v_r + v_l}{2}$ and $\omega = \frac{v_r - v_l}{a}$, where $a = 0.5$ m is the distance between the drive wheels. Thus, the standard deviations of the linear and rotational velocity measurements are $\sigma_v = \frac{\sqrt{2}}{2}\sigma$ and $\sigma_\omega = \frac{\sqrt{2}}{a}\sigma$, respectively.

Each robot records distance and bearing measurements to all other robots. Note that for simplicity we assume that each robot can observe all others at every time step. However, this is not a necessary assumption, as the analysis can easily be extended to the case where multiple propagation steps occur between updates (e.g., limited sensing range, or different sampling frequencies between proprioceptive and exteroceptive sensors). The standard deviation of the distance measurement noise is equal to $\sigma_d = 0.1$ m, while the standard deviation of the bearing measurement noise is set to $\sigma_\theta = 5$ deg. It should be pointed out that the sensor-noise levels selected for the simulations are larger than what is typically encountered in practice. This was done purposefully, since larger noise levels lead to higher estimation errors, which make the effects of inconsistency more apparent.

The results of all filters for one of the robots are presented in Fig. 1 (the results for the other three robots are very similar and are omitted due to space limitations). Specifically, Fig. 1(a) shows the orientation estimation errors, obtained from a typical simulation. Clearly, the standard-EKF errors grow significantly faster than those of the ideal and the OC-EKF, which indicates that the standard EKF tends to diverge. Note also that although the orientation errors of the ideal and the OC-EKF remain well within their corresponding 3σ bounds, those of the standard EKF exceed them. Most importantly, in contrast to those of the

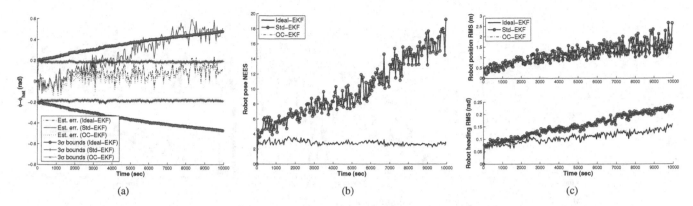

Fig. 1. Simulation results for one robot of a 4-robot team. The results for the other robots are similar to the ones presented here. (a) Orientation estimation errors vs. 3σ bounds obtained from one typical realization of the Monte Carlo simulations. The σ values are computed as the square-root of the corresponding element of the estimated covariance matrix. (b)-(c) Monte Carlo results of average NEES and RMS errors, respectively. In these two plots, the solid lines correspond to the ideal EKF, the solid lines with circles to the standard EKF, and the dash-dotted lines to the OC-EKF. Note that the NEES and RMS errors as well as the estimation errors of the ideal and the OC-EKF are almost identical, which makes the corresponding lines difficult to distinguish.

OC-EKF, the 3σ bounds of the standard EKF remain almost *constant* as if the orientation of the robot was observable. However, as discussed in Section IV, the robots have no access to absolute orientation information and thus the orientation covariance should continuously grow (as is the case for the ideal and the OC-EKF). The results of Fig. 1(a) clearly demonstrate that the incorrect observability properties of the standard EKF cause an unjustified reduction of the orientation uncertainty.

Figs. 1(b)-1(c) show the average NEES and RMS errors. As evident, the performance of the OC-EKF is *almost identical* to that of the ideal EKF, and substantially better than the standard EKF, both in terms of RMS errors and NEES. This occurs even though the OC-EKF Jacobians are less accurate than those of the standard EKF, as explained in the preceding section. This fact indicates that the errors introduced by the use of inaccurate linearization points (for computing the Jacobians) have a less detrimental effect on consistency than the use of an error-state system model with observable subspace of dimension larger than that of the actual CL system.

VII. EXPERIMENTAL RESULTS

A real-world experiment was performed to further validate the presented analysis and the OC-EKF algorithm. During the test, a team of four Pioneer I robots moves in a rectangular area of 2.5 m × 4.5 m, within which the positions of the robots are tracked by an overhead camera. The vision system provides measurements of the robot poses in a global coordinate frame, which serve as the ground truth for evaluating the estimators' performance in terms of NEES and RMS errors. The standard deviation of the noise in these measurements is approximately 0.5 deg for orientation and 0.01 m, along each axis, for position. The robots were commanded to move at a constant velocity of $v = 0.1$ m/sec while avoiding collision with the boundaries of the arena as well as with their teammates. In order to preserve legibility, only the first 200 sec of the four robot trajectories are shown in Fig. 2(a).

Although four robots of the same model were used, their wheel-encoders are not equally accurate. Specifically, during calibration, the velocity measurement errors, modeled as zero-mean white Gaussian noise processes, exhibited standard deviations ranging from $\sigma_{v_{\min}} = 3.8\%v$ for the most accurate odometer to $\sigma_{v_{\max}} = 6.9\%v$ for the robot with the highest noise levels. Similarly, the standard deviations of the rotational velocity measurements assumed values between $\sigma_{\omega_{\min}} = 0.0078$ rad/sec and $\sigma_{\omega_{\max}} = 0.02$ rad/sec for the four robots. We observe that as a result of this variability of sensor characteristics, the experiment can be considered as involving a *heterogeneous* robot team.

Distance-bearing measurements are produced synthetically using the differences in the positions of the robots, as these are recorded by the overhead camera, expressed in the measuring robot's coordinate frame, with the addition of noise. For the experimental results shown in this section, the distance and bearing measurements are corrupted by zero-mean white Gaussian noise, with standard deviation $\sigma_d = 0.05$ m for distance and $\sigma_\theta = 2$ deg for bearing measurements, respectively. Pose estimation was run offline, and two filters were compared: (i) the standard EKF, and (ii) the proposed OC-EKF. Comparative results of the standard EKF and the proposed OC-EKF for one of the robots are presented in Figs. 2(b) and 2(c). The results for the other robots are similar to the ones presented here. From the experimental results it becomes clear that, just as in simulation, also in the real-world experiment the OC-EKF outperforms the standard EKF, in terms of both accuracy and consistency. Moreover, these results further support our conjecture that the mismatch in the dimension of the unobservable subspace between the linearized CL system and the underlying nonlinear system is a fundamental cause of filter inconsistency.

VIII. CONCLUSIONS

In this paper, we have studied in depth the issue of consistency in EKF-based CL, from an observability perspective. By

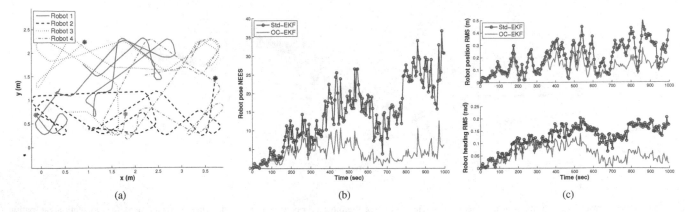

(a)   (b)   (c)

Fig. 2. Experimental results: (a) Four Pioneer I robots move inside a 2.5 m × 4.5 m arena, with starting positions marked by ∗. To maintain clarity, only the first 200 sec of the trajectories are plotted. (b)-(c) NEES and RMS errors respectively for one robot. The results for the other robots are similar to the ones presented here. As in the simulations, the OC-EKF outperforms the standard EKF in terms of both accuracy (RMS errors) and consistency (NEES).

comparing the observability properties of the linearized error-state model employed in the EKF with those of the underlying nonlinear CL system, we proved that the observable subspace of the standard EKF-CL is always of higher dimension than that of the actual CL system. As a result, the covariance estimates of the EKF undergo reduction in directions of the state space where no information is available, and thus the standard EKF-CL is always inconsistent. Moreover, we proposed a new OC-EKF, which improves the consistency of EKF-based CL. The design methodology followed for deriving the OC-EKF requires appropriate selection of the linearization points at which the Jacobians are evaluated, which ensures that the observable subspace of the linearized error-state system model is of the same dimension as that of the underlying actual system. Simulations and experiments have verified that the proposed OC-EKF performs better, in terms of both accuracy and consistency, than the standard EKF.

ACKNOWLEDGEMENTS

This work was supported by the University of Minnesota (DTC), the National Science Foundation (IIS-0643680, IIS-0811946, IIS-0835637), and the University of California, Riverside (BCOE).

REFERENCES

[1] R. Kurazume and S. Hirose, "An experimental study of a cooperative positioning system," *Auto. Rob.*, vol. 8, no. 1, pp. 43–52, Jan. 2000.

[2] I. M. Rekleitis, G. Dudek, and E. E. Milios, "Multi-robot cooperative localization: a study of trade-offs between efficiency and accuracy," in *Proc. of the IEEE/RSJ Int. Conf. on Intel. Rob. and Sys.*, Lausanne, Switzerland, Sept. 30–Oct. 4, 2002, pp. 2690–2695.

[3] S. I. Roumeliotis and G. A. Bekey, "Distributed multirobot localization," *IEEE Trans. Rob. Autom.*, vol. 18, no. 5, pp. 781–795, Oct. 2002.

[4] A. Howard, M. J. Mataric, and G. S. Sukhatme, "Localization for mobile robot teams using maximum likelihood estimation," in *Proc. of the IEEE/RSJ Int. Conf. on Intel. Rob. and Sys.*, Lausanne, Switzerland, Sept. 30–Oct. 4, 2002, pp. 434–439.

[5] E. D. Nerurkar, S. I. Roumeliotis, and A. Martinelli, "Distributed maximum a posteriori estimation for multi-robot cooperative localization," in *Proc. of the IEEE Int. Conf. on Rob. and Autom.*, Kobe, Japan, May 12-17, 2009, pp. 1402–1409.

[6] D. Fox, W. Burgard, H. Kruppa, and S. Thrun, "A probabilistic approach to collaborative multi-robot localization," *Auto. Rob.*, vol. 8, no. 3, pp. 325–344, June 2000.

[7] S. Panzieri, F. Pascucci, and R. Setola, "Multirobot localization using interlaced extended Kalman filter," in *Proc. of the IEEE/RSJ Int. Conf. on Intel. Rob. and Sys.*, Beijing, China, Oct. 9-15, 2006, pp. 2816–2821.

[8] N. Karam, F. Chausse, R. Aufrere, and R. Chapuis, "Localization of a group of communicating vehicles by state exchange," in *Proc. of the IEEE/RSJ Int. Conf. on Intel. Rob. and Sys.*, Beijing, China, Oct. 9-15, 2006, pp. 519–524.

[9] A. Martinelli, "Improving the precision on multi robot localization by using a series of filters hierarchically distributed," in *Proc. of the IEEE/RSJ Int. Conf. on Intel. Rob. and Sys.*, San Diego, CA, Oct. 29–Nov. 2, 2007, pp. 1053–1058.

[10] Y. Bar-Shalom, X. Li, and T. Kirubarajan, *Estimation with applications to tracking and navigation.*   New York: Wiley, 2001.

[11] S. Julier and J. Uhlmann, "A counter example to the theory of simultaneous localization and map building," in *Proc. of the IEEE Int. Conf. on Rob. and Autom.*, Seoul, Korea, May 21-26, 2001, pp. 4238–4243.

[12] J. Castellanos, J. Neira, and J. Tardos, "Limits to the consistency of EKF-based SLAM," in *Proc. of the 5th IFAC Symp. on Intel. Auto. Veh.*, Lisbon, Portugal, July 5-7, 2004, pp. 1244–1249.

[13] T. Bailey, J. Nieto, J. Guivant, M. Stevens, and E. Nebot, "Consistency of the EKF-SLAM algorithm," in *Proc. of the IEEE/RSJ Int. Conf. on Intel. Rob. and Sys.*, Beijing, China, Oct. 9-15, 2006, pp. 3562–3568.

[14] S. Huang and G. Dissanayake, "Convergence and consistency analysis for extended Kalman filter based SLAM," *IEEE Trans. Rob.*, vol. 23, no. 5, pp. 1036–1049, Oct. 2007.

[15] G. P. Huang, A. I. Mourikis, and S. I. Roumeliotis, "Analysis and improvement of the consistency of extended Kalman filter-based SLAM," in *Proc. of the IEEE Int. Conf. on Rob. and Autom.*, Pasadena, CA, May 19-23, 2008, pp. 473–479.

[16] A. Bahr, M. R. Walter, and J. J. Leonard, "Consistent cooperative localization," in *Proc. of the IEEE Int. Conf. on Rob. and Autom.*, Kobe, Japan, May 12-17, 2009, pp. 3415–3422.

[17] A. Howard, M. J. Mataric, and G. S. Sukhatme, "Putting the 'i' in 'team': an ego-centric approach to cooperative localization," in *Proc. of the IEEE Int. Conf. on Rob. and Autom.*, Taipei, Taiwan, Sept. 14-19, 2003, pp. 868–874.

[18] A. Martinelli and R. Siegwart, "Observability analysis for mobile robot localization," in *Proc. of the IEEE/RSJ Int. Conf. on Intel. Rob. and Sys.*, Alberta, Canada, Aug. 2-6, 2005, pp. 1471–1476.

[19] P. Maybeck, *Stochastic Models, Estimation and Control.*   Academic Press, 1979, vol. 1.

[20] G. P. Huang and S. I. Roumeliotis, "On the consistency of multi-robot cooperative localization," MARS Lab, University of Minnesota, Minneapolis, MN, Tech. Rep., Jan. 2009. [Online]. Available: www.cs.umn.edu/~ghuang/paper/TR_CL_Consistency.pdf

[21] C. Meyer, *Matrix Analysis and Applied Linear Algebra.*   SIAM, 2001.

Inner Sphere Trees
for Proximity and Penetration Queries

Rene Weller
Clausthal University, Germany
Email: rwe@tu-clausthal.de

Gabriel Zachmann
Clausthal University, Germany
Email: zach@tu-clausthal.de

Abstract—**We present a novel geometric data structure for approximate collision detection at haptic rates between rigid objects. Our data structure, which we call *inner sphere trees*, supports different kinds of queries, namely, proximity queries and a new method for interpenetration computation, the penetration *volume*, which is related to the water displacement of the overlapping region and, thus, corresponds to a physically motivated force.**

The main idea is to bound objects from the *inside* with a set of *non-overlapping* spheres. Based on such sphere packings, a "inner bounding volume hierarchy" can be constructed. In order to do so, we propose to use an AI clustering algorithm, which we extend and adapt here.

The results show performance at haptic rates both for proximity and penetration volume queries for models consisting of hundreds of thousands of polygons.

I. INTRODUCTION

Collision detection between rigid objects is important for many fields of robotics and computer graphics, e.g. for path-planning, haptic rendering, physically-based simulations, and medical applications. Today, there exist a wide variety of freely available collision detection libraries and nearly all of them are able to work at visual interactive rates, even for very complex objects [1]. Most collision detection algorithms dealing with rigid objects use some kind of bounding volume hierarchy (BVH). The main idea behind a BVH is to subdivide the primitives of an object hierarchically until there are only single primitives left at the leaves. BVHs guarantee very fast responses at query time, so long as no further information than the set of colliding polygons is required for the collision response. However, most applications require much more information in order to resolve or avoid the collisions.

One way to do this is to compute the exact time of contact of the objects. This method is called continuous collision detection. Another approach, called penalty methods, is to compute repelling forces based on the penetration depth. However, there is no universally accepted definition of the penetration depth between a pair of polygonal models [2], [3]. Mostly, the minimum translation vector to separate the objects is used, but this may lead to discontinuous forces.

Another approach is to avoid penetrations or contacts before they really happen. In this case, the minimum distance between the objects can be used to compute repelling forces.

Haptic rendering requires update rates of at least 200 Hz, but preferably 1 kHz to guarantee a stable force feedback. Consequently, the collision detection time should never exceed 5 msec.

One example of an approach that offers fairly constant query times are voxel-based methods like the Voxmap Pointshell algorithm (VPS), where objects, in general, have to be voxelized (the "voxmap") and covered by a point cloud (the "point shell"). This can be very memory consuming and produce aliasing artifacts due to the discretization errors.

A. Main Contributions

This paper contributes the following novel ideas to the area of collision detection:

- a novel geometric data structure, the *Inner Sphere Trees (IST)*, that provides hierarchical bounding volumes from the *inside* of an object;
- a method to compute a dense sphere packing inside a polygonal object;
- we propose to utilize a clustering algorithm to construct a sphere hierarchy;
- a unified algorithm that can compute for a pair of objects, based on their ISTs, both an approximate minimal distance and the approximate penetration volume; the application does not need to know in advance which situation currently exists between the pair of objects;

Our ISTs and, consequently, the collision detection algorithm are independent of the geometry complexity; they only depend on the approximation error.

The main idea is that we do not build an (outer) hierarchy based on the polygons on the boundary of an object. Instead, we fill the interior of the model with a set of non-overlapping simple volumes that approximate the object's volume closely. In our implementation, we used spheres for the sake of simplicity, but the idea of using inner BVs for lower bounds instead of outer BVs for upper bounds can be extended analogously to all kinds of volumes. On top of these inner BVs, we build a hierarchy that allows for fast computation of the approximate proximity and *penetration volume*.

The penetration volume corresponds to the water displacement of the overlapping parts of the objects and, thus, leads to a physically motivated and continuous repulsion force. According to [4, Sec. 5.1], it is "the most complicated yet accurate method" to define the extent of intersection, which was also reported earlier by [5, Sec. 3.3]. However, to our

Fig. 1. These images show the different stages of our sphere packing algorithm. First, we voxelize the object (left) and compute distances from the voxels to the closest triangle (second image; transparency = distance). Then, we pick the voxel with the largest distance and put a sphere at its center. We proceed incrementally and, eventually, we obtain a dense sphere packing of the object (right).

knowledge, there are no algorithms to compute it efficiently as yet.

Our data structure can support all kinds of object representations, e.g. polygon meshes or NURBS surfaces. The only precondition is that they be watertight. In order to build the hierarchy on the inner spheres, we utilize a recently proposed clustering algorithm that allows us to work in an adaptive manner.

The results shows that our new data structure can answer both kinds of queries at haptic rates with a negligible loss of accuracy.

II. PREVIOUS WORK

Collision detection has been extensively investigated by researchers in the past decades. There already exist a large variety of techniques and software packages for collision detection queries. They vary greatly in their techniques and computational method. Here we concentrate on those approaches, that are able to compute extended contact informations like a penetration depth or the separation distance between a pair of objects.

Most of them are not designed to work at haptic refresh rates, or they are restricted to simple point probes, or require special objects, such as convex objects as input. In the following, we will give a short overview of classical and also state of the art approaches to manage these tasks.

A. BVH Based Methods

A classical approach for proximity queries is the GJK algorithm [6], [7]. It derives the distance between a pair of convex objects, by utilizing the Minkowski sum. It is also possible to extend GJK in order to compute the penetration depth [8].

In [9] a generalized framework for minimum distance computations that depends on geometric reasoning is presented. It also includes time-critical properties.

PQP [10] creates a hierarchy of rectangle swept spheres. Distance computations are performed between these volumes on the hierarchical tree. Moreover, it uses specialized algorithms to improve the efficiency of distance calculations. We used it in this paper to compute the ground truth for the proximity queries.

Another package, supporting proximity queries between any convex quadrics is SOLID [11]. It uses axis aligned bounding boxes and the Minkowski difference between convex polytopes together with several optimization techniques.

SWIFT++ [12] provides a convex surface decomposition scheme and a modified Lin-Canny closest feature algorithm to compute approximate as well as exact distance between general rigid polyhedra.

Sphere trees have also been used for distance computation in the past [13]–[15]. The algorithms presented there are interruptible and they are able to deliver approximative distances. Moreover, they all compute a lower bound on the distance, because of using outer hierarchies, while our ISTs derive an upper bound. Thus, a combination of these approaches with our ISTs could deliver good error bounds in both directions.

[2] presented an extended definition of the penetration depth that also takes the rotational component into account, called the generalized penetration depth. However, this approach is computationally very expensive and, thus, might currently not be fast enough for haptic interaction rates.

[16] approximate a local penetration depth by first computing a local penetration direction and then use this information to estimate a local penetration depth on the GPU.

The literature on penetration volume computation is sparse. One approach, proposed by [17], constructs the intersection volume of convex polyhedra explicitly. For this reason, it is applicable only to very simple geometries at interactive rates.

Another interesting method is given by [18]. They compute an approximation of the intersection volume from layered depth images on the GPU. This approach is applicable to deformable geometries but restricted to image space precision.

B. Voxel Based Data Structures

Classical Voxmap Pointshell approaches [19] divide the environment into a dynamic object, that can move freely through space and static objects that are fixed in the world. The static world is discretized into a set of discrete volume elements, the voxels, while the dynamic object is described by a set of points that represent its surface, the pointshell. At query time, for each of these points in the pointshell, it is determined with a simple boolean test, whether it is located in a filled voxel or not.

There also exist extensions to VPS, including optimizations to the force calculation in order to increase its stability [20]. However, even such optimizations cannot completely remedy the limits of VPS, namely aliasing effects and the high memory consumption.

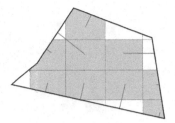 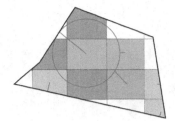 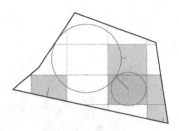

Fig. 2. This figure shows the first steps of the creation of the inner spheres. First, the object is voxelized (left). Then, we compute the shortest distance to the surface (gray lines) for interior voxel centers (light gray), i.e., a discretization of the interior distance field. Next, we place a maximal sphere at the voxel center with the largest radius (gray sphere). Then, the dark gray colored voxels are deleted, and the shortest distances of some voxels are updated, because they are closer now to an inner sphere (medium gray). This procedure continues greedily as shown in the drawing to the right.

Closely related to voxel-based approaches are distance field based methods. [21] use a pointshell of reduced deformable models in combination with distance fields in order to guarantee continuous contact forces.

III. CREATION OF THE INNER SPHERE TREE

In this section we describe the construction of our data structure. In a fist step, we want a watertight object to be densely filled with a set of non-overlapping spheres. The volume of the object should be approximated well by the spheres, while their number should be small. In a second step, we create a hierarchy over this set of spheres.

For squared objects, spheres seem not to be a good choice as filling volumes. However, they compensate this disadvantage because of the trivial overlap test and their rotationally invariance. Moreover, it is easy, in contrast to AABBs or OBBs, to compute the exact intersection volume.

A. The Sphere Packing

Filling objects densely with smaller volumes is a highly non-trivial task and still an active field of research, even when restricted to spheres [22], [23]. We present a simple heuristic that offers a good trade-off between accuracy and speed in practice.

This heuristic is currently based on a distance field. We start with a flood filling voxelization, but instead of simply storing whether or not a voxel is filled, we additionally store the distance d from the center of the voxel to the nearest point on the surface, together with the triangle that realizes this distance.

After this initialization, we use a greedy algorithm to generate the inner spheres. All voxels are stored in a priority queue, sorted by their distance to the surface. Until the p-queue is empty, we extract the maximum element, i.e. the voxel V^* with the largest distance d^*. We create an inner sphere with radius d^* and centered on the center of the voxel V^*. Then, all voxels whose centers are contained in this sphere are deleted from the p-queue. Additionally, we have to update all voxels V_i with $d_i < d^*$ and distance $d(V_i, V^*) < 2d^*$. This is because they are now closer to the sphere around V^* than to a triangle on the hull (see Figure 2). Their d_i must now be set to the new free radius.

After this procedure, the object is filled densely with a set of non-overlapping spheres. The density, and thus the accuracy, can be controlled by the number of voxels.

B. Building the IST

Our sphere hierarchy is based on the notion of a *wrapped hierarchy* [24], where inner nodes are tight BVs for all their leaves, but they do not necessarily bound their direct children. Compared to layered hierarchies, the big advantage is that the inner BVs are tighter. We use a top-down approach to create our hierarchy, i.e., we start at the root node that covers all inner spheres and divide these into several subsets.

The partitioning of the inner spheres has significant influence on the performance during runtime. Previous algorithms for building ordinary sphere trees, like the medial axis approach [14], [25] work well if the spheres constitute a *covering* of the object and have similar size, but in our scenario we use disjoint inner spheres that exhibit a large variation in size. Other approaches based on the *k-center problem* work only for sets of points and do not support spheres.

So, we decided to use the *batch neural gas* clustering algorithm (BNG) known from artificial intelligence [26]. BNG is a very robust clustering algorithm, which can be formulated as stochastic gradient descent with a cost function closely connected to quantization error. Like *k-means*, the cost function minimizes the mean squared euclidean distance of each data point to its nearest center. But unlike k-means, BNG exhibits very robust behavior with respect to the initial cluster center positions (the *prototypes*): they can be chosen arbitrarily without affecting the convergence. Moreover, BNG can be extended to allow the specification of the *importance* of each data point; below, we will describe how this can be used to increase the quality of the ISTs.

In the following, we will give a quick recap of the basic batch neural gas and then describe our extensions and application to building the inner sphere tree.

Given points $x_j \in \mathbb{R}^d, j = 0, \dots, m$ and prototypes $w_i \in \mathbb{R}^d, i = 0, \dots, n$ initialized randomly, we set the rank for every prototype w_i with respect to every data point x_j as

$$k_{ij} := |\{w_k : d(x_j, w_k) < d(x_j, w_i)\}| \in \{0, \dots, n\} \quad (1)$$

In other words, we sort the prototypes with respect to every data point. After the computation of the ranks, we compute

Fig. 3. This figure shows the results of our hierarchy building algorithm based on batch neural gas clustering with magnification control. All of those inner spheres that share the same shade of gray are assigned to the same bounding sphere. The left image shows the clustering result of the root sphere, the right images the partitioning of its four children.

the new positions for the prototypes:

$$w_i := \frac{\sum_{j=0}^{m} h_\lambda(k_{ij}) x_j}{\sum_{j=0}^{m} h_\lambda(k_{ij})} \tag{2}$$

These two steps are repeated until a stop criterion is met. In the original paper, a fixed number of iterations is proposed. We propose to use an adaptive version and stop the iteration if the movement of the prototypes is smaller than some ε. In our examples, we chose $\varepsilon \approx 10^{-5} \times \text{BoundingBoxSize}$, without any differences in the hierarchy compared to the non-adaptive, exhaustive approach. This improvement speeds up the creation of the hierarchy significantly.

The convergence rate is controlled by a monotonically decreasing function $h_\lambda(k) > 0$ that decreases with the number of iterations t. We use the function proposed in the original paper: $h_\lambda(k) = e^{-\frac{k}{\lambda}}$ with initial value $\lambda_0 = \frac{n}{2}$, and reduction $\lambda(t) = \lambda_0 \left(\frac{0.01}{\lambda_0} \right)^{\frac{t}{t_{\max}}}$, where $t_{\max}$ is the maximum number of iterations. These values have been taken according to [27].

Obviously, the number of prototypes defines the arity of the tree. Experiments with our data structure have shown that a branching factor of 4 produces the best results. Additionally, this has the benefit that we can use the full capacity of SIMD units in modern CPUs.

So far, the BNG only utilizes the location of the centers of the spheres. In our experience, this already produces much better results than other, simpler heuristics, such as greedily choosing the biggest spheres or the spheres with the largest number of neighbors. However, it does not yet take the extent of the spheres into account. As a consequence, the prototypes tend to avoid regions that are covered with a very large sphere, i.e., centers of big spheres are treated as outliers and they are thus placed on very deep levels in the hierarchy. However, it is better to place big spheres at higher levels of the hierarchy in order to get early lower bounds during distance traversal (see Section IV-A for details).

Therefore, we use an extended version of the classical batch neural gas, that also takes the size of the spheres into account. Our extension is based on an idea of [28], where *magnification control* is introduced. The idea is to add weighting factors in order to "artificially" increase the density of the space in some areas.

With weighting factors $v(x_j)$, Eq. 2 becomes

$$w_i := \frac{\sum_{j=0}^{m} h_\lambda(k_{ij}) v(x_j) x_j}{\sum_{j=0}^{m} h_\lambda(k_{ij}) v(x_j)} \tag{3}$$

In our scenario, we already know the density, because our spheres are disjoint. Thus, we can directly use the volumes of our spheres to let $v(x_j) = \frac{4}{3} \pi r^3$.

Summing up the hierarchy creation algorithm: we first compute a bounding sphere for all inner spheres (at the leaves), which becomes the root node of the hierarchy. To do that, we use the fast and stable smallest enclosing sphere algorithm proposed in [29]. Then, we divide the set of inner spheres into subsets in order to create the children. To do that, we use the extended version of batch neural gas with magnification control. We repeat this scheme recursively (See Figure 3 for some clustering results).

In the following, we will call the spheres in the hierarchy that are not leaves *hierarchy spheres*. Spheres at the leaves, which were created in Section III-A, will be called *inner spheres*. Note that hierarchy spheres are not necessarily contained completely within the object.

IV. BVH TRAVERSAL

Our new data structure supports different kinds of queries, namely proximity queries, which report the separation distance between a pair of objects, and penetration volume queries, which report the common volume covered by both objects. As a by-product, the proximity query can return a witness realizing the distance, and the penetration algorithm can return a partial list of intersecting polygons.

Fig. 4. After constructing the sphere packing (see Section III-A), every voxel can be intersected by several non-overlapping spheres (left). These do not necessarily account for the whole voxel space (light gray space in the left picture). In order to account for these voids, too, we simply increase the radius of the sphere that covers the center of the voxel (right).

Algorithm 1: computeVolume(A, B, totalOverlap)

---

input : A, B = spheres in the inner sphere tree

in/out: totalOverlap = overall volume of intersection

if A *and* B *are leaves* **then**
    // end of recursion
    totalOverlap += overlapVolume(A, B)
else
    // recursion step
    forall *children* a[i] *of* A **do**
        forall *children* b[j] *of* B **do**
            if overlap(a[i], b[j]) > 0 **then**
                checkVolume(a[i], b[j], totalOverlap)

In this section, we concentrate on these two types of queries, because they are more interesting for physically-based simulations that need some contact information to compute proper forces. But it should be obvious that the traversal can be easily modified in order to provide also approximate yes-no answer, which would further increase the speed of collision detection.

We start with a separate discussion of the two query types in order to point out their specific requirements. In Section IV-C we describe how to combine these traversal schemes to a unified algorithm that is able to provide distance and overlap volume informations, without the user has to know in advance, whether the objects overlap or not.

A. Proximity Queries

Our proximity query algorithm works like most other classical BVH traversal algorithms: We check whether two bounding volumes overlap or not. If this is the case, we recursively step to their children. In order to compute lower bounds for the distance, we simply have to add an appropriate distance test at the right place. This has to be done, when we reach a pair of inner spheres (i.e., leaves) during traversal. Due to Section III-A, these inner spheres are located completely inside the object and provide, thus, a lower bound on the sought-after distance. During traversal, there is no need to visit bounding volumes in the hierarchy that are farther away than the current minimum distance, because of the bounding property. This results in a high culling efficiency.

It should be obvious, that this algorithm can be easily extended to triangle accuracy by additionally taking the triangles

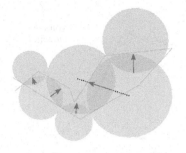

Fig. 5. The direction of the penalty force can be derived from the weighted average of all vectors between the centers of colliding pairs of spheres, weighted by their overlap.

into account that are closest to an inner sphere.

B. Penetration Volume Queries

As stated before, our data structure does not only support proximity queries, but also a new kind of penetration query, namely the *penetration volume*. This is the volume of the intersection of the two objects, which can be interpreted directly as the amount of the repulsion force, if it is considered as the amount of water being displaced.

The algorithm to compute the penetration volume (see Algorithm 1) does not differ very much from the proximity query test: we simply have to replace the distance test by an overlap test and maintain an accumulated overlap volume during the traversal. The overlap volume of a pair of spheres can be easily derived by adding the volumes of the spherical caps.

Due to the non-overlapping constraint of the inner spheres, the accumulated overlap volumes provides a lower bound on the real overlap volume of the objects.

1) Filling the gaps: The algorithm described in Section III-A results in densely filled objects. However, there still remain small voids between the spheres that cannot be completely compensated by increasing the number of voxels. This results in bad lower bounds.

As a remedy, we propose a simple heuristic to compensate this problem: We additionally assign a *secondary radius* to each inner sphere, such that the volume of the secondary sphere is equal to the volume of all voxels whose centers are contained within the radius of the primary sphere (see Figure 4). This guarantees that the total volume of all secondary spheres equals the volume of the object, within the accuracy of the voxelization, because each voxel volume is accounted for exactly once.

Certainly, these secondary spheres may slightly overlap, but this simple heuristic leads to acceptable estimations of the penetration volume. (Note, however, that the secondary spheres are not necessarily larger than the primary spheres.)

2) Collision response: In order to apply penalty forces in haptic environments or simulations, we also need the direction of the force in addition to its amount.

This can be derived easily from our ISTs by considering all overlapping pairs of spheres (R_i, S_j) separately. Let $\mathbf{c}_i, \mathbf{c}_j$

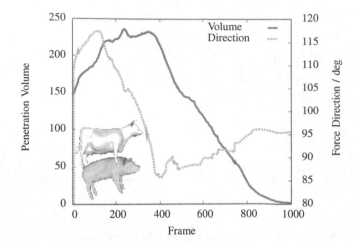

Fig. 6. Penetration volume (continuous line) and direction of the force vector (dotted line) during the path of a cow scraping alongside a pig.

be their sphere centers and $\mathbf{n}_{ij} = \mathbf{c}_i - \mathbf{c}_j$. Then, we compute the overall direction of the penalty force as the weighted sum $\mathbf{n} = \sum_{i,j} \mathrm{Vol}(R_i \cap S_j) \cdot \mathbf{n}_{ij}$ (see Figure 5). Obviously, this direction is continuous, provided the path of the objects is continuous (see Figure 6).

In case of deep penetrations, it can be necessary to flip some of the directions $\mathbf{n}_{ij}$. Computing normal cones for all spheres throughout the hierarchy can help to identify these pairs. It is also possible to extend this approach to provide continuous torques. Details are omitted here due to space constraints.

C. The Unified Algorithm

In the previous sections, we introduced the proximity and the penetration volume computation separately. However, it is quite easy to combine both algorithms. This yields a unified algorithm that can compute both the distance and the penetration volume, without the user having to know in advance, whether the objects overlap or not.

We start with the distance traversal. If we find the first pair of intersecting inner spheres, then we simply switch to the penetration volume computation.

The correctness is based on the fact, that all pairs of inner spheres we visited so far during distance traversal do not overlap and thus do not extend the penetration volume. Thus, we do not have to visit them again and can continue with the traversal of the rest of the hierarchies using the penetration volume algorithm. If we do not meet an intersecting pair of inner spheres, the unified algorithm still reports the minimal separating distance.

V. RESULTS

We have implemented our new data structure in C++. The testing environment consists of a PC running Windows XP with an Intel Pentium IV 3GHz dual core CPU and 2GB of memory. The initial distance field was computed using a slightly modified version of Dan Morris' *Voxelizer* [30].

The benchmark includes hand recorded object paths with distances ranging from about 0–20% of the object's BV size

for the proximity queries. We concentrated on very close configurations, because they are more interesting in real world scenarios and more challenging regarding the running time. The paths for the penetration volume queries concentrate on light to medium penetrations of about 0–10% of the object's volume. This scenario resembles the usage in haptic applications best, because the motive for using collision detection algorithms is to avoid heavy penetrations. However, we also included some heavy penetrations of 50% of the object's volume to stress our algorithm.

We used highly detailed objects with a polygon count ranging up to 370k to test the performance and the quality of our algorithm.[1] The quality of the resulting distances and penetration volumes is closely related to the quality of the underlying voxelization. Consequently, we voxelized each object in different resolutions in order to evaluate the trade-off between the number of spheres and the accuracy.

We computed the ground truth data for the proximity queries with the PQP library. We also included the running time of PQP in our plots, even if the comparison seems to be somewhat unfair, because PQP computes exact distances. However, it shows the impressive speed-up that is achievable when using approximative approaches. Moreover, it is possible to extend ISTs to support exact distance calculations, too.

To our knowledge, there are no implementations available to compute the exact penetration volume efficiently. In order to still evaluate the quality of our penetration volume approximation, we used a tetrahedralization to compute the exact volume. Even though we speed it up by a hierarchy built on the tetrahedra, the running time of this approach is in the order of 0.5 sec/frame.[2]

The results of our benchmarking show that our ISTs with the highest sphere resolution have an average speed-up of 50 compared to PQP, while the average error is only 0.15%. Even in the worst case, they are suitable for haptic rendering with response rates of less than 2 mesc in the highest resolution (see Figure 7).

Our penetration volume algorithm is able to answer queries at haptic rates between 0.1 msec and 2.5 msec on average, depending on the voxel resolution, even for very large objects with hundreds of thousands of polygons (see Figure 8). The average accuracy using the highest sphere resolution is around 0.5%.

The per-frame quality displayed in Figure 9 re-emphasizes the accuracy of our approach and, additionally, shows the continuity of the distance and the volume.

VI. CONCLUSIONS AND FUTURE WORK

We have presented a novel hierarchical data structure, the *inner sphere trees*. The ISTs support different kinds of collision detection queries, including proximity queries and penetration volume computations with one unified algorithm.

---

[1] Please visit cg.in.tu-clausthal.de/research/ist to watch some videos of our benchmarks.
[2] This is due to bad BV tightness and the costly tetrahedron-tetrahedron overlap volume calculation.

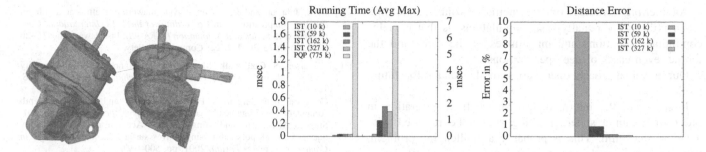

Fig. 7. Left: Snapshot from our oil pump scene (330k triangles). The dark gray spheres show the closest pair of spheres. Center: average and maximum time/frame. Left: relative error compared to accurate distance.

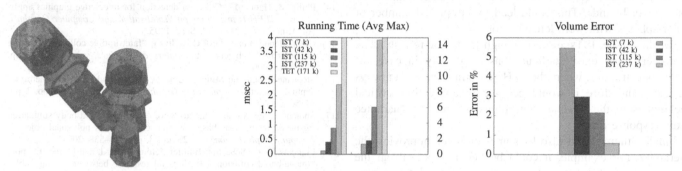

Fig. 8. Left: Snapshot from our bolt scene (171k triangles). The internal gray spheres show the overlapping inner spheres. Center: average and maximum time/frame. Right: relative error compared to accurate penetration volume.

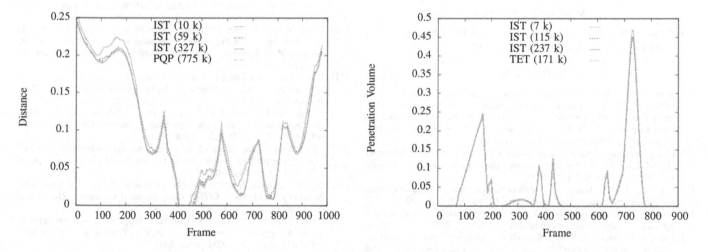

Fig. 9. Left: Distance per frame in the oil pump scene. Right: Penetration volume per frame in the bolt scene.

All types of queries can be answered at rates of about 1 kHz (which makes the algorithm suitable for haptic rendering) even for very complex objects with several hundreds of thousands of polygons.

For proximity situations, typical average running times are in the order of 0.05 msec with 327 000 spheres per object and an error of about 0.1%. In penetration situations, the running times depend, obviously, much more on the intersection volume; here, we are in the order of around 2.5 msec on average with 237 000 spheres and an error of about 0.5%.

The balance between accuracy and speed can be defined by the user. Moreover, the speed is independent of the object

complexity, because the number of leaves of our hierarchy is mostly independent of the number of polygons.

Our algorithm for both kinds of queries can be integrated into existing simulation software very easily, because there is only a single entry point, i.e., the application does not need to know in advance whether or not a given pair of objects will be penetrating each other.

Memory consumption of our inner sphere trees is similar to other bounding volume hierarchies, depending on the predefined accuracy (in our experiments, it was always in the order of a few MB). This is very modest compared to voxel-based approaches.

Another big advantage of our penetration volume algorithm, when utilized for penalty-based simulations, is that it yields continuous directions and magnitudes of the force and the torque, even cases of deep penetrations.

Our novel approach opens up several avenues for future work.

First of all, the intermediate distance field generation in order to obtain a sphere packing should be replaced with a better algorithm. This is probably a challenging problem, because several goals must be met: accuracy, query efficiency, and small build times.

An interesting question is the analytical determination of exact error bounds. This could lead to an optimal number of inner spheres with well-defined errors.

On the whole, ISTs are fast enough for haptic refresh rates. However, there exist configurations, especially in cases of heavy penetrations, where the 1 kHz constraint may not always be met. Therefore it would be nice to apply time critical techniques to the traversal algorithms in order to guarantee fixed response times.

Finally, there might also be some room for improving the hierarchy. For example, it could be better, especially at the borders of an object, to minimize the volume of those parts of hierarchy spheres that are outside of the object, instead of minimizing their volume.

ACKNOWLEDGMENT

This work was partially supported by DFG grant ZA292/1-1 and BMBF grant Avilus / 01 IM 08 001 U.

REFERENCES

[1] Sven Trenkel, René Weller, and Gabriel Zachmann, "A benchmarking suite for static collision detection algorithms", in *International Conference in Central Europe on Computer Graphics, Visualization and Computer Vision (WSCG)*, Václav Skala, Ed., Plzen, Czech Republic, 29 January–1 February 2007, Union Agency.

[2] Liangjun Zhang, Young J. Kim, Gokul Varadhan, and Dinesh Manocha, "Generalized penetration depth computation", *Computer-Aided Design*, vol. 39, no. 8, pp. 625–638, 2007.

[3] C.J. Ong and E.C. Gilbert, "Growth distances: New measures for object separation and penetration", *T-RA*, vol. 12, pp. 888–903, 1996.

[4] Susan M. Fisher and Ming C. Lin, "Fast penetration depth estimation for elastic bodies using deformed distance fields", 2001.

[5] James F. O'Brien and Jessica K. Hodgins, "Graphical modeling and animation of brittle fracture", in *SIGGRAPH '99: Proceedings of the 26th annual conference on Computer graphics and interactive techniques*, New York, NY, USA, 1999, pp. 137–146, ACM Press/Addison-Wesley Publishing Co.

[6] E. G. Gilbert, D. W. Johnson, and S. S. Keerthi, "A fast procedure for computing the distance between complex objects in three-dimensional space", *IEEE Journal of Robotics and Automation*, vol. 4, pp. 193–203, 1988.

[7] Gino van den Bergen, "A fast and robust GJK implementation for collision detection of convex objects", *Journal of Graphics Tools: JGT*, vol. 4, no. 2, pp. 7–25, 1999.

[8] Stephen Cameron, "Enhancing GJK: Computing minimum and penetration distances between convex polyhedra", in *Proceedings of International Conference on Robotics and Automation*, 1997, pp. 3112–3117.

[9] D. E. Johnson and E. Cohen, "A framework for efficient minimum distance computations", in *Proceedings of the IEEE International Conference on Robotics and Automation (ICRA-98)*, Piscataway, May 16–20 1998, pp. 3678–3684, IEEE Computer Society.

[10] E. Larsen, S. Gottschalk, M. Lin, and D. Manocha, "Fast proximity queries with swept sphere volumes", in *Technical Report TR99-018*, 1999.

[11] Gino van den Bergen, "Proximity queries and penetration depth computation on 3d game objects", *Game developers conference*, 2001.

[12] Stephen A. Ehmann and Ming C. Lin, "Accurate and fast proximity queries between polyhedra using convex surface decomposition", in *in Computer Graphics Forum*, 2001, pp. 500–510.

[13] Sean Quinlan, "Efficient distance computation between non-convex objects", in *In Proceedings of International Conference on Robotics and Automation*, 1994, pp. 3324–3329.

[14] Philip M. Hubbard, "Collision detection for interactive graphics applications", *IEEE Transactions on Visualization and Computer Graphics*, vol. 1, no. 3, pp. 218–230, Sept. 1995.

[15] Cesar Mendoza and Carol O'Sullivan, "Interruptible collision detection for deformable objects", *Computers & Graphics*, vol. 30, no. 3, pp. 432–438, 2006.

[16] Stephane Redon and Ming C. Lin, "A fast method for local penetration depth computation", *Journal of Graphics Tools: JGT*, vol. 11, no. 2, pp. 37–50, 2006.

[17] Shoichi Hasegawa and Makoto Sato, "Real-time rigid body simulation for haptic interactions based on contact volume of polygonal objects", *Comput. Graph. Forum*, vol. 23, no. 3, pp. 529–538, 2004.

[18] François Faure, Sébastien Barbier, Jérémie Allard, and Florent Falipou, "Image-based collision detection and response between arbitrary volumetric objects", in *ACM Siggraph/Eurographics Symposium on Computer Animation, SCA 2008, July, 2008*, Dublin, Irlande, July 2008.

[19] William A. McNeely, Kevin D. Puterbaugh, and James J. Troy, "Six degrees-of-freedom haptic rendering using voxel sampling", in *Siggraph 1999*, Alyn Rockwood, Ed., Los Angeles, 1999, ACM Siggraph, Annual Conference Series, pp. 401–408, Addison Wesley Longman.

[20] Matthias Renz, Carsten Preusche, Marco Ptke, Hans peter Kriegel, and Gerd Hirzinger, "Stable haptic interaction with virtual environments using an adapted voxmap-pointshell algorithm", in *In Proc. Eurohaptics*, 2001, pp. 149–154.

[21] Jernej Barbič and Doug L. James, "Six-dof haptic rendering of contact between geometrically complex reduced deformable models", *IEEE Transactions on Haptics*, vol. 1, no. 1, pp. 39–52, 2008.

[22] Ernesto G. Birgin and F. N. C. Sobral, "Minimizing the object dimensions in circle and sphere packing problems", *Computers & OR*, vol. 35, no. 7, pp. 2357–2375, 2008.

[23] Achill Schuermann, "On packing spheres into containers (about kepler's finite sphere packing problem)", in *Documenta Mathematica*, Sept. 09 2006, vol. 11, pp. 393–406.

[24] Agarwal, Guibas, Nguyen, Russel, and Zhang, "Collision detection for deforming necklaces", *CGTA: Computational Geometry: Theory and Applications*, vol. 28, 2004.

[25] Gareth Bradshaw and Carol O'Sullivan, "Adaptive medial-axis approximation for sphere-tree construction", in *ACM Transactions on Graphics*, vol. 23(1), pp. 1–26. ACM press, 2004.

[26] M. Cottrell, B. Hammer, A. Hasenfuss, and T. Villmann, "Batch and median neural gas", *Neural Networks*, vol. 19, pp. 762–771, jul 2006.

[27] Thomas M. Martinetz, Stanislav G. Berkovich, and Klaus J. Schulten, "'Neural-gas' network for vector quantization and its application to time-series prediction", *IEEE Trans. on Neural Networks*, vol. 4, no. 4, pp. 558–569, 1993.

[28] Barbara Hammer, Alexander Hasenfuss, and Thomas Villmann, "Magnification control for batch neural gas", in *ESANN*, 2006, pp. 7–12.

[29] Bernd Gärtner, "Fast and robust smallest enclosing balls", in *ESA*, Jaroslav Nesetril, Ed. 1999, vol. 1643 of *Lecture Notes in Computer Science*, pp. 325–338, Springer.

[30] Dan Morris, "Algorithms and data structures for haptic rendering: Curve constraints, distance maps, and data logging", in *Technical Report 2006-06*, 2006.

Using the Distribution Theory to Simultaneously Calibrate the Sensors of a Mobile Robot

Agostino Martinelli

INRIA Rhone Alpes, Grenoble, FRANCE

email: agostino.martinelli@ieee.org

Abstract—This paper introduces a simple and very efficient strategy to extrinsically calibrate a bearing sensor (e.g. a camera) mounted on a mobile robot and simultaneously estimate the parameters describing the systematic error of the robot odometry system. The paper provides two contributions. The first one is the analytical computation to derive the part of the system which is observable when the robot accomplishes circular trejectories. This computation consists in performing a local decomposition of the system, based on the theory of distributions. In this respect, this paper represents the first application of the distribution theory in the frame-work of mobile robotics. Then, starting from this decomposition, a method to efficiently estimate the parameters describing both the extrinsic bearing sensor calibration and the odometry calibration is derived (second contribution). Simulations and experiments with the robot e-Puck equipped with encoder sensors and a camera validate the approach.

I. INTRODUCTION

A sensor calibration technique is a method able to estimate the parameters characterizing the systematic error of the sensor. In the last decades, this problem has been considered with special emphasis in the field of computer vision. The problem of camera calibration consists in estimating the intrinsic and extrinsic parameters based on a number of points whose object coordinates in a global frame are known and whose image coordinates are measured [19]. In robotics, when a camera is adopted, the calibration only regards the estimation of the extrinsic parameters, i.e. the parameters describing the configuration of the camera in the robot frame. In the case of robot wrists, very successful approaches are based on the solution of a homogeneous transform equation of the form $AX = XB$ which is obtained by moving the robot wrist and observing the resulting motion of the camera [15], [18]. In particular, in the previous equation, A describes the configuration change of the wrist, B the configuration change of the camera and X the unknown camera configuration in the wrist reference frame. A and B are assumed to be known with high accuracy: A is obtained by using the encoder data and B by using the camera observations of a known object before and after the wrist movement [18]. In mobile robotics the situation changes dramatically and the previous methods cannot be applied. Unfortunately, the displacement of a mobile robot obtained by using the encoder data is not precise as in the case of a wrist. In other words, the previous matrix A is roughly estimated by the encoders. A possible solution to this inconvenient could be obtained by adopting other sensors to estimate the robot movement (i.e. the matrix A).

However, most of times the objective is to estimate the camera configuration in the reference frame attached on the robot odometry system. Therefore, it is important to introduce a new method able to perform simultaneously the extrinsic camera calibration and the calibration of the robot odometry system. So far, the two problems have been considered separately.

A. Previous Works

Regarding the odometry, a very successful strategy has been introduced in 1996 by Borenstein and Feng [3]. This is the UMBmark method. It is based on absolute robot position measurements after the execution of several square trajectories. In [10] the problem of also estimating the non-systematic odometry errors was considered. More recently, a method based on two successive least-squares estimations has been introduced [2]. Finally, very few approaches calibrate the odometry without the need of an a priori knowledge of the environment and/or of the use of global position sensors (like a GPS) [6], [17], [20].

Regarding the problem of sensor to sensor calibration, several cases have recently been considered (e.g. IMU-camera [16], laser scanner-camera [4], [21], [22] and odometry-camera [12]). In [12], an observability analysis taking into account the system nonlinearities was also provided to understand whether the system contains the necessary information to perform the self calibration. Indeed, a necessary condition to perform the estimation of a state, is that the state is observable. In [12] it was investigated whether the state containing the extrinsic parameters of the vision sensor is or is not observable. The *observability rank criterion* introduced by Hermann and Krener [8] was adopted to this scope. The same observability analysis was later applied to the case of the odometry self-calibration [13]. However, in these works, what it was determined is only whether the state containing the parameters defining the sensor systematic error is observable or not. On the other hand, when a state is not observable, suitable functions of its components could be observable and therefore could be estimated. The derivation of these functions is very important in order to properly exploit the information contained in the sensor data to estimate a given set of parameters. This derivation requires to perform a local decomposition [9]. While in the linear case this decomposition is easy to be performed, in the non linear case it is often troublesome and requires to apply the theory of distributions developed in [9].

B. Paper Contributions

In this paper we consider simultaneously the problems of odometry calibration and the extrinsic calibration of a bearing sensor. Furthermore, the calibration is carried out by only using a single point feature. To the best of our knowledge this problem has never been investigated before. The paper provides two contributions:

- A local decomposition of the considered system based on the theory of distributions developed in [9];
- A new strategy to robustly, efficiently and accurately calibrate both the bearing sensor and the odometry system.

The first contribution was preliminary discussed in [14] where the local decomposition has been performed only for the special cases of straight trajectory and pure rotation. To the best of our knowledge this contribution represents the first application of the distribution theory in the field of mobile robotics and the first non-trivial application of this theory to face a real estimation problem. In the specific case, it allows us to detect for circular trajectories which functions of the original calibration parameters are observable. Then, by performing at least three independent circular trajectories, it is possible to evaluate all the parameters describing our calibration problem.

Section II defines the calibration problem and provides the basic equations to characterize the system dynamics and the observation. In section III we remind some results from the theory developed in [8] (III-A) and [9] (III-B). In section IV we perform the decomposition for circular trajectories. This allows us to detect the functions of the parameters which are observable. Then, in section V, we summarize some important analytical properties for the observation function whose derivation is available in [11]. Based on these properties, we introduce the calibration strategy in section VI whose validation is provided in VII through simulations and experiments. Conclusions are provided in section VIII.

II. THE CONSIDERED SYSTEM

We consider a mobile robot moving in a $2D$-environment. The configuration of the robot in a global reference frame can be characterized through the vector $[x_R, y_R, \theta_R]^T$ where x_R and y_R are the cartesian robot coordinates and θ_R is the robot orientation. The dynamics of this vector are described by the following non-linear differential equations:

$$\begin{bmatrix} \dot{x}_R = v \cos \theta_R \\ \dot{y}_R = v \sin \theta_R \\ \dot{\theta}_R = \omega \end{bmatrix} \tag{1}$$

where v and ω are the linear and the rotational robot speed, respectively. The link between these velocities and the robot controls (u) depends on the considered robot drive system. We will consider the case of a differential drive. In order to characterize the systematic odometry error we adopt the model introduced in [3]. We have:

$$v = \frac{\delta_R v_R + \delta_L v_L}{2} \qquad \omega = \frac{\delta_R v_R - \delta_L v_L}{\delta_B B} \tag{2}$$

where v_R and v_L are the control velocities (i.e. $u = [v_R, v_L]^T$) for the right and the left wheel, B is the nominal value for the distance between the robot wheels and δ_R, δ_L and δ_B characterize the systematic odometry error due to an uncertainty on the wheels diameters and on the distance between the wheels.

Furthermore, a bearing sensor (e.g. a camera) is mounted on the robot. We assume that its vertical axis is aligned with the $z-$axis of the robot reference frame and therefore the transformation between the frame attached to this sensor and the one of the robot is characterized through the three parameters ϕ, ρ and ψ (see fig. 1).

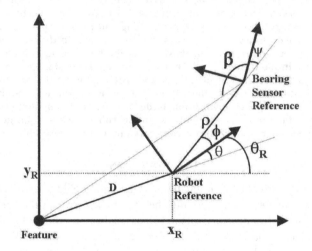

Fig. 1.   The two frames attached to the robot and to the bearing sensor.

The available data are the control $u = [v_R, v_L]^T$ and the bearing angle of a single feature (β in fig. 1) at several time steps during the robot motion.

We introduce the following quantities:

$$\mu \equiv \frac{\rho}{D} \equiv \frac{\rho}{\sqrt{x_R^2 + y_R^2}}; \qquad \begin{array}{l} \theta \equiv \theta_R - atan2(y_R, x_R); \\ \gamma \equiv \theta + \phi \end{array} \tag{3}$$

By using simple trigonometry algebra we obtain (fig. 1):

$$\beta = \begin{cases} -atan\left(\dfrac{\sin \gamma}{\mu + \cos \gamma}\right) - \psi + \pi & \gamma_- \leq \gamma \leq \gamma_+ \\ -atan\left(\dfrac{\sin \gamma}{\mu + \cos \gamma}\right) - \psi & otherwise \end{cases} \tag{4}$$

where γ_- and γ_+ are the two solutions (in $[-\pi, \pi)$) of the equation $\cos \gamma = -\mu$ with $\gamma_+ = -\gamma_-$ and $\gamma_+ > 0$. We made the assumption $0 < \mu < 1$ since we want to avoid collisions between the robot and the feature ($D > \rho$).

By using (1) and the definitions in (3) the dynamics of our system are described by the following equations:

Original Calibration Parameters
Camera: ϕ, ρ, ψ Odometry: δ_R, δ_L, δ_B
Observable Parameters
ϕ, ψ, $\eta \equiv \frac{\delta_R}{2\rho}$, $\delta \equiv \frac{\delta_L}{\delta_R}$, $\xi \equiv \frac{1}{B}\frac{\delta_R}{\delta_B}$
Parameters observable in a single q-trajectory
$A^q \equiv \frac{\Psi_1^q - \Psi_3}{1 + \Psi_1^q \Psi_3}$, $V^q \equiv \Psi_2^q \frac{1 + \Psi_1^q \Psi_3}{1 + \Psi_3^2}$, $L^q \equiv \psi - atan\Psi_1^q$, $\xi_q \equiv \xi(1 - q\delta)$
where: $\Psi_1^q \equiv \frac{\xi_q - \eta_q \sin\phi}{\eta_q \cos\phi}$, $\Psi_2^q \equiv \frac{\mu\eta_q \cos\phi}{\sin\gamma}$,
$\Psi_3 \equiv \frac{\mu + \cos\gamma}{\sin\gamma}$, $\eta_q \equiv \eta(1 + q\delta)$

TABLE I
VARIABLES ADOPTED IN THIS PAPER

$$\begin{cases} \dot{\mu} = -\mu^2 \frac{v}{\rho}\cos(\gamma - \phi) \\ \dot{\gamma} = \omega - \mu\frac{v}{\rho}\sin(\gamma - \phi) \\ \dot{\phi} = \dot{\rho} = \dot{\psi} = \dot{\delta}_R = \dot{\delta}_L = \dot{\delta}_B = 0 \end{cases} \quad (5)$$

The goal is to simultaneously estimate the parameters ϕ, ρ, ψ, δ_R, δ_L and δ_B using the available data (i.e. v_R, v_L and β in a given time interval). Since these data consists of angle measurements (the wheel diameters are not known and in fact are to be estimated), the best we can hope is the possibility to estimate these parameters up to a scale factor. In particular, we will refer to the following parameters:

$$\phi, \quad \psi, \quad \eta \equiv \frac{\delta_R}{2\rho}, \quad \delta \equiv \frac{\delta_L}{\delta_R}, \quad \xi \equiv \frac{1}{B}\frac{\delta_R}{\delta_B} \quad (6)$$

whose dynamics are:

$$\begin{cases} \dot{\mu} = -\mu^2\eta(v_R + \delta v_L)\cos(\gamma - \phi) \\ \dot{\gamma} = \xi(v_R - \delta v_L) - \mu\eta(v_R + \delta v_L)\sin(\gamma - \phi) \\ \dot{\phi} = \dot{\psi} = \dot{\eta} = \dot{\delta} = \dot{\xi} = 0 \end{cases} \quad (7)$$

In section IV we derive for circular trajectories, which functions of μ, γ, ϕ, ψ, η, δ and ξ are observable and hence can be estimated. Then, in section VI we introduce a very efficient strategy to estimate these parameters by considering three independent circular trajectories. Finally, by adding a simple metric measurement (e.g. the initial distance between the robot and the feature) the original parameters ϕ, ρ, ψ, δ_R, δ_L and δ_B can also be estimated.

For the sake of clarity we report all the variables appearing in this paper in table I.

III. OBSERVABILITY PROPERTIES AND LOCAL DECOMPOSITION

A general characterization for systems in the frame work of autonomous navigation is provided by the following equations:

$$\begin{cases} \dot{S} = \sum_{i=1}^{M} f_i(S)u_i \\ y = h(S) \end{cases} \quad (8)$$

where $S \in \Sigma \subseteq \Re^n$ is the state, u_i are the system inputs, $y \in \Re$ is the output (we are considering a scalar output for the sake of clarity, the extension to a multi dimensions output is straightforward). The system defined by (4) and (7) can be characterized by (8). We have: $S = [\mu, \gamma, \phi, \psi, \eta, \delta, \xi]^T$, $M = 2$, $u_1 = v_R$, $u_2 = v_L$, $f_1(S) = [-\mu^2\eta\cos(\gamma - \phi), \xi - \mu\eta\sin(\gamma - \phi), 0, 0, 0, 0, 0]^T$, $f_2(S) = -[\mu^2\eta\delta\cos(\gamma - \phi), \xi\delta + \mu\eta\delta\sin(\gamma - \phi), 0, 0, 0, 0, 0]^T$ and $y = \beta$.

A. Observability Rank Criterion

We want to remind some concepts in the theory by Hermann and Krener in [8]. We will adopt the following notation. We indicate the kth order Lie derivative of a scalar field Λ along the vector fields $v_{i_1}, v_{i_2}, ..., v_{i_k}$ with $L_{v_{i_1}, v_{i_2}, ..., v_{i_K}}^k \Lambda$. We remind the definition of the Lie derivative. It is provided by the following two equations:

$$L^0\Lambda = \Lambda, \quad (9)$$

$$L_{v_{i_1}, ..., v_{i_k}, v_{i_{k+1}}}^{k+1}\Lambda = \nabla_S \left(L_{v_{i_1}, ..., v_{i_k}}^k \Lambda\right) \cdot v_{i_{k+1}}$$

where the symbol ".." denotes the scalar product. Now, let us refer to the system in (8) and let us indicate with Ω the space of all the Lie derivatives $L_{f_{i_1}, ..., f_{i_k}}^k h$, $(i_1, ..., i_k = 1, ..., M)$ where the functions f_{i_j} $(j = (1, ..., M))$ are the ones appearing in (8). Furthermore, we denote with $dL_{f_{i_1}, ..., f_{i_k}}^k h$ the gradient of the corresponding Lie derivative (i.e. $dL_{f_{i_1}, ..., f_{i_k}}^k h \equiv \nabla_S L_{f_{i_1}, ..., f_{i_k}}^k h$) and we denote with $d\Omega$ the space spanned by all these gradients.

In this notation, the observability rank criterion can be expressed in the following way: *The dimension of the largest observable sub-system at a given S_0 is equal to the dimension of $d\Omega$. Therefore, the problem of understanding wheter a system is observable or not can be solved by computing the rank of the matrix whose lines are the gradients of the Lie derivatives.*

B. Local Decomposition

Let us suppose that the system in (8) is not observable and that the dimension of the largest observable subsystem is n_{obs}. According with the theory of distributions developed in [9], we can find n_{obs} independent functions of the components of the original state S which are observable and $n - n_{obs}$ independent functions of the components of S which are not observable. More precisely, if we include the n_{obs} observable functions in the vector S_b and the other $n - n_{obs}$ functions in the vector S_a, we have the following decomposition for the original system:

$$\begin{cases} \dot{S}_a = \sum_{i=1}^{M} f_i^a(S_a, S_b)u_i \\ \dot{S}_b = \sum_{i=1}^{M} f_i^b(S_b)u_i \\ y = h_b(S_b) \end{cases} \quad (10)$$

In particular, the subsystem defined by the last two equations in (10) is independent of the value of S_a and it is observable. Therefore, by performing this decomposition, we can use the information coming from the dynamics (i.e. the knowledge of $u(t)$) and the observations ($y(t)$) in order to estimate the observable quantities (S_b). This decomposition is very important in every estimation problem when the state is not observable. Indeed, estimating directly the original state S results in an erroneous evaluation.

In section II, when we introduced μ and γ (defined in (3)) and the three parameters η, δ, ξ, we performed such a decomposition for the state $[x_R, \; y_R, \; \theta_R, \; \phi, \; \rho, \; \psi, \; \delta_R, \; \delta_L, \; \delta_B]^T$: indeed, the new state $[\mu, \; \gamma, \; \phi, \; \psi, \; \eta, \; \delta, \; \xi]^T$ is observable as proven in [11] and its components are non linear functions of the components of the original state (which is not observable). On the other hand, in most of cases it is very troublesome to perform such a decomposition. In the next section we perform such a decomposition for the same state (i.e. $[\mu, \; \gamma, \; \phi, \; \psi, \; \eta, \; \delta, \; \xi]^T$) but when we only allow the robot to move along circular trajectories. We apply the distributions theory developed in [9].

IV. LOCAL DECOMPOSITION FOR CIRCULAR TRAJECTORIES

We consider the one-degree of freedom motion obtained by setting $v_R = \nu$, $v_L = q\nu$. Let us define:

$$\eta_q \equiv \eta(1 + q\delta) \qquad \xi_q \equiv \xi(1 - q\delta) \qquad (11)$$

From the dynamics in (7) we obtain the following dynamics:

$$\begin{cases} \dot{\mu} = -\mu^2 \eta_q \nu \cos(\gamma - \phi) \\ \dot{\gamma} = \xi_q \nu - \mu \eta_q \nu \sin(\gamma - \phi) \\ \dot{\eta}_q = \dot{\xi}_q = \dot{\phi} = \dot{\psi} = 0 \end{cases} \qquad (12)$$

In the next we provide the steps necessary to perform the decomposition of the system whose dynamics are given in (12) and whose output is the observation in (4). We proceed in two separate steps. We first consider the following simpler system:

$$\begin{cases} \dot{\mu} = -\mu^2 \eta_q \nu \cos(\gamma - \phi) \\ \dot{\gamma} = \xi_q \nu - \mu \eta_q \nu \sin(\gamma - \phi) \\ \dot{\eta}_q = \dot{\xi}_q = \dot{\phi} = 0 \\ y = \dfrac{\sin\gamma}{\mu + \cos\gamma} \end{cases} \qquad (13)$$

where we removed the variable ψ. We apply the method developed in [9] in order to find the local decomposition for this simplified system.

The associated partial differential equation is in this case:

$$\frac{\mu\cos\gamma + 1}{\sin\gamma}\frac{\partial\Psi}{\partial\mu} + \frac{\partial\Psi}{\partial\gamma} + \frac{\xi_q\cos\phi}{\eta_q\mu\sin\gamma}\frac{\partial\Psi}{\partial\phi} + \frac{\xi_q\sin\phi - \eta_q}{\mu\sin\gamma}\frac{\partial\Psi}{\partial\eta_q} = 0$$

namely, every solution $\Psi(\mu, \gamma, \phi, \eta_q, \xi_q)$ of the previous partial differential equation is an observable quantity for the

system in (13). We found the following four independent solutions:

$$\Psi_1^q \equiv \frac{\xi_q - \eta_q\sin\phi}{\eta_q\cos\phi}, \quad \Psi_2^q \equiv \frac{\mu\eta_q\cos\phi}{\sin\gamma},$$

$$\Psi_3 \equiv \frac{\mu + \cos\gamma}{\sin\gamma}, \quad \xi_q \qquad (14)$$

The local decomposition of (13) is:

$$\begin{cases} \dot{\Psi}_1^q = 0 \\ \dot{\Psi}_2^q = \nu\Psi_2^q(\Psi_1^q\Psi_2^q - \xi_q\Psi_3) \\ \dot{\Psi}_3 = \nu(\Psi_2^q + \Psi_1^q\Psi_2^q\Psi_3 - \xi_q - \xi_q\Psi_3^2) \\ \dot{\xi}_q = 0 \\ y = \dfrac{1}{\Psi_3} \end{cases} \qquad (15)$$

Let us proceed with the second step. We add to the system in (15) the parameter ψ (with $\dot{\psi} = 0$) and we consider the output $y = \beta$ instead of $y = \frac{1}{\Psi_3}$. We apply again the method in [9] on the resulting system. The associated partial differential equation is in this case:

$$(\Psi_1^{q^2}+1)\frac{\partial G}{\partial\Psi_1^q} + (\Psi_2^q(\Psi_3 - \Psi_1^q))\frac{\partial G}{\partial\Psi_2^q} + (\Psi_3^2+1)\frac{\partial G}{\partial\Psi_3} + \frac{\partial G}{\partial\psi} = 0$$

namely, every solution $G(\Psi_1^q, \Psi_2^q, \Psi_3, \xi_q, \psi)$ of the previous partial differential equation is an observable quantity for this system. We found the following four independent solutions:

$$A^q \equiv \frac{\Psi_1^q - \Psi_3}{1 + \Psi_1^q\Psi_3}, \qquad V^q \equiv \Psi_2^q\frac{1 + \Psi_1^q\Psi_3}{1 + \Psi_3^2},$$

$$L^q \equiv \psi - atan\Psi_1^q, \qquad \xi_q \qquad (16)$$

and the local decomposition is:

$$\begin{cases} \dot{A}^q = \nu(1 + A^{q^2})(\xi_q - V^q) \\ \dot{V}^q = \nu A^q V^q(2V^q - \xi_q) \\ \dot{L}^q = \dot{\xi}_q = 0 \\ \beta = -atanA^q - L^q + S_p\dfrac{\pi}{2} \end{cases} \qquad (17)$$

where S_p can be ± 1 depending on the values of the parameters. We do not provide here this dependence. In [11] we derive some important properties relating S_p to the robot motion.

Deriving this decomposition is very hard. As shown, it is based on the solutions of two partial differential equations. However, to check the validity of this decomposition is very simple since it only requires to compute derivatives (e.g. this can be done by using the matlab symbolic computation). Furthermore, also the solution has a simple analytical expression.

This decomposition has a very practical importance. It tells us that, when the robot accomplishes circular trajectories, the information contained in the sensor data (i.e. the information contained in the function $\nu(t)$ and $\beta(t)$) allows us to estimate only the state $\Theta \equiv [A^q, \; V^q, \; L^q, \; \xi_q]^T$ and not the original state $[\mu, \ \gamma, \ \phi, \ \psi, \ \xi, \ \delta, \ \eta]^T$. In the next sections we will provide a powerful strategy to estimate the initial value $\Theta_0 \equiv [A_0^q, \ V_0^q, \ L^q, \ \xi_q]^T$ for a given circular trajectory. We remark that Θ_0 depends on the calibration parameters and on the initial values of μ and γ. Our goal is the estimation of $\phi, \ \rho, \ \psi, \ \eta, \ \delta, \ \xi$, which are five parameters. By performing the estimation of Θ_0 we obtain four independent equations on these parameters. On the other hand, we also add two new unknowns (the values of the initial μ and γ). Therefore, in order to have enough equations, we must combine at least three independent circular trajectories (i.e. with a different q).

V. ANALYTICAL PROPERTIES OF THE OBSERVATION FUNCTION

In this section we summarize some important properties of the observation β obtained when the robot accomplishes circular trajectories. These properties are fundamental to introduce our calibration strategy. For the sake of conciseness, we cannot derive these properties here. However, a detailed derivation is available in [11]. Furthermore, it is possible to directly verify the validity of these properties with a simple substitution. For the sake of simplicity in the next two sections we neglect the suffix q on the three parameters A, V and L. On the other hand, to distinguish ξ_q from ξ previously defined, we still maintain q in ξ_q.

First of all, it is possible to directly integrate the dynamics in (17) to get an analytical expression for β vs the time or vs the curve length s defined by the following equation:

$$s = s(t) = \int_0^t \nu(\tau)d\tau \qquad (18)$$

The expression of β vs s is given by the following equations:

$$w = \xi_q \tan(c + S_w \xi_q s) \qquad (19)$$

$$V = \frac{\xi_q k(2k - \xi_q) + kw^2 + S_V w\sqrt{k(k - \xi_q)(w^2 + \xi_q^2)}}{(2k - \xi_q)^2 + w^2} \qquad (20)$$

$$A = S_y \sqrt{\frac{k(2V - \xi_q) - V^2}{V^2}} \qquad (21)$$

$$\beta = atan(A) - L + S_p \frac{\pi}{2} \qquad (22)$$

where c and k are two time-independent parameters whose value depends on the initial robot configuration with respect to the feature and S_w, S_V, and S_y are three sign variables, namely, as for S_p, they can assume the value of $+1$ or -1. The validity of this expression for β can be checked by directly computing the time derivatives of V and A respectively in (20) and (21) and by verifying that they satisfy (17).

The variables S_V, S_y and S_p flip in correspondence of special points whose value can be determined by imposing the continuity of the expressions (19-22) and their first derivatives (see [11] for a detailed discussion of this continuity analysis).

Among these points, there are the ones where the function w in (19) diverges. We call them nodes. They are:

$$s_n = -S_w \frac{c}{\xi_q} + j\frac{\pi}{\xi_q} + S_w \frac{\pi}{2\xi_q} \qquad (23)$$

j being an integer. In other words, there are infinite nodes at the distance of $\frac{\pi}{|\xi_q|}$ one each other.

For every point s we introduce an operation which associates to s the point $\hat{s}$ defined by the following equation:

$$\hat{s} \equiv 2s_n^R - s \qquad (24)$$

where s_n^R is the closest node to s on the right. In [11] we derive the following fundamental theorem whose validity can in any case be checked by a direct substitution:

Theorem 1 (Reflection of the Observation Function) *The observation function satisfies the following fundamental equation $\forall s$:*

$$\beta(s) + \beta(\hat{s}) = -2L \ (\mod \pi) \qquad (25)$$

This theorem is fundamental for estimating the parameters ξ_q, L, k and c related to a given trajectory (see the algorithm 1 in the next section and the videos in [23]). The name reflection comes from the fact that according with (24) $\hat{s}$ is the reflection of s with respect to the node between them (s_n^R).

VI. THE STRATEGY TO ESTIMATE THE SYSTEM PARAMETERS

It is possible to verify that the observation function is a periodic function whose period is:

$$T_S = \frac{2\pi}{|\xi_q|} \qquad (26)$$

Equation (26) allows us to estimate the parameter ξ_q by evaluating the period of the observation function. Actually, this equation does not provide the sign of ξ_q. However, this sign is positive for all the values of $q < \frac{1}{\delta}$ (i.e. when the robot accomplishes counter clock-wise circular trajectories).

Once ξ_q is estimated, the next step consists in the evaluation of the position of one node. Indeed, once we know the position of one node, we can determine c (or better few candidates of c) by using (23). On the other hand, the position of one node can be evaluated by using the previous theorem (i.e. the equation (25). The algorithm 1 describes the procedure to perform this evaluation. In [23] it is possible to get some videos which illustrate how this algorithm works in simulated and real scenarios. The algorithm computes the left hand side of equation (25), called $\theta(s_c, s)$, for every possible node candidate (s_c), which is in the interval $[0, \frac{T_S}{2}]$. The function $\theta(s_c, s)$ is independent of the second argument s when $s_c = s_n$. Indeed, $\theta(s_n, s) = -2L \ \forall s$. This means that the standard deviation of $\theta(s_c, s)$ respect to the second argument (s) is zero when computed in s_n (i.e. $\sigma(s_n) = 0$). When the robot sensors are affected by measurement errors, the function $\sigma(s_c)$ attains its minimum on s_n (see figure 4e).

Algorithm 1 (Returns one Node)

for $s_c = 0$ *to* $\frac{T_S}{2}$ *do*
    for $s = 0$ *to* $\frac{T_S}{2}$ *do*
        $\hat{s} = 2s_c - s$
        $\theta(s_c, s) = \beta(s) + \beta(\hat{s})(\ \mod \pi)$
    end for
end for
for $s_c = 0$ *to* $\frac{T_S}{2}$ *do*
    $\sigma(s_c) = $ *standard deviation of* $\theta(s_c, s)$
end for
$s_n = \arg\min_{s_c} \sigma(s_c)$

Once s_n is determined, equations (24) and (25) allow us to immediately evaluate the parameter L. Furthermore, as said before, equation (23) allows us to evaluate c. In both cases few possible values for these parameters are actually provided. The correct ones can be selected by combining more than one circular trajectory. On the other hand, combining at least three trajectories (with different q) is necessary also to estimate our original parameters ϕ, ψ, η, δ and ξ once the parameters ξ_q and L are evaluated for each trajectory.

Once the parameters ξ_q and L are estimated for at least three independent trajectories (i.e. corresponding to three different values of q), the calibration parameters ϕ, ψ, η, δ and ξ can be found by using (11), the first equation in (14) and the third equation in (16). The method to get this estimation is simple and is explained in [11]. Finally, in [11], we also explain how to estimate the original calibration parameters (i.e. ϕ, ρ, ψ, δ_R, δ_L and δ_B) when it is available a supplementary metric measurement consisting in the initial distance of the robot from the feature for just one among the three considered trajectories.

VII. PERFORMANCE EVALUATION

We evaluate the performance of the proposed strategy by carrying out both simulations and real experiments. In particular, since in the simulations the ground truth is available, we compare our strategy with a method based on an Extended Kalman Filter (EKF). This method uses the EKF to integrate the encoders data and the data provided by the camera. The matrices defining this filter can be obtained starting from the analytical expressions in (4) and (7). From now on we will refer to this calibration method with $CEKF$ (Calibration based on the EKF).

A. Simulations

We simulate a mobile robot equipped with encoder sensors and an omnidirectional camera.

Regarding the encoders we simulate their data according with the model introduced in [3]. In other words the measurements are affected by zero mean Gaussian errors independent among them. Furthermore, the variance of each measurement is proportional to the value provided by the sensor. Let us consider for instance the right wheel and let us suppose that the true rotation occurred at a given time step is equal to $\delta\alpha_R^{true}$. We generate the following measurement: $\delta\alpha^R = N\left(\delta\alpha_R^{true}, \frac{K^2}{r}|\delta\alpha_R^{true}|\right)$, where $N(m, \sigma^2)$ indicates

the normal distribution with mean value m and variance σ^2, r is the nominal value of the radius of the wheel and K characterizes the non systematic odometry error. The precision of our strategy is always excellent, even when we considered values of K much larger (hundred times) than the values obtained through real experiments (see [5] where $K \simeq 5\ 10^{-4}m^{\frac{1}{2}}$). In this section we provide a part of the results obtained with our simulations. In particular, we set $r = 0.3m$, $K = 0.001m^{\frac{1}{2}}$ and the distance between the wheels $B = 0.5m$.

The simulated exteroceptive sensor provides the bearings of a single feature at the origin. Furthermore, we assume that these bearing measurements are affected by a zero mean Gaussian error with a variance equal to $(0.5deg)^2$. The small value of this variance was necessary in order to guarantee the convergence of the $CEKF$. However, in [11] we consider larger values for this variance (up to $(5deg)^2$) and the performance of our strategy is always very good.

We also investigate the impact of having a path not perfectly circular. To this goal we divide the circular path in ten thousands segments. For each one we compute the corresponding displacement for the right and the left wheel (Δs_R^c and Δs_L^c). Then, we generate the displacements of the right and left wheel (Δs_R and Δs_L) randomly with a Gaussian distribution: $\Delta s_R = N(\Delta s_R^c, (\imath \times \Delta s_R^c)^2)$ and $\Delta s_L = N(\Delta s_L^c, (\imath \times \Delta s_L^c)^2)$. We consider three cases: Model 1 where $\imath = 0$ (i.e. perfect circular trajectory), model 2 where $\imath = 0.02$ and model 3 where $\imath = 0.04$. In fig. 2 we plot the ratio $\frac{\Delta s_L}{\Delta s_R}$ vs time when the robot accomplishes a circular trajectory with $q = 0.7$. On the left it is displayed the real case obtain by using the robot AMV-1 from the BlueBotics company (see [1] for a description of the robot). On the right it is considered the simulated robot when $\imath = 0.02$. It is possible to realize that the real case satisfies the circular hypothesis better than the case when $\imath = 0.02$.

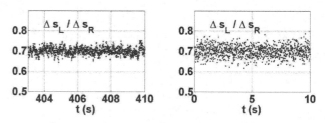

Fig. 2. The ratio $\frac{\Delta s_L}{\Delta s_R}$ vs time when the robot accomplishes a circular trajectory ($q = 0.7$). On the left the result obtained with the real robot AMV-1 and on the right our simulated robot when $\imath = 0.02$.

In fig. 3 we display the precision on the parameters vs the number of camera observations when the estimation is performed by the $CEKF$. In contrast with our strategy, this filter requires to initialize the parameters. We obtained that, in order to have the convergence, the initial relative error on the parameter η must be smaller than 10%. Regarding δ and ξ the relative error must be smaller than 20%. Finally, regarding ϕ and ψ the initial error must be smaller than $10deg$. From fig. 3 we remark that the convergence is very slow for the parameters ϕ, ψ and η.

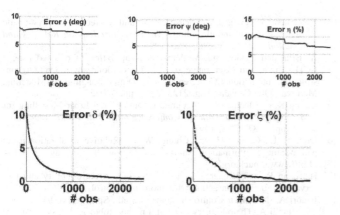

Fig. 3. The precision on the parameters estimated through $CEKF$ vs the number of exteroceptive observations

Method	#Obs.	$\Delta\phi$ deg	$\Delta\psi$ deg	$\frac{\Delta\eta}{\eta}$ %	$\frac{\Delta\delta}{\delta}$ %	$\frac{\Delta\xi}{\xi}$ %
Our	250	0.06	0.05	0.5	0.02	0.07
EKF	250	7.8	7.6	10	3.2	3.6
Our	500	0.06	0.05	0.4	0.02	0.05
EKF	500	7.7	7.5	9.8	1.9	2.5
Our	1000	0.06	0.05	0.3	0.02	0.05
EKF	1000	7.5	7.3	9.2	1.0	1.2

TABLE II
THE ERRORS ON THE PARAMETERS AVERAGED ON 100 SIMULATIONS
ESTIMATED BY THE $CEKF$ AND OUR STRATEGY.

Table II compares the performance of our approach with respect to $CEKF$ when the number of exteroceptive observations are 250, 500 and 1000. Even for model 3 (i.e. when $\iota = 0.04$) our method significantly outperforms $CEKF$.

B. Real Experiments

We used the mobile robot e-puck (see [7] for a detailed description of this robot and its sensors). In our experiments we only used the camera and the odometry sensors. Actually, our strategy has been developed to calibrate an omnidirectional bearing sensor. In contrast, the available camera, has a very limited angle of view ($\simeq 38deg$). In practice, it is in general not possible to observe a single feature during the entire circular trajectory accomplished by the robot. The only situation where this is possible occurs when the feature is inside the circular trajectory and close to the center. Furthermore, the camera must look towards the center of the circumference. This is the case when the angle ϕ is close to $0deg$ and ψ is close to $90deg$). Since the available camera looks ahead, we fixed in front of the robot a small mirror (see figure 4a). Obviously, in these conditions our strategy cannot estimate the extrinsic calibration parameters related to the real camera. However, it estimates the parameters related to the virtual camera, i.e. the one mirrored. We remark that the goal of this experiment is not to estimate the configuration of the true camera but to validate our strategy. Therefore, we never mind whether the camera we are considering is the virtual camera.

An issue which arises when the feature is inside the trajectory is the possibility to have collisions with the feature.

Tape	ϕ (deg)	ψ (deg)	η (m^{-1})	δ	ξ (m^{-1})
No	−5.80	117.18	10.21	0.9987	18.99
Yes	−5.67	116.91	10.40	0.9959	18.97

TABLE III
THE CALIBRATION PARAMETERS WITH AND WITHOUT TAPE ESTIMATED
IN OUR EXPERIMENTS.

In order to avoid this, the circumference has to be large. In practice we could not consider trajectories with values of q smaller than 0.4.

The robot camera provides images with resolution 60×60. Figure 4b shows the robot e-puck together with the source of light we adopted as a point feature in our experiments. Figure 4c is an image of the feature taken by the e-puck camera during our experiments. The images were provided at a frequency in the range $[0.5, 3]Hz$.

We carried out two complete experiments. In the latter we increased the radius of the right wheel by $0.062mm$ with a piece of tape. Each experiment consists of four independent trajectories with the following values q: 0.9, 0.7, 0.6, 0.4. Regarding the estimation of the parameters ξ_q, L, c and k we show in figures 4d, 4e and 4f only the results for $q = 0.6$ without tape. In particular, we show the observation function with the estimated nodes (4d), the function $\sigma(s_c)$ (4e) and the observation functions as observed and as estimated by our strategy (4f).

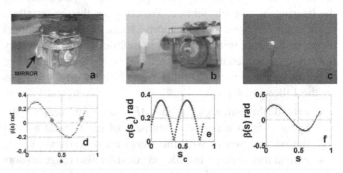

Fig. 4. The experiment performed with the robot e-puck. (a) displays the robot e-puck with a small mirror in front of the camera. The robot together with the source of light used as the feature (b) and the feature observed by the robot camera (c). The images below refer to an experiment with $q = 0.6$ and without tape. d displays the observation function with the estimated nodes, e displays the function $\sigma(s_c)$ attaining its minima in correspondence of the nodes and f displays the observation function as observed and as estimated by our strategy.

The calibration parameters with and without tape are reported in table III. Regarding the angles ϕ and ψ, we remark that the difference between the case with tape and without tape is smaller than $0.3deg$, which is roughly the mean errors obtained in our simulations. This is consistent with the fact that the tape does not affect these angle parameters. Regarding η the difference is $\simeq 2\%$ which is a bit larger than the mean error obtained in our simulations. Also regarding δ the difference is $\simeq 0.3\%$ which is definitely larger than the mean error obtained in our simulations. This is consistent with the increased radius of the right wheel due to the tape ($0.062mm$). In particular, since the wheel radius is $\simeq 2cm$, we obtain $\simeq 0.06mm$ for the radius change. The variation in the parameter ξ is very small and not in the expected direction. In particular, it is $\simeq 0.1\%$ which is roughly the mean error obtained in our simulations. This parameter should be affected by the tape since it depends on δ_R. We believe that the tape also increased the effective distance between the wheels (i.e. the parameter δ_B) making $\xi = \frac{1}{B}\frac{\delta_R}{\delta_B}$ unaffected by the tape.

VIII. Conclusions

In this paper we considered the problem of simultaneously calibrating an odometry system and the extrinsic parameters of a bearing sensor. The calibration only uses a single point feature.

Two new contributions were provided:

- A local decomposition of the considered system based on the theory of distributions developed in [9];
- A simple strategy to robustly, efficiently and accurately estimate the parameters describing both the extrinsic bearing sensor calibration and the odometry calibration.

The performance of the proposed strategy was carefully evaluated by carrying out both simulations and real experiments with the mobile robot e-puck. We found excellent results in terms of precision, robustness and facility of implementation. We remark that the proposed strategy was able to detect a variation in the wheel diameter of less than $0.1mm$ although the robot moved along very short paths and by just observing the bearing of a single feature. We consider this an excellent result and we believe that this approach can have a huge impact on the problem of self-calibration in mobile robotics especially if it can be extended to the $3D$ case and hence it can be applied in outdoor environment. At this regard, future works will focus on the following topics:

- extend this strategy to the $3D$ case, i.e. remove the hypothesis of a perfect alignment of the vertical axis of the bearing sensor with the world frame vertical axis;
- extend this strategy in order to consider also range sensors and other kind of features.

Finally, we believe that the theory of distributions is a very powerful tool to face many estimation problems in the frame-work of mobile robotics and this paper is the first very successful application of this sophisticated theory.

Acknowledgment

This work was supported by the European Commission FP7-ICT-2007-3.2.2 Cognitive Systems, Interaction, and Robotics under the contract #231855 (sFLY) and the EU Integrated Projects BACS - FP6-IST-027140. The author wishes also to thank Jean-Marc Bollon for very useful discussions and suggestions regarding the experiment.

References

[1] http://www.bluebotics.com/automation/AMV-1/
[2] G. Antonelli and S. Chiaverini, Linear estimation of the physical odometric parameters for differential-drive mobile robots, Autonomous Robot, July 2007, Vol 23, pages 59-68
[3] Borenstein J., Feng L., "Measurement and correction of systematic odometry errors in mobile robots," *IEEE Transactions on Robotics and Automation*, vol. 12, pp. 869–880, 1996.
[4] X. Brun and F. Goulette, Modeling and calibration of coupled fisheye CCD camera and laser range scanner for outdoor environment reconstruction, International Conference on 3D Digital Imaging and Modeling, Montreal, QC, Canada, Aug. 2123, 2007, pp. 320327.
[5] Chong K.S., Kleeman L., "Accurate Odometry and Error Modelling for a Mobile Robot," *International Conference on Robotics and Automation*, vol. 4, pp. 2783–2788, 1997.
[6] Doh, N. L., Choset, H. and Chung, W. K., Relative localization using path odometry information, Autonomous Robots, Vol 21, pages 143-154
[7] http://www.e-puck.org/
[8] Hermann R. and Krener A.J., 1977, Nonlinear Controllability and Observability, IEEE Transaction On Automatic Control, AC-22(5): 728-740
[9] Isidori A., Nonlinear Control Systems, 3rd ed., Springer Verlag, 1995.
[10] Martinelli A, The odometry error of a mobile robot with a synchronous drive system, *IEEE Trans. on Rob. and Aut.* Vol 18, NO. 3 June 2002
[11] Martinelli A, Using the Distribution Theory to Simultaneously Calibrate the Sensors of a Mobile Robot, Internal Research Report, INRIA, http://hal.inria.fr/docs/00/35/30/79/PDF/RR-6796.pdf
[12] A. Martinelli, D. Scaramuzza and R. Siegwart, Automatic Self-Calibration of a Vision System during Robot Motion, International Conference on Robotics and Automation, Orlando, Florida, April 2006.
[13] A. Martinelli and R. Siegwart, Observability Properties and Optimal Trajectories for On-line Odometry Self-Calibration, International Conference on Decision and Control, San Diego, California, December 2006.
[14] A. Martinelli, Local Decomposition and Observability Properties for Automatic Calibration in Mobile Robotics, International Conference on Robotics and Automation, Kobe, Japan, May 2009.
[15] Park F. C. and B. J. Martin, Robot Sensor Calibration: Solving AX=XB on the Euclidean Group, IEEE Trans. on Rob. and Aut., Vol 10, No 5 Oct 1994.
[16] F. M. Mirzaei and S. I. Roumeliotis, A Kalman filter-based algorithm for IMU-camera calibration: Observability analysis and performance evaluation, IEEE Trans. on Rob., 2008, Vol. 24, No. 5, October 2008
[17] Roy N., and Thrun S., Online Self-calibration for Mobile Robots, proceedings of the 1999 IEEE International Conference on Robotics and Automation, 19 May 1999 Detroit, Michigan, pp. 2292-2297.
[18] Shiu Y. C. and S. Ahmad, Calibration of Wrist-Mounted Robotic Sensors by Solving Homogeneous Transform Equations of the Form AX=XB, IEEE Trans on Rob. and Aut. Vol 5 No 1 Feb 1989
[19] R. Y. Tsai, A versatile camera calibration technique for high-accuracy 3d machine vision metrology using off-the-shelf tv cameras and lenses. IEEE J. Robotics Automat, 3(4), 323-344, 1987.
[20] H.J. von der Hardt, R. Husson, D. Wolf, An Automatic Calibration Method for a Multisensor System: Application to a Mobile Robot Localization System, Interbational Conference on Robotics and Automation, Leuven, Belgium, May 1998.
[21] S. Wasielewski and O. Strauss, Calibration of a multi-sensor system laser rangefinder/camera, in Proceedings of the Intelligent Vehicles Symposium, Detroit, MI, Sept. 2526, 1995, pp. 472477.
[22] Q. Zhang and R. Pless, Extrinsic calibration of a camera and laser range finder (improves camera calibration), in Proceedings of the IEEE/RSJ International Conference on Intelligent Robots and Systems, Sendai, Japan, Sept. 28Oct. 2, 2004, pp. 23012306.
[23] http://emotion.inrialpes.fr/people/martinel/Video.zip

Planning Motion in Environments with Similar Obstacles

Jyh-Ming Lien and Yanyan Lu

George Mason University, Fairfax, Virginia 22030

{jmlien, ylu4}@gmu.edu

Abstract—In this work, we investigate solutions to the following question: Given two motion planning problems $\mathcal{W}_1$ and $\mathcal{W}_2$ with the same robot and similar obstacles, can we reuse the computation from $\mathcal{W}_1$ to solve $\mathcal{W}_2$ more efficiently? While the answer to this question can find many practical applications, all current motion planners ignore the correspondences between similar environments. Our study shows that by carefully storing and reusing the computation we can gain significant efficiency.

I. INTRODUCTION

In our everyday life, we face motion planning problems. We know how to navigate in our environment from the experiences learned since our childhood. The learning process may be complex but one of the reasons that we can learn such tasks is that most objects we encounter today are identical or similar to the objects we encountered yesterday or even years ago. That is, we as human beings, remember how to navigate around or manipulate similar objects using similar strategies.

 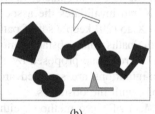

(a)	(b)

Fig. 1.   Two similar workspaces sharing the same robot and several similar obstacles. The white and gray objects are the start and goal configurations. Similar workspaces are usually viewed as completely different problems.

In this work, we propose a method that mimics this simple observation. The goal of our work is to solve a motion planning problem, e.g., the problem in Fig. 1(b), more efficiently by reusing the computation from other similar problems, e.g., the problem in Fig. 1(a). It is important to note that the similarity in this paper is defined by the shapes of the (C-space) obstacles, not by the similarity of the free C-space.

Similar environments are common. For example, desks and chairs in a classroom or in an office may be moved around from one place to another frequently, but new items are seldom introduced. Even different environments, such as two apartments or a manufacturing factory and an airport garage, may share many similar items. The main differences are usually the arrangements. Similar environments can also be found in simulated reality, e.g., in different levels of a video game or in the different regions of a VR world, where many objects

are intentionally duplicated to reduce the (e.g., modeling and rendering) complexity. A planner that exploits the similarity between its workspaces, such as the one proposed in this paper, can provide significant efficiency.

This paper also attempts to address a closely related but slightly different question: how much pre-processing can be done to solve a family of motion planning problems with similar environments? If a significant portion of the computation can be done offline, we can pre-process a very large set of geometric models (e.g., all the public 3-d models on the internet) and store the computation in a database for fast retrieval. An important consequence of this is that almost all (either known or unknown) motion planning problems can be solved more efficiently.

In this paper, we propose a motion planner (ReUse-based PRM or simply Ru-Prm) that constructs local roadmaps around geometric models and stores the local roadmaps in a database. When an environment is given, the planner will match the obstacles in the environment to the existing models in the database. Then the roadmap associated with the matched models is transformed and merged into a global roadmap. In essence, our planning method solves the motion problems by reconstructing the given environment from its "memory" (i.e., the existing computation in the database).

Although the proposed method is simple and extends from the existing PRM-based frameworks, Ru-Prm is unique in several ways. The most critical distinction is that much "reusable" computation is ignored by the existing planners. Given two workspaces with identical models (obstacles) but with different arrangements of these obstacles (e.g., Fig. 1), all existing planners treat these two workspaces as two distinct problems and completely ignore the correspondences between them. More detailed comparison to the related work is in Section II. In addition to the Ru-Prm planner, we also propose and develop a new shape matching method (in Section V-D and the Appendix) that allows Ru-Prm to perform sub-part matching and roadmap transformation more easily.

To the best of our knowledge, this is the first work dealing with similar environments. Although we consider this paper as preliminary work that provides a proof of concept in this new research direction, our experimental results are very encouraging. Our experimental results are very encouraging and show significant efficiency improvement (by 1~3 orders of magnitude faster in most of the studied environments) over the existing PRM planners.

II. RELATED WORK

We are not aware of any planners that consider similarities among motion planning problems. There exist PRM planners that identify and learn *features* in C-free but their goal is to determine sampling strategies [1, 2, 3, 4, 5] using machine learning techniques. Given similar environments, these methods still build the roadmaps from scratch. There are also methods that pre-compute and reuse configurations for highly constrained systems (e.g., closed-chain [6]) in which feasible configurations are usually difficult to obtain. These methods do not consider similarities among motion planning problems.

Our method can also be viewed as a type of "self-improving" algorithm [7, 8]. Existing self-improving planners consider the performance for a single environment with multiple queries, e.g., [9, 10], but do not consider the performance improvement across different environments.

In the rest of this section, we will review related work in motion planning. We classify these work into methods that deal with static and dynamic environments, respectively. In our discussion, we will focus on the difficulty of applying the existing work directly to efficiently plan motions in similar environments. From our reviews, we notice that all existing methods cannot efficiently handle the problem of similar environments. In similar environments, even though the robot and the obstacles may remain the same, the arrangement of the obstacles can be totally different, all existing methods essentially treat them as distinct problems.

A. Static Environments

In this paper, we focus on the problems with static environments. The problem of motion planning in static environments has been studied extensively. It is well known that any *complete motion planners* [11], which always return a solution as long as there is a path or no if there is not, are unlikely to be practical. During the two last decades, researchers have focused on *probabilistically complete* planners, e.g., PRM [12, 13, 14], EST [15], and RRT [16], which are easy to implement and are capable of solving challenging problems. A complete review of these methods can be found in [17, 18].

Briefly, the classic PRM works by randomly generating collision-free configurations and then connecting them using local planners to form a roadmap in C-space [12]. Due to the uniform sampling strategy used by the classic PRM, the roadmap may not capture the connectivity of C-free. This is known as the "narrow passage problem." Therefore, many of the PRM variants that deal with static environments focus on problems with narrow passages. Strategies have been proposed to increase the chance of creating samples inside the narrow passage [13, 19, 20, 21]. Difficult problems, such as the alpha puzzle, become the center of study in these works. Nevertheless, these difficult problems are usually created artificially for testing purposes, and in many real-life problems, such as planning motion in a factory or in a virtual world, are usually easier and do not contain very narrow passages. Therefore, instead of focusing on these difficult but rare problems, our work attempts to increase the planner efficiency for the easier but more commonly seen problems. Our method is based on extending the existing work of PRMs with the functionality of reusing computations among similar environments.

B. Changing Environments

In changing environments, obstacles become movable [22] or even deformable [23]. The problem of changing environments can be considered as a 'continuous' version of the similar environments. If we take snapshots of the changing environments, the consecutive shots are a set of similar environments with small changes.

Many methods have been proposed to handle changing environments. These methods usually combine known techniques to capture both globally static and locally dynamic connectivity of C-free and can be categorized into two main frameworks: state-time space and hybrid methods. The state-time space framework is (probabilistically) complete but requires perfect knowledge of obstacle velocities and trajectory. On the other hand, the hybrid methods are more suitable for on-line planning. For example, Fiorini and Shiller [24] combine the idea of C-space with velocity obstacle and Petti and Fraichard [25] incorporate the idea of inevitable collision zone. More recent work usually uses PRMs to compute the global roadmap. In order to quickly reflect the dynamic changes in the global roadmap, Jaillet and Simeon [26] incorporate RRT to connect disconnected regions, van den Berg et al. [27] propose the idea of D*, and Leven and Hutchinson [28] construct a regular grid in workspace that maps each of its cell to the roadmap nodes and edges. Once obstacles move, [28] quickly checks occupied cells and invalidates the associated nodes and edges. Vannoy and Xiao [29] keep a set of paths and update the fitness values of the paths when obstacles move. Our method is closer to the Elastic roadmap proposed by Yang and Brock [30], where the configurations are sampled around obstacles and are moved along with the obstacles.

Most of these methods, either explicitly or implicitly, depend on the idea that the moving obstacles do not significantly affect the roadmap connectivity (at least for a short period of time), thus the planner should be able to quickly repair and replan. This assumption becomes unrealistic when obstacles are totally rearranged.

III. OUR METHOD

In either static or dynamic environments, all existing methods compute the entire roadmap from scratch once the robot is placed in a new workspace. This is because, even though similar environments are composed of similar obstacles, a roadmap capturing the connectivity of a particular environments is unlikely to be reusable in other environments.

On the contrary, RU-PRM represents its "mental image" as a roadmap constructed for each obstacle (called ob-map) independently. For an unknown environment, RU-PRM also provides a mechanism to *transform* the ob-maps to reflect the new arrangement of the obstacles. Finally, RU-PRM solves motion planning problems by *composing* a global roadmap from all ob-maps in the new environment.

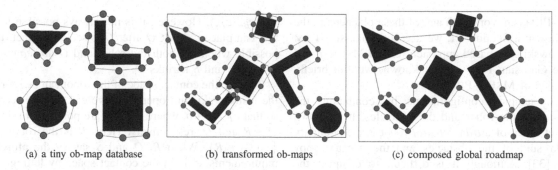

(a) a tiny ob-map database   (b) transformed ob-maps   (c) composed global roadmap

Fig. 2. (a) A database of pre-computed ob-maps. (b) Transformed ob-maps in a new environment. (c) A global roadmap is composed from the ob-maps.

More specifically, given a motion planning problem (O, R, Q), where O is a set of obstacles, R is the robot, and Q is a query, RU-PRM first looks for an existing computation (ob-map) for each obstacle $O_i \in O$. If the same or a similar obstacle O'_i for O is found in the database, we reuse the computation by transforming the ob-map constructed for O'_i. If there is no such an obstacle found in the database, we compute an ob-map for O_i and store it in the database. To construct ob-maps, we use the idea of obstacle-based PRMs [19, 13], which sample configurations on or near the surface of the C-space obstacles (C-obst). Once all the ob-maps are either loaded and transformed or constructed for all the obstacles in the workspace, these ob-maps are merged into a global roadmap, which then is used to answer the query Q. Fig. 2 illustrates how RU-PRM works. Algorithm III.1 summarizes the main steps of RU-PRM.

Algorithm III.1: RU-PRM(O, R, Q)

comment: Obstacles O, Robot R and query Q

for each $O_i \in O$

$\mathbf{do} \begin{cases} \textbf{if } \nexists M_{O_i}, \text{ an ob-map of } O_i \\ \quad \textbf{then } \begin{cases} \text{Create}(M_{O_i}) \\ \text{Store}(M_{O_i}) \end{cases} \\ \text{Read}(M_{O_i}) \\ \text{Transform}(M_{O_i}) \end{cases}$

$M_O \leftarrow \text{Merge}(\cup_i \{M_{O_i}\})$

$\text{Query}(M_O, Q)$

In Algorithm III.1, the sub-routines Store($\cdot$), Read($\cdot$), and Query($\cdot$) are straightforward. The other three main sub-routines: Create($\cdot$), Transform($\cdot$) and Merge($\cdot$) will be discussed in detail in Section IV.

For the rest of this section, we will provide a short discussion on the necessary conditions and assumptions that make RU-PRM efficient. When a given problem does not satisfy these assumptions, RU-PRM degrades to the traditional PRM.

A. Assumptions

The benefits provided by our method is based on the following assumptions and conditions.

1) The robot remains the same in the similar environments.

2) The unknown workspace $\mathcal{W}_i$ has high correspondences to other known workspaces $\mathcal{W}_1 \cdots \mathcal{W}_{i-1}$.

3) A large storage space is available to store all ob-maps.

4) Pre-processing time is available.

These assumptions are general enough to cover many practical situations. As we have pointed out earlier, the first two assumptions are from the observation that many motion planning problems in real life and in virtual worlds share many similar items in their workspaces. Moreover, the type and the number of the robots used in these environments, e.g., characters in a game or robot arms in a factory, are usually limited and do not change often.

Other assumptions are also supported by the current technologies. Most off-the-shelf hard-drive disc with Tera-byte capacity can be obtained for just a few hundred US dollars. Multi-core processors are becoming cheaper and allow more background computation for creating a database of ob-maps. In addition, due to the advances of modeling software and digitizing techniques, constructing complex geometric models becomes easier than ever. Several 3-d geometric databases, e.g., [31], that contain thousands of geometric models, are also available to the public and provide us a set of common objects, e.g., desks, tables, and chairs, to bootstrap the proposed planner.

RU-PRM combines all these ingredients and reuses the computation from the pre-processed geometric models. We envision that our method will allow these computation to be shipped with a new robot or a new character (e.g., in a video game). As far as we know this is the first work dealing with this types of problems. In this paper, we present a preliminary work to provide a proof of the concept in this research direction.

IV. CREATE, TRANSFORM AND MERGE OB-MAPS

In this section, we present three main sub-routines for RU-PRM in their basic forms. Advanced techniques for optimizing the efficiency of these operations will be discussed in the next section (Section V).

A. Create Ob-maps

Creating roadmaps around C-obst allows us to reuse the computation when the same or similar obstacles are given. We call these roadmaps ob-maps. There exist several planners creating ob-maps, e.g., OBPRM [13] and Gaussian PRM [19].

Although RU-PRM can work with any of these planners, in the experiment shown in Section VI, we use both Gaussian PRM and the Minkowski sum based approach (called MSUM-PRM) [32]. Because Gaussian PRM is well known, we will briefly describe the idea of MSUM-PRM.

MSUM-PRM samples configurations by computing the *Minkowski sums* of the robot and the obstacles. It is known that the contact space of a *translational* robot is the boundary of Minkowski sum of the obstacles and the negated copy of the robot [33]. Although it is difficult to compute the exact Minkowski sums of polyhedra [34], we have shown that sampling points from the Minkowski sum boundary without explicitly generating its mesh boundary, can be done much more easily and efficiently [35]. To handle the case of non-translational robots, we simply draw samples from a set of Minkowski sums, each of which is constructed by randomly assigning different orientations and joint angles to the robot.

In [32], we have shown that MSUM-PRM generates a free configuration significantly faster than OBPRM and Gaussian PRM. We have also demonstrated that these configurations can be connected into roadmap using more powerful local planners based on the geometric properties of the Minkowski sum. We refer the interested reader to [32] for details regarding this sampling strategy.

MSUM-PRM provides an important advantage for RU-PRM that is missing from the other obstacle-based samplers. A configuration created by MSUM-PRM can be transformed easily because the configuration is represented by a pair of points from the robot and an obstacle. Details of the transformation operation will be discussed in the next section.

B. Transform Ob-maps

When an unknown obstacle X is matched to an existing model O by translating, rotating, and scaling O, we consider how these transformations can affect O's ob-map (denoted as M_O) for free-flying robots. More precisely, we seek methods that transform M_O to approximate the ob-map of X.

We let CO be the C-obst of O. We further let T, R, and S be the translation, rotation and uniform scale applied to O, respectively. Given a configuration c from ∂CO, our goal is to obtain the corresponding configuration c' of c so that c' is in ∂CO of O transformed using T, R and S.

To ease our discussion, we assume that (1) when we generate M_O, the center of O is placed at the origin of the coordinate system and (2) the robot is a free-flying articulated robot. The configuration c is composed of three components (c_T, c_{BR}, c_{JR}), where c_T and c_{BR} are the values of the translational and rotational degrees of freedom (dof) for the base, respectively. and c_{JR} represents the rotational dof of the joints.

When we consider only the translation T and rotation R, obtaining the corresponding c' is straightforward, i.e.,

$$c' = (T + R \cdot c_T, R \cdot c_{BR}, c_{JR}) . \tag{1}$$

Note that T and R have no effect on c_{JR}.

When we consider the scale S, the only component of c affected by S is c_T. Therefore, c' will have the form of (c'_T, c_{BR}, c_{JR}). However, it is not always possible to obtain c'_T that can place c' in ∂CO after O is scaled. In fact, this is only possible when (1) both the robot and O are *convex* or (2) O is convex and S shrinks O.

Because the contact space of the robot can be represented by the Minkowski sum operation, we can always decompose c_T so that $c_T = r + o$, where r and o are points from the boundary of $-R$ and O, respectively. Here, R is the robot and $-R = \{-x \mid x \in R\}$. When R, O and S satisfy the aforementioned requirements, c' is in the contact space by letting

$$c'_T = r + S \cdot o . \tag{2}$$

If O and R are not convex or if O is convex but S enlarges O, we can work around this problem using convex decomposition. However, convex decomposition can slow down the computation significantly and can generate a lot of components. Therefore, instead of decomposing exactly, we use approximate convex decomposition [36] and represent the model using the convex hulls of the components in the decomposition. Details of this approach is discussed in the Appendix when we describe our shape matching method.

Alternatively, we can apply Eq. 2 to non-convex shapes. The consequence of this is that c' may make the robot collide with $S(O)$. Therefore, collision detection is used to check the feasibility of every transformed configuration.

C. Merge Ob-maps

So far, we have treated each obstacle independently. After the pre-processing step that either loads or generates ob-maps, we proceed to compose a global roadmap.

The configurations in a (transformed) ob-map, although are near the surface of the associated obstacle, may be outside the bounding box or colliding with other obstacles. We validate each configuration in an ob-map using a collison detector. For edges connecting two collision-free configurations, we evaluated them in a lazy manner [37], i.e., we only check the feasibility of the edges in the extracted paths during the query phase.

After all configurations are verified, we merge the ob-maps pairwisely. For every pair of obstacles O_i and O_j in the workspace, we connect their ob-maps by simply adding edges for k pairs of the closest configurations. Each pair consists of a configuration from the ob-map of O_i and a configuration from the ob-map of O_j. Again, we do not check the feasibility of these edges, although some edges may collide with the obstacles.

Once the global roadmap is constructed, we iteratively extract and evaluate new paths and delete invalid edges from the roadmap until a collision-free path is found, otherwise "no solution" will be reported.

V. PLANNER OPTIMIZATION

In the previous section, the basic framework of RU-PRM is described. In this section, we will present several optimization techniques to improve the efficiency of the framework. We also present a new shape matching method using approximate convex decomposition.

A. Create Ob-maps with Quality Control

Like most PRM-based methods, the quality of the roadmap generated by MSUM-PRM also depends on the number n of configurations sampled. The value of n is usually provided by the user based on the "difficulty of the problem." However, when we consider similar workspaces, their difficulties may not be known at the ob-map creation time. Therefore, requiring users to specify the value of n not only puts too much burden on the users, but it is also unrealistic and impractical.

Xie et al. [38] have attempted to remove this limitation from PRMs by incrementally increasing the resolution of a roadmap. Their method adds and connects additional nodes to the roadmap until certain criteria are met. We follow the same approach to avoid specifying the value of n. In each iteration of our map generation process, we add and connect n_i new configurations to the existing ob-map (which is initially empty) using MSUM-PRM. Then, we check the number n_{CC} of connected components that comprise $m\%$ of the configurations in the map. If n_{CC} is decreasing for the last z consecutive iterations, we stop. Note that, although it seems that we introduce more parameters (n_i, m and z) by constructing ob-maps incrementally, these parameters are easier to set and usually remain fixed regardless the workspace.

We realize that this simple heuristic method may not handle all possible scenarios properly. However, from the results of our experiments, the ob-maps generated by this method cover the obstacles well, and, in most examples, ob-maps usually contain a single connected component.

B. Transform Ob-maps Hierarchically

An ob-map may still contain many nodes. Since we do not know in advance if all these nodes are necessary. For example, in an open workspace, only few of these nodes are needed to solve the problem. Transforming and validating all the nodes in the ob-map significantly slows down the computation. However, in a cluttered workspace, it is necessary to keep all the nodes in order to capture the global connectivity.

To address this issue, we organize the roadmap nodes into a hierarchy and process these nodes in the same manner. More specifically, we compute the core of an ob-map, which is a sub-graph that maintains the same connectivity as the ob-map. In a new workspace, all the nodes in the core will be transformed and their feasibility will be checked. When the core does not have the desired connectivity, we repair the core by transforming additional nodes from the ob-map.

The core is constructed by determining and removing the redundant nodes in the ob-map. Given a pair of adjacent nodes u and v in the ob-map, we say that v is redundant if the roadmap contains the same number of connected components after (1) removing v and (2) connecting u to v's adjacent nodes without colliding with the associated obstacle. Finally, the core is simply a sub-graph of the ob-map without redundant nodes. We store the core with the ob-map in the database.

In a new workspace, the core is loaded from the database first. After transformation and evaluation, the core may be split into many connected components due to the removal of

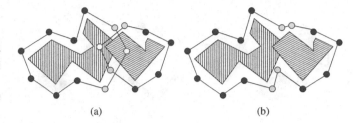

Fig. 3. Merge the ob-maps of two overlapping obstacles. (a) Before merging. Free, collision, and boundary nodes are shown in dark gray, white and light gray, respectively. (b) After merging, the boundary nodes are connected.

in-collision nodes. When this happens, more nodes are added to the core from the original ob-map. Let CC_1 and CC_2 be two connected components in the core, we improve the core's connectivity by finding a path in the ob-map that connects a pair of nodes from CC_1 and CC_2. This process repeats until both the core and the ob-map have the same number of connected components or when no path in the ob-map can be found.

C. Merge Ob-maps Using Boundary Nodes

The merging operation can also be optimized by classifying the relationships of the C-space obstacles. Let O_i and O_j be a pair of obstacles in the workspace, we can classify their relationships in C-space based on the feasibility of the configurations in their ob-maps, M_{O_i} and M_{O_j}. More specifically, we identify the *boundary nodes* in the ob-maps. Let $B_{\{O_i,O_j\}}$ be a set of boundary nodes in M_{O_i} w.r.t. O_j. For each configuration $c \in M_{O_i}$, we say $c \in B_{\{O_i,O_j\}}$ if an adjacent node of c makes the robot collide with O_j. Fig. 3 shows an example of the boundary nodes. Note that $B_{\{O_i,O_j\}} \neq B_{\{O_j,O_i\}}$ unless they are both empty. We use the boundary nodes to classify and connect the ob-maps. There are three cases to consider:

1) M_{O_i} and M_{O_j} are far away from each other, i.e., $B_{\{O_i,O_j\}} = B_{\{O_j,O_i\}} = \emptyset$.
2) M_{O_i} and M_{O_j} overlap, i.e., $B_{\{O_i,O_j\}} \neq \emptyset$ and $B_{\{O_j,O_i\}} \neq \emptyset$.
3) M_{O_i} and M_{O_j} are near each other, i.e., $B_{\{O_i,O_j\}} \neq \emptyset$ and $B_{\{O_j,O_i\}} = \emptyset$ or vice versa.

Case 1. When the C-obsts of O_i and O_j are far away from each other, we connect their ob-maps using the method described in Section IV-C, which simply connects the k-closest pairs between the ob-maps.

Case 2. This is the situation that the C-obsts of O_i and O_j overlap. The ob-maps are connected by adding edges between $B_{\{O_i,O_j\}}$ and $B_{\{O_j,O_i\}}$. Fig. 3 shows an example in this case.

Case 3. This situation requires us to combine the techniques for **Case 1** and **Case 2**. For $B_{\{O_i,O_j\}} \neq \emptyset$ and $B_{\{O_j,O_i\}} = \emptyset$, ob-maps are connected by adding edges between $c_i \in B_{\{O_i,O_j\}}$ and its k closest nodes $c_j \in M_{O_j}$. For $B_{\{O_i,O_j\}} = \emptyset$ and $B_{\{O_j,O_i\}} \neq \emptyset$, ob-maps are merged in the same manner.

D. Shape Matching Using Approximate Convex Decomposition

So far, we have assumed that the same obstacle can be found in the database. To reuse ob-maps for different but

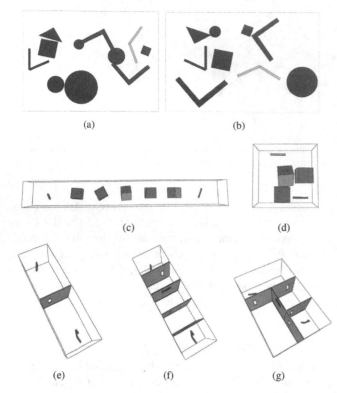

(a)   (b)

(c)   (d)

(e)   (f)   (g)

Fig. 4. Environments used in the experiments. (a, b) 2-d workspaces with an articulated robot. (c, d) Similar workspaces with a rigid robot and cubes. (e, f, g) Similar workspaces with a U-shaped rigid robot and walls.

similar obstacles, we propose a new shape matching method. Essentially, our matching method estimates the dissimilarity of two objects by measuring their morphing distance. The proposed matching method is composed of two main steps: (1) decomposing shapes using approximate convex decomposition (ACD), and (2) computing dissimilarity using morphing and bipartite matching. Details of the shape matching method and some preliminary results are discussed in the Appendix.

VI. IMPLEMENTATION AND EXPERIMENTAL RESULTS

In this section, we show experimental results. All the experiments are performed on a PC with two Intel Core 2 CPU at 2.13 GHz with 4 GB RAM. Our implementation is coded in C++. Our current implementation only supports free-flying robots and 2-d shape matching.

We test RU-PRM in three sets of similar environments (see Fig. 4). We assume that the robots and the obstacles in these environments are known, therefore we can pre-process them and store their ob-maps in a database. We use two types of RU-PRMs in our experiments: (1) RU-PRM with Gaussian PRM and (2) RU-PRM with MSUM-PRM. During the pre-processing step, RU-PRM with Gaussian PRM takes about 4.9, 96, and 382 seconds to create and store the ob-maps for the obstacles in Env. (a-b), Env. (c-d), and Env. (e-g), respectively. RU-PRM with MSUM-PRM takes about 35 and 131 seconds to create and store the ob-maps for the obstacles in Env. (c-d) and Env. (e-g), respectively. Because currently MSUM-PRM has not been implemented to handle 2-d workspaces, RU-PRM with

MSUM-PRM is not used in Env. (a-b). In these experiments, we compare RU-PRMs against several well-known planners: MSUM-PRM [32], Uniform [12], Gaussian [19] and Visibility PRMs [39].

RU-PRM is significantly faster The running times for all three environments are shown in Fig. 5. Notice that the y-axes of the charts in Fig. 5 are in logarithmic scale. The running time for RU-PRMs includes the time for reading the ob-maps from the hard disk drive, as well as the time for transforming, evaluating and merging the ob-maps and the time for performing the query in the global roadmap.

From the results, RU-PRM shows significant efficiency improvement in *all* studied environments. Env. (a-b) and Env. (c-d) are simple environments where the Uniform PRM usually solves the problems with a few hundred nodes and outperforms Gaussian and Visibility PRMs and MSUM-PRM. In these simple environments, RU-PRM with MSUM-PRM or Gaussian PRM still provides noticeable improvements (up to 100 times faster) over the Uniform PRM (and therefore over Gaussian PRM only and MSUM-PRM only). This evidence demonstrates the strength of reusing computation. These plots also shows that combining RU-PRM with either MSUM-PRM or Gaussian PRM does not seem to affect its performance in simple environment, however, the difference becomes more noticeable in more difficult environments.

In more difficult environments, e.g., Env. (e-g) that contains narrow passages, RU-PRM with MSUM-PRM or Gaussian is significantly faster (by 5∼8 orders of magnitude) than Uniform and Gaussian PRMs and is still significantly faster than using MSUM-PRM or Gaussian only (by 2∼5 orders of magnitude). A possible source of the improvement is from MSUM-PRM, which has been shown to be better than the classic PRMs [32], however, our results also show that combining RU-PRM with MSUM-PRM further improves MSUM-PRM. The main reason for this performance improvement is obvious. Obtaining connectivity in the free C-space is usually the most time-consuming step in PRMs, RU-PRM saves significant amount of time by reusing the connectivity provided by the ob-maps.

VII. CONCLUSION

In this paper, we proposed the first work considering motion planning in similar environments. We developed a new method called RU-PRM that reuses the computation from the previously solved problems. Essentially, RU-PRM stores the local roadmap built around each C-obst. When a new environment is given, RU-PRM matches the obstacles and loads the matched roadmaps. These roadmaps are transformed, evaluated and finally merged to solve the queries in the new environments. We discussed several optimization techniques that improve the efficiency of the basic RU-PRM framework. We also proposed a new shape matching method that allows RU-PRM to do sub-part matching and roadmap transformation more easily.

Although we consider this paper as a preliminary work that provides a proof of the concept in this new research direction, our experimental results are very encouraging and

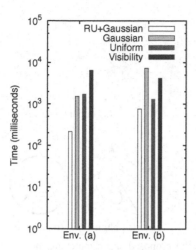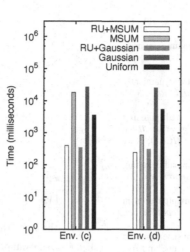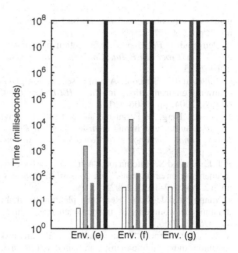

Fig. 5. Experimental results for the environments in Fig.4. Notice that the y-axes are all in logarithmic scale and Envs. (c)-(g) share the same legend. We stop the planner when it takes more than 10^8 milliseconds. The running times for all PRMs are collected over 30 runs.

show significant efficiency improvement (by 5~8 orders of magnitude faster) over the classic PRM planners (including Uniform, Gaussian, and Visibility PRMs) and is faster then MSUM-PRM by 3~5 orders of magnitude.

Limitations and future work. Many aspects in the proposed work can be improved. One of the most critical bottlenecks of our current implementation is the efficiency of the shape matching method. We also plan to investigate better criteria for evaluating the connectivity of the ob-map and its core.

REFERENCES

[1] M. Morales, L. Tapia, R. Pearce, S. Rodriguez, and N. M. Amato, "A machine learning approach for feature-sensitive motion planning," in *Proc. Int. Workshop Alg. Found. Robot.(WAFR)*, Utrecht/Zeist, The Netherlands, July 2004, pp. 361–376.

[2] B. Burns and O. Brock, "Toward optimal configuration space sampling," in *Proc. Robotics: Sci. Sys. (RSS)*, 2005.

[3] A. Yershova, L. Jaillet, T. Simeon, and S. M. Lavalle, "C-space subdivision and integration in feature-sensitive motion planning," in *Proc. of IEEE Int. Conf. on Robotics and Automation*, April 2005, pp. 3867–3872.

[4] S. Rodriguez, S. Thomas, R. Pearce, and N. M. Amato., "(resampl): A region-sensitive adaptive motion planner," in *Proc. Int. Workshop Alg. Found. Robot.(WAFR)*, July 2007.

[5] S. Finney, L. Kaelbling, and T. Lozano-Perez, "Predicting partial paths from planning problem parameters," in *Proceedings of Robotics: Science and Systems*, Atlanta, GA, USA, June 2007.

[6] L. Han and N. M. Amato, "A kinematics-based probabilistic roadmap method for closed chain systems," in *Robotics:New Directions*. Natick, MA: A K Peters, 2000, pp. 233–246, book contains the proceedings of the International Workshop on the Algorithmic Foundations of Robotics (WAFR), Dartmouth, March 2000.

[7] N. Ailon, B. Chazelle, S. Comandur, and D. Liu, "Self-improving algorithms," in *SODA '06: Proceedings of the seventeenth annual ACM-SIAM symposium on Discrete algorithm*. New York, NY, USA: ACM, 2006, pp. 261–270.

[8] K. L. Clarkson and C. Seshadhri, "Self-improving algorithms for delaunay triangulations," in *SCG '08: Proceedings of the twenty-fourth annual symposium on Computational geometry*. New York, NY, USA: ACM, 2008, pp. 148–155.

[9] T.-Y. Li and Y.-C. Shie, "An incremental learning approach to motion planning with roadmap management," in *Proc. of IEEE Int. Conf. on Robotics and Automation*, 2002, pp. 3411–3416.

[10] R. Gayle, K. R. Klingler, and P. G. Xavier, "Lazy reconfiguration forest (LRF): An approach for motion planning with mulitple tasks in dynamic environments," in *Proc. of IEEE Int. Conf. on Robotics and Automation*, 2007, pp. 1316–1323.

[11] J. F. Canny, *The Complexity of Robot Motion Planning*. Cambridge, MA: MIT Press, 1988.

[12] L. E. Kavraki, P. Svestka, J. C. Latombe, and M. H. Overmars, "Probabilistic roadmaps for path planning in high-dimensional configuration spaces," *IEEE Trans. on Robotics and Automation*, vol. 12, no. 4, pp. 566–580, August 1996.

[13] N. M. Amato, O. B. Bayazit, L. K. Dale, C. V. Jones, and D. Vallejo, "OBPRM: An obstacle-based PRM for 3D workspaces," in *Robotics: The Algorithmic Perspective*. Natick, MA: A.K. Peters, 1998, pp. 155–168, proc. Third Workshop on Algorithmic Foundations of Robotics (WAFR), Houston, TX, 1998.

[14] N. M. Amato and Y. Wu, "A randomized roadmap method for path and manipulation planning," in *Proc. of IEEE Int. Conf. on Robotics and Automation*, 1996, pp. 113–120.

[15] D. Hsu, J.-C. Latombe, and R. Motwani, "Path planning in expansive configuration spaces," *Int. J. Comput. Geom. & Appl.*, pp. 2719–2726, 1997.

[16] S. M. LaValle and J. J. Kuffner, "Rapidly-Exploring Random Trees: Progress and Prospects," in *Proc. Int. Workshop Alg. Found. Robot.(WAFR)*, 2000, pp. SA45–SA59.

[17] R. Geraerts and M. H. Overmars, "Sampling techniques for probabilistic roadmap planners," in *Intelligent Autonomous Systems (IAS)*, 2004, pp. 600–609.

[18] S. M. LaValle, *Planning Algorithms*, 6th ed. Cambridge University Press, 2006.

[19] V. Boor, M. H. Overmars, and A. F. van der Stappen, "The Gaussian sampling strategy for probabilistic roadmap planners," in *Proc. of IEEE Int. Conf. on Robotics and Automation*, vol. 2, May 1999, pp. 1018–1023.

[20] S. A. Wilmarth, N. M. Amato, and P. F. Stiller, "MAPRM: A probabilistic roadmap planner with sampling on the medial axis of the free space," in *Proc. of IEEE Int. Conf. on Robotics and Automation*, vol. 2, 1999, pp. 1024–1031.

[21] D. Hsu, T. Jiang, J. Reif, and Z. Sun, "Bridge test for sampling narrow passages with proabilistic roadmap planners," in *Proc. of IEEE Int. Conf. on Robotics and Automation*, 2003, pp. 4420–4426.

[22] J. van den Berg and M. Overmas, "Roadmap-based motion planning in dynamic environments," in *Proc. IEEE Int. Conf. Intel. Rob. Syst. (IROS)*, 2004, pp. 1598–1605.

[23] S. Rodriguez, J.-M. Lien, and N. M. Amato, "Planning motion in completely deformable environments," in *Proc. of IEEE Int. Conf. on Robotics and Automation*, May 2006, pp. 2466–2471.

[24] P. Fiorini and Z. Shiller, "Motion planning in dynamic environments

using velocity obstacles," *Int. Journal of Robotics Research*, vol. 17, no. 7, pp. 760–772, 1998.

[25] S. Petti and T. Fraichard, "Safe motion planning in dynamic environments," in *Proc. IEEE Int. Conf. Intel. Rob. Syst. (IROS)*, 2005, pp. 2210–2215.

[26] L. Jaillet and T. Simeon, "A prm-based motion planner for dynamically changing environments," in *Proc. IEEE Int. Conf. Intel. Rob. Syst. (IROS)*, 2004, pp. 1606–1611.

[27] J. van den Berg, D. Ferguson, and J. Kuffner, "Anytime path planning and replanning in dynamic environments," in *Proceedings of the IEEE International Conference on Robotics and Automation (ICRA)*, May 2006, pp. 2366 – 2371.

[28] P. J. Leven and S. Hutchinson, "A framework for real-time path planning in changing environments," *Int. Journal of Robotics Research*, vol. 21, no. 12, pp. 999–1030, 2002.

[29] J. Vannoy and J. Xiao, "Real-time planning of mobile manipulation in dynamic environments of unknown changes," in *Proc. Robotics: Sci. Sys. (RSS)*, 2006.

[30] Y. Yan and O. Brock, "Elastic roadmaps: Globally task-consistent motion for autonomous mobile manipulation in dynamic environments," in *Proc. Robotics: Sci. Sys. (RSS)*, 2006.

[31] M. K. Philip Shilane, Patrick Min and T. Funkhouser, "The princeton shape benchmark," in *Proceedings of the Shape Modeling International*, 2004.

[32] J.-M. Lien, "Hybrid motion planning using Minkowski sums," in *Proc. Robotics: Sci. Sys. (RSS)*, Zurich, Switzerland, 2008.

[33] T. Lozano-Pérez, "Spatial planning: A configuration space approach," *IEEE Trans. Comput.*, vol. C-32, pp. 108–120, 1983.

[34] G. Varadhan and D. Manocha, "Accurate Minkowski sum approximation of polyhedral models," *Graph. Models*, vol. 68, no. 4, pp. 343–355, 2006.

[35] J.-M. Lien, "Point-based minkowski sum boundary," in *PG '07: Proceedings of the 15th Pacific Conference on Computer Graphics and Applications*. Washington, DC, USA: IEEE Computer Society, 2007, pp. 261–270.

[36] J.-M. Lien and N. M. Amato, "Approximate convex decomposition," in *SCG '04: Proceedings of the twentieth annual symposium on Computational geometry*. New York, NY, USA: ACM Press, 2004, pp. 457–458, video Abstract.

[37] R. Bohlin and L. E. Kavraki, "A randomized algorithm for robot path planning based on lazy evaluation," in *Handbook on Randomized Computing*, P. Pardalos, S. Rajasekaran, and J. Rolim, Eds. Kluwer Academic Publishers, 2001, pp. 221–249.

[38] D. Xie, M. A. Morales, R. Pearce, S. Thomas, J.-M. Lien, and N. M. Amato, "Incremental map generation (IMG)," in *Proc. Int. Workshop Alg. Found. Robot.(WAFR)*, July 2006.

[39] C. Nissoux, T. Simeon, and J.-P. Laumond, "Visibility based probabilistic roadmaps," in *Proc. IEEE Int. Conf. Intel. Rob. Syst. (IROS)*, 1999, pp. 1316–1321.

[40] R. C. Veltkamp and M. Hagedoorn, "State of the art in shape matching," *Principles of visual information retrieval*, pp. 87–119, 2001.

APPENDIX

Shape approximation using ACD In the first step, we decompose the polygon P into several approximately convex pieces [36]. A component C in the decomposition is approximately convex if the concavity of C is less than a user pre-defined tolerance. It has been shown that ACD contains much fewer components compared to the exact decomposition and, more importantly, it maintains the key structural features. Because of these advantages, we can represent P using the convex hulls of the components in its ACD. An example of ACD is shown in Fig. 6(a) and (b).

We first compute the dissimilarity between a pair of *convex* components by estimating their *morphing* distance. Let $\{C_i\}$ and $\{D_j\}$ be two sets of convex components obtained from the ACDs of the polygons P and Q. Given two convex components C_i with m vertices (colored in green in Fig. 6(d)) and D_j with n vertices (colored in blue in Fig. 6(d)), we compute the morphing distance as the follows:

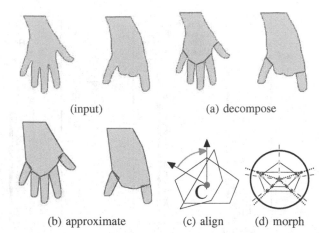

(input)   (a) decompose

(b) approximate   (c) align   (d) morph

Fig. 6.   An example of the proposed shape matching method.

1) Align C_i and D_j by overlapping their centers and their principal axes, (see Fig. 6(c)).
2) Draw m rays from the common center to the vertices of C_i and n rays to those of D_j, (see Fig. 6(d)).
3) Compute the intersections of the rays with vertices C_i and D_j. For each ray, a line segment between the intersection and the vertex is identified.
4) Let the morphing distance of C_i and D_j be the sum of the lengths of all $m + n$ segments.

Once the morphing distances between all pairs of components in $\{C_i\}$ and $\{D_i\}$ are computed, we arrange the components $\{C_i\}$ and $\{D_i\}$ into a bipartite graph and solve the minimum bipartite matching problem.

Note that although we only demonstrate using polygons in this section, the proposed matching method can be extended to handle polyhedra without modification.

Although there exist many shape matching methods (e.g., see survey [40]), this new method provides unique functionalities for Ru-Prm. In particular, it represents shape using a set of convex objects. This not only allows the scale transformation discussed in Section IV-B, but also allows sub-part matching. For example, the ob-map for a hand gesture can be "deformed" to fit around another gesture by transforming the fingers even if the gestures are very different. Due to the space limitation, we will skip the technical details of the proposed method.

The proposed method correctly returns its best match for every test case using 10 randomly selected polygons (shown in Fig.7). Moreover, it takes 70 ms on average for a polygon to find the best match. Therefore, even when we consider the time spent on shape matching, Ru-Prm still outperforms the other three planners.

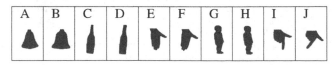

A	B	C	D	E	F	G	H	I	J

Fig. 7.   Matching Results. 10 models used in this shape matching experiments.

96

Robust Visual Homing with Landmark Angles

John Lim[1,2]
[1]Department of Information Engineering, RSISE
Australian National University
john.lim@rsise.anu.edu.au

Nick Barnes[1,2]
[2]Canberra Research Laboratory,
NICTA, Australia
nick.barnes@nicta.com.au

Abstract—**This paper presents a novel approach to visual homing for robot navigation on the ground plane, using only the angles of landmark points. We focus on a robust approach, leading to successful homing even in real, dynamic environments where significant numbers of landmark points are wrong or missing. Three homing algorithms are presented, two are shown to be provably convergent, and the other shown to converge empirically. Results from simulations under noise and robot homing in real environments are provided.**

I. INTRODUCTION

Visual homing is the problem of using information from visual images to return to some goal or home location after an agent has been displaced from that initial location.

Many approaches are possible. If some metric map of the environment was constructed (eg: using laser, sonar, stereo within a SLAM framework [1, 2]), the agent can plan its return path using the map. Alternatively, the estimated camera motion from two views can give the homing direction using computer vision algorithms such as the 8-point algorithm [3]. Visual servoing [4] and other approaches [5, 6] are also possible.

Here, we focus on *local visual homing* methods which are typically used within graph-like *topological maps* [7, 8, 9] where connected nodes represent nearby locations in the environment. A local homing algorithm is then used to drive the agent from one node to the next. Various local homing methods exist, including [10, 11, 12, 13, 14, 15, 16, 17] and many others, which can be found in reviews such as [18]. The elegance and simplicity of these methods make them attractive not only for robotics, but also for modeling homing behavior in insects. (Indeed, many of these algorithms were directly inspired by or aimed at explaining insect homing.)

Using only the angles of landmark or feature points, we follow in the footsteps of other biologically inspired homing methods. These include the snapshot model [16], which matches sectors of the image snapshot taken at the goal position with the sectors in the current image to obtain 'tangential' and 'radial' vectors used to perform homing. The average landmark vector method [11] computes the average of several unit *landmark vectors* (a vector pointing from the agent to some landmark point) and obtains a homing vector by subtracting the average landmark vectors at the current and goal positions. The average displacement model of [15] and the work of [19] are examples of other such methods. These rely on some globally known compass direction in order to perform a correct comparison of landmark vectors

observed at the current and goal positions. More recently, [10, 20] proposed a *compass-free* framework which works for any combination of start, goal and landmark positions on the entire plane. This approach uses a 'basic' and a 'complementary' control law that performs homing with three or more landmarks.

In our work, we assume that correspondences have been found between the landmark points in an image taken at the current position with those in an image taken at the goal location. The robot starts at some point, calculates the homing direction and moves in that direction. This is iterated and the robot moves step by step, until it arrives at the goal location. The landmark points used are image features obtained, matched or tracked with methods such as Scale Invariant Feature Transform (SIFT) matching [22], Harris corners [23], KLT [24] and others. These are 'naturally' occurring landmark points, rather than objects placed in the scene for the sole purpose of being an easily tracked and unmoving landmark. While we focus on local, short-range homing, this can be integrated with a topological map framework [7, 8, 9, 20] to perform long-range homing and localization tasks.

Section II introduces three algorithms. Firstly, a homing algorithm with both the robot and landmark points lying on the same plane is presented and shown to converge theoretically. Next, we investigate the case of planar homing with general, 3D landmark points and present two algorithms - the *conservative* method, which has theoretically guaranteed convergence and a *non-conservative* method which is demonstrated to converge empirically. Implementation details are in Section III and experimental results are in Section IV.

Motivation. We aim to address the issues of:

Robustness to outliers. In general, existing methods assume that a correct correspondence has been found between landmarks seen in the goal and current positions. However, if some of these landmarks have moved, are moving, were occluded, or were erroneously matched, there is often no provision to ensure successful homing. We term such moved or mismatched landmarks as *outliers* in the observed data. We propose a *voting* framework that efficiently integrates information from all landmarks to obtain homing cues that are extremely robust to outliers and other types of measurement noise.

Provable convergence. Homing is not provably successful in most of the cited local homing methods. The work of [10] showed successful homing through extensive simulations. [12] provided the first planar homing algorithm with a proof of convergence. This paper differentiates itself from [12], by proposing provably convergent algorithms for the more general case where the relative rotation between current and goal positions need not be known (compass-free), and where the landmarks need not lie on the same plane as the robot.

II. THEORY

Let the goal or home be a point G and the current location of the robot be the point C. Images of the environment are taken at both points. At each step, the robot uses this image pair to compute and move in some homing direction, $\mathbf{h}$. Hence, C changes location with each iteration and homing is successful when C converges on G. Define convergence as moving the robot to within some neighbourhood of G and having it remain in that neighbourhood. We begin with the observation that:

Observation 1: The robot converges, if at each iteration, it takes a small step in the computed direction $\mathbf{h}$, such that the angle between $\overrightarrow{CG}$ and $\mathbf{h}$ is less than $90°$.

C lies on a circle centered on G with radius $|\overrightarrow{CG}|$. If the movement of the robot is small, it will always move *into* the circle if homing direction $\mathbf{h}$ deviates from $\overrightarrow{CG}$ by less than $90°$ (At exactly $90°$ from $\overrightarrow{CG}$, the robot moves off the circle in a tangent direction). This ensures that each step the robot makes will take it a little closer to the goal (since it is moving into the circle). Hence the robot-goal distance decreases monotonically, and at the limit, it will reach G. (In practice, robot motion is not infinitesimally small, but as long as the angle between $\mathbf{h}$ and $\overrightarrow{CG}$ is not too close to $90°$, the above still holds true.)

A. The Case of the Planar World

Consider the case in which both robot and landmark points lie on a plane. Given 2 landmarks L_1 and L_2 (Figure 1(a)), and the current position at point C, the *inter-landmark angle* is the angle between the two rays $\overrightarrow{CL_1}$ and $\overrightarrow{CL_2}$. All points on circular arc L_1CL_2 observe the same angle $\angle L_1CL_2$. Following [10], we define a convention where angle $\angle L_1CL_2$ is consistently measured, going from $\overrightarrow{CL_1}$ to $\overrightarrow{CL_2}$, in a anticlockwise (or clockwise) direction. Then any point on arc L_1CL_2 will have acute $\angle L_1CL_2$ (or obtuse, if measuring in the clockwise direction). The inter-landmark angle observed at the current position, $\angle L_1CL_2$, is the current angle, whilst that observed at the goal position, $\angle L_1GL_2$, is the goal angle.

The circular arc L_1CL_2 is termed a *horopter*. The dashed landmark line L_1L_2 splits the plane into upper and lower half-planes. Let the set of points on the horopter be R_0; let the region in the upper half-plane and within the horopter be R_1; the region outside the horopter (shaded region in Figure 1(a)) be R_2; and the region in the lower half-plane be R_3.

In the configuration of Figure 1(a) and using the anticlockwise angle measurement convention, if C lies on the horopter, and G lies within R_1, that is $G \in R_1$, then $\angle L_1GL_2 > \angle L_1CL_2$. However, if $G \in R_2$, then $\angle L_1GL_2 < \angle L_1CL_2$. Furthermore, the anticlockwise convention for measuring angles implies that if point $P \in R_1 \cup R_2$, then $\angle L_1PL_2$ is acute, but if $P \in R_3$, then $\angle L_1PL_2$ is obtuse. Therefore,

$C \in R_0 \cup R_1 \cup R_2$ and $G \in R_3$, implies $\angle L_1CL_2 < \angle L_1GL_2$ since acute angles are always smaller than obtuse ones.

The acute-obtuse cases are reversed if C lies on the other side of the line L_1L_2 (the lower half-plane). Angle $\angle L_1CL_2$ is now acute when measured in the clockwise direction and obtuse if measured in the anticlockwise. An analysis will yield relationships symmetrical to the above.

With this, given knowledge of whether the current or the goal angle is larger, one can constrain the location of G to one of regions R_1, R_2, R_3. However, since distance from the current point to the landmarks is unknown, we know neither the structure of the horopter nor that of the line L_1L_2. Even so, one can still obtain valid but weaker constraints on the location of G purely from the directions of the landmark rays.

These constraints on G come in two types. Type 1: Define a region R_{A1} that lies between vectors $\overrightarrow{CL_1}$ and $\overrightarrow{CL_2}$ (shaded region in Figure 1(b)). Let the complement of R_{A1} be $R_{B1} = R_{A1}^c = \Pi \setminus R_{A1}$, where Π is the entire plane and $\setminus$ denotes set difference. R_{B1} is the Type 1 constraint region (the entire unshaded region in Figure 1(b)) and G must lie within R_{B1}.

Type 2: Let R_{A2} lie between the vectors $-\overrightarrow{CL_1}$ and $-\overrightarrow{CL_2}$ (shaded region in Figure 1(c)). The Type 2 constraint region is the complement set, $R_{B2} = R_{A2}^c$ (the unshaded region).

Lemma 1: If an acute current angle is greater than the goal angle, then $G \in R_{B1}$ but if it is less than the goal angle, then $G \in R_{B2}$. If an obtuse current angle is greater than the goal angle, then $G \in R_{B2}$ but if it is less, then $G \in R_{B1}$.

Proof: If $\angle L_1CL_2$ is acute and $\angle L_1CL_2 > \angle L_1GL_2$, the goal, G, must lie in R_2. R_{A1} is the shaded area in Figure 1(b), which does not intersect R_2. One can see that for any C lying on the horopter, R_{A1} never intersects R_2. R_2 is a subset of R_{B1}. Hence, $G \in R_2 \Rightarrow G \in R_{B1}$.

Conversely, if $\angle L_1CL_2$ is acute and $\angle L_1CL_2 < \angle L_1GL_2$, then $G \in R_1 \cup R_3$ and $G \notin R_2$. R_{A2} is the shaded area in Figure 1(c) and $R_{A2} \cap (R_1 \cup R_3) = \emptyset$ for any point C on the horopter. So, $G \in (R_1 \cup R_3) \subset R_{B2}$.

For obtuse $\angle L_1CL_2$ (when using the clockwise convention), the proof is symmetrical to the above (that is, if $\angle L_1CL_2 > \angle L_1GL_2$, then $G \in R_1 \cup R_3$ and $(R_1 \cup R_3) \subset R_{B2}$ so $G \in R_{B2}$. Also, if $\angle L_1CL_2 < \angle L_1GL_2$, then $G \in R_2$ and $R_2 \subset R_{B1}$ so $G \in R_{B1}$). ∎

From the constraints on G, we can now define constraints on $\overrightarrow{CG}$. Some region, R, consists of a set of points, $\{P_1, P_2 \cdots P_i, \cdots\}$. We define a set of direction vectors:

$$D = \{\mathbf{d_j} \mid \mathbf{d_j} = \frac{\overrightarrow{CP_i}}{|\overrightarrow{CP_i}|}, \ \forall \ i > 0, j \leqslant i\} \quad (1)$$

such that for every point in set R there exists some $k > 0$ and some $\mathbf{d_j} \in D$ such that $P_i = C + k\mathbf{d_j}$. (A robot starting at C and moving k units in direction $\mathbf{d_j}$ will arrive at P_i). As an example, consider the case in Figure 1(b). Let $\overrightarrow{CL_1}$ correspond to the polar coordinate angle, θ, of $\theta_{\overrightarrow{CL_1}} = 0$, and let the direction of θ vary in the anticlockwise direction. Then, region R_{A1} maps to the fan of vectors with polar angle $0 < \theta < \theta_{\overrightarrow{CL_2}}$ whilst region R_{B1} maps to the vectors with polar angle $\theta_{\overrightarrow{CL_2}} < \theta < 360°$.

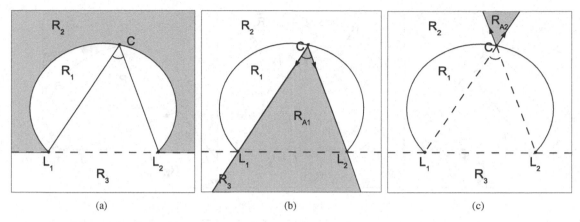

Fig. 1. (a) Horopter L_1CL_2 and line L_1L_2 splits the plane into 3 regions. (b-c) Regions $R_{A1,A2}$ (shaded) and $R_{B1,B2}$ (everything that is unshaded).

$\overrightarrow{CG}$ is the vector pointing from current to goal location. Recovering it guarantees convergence (moving in the direction $\overrightarrow{CG}$ repeatedly will, inevitably, bring the robot to G). It also gives the most efficient path (the straight line) to the goal. In the following, we are interested only in the direction of $\overrightarrow{CG}$ and we will use $\overrightarrow{CG}$, and the unit vector in the direction $\overrightarrow{CG}$, interchangeably. Then:

Theorem 1: From one landmark pair, Lemma 1 constrains G within regions R_{B1} or R_{B2}. Equation 1 maps this to a set of directions, D, where $\overrightarrow{CG} \in D$. For N landmarks, we have up to $^{N}C_2$ vector sets $\{D_1, D_2, \cdots D_{N_{C_2}}\}$. $\overrightarrow{CG}$ lies in their intersection, $D_{res} = D_1 \cap D_2 \cap \cdots \cap D_{N_{C_2}}$.

Assuming isotropically distributed landmarks, as $N \to \infty$, $D_{res} = \overrightarrow{CG}$. In practice, successful homing only requires N to be large enough, such that a homing direction can be chosen, that is less than 90° from *every* vector in D_{res}. It will then be less than 90° from $\overrightarrow{CG}$ and from Observation 1, the robot will converge. A stricter condition is to have the maximum angle between any two vectors in D_{res} less than 90°. Then all vectors in D_{res} will be less than 90° from $\overrightarrow{CG}$.

B. Robot on the Plane and Landmarks in 3D

Whilst the robot can move on some plane with normal vector, $\mathbf{n}$, the observed landmark points will, in general, not lie on that plane. Here, we extend the previous results to this more general situation. It is not unreasonable to assume that the robot knows which direction is 'up', that is, the normal to the ground plane. Then, we can measure angles according to the same clockwise or anticlockwise conventions as before. (Note that this is *not* equivalent to finding the ground plane, which involves segmenting ground pixels from non-ground pixels, and is a non-trivial task in the most general situations.)

Let C, G lie on the x-y plane (so normal vector $\mathbf{n}$ is the z-axis) and suppose C lies in the negative-x region. There always exists some plane passing through two 3D points such that this plane is orthogonal to the x-y plane. So, without loss of generality, we can let the landmark pair L_1, L_2 lie on the y-z plane. There exists a circular arc passing through L_1, C and L_2. Revolving the arc about line L_1L_2 sweeps out a surface of

revolution. The portion of the surface that lies in the negative-x region of $\Re^3$ is the horopter surface. All points on the horopter have inter-landmark angles equal to $\angle L_1CL_2$.

The intersection of this 3D horopter surface with the x-y plane containing C and G is a 2D horopter curve (some examples shown in Figure 2 (a-f)). Rather than a circular arc, the curve is elliptical. Let region R_1 be the set of all points inside the horopter curve and in the negative-x region; let R_2 be the set of points in the negative-x region and outside the horopter; let R_3 be the set of points in the positive-x region. Then, the inequality relationships between $\angle L_1CL_2$ and $\angle L_1GL_2$ if C lay on the horopter and G in one of regions R_1, R_2 or R_3, are similar to those discussed in Section II-A.

Since landmark distances are unknown, once again, we attempt to constrain G using only landmark directions. The difference is that we now use the projected landmark vectors $\overrightarrow{C\tilde{L}_1}$ and $\overrightarrow{C\tilde{L}_2}$ instead of the landmark vectors as was previously the case. The projection of a landmark ray vector, $\overrightarrow{CL_1}$, onto the plane is $\overrightarrow{C\tilde{L}_1} = \mathbf{n} \times (\overrightarrow{CL_1} \times \mathbf{n})$ where $\tilde{L}_1$ is the projection of L_1 onto the plane and $\times$ is the cross product.

The horopter curve intersects the y-axis at H_1, H_2 and the projected landmark rays, $\overrightarrow{C\tilde{L}_1}$ and $\overrightarrow{C\tilde{L}_2}$ intersect the y-axis at $\tilde{L}_1, \tilde{L}_2$. However, whilst L_1, L_2 lie on the horopter, the projected landmark rays are such that $\tilde{L}_1, \tilde{L}_2$ do not lie on the horopter in general. (If they did, then the problem is reduced to that of Section II-A.) Let P_y be the y-coordinate of point P. Three cases arise: (Case 1) both $\tilde{L}_{1y}, \tilde{L}_{2y}$ lie in the interval $[H_{1y}, H_{2y}]$; (Case 2) one of $\tilde{L}_{1y}$ or $\tilde{L}_{2y}$ lies outside that interval; (Case 3) both $\tilde{L}_{1y}, \tilde{L}_{2y}$ lie outside $[H_{1y}, H_{2y}]$.

For Case (1), the results of Lemma 1 hold except that the Type 1 and 2 constraint regions are now bounded by projected landmark rays. Figure 2(a) illustrates the case of $G \in R_2$. The shaded region is $R_{B1} = R_{A1}^c$, which is a Type 1 constraint bounded by $\overrightarrow{C\tilde{L}_1}, \overrightarrow{C\tilde{L}_2}$ such that $G \in R_{B1}$ (since $R_2 \subset R_{B1}$) for any C on the horopter curve. Meanwhile, Figure 2(b) illustrates the case of $G \in R_1 \cup R_3$ and the shaded region is the Type 2 constraint region, $R_{B2} = R_{A2}^c$ bounded by $-\overrightarrow{C\tilde{L}_1}, -\overrightarrow{C\tilde{L}_2}$ such that $G \in (R_1 \cup R_3) \subset R_{B2}$. Using

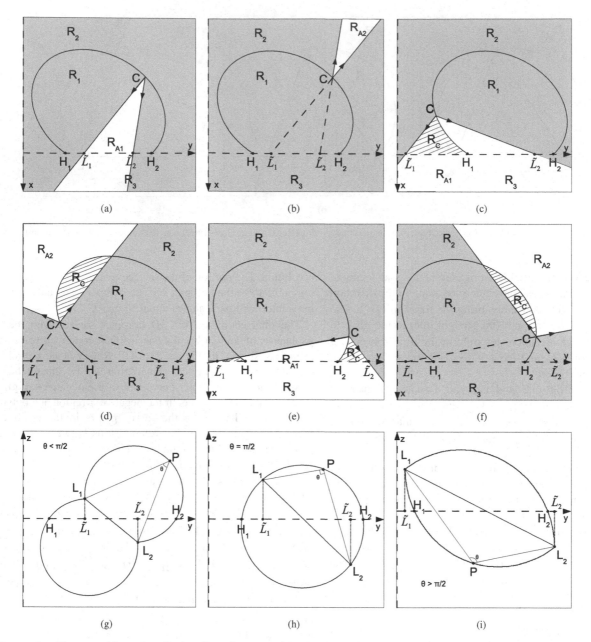

Fig. 2. Intersection of horopter with x-y plane. Regions R_{B1}, R_{B2} constraining location of G are shaded. The complement regions, R_{A1}, R_{A2} are unshaded. (a-b) Case1: both $\tilde{L}_{1y}$, $\tilde{L}_{2y}$ in $[H_{1y}, H_{2y}]$. (c-d) Case 2: one of $\tilde{L}_{1y}$ or $\tilde{L}_{2y}$ outside $[H_{1y}, H_{2y}]$. (e-f) Case 3: both $\tilde{L}_{1y}$, $\tilde{L}_{2y}$ outside $[H_{1y}, H_{2y}]$. (g-i) Intersection of horopter surface with y-z plane. 3 cases arising from $\theta < 90°$, $\theta = 90°$ and $\theta > 90°$.

Equation 1, these regions can be mapped to sets of possible homing directions.

Unfortunately, Lemma 1 does not always hold for Cases (2) and (3) such as in the example configurations in Figures 2(c, d) for Case (2) and (e, f) for Case (3). Regions R_C are marked with diagonal hatching. In Figure 2(c), the Type 1 constraint, R_{B1} should contain R_2. However, $R_C \subset R_2$ but $R_C \not\subset R_{B1}$. Likewise, in Figure 2(d), the Type 2 constraint, R_{B2} should contain $R_1 \cup R_3$ but it misses out on $R_C \subset R_1$. This means that if G lay in R_C, there would exist configurations of L_1, L_2, C where G would not lie in the constraint region.

Two approaches are possible - the first is a non-conservative approach that uses all constraints arising from all landmark pairs. This method will very probably converge and in all (thousands of) experiments (Section IV), was indeed observed to converge. A second approach uses only a subset of all the constraints and its convergence is theoretically guaranteed.

1. Non-conservative Method - Likely Convergence

As a result of the above problem, Theorem 1 will no longer hold. D_{res} was previously defined as the intersection of *all* sets of possible homing directions arising from all the landmark pairs. This may not exist, so we will instead define $\bar{D}_{res}$, which is the intersection of the *largest number* of such sets.

One observes that the size of R_C tends to be small relative to the total region in which G can lie. For example, in the case of $G \in R_2$ and R_{B1} should contain R_2 but misses out the R_C regions (Figures 2 (c, e)), R_C is largest when C is at the highest point on the horopter and $\tilde{L}_{1y}, \tilde{L}_{2y}$ approach $\pm\infty$. Even then, R_C is small compared to all of region R_2.

Therefore, the probability of $G \in R_C$ is actually rather small. Assuming N randomly scattered landmarks, out of $^N C_2$ constraints, some proportion of constraints would give rise to these 'missed' regions, R_C, where there is potential for error. Of these, an even smaller number will actually be erroneous constraints, i.e. $G \in R_C$. These erroneous constraints map to a set of direction vectors that is 'wrong' in the sense that $\overrightarrow{CG}$ is not in this set.

The direction $\overrightarrow{CG}$ might not lie in $\bar{D}_{res}$. However, there is a good chance that $\overrightarrow{CG}$ lies close to it, and in fact, there is a very high probability that $\overrightarrow{CG}$ is less than 90° from a homing direction chosen from the set $\bar{D}_{res}$ (eg: the average of all directions in $\bar{D}_{res}$). In the highly unlikely event of a wrong estimate, that is, $\overrightarrow{CG}$ being greater than 90° from the homing direction, the robot will make a step that takes it further away from the goal than it was at its last position.

However, this alone is insufficient to cause non-convergence. The robot will move to a new position, leading to a change in the directions of the landmark rays, and a new estimate of the new homing direction is found. In order for the robot to *not* converge to G, it would have to obtain so many wrong estimates that it was going in the wrong direction most of the time. This requires a stacking of many improbable odds and indeed, in no experiment was non-convergence ever observed. Even so, there is a remote but finite chance of failure, which leads us to the next method.

2. Conservative Method - Guaranteed Convergence

Assume the camera is mounted some distance above the ground and the x-y plane (we are working in the camera coordinate frame) is parallel to the ground. Homing motion is then restricted to this x-y plane. As before, landmark pair L_1, L_2 lies on the y-z plane and the horopter surface is the surface swept out by revolving the arc L_1CL_2 about the line L_1L_2, restricted to the half-space that has x-coordinates with the same sign as the sign of C_x.

There is a strategy for picking landmark pairs so that $\tilde{L}_{1y}, \tilde{L}_{2y}$ lie in the interval $[H_{1y}, H_{2y}]$. This is the Case 1 configuration which is free of erroneous constraints:

Lemma 2: If a landmark pair, L_1, L_2 is chosen such that one lies above the x-y plane and one lies below it, and such that $\theta = cos^{-1}(\overrightarrow{CL_1} \cdot \overrightarrow{CL_2}) \leqslant 90°$, then $\tilde{L}_{1y}, \tilde{L}_{2y} \in [H_{1y}, H_{2y}]$.

Here, θ is always acute; it differs from the inter-landmark angle which can be acute or obtuse depending on how it is measured. The intersection of the horopter surface with the y-z plane is as shown in Figures 2(g-i) which depict the three cases arising when the angle, θ, between two landmark rays as observed at any point C on the horopter, is greater than, equal to or less than 90°. With one landmark above and one below the x-y plane, the line segment lying between L_1 and

L_2 intersects the y-axis. $\tilde{L}_1, \tilde{L}_2$ are the projections of L_1 and L_2 onto the x-y plane, whilst H_1 and H_2 are the intersections of the horopter with the plane.

If $\theta \leqslant 90°$ (Figures 2(g) and (h)), it can be shown that $\tilde{L}_{1y}, \tilde{L}_{2y} \in [H_{1y}, H_{2y}]$ is always true for any L_1, L_2 satisfying these conditions. However, if $\theta > 90°$, this is not always true, as the counterexample of Figure 2(i) demonstrates.

Using only landmark pairs that have $\theta \leqslant 90°$ is a conservative method that ensures only correct constraints are used, in the noise-free case. However, most of the discarded constraints will in fact be correct, according to the earlier argument that the probability that $G \in R_C$ is low. (Note that if insufficient pairs of landmarks meet the conditions of Lemma 2, one can still use the earlier, non-conservative method to home.)

The conservative method ensures Theorem 1 holds and $\overrightarrow{CG} \in D_{res}$. Hence, if there are sufficient landmarks such that a vector **h** which is within 90° of all direction vectors in D_{res} can be found, then moving in the direction of **h** will bring the robot closer to the goal. If this condition is met at each step of the robot's estimate and move cycle, convergence on G is guaranteed as per Observation 1.

III. Algorithms and Implementation

Algorithm 1 Planar Robot, 3D Landmarks

1: **while** angular error, $\alpha_{ave} > \alpha_{thres}$ **do**
2:     **for** $j = 1$ to K **do**
3:         Select a pair L_1^j, L_2^j from set of landmark pairs with known correspondence.
4:         **if** (Conservative) and ((L_1^j, L_2^j are both above the plane or both below) or ($cos^{-1}(\overrightarrow{CL_1^j} \cdot \overrightarrow{CL_2^j}) > 90°$)) **then**
5:             continue to next j.
6:         **end if**
7:         $\alpha_j = |\angle L_1^j C L_2^j - \angle L_1^j G L_2^j|$.
8:         Find regions R_{B1}, R_{B2} as per Lemma 1 using projected landmark rays, $C\tilde{L}_1^j, C\tilde{L}_2^j$ and $G\tilde{L}_1^j, G\tilde{L}_2^j$
9:         Find the set of possible homing directions D_j.
10:         Cast votes for D_j.
11:     **end for**
12:     Find bin(s) with votes > $vote_{thres}$. Mean direction is homing vector, **h**. Calculate average α_{ave}.
13:     Move robot by $StepSz * \mathbf{h}$.
14: **end while**

The homing approach proposed in Section II involves finding sets of possible homing directions and obtaining the intersection of all or of the largest number of these sets. A voting framework accomplishes this quickly and robustly. The table of votes divides the space of possible directions of movement into voting bins. For planar motion, voting is done in the range $\theta = [0, 360°)$. From the sets of possible homing directions $\{D_1, D_2, \cdots, D_i, \cdots\}$, votes for bins corresponding to the directions in each D_i are incremented, and the bin(s) with maximum vote (or with votes exceeding a threshold, $vote_{thres}$) gives D_{res} or $\bar{D}_{res}$. When more than one bin has maximum vote, we take the average direction to be the homing vector, $\mathbf{h}$. The non-conservative and conservative approaches for homing with planar robots and 3D landmarks are implemented in Algorithm 1.

In the algorithm, K can be all $^N C_2$ combinations of landmark pairs, or some random sample thereof. To sense whether the robot is far from or close to the goal, we average the inter-landmark angular error, $\alpha_j = |\angle L_1^j C L_2^j - \angle L_1^j G L_2^j|$ over all j. α_{ave} gives a measure of how similar the current and goal images are. Ideally, as the robot approaches the goal position, $\alpha_{ave} \to 0$. We specify that when $\alpha_{ave} < \alpha_{thres}$, homing is completed and the robot stops. α_{ave} may be used to control the size of robot motion, the variable $StepSz$.

Note that Algorithm 1 is more efficiently implemented by using the directions mapped from R_{A1}, R_{A2} instead of R_{B1}, R_{B2}, since the former regions are typically smaller. This way, less computations (incrementing of votes) occur. $\mathbf{h}$, is then found from the *minimum* vote instead of the maximum.

IV. EXPERIMENTS AND RESULTS

A. Simulations

In the Matlab simulations, landmarks were randomly scattered within a cube of $90 \times 90 \times 90$ units. The robot moves between randomly generated start and goal positions by iterating the homing behaviour. $\mathbf{h}$ was taken to be the average of the set of directions with maximum votes.

Firstly, for the noise-free case, 1000 trials were conducted for each of the planar conservative and planar non-conservative cases. Secondly, a series of experiments involving 100 trials each, investigated homing under outliers and Gaussian noise. Outliers were simulated by randomly replacing landmark rays with random vectors. The circle of possible directions of movement was divided into 360 voting bins (1° per bin). We tested for up to 40% outliers and for Gaussian noise with standard deviation up to 9°.

The robot successfully homed in all experiments. This confirms the theoretical convergence proof for the conservative case, and gives empirical evidence supporting the statistical argument for likely convergence in the non-conservative case.

In order to examine the quality of homing, we compute the directional error (angle between $\overrightarrow{CG}$ and $\mathbf{h}$), which measures how much the homing direction deviates from the straight line home. The histograms in Figure 3(a-d) summarize the directional error in the noise and outlier experiments.

Recall, from Observation 1 that a homing step will take the robot closer to the goal if $\mathbf{h}$ is less than 90° from $\overrightarrow{CG}$. For no noise and no outliers, all homing directions computed were indeed less than 90° from $\overrightarrow{CG}$ (even for the non-conservative method). However, even with up to 40% of landmark rays being outliers, the number of homing estimates that were more than 90° from $\overrightarrow{CG}$ was insignificant for the non-conservative case (Figure 3(b)), and less than 8% for the conservative case (Figure 3(a)). This means the robot was heading in the correct direction the vast majority of the time in spite of the outliers, hence homing was successful.

Successful convergence was observed in trials with Gaussian noise. Figures 3(c-d) illustrate the performance as the Gaussian noise standard deviation is varied from 0° to 9°. Performance degrades gracefully with noise but the proportion of homing vectors with directional error greater than 90° was once again insufficient to prevent convergence.

It is interesting to note that the planar non-conservative case outperformed the planar conservative case. Voting proved particularly robust to the incorrect constraints arising in the non-conservative case. Since correct constraints dominate in number, these form a robust peak in the vote space which requires large numbers of incorrect constraints voting *in consistency* with each other, to perturb from its place. However, the incorrect constraints are quite random and do not generally vote to a consistent peak at all; and the 'missed' R_C regions also tend to be small, so the total effect on the performance of the non-conservative case is quite minimal.

Conversely, although the conservative case guarantees convergence, that guarantee comes at the cost of throwing away all constraints that are not certain to be correct. In the process, many perfectly fine constraints are discarded as well. Thus, the set of directions with maximum votes was larger, and the average direction (taken as the homing direction) deviated further from $\overrightarrow{CG}$, compared to the non-conservative method.

B. Real Experiments

Grid Trials: A camera captured omnidirectional images of some environment at every 10 cm in a 1 m x 1 m grid. With some point on the grid as the home position, the homing direction was calculated at every other point on the grid. The result is summarized in a vector-field representation where each vector is the direction a robot would move in if it was at that point in the field. By taking each point on the grid in turn to be the home position, we can examine how the algorithm performs in that environment for all combinations of start and goal positions (marked with a '+') within the grid.

All experiments were successful. Figure 3(e-h) shows sample results from the 3 different environments. Landmarks were SIFT features [22] matched between current and goal images.

The high-resolution (3840 x 1024) Ladybug camera [25] captured images of a room (Figure 3(k)) and an office cubicle (l). Figure 3(e,f) are results for the room using conservative and non-conservative algorithms respectively. Room images were taken with the camera 0.85 m above the ground. This is necessary for the conservative method, which requires landmarks both above and below the image horizon. (In contrast, cubicle images were taken with the camera lying on the floor and were unsuitable for the conservative algorithm since everything below the horizon is featureless ground.)

Both conservative and non-conservative methods homed successfully in Figure 3(e,f). Since the conservative method discards many constraints, the homing direction deviated further from $\overrightarrow{CG}$, compared to the non-conservative method which gave more direct homing paths. Nevertheless, it is clear that a robot placed anywhere within the grid will follow the vectors and eventually reach the goal position.

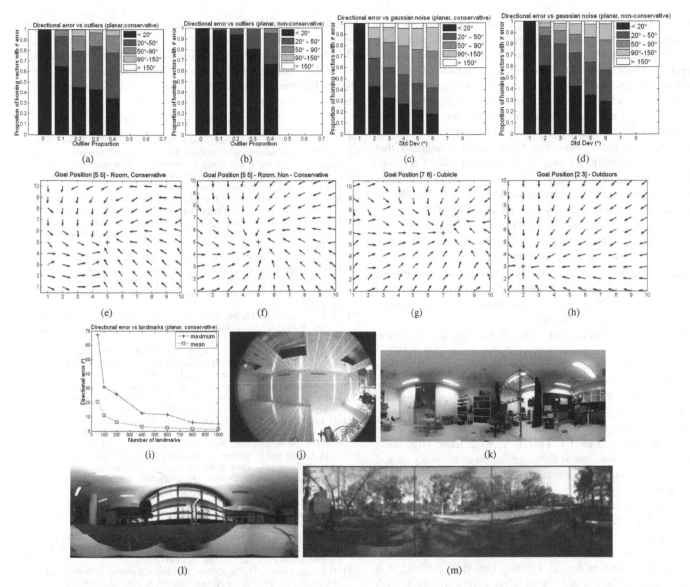

Fig. 3. (a-d) Homing direction under outliers and Gaussian noise. (e-f) Grid trials for room, comparing conservative and non-conservative methods. (g-h) Grid trials in cubicle and outdoors. (i) Effect of varying number of landmarks (j) Robot's view in atrium experiment. (k) High-resolution panorama taken in a room with an elevated camera and (l) taken in a cubicle with camera on the floor. (m) Low-resolution panorama taken outdoors.

Figure 3(g) demonstrates successful homing in the cubicle environment using the non-conservative method. We also tested the algorithm on a low-resolution (360 x 143) outdoor image set supplied by authors of [17, 21]. Figure 3(m) is an example image from the set and 3(h) is a sample result. Interestingly, the low-resolution outdoor images gave smoother vector fields than the high-resolution, indoor ones. We believe this is due to a more even distribution of landmarks outdoors, whereas indoors, features tend to be denser in some areas of the image but are very sparse in other areas, leading to a bias in some directions. However, convergence is unaffected and homing paths remain fairly straight.

Robot Trials: Videos at [26]. A holonomic wheeled robot [27] homed in various environments including an office and a building atrium. The robot performed local homing multiple times to move along a chain of goal positions (simplest instance of the topological maps mentioned in Section I) in order to get from one part of the environment to another. Snapshots were captured at intermediate goal positions and stored in the chain. An Omnitech Robotics fish-eye lens camera [28] was mounted pointing upwards, giving views such as Figure 3(j), where the image rim corresponds to the horizon.

Homing was successful to within the order of a few centimeters, in the presence of mismatched landmarks, moving objects (leading to bad landmark points) and illumination changes.

V. DISCUSSION

Robustness to Outliers: In methods such as [11, 12, 10, 20], a series of vectors, $\{M_1, M_2 \cdots M_N\}$, are derived from landmark bearings by some rule, and the average direction is taken as the homing vector, $M_{res} = M_1 + M_2 + \cdots + M_N$. Unfortunately, this average is sensitive to outliers which can potentially cause the homing vector to point in any random direction. The experiments demonstrated that our voting-based method was able to efficiently incorporate information from a set of landmarks while being resistant to significant numbers of mismatched and moved landmarks in the set.

Number of Landmarks: More landmarks lead to more direct homing paths. However, Observation 1 suggests that homing is successful as long as homing directions are consistently less than 90° from $\overrightarrow{CG}$. This condition is satisfied even with as few as 50 landmark matches (Figure 3(i)). In the experiments, images typically gave hundreds of SIFT matches.

Speed: Our algorithm runs in 35 msec (28 fps) on a Linux OS, Pentium4, 3GHz, 512Mb RAM machine (~200 landmarks). The inputs to the algorithm were SIFT matches in the experiments, but these can take several seconds to compute. However, real-time SIFT implementations (for example, on a GPU) do exist. Alternatively, faster feature tracking methods such as KLT are also possible.

Poor Image Resolutions: The trials with low-resolution (360 x 143 pixels) outdoor images demonstrate that the algorithm works well even with poor image resolutions. This is in agreement with the results of the simulations under increasing Gaussian noise. The image noise in most cameras is generally in the order of a couple of pixels (typically less than a degree), which is far less than the amounts of Gaussian noise used in the simulations (up to 9°). Therefore, what these large noise trials do in fact simulate is the degradation of accuracy caused by using very coarse image resolutions causing uncertainty in the landmark ray.

Voting Resolution: Lower voting resolutions (currently 1° per bin) could achieve even greater speeds without affecting convergence (it will, however, lead to a less direct path home). As an extreme example, the conditions of Observation 1 are satisfied even with as few as 4 voting bins (90° per bin), if h is the center of the bin and conservative voting was used. Such a minimalist version of the method is well-suited for small, resource limited robots.

Flexible Probability Maps: Whilst our experiments (and most existing methods) obtained a single homing vector, the voting table is in fact a weighted map of likely directions for homing. This leads to a flexible and natural framework for incorporating additional navigational constraints such as obstacle avoidance, kinematic constraints (in non-holonomic robots), centering behaviour and path optimality planning. For example, directions blocked by obstacles can be augmented with negative votes so that the robot is discouraged from moving in that direction.

VI. CONCLUSION

We investigated the problem of planar visual homing using landmark angles and presented three novel algorithms and demonstrated their performance in simulations and real robot trials. **Videos available on the author's website [26].**

Acknowledgments: We thank Jochen Zeil for providing the outdoor image set and Luke Cole for help with the robot. NICTA is funded by the Australian Government as represented by the Department of Broadband, Communications and the Digital Economy and the ARC through the ICT Centre of Excellence program.

REFERENCES

[1] S. Thrun, W. Burgard, D. Fox, *Probabilistic Robotics*, MIT Press, 2005.
[2] A. Davison, *Real-time simultaneous localisation and mapping with a single camera*, Proc. International Conference on Computer Vision, 2003.
[3] R. Hartley and A. Zisserman, *Multiple View Geometry in Computer Vision*, Cambridge University Press, 2000.
[4] S. Hutchinson, G. Hager and P. Corke, *A tutorial on visual servo control*, IEEE Trans. Robotics and Automation, vol. 12, no. 5, pp. 651-670, 1996.
[5] D.T. Lawton and W. Carter, *Qualitative spatial understanding and the control of mobile robots*, IEEE Conf. on Decision and Control, vol. 3, pp. 1493-1497, 1990.
[6] J. Hong, X. Tan, B. Pinette, R. Weiss and E.M. Riseman, *Image-based homing*, IEEE Control Systems Magazine, vol. 12, pp. 38-45, 1992.
[7] M.O. Franz, B. Schölkopf, H.A. Mallot, H.H. Bülthoff, *Learning view graphs for robot navigation*, Autonomous Robots, vol. 5, no. 1, 1998, pp. 111-125.
[8] M.O. Franz, H.A. Mallot, *Biomimetic robot navigation*, Biomimetic Robots, Robotics and Autonomous Systems, vol. 30 no. 1, 2000, pp. 133-153.
[9] B. Kuipers and Y.T. Byun, *A robot exploration and mapping strategy based on a semantic hierarchy of spatial representations*, Robotics and Autonomous Systems, vol. 8 no. 12, 1991, pp. 47-63.
[10] K.E. Bekris, A.A. Argyros, L.E. Kavraki, *Angle-Based Methods for Mobile Robot Navigation: Reaching the Entire Plane*, in Proc. Int. Conf. on Robotics and Automation, pp. 2373-2378, 2004.
[11] D. Lambrinos, R. Möller, T. Labhart, R. Pfeifer and R. Wehner, *A mobile robot employing insect strategies for navigation*, Biomimetic Robots, Robotics and Autonomous Systems, vol. 30, no. 1, 2000, pp. 39-64.
[12] S.G. Loizou and V. Kumar , *Biologically Inspired Bearing-Only Navigation and Tracking*, IEEE Conf. on Decision and Control, 2007.
[13] K. Weber, S. Venkatesh and M. Srinivasan, *Insect-inspired robotic homing*, Adaptive Behavior, vol. 7, no. 1, 1999, pp. 65-97.
[14] S. Gourichon, J.A. Meyer and P. Pirim, *Using colored snapshots for short range guidance in mobile robots*, Int. J. Robot. Automation, vol. 17, pp. 154-162, 2002.
[15] M.O. Franz, B. Schölkopf, H.A. Mallot, H.H. Bülthoff, *Where did I take that snapshot? Scene-based homing by image matching*, Biological Cybernetics, vol. 79, no. 3, 1998, pp. 191-202.
[16] B.A. Cartwright and T.S. Collett, *Landmark learning in bees*, Journal of Comparative Physiology A, vol. 151, no. 4, 1983, 521-543.
[17] J. Zeil, M.I. Hoffmann and J.S. Chahl, *Catchment areas of panoramic images in outdoor scenes*, Journal of the Optical Society of America A, vol. 20, no. 3, 2003, pp. 450-469.
[18] A. Vardy, R. Möller, *Biologically plausible visual homing methods based on optical flow techniques*, Connection Science, vol. 17, no. 1, 2005, pp. 47-89.
[19] R. Möller, *Insect visual homing strategies in a robot with analog processing*, Biological Cybernetics: special issue in Navigation in Biological and Artificial Systems, vol. 83, no. 3 pp. 231-243, 2000.
[20] A.A. Argyros, C. Bekris, S.C. Orphanoudakis, L.E. Kavraki, *Robot Homing by Exploiting Panoramic Vision*, Journal of Autonomous Robots, Springer, vol. 19, no. 1, pp. 7-25, July 2005.
[21] W. Stürzl, J. Zeil, *Depth, contrast and view-based homing in outdoor scenes*, Biological Cybernetics, vol. 96, pp. 519-531, 2007.
[22] D.G. Lowe, *Object Recognition from local scale-invariant features*, in Proc. of Int. Conf. on Computer Vision, pp. 1150-1157, September, 1999.
[23] C. Harris and M. Stephens, *A combined corner and edge detector*, Proceedings of the 4th Alvey Vision Conference, pp. 147-151, 1988.
[24] J. Shi and C. Tomasi, *Good Features to Track*, IEEE Conference on Computer Vision and Pattern Recognition, pp. 593-600, 1994.
[25] Point Grey Research, http://www.ptgrey.com , 2009.
[26] J.Lim, http://users.rsise.anu.edu.au/%7Ejohnlim/ , 2009.
[27] L. Cole and N. Barnes, *Insect Inspired Three Dimensional Centering*, In Proc. Australasian Conf. on Robotics and Automation, 2008.
[28] Omnitech Robotics, http://www.omnitech.com/ , 2009.

Rut Detection and Following
for Autonomous Ground Vehicles

Camilo Ordonez *, Oscar Y. Chuy Jr. †, Emmanuel G. Collins Jr. ‡, and Xiuwen Liu §

Center for Intelligent Systems, Control and Robotics (CISCOR)

* † ‡ Department of Mechanical Engineering

Florida A&M - Florida State University

§ Department of Computer Science

Florida State University

Tallahassee, FL 32310, United States

Email: camilor@eng.fsu.edu chuy@eng.fsu.edu ecollins@eng.fsu.edu liux@cs.fsu.edu

Abstract—Expert off road drivers have found through experience that ruts formed on soft terrains as a result of vehicular transit can be used to improve vehicle safety and performance. Rut following improves vehicle performance by reducing the energy wasted on compacting the ground as the vehicle traverses over the terrain. Furthermore, proper rut following can improve vehicle safety on turns and slopes by utilizing the extra lateral force provided by the ruts to reduce lateral slippage and guide the vehicle through its path. This paper presents a set of field experiments to show the relevance of rut following for autonomous ground vehicles and proposes a reactive based approach based on knowledge of the width of the tires and the vehicle body clearance to provide mobile robots with rut detection and following abilities. Experimental results on a Pioneer 3AT robot show that the proposed system was able to detect and follow S-shaped ruts, and ruts that are not directly in front of the robot.

I. INTRODUCTION

Autonomous ground vehicles (AGVs) are increasingly being considered and used for challenging outdoor applications. These tasks include fire fighting, agricultural applications, search and rescue, as well as military missions. In these outdoor applications, ruts are usually formed in soft terrains like mud, sand, and snow as a result of habitual passage of wheeled vehicles over the same area. Fig. 1 shows a typical set of ruts formed by the traversal of manned vehicles on off road trails.

Expert off road drivers have realized through experience that ruts can offer both great help and great danger to a vehicle [1].

Fig. 1. Typical Off Road Ruts Created by Manned Vehicles

On soft terrains ruts improve vehicle performance by reducing the energy wasted on compacting the ground as the vehicle traverses over the terrain [2]. Furthermore, when traversing soft and slippery terrains, proper rut following can improve vehicle safety on turns and slopes by utilizing the extra lateral force provided by the ruts to reduce lateral slippage and guide the vehicle through the desired path [1, 3, 4, 5, 6]. On the other hand, a vehicle moving at high speed that hits a rut involuntarily can lose control and tip over. An AGV provided with rut detection and rut following abilities can benefit from the correct application of this off road driving rule, and thereby improve its efficiency and safety in challenging missions.

Besides the benefits of rut following already explained, proper rut detection and following can be applied in diverse applications. Rut detection can signal the presence of vehicles in the area, and also can help in the guidance of loose convoy operations. In planetary exploration, ruts can play an important role; due to the high cost of these missions, it is desirable, in some situations, for a rover to retrace its path after a successful exploration mission and minimize the risk of getting stuck in terrain that is difficult to traverse. A rut detection system can be used as a robot sinkage measurement system, which is key in the prediction of center high situations. Automatic rut detection can also be employed to determine the coefficient of rolling resistance [7] (a vital parameter in robot dynamic models), and in general can be used to learn different properties of the terrain being traversed.

Prior to the research in [8], work on rut detection focused exclusively on paved surfaces in a road surface inspection application [9, 10]. However, these approaches are not concerned with the continuity of the ruts, something achieved in the proposed approach by using local rut models in the vicinity of the vehicle. In contrast to the method presented in [8], the rut detection method presented here incorporates domain knowledge regarding tire width and vehicle body clearance into the rut detection problem. By doing so, the detection process becomes more efficient because the search for ruts can be performed on specific candidate points over the laser scan instead of at every point as in [8]. Besides that, by incorporating geometric constraints on the rut depth and width, center high situations can be reduced and ruts that are too wide or too narrow can be avoided. Another important difference between the current rut detection implementation and the one of [8] is that this new approach uses a polynomial representation of the left and right ruts in the local vicinity of the robot. By doing this, the robot can differentiate between the left and the right rut, which is necessary for waypoint assignment during rut following.

Additional research that is related to rut detection is the development of a seed row localization method using machine vision to assist in the guidance of a seed drill [11]. This system was limited to straight seed rows and was tested in agricultural type environments, which are relatively structured. The work presented on [12], presents a vision-based estimation system for slip angle based on the visual observation of the trace produced by the wheel of the robot. However, it only detects the wheel traces being created by the robot. An important result was shown in [7], where a correlation between the rut depth and the rolling resistance was presented. However, this work did not deal with the rut detection problem. As previously mentioned, a rut detection method for mobile robots was developed in [8]. However, that paper did not present any approach to the rut following problem. Two successful systems of road lane detection and tracking are presented in [13, 14]. However, these approaches are tested on flat ground and are mainly concerned with keeping the vehicle inside the road and not with keeping the wheels inside specific regions of the terrain as is the case for rut following.

The main contributions of this paper are the conception, design and performance of field experiments to show the relevance of rut detection and following for autonomous vehicles. In addition, the paper proposes, implements and performs an experimental validation of a solution to provide mobile robots with rut detection and following capabilities.

The remainder of the paper is organized as follows. Section II presents a series of motivational experiments with two different robotic platforms and two different terrains. Section III describes the proposed approach to rut detection and following. Section IV introduces the experimental setup and shows experimental results. Section V provides a set of improvements to the proposed approach. Finally, Section VI presents concluding remarks, including a discussion of future research.

II. MOTIVATIONAL EXPERIMENTS

To show some of the benefits of rut following three controlled experiments were performed using two different robotic platforms on two different soft terrains. It is important to note that the motivational experiments do not use the proposed algorithm, but do experimentally show the relevance of rut following for off road robot navigation.

During the motivational experiments, the robot stays in the ruts by following a set of preassigned waypoints. In the case of the Pioneer 3-AT, which has less accurate localization capabilities than the XUV, the runs were performed on short and straight ruts and the vehicle was carefully placed and

(a)   (b)

Fig. 2.   (a) Pioneer 3-AT Robotic Platform on Sand. (b) XUV Robotic Platform on Mud

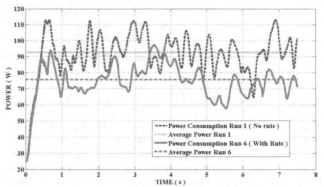

Fig. 3.   Decrease in Power Consumption by Following Ruts (Pioneer 3 on Sand)

aligned at the starting point of the ruts. In the case of the XUV the experiments were performed over longer ruts because it counts with a localization system comprised of a differential GPS and a high cost IMU. Fig. 2(a) shows the Pioneer 3-AT robot following ruts in sandy terrain, and Fig. 2(b) shows the XUV robot following ruts in muddy terrain.

In the motivational experiments, power consumption and velocity tracking are used as performance metrics. The power consumption (P_c) is computed as the RMS value of $F_r V_r$, where F_r is the force required to overcome the rolling resistance when the vehicles is moving at constant velocity V_r. The velocity tracking performance is computed as the RMS value of the velocity error $E_v(t) = V_r(t) - V_c(t)$, where V_r is the robot velocity and V_c is the commanded velocity.

First, a Pioneer 3-AT robotic platform was commanded to follow a set of ruts over sandy terrain at 0.8 m/s. Six trials were performed; the first run was used as a baseline because it corresponds to the no-rut case (i.e., the robot is beginning the first creation of ruts). Fig. 3 shows a comparison of the power consumption for the first (no ruts) pass and the sixth pass. Notice that by following the ruts, there is an average reduction in power consumption of 18.3%. Furthermore, the experiments revealed that as early as the second pass, there is an average reduction in power consumption of 17.9%.

A second experiment was performed on mud with the XUV robotic platform. The robot was commanded to follow a set of waypoints along a straight line at a speed of 5 mph. Fig. 4 shows the reduction of the rolling resistance coefficient μ_ρ and power consumption for 4 successive trials. Notice that in the second pass, there is a reduction in power consumption of 12.6%.

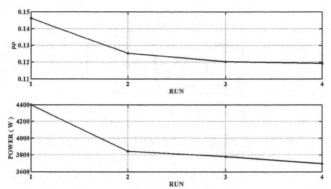

Fig. 4. Decrease in Power Consumption by Following Ruts (XUV on Mud)

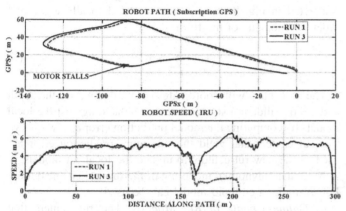

Fig. 5. Velocity Tracking Improvement by Following Ruts (XUV on Mud)

A third experiment was performed on mud with the XUV robotic platform. The robot was commanded to follow a set of waypoints along a curved path at 11 mph and three trials were performed. Fig. 5 shows the robot path and velocity profiles for the first and third run. Notice that on the first run, when there were no ruts, the vehicle was not capable of generating enough torque to track the commanded speed. This caused the motor to stall and the vehicle was not able to complete its mission. On the contrary, in the 3rd trial the robot was able to complete its mission by using the ruts created during the first two passes. The velocity tracking error reduced from 46.2% for the first run to 19.3% for the third run.

It is also worth mentioning that the robot finished the mission successfully on the second pass and exhibited a velocity tracking error of 20%. In the above experimental results it is clear that rut following improved the vehicle performance. This is important from a practical stand point because it means that a robot in the field can benefit from detecting and following ruts, even those that are freshly formed.

III. PROPOSED APPROACH TO RUT DETECTION AND FOLLOWING

The proposed approach assumes that the AGV is equipped with a laser range finder that observes the terrain in front of the vehicle. The proposed approach is divided into two subsystems as shown in Fig. 6: 1) a reactive system in charge of generating fine control commands to place the robot wheels in the ruts, and 2) a local planning system conceived to select the best

rut to follow among a set of possible candidates based on a predefined cost function. Once the planner has selected a rut to follow, the reactive system is engaged. This paper focuses on the reactive system, which is a very important component of the proposed approach because it allows precise rut following. A reactive system is selected because it can handle situations for which a system based only on global information would fail. As shown in Fig. 6, the reactive system is composed of the stages described below.

A. Rut Detection

The rut detection stage is in charge of analyzing the laser scans to find a set of possible rut candidates. These rut candidates are then passed through a two stage validation process, which efficiently removes the candidates that don't satisfy the validation criteria. First, the candidate ruts are validated both in depth and width using important domain knowledge regarding the width of the tires and the vehicle body clearance. Second, the candidate ruts are validated using a set of current local models of the ruts in the vicinity of the vehicle.

1) Rut Center Candidate Generation: Fig. 7 illustrates all the relevant coordinate systems used in this work: the inertial system N, the sensor frame S, the sensor head frame H and the vehicle frame B. This stage starts by transforming the laser scan from sensor coordinates to the B_p frame coordinates, which is coincident with the the vehicle kinematic center (B) and has the X_{bp} axis oriented with the robot and the Z_{bp} axis perpendicular to the terrain. This is a convenient transformation because it compensates for the vehicle roll and pitch.

The rut candidates are the local minima of the function $Z(\theta)$, where θ is the angle of the laser beam with respect to the X_s axis, and $Z(\theta)$ is the elevation of a laser point in the B_p frame. The current laser has a coverage of $\approx 140°$ and an angular resolution $\theta_{res} \approx 0.3515°$. Therefore, θ is given by

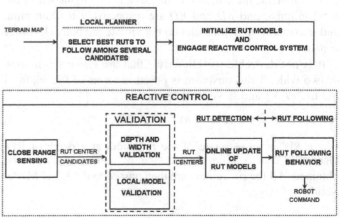

Fig. 6. Schematic of the Proposed Approach to Rut Detection and Following

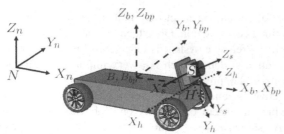

Fig. 7.   Coordinate Systems

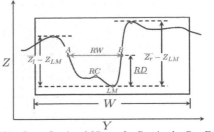

Fig. 8.   Cross Sectional View of a Rut in the Rut Frame

$$\theta = 20° + (i-1)\theta_{res}, \; i = 1, 2, ..., 399. \tag{1}$$

Given two laser beams with angles θ_a and θ_b with respect to the X_s axis, $\theta_i \in [\theta_a, \theta_b]$[1], is a local minima of $Z(\theta)$ if the following three conditions are satisfied:

1) $Z(\theta_i) < Z(\theta) \; \forall \; \theta \in [\theta_a, \theta_b]$,
2) $Z(\theta_i) \leq Z(\theta_a) - \underline{RD}$,
3) $Z(\theta_i) \leq Z(\theta_b) - \underline{RD}$,

where $\underline{RD}$ is the minimum depth that a rut should have to be considered a rut. As explained before, these local minima are only rut candidates, which need to be validated in two stages as now discussed. Note that the validation stages are implemented in cascade and therefore if a rut candidate doesn't pass the first stage is immediately removed from the candidate list and doesn't have to go through the second stage.

2) Depth and Width Validation: Once a set of rut candidates has been selected as described in III-A.1, a local window W is constructed around each candidate. The size of this window is a design parameter. In the proposed approach W is selected so that the widest ruts to be detected are covered by W when the relative orientation between the vehicle and the rut is 30°.

As explained in III-A, it is important to verify that the rut does not violate the vehicle body clearance. This constraint is checked by using the following two rules:

$$max(\overline{Z_r} - Z_{LM}, \overline{Z_l} - Z_{LM}) \leq \overline{RD}, \tag{2}$$
$$min(\overline{Z_r} - Z_{LM}, \overline{Z_l} - Z_{LM}) \geq \underline{RD}, \tag{3}$$

where as shown in Fig. 8, Z_{LM} is the elevation of the local minima, $\overline{Z_r}$ and $\overline{Z_l}$ are respectively the points with maximum elevation inside the window W to the right and to the left of the local minima, and $\underline{RD}$ and $\overline{RD}$ are respectively the minimum and maximum rut depths that do not violate the body clearance constraint.

It is not desirable to follow ruts that are either too narrow or two wide. This constraint is posed in terms of the width of the tire (TW) and is verified using

$$TW \leq RW \leq 1.5 \; TW, \tag{4}$$

where RW is an estimate of the rut width at a depth of $\underline{RD}$. Once RW has been estimated, the rut center (RC) can be obtained. All the rut centers are then passed to the Local Model Validation stage.

[1]Note that θ takes on multiple values of the angular resolution of the sensor (0.3515°) in $[\theta_a, \theta_b]$.

3) Local Model Validation: The robot keeps local models of the right and left ruts in the vicinity of the vehicle. As illustrated in Fig. 9, the ruts are modeled locally as second order polynomials of the form,

$$y(x) = \sum_{k=1}^{3} a_k x^{k-1}. \tag{5}$$

The rut centers with coordinates (x_i, y_i) that passed the depth and width validation stage are then validated against the local models by computing the model prediction error $e_i = y(x_i) - y_i$. The rut centers that yield the minimum prediction error are used as the new rut centers to update the local rut models.

Note that polynomial modeling of the ruts is just one option. For example, a clothoid model can be used.

4) Online Update of Rut Models: The rut centers that passed the two stage validation process are then used to update the rut local models given by (5). The model parameters a_k are found using a least squares minimization approach.

In the current implementation the laser has a fixed pitch, and therefore the robot has to move to initialize the models. It does so by looking for 10 pairs of ruts centers that have a separation similar to the track width of the robot. However, in the future implementation, this constraint will be removed by the inclusion of a tilt platform. In addition, the model initialization will be performed by the local path planning subsystem, which uses a predefined cost function to select the best rut to follow among several possible candidates. By doing this, the possibility of following random tracks can be minimized.

B. Rut Following

The rut that exhibits the minimum prediction error is used to generate a new waypoint for the robot as shown in Fig. 10.

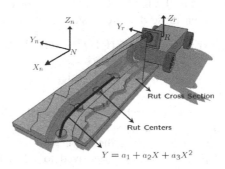

Fig. 9.   Rut Model

This waypoint takes into consideration the vehicle geometry so that the wheels of the robot can be placed in the rut. Assume that the right rut presents the minimum model prediction error. Then, the waypoint $W_p = (X_w, Y_w)$ is chosen using the rut center $R_{c1} = (X_{c1}, Y_{c1})$ as follows:

$$X_w = X_{c1}, \tag{6}$$
$$Y_w = Y_{c1} + RobW/2, \tag{7}$$

where $RobW$ is the vehicle track width and W_p and R_{c1} are expressed in the body frame B. It is important to clarify that R_{c1} is located at the intersection of the laser plane with the rut. In the current implementation R_{c1} is located at ($\approx 15cm$) from the front of the robot to allow a maximum traversal speed of $75cm/s$.

After a waypoint has been generated, a command for the angular velocity ω is generated using

$$l^2 = X_w^2 + Y_w^2, \quad r^2 = (r - Y_w)^2 + X_w^2, \tag{8}$$
$$r = \frac{l^2}{2\,Y_w}, \tag{9}$$
$$\omega = \frac{v}{r}, \tag{10}$$

where r is the turning radius and v is the linear velocity of the robot, which is kept low and constant as is recommended for off road driving [4]. Equations (8)-(10) define an algorithm similar to the Pure Pursuit algorithm [15].

IV. EXPERIMENTAL SETUP AND RESULTS

The experiments were conducted on a Pionner 3-AT robotic platform. It was equipped with a laser range finder URG-04LX [16]. This laser has an angular resolution of $0.36°$, a scanning angle of $240°$, and a detection range of 0.02m-4m. In the current implementation, the laser readings were taken at 5 Hz. In addition, a tilt sensor was employed to obtain pitch and roll information with an accuracy of $\pm 0.2°$ (static measurements) and a sampling rate of 8 Hz [17].

The experimental evaluation was performed on soft dirt. It is important to note that the ruts created in this terrain type are structured similarly to the ruts typically encountered in off road trails as illustrated in Fig. 1. The evaluation of the algorithm on less structured ruts and different terrains is part of our current research. The depth of the ruts was in the range of $3 - 6cm$ which is comparable to the changes in elevation of the non-compacted terrain (i.e., the terrain that is not part of the ruts.)

A. Rut Following of an S-shaped Rut

An S-shaped rut, shown in Fig. 11, was chosen to evaluate the tracking performance of the algorithm. This particular shape was chosen because it includes both straight and curved regions.

Fig. 12 shows the raw laser readings corresponding to the scenario with the S-shaped rut. The figure also shows the rut detection results (filled circles), false alarms (filled stars) and two regions were the algorithm fails to detect the ruts. These false negatives can be caused by occlusions, excessive pitch of

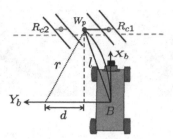

Fig. 10. Waypoint Assignment

TABLE I
RUT DETECTION PERFORMANCE S-SHAPED RUT

No of Rut Cross Sections	Detection Rate	False Alarm Rate
612	89%	16.67%

the robot, and in some situations (see, for example, Region 2 in Fig. 12) they are mainly caused by the relative orientation of the robot and the rut. However, it is important to notice that due to the online models of the left and right ruts, the robot was able to remain inside the ruts, despite the missed detections and the false alarms. Table I summarizes the rut detection results.

In order to quantify the tracking performance, define the cross-track error as $e_{ct}(l_p) \triangleq y(l_p) - y_{des}(l_p)$, where y_{des} is the desired path for the rear right wheel as a function of the path length (l_p) and y corresponds to the actual path followed by the rear right wheel. The RMS value of the cross-track error computed for two different trials was approximately $2cm$. The actual path followed by the wheel was manually measured by using a distinct mark left by the rear right wheel.

B. Rut Following with an Initial Position Offset

To test the ability of the proposed approach to track ruts that are not directly in front of the robot, the following experiment was performed. As shown in Fig. 13, the robot started its mission with an offset. This offset is a non dimensional quantity computed as the distance from the center of the right rut to the center of the front right wheel and normalized by the track width of the vehicle. Three experiments were conducted for offsets of 0.5, 1.0, and 1.5.

Fig. 13 shows the trajectory followed by the rear right wheel for the three different offsets. In all the trials the robot was able to find the ruts and position itself in the right location to follow the ruts. Table II summarizes the rut detection results for this experiment.

TABLE II
RUT DETECTION PERFORMANCE UNDER INITIAL POSITION OFFSET

No of Rut Cross Sections	Detection Rate	False Alarm Rate
328	82.9%	1.83%

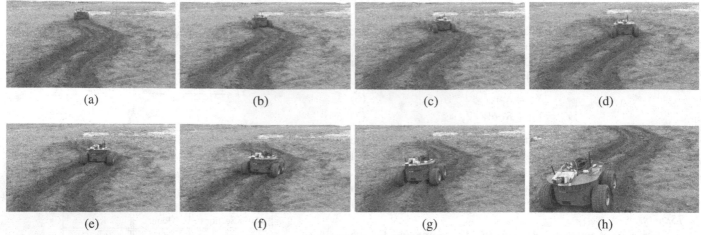

Fig. 11.   Pioneer 3-AT Following an S-Shaped Rut

V. IMPROVEMENTS TO THE PROPOSED APPROACH

This section presents a brief description of a set of improvements to the rut detection and following approach. The new features of the approach are introduced with the objective of improving the robustness of the algorithm by using a probabilistic framework to perform the rut detection and a tracking module, based on an Extended Kalman Filter (EKF), that exploits the spatio-temporal coherence that exists between consecutive rut detections and generates state estimates that directly feed a steering control system to follow the ruts.

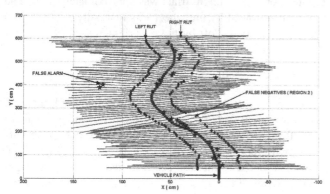

Fig. 12.   Terrain Map and Rut Detection Results

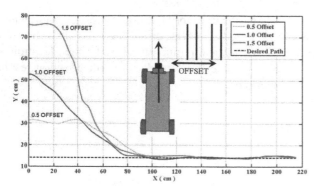

Fig. 13.   Wheel Path for Different Initial Position Offsets

A. *Probabilistic Based Rut Detection*

A probabilistic framework is selected because it provides the means to account for the uncertainty that arises in the rut detection process due to sensor noise and modeling approximations. Since ruts are expected to vary in shape depending on the terrain and vehicle, we propose to experimentally generate a set of rut templates obtained using rut samples from the range of traversable ruts. That is, ruts with a width in the range $[TW, 1.5TW]$ and with a depth in the range $[0.5BC, 0.8BC]$, where TW is the tire width and BC represents the body clearance. To improve computational efficiency, only 4 rut templates were used in the current implementation.

Once a laser scan is obtained, the rut templates are passed point by point through a search region (designated by the EKF, see subsection V-B) of the laser scan and the sum of squared errors between each of the templates and the laser points are computed for each position. Then, the minimum of these errors e_{min} is used as the feature to estimate the probability of the laser point being a rut center. These probabilities are computed using Bayes' theorem as follows:

$$p(w_j/e_{min}) = \frac{p(e_{min}/w_j)p(w_j)}{\sum_{j=1}^{2} p(e_{min}/w_j)p(w_j)}, \qquad (11)$$

where w_j represents the class of the measurement (rut or not rut) and $p(w_j)$ are the prior probabilities of each class, which are assumed equal to 0.5. The likelihoods ($p(e_{min}/w_j)$) are estimated using a maximum likelihood approach [18] and a training set which contains 100 rut samples. Fig. 14 illustrates the posterior probability estimates $p(Rut/e_{min})$ for each point of a laser scan that contains two ruts.

B. *Rut Tracking*

The rut tracking relies on an EKF that recursively estimates the lateral offset (y_{off}), the relative angle between the vehicle and the rut (θ_{vr}), and the parameters of the rut, which motivated by the work of [19] is here modeled locally as a curve of constant curvature (κ). The rut is modeled using frame R as illustrated in Fig. 15, which makes an angle θ_r

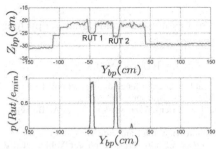

Fig. 14. Laser Data Containing Two Ruts (top) and Corresponding Probability Estimates of $p(RUT/e_{min})$ (bottom)

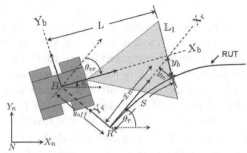

Fig. 15. Rut Frame Coordinates used by the Process and Measurement Models

with the inertial frame N and moves with the vehicle having the X_r axis tangent to the rut at all times.

Assuming that the vehicle moves with forward velocity v and angular velocity $\omega = \frac{d\theta_v}{dt}$, the evolution of θ_r, θ_{vr}, and y_{off} are computed using

$$\dot{\theta}_r = v\cos(\theta_{vr})\kappa, \tag{12}$$

$$\dot{\theta}_{vr} = \omega - v\sin(\theta_{vr})\kappa, \tag{13}$$

$$\dot{y}_{off} = v\sin(\theta_{vr}). \tag{14}$$

Using the backward Euler rule with sampling time δ_t and assuming that the evolution of the curvature is driven by white and Gaussian noise, it is possible to express the process model as

$$\begin{bmatrix} \theta_{vr_k} \\ \kappa_k \\ y_{off_k} \end{bmatrix} = \begin{bmatrix} \theta_{vr_{k-1}} - \kappa_{k-1}v\cos(\theta_{vr_{k-1}})\delta_t \\ \kappa_{k-1} \\ y_{off_{k-1}} + v\sin(\theta_{vr_{k-1}})\delta_t \end{bmatrix} + \begin{bmatrix} 1 \\ 0 \\ 0 \end{bmatrix}\delta\theta_{v_{k-1}} + w_{k-1}, \tag{15}$$

where $\delta\theta_{v_{k-1}}$ is the model input (the commanded change in vehicle heading and w represents the process noise, which is assumed white and with normal probability distribution with zero mean, and covariance Q $(p(w) \sim N(0,Q))$.

The measurement model corresponds to the lateral distance y_b from the vehicle X_b axis to the rut center, which is located at the intersection of the laser and the rut (see Fig. 15). Using geometry, it is possible to express y_b as

$$y_{b_k} = -\sin(\theta_{vr_k})x_m + \frac{1}{2}\kappa_k x_m^2\cos(\theta_{vr_k}) - y_{off_k}\cos(\theta_{vr_k}) + v_k, \tag{16}$$

where v is a white noise with normal probability distribution $(p(v) \sim N(0,R))$. As shown in Fig. 15, x_m is a function of the state $x_k = [\theta_{vr_k}, \kappa_k, y_{off_k}]^T$ and the lookahead distance (L) of the laser and satisfies

$$\frac{1}{2}x_m^2\kappa_k\sin(\theta_{vr_k}) + \cos(\theta_{vr_k})x_m - (L + y_{off_k}\sin(\theta_{vr_k})) = 0, \tag{17}$$

where (17) is obtained as a result of a coordinate transformation from the rut frame R to the vehicle frame B. As mentioned in subsection V-A, the measurement model of the EKF (16) is used to generate a prediction of the rut location for the next iteration and therefore allows the rut detection module to limit the search to a small region around the predicted value.

C. Steering Control for Rut Following

The state estimates generated by the EKF are then used by a nonlinear steering control law, which is an adaptation of the controller proposed in [20], which was designed for an Ackerman steered vehicle. In this work, we approximate the vehicle kinematics using a differential drive model.

The main objective of the controller is to drive the relative angle between the vehicle and the rut θ_{vr} to zero and the lateral offset y_{off} to a desired offset $y_{off_{des}} = \frac{RobW+TW}{2}$, where $RobW$ is the width of the robot and TW is the width of the tire. To achieve this, a desired angle for the vehicle $\theta_{v_{des}}$ is computed using a nonlinear steering control law as follows

$$\theta_{v_{des}} = \theta_r + \arctan(\frac{k_1(y_{off_{des}} - y_{off})}{v}), \tag{18}$$

where θ_r is the angle of the rut with respect to the global frame N, v is the robot velocity, and k_1 is a gain that controls the rate of convergence towards the desired offset. The desired angle $(\theta_{v_{des}})$ is then tracked using the proportional control law

$$\omega = k_2(\theta_v - \theta_{v_{des}}) = k_2(\arctan(\frac{k_1(y_{off_{des}} - y_{off})}{v}) - \theta_{vr}), \tag{19}$$

where ω is the commanded angular velocity for the robot. Notice that (19) takes as inputs the state estimates generated by the EKF.

D. Simulation Evaluation

To test the proposed approach, a computer simulation using Matlab was developed. A theoretical rut was simulated using a curved path with constant curvature $\kappa = 0.25m^{-1}$. The sensor measurements were simulated by finding the intersection of the laser L_1 with the rut as illustrated in Fig. 16. The lookahead distance was set to $45cm$ and the robot linear velocity was maintained constant at $20cm/s$. The process noise covariance Q was set to $Q = diag(1e-5, 2e-4, 1e-5)$ and the measurement noise covariance was set to $R = 1e-3$, which is 10 times larger than the typical variance for a laser sensor. The initial covariance estimate P_o was set equal to Q and the robot was originally placed parallel to the rut but with a lateral offset of $1m$. Notice that the desired offset is $y_{off_{des}} = \frac{RobW+TireW}{2} = 25cm$.

The first performance metric RMS_{TvsE} is the RMS error between the true and estimated offsets, where the true offset

111

is defined as the distance between the kinematic center of the vehicle B and the closest point on the rut and the estimated offset is the one estimated by the EKF. The second performance metric RMS_{EvsD} is the RMS error between the estimated offset and the desired offset at steady state . The average RMS values for 10 runs were $RMS_{TvsE} = 0.33cm$ and $RMS_{EvsD} = 0.9cm$. Notice that both of the RMS errors are very small. However, these errors are expected to increase in the physical experiments because the curvature of actual ruts changes continuously and there will be more uncertainty in the initial state estimates of the filter $x_0 = [\theta_{vr_0}, \kappa_0, y_{off_0}]^T$.

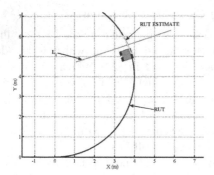

Fig. 16. Robot Following a Rut of Constant Curvature Using an EKF and the Proposed Steering Control

VI. CONCLUSIONS AND FUTURE WORK

A set of experiments on different robotic platforms and terrains were conducted to show the value of rut following for off road navigation. Then, the first stage of a rut detection and following system was designed, implemented and experimentally evaluated. The experimental results showed that the proposed system was able to detect and follow S-shaped ruts and it also showed its ability to follow ruts that have a lateral offset with respect to the robot. To increase the robustness of the proposed reactive system, a set of improvements including a probabilistic based rut detection approach and a tracking module based on an EKF were suggested and tested in simulation with promising results for future implementation.

A planner based subsystem needs to be developed to select the best rut to follow among several candidates and to provide a mechanism to initialize the EKF (i.e., provide the initial state values). In addition, a vision based approach to rut detection should be investigated because it would provide long range information to complement the current local information obtained with the laser range finder and open the possibility of detecting ruts based on different features (e.g., texture) and not only range. Therefore, shallower ruts could be detected.

ACKNOWLEDGMENT

Prepared through collaborative participation in the Robotics Consortium sponsored by the U. S. Army Research Laboratory under the Collaborative Technology Alliance Program, Cooperative Agreement DAAD 19-01-2-0012. The U. S. Government is authorized to reproduce and distribute reprints for Government purposes notwithstanding any copyright notation thereon.

REFERENCES

[1] W. Blevins. Land rover experience driving school. Class notes for Land Rover experience day, Biltmore, NC, 2007.
[2] T. Muro and J. O'Brien. *Terramechanics*. A.A. Balkema Publishers, 2004.
[3] Land rover lr3 overview mud and ruts. Available: http://www.landroverusa.com/us/en/Vehicles/LR3/Overview.htm,[Accesed: Aug. 12 2008].
[4] J. Allen. *Four-Wheeler's Bible*. Motorbooks, 2002.
[5] N. Baker. Hazards of mud driving. Available: http://www.overland4WD.com/PDFs/Techno/muddriving.pdf, [Accesed: Aug. 12 2008].
[6] 4x4 driving techniques. Available: http://www.ukoffroad.com /tech/driving.html [Accesed: Aug. 12 2008].
[7] M. Saarilahti and T. Anttila. Rut depth model for timber transport on moraine soils. In *Proceedings of the 9th International Conference of International Society for Terrain-Vehicle Systems, Munich, Germany*, 1999.
[8] C. Ordonez and E. Collins. Rut Detection for Mobile Robots. In *Proceedings of the IEEE 40th Southeastern Symposium on System Theory*, pages 334–337, 2008.
[9] J. Laurent, M. Talbot, and M. Doucet. Road Surface Inspection Using Laser Scanners Adapted for the High Precision 3D Measurements on Large Flat Surfaces. In *Proceedings of the IEEE International Conference on Recent Advances in 3D Digital Imaging and Modeling*, 1997.
[10] W. Ping, Z. Yang, L. Gan, and B. Dietrich. A Computarized Procedure for Segmentation of Pavement Management Data. In *Proceedings of Transp2000, Transportation Conference*, 2000.
[11] V. Leemand and M. F. Destain. Application of the hough transform for seed row localization using machine vision. *Biosystems Engineering*, 94:325–336, 2006.
[12] K. Nagatani G. Reina, G. Ishigami and K. Yoshida. Vision-based Estimation of Slip Angle for Mobile Robots and Planetary Rovers. In *Proceedings of the IEEE International Conference on Robotics and Automation*, 2008.
[13] Z. Kim. Realtime Lane Tracking of Curved Local Road. In *Proceedings of the IEEE Intelligent Transportation Systems Conference*, 2006.
[14] X. Hu Y. Zhou, R. Xu and Q. Ye. A robust lane detection and tracking method based on computer vision. *Measurement Science and Technology*, 17:736–745, 2006.
[15] R. C. Coulter. Implementation of the pure pursuit path tracking algorithm. Technical Report CMU-RI-TR-92-01, Robotics Institute, Carnegie Mellon University, 1992.
[16] LTD Hokuyo Automatic Co. *Range Finder Type Laser Scanner URG-04LX Specifications*.
[17] PNI Corporation. *TCM2 Electronic Compass Module User's Manual*.
[18] R. Duda, P. Hart, and D. Stork. *Pattern Classification*. John Wiley & Sons, INC., 2001.
[19] L.B. Cremean and R.M. Murray. Model-based estimation of off-highway road geometry using single-axis ladar and inertial sensing. pages 1661–1666, May 2006.
[20] Sebastian Thrun, Mike Montemerlo, Hendrik Dahlkamp, David Stavens, Andrei Aron, James Diebel, Philip Fong, John Gale, Morgan Halpenny, Gabriel Hoffmann, Kenny Lau, Celia Oakley, Mark Palatucci, Vaughan Pratt, Pascal Stang, Sven Strohband, Cedric Dupont, Lars-Erik Jendrossek, Christian Koelen, Charles Markey, Carlo Rummel, Joe van Niekerk, Eric Jensen, Philippe Alessandrini, Gary Bradski, Bob Davies, Scott Ettinger, Adrian Kaehler, Ara Nefian, and Pamela Mahoney. Stanley: The robot that won the darpa grand challenge: Research articles. *J. Robot. Syst.*, 23(9):661–692, 2006.

Unsupervised Discovery of Object Classes from Range Data Using Latent Dirichlet Allocation

Felix Endres[1]   Christian Plagemann[2]   Cyrill Stachniss[1]   Wolfram Burgard[1]

[1] University of Freiburg, Dept. of Computer Science, Georges Koehler Allee 79, 79110 Freiburg, Germany
[2] Stanford University, 353 Serra Mall, Gates Building – Office 244, Stanford, CA 94305-9010, USA

Abstract— Truly versatile robots operating in the real world have to be able to learn about objects and their properties autonomously, that is, without being provided with carefully engineered training data. This paper presents an approach that allows a robot to discover object classes in three-dimensional range data in an unsupervised fashion and without a-priori knowledge about the observed objects. Our approach builds on Latent Dirichlet Allocation (LDA), a recently proposed probabilistic method for discovering topics in text documents. We discuss feature extraction, hypothesis generation, and statistical modeling of objects in 3D range data as well as the novel application of LDA to this domain. Our approach has been implemented and evaluated on real data of complex objects. Practical experiments demonstrate, that our approach is able to learn object class models autonomously that are consistent with the true classifications provided by a human. It furthermore outperforms unsupervised method such as hierarchical clustering that operate on a distance metric.

I. INTRODUCTION

Home environments, which are envisioned as one of the key application areas for service robots, typically contain a variety of different objects. The ability to distinguish objects based on observations and to relate them to known classes of objects therefore is important for autonomous service robots. The identification of objects and their classes based on sensor data is a hard problem due to the varying appearances of the objects belonging to specific classes. In this paper, we consider a robot that can observe a scene with a 3D laser range scanner. The goal is to perform

- unsupervised learning of a model for object classes,
- consistent classification of the observed objects, and
- correct classification of unseen objects belonging to one of the known object classes.

Figure 1 depicts a typical point cloud of a scene considered in this paper. It contains four people, a box, and a balloon-like object. The individual colors of the 3D data points illustrate the corresponding object classes that we want our algorithm to infer.

An important distinction between different approaches to object detection and recognition is the way the objects or classes are modeled. Models can be engineered manually, learned from a set of labeled training data (supervised learning) or learned from unlabeled data (unsupervised learning). While the former two categories have the advantage that detailed prior knowledge about the objects can be included easily, the effort for manually building the model or labeling

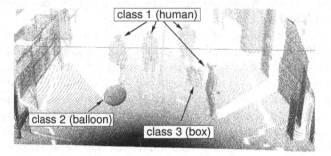

Fig. 1: Example of a scene observed with a laser range scanner mounted on a pan-tilt unit. Points with the same color resemble objects belonging to the same class.

a significant amount of training data becomes infeasible with increasing model complexity and larger sets of objects to identify. Furthermore, in applications where the objects to distinguish are not known beforehand, a robot needs to build its own model, which can then be used to classify the data.

The contribution of this paper is a novel approach for discovering object classes from range data in an unsupervised fashion and for classifying observed objects in new scans according to these classes. Thereby, the robot has no a-priori knowledge about the objects it observes. Our approach operates on a 3D point cloud recorded with a laser range scanner. We apply Latent Dirichlet Allocation (LDA) [2], a method that has recently been introduced to seek for topics in text documents [9]. The approach models a distribution over feature distributions that characterize the classes of objects. Compared to most popular unsupervised clustering methods such as k-means or hierarchical clustering, no explicit distance metric is required. To describe the characteristics of surfaces belonging to objects, we utilize spin-images as local features that serve as input to the LDA. We show in practical experiments on real data that a mobile robot following our approach is able to identify similar objects in different scenes while at the same time labeling dissimilar objects differently.

II. RELATED WORK

The problem of classifying objects and their classes in 3D range data has been studied intensively in the past. Several authors introduced features for 3D range data. One popular free-form surface descriptor are spin-images, which have been applied successfully to object recognition problems [13; 12; 14; 15]. In this paper, we propose a variant of spin-images that—instead of storing point distributions of the surface—

stores the angles between the surface normals of points, which we found to yield better results in our experiments. An alternative shape descriptor has been introduced by [18]. It relies on symbolic labels that are assigned to regions. The symbolic values, however, have to be learned from a labeled training set beforehand. Stein and Medioni [19] present a point descriptor that, similar to our approach, also relies on surface orientations. However, it focuses on the surface normals in a specific distance to the described point and models their change with respect to the angle in the tangent plane of the query point. Additional 3D shape descriptors are described in [5] and [6].

A large amount of work has focused on supervised algorithms that are trained to distinguish objects or object classes based on a labeled set of training data. For example, Anguelov *et al.* [1] and Triebel *et al.* [20] use supervised learning to classify objects and associative Markov networks to improve the results of the clustering by explicitly considering relations between the class predictions. In a different approach, Triebel *et al.* [21] use spin-images as surface descriptors and combine nearest neighbor classification with associative Markov networks to overcome limitations of the individual methods. Another approach using probabilistic techniques and histogram matching has been presented by Hetzel *et al.* [10]. It requires a complete model of the object to be recognized, which is an assumption typically not fulfilled when working on 3D scans recorded with a laser range finder. Ruhnke *et al.* [17] proposed an approach to reconstructing full 3D models of objects by registering several partial views. The work operates on range images from which small patches are selected based on a region of interest detector.

In addition to the methods that operate on 3D data, much research has also focused on image data as input. A common approach to locate objects in images is the sliding window method [4; 7]. Lampert *et al.* [16] proposed a new framework that allows to efficiently find the optimal bounding box without applying the classification algorithm explicitly to all possible boxes. Another prominent supervised detector is the face detector presented by Viola and Jones [22]. It computes Haar-like features and applies AdaBoost to learn a classifier.

In the domain of unsupervised classification of text documents, several models that greatly surpass mere counting of words have been proposed. These include probabilistic latent semantic indexing (PLSI) [11] and Latent Dirichlet Allocation [2], which both use the co-occurrence of words in a probabilistic framework to group words into topics. In the past, LDA has also been applied successfully to image data. In contrast to text documents [9], images often contain data of many different categories. Wang and Grimson [23], therefore, first perform a segmentation before applying LDA. Bosch *et al.* [3] used PLSI for unsupervised discovery of object distributions in image data. As shown in [8], LDA supersedes PLSI and it has been argued that the latter can be seen as a special case of LDA, using a uniform prior and maximum a posteriori estimation for topic selection. Fritz and Schiele [7] propose the sliding window approach on a grid of

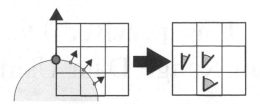

Fig. 2: Variant of spin-images used to compute a surface signature: the 3D object structure is rotated around the surface normal of a query point (large point) and a grid model accumulates the average angular distances between the surface normal at the query point and those of the points falling into the grid cells (small points).

edge orientations to evaluate topic probabilities on subsets of the whole image. While the general approach of these papers is related to ours, to the best of our knowledge the algorithm described in this paper is the first to apply LDA on laser range data and which addresses the specific requirements of this domain.

III. DATA PRE-PROCESSING AND LOCAL SHAPE FEATURES

As most approaches to object detection, identification, and clustering, we operate on local features computed from the input data. Our primary focus lies on the description of *shape* as this is the predominant feature captured in 3D range data. However, real-world objects belonging to the same class do not necessarily have the same shape and vice versa. Humans, for example, have a significant variability in shape. To deal with this problem, we model classes of objects as distributions of local shape features.

In the next sections, we first describe our local feature used to represent the characteristics of surfaces and after than, we address the unsupervised learning problem to estimate the distributions over local features.

A. Representation and Data Pre-processing

Throughout this work, we assume our input data to be a point cloud of 3D points. Such a point cloud can be obtained with a 2D laser range finder mounted on a pan-tilt unit, a standard setting in robotics to acquire 3D range data. An example point cloud recorded with this setup is shown in the motivating example in Figure 1 on the first page of this paper.

As in nearly all real world settings, the acquired data is affected by noise and it is incomplete due to perspective occlusions. The segmentation of range scans into a set of objects and background structure is not the key focus of this work. We therefore assume a ground plane as well as walls that can be easily extracted and assume the objects to be spatially disconnected. This allows us to apply a spatial clustering algorithm to create segments containing only one object.

B. Local Shape Descriptors

For characterizing the local shape of an object at a query point, we propose to use a novel variant of spin-images [12]. Spin-images can be seen as small raster images that are aligned to a point such that the upwards pointing vector of the raster image is the surface normal of the point. The image is then virtually rotated around the surface normal, "collecting" the neighboring points it intersects. To account for the differences in data density caused by the distance between sensor and object, the spin-images are normalized.

To actually compute a normal for each data point, we compute a PCA using all neighboring points in a local region of 10cm. Then, the direction of the eigenvector corresponding to the smallest eigenvalue provides a comparably stable but smoothed estimate of the surface normal.

We have developed a variant of spin-images that does not count the points "collected" by the pixels of the raster image. Instead, we compute the average angle between the normal of the query point for which the spin-image is created and the normals of all collected points. See Figure 2 for an illustration. The average between the normals is then discretized to obtain a discrete feature space, as required in the LDA approach. As we will show in our experiments, this variant of spin-images provides better results, since they contain more information about the shape of the object.

IV. PROBABILISTIC TOPIC MODELS FOR OBJECT SHAPE

After segmenting the scene into a finite set of scan segments and transforming the raw 3D input data to the discrete feature space, the task is to group similar segments to classes and to learn a model for these classes. Moreover, we aim at solving the clustering and modeling problems simultaneously to achieve a better overall model. Inspired by previous work on topic modeling in text documents, we build on Latent Dirichlet Allocation for the unsupervised discovery of object classes from feature statistics.

Following this model, a multinomial distribution is used to model the distribution of discrete features in an object class. Analogously, another multinomial distribution is used to model the mixture of object classes which contribute to a scan segment. In other words, we assume a generative model, in which (i) segments generate mixtures of classes and (ii) classes generate distributions of features.

Starting from a prior distribution about these latent (i.e., hidden) mixtures, we update our belief according to the observed features. To do this efficiently, we express our prior $P(\theta)$ as a distribution that is conjugate to the observation likelihood $P(y \mid \theta)$. $P(\theta)$ being a conjugate distribution to $P(y \mid \theta)$ means that

$$P(\theta \mid y) = \frac{P(y \mid \theta)P(\theta)}{\int P(y \mid \theta)P(\theta)\,d\theta} \qquad (1)$$

is in the same family as $P(\theta)$ itself. For multinomial distributions, the conjugate prior is the Dirichlet distribution, which we explain in the following.

A. The Dirichlet Distribution

The Dirichlet distribution is a distribution over multivariate probability distributions, i.e., a distribution assigning a probability density to every possible multivariate distribution. For the multinomial variable $\mathbf{x} = \{x_1, \ldots, x_K\}$ with K exclusive states x_i, the Dirichlet distribution is parameterized by a vector

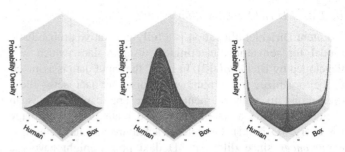

Fig. 3: Three Dirichlet distributions. On the left for the parameter vector $\alpha = \{2,2,2\}$, in the middle for $\alpha = \{3,6,3\}$ and on the right for $\alpha = \{0.1,0.1,0.1\}$.

$\alpha = \{\alpha_1, \ldots, \alpha_K\}$. If $\alpha_i = 1$ for all i, the Dirichlet distribution is uniform. One can think of $(\alpha_i - 1)$ for $\alpha_i \in \mathbb{N}^{>0}$ as the number of observations of the state i. The Dirichlet distribution can be calculated as

$$f(\mathbf{x}) = \underbrace{\frac{\Gamma(\sum_{i=1}^{K} \alpha_i)}{\prod_{i=1}^{K} \Gamma(\alpha_i)}}_{\text{Normalization}} \prod_{i=1}^{K} x_i^{\alpha_i - 1}, \qquad (2)$$

where $\Gamma(\cdot)$ is the Gamma function and where the elements of $\mathbf{x}$ have to be positive and sum up to one.

Consider the following example: let there be three object classes "human", "box", and "chair" with a Dirichlet prior parameterized by $\alpha = \{2,2,2\}$. This prior assigns the same probability to all classes and hence results in a *symmetric* Dirichlet distribution. A 3D Dirichlet distribution $Dir(\alpha)$ can be visualized by projecting the the manifold where $\sum \alpha_i = 1$ to the 2D plane, as depicted in the left plot of Figure 3. Here the third variable is given implicitly by $\alpha_3 = 1 - \alpha_1 - \alpha_2$. Every corner of the depicted triangle represents the distributions where only the respective class occurs and the center point represents the uniform distribution over all classes. Now consider an observation of one human, four boxes, and a chair. By adding the observation counts to the elements of α, the posterior distribution becomes $Dir(\{5,8,5\})$ which is shown in the middle plot in Figure 3. The same result would of course occur when calculating the posterior using Eq. (1).

However choosing the values of α_i larger than 1 favors distributions that represent mixtures of classes, i.e. we expect the classes to occur together. To express a prior belief that either one or the other dominates we need to choose values smaller than 1 for all α_i. The shape of the distribution then changes in a way that it has a "valley" in the middle of the simplex and peaks at the corners. This is depicted in the right plot in Figure 3. In our setting, where a Dirichlet distribution is used to model the distribution of object classes, such a prior would correspond to the proposition that objects are typically assigned to one (or only a few) classes.

The calculation of the *expected probability distribution* over the states and can be performed easily based on α. The expected probability for x_i is given by

$$\mathbb{E}[x_i] = \frac{\alpha_i}{\sum_{i'} \alpha_{i'}}. \qquad (3)$$

B. Latent Dirichlet Allocation

Latent Dirichlet allocation is a fully generative probabilistic model for semantic clustering of discrete data, which was developed by Blei *et al.* [2]. In LDA, the input data is assumed to be organized in a number of discrete data sets—these correspond to scan segments in our application. The scan segments contain a set of discretized features (a spin image for every 3D point). Obviously, a feature can have multiple *occurrences* since different 3D data points might have the same spin image. Often, the full set of data (from multiple scans) is referred to as "corpus". A key feature of LDA is that it does not require a distance metric between features as most approaches to unsupervised clustering do. Instead, LDA uses the co-occurrence of features in scan segments to assign them probabilistically to classes—called *topics* in this context.

Being a generative probabilistic model, the basic assumption made in LDA is that the scan segments are generated by random processes. Each random process represents an individual topic. In this work, we distinguish topics using the index j and scan segments are indexed by d. A random process generates the features in the segments by sampling them from its own specific discrete probability distribution $\phi^{(j)}$ over the features. A segment can be created by one or more topics, each topic having associated a distinct probability distribution over the features.

To represent the mixture of topics in a segment d, a multinomial distribution $\theta^{(d)}$ is used. For each feature in the segment, the generating topic is selected by sampling from $\theta^{(d)}$. The topic mixture $\theta^{(d)}$ itself is drawn from a Dirichlet distribution once for every segment in the corpus. The Dirichlet distribution represents the prior belief about the topic mixtures that occur in the corpus, i.e., whether the segments are generated by single topics or from a mixture of many topics. We express the prior belief with respect to the topic distribution using the Dirichlet parameter vector α.

Griffiths and Steyvers [9] extended LDA by additionally specifying a Dirichlet prior $Dir(\beta)$ on the conditional distributions $\phi^{(j)}$ over the features. This prior is useful in our application since it enables us to model a preference for selecting few characteristic features of a topic.

C. Learning the Model

In this section, we describe how to find the assignments of topics to 3D data points in range scans following the derivation of Griffiths and Steyvers [9]. Given the corpus $\mathbf{w} = \{w_1, w_2, ...w_n\}$ as the set of all feature occurrences, where each occurrence w_i belongs to exactly one scan segment. We are then looking for the most likely topic assignment vector $\mathbf{z} = \{z_1, z_2, ...z_n\}$ for our data $\mathbf{w}$. Here, each z_i is an index referring to topic j that generated w_i. Hence, we seek to estimate the probability distribution $P(\mathbf{z} \mid \mathbf{w})$. Based on $P(\mathbf{z} \mid \mathbf{w})$, we can then obtain the most likely topic assignment for each 3D data point. Using Bayes rule, we know that

$$P(\mathbf{z} \mid \mathbf{w}) = \frac{P(\mathbf{w} \mid \mathbf{z})P(\mathbf{z})}{P(\mathbf{w})}. \tag{4}$$

Unfortunately, the partition function $P(\mathbf{w})$ is not known and cannot be computed directly because it involves T^N terms, where T is the number of topics and N is the number of feature occurrences.

A common approach to approximate a probability distribution, for which the partition function $P(\mathbf{w})$ is unknown, is Markov chain Monte Carlo (MCMC) sampling. MCMC approximates the target distribution $P(\mathbf{z} \mid \mathbf{w})$ by randomly initializing the states of the variables—here the topic assignments. Subsequently, it samples new states using a Monte Carlo transition function leading to the target distribution. Therefore, the target distribution has to be the equilibrium distribution of the transition function. The transition function obeys the Markov property, i.e., it is independent of all states but the last. In our approach, we use Gibbs sampling as the transition function where the new state (the topic assignment) for each feature occurrence is sampled successively.

Gibbs sampling requires a *proposal distribution* to generate new states. Therefore, the next section describes how to obtain an appropriate proposal distribution for our problem.

D. Computing the Proposal Distribution for Gibbs Sampling

The proposal probability distribution over the possible topic assignments of a feature occurrence is calculated conditioned on the current assignments of the other feature occurrences. A new topic assignment is then sampled from this proposal distribution.

For estimating $P(\mathbf{z} \mid \mathbf{w})$, we successively sample from the distribution in the numerator on the right hand side of Eq. (4) the topic assignment z_i for each feature occurrence w_i given the topics of all other features. The distribution over the topics for sampling z_i is given by

$$P(z_i = j \mid \mathbf{z_{-i}}, \mathbf{w}) = \frac{\overbrace{P(w_i|z_i = j, \mathbf{z}_{-i}, \mathbf{w}_{-i})}^{\text{likelihood of } w_i}\overbrace{P(z_i = j|\mathbf{z}_{-i})}^{\text{prior of } z_i}}{\sum_{j=1}^{T} P(w_i|z_i = j, \mathbf{z}_{-i}, \mathbf{w}_{-i})P(z_i|\mathbf{z}_{-i})}. \tag{5}$$

In Eq. (5), $\mathbf{w}_{-i}$ denotes the set $\mathbf{w}$ without w_i and $\mathbf{z}_{-i}$ the corresponding assignment vector. We can express the conditional distributions in the nominator of Eq. (5) by integrating over ϕ and θ, where ϕ denotes the feature distribution of all topics and θ denotes the topic distribution for each scan segment.

The likelihood of w_i in Eq. (5) depends on the probability of the distribution of topic j over features, so we need to integrate over all these distributions $\phi^{(j)}$:

$$P(w_i = w \mid z_i = j, \mathbf{z}_{-i}, \mathbf{w}_{-i}) =$$
$$\int \underbrace{P(w_i = w \mid z_i = j, \phi^{(j)})}_{\phi_w^{(j)}} \underbrace{P(\phi^{(j)} \mid \mathbf{z}_{-i}, \mathbf{w}_{-i})}_{\text{posterior of } \phi^{(j)}} d\phi^{(j)} \tag{6}$$

Since the Dirichlet distribution is conjugate to the multinomials (to which $\phi^{(j)}$ belongs to), this posterior can be computed easily from the prior and the observations by adding the observations to the respective elements of the parameter vector β of the prior (see also Section IV-A). As a result, we obtain a Dirichlet posterior with parameter vector $\beta + n_{-i,j}^{(w)}$

where the elements of $n_{-i,j}^{(w)}$ are the number of occurrences of feature w assigned to topic j by the assignment vector $\mathbf{z}_{-i}$.

The first term on the right hand side of Eq. (6) is the probability for feature w under the multinomial $\phi^{(j)}$ and the second term denotes the probability of that multinomial. Therefore, solving this integral results in computing the expectation of $\phi_w^{(j)}$ which is the probability of w under $\phi^{(j)}$. According to Eq. (3), this expectation can be easily computed. The probability that an occurrence w_i is feature w is

$$P(w_i = w \mid z_i = j, \mathbf{z}_{-i}, \mathbf{w}_{-i}) = \mathbb{E}(\phi_w^{(j)}) = \frac{n_{-i,j}^{(w)} + \beta_w}{\sum_{w'} n_{-i,j}^{(w')} + \beta_{w'}}. \quad (7)$$

In the same way, we integrate over the multinomial distributions over topics θ, to find the prior of z_i from Eq. (5). With d_i being the index of the scan segment to which w_i belongs, we can compute the probability of a topic assignment for feature occurrence w_i as:

$$P(z_i = j \mid \mathbf{z}_{-i}) = \int \underbrace{P(z_i = j \mid \theta^{(d_i)})}_{\theta_j^{(d_i)}} \underbrace{P(\theta^{(d_i)} \mid \mathbf{z}_{-i})}_{\text{posterior of } \theta^{(d_i)}} d\theta^{(d_i)} \quad (8)$$

Let $n_{-i,j}^{(d_i)}$ be the number of features in the scan segment d_i that are assigned to topic j. Then, analogous to Eq. (7), the expected value of $\theta_j^{(d_i)}$ can be calculated by adding $n_{-i,j}^{(d_i)}$ to the elements of the parameter vector α of the prior:

$$P(z_i = j \mid \mathbf{z}_{-i}) = \mathbb{E}(\theta_j^{(d_i)}) = \frac{n_{-i,j}^{(d_i)} + \alpha_j}{\sum_{j'} n_{-i,j'}^{(d_i)} + \alpha_{j'}} \quad (9)$$

Combining the results of Eq. (7) and (9) in Eq. (5), we obtain the *proposal distribution* for the sampling of z_i as

$$P(z_i = j \mid \mathbf{z}_{-i}, \mathbf{w}) \propto \frac{n_{-i,j}^{(w)} + \beta_w}{\sum_{w'} n_{-i,j}^{(w')} + \beta_{w'}} \frac{n_{-i,j}^{(d_i)} + \alpha_j}{\sum_{j'} n_{-i,j'}^{(d_i)} + \alpha_{j'}} . \quad (10)$$

Eq. (10) is the proposal distribution used in Gibbs sampling to obtain next generation of assignments.

After a random initialization of the Markov chain, a new state is generated by drawing the topic for each feature occurrence successively from the proposal distribution. From these samples, the distributions θ and ϕ can be estimated by using the sampled topic assignments $\mathbf{z}$.

Note that in our work, we restrict the Dirichlet priors to be symmetric. This implies that all topics and all features have the same initial prior occurrence probability. As a result, we only have to specify only value for the elements of the parameter vectors α and β which we denote by $\hat{\alpha}$ and $\hat{\beta}$. This leads to:

$$\phi_j^{(w)} \sim \frac{n_j^{(w)} + \hat{\beta}}{\left(\sum_{w'} n_j^{(w')}\right) + W\hat{\beta}} \qquad \theta_j^{(d)} \sim \frac{n_j^{(d)} + \hat{\alpha}}{\left(\sum_{j'} n_{j'}^{(d)}\right) + T\hat{\alpha}} \quad (11)$$

where T is the number of topics and W the number of features.

To summarize, we explained how to compute the proposal distribution in Eq. (10) used in Gibbs sampling during MCMC. The obtained samples can then be used to estimate the distributions ϕ and θ. Due to our restriction to symmetric priors, only two parameters ($\hat{\alpha}, \hat{\beta} \in \mathbb{R}$) have to be specified.

E. Unsupervised Topic Discovery and Classification of Newly Observed Objects

This section briefly summarizes how the components presented so far are integrated to perform the unsupervised discovery of object classes and the classification when new observations are made.

First of all, we preprocess the data according to Section III-A to extract the scan segments which correspond to objects in the scene and for which we aim to learn a topic model. For each data point in a scan segment, we compute our feature, a variant of the spin-image, according to Section III-B to describe the surfaces characteristics.

For the discovery of topics, we then compute the feature distributions ϕ of the object classes as well as the topic mixtures θ for the scan segments using MCMC as described in the previous section. The learned distributions θ denote a probabilistic assignment of objects to topics.

Class inference, that is, the classification of objects contained in *new* scenes can be achieved using the feature distribution ϕ. In this case, ϕ and θ can be used to compute the proposal distribution directly and are not updated.

Note that the approach presented here does not automatically determine the number of object classes. This is similar to other unsupervised techniques such as k-means clustering or EM-based Gaussian mixture models in which the number of object classes is assumed to be known. We experimentally evaluated settings in which the number of topics was higher or lower than the number of manually assigned classes in the data set. Our observation was that a higher number of topics leads to the detection of shape classes such as "corner", "edge", or "flat surface" and that the objects are modeled as mixtures of those.

F. The Influence of the Dirichlet Priors $\hat{\alpha}$ and $\hat{\beta}$

Two hyperparameters $\hat{\alpha} \in \mathbb{R}$ and $\hat{\beta} \in \mathbb{R}$ need to be provided as the input to the presented approach. They define the prior distributions for the mixture of object classes in a data set and for the mixture of features in an object class respectively.

As briefly discussed in Section IV-A, choosing $\hat{\alpha}$ larger than one favors the occurrence of many topics in each scan segment, while lower values result in less topics per scan segment. Similarly, the lower the hyperparameter $\hat{\beta}$ for the Dirichlet distribution over the features, the stronger the preference for fewer features per topic and unambiguous ones. Due to the segmentation in the preprocessing step, we assume that there are only few topics per scan segment and thus a low value for the hyperparameter is favored in this setting. For $\hat{\beta}$ holds: On the one hand different objects can yield the same individual features (yet in distinct distributions). On the other hand, we expect features to be related to specific topics.

From this intuitions about the Dirichlet parameters, a high performance can be expected if both parameters are selected between zero and one. This could be confirmed experimentally and the results are given in Section V-D, where we analyze the influence of the hyperparameters on manually labeled data sets.

Fig. 4: Example point cloud segments of Corpus-A (box, balloon) and Corpus-B (box, balloon, human, swivel chair, chair)

V. EXPERIMENTAL EVALUATION

In this section, we present experiments carried out to evaluate our approach on recorded data. All results are based on scans of real scenes acquired with an ActivMedia pioneer robot equipped with a SICK LMS range finder mounted on a Schunk pant-tilt unit. No simulator was involved in the evaluation.

The goal of the evaluation is to answer the following questions: (i) Are the proposed local shape features in conjunction with the topic model approach expressive enough to represent real-world objects? (ii) Is the approach able to discover object classes from unlabeled point clouds and are these classifications consistent with human-provided class labels? (iii) How does our LDA-based approach compare to conceptually simpler approaches for unsupervised clustering? (iv) How sensitive is the proposed algorithm w.r.t to the choice of parameters for the feature extraction step as well as of the Dirichlet priors?

A. Test Data

For the experimental evaluation, we prepared and re-arranged indoor scenes containing five different object types: balloons, boxes, humans, and two types of chairs. In total, we recorded 51 full laser-range scans containing 121 object instances. The first part of this data set is termed *Corpus-A*. It contains 31 object instances of low geometric complexity (different boxes and balloons). The second and larger part comprising of 82 object instances, *Corpus-B*, additionally contains complex and variable shapes of chairs and humans. See Figure 4 for examples of such object segments represented as 3D point clouds.

The data was acquired and pre-processed as described in Section III-A. Some difficulties, inherent in 3D data recorded in this way, should be pointed out: Only one side of an object can be recorded and non-convex objects typically occlude themselves partially. Objects were scanned from different view points and thus different parts are observed. Different objects of the same class were scanned (different humans, different chairs, etc.). Metal parts, such as the legs of chairs, reflect the laser beams and, thus, are invisible to the sensor. Finally, local shape features extracted from the scans of humans are highly diverse compared to the simpler objects.

Figure 5 shows typical classification results achieved by our algorithm when applied to entire scans in three example

scenes. Here, the points are color-coded according to their class assignments (elements of Corpus-A on the left and Corpus-B in the middle and on the right). The labels assigned to the individual points are taken from a sample of the posterior distribution $P(\mathbf{z} \mid \mathbf{w})$ as generated during the clustering process. It can be seen that the point labels are almost perfectly consistent within each object segment and, thus, the maximum likelihood class assignment per segment is unambiguous.

In addition to that, Figure 6 gives a visual impression of the topics assigned by our approach to the 82 scan segments of Corpus-B. The labels in this diagram show the true object class. Each color in the diagram denotes one topic and the ratios of colors denote for each object segment the class assignment weight. As the diagram shows, except of one chair, all objects are grouped correctly when using the maximum likelihood assignment.

We furthermore analyzed the runtime requirements of our approach, disregarding the time for pre-processing and the computation of the spin images. In Corpus-B (82 objects from 39 different 3D scans, 300 000 spin image in total), it took less than 20 s to learn the topic distributions via MCMC and to classify the objects. Thus, the computation time per 3D scan is around 500 ms which is faster than the time needed to record a 3D scan.

B. Clustering by Matching Shape Histograms

In order to compare our LDA-based approach to an unsupervised clustering technique, we implemented hierarchical clustering (HC) using the similarity between spin-image histograms as the distance metric. In this implementation, we build a feature histogram for each object segment by counting the occurrences of the individual spin-images from the (finite) spin-image dictionary (see. Section III-B). To compare two scan segments, we first normalize their histograms to sum up to one over all bins. Among the popular measures for comparing histograms, namely histogram intersection [10], χ^2 distance, and the Kullback Leibler divergence (KL-D), histogram intersection appeared to provide the best results in our domain. This is due to the fact that the χ^2 distance and the KL-D are heavily influenced by features with few or no occurrences—an effect that can be observed frequently in our data sets. The quantitative results comparing LDA to HC are given in Table I. As can be seen for the simpler setting of Corpus-A, HC gives acceptable results but is still outperformed by LDA. In the more complex setting of Corpus-B, however, HC was not able to find a good clustering of the scene. In multiple runs using different setups, we found that the difference is statistically significant.

Figure 7 visualizes the similarity matrix between scan

TABLE I: Summary of the classification results on the test data sets. The percentages give the average correct classifications achieved by hierarchical clustering (HC) and the proposed model based on LDA.

Data set	No. of scenes	No. of segments	HC	LDA
Corpus-A	12	31	94.84%	99.89%
Corpus-B	39	82	71.19%	90.38%

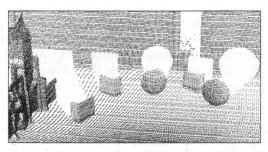  

Fig. 5: Example classification results on test scans from Corpus-A (left) and Corpus-B (middle and right). The detected object classes are colored according to the LDA-assigned shape model.

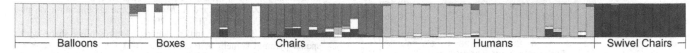

Fig. 6: Resulting topic mixtures θ for 82 segments of Corpus-B computed via LDA (the labels were not provided to the system).

segments obtained using histogram intersection. Due to their rather uniform shape, balloons can be well distinguished from other objects. Objects with a more complex shape, however, are confused easily. This indicates that approaches working only based on such a distance metric are likely operate less accurately in more complex scenes. In contrast to that, LDA considers distributions of features and their dependencies and therefore perform substantially better.

C. Parameters of the Spin-Image Features

In this experiment, we analyzed the difference of the clustering performance when the regular spin-images (referred to as "Type 1") and our variant (referred to as "Type 2") is used. We also investigated the influence of the parameters used to create the features. These parameters are (i) the support distance, i.e., the size of the spinning image, (ii) the grid resolution, and (iii) the discretization of the stored values.

To compare the two alternative types of spin images, we collected statistics measuring the LDA clustering performance on a labeled test set, integrating over the three feature parameters. That way, we analyzed 10 780 different parameter settings—each for regular spin-images and for our variant. Figure 8 shows the results of this experiment as a histogram. The higher the bars on the right hand side of the histogram, the better the results. As can be seen, our approach outperforms regular spin-images.

In addition to that, we computed the clustering performance of our approach and HC for a wide variety of feature parameters using Corpus-B. Figure 9 shows the results for HC and LDA. Again, our approach clearly outperforms HC. The broad spread of high classification rates over the range of parameters demonstrates that the results presented in the previous section were not caused by selecting feature parameters that were suboptimal for HC.

We observe that for smaller support distances, a higher discretization resolutions work well and vice versa. The intuition for this finding is that feature distributions with a large support and a very accurate discretization have overly detailed features, that do not match the distributions of other segments well.

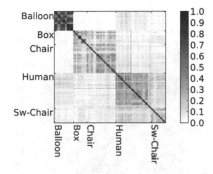

Fig. 7: Visualization of the confusion matrix of classification based on matching spin-image histograms.

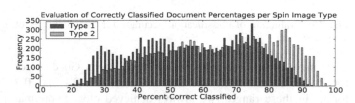

Fig. 8: Classification using standard spin-image features ("Type 1") generally labels less documents correctly than classification upon the features we proposed ("Type 2").

The best results in our setting are obtained for features with a discretization resolution between 5 and 27 and a rather short support distance. In conclusion we see, that choosing such parameters for the feature generation, we can achieve over 90 % correct classifications (compare lower plot in Figure 9).

D. Sensitivity of the Dirichlet Priors

We furthermore evaluated how sensitive our approach is with respect to the choice of the parameters $\hat{\alpha}$ and $\hat{\beta}$ for the Dirichlet priors. Figure 10 depicts the average classification rates for varying parameters. In this plot, we integrate over the three feature parameters in a local region around the values determined in the previous experiment to illustrate how robust LDA performs. As can be seen from Figure 10, determining the hyperparameters is not a critical task since the performance

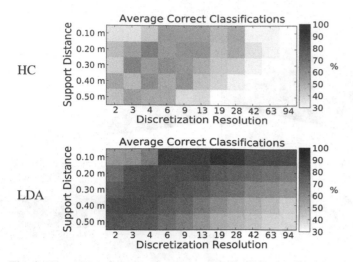

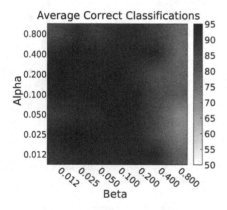

Fig. 9: Classification accuracy on Corpus-B for different discretization resolutions and respect to support distances for HC (top) and LDA (bottom).

Fig. 10: Evaluation of classification accuracy for various values of alpha and beta.

stays more or less constant when varying them. Good values for $\hat{\alpha}$ lie between 0.1 and 0.8 and between 0.1 and 0.3 for $\hat{\beta}$. In these ranges, we always achieved close-to-optimal classification accuracies on labeled test sets.

VI. CONCLUSION

In this paper, we presented a novel approach for discovering object classes from laser range data in an unsupervised fashion. We use a feature-based approach that applies a novel variant of spin-images as surfaces representations but is not restricted to this kind of features. We model object classes as distributions over features and use Latent Dirichlet Allocation to learn clusters of 3D objects according to similarity in shape. The learned feature distributions can subsequently be used as models for the classification of unseen data. An important property of our approach is that it is unsupervised and does not need labeled training data to learn the partitioning.

We carried out experiments using 3D laser range data acquired with a mobile robot. Even for datasets containing complex objects with varying appearance such as humans, we achieve a robust performance with over 90% correctly

grouped objects. We furthermore demonstrate that our approach clearly outperforms unsupervised clustering approaches such as hierarchical clustering. Not only does LDA achieve higher classification accuracy throughout the entire parameter range, it is also less sensitive to the choice of parameters.

REFERENCES

[1] D. Anguelov, B. Taskar, V. Chatalbashev, D. Koller, D. Gupta, G. Heitz, and A. Ng. Discriminative learning of markov random fields for segmentation of 3d scan data. In *Proc. of the Conf. on Comp. Vision and Pattern Recognition (CVPR)*, pages 169–176, 2005.
[2] D.M. Blei, A.Y. Ng, M.I. Jordan, and J. Lafferty. Latent dirichlet allocation. *Journal of Machine Learning Research*, 3, 2003.
[3] A. Bosch, A. Zisserman, and X. Munoz. Scene classification via plsa. In *In Proc. ECCV*, pages 517–530, 2006.
[4] A. Bosch, A. Zisserman, and X. Munoz. Representing shape with a spatial pyramid kernel. In *Proc. of the ACM Int. Conf. on Image and Video Retrieval*, pages 401–408, 2007.
[5] B. Bustos, D.A. Keim, D. Saupe, T. Schreck, and D.V. Vranić. Feature-based similarity search in 3d object databases. *ACM Comput. Surv.*, 37(4):345–387, 2005.
[6] R.J. Campbell and P.J. Flynn. A survey of free-form object representation and recognition techniques. *Computer Vision and Image Understanding*, 81(2):166–210, 2001.
[7] M. Fritz and B. Schiele. Decomposition, discovery and detection of visual categories using topic models. In *Proc. of the Conf. on Comp. Vision and Pattern Recognition (CVPR)*, pages 1–8, 2008.
[8] M. Girolami and A. Kabán. On an equivalence between PLSI and LDA. In *Proc. of the Int. ACM SIGIR Conf. on Research and Development in Information Retrieval*, pages 433–434, 2003.
[9] T. L. Griffiths and M. Steyvers. Finding scientific topics. *Proc Natl Acad Sci U S A*, 101 Suppl 1:5228–5235, 2004.
[10] G. Hetzel, B. Leibe, P. Levi, and B. Schiele. 3d object recognition from range images using local feature histograms. In *Proc. of the Conf. on Comp. Vision and Pattern Recognition (CVPR)*, pages 394–399, 2001.
[11] T. Hofmann. Probabilistic latent semantic indexing. In *Proc. of the Int. ACM SIGIR Conf. on Research and Development in Information Retrieval*, pages 50–57, 1999.
[12] A. Johnson. *Spin-Images: A Representation for 3-D Surface Matching*. PhD thesis, Carnegie Mellon University, Pittsburgh, PA, 1997.
[13] A. Johnson and M. Hebert. Recognizing objects by matching oriented points. Technical Report CMU-RI-TR-96-04, Robotics Institute, Carnegie Mellon University, Pittsburgh, PA, May 1996.
[14] A.E. Johnson and M. Hebert. Surface matching for object recognition in complex three-dimensional scenes. *Image and Vision Computing*, 16:635–651, 1998.
[15] A.E. Johnson and M. Hebert. Using spin images for efficient object recognition in cluttered 3d scenes. *IEEE Transactions on Pattern Analysis and Machine Intelligence*, 21:433–449, 1999.
[16] C.H. Lampert, M.B. Blaschko, and T. Hofmann. Beyond sliding windows: Object localization by efficient subwindow search. In *Proc. of the IEEE Conf. on Computer Vision and Pattern Recognition (CVPR)*, pages 1–8, 2008.
[17] M. Ruhnke, B. Steder, G. Grisetti, and W Burgard. Unsupervised learning of 3d object models from partial views. In *Proc. of the IEEE Int. Conf. on Robotics & Automation (ICRA)*, 2009. To appear.
[18] S. Ruiz-Correa, L.G. Shapiro, and M. Meila. A new paradigm for recognizing 3-d object shapes from range data. *Computer Vision, IEEE International Conference on*, 2:1126, 2003.
[19] F. Stein and G. Medioni. Structural indexing: Efficient 3-d object recognition. *IEEE Transactions on Pattern Analysis and Machine Intelligence*, 14(2):125–145, 1992.
[20] R. Triebel, K. Kersting, and W. Burgard. Robust 3d scan point classification using associative markov networks. In *Proc. of the IEEE Int. Conf. on Robotics & Automation (ICRA)*, 2006.
[21] R. Triebel, R. Schmidt, O. Martinez Mozos, and W. Burgard. Instace-based amn classification for improved object recognition in 2d and 3d laser range data. In *Proc. of IJCAI*, pages 2225–2230, 2007.
[22] P. Viola and M.J. Jones. Robust real-time object detection. In *Proc. of IEEE Workshop on Statistical and Theories of Computer Vision*, 2001.
[23] X. Wang and E. Grimson. Spatial latent dirichlet allocation. In *Advances in Neural Information Processing Systems*, volume 20, 2007.

Towards Cyclic Fabrication Systems for Modular Robotics and Rapid Manufacturing

Matthew S. Moses*, Hiroshi Yamaguchi†, Gregory S. Chirikjian‡

Department of Mechanical Engineering
Whiting School of Engineering,
Johns Hopkins University,
Baltimore, Maryland 21218-2682
Email: matt.moses*, hyamaguchi†, gregc‡ @ jhu.edu

Abstract—A cyclic fabrication system (CFS) is a network of materials, tools, and manufacturing processes that can produce all or most of its constituent components. This paper proposes an architecture for a robotic CFS based on modular components. The proposed system is intended to self-replicate via producing necessary components for replica devices. Some design challenges unique to self-replicating machines are discussed. Results from several proof-of-principle experiments are presented, including a manipulator designed to handle and assemble modules of the same type it is constructed from, a DC brush motor fabricated largely from raw materials, and basic manufacturing tools made with a simple CFS.

I. INTRODUCTION

In the most general terms, we are interested in "making machines that make others of their kind" [1]. This topic, first formally introduced by von Neumann and Burks [2] has seen sporadic interest over ensuing decades with a recent increase in research activity. Applications of self-replicating machines have been proposed ranging from space exploration [3] [4], construction of solar power arrays [5], desktop rapid manufacturing [6] [7] [8], and nanotechnology. Reviews of much of this research are presented in [9] [10].

For the purpose of discussion it is helpful to classify the process of "making" into "assembly" and "fabrication". Assembly involves the formation of kinematic constraints between prefabricated components (e.g. nuts and bolts, keyways, snap-connectors, magnetic latches), while fabrication involves forming useful components from raw materials via changes in molecular structure (e.g. casting, welding, soldering, gluing, cutting, bending). Most recent research falls nicely into one category or the other. In the assembly category, researchers have demonstrated general processes of shape change through reconfiguration [11] [12] [13], stochastic assembly of random components [14] [15], prototype assembly systems for MEMS [16], and self-replication through centrally directed assembly of modular robots [17] [18] [19]. Within fabrication, two groups are developing desktop printers with the potential to replicate their constituent components [6] [7], and other groups are using rapid manufacturing machines to produce fully formed robotic devices [20] [21]. Research on self-assembly of meso-scale mechanical components falls at the boundary between fabrication and assembly [22]. There is ongoing activity in the simulation of self-replicating machines, using

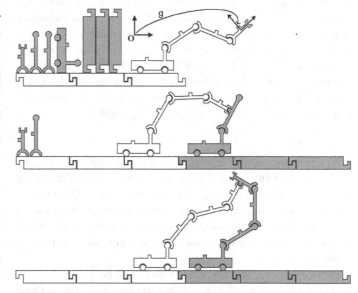

Fig. 1. Schematic representation of assembly during self-replication. Prefabricated components are stacked in a known arrangement. The task of the mobile manipulator is to extend the structural lattice and assemble a new manipulator from the stacked components.

both cellular automata [23] and physics based simulations [24] [25]. There is also work on theoretical aspects of self-replicating machines [26] [27] [28].

This paper proposes an architecture for a self-reconfigurable, self-replicating manufacturing system based on a set of "universal" electro-mechanical components. Section II discusses some of the challenges that generally arise in self-replicating machine design. Section III illustrates one particular approach to the problem and presents a conceptual vision for what a completed solution would look like. Section IV presents results from several proof-of-principle experiments performed to verify elements of the overall system design.

II. PROBLEM FORMALIZATION

While the topics of robot assembly planning and automated manufacturing are well-studied, there are certain complications introduced when a robot must build a replica of itself. The common theme across these complications is that the robot

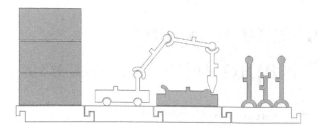

Fig. 2. Schematic representation of fabrication. A robot manipulator equipped with material deposition/removal nozzles fabricates components from a source of raw materials. Fabricated components are then made available to the assembly system.

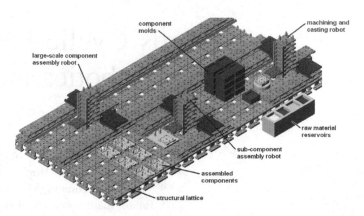

Fig. 3. Concept of a large-scale automated cyclic fabrication system. Mobile 3-axis manipulators operate on a structural grid, fabricating components from raw materials and then adding them to the grid and building additional machines.

and the thing it builds cannot be "decoupled". In industrial practice, a product may be designed to suit the capabilities of a certain robot, or a robot may be modified to perform a specialized task. In the case of self-replication, the robot and the product cannot be independently modified - a change to one necessitates a change to the other. This section discusses some of these complications as they arise in assembly and fabrication processes in machine self-replication.

A. Assembly

A schematic of the assembly process is shown in Figure 1. A mobile manipulator has access to a set of prefabricated components arranged in known configurations. The task of the robot is to obtain components and assemble them into a duplicate robot. The robot begins by constructing an extension to the platform, and then assembles a duplicate manipulator in the new space.

1) Workspace: The assembly robot must be able to build a device as large as itself. For each component in the robot, there must be a collision-free path between the part storage location and the part's destination on a growing replica. Numerous approaches have been reported for dealing with the workspace problem. For example, [19] and [29] actuate the partially-constructed replica during assembly so that the original machine may access necessary areas. In contrast [17] and [18] use mobile manipulators operating in an environment common to original and replica. The environment guides the original robot around the growing replica, effectively increasing the workspace.

2) Assembly Error Tolerance: A physical robot manipulator will have some error in the position of the end-effector, due to imperfections in measurement and manufacturing. For a single link, this positioning error can be represented as a probability density function $\rho(g)$ on the group of rigid-body transformations $g \in SE(3)$. A pdf for the entire manipulator, representing positional uncertainty of the end-effector, can be generated by "concatenating" the error distributions of individual links, using convolutions of the form [30]

$$\rho_{12}(g) = (\rho_1 * \rho_2)(g) = \int_G \rho_1(h)\rho_2(h^{-1} \circ g)dh.$$

The allowable assembly tolerance between two mechanical parts can be represented as a function $\alpha(g)$ on $g \in SE(3)$.

This function varies between 0 and 1 and represents the likelihood of a successful connection when the two components are placed with relative configuration g. The overall likelihood that a given manipulator can successfully assemble two components is given by the integral

$$\gamma = \int_G \alpha(g)\rho(g)dg.$$

An important design goal is to maximize γ.

Intuitively, it is good for the manipulator to have a tight distribution, the optimal being the delta function $\rho(g) = \delta(g)$. It is also desirable for the mechanical parts to tolerate large relative displacements. One measure of the uncertainty tolerance in the relative part configurations is parts entropy [31]

$$S = -\int_G \bar{\alpha}(g) \log \bar{\alpha}(g)dg,$$

where $\bar{\alpha}(g)$ indicates $\alpha(g)$ normalized to a pdf. A large S corresponds to a greater tolerance of uncertainty and is hence desirable. Note that $\rho(g)$ is conditioned on design parameters of the parts themselves, which can be represented as a vector $\mathbf{a}$, and on the arrangement of parts in the manipulator, which can be represented as a vector of parameters $\mathbf{b}$. The assembly tolerance function $\alpha(g)$ is itself a function of the part parameters $\mathbf{a}$. The dependence of γ on $\mathbf{a}$ and $\mathbf{b}$ can be written as

$$\gamma(\mathbf{a}, \mathbf{b}) = \int_G \alpha(g, \mathbf{a})\rho(g \,|\, \mathbf{a}, \mathbf{b})dg.$$

This illustrates the "coupling" between the robot and what it builds, as both ρ and α depend on the same parameter vector $\mathbf{a}$. Challenges arise because changing $\mathbf{a}$ to induce a desirable change in α can induce undesirable changes in ρ, resulting in minimal or detrimental impact to the overall function γ, which is what should actually be maximized.

The set of components described in Section IV-A are designed largely by engineering intuition. However, this formalism provides a way to approach the design process in a more systematic way. A topic of current study is the quantification

122

of $\rho(g\,|\,\mathbf{a},\mathbf{b})$, $\alpha(g,\mathbf{a})$, and $\gamma(\mathbf{a},\mathbf{b})$, using data collected over many assembly trials with various part designs.

3) Connector Mechanism Design: Connector design is a challenging problem in all types of reconfigurable modular robot systems. A standard connection mechanism must be devised that provides mechanical strength and electrical interconnection between modules. Additionally, the connector usually also functions as a "handle" so that modules may be grasped and manipulated using a standard end-effector. It is important to keep the connector as simple as possible, because the connector must be built by other parts of a self-replicating system. A more complex connector may solve certain assembly problems, but it introduces other problems during the fabrication steps.

B. Fabrication

A schematic representation of fabrication is shown in Figure 2. The fabricating machine takes some form of refined raw materials as input (blocks, liquid resins, etc) and produces finished components as output.

1) Selection of Materials and Processes: Materials and processes must be carefully chosen so that the system can fabricate its constituent components. In industrial practice, cutting tools (for example) can be produced on special purpose machines able to form hard materials. Hardened tools are then used to form more generic products from softer materials. In order to "close the loop" in self-replication, some process must be selected in which machines made of soft materials can form hardened cutting tools.

2) Force and Temperature Limitations: Fabrication processes often require the use of large forces and temperatures, in turn requiring heavy, high-precision machinery. In general, it is desirable to select manufacturing processes that use forces and temperatures that are as low as possible, as this simplifies the associated machinery, which of course must be built by some part of the self-replicating system.

3) Component Sub-assembly: Ideally the fabricating robot could simply print entire functional components to be used in assembly. However in many cases it is necessary for several fabricated components to be sub-assembled into a component suitable for use by the assembly system.

4) Resolution Limitations: The problem of resolution is somewhat analogous to the assembly tolerance problem (Section II-A.2). The components produced by the fabricating system must be functional even though they are produced to rather crude tolerances. Because the system self-replicates, the range of acceptable part tolerances must be the same for both the fabricating machine and the products it fabricates.

5) Generational Error Correction: Generational error correction really applies to the entire system. We place it in the fabrication category because it is assumed that given suitably fabricated components, the assembly error tolerance built-in to the modules allows the assembly system to produce functional robots at some guaranteed success rate. Only a few studies have looked at this topic [27]. It is clear that some type of error correction must be built in to the fabrication system to avoid

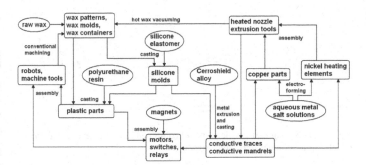

Fig. 4. Materials, tools, and manufacturing processes comprise a cyclic fabrication system. Raw materials are shown in ellipses, products are shown in boxes, and processes are indicated by labeled arrows.

accumulation of errors over successive replications. This could be performed by some intergenerational calibration methods, or by calibration to some external standard.

III. CONCEPT OVERVIEW

The model system we use for inspiration is something most engineers are familiar with: LEGO™. Modern LEGOs can be made into many kinds of robots and automated devices. They are mentioned by name in some of the early papers on modular robotics [32]. Several LEGO systems have been demonstrated that assemble simple devices also made from LEGOs [33] [34]. It is a much harder task to build a LEGO machine that can fabricate LEGO blocks. To state the goal of this work in simple terms, we seek to build a set of "LEGO-like" blocks, such that a machine made from them can 1) assemble complex devices from similar blocks, and 2) produce new blocks from a stock of raw materials. Ultimately, a fabricating machine built from this set of blocks should be able to fabricate all the essential components in the set, including sensors, actuators, and control devices.

A. Lattice Architecture

Figure 3 shows a concept of how the fabrication system might be implemented. Multiple mobile manipulator robots are installed on a structural grid. Each robot is based on a standardized 3-axis Cartesian manipulator with a general purpose end effector. An external system provides power and control. Tasks are classified into fabrication, mid-level component assembly, and large-scale structural assembly. The fabrication sub-system obtains raw materials from reservoirs and produces basic parts. These parts are then assembled into larger components by the mid-level assembly robot. The large-scale assembly robot assembles components into new devices (e.g. new robots and new areas of structural lattice).

The system as a whole can self-replicate through a gradual process of growth. An initial system with a given lattice structure and a certain number of robots replicates by increasing the lattice size and doubling the number of robots. Manufacturing output is not limited to strict replication; the system can grow and reconfigure in a flexible manner. For example, the large-scale assembly robots can also reconfigure

simple polyurethane castings → castings bonded with adhesive into functional components → components assembled into structural and mechanical devices

Fig. 5. Individual cast plastic parts are combined into mid-level components, which are assembled into larger structures.

Fig. 6. The base component has four tapered compression pins, two threaded tension pins, and two captured-nut fasteners. A common end-effector is used to grasp components, and tighten/untighten them from an assembly.

their working envelope by adding and removing components in the structural grid. Depending on the application, the useful output of the system may be some product made by the robots (gears, motors, etc.), or it may be the structural grid itself. For example, some proposed applications of modular robots call for a large reconfigurable lattice [35]. A large lattice composed primarily of passive components, but reconfigured by a small number of active robots moving within the lattice, might be a feasible alternative to reconfigurable lattices in which every element is an active modular robot.

B. Cyclic Fabrication Systems

A cyclic fabrication system (CFS) is a network of materials, tools, and manufacturing processes that can produce all or most of its constituent components. It is cyclic in the way the game "rock-paper-scissors" is cyclic: tools, materials, and fabrication processes are chosen such that one process creates tools used in the next process, which is used in the next, and so on until a final process produces tools needed to perform the initial process. Complex CFSs have been proposed in the context of self-replicating machines. Processes for producing machine components from lunar soil are proposed in [3] and [4]. A cyclic process for separating common terrestrial soil into useful materials is proposed in [36]. A comparatively simple system is outlined in [37] based on UV-catalyzed polymer resin.

A proposed cyclic fabrication system is shown in Figure 4, based on polyurethane and silicone resins, wax, low-temperature-melting alloys, and solutions of metal salts. The primary benefit of these materials is that they are readily available and fairly safe and easy to work with in a small laboratory environment. They also serve as proxy materials for performing proof-of-principle experiments that can later be extended to materials that are harder to work with. The processes used in the system are conventional machining and assembly operations, casting, hot wax vacuuming, and electroforming.

The polyurethane resin is a two-part liquid formulation. Upon mixing, the liquid polyurethane hardens within minutes into a machineable plastic with good mechanical properties. Solid polyurethane can form a wide range of components including gears, bushings, springs, and structural members. The wax has a low melting point, is dimensionally stable and easily machined. The key features of these materials are that the polyurethane in liquid form can easily be cast

in wax molds, while the wax can easily be machined by solid polyurethane tools. Master patterns made from wax can be used to form polyurethane and silicone molds, which in turn can be used to cast metal parts from low-temperature melting alloys. The resultant metal parts may be used directly in machines or as conductive mandrels for electroforming. Copper, nickel, and many other metals can be electroformed over a mandrel made of low-temperature melting alloy. The alloy can then be removed from the metal part by melting and reused. Copper can form electrical contacts, motor coils, conductive traces, and nozzles for material deposition/removal. Nickel may be used in magnetic circuits and in resistive heaters.

IV. EXPERIMENTS

A. Modular Components and Assembly Robot

This Section presents a "universal" set of electromechanical modules. Design goals for the modules are: versatility and usefulness in assembly, ease of assembly, ease of fabrication using CFS processes, tolerance to misalignment during assembly, and strong, reversible intermodule connections. The modular components are based on an earlier design [29]. The components are made of polyurethane. Each of the castings that make up a component are designed to be produced in a single-piece wax or silicone mold with minimal undercuts. This greatly simplifies the design and fabrication of the mold, simplifies the demolding process, and eliminates difficulties that arise with removing sprues and flash. However, the exposed top surface of a part cast in a single-piece mold is not suitable for precise assembly. This problem can be solved by combining multiple parts with adhesive, so that precise dimensions between critical surfaces are maintained. With these components it is primarily achieved through the use of pins and plates. Each plate contains four compression pins, two tension pins, and two captured nuts. The compression pins are tapered and mate with tapered holes in other plates, while the tension pins mate with captured nuts.

A minimum connection between two components consists of a tension pin connected to a nut, in between two compression pins. This form of mechanical connection was chosen over other methods (magnets, snap-fittings, etc) because of its high strength and ease of reversibility. When a new component is placed on a growing assembly the tapered mating surfaces engage, locating the component. The captured nuts are then tightened onto the mating tension pins, providing a solid

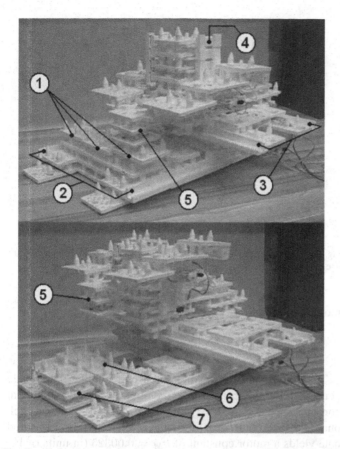

Fig. 7. A 3-axis manipulator performing a simple assembly task; (2) - X axis, (3) - Y axis, (4) - Z axis, (5) - end-effector. Three separate components (1) are initially placed on the assembly station (6). The robot assembles the three components and removes them from the assembly station (7).

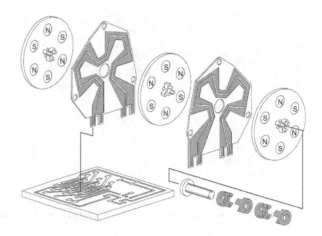

Fig. 8. Diagram of motor construction.

mechanical connection between the new component and the assembly. Example assemblies shown in Figure 5 include a structural platform and a linear motion slide. Variations on these components can form other mechanisms, including revolute joints, gear trains, 3-axis manipulators. The threaded fasteners provide a reversible assembly mechanism, so that an existing structure may be disassembled and its parts reused in a new structure. Additional small holes in the plates are accommodations for future electrical connectors.

A single end-effector is used for grasping components and to tighten and un-tighten the threaded fasteners (see Figure 6). The end-effector consists of a spring loaded tool-piece with a slot and internal thread. For grasping a part, the internal thread of the tool mates with the threaded tension pin on a component. After a component is placed, the tool is unscrewed from the tension pin and placed on the component's captured-nut fastener. The slot on the tool-piece self aligns with the fastener and tightens or loosens it in the manner of socket and nut. Apart from the spring and motor, the entire end-effector assembly is built from polyurethane components.

A 3-axis Cartesian manipulator was constructed from about 80 modular components. The manipulator is driven by seven small gearmotors (Solarbotics GM2). Two motors drive each of the axes in parallel, and the seventh is used to drive the

end-effector. Low resolution optical encoders are used on one motor on each axis for position feedback. Resolution is approximately 1 count per mm. Power is supplied to the motors through ordinary hookup wire. The manipulator is controlled by a simple microcontroller (PIC16F690). The controller simply follows a sequence of position commands, driving each axis independently for a predetermined number of encoder counts. Due to the high tolerance for misalignment of the parts, even this crude form of control is able to automatically assemble components.

Figure 7 shows two snapshots in a test assembly sequence of the 3-axis manipulator. Three unassembled components are placed (unconnected) on an assembly station in front of the manipulator. The robot then retrieves each of the outside parts and assembles them in sequence to the center component. After the three parts are connected together, the robot lifts the assembly and removes it from the station. This is a simple example that demonstrates the feasibility of using machines made from these components to perform assembly tasks.

B. Actuator Production

This section describes a permanent magnet axial-gap Lorenz-force motor, fabricated almost entirely using techniques from the cyclic fabrication system described in Section III-B. Motors of this general design are well known, including "printed coil" motors used in many applications. The challenge here is to build such a motor using the constrained set of materials, tools, and methods that comprise the cyclic fabrication system. The Fab@home group has demonstrated printing of batteries, circuit elements, and actuators using variations on rapid manufacturing processes [38]. The Reprap group has demonstrated formation of circuit wiring by extruding metal into plastic parts [39] [40]. Inspired by their success, we designed a metal extruding mechanism with the intention of installing it on a 3-axis manipulator similar to the one in Section IV-A. The extruder consists of a heated copper nozzle and a motorized syringe pump (see Figure 9). A heated cup is used in conjunction with the nozzle device. Thermostat control of the heated nozzle and cup was performed with Reprap

Fig. 9. Heated copper nozzle for extrusion of low-melting alloy into plastic parts.

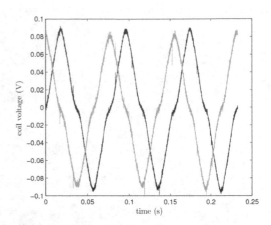

Fig. 10. Motor back-emf measured at each coil circuit. Data shown is for one revolution, $\omega = 27 rad/s$.

electronics. The cup contains a small amount of molten metal, which is retrieved by the nozzle using the syringe pump. The nozzle is moved to a plastic part, where the molten metal is then pumped out into channels in the part.

1) Method of Construction: The main components of the motor are illustrated in Figures 8 and 11. The motor consists of 15 cast plastic parts - 6 coil plates, 7 magnet disks, 1 commutator shaft, 1 base plate (Smooth-On Smooth-Cast 300), 4 cast metal commutator rings (Cerroshield alloy, melting point 95 C, McMaster P/N 8921K23), 8 metal brushes (brass strip), 6 magnetic yoke pieces (mild steel strips), and 42 NdFeB magnets (1/2" diameter by 1/8" thick, Digikey P/N 496-1002-ND). The part masters were fabricated using conventional machining and laser cutting, and the plastic and cast-metal parts were produced manually from silicone molds. To form the magnet disks, magnets were positioned in a disk mold with a plastic fixture. The fixture was then removed and liquid resin cast around the magnets to embed them in the final part. Wiring channels were cast in place as part of the coil plates and base plate. These channels were manually filled with molten Cerroshield using the heated nozzle device. During assembly, the base plate was clamped to a hot plate and coil plates with filled channels were then lowered into place onto the base plate.

2) Motor Performance: The motor can operate as a stepper motor, DC brush motor, or generator. When configured as a DC brush motor, the commutator excites the coils with a quadrature square wave. The steady-state behavior of the motor in DC operation can be described using a simple model [41]

$$\tau = k_M i - \tau_f,$$

$$i = \frac{V - k_M \omega}{R},$$

where τ_f is a constant torque representing sliding friction, V is constant DC voltage applied to the motor terminals, R is lumped resistance of the motor windings, brushes, and commutator. The model is valid for $i > \tau_f / k_M$. For this simple motor, the characteristic we are most interested in is mechanical output power. The maximum power output in DC

motor configuration is given by

$$P_{max} = \frac{1}{4R} \left(V - \frac{\tau_f R}{k_M} \right)^2.$$

The motor constant was measured by recording the voltage generated by the coils while the motor shaft was spun at constant speed. Figure 10 shows the voltage output of each phase in the motor over a single revolution. The measured data yields a motor constant of $k_M = 0.00325$ (in units of Vs or Nm/A).

The motor was tested in DC brush configuration and operated continuously for several minutes. At a terminal voltage of $6V$ it drew about $7.5A$ and had a no load speed of approximately $400 rpm (42 rad/s)$. The winding resistance was estimated to be $R = 0.78\Omega$, the friction torque $\tau_f = 0.024 Nm$, and the maximum available power output $P_{max} = 6mW$. When compared with the commercial motor used in the assembly robot in Section IV-A, which has a power output of about $100mW$ at $6V$, it is clear there is much room for improvement in the motor design.

Assembly of the motor relies extensively on human intervention, so this experiment should be regarded as a proof-of-principle that a subset of the CFS (Figure 4) can build working actuators. Topics of current study related to this experiment are 1) incorporating the metal-deposition nozzle into a Reprap machine to print entire coil plates, 2) construction of a motor assembly machine using the modular electro-mechanical components, and 3) increasing power output and eliminating need for permanent magnets.

C. Fabrication Tool Production

Figure 13 illustrates a closed manufacturing process for replicating polyurethane cutting tools, which is a subset of the CFS shown in Figure 4. Three types of cutting tool are necessary for the process: a standard endmill, a tapered endmill, and a lathe cutting tool. The process begins with bulk wax material, which is turned and faced into two cylinders using lathe operations. One cylinder receives a conical bore

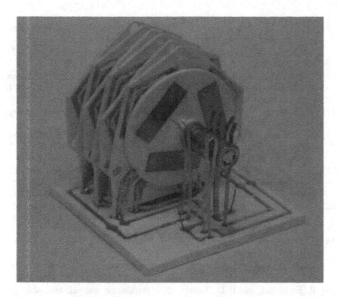

Fig. 11. A DC brush motor fabricated using materials and processes from the cyclic fabrication system. For size reference, the base plate is about $12cm$ square.

Fig. 12. Clockwise from top left: wax mold, polyurethane part; polyurethane mold, wax and low-melt alloy parts; wax master, polyurethane mold, Cerroshield alloy mandrel, electroformed copper nozzle; wax mold, polyurethane lathe tool; silicone molds and polyurethane end-mills.

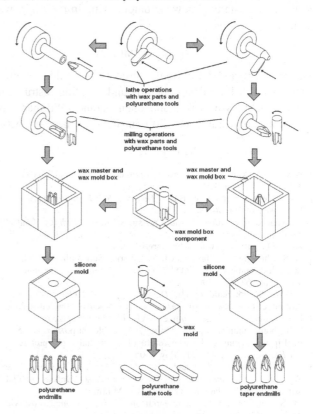

Fig. 13. Process flowchart for replicating cutting tools.

and the other a tapered surface. Flutes (cutting surfaces) are cut in the cylinders using milling operations. This results in a wax "master pattern" replica of a standard and tapered endmill. Additional milling operations generate parts for a wax mold container. The master pattern together with the mold container form a pattern from which silicone molds can be cast. The silicone molds contain a negative replica of the wax masters in which polyurethane parts can be cast. The process for replicating lathe cutting tools is simpler - a negative pattern can be machined directly into a wax block using a polyurethane milling tool. This process was verified by using a manually operated milling base (Sherline) to replicate polyurethane cutting tools. An additional subset of the CFS was tested by producing a copper nozzle similar to the one used in the assembly of Figure 9. Results are shown in Figure 12. Materials used in the experiments are: wax (melting point 68C, McMaster P/N 93955K77), polyurethane resin (Smooth-On Smooth-Cast 300), silicone elastomer resin (Smooth-On Mold-Max 20), Cerroshield alloy (melting point 95C, McMaster P/N 8921K23), and aqueous copper sulfate solution.

V. CONCLUSION

We have proposed an architecture for a modular robotic system that implements a cyclic fabrication system. The system is intended for use as a general-purpose manufacturing system, with the special feature that it can self-reconfigure and self-replicate. Some of the unique problems encountered in design of self-replicating machines were addressed. All of these problems arise due to the coupling between robot and what it makes - the robot cannot be designed independently of its "product".

Several proof-of-principle experiments supported the feasibility of the cyclic fabrication system. A primary factor in selecting the processes for the CFS is ease of use in the laboratory. While the exact processes chosen may not be practical for making machines that operate in the "real-world" we argue that the results are useful for several reasons. First, the processes investigated serve as proxies for industrial processes. Design principles uncovered during investigation may be applied to more common methods of manufacturing, for example machineable ceramics and aluminum casting might be used in place of wax and Cerroshield with minor modifications to the basic design of the production cycle. Second, the products generated by the production system as-is may be useful in certain applications, just as the objects built by commercial rapid prototyping devices have found use in certain applications, despite a limited selection of materials to work with.

Finally, the most exciting application for a compact automated production system may be for building things small. Silicone elastomer casting and electroforming are routinely used in microfabrication. The physicist Richard Feynman famously suggested using a remotely operated machining system to replicate itself in incrementally smaller and smaller iterations. Admittedly, he called the idea "one weird possibility" and "a very long and very difficult program" [42], but perhaps it may not be so weird or so difficult if approached in the right way.

Topics for future research include: quantifying assembly tolerance functions, incorporating electrical conductors into the modular components, designing new robot assemblies with increased workspace, automating the motor fabrication process, and designing new actuators with improved power output.

ACKNOWLEDGMENT

The RepRap and Fab@home projects have done a great service by publishing design details and building instructions for their machines in a free and open manner. We thank the large number of volunteers who contribute to these projects.

REFERENCES

[1] A. Cho, "Making machines that make others of their kind," *Science*, vol. 318, no. 5853, pp. 1084–1085, 2007.
[2] J. von Neumann and A. W. Burks, *Theory of Self-Reproducing Automata.* University of Illinois Press, 1966.
[3] "Advnaced automation for space missions," in *Proceedings of the 1980 NASA/ASEE Summer Study*, R. A. Freitas, Jr., Ed., 1980. [Online]. Available: http://www.islandone.org/MMSG/aasm/
[4] G. S. Chirikjian, Y. Zhou, and J. Suthakorn, "Self-replicating robots for lunar development," *IEEE/ASME Transactions on Mechatronics*, vol. 7, no. 4, pp. 462–472, 2002.
[5] T. Bass, "Robot, build thyself," *Discover*, pp. 64–72, Oct. 1995. [Online]. Available: http://discovermagazine.com/1995/oct/robotbuildthysel569
[6] Reprap main web page. [Online]. Available: http://reprap.org
[7] Fab@home main web page. [Online]. Available: http://fabathome.org
[8] H. Lipson, "Homemade: The future of functional rapid prototyping," *IEEE Spectrum*, pp. 24–31, May 2005.
[9] R. A. Freitas, Jr. and R. C. Merkle, *Kinematic Self-Replicating Machines.* Landes Bioscience, 2004. [Online]. Available: http://www.molecularassembler.com/KSRM.htm
[10] M. Sipper, "Fifty years of research on self-replication: An overview," *Artificial Life*, vol. 4, no. 3, pp. 237–257, 1998.
[11] M. H. Yim, Y. Zhang, and D. G. Duff, "Modular robots," *IEEE Spectrum*, pp. 30–34, Feb. 2002.
[12] H. Kurokawa, K. Tomita, A. Kamimura, S. Kokaji, T. Hasuo, and S. Murata, "Distributed self-reconfiguration of M-TRAN III modular robotic system," *International Journal of Robotics Research*, vol. 27, pp. 373–386, 2008.
[13] Z. Butler, K. Kotay, D. Rus, and K. Tomita, "Generic decentralized locomotion control for lattice-based self-reconfigurable robots," *International Journal of Robotics Research*, vol. 23, no. 9, pp. 919–937, 2004.
[14] S. Griffith, D. Goldwater, and J. M. Jacobson, "Self-replication from random parts," *Nature*, vol. 437, p. 636, Sep. 2005.
[15] E. Klavins, "Programmable self-assembly," *Control Systems Magazine*, vol. 24, no. 4, pp. 43–56, Aug. 2007.
[16] K. Tsui, A. A. Geisberger, M. Ellis, and G. D. Skidmore, "Micromachined end-effector and techniques for directed MEMS assembly," *Journal of Micromechanics and Microengineering*, vol. 14, no. 4, p. 542, 2004.

[17] K. Lee, M. Moses, and G. S. Chirikjian, "Robotic self-replication in structured environments: Physical demonstrations and complexity measures," *International Journal of Robotics Research*, vol. 27, pp. 387–401, 2008.
[18] J. Suthakorn, A. B. Cushing, and G. S. Chirikjian, "An autonomous self-replicating robotic system," in *Proceedings of 2003 IEEE/ASME International Conference on Advanced Intelligent Mechatronics*, 2003.
[19] V. Zykov, E. Mytilinaios, B. Adams, and H. Lipson, "Self-reproducing machines," *Nature*, vol. 435, no. 7038, pp. 163–164, 2005.
[20] K. J. De Laurentis, C. Mavroidis, and F. F. Kong, "Rapid robot reproduction," *IEEE Robotics & Automation Magazine*, vol. 11, no. 2, pp. 86–92, Jun. 2004.
[21] J. G. Cham, S. A. Bailey, J. E. Clark, R. J. Full, and M. R. Cutkosky, "Fast and robust: Hexapedal robots via shape deposition manufacturing," *International Journal of Robotics Research*, vol. 21, no. 10, Oct. 2002.
[22] D. H. Gracias, J. Tien, T. L. Breen, C. Hsu, and G. M. Whitesides, "Forming electrical networks in three dimensions by self-assembly," *Science*, vol. 289, no. 5482, pp. 1170–1172, Aug. 2000.
[23] U. Pesavento, "An implementation of von neumann's self-reproducing machine," pp. 337–354, 1995.
[24] W. M. Stevens, "Simulating self-replicating machines," *Journal of Intelligent and Robotic Systems*, vol. 49, no. 2, pp. 135–150, Jun. 2007.
[25] R. Ewaschuk and P. D. Turney, "Self-replication and self-assembly for manufacturing," *Artificial Life*, vol. 12, no. 3, pp. 411–433, 2006.
[26] A. Menezes and P. Kabamba, "A combined seed-identification and generation analysis algorithm for self-reproducing systems," in *Proceedings of the 2007 American Control Conference*, Jul. 2007, pp. 2582–2587.
[27] P. D. Owens and A. G. Ulsoy, "Self-reproducing machines: preventing degeneracy," in *Proceedings of ASME International Mechanical Engineering Congress and Exposition*, Nov. 2006.
[28] W. R. Buckley, "Computational ontogeny," *Biological Theory*, vol. 3, no. 1, pp. 3–6, 2008.
[29] M. Moses, "Physical prototype of a self-replicating universal constructor," Master's thesis, University of New Mexico, 2001. [Online]. Available: http://home.earthlink.net/ mmoses152
[30] G. S. Chirikjian, *Stochastic Models, Information Theory, and Lie Groups.* Birkhäuser, 2009.
[31] A. C. Sanderson, "Parts entropy methods for robotic assembly system design," in *Proceedings of the IEEE International Conference on Robotics and Automation*, 1984, pp. 600–608.
[32] S. Murata, H. Kurokawa, and S. Kokaji, "Self-assembling machine," in *Proceedings of the IEEE International Conference on Robotics and Automation*, 1994, pp. 441–448.
[33] (2006, Dec.) Mindstorms autofabrik. [Online]. Available: http://www.youtube.com/watch?v=GQ3AcPEPbH0
[34] D. Esterman, M. Sullivan, J. Bergendahl, and C. G. Cassandras, "Computer-controlled lego factory," University of Massachusetts/Amhherst, Tech. Rep., Jun. 1995. [Online]. Available: http://vita.bu.edu/cgc/newlego/index.html
[35] B. Kirby, B. Aksak, S. C. Goldstein, J. F. Hoburg, T. C. Mowry, and P. Pillai, "A modular robotic system using magnetic force effectors," in *Proceedings of the IEEE International Conference on Intelligent Robots and Systems*, Oct. 2007, pp. 2787–2793.
[36] K. S. Lackner and C. H. Wendt, "Exponential growth of large self-reproducing machine systems," *Mathematical and Computer Modelling*, vol. 21, no. 10, pp. 55–81, 1995.
[37] C. J. Phoenix. (1998, Mar.) Partial design for macro-scale machining self-replicator. [Online]. Available: http://groups.google.com/group/sci.nanotech/msg/96a67c84809c9a5d
[38] E. Malone, K. Rasa, D. L. Cohen, T. Isaacson, H. Lashley, and H. Lipson, "Freeform fabrication of 3d zinc-air batteries and functional electromechanical assemblies," *Rapid Prototyping Journal*, vol. 10, no. 1, pp. 58–69, 2004.
[39] E. Sells and A. Bowyer, "Rapid prototyped electronic circuits," University of Bath, Tech. Rep., Nov. 2004. [Online]. Available: http://staff.bath.ac.uk/ensab/replicator/Downloads/report-01-04.doc
[40] R. Jones. (2009, Apr.) Reprap solder extruder. [Online]. Available: http://www.youtube.com/watch?v=Wnu4lmOnX00
[41] A. E. Fitzgerald, C. Kingsley, and S. D. Umans, *Electric Machinery*, 6th ed. McGraw-Hill, 2003, pp. 30–42.
[42] R. P. Feynman. (1959) There's plenty of room at the bottom. [Online]. Available: http://www.zyvex.com/nanotech/feynman.html

Setpoint Regulation for Stochastically Interacting Robots

Nils Napp

nnapp@u.washington.edu

Electrical Engineering
University of Washington
Seattle WA 98195

Samuel Burden

sburden@eecs.berkeley.edu

Electrical Engineering
University of California at Berkeley
Berkeley CA 94720

Eric Klavins

klavins@u.washington.edu

Electrical Engineering
University of Washington
Seattle WA 98195

Abstract—We present an integral feedback controller that regulates the average copy number of a particular assembly in a system of stochastically interacting robots. The mathematical model for the stochastic system is a *tunable reaction network*, which makes this approach applicable to a large class of other systems, including ones that exhibit stochastic self assembly at various length scales. We prove that this controller works for a range of set-points, and how to compute this range. Finally, we demonstrate the approach on a physical testbed.

I. INTRODUCTION

Self-assembly of complex systems and structures promises many new applications, such as easily combining different micro-fabrication technologies [1] or building arbitrary, complex nano-structures [2]. While many natural systems are reliably self-assembled at vastly different length and time scales, engineered self-assembled systems remain comparatively simple. The difficulties of engineering complex self-assembling systems are associated with large configuration spaces, our lack of understanding the relationship between local and global dynamics, and the stochastic or uncertain nature of their dynamic models.

In the context of engineering, the interplay between uncertainty and sensitivity of global to local behavior can often lead to a profound lack of modularity as small unintended local interactions can drastically alter the behavior from what is expected by composition.

In this paper we partially address this problem by designing a feedback controller that can regulate the expected value of the number of an arbitrary component type. This approach could be used for composition in the sense that other subsystems can rely on the presence of these regulated quantities.

We are guided by the application of *stochastic self-assembly*, in which self-assembling particles interact randomly. Such systems abound in engineered settings, such as in DNA self-assembly [2], micro and meso-scale self-assembly [1, 3, 4], and robotic self-assembly [5, 6].

Self-assembly can be either *passive* or *active*. Designing systems that passively self-assemble is a problem of engineering a favorable free energy landscape in configuration space. Passive self-assembling systems often lack flexibility since a specific design of the energy landscape can be difficult to adapt to new tasks. In addition, there are physical limitations to how much the energy landscape can be manipulated. The yield

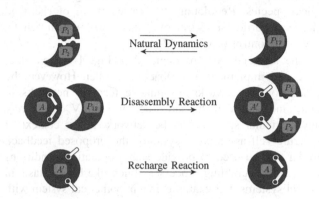

Fig. 1. Schematic representation of the different robot interactions. The passive robots P_1 and P_2 can form heterodimers, which can disassemble spontaneously. The active robot A can expend energy to undo bonds. When the arms of an active robot are retracted, it is charged and can actively disassemble a dimer. If the arms of an active robot are extended (denoted A') then it is not charged, but may become charged via the recharge reaction, the rate of which can be controlled.

of a desired output structure is a function of the shape and depth of energy wells, as a result the limits in manipulating the energy landscape in passive self-assembly generally lead to low yields.

In *active self-assembly*, energy can be locally injected into the system. In particular, we focus on the situation when we have the ability selectively undo bonds that are formed by passive dynamics. Active self-assembly can overcome the lack of flexibility of passive self-assembling system by making aspects of the system re-programmable while leaving other areas in the energy landscape untouched. As a result, the changes in the global dynamics remain tractable.

The particular model for active self-assembly we investigate is that of a *tunable reaction network*. We present a system of simple stochastically interacting robots that are well modeled as a tunable reaction network and demonstrate the feedback setpoint regulation scheme. Fig. 1 shows a pictorial representation of the tunable reaction network investigated in this paper. There are three robot types and several instances of each (see Fig. 2a). The passive robots P_1 and P_2 are able to bind and form heterodimer complexes P_{12}, which in turn can spontaneously disassemble. The active robots A can dock with heterodimers and disassemble them. The disassembly reaction leaves active robots in an uncharged state, denoted by A'. The last reaction in Fig. 1 recharges uncharged robots at a rate that is controlled externally. *The control problem for this system is to regulate the number of heterodimers P_{12} in the system by adjusting the recharge rate.* (This problem is re-stated formally in Sec. IV.) While the tunable reaction network shown in Fig. 1 is comparatively simple, tunable reaction networks in general can describe much more complicated systems.

For example, many biological systems can be viewed as tunable reaction networks. Inside cells, enzymes are expressed to control the rates of various metabolic reactions. Similar to the problem solved here, one of the many functions of the biochemical processes inside cells is maintaining equilibria of chemical species. Regulating the concentration of chemical species is a particular aspect of *homeostasis*, which can be viewed as a control problem [7].

For the artificial systems depicted in Fig. 1 we propose, analyze, and implement a feedback controller. However, the proposed controller works for tunable reaction networks in general, since the analysis and proof in Sec. IV do not rely on any particular structure of the network. In the context of engineering self-assembling systems, the proposed feedback controller can be used to provide stable operating conditions for other self-assembling processes, much like homeostasis in biological systems. For example, in a hypothetical system with a vat of self-assembling miniature robots, we might care that the relative concentration of robot feet and robot legs is fixed in order to maximize the yield of functioning miniature robots. In general, we envision the self-assembling systems of the future as having metabolisms of their own that regulate the various species of partially assembled objects in the system to maximize the yield of the desired final assembly.

II. EXPERIMENTAL ROBOTIC CHEMISTRY

The robots described here interact stochastically as in [5, 6], however, they are much simpler both mechanically and electronically. Also, while other robotic platforms consist of a homogeneous group of robots, the robotic testbed described here is a heterogeneous mixture of three different robot types, Fig. 2bc. The assembly of the two passive robot types P_1 and P_2 is driven by complementary shape and embedded magnets. The magnetic force creates an energy well that tends to pull two robots together and form a heterodimer. The third, active robot type can expend energy to disassemble a heterodimer into its constituents.

The energy for this disassembly is supplied to the active robots via solar panels. Each active robot stores energy from its solar panel in a capacitor, if the charge in the capacitor reaches a threshold and an active robot A is bound to a heterodimer it activates a motor and disassembles the heterodimer. Disassembling heterodimers depletes the on-board energy storage of active robots requiring more energy from the solar cells to disassemble additional heterodimers. Adjusting the amount of incident light changes the recharge rate of active robots and thus indirectly affects the rate at which heterodimers are disassembled.

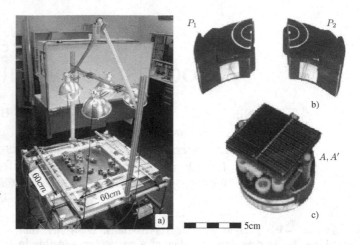

Fig. 2. Hardware of test-bed. a) Picture of the air-table showing the robots, the air-jets, the overhead lamps, and the overhead camera. b) Picture of the two passive component types showing the tracking targets and the complementary shapes. c) The active robot showing solar cells, contact sensors, the spinning levers that pull bound passive complexes apart.

Although this indirect approach may seem unnecessarily complicated, it possesses a key design feature that we believe justifies the added complexity: the structural, energy delivery, and computational functions reside on separate components of the overall system. We think of P_1 and P_2 as the structural components we want to control, the active robots as agents of energy delivery, and the controller implemented on a computer as the computational component. This division of labor is analogous to many biological systems where different cellular functions are largely separated into different types of molecules. We believe that such a separation of functionality in self-organization is essential to engineering large scale complex systems. Distributing the functionality in this way can yield much simpler individual components on average. For example, the passive robots contain no electronic components whatsoever, and the active robots only contain a simple circuit made from discrete electrical components, a motor, and a solar panel.

A. Physical Characteristics of Testbed

The body of each robot is machined from polyurethane prototyping foam and painted black to aid the vision system. This material is easy to machine, light, and stiff.

The robots float on an air-table shown in Fig. 2a), which has a large HVAC blower attached to the bottom of a perforated board (blower not visible in image). The blower is able to maintain a high flow-rate of air through the table surface and allows us to float relatively heavy pieces $\approx 2.5 \frac{g}{cm^2}$. The active area of the table is 60cm × 60cm. Mounted along the perimeter of the table are computer controlled solenoid valves. These valves can deliver short bursts of pressurized air from a compressor (30psi). By randomly activating these air-jets robots on the air-table are driven to perform a random walk. The bursts are randomized and controlled via a MATLAB script, which also updates the state of the controller and adjust the intensity of four overhead lamps. These lamps determine the amount of incident light to the solar panels, thereby setting the recharge reaction rate.

Images from the overhead camera are used to extract the number and position of *targets*, consisting of small, circular disks with a pattern of concentric light and dark rings, see Fig. 2b. We detect targets in real time and use the data both in the feedback loop to exert control and open loop to estimate the system reaction rates and diffusion constants.

We determine the number of heterodimers by adjusting the image processing parameters so that only whole targets register in the vision system. Half of a target is attached each passive robot in such a way that when a heterodimer forms the two halves from a complete target that is picked up by the vision system. The rotational symmetry of the targets simplifies the image processing by reducing the convolution of the target kernel from three to two dimensions, allowing sample rates of ≈ 1 Hz.

III. MATHEMATICAL MODEL

This section describes *stochastic chemical kinetics* [8] and the associated *chemical master equation* (CME), used to model the discrete configurations of the robotic testbed. This section also describes a *stochastic hybrid system* (SHS) model that extends stochastic chemical kinetics to include continuous state variables, needed to model the closed loop feedback system.

A. Model for Stochastic Chemical Kinetics

The idea is to create a stochastic model that reflects our understanding of how chemical reactions occur at a microscopic level, as opposed to *mass action kinetics*, which is a deterministic model of the evolution of chemical concentrations. When the number of molecules involved in a set of chemical reactions grows, the approximations of mass action kinetics become very good. The large number of molecules averages stochastic effects away [9, Ch. 5.8]. However, when only a few molecules are involved, the stochastic nature of chemical reactions dominates the dynamics and requires explicit modeling.

Let

$$S = \{A, A', P_1, P_2, P_{12}\},$$

denote the set of chemical species, in this case the robot types of the testbed. The symbol A stands for an active robot that is charged, A' is an uncharged active robot. The symbol P_1 and P_2 are the two different types of passive robots and P_{12} is a heterodimer of passive robots, see Fig. 1 and 2. The *copy number* of each species is the number of instances of that particular species and is denoted by a capital N subscripted with the appropriate symbol, i.e. N_A specifies the copy number of species A. The *state* q of the system is described by the vector of copy numbers $\mathbf{q} = (N_A, N_{P_{12}}, N_{A'}, N_{P_1}, N_{P_2})^T$. The set of all possible states is denoted by Q.

Events that affect the state $\mathbf{q}$ are called *reactions*. This paper considers the set of reactions in Fig. 1. In General, if reactions are indexed by a set L and the state of the system is $\mathbf{q}$ before a reaction l and $\mathbf{q}'$ after the reaction, then we have

$$\mathbf{q}' = \mathbf{q} + \mathbf{a}_l,$$

where $\mathbf{a}_l$ is a vector that is specific to the reaction type. The chemical species that correspond to negative entries in $\mathbf{a}_l$ are called *reactants* and those that correspond to positive entries are called *products*. For example, the reaction

$$P_1 + P_2 \xrightarrow{\quad k \quad} P_{12}$$

where two different passive robots form a dimer has the associated **a** vector

$$\mathbf{a} = (0, 0, -1, -1, 1)^T.$$

Both P_1 and P_2 are reactants and P_{12} is a product. The *multiplicity* of a reaction from a given state $\mathbf{q}$, denoted $M(\mathbf{a}, \mathbf{q})$, specifies the number of different ways the reactants of **a** can be chosen from state $\mathbf{q}$. In addition to the **a** vector each reaction has associated with it a *rate constant* k_l, that depends on the underlying stochastic behavior of the interacting species. Determining these rate constants for the system of robots is the topic of Sec. III-D.

Stochastic chemical kinetics defines a discrete state, continuous time *Markov process* with state space Q and the following transitions rates. The transition rate between $\mathbf{q}$ and $\mathbf{q}'$ is given by

$$k_l M(\mathbf{a}_l, \mathbf{q}),$$

when $\mathbf{q}' = \mathbf{q} + \mathbf{a}_l$ and $\mathbf{a}_l$ is applicable in $\mathbf{q}$ (i.e. $\mathbf{q}'$ is non-negative). Given that the process is in state $\mathbf{q}$ at time t, the probability of transitioning to state $\mathbf{q}'$ within the next dt seconds is

$$k_l M(\mathbf{a}_l, \mathbf{q}) dt.$$

This property suffices to define the conditional transition probabilities of the stochastic process and together with an initial distribution over the states defines the Markov process that comprises the stochastic chemical kinetics model. This model is applicable to a set of interacting molecules if the system is *well mixed* [9, 10]. In practice this assumption is difficult to verify. However, in our system of robots we can explicitly check the assumptions, since we can observe the position of all involved particles. A description of the procedures used to verify the well-mixed assumption is given in Sec. III-C.

Conveniently, discrete state Markov Processes can be expressed as linear algebra in the following way. Fix an enumeration of Q and let $\mathbf{p}_i$ denote the probability of being in the ith state $\mathbf{q} \in Q$. The enumeration is arbitrary but assumed fixed for the remainder of this paper. The dynamics of the probability vector $\mathbf{p}$ are governed by the *infinitesimal generator* $\mathbf{A}$ defined as follows: All entries of $\mathbf{A}$ are zero unless

- If $i \neq j$ and $\mathbf{q}_i + \mathbf{a}_l = \mathbf{q}_j$: $\mathbf{A}_{ij} = k_l M(\mathbf{a}_l, \mathbf{q}_i)$
- If $i = j$: $\mathbf{A}_{ii} = -\sum_m \mathbf{A}_{im}$.

By construction the rows of $\mathbf{A}$ sum to zero and all off-diagonal entries are non-negative. Probability mass functions over Q are expressed as row vectors and real functions on Q, $y : Q \to \mathbb{R}$ as column vectors. The dynamics an arbitrary probability mass function $\mathbf{p}$ is governed by

$$\dot{\mathbf{p}} = \mathbf{p}\mathbf{A}, \qquad (1)$$

the CME.

B. A Reaction Network for the Testbed

The reaction network description for our robotic testbed consists of four distinct reactions: two describe the spontaneous association and dissociation of passive robots P_1 and P_2, one describes the disassembly of P_{12} by active robots, and the last reaction describes recharging of active robots. Denote the rate constant for association and dissociation by the natural dynamics by k_1 and k_{-1}, for the disassembly reaction by k_2, and for the tunable recharge reaction by k_3. The rate constant for the tunable recharge reaction corresponds to the maximal physically possible rate, in this case highest operating intensity of the overhead lamps. These reactions are summarized in (2)-(4).

$$P_1 + P_2 \xrightarrow[k_{-1}]{k_1} P_{12} \qquad (2)$$

$$P_{12} + A \xrightarrow{k_2} P_1 + P_2 + A' \qquad (3)$$

$$A' \xrightarrow{uk_3} A. \qquad (4)$$

Note that the rate constant in (4) depends on u. As a result the infinitesimal generator matrix $\mathbf{A}$ is a function of u.

The discrete state space Q is finite and obeys the conservation equations

$$N_{P_1} + N_{P_{12}} = N_{P_2} + N_{P_{12}} \doteq C_1, \qquad (5)$$

$$N_A + N_{A'} \doteq C_2. \qquad (6)$$

The first relation (5) holds when the system has the same number of both types of passive robots (C_1 of each, which we ensure in our experiments), while (6) asserts that there are C_2 active robots that can either be in a charged or discharged state. As a consequence of (5) and (6), N_{P_1}, N_{P_2}, and A' can be expressed in terms of $N_{P_{12}}$, A, and the constants C_1 and C_2. Instead of five different species we can keep track of only two. For the remainder of this paper we will assume that

$$\mathbf{q} = \begin{pmatrix} N_A \\ N_{P_{12}} \end{pmatrix} \in \mathbb{N}^2$$

and note that the copy number for the missing species can be reconstructed from this reduced state.

C. Checking the Well-Mixed Condition

There are several equivalent definitions of what it means for a system to be well-mixed. Basically, all definitions are sufficient conditions for guaranteeing that a process is Markov and that each *possible* combination of reactants for a particular reaction $\mathbf{a}_l$ is equally likely to be involved in the next reaction.

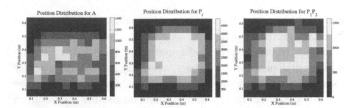

Fig. 3. Observed distribution of robots on air-table. The left figure corresponds to active robots, A or A', the middle plot correspond to passive robots P_1/P_2, and the right figure to heterodimers P_{12}. These plots demonstrate that the occupancy of parts on the air-table is roughly uniform on the table. The area of low occupancy around the perimeter is due to the geometry of the components interacting with the boundary of the air-table.

While being well-mixed in this sense is a strong assumption, it allows for the characterization of a reaction by a single parameter, the rate constant k_l. For the remainder of this section we use the definition of well-mixedness from [10]. For alternative conditions see [9, Ch. 7.2]. The two conditions that must be checked are that: (a) the reactants are uniformly distributed throughout the environment and (b) that the reactants diffuse through the reaction domain faster than they react.

To estimate the distribution of the different types of robots on the air-table we decomposed it into a 11×11 grid and extracted the occupancy statistics for each grid box from video data. Fig. 3 shows the resulting distributions. The red area in the center of each plot is roughly at the same level and indicates a uniform distribution. The area of low occupancy around the perimeter results from the fact the position of each robot is estimated at its center yet geometric constraints keep the center away from from the air-table border.

The diffusion coefficient for a robot is defined as

$$D = \frac{\mathbb{E}\, r^2(t)}{4t},$$

where $r(t)$ denotes the displacement of the robot as a function of time. We used the targets described in Sec. II to track the position of different robot types. We averaged over multiple experiments as well as the instances of of each robot type to compute the expected value. The resulting estimates for the diffusion coefficient are given in Tab. I. The subscripts of D indicates what robot type the diffusion coefficient was calculated for. For example, $D_{P_{12}}$ is the diffusion coefficient of heterodimers.

Combined with the rate constants measured in Sec. III-D we conclude that condition (a) and (b) are approximately met. The testbed is well-mixed and the stochastic chemical kinetic model is appropriate.

D. Characterizing Rate Constants

One method to determine rate constants is to measure the average waiting time between reactions from a known state. This quantity, together with the known inverse relationship between the reaction rate and average waiting time, yields an estimate of the rate [5]. Although useful in simulation, one drawback of this method is that one needs to repeatedly re-initialize the system to gather statistical data, which is tedious and time consuming. An exception is k_3, which was measured

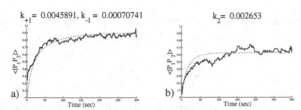

$k_{+1} = 0.0045891, k_{-1} = 0.00070741$

$k_2 = 0.002653$

a)

b)

Fig. 4. Curve fitting results used to determine rate constants. a) The lights are off ($u = 0$) and the system starts in $N_A = N_{P_{12}} = 0$. b) The light are on ($u = 1$) the system also starts in $N_A = N_{P_{12}} = 0$.

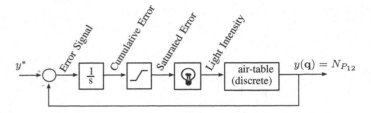

Fig. 5. Block diagram of the proposed control system. Only the air-table state and output signal are discrete, all other signals are continuous.

in this way. The reason is that the recharge reaction represents a change in internal state, which is easy to re-initialize.

For the other rate constants we take a different approach. We average multiple longer trajectories all *starting* from the same initial condition. However, the system is allowed to continue evolving for a set amount of time, possibly undergoing many reactions. This has the advantage that each re-initialization gives much more information than a single waiting time. We then fit this empirical average to solutions of the CME (1).

We determined the remaining rate constants k_1, k_{-1} and k_2 in two steps. First, we gathered trajectories starting from $N_A = N_{P_{12}} = 0$ with $u = 0$ (lights off). This way the disassembly reaction and recharge reaction do not influence the natural dynamics. We then used MATLAB to numerically fit the CME solution with the two free parameters k_1 and k_{-1} to the empirical average, minimizing the mean squared error, see Figure 4a.

Using the values previously determined for k_3, k_1, and k_{-1} we then used the same approach (this time with $u = 1$) for determining the only remaining parameter in the CME solution, k_2. The resulting curve fit is shown in Fig. 4b.

Parameter	Estimate	Uncertainty	Units
k_1	0.0046		$\frac{reaction}{sec\ number^2}$
k_{-1}	0.00071		$\frac{reaction}{sec\ number}$
k_2	0.0027		$\frac{reaction}{sec\ number^2}$
k_3	0.08		$\frac{reaction}{sec\ number}$
$D_A/D_{A'}$	0.0018	0.0002	$\frac{m^2\ sec}{sec}$
D_{P_1}/D_{P_2}	0.0015	0.0001	$\frac{m^2\ sec}{sec}$
$D_{P_{12}}$	0.00083	0.00001	$\frac{m^2\ sec}{sec}$

TABLE I

ESTIMATES OF RATE CONSTANTS AND DIFFUSION COEFFICIENTS.

IV. CONTROLLER DESIGN AND ANALYSIS

This section describes an integral feedback controller for the reaction network (2)-(4) and a stochastic hybrid system (SHS) that describes the closed loop system.

The idea of the feedback controller is simple, increasing light intensity recharges the active robots more quickly and disassemble more heterodimers. An increase in u will decrease $N_{P_{12}}$. Decreasing the light intensity results in fewer charged robots to disassemble heterodimers. A decrease in u increases the average number $N_{P_{12}}$. With the mathematical model for the testbed in place we can state the problem formally.

Design a control system that measures $N_{P_{12}}$ and adjusts intensity of the overhead lamps such that $\mathbb{E}N_{P_{12}} = y^*$ for a given reference value y^*.

A. Integral Control

The discrete state $\mathbf{q}$ of the closed loop system develops according to (2)-(4). Let $\mathbf{y}$ be the a vector corresponding to an output function $y : Q \rightarrow \mathbb{R}$, in this case $y(\mathbf{q}) = N_{P_{12}}$. Define a new continuous part of the state that models the cumulative error from a setpoint y^* as

$$\dot{x} = f(\mathbf{q}, x) = \gamma(y(\mathbf{q}) - y^*) = \gamma(N_{P_{12}} - y^*). \quad (7)$$

In order to express saturation of the input, here the physical limitations of the overhead lamps, define $h : \mathbb{R} \rightarrow \mathbb{R}$ by

$$h(x) = \begin{cases} 0, & x \leq 0 \\ x, & 0 < x \leq 1 \\ 1, & 1 < x. \end{cases}$$

With this notation we can define an integral feedback controller by

$$u = h(x). \quad (8)$$

A block diagram of the control system is shown in Fig. 5. The remainder of this section is dedicated to analyzing the closed loop system.

B. Stochastic Hybrid System

Adding a continuous random variable whose dynamics depend on the discrete state $\mathbf{q}$ of a Markov process results in a *stochastic hybrid system* (SHS). This section is a brief description of the notation and some specific mathematical tools available for SHSs, for more information see [11, 12].

The key feature of an SHS is that the dynamics of the system are stochastic and that the state are hybrid, meaning the state space of the system has the form $Q \times X$ where Q is some discrete set and $X \subseteq \mathbb{R}$ is continuous. The set of possible discrete states Q is typically finite or countably infinite. We use $z \in Z = Q \times X$ as shorthand for the pair (q, x). Let $\mathcal{Q}$, $\mathcal{X}$, and $\mathcal{Z}$ denote the stochastic processes on the various components of the state space.

In each discrete state, the dynamics of $\mathcal{X}$ are governed by a differential equation,

$$\dot{x} = f(\mathbf{q}, x) \qquad f : Q \times X \rightarrow TX. \quad (9)$$

The dynamics of the discrete state $\mathcal{Q}$ are governed by a set of transitions, indexed by a finite set L. Each transition $l \in L$

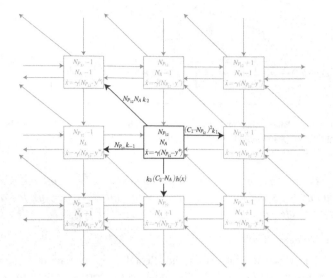

Fig. 6. A schematic representation of the closed loop SHS. The boxes represent discrete states and the arrows represent transitions. Each box shows both the discrete state it represents and the ODE describing the continuous states. An arbitrary state $(N_A, N_{P_{12}})^T$ is highlighted in black. The transition intensities for all transitions leaving $(N_A, N_{P_{12}})^T$ are shown next to the arrows.

has associated with it an intensity function

$$\lambda_l(\mathbf{q}, x) \qquad \lambda_l : Q \times X \to [0, \infty), \qquad (10)$$

and a reset map

$$(\mathbf{q}, x) = \phi_l(\mathbf{q}^-, x^-) \qquad \phi_l : Q \times X \to Q \times X. \qquad (11)$$

The intensity function is the instantaneous rate of the transition l occurring, so that $P(l$ occurs during$(t, t+dt)|\mathcal{Q} = \mathbf{q}, \mathcal{X} = x) = \lambda_l(q, x, t)dt$. The reset map ϕ_l determines where the process jumps after a transition is triggered at $(\mathbf{q}^-, x^-)$ at time t. The minus in the superscript denotes the left hand limit of $\mathbf{q}$ and x at time t. We think of this limit as the state of the process immediately before the jump. Fig. 6 shows part of the system representing the closed loop SHS. The boxes represent discrete states and the arrows represent transitions.

C. Extended Generator

This section describes the *extended generator* $\mathcal{L}$ associated with an SHS. This operator is analogous to the *generator matrix* of a discrete state Markov process but in the hybrid case is a partial differential equation describing the dynamics of the expected value of arbitrary *test functions* on the state space. In particular, the extended generator allows us to derive ordinary differential equations (ODEs) that govern the dynamics of the statistical moments of the state variables of an SHS.

Operator $\mathcal{L}$ in (12) is the extended generator for an SHS described by (9)-(11). Let ψ be a real valued function on $Q \times X$ and define

$$\mathcal{L}\psi(z) = \frac{\partial \psi(z)}{\partial x} f(z) + \sum_{l \in L} (\psi(\phi_l(z)) - \psi(z)) \lambda_l(z). \quad (12)$$

The operator $\mathcal{L}$ has the following useful property relating the time derivative of the expected value of a test function ψ to $\mathcal{L}\psi$

$$\frac{d\,\mathbb{E}\psi}{dt} = \mathbb{E}\,\mathcal{L}\psi \qquad (13)$$

[11]. The extended generator for the closed loop system is given by

$$
\begin{aligned}
\mathcal{L}\quad & \psi(N_{P_{12}}, N_A, x) \qquad\qquad\qquad\qquad (14)\\
= \quad & \frac{\partial \psi(N_{P_{12}}, N_A, x)}{\partial x}\gamma(N_{P_{12}} - y^*)\\
+ \quad & (\psi(N_{P_{12}}+1, N_A, x) - \psi(N_{P_{12}}, N_A, x))k_1(C_1 - N_{P_{12}})^2\\
+ \quad & (\psi(N_{P_{12}}-1, N_A, x) - \psi(N_{P_{12}}, N_A, x))k_{-1}N_{P_{12}}\\
+ \quad & (\psi(N_{P_{12}}-1, N_A-1, x) - \psi(N_{P_{12}}, N_A, x))k_2 N_{P_{12}}N_A\\
+ \quad & (\psi(N_{P_{12}}, N_A+1, x) - \psi(N_{P_{12}}, N_A, x))x(C_2 - N_A).
\end{aligned}
$$

It can be used to find ODEs describing the evolution of the statistical moments of the SHS. Specifically, letting $\psi = x$ we obtain

$$\frac{d\,\mathbb{E}x}{dt} = \mathbb{E}\,\gamma(N_{P_{12}} - y^*). \qquad (15)$$

If the closed loop system is stochastically stable, in the sense that the probability distribution of states approaches a fixed invariant distribution, then by (15) we can conclude that

$$\mathbb{E}N_{P_{12}} = y^*.$$

The controller works in expected value when the system is in steady state. Now, the problem of showing correctness of the controller reduces to showing that the system is stochastically stable or *ergodic*, i.e. that the system always approaches a unique steady state distribution.

D. Ergodicity

We use a Lyapunov function argument [13, THM 5.1] to show that the closed loop SHS is ergodic. This allows us to set the LHS in (15) to zero and argue that the controller works in steady state. We show that the system is ergodic for some reference values y^* and give sufficient conditions for ergodicity for a range of y^*.

Denote the generator matrices of minimum and maximum input by $\mathbf{A}_m = \mathbf{A}(0)$, $\mathbf{A}_M = \mathbf{A}(1)$ and the corresponding steady state probability mass functions by $\mathbf{p}_m$ and $\mathbf{p}_M$ respectively.

Theorem: Let $\mathbf{A}(u)$ be the generator of a tunable reaction network and $\mathbf{y}$ the vector corresponding to an output function $y : Q \to \mathbb{R}$ of the discrete state. The feedback controller proposed in (8) results in a closed loop system with a stationary distribution that has $\mathbb{E}y = y^*$ when y^* is in the *controllable region*, $\mathbf{p}_M\mathbf{y} < y^* < \mathbf{p}_m\mathbf{y}$.

Note: If $\mathbf{p}_M\mathbf{y} > \mathbf{p}_m\mathbf{y}$, then the theorem applies with the sign in (7), and the upper and lower limits of the controllable region reversed.

Proof: Let $\mathcal{Z}$ be the SHS corresponding to the closed loop system. By [13, THM 5.1], $\mathcal{Z}$ is ergodic when there exists a function $V : Z \to \mathbb{R}^+$ with the property that $V(z) \to \infty$ as $|z| \to \infty$ and

$$\mathcal{L}V(z) \leq -f(z) \quad \forall z \notin C \qquad (16)$$

for some compact region C and positive function f^1.

For our system, we define the function $\widehat{V}$ to be

$$\widehat{V}(q, x) = \begin{cases} x + c^+(q) & \text{for } x > 0 \\ -x + c^-(q) & \text{for } x < 0, \end{cases}$$

where c^+ and c^- depend on q. Note that the function $\widehat{V}$ is neither differentiable (required to apply $\mathcal{L}$) nor positive (required by theorem) since the offsets can be negative. To address this problem, let V be a function that agrees with $\widehat{V}$ when x is outside some interval $[v_{min}, v_{max}]$ for all $q \in Q$, and is both non-negative and twice differentiable. This function always exists since Q is finite and $\widehat{V}$ increases with $|x|$ so that $\widehat{V}$ is positive for sufficiently large $|x|$.

Let the compact region required by the theorem be $C = Q \times [\min(v_{min}, 0), \max(v_{max}, 1)]$. Since we are only interested in V outside C, we look at the cases when the feedback input is saturated at either $u = 0$ or $u = 1$. This situation simplifies the analysis, since the transition intensities $\lambda(q, x)$ are independent of x in the saturated regions. We now argue that for some range of set points y^* we can find c^+ and c^- to make V a Lyapunov function in the sense of (16).

Choosing $f = \epsilon$ and considering saturation at $u = 1$ first, we can rewrite the conditions of (16) in vector from,

$$\mathbf{y} - y^*\mathbf{1} + \mathbf{A}_M \mathbf{c}^+ \leq -\epsilon \mathbf{1}. \qquad (17)$$

Let $\tilde{\epsilon}$ be an arbitrary vector with strictly positive entries, then we can rewrite (17) as

$$\mathbf{y} - y^*\mathbf{1} + \mathbf{A}_M \mathbf{c}^+ = -\tilde{\epsilon}. \qquad (18)$$

We want to determine when this equation has a solution for $\mathbf{c}^+$. Note that

$$\mathbf{A}_M \mathbf{c}^+ = -\tilde{\epsilon} + y^*\mathbf{1} - \mathbf{y}$$

has a solution only if $(-\tilde{\epsilon} + y^*\mathbf{1} - \mathbf{y})$ is in the column space of $\mathbf{A}_M$, which we write $(-\tilde{\epsilon} + y^*\mathbf{1} - \mathbf{y}) \in \mathrm{Col}\mathbf{A}_M$. Equivalently

$$(\mathrm{Col}\mathbf{A}_M)^\perp \perp (-\tilde{\epsilon} + y^*\mathbf{1} - \mathbf{y}) \qquad (19)$$
$$(\mathrm{Nul}\mathbf{A}_M^T) \perp (-\tilde{\epsilon} + y^*\mathbf{1} - \mathbf{y}) \qquad (20)$$
$$(\mathbf{p}_M^*)^T \perp (-\tilde{\epsilon} + y^*\mathbf{1} - \mathbf{y}) \qquad (21)$$
$$0 = \mathbf{p}_M^*(-\tilde{\epsilon} + y^*\mathbf{1} - \mathbf{y}) \qquad (22)$$
$$0 = -\mathbf{p}_M^*\tilde{\epsilon} + y^* - \mathbf{p}_M^*\mathbf{y}, \qquad (23)$$

where Nul denotes the null space and $\perp$ the orthogonal complement. Since $\tilde{\epsilon}$ has arbitrary, strictly positive entries, a solution for $\mathbf{c}^+$ exists when

$$\mathbf{p}_M^*\mathbf{y} < y^*.$$

---

[1]The theorem has some technical preconditions, which are fulfilled in our case, namely that all compact sets are petite see [13]. This follows from [14, THM 4.1], [12, THM 27.6] and the fact that every Feller process is also a T-process.

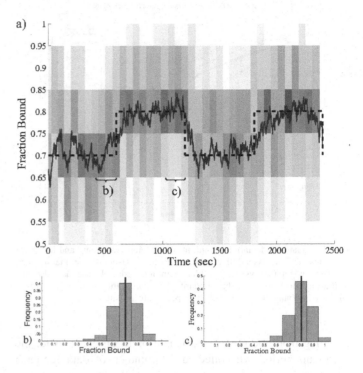

Fig. 7. Tracking data from robotic test-bed. This is the average of 25 different experiments. Each experiment has $C_1 = 10$ P_1 and P_2 each, and $C_2 = 4$ active robots. The grey shading in the background corresponds to the fraction of trajectories with that number of pairs. Darker regions correspond to higher occupancy. The two histograms on the bottom show the fraction of possible dimers taken for the last half of a step (b,c). The vertical line is the computed mean demonstrating the correct behavior for the controller.

Similarly, for saturation with $u = 0$ we get

$$\mathbf{p}_m^*\mathbf{y} > y^*.$$

Thus the system is ergodic if

$$\mathbf{p}_M^*\mathbf{y} < y^* < \mathbf{p}_m^*\mathbf{y}.$$

Furthermore, by (15) the expected value of $N_{P_{12}}$ tracks the reference value y^* when it is in the controllable region. ∎

The proof does not rely on any special structure of $\mathbf{A}(u)$ nor the value of γ, as a result the theorem is generally applicable to tunable reaction networks with saturating inputs.

V. EXPERIMENTAL RESULTS

We implemented the proposed controller on the robotic test-bed described in Sec. II. The generator matrix $\mathbf{A}(u)$ is defined by (2)-(4), and the output function is $y(\mathbf{q}) = N_{P_{12}}$. To show its the tracking capability we tracked two periods of a square wave. The low and high set-points were 0.7 and 0.8 (corresponding to 7 and 8 P_{12}). Both of the setpoints are inside the empirically determined controllable region for this system, 0.60-0.86.

The combined results of 25 trajectories are shown in Fig. 7. We let each trajectory run with a set point of 0.7 for 5 minutes (a half period) before recording data, which allowed transients

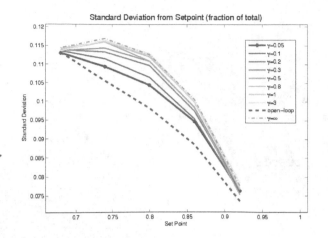

Fig. 8. Standard deviation of output for different set points and integrator constants γ. The dashed line at the bottom corresponds the the standard deviation when the system is under open loop control. The dash-dot line at the top corresponds to the standard deviation of the system when driven with a bang-bang controller and the input is always saturated.

resulting from the manual initialization to dissipate. After the warm up period we collected 20 minutes of data for each trajectory.

This experiment demonstrates the controller tracking a reference signal in mean (Fig. 7bc). This experiment also demonstrates the fundamental stochasticity in the system. The spread in Fig. 7bc is not due to measurement uncertainty or noise, but a fundamental property of the stochastic system we are controlling.

We also present some preliminary simulation experiments exploring how the variance of the copy number relates to the integrator gain γ. The proof for tracking in mean did not depend on the value of γ, so the proposed controller will always yield the desired mean steady-state copy number. However, it might differ in the degree of fluctuation around the correct mean.

The relation between set-point, integrator gain, and standard deviation at steady state are shown in Fig. 8. Each data point was collected by setting γ and estimating the standard deviation at steady state. This approach limits smallest value of γ we can feasibly simulate, since small values slow down the system dynamics and make simulations prohibitively computationally expensive.

We observe that less aggressive values of γ result in a smaller standard deviation of the output. The upper and lower limits of the standard deviation correspond to open-loop and bang-bang control. Another interesting feature of Fig. 8 is that the standard deviation of the output seems to be less sensitive to γ if the reference y^* is close to the edge of the controllable region.

VI. CONCLUSIONS AND FUTURE WORK

We proposed an integral feedback controller for controlling the average copy number of an arbitrary species in a system modeled by stochastic chemical kinetics. We proved that the controller tracks a reference in mean and demonstrated the approach on an robotic experimental platform. We also presented some preliminary simulation results regarding the variance of the the copy number as a function of the integrator gain. We are currently working on analytical results describing the steady state variance of the control scheme. We are also investigating ways to decentralize the controller by using local estimates of the global output.

Finally, we would like to emphasize the generality of our approach. The functionality of the controller requires no tuning of the integrator gain γ as long as the reference is in the controllable region, which is easy to measure experimentally. The detailed structure of the underlying Markov process does not matter.

ACKNOWLEDGEMENTS

We would like to thank Sheldon Rucker, Michael McCourt, and Yi-Wei Li for their dedication and hard work in designing and building the robots. We would also like to thank Alexandre Mesquita and João Hespanha for their helpful discussions about proving ergodicity.

REFERENCES

[1] E. Saeedi, S. Kim, H. Ho, and B. A. Parviz, "Self-assembled single-digit micro-display on plastic," in *MEMS/MOEMS Components and Their Applications V. Special Focus Topics: Transducers at the Micro-Nano Interface*, vol. 6885, p. 688509, SPIE, 2008.
[2] P. W. K. Rothemund, "Folding dna to create nanoscale shapes and patterns," *Nature*, vol. 440, pp. 297–302, Mar. 2006.
[3] M. Boncheva, D. A. Bruzewicz, and W. G. M., "Millimeter-scale self-assembly and its applications," *Pure and Applied Chemistry*, vol. 75, pp. 621–630, 2003.
[4] H. Onoe, K. Matsumoto, and I. Shimoyama, "Three-dimensional micro-self-assembly using hydrophobic interaction controlled by self-assembled monolayers," *Journal of Microelectromechanical Systems*, vol. 13, no. 4, pp. 603– 611, 2004.
[5] S. Burden, N. Napp, and E. Klavins, "The statistical dynamics of programmed robotic self-assembly," in *Conference Proceedings ICRA 06*, pp. pp. 1469–76, May 2006.
[6] P. J. White, K. Kopanski, and H. Lipson, "Stochastic self-reconfigurable cellular robotics ieee international conference on robotics and automation (icra04), pp. 2888-2893," *IEEE International Conference on Robotics and Automation (ICRA04)*, pp. 2888–2893, 2004.
[7] H. El-Samad, J. P. Goff, and M. Khammash, "Calcium homeostasis and parturient hypocalcemia: An integral feedback perspective," *Journal of Theoretical Biology*, vol. 214, pp. 17–29, Jan. 2002.
[8] D. A. McQuarrie, "Stochastic approach to chemical kinetics," *Journal of Applied Probability*, vol. 4, pp. 413–478, Dec 1967.
[9] N. V. Kampen, *Stochastic Processes in Physics and Chemistry*. Elsevier, 3 ed., 2007.
[10] D. T. Gillespie, "Exact stochastic simulation of coupled chemical reactions," *Journal of Physical Chemistry*, vol. 81, no. 25, pp. 2340–2361, 1977.
[11] J. P. Hespanha, "Modeling and analysis of stochastic hybrid systems," *IEE Proc — Control Theory & Applications,* Special Issue on Hybrid Systems, vol. 153, no. 5, pp. 520–535, 2007. Available at http://www.ece.ucsb.edu/ hespanha/published.
[12] M. Davis, *Markov Processes and Optimization*. Chapman & Hall, 1993.
[13] S. P. Meyn and R. L. Tweedie, "Stability of markovian processes *III*: Foster-lyapunov criteria for continuous-time processes," *Advances in Applied Probability*, vol. 25, no. 3, pp. 518–548, 1993.
[14] S. P. Meyn and R. L. Tweedie, "Stability of markovian processes *II*: Continuous-time processes and sampled chains," *Advances in Applied Probability*, vol. 25, no. 3, pp. 487–517, 1993.

Centralized Path Planning for Multiple Robots: Optimal Decoupling into Sequential Plans

Jur van den Berg   Jack Snoeyink   Ming Lin   Dinesh Manocha

Department of Computer Science, University of North Carolina at Chapel Hill, USA.

E-mail: {berg, snoeyink, lin, dm}@cs.unc.edu

Abstract— We develop an algorithm to decouple a multi-robot path planning problem into subproblems whose solutions can be executed sequentially. Given an external path planner for general configuration spaces, our algorithm finds an execution sequence that minimizes the dimension of the highest-dimensional subproblem over all possible execution sequences. If the external planner is complete (at least up to this minimum dimension), then our algorithm is complete because it invokes the external planner only for spaces of dimension at most this minimum. Our algorithm can decouple and solve path planning problems with many robots, even with incomplete external planners. We show scenarios involving 16 to 65 robots, where our algorithm solves planning problems of dimension 32 to 130 using a PRM planner for at most eight dimensions.[1]

I. INTRODUCTION

In this paper, we discuss the problem of path planning for multiple robots, which arises in different applications in Robotics. The objective is to move multiple robots in a common workspace from a given start configuration to a given goal configuration without mutual collisions and collisions with obstacles. We assume that an exact representation of the geometry of the robots and the workspace is given.

This problem has been studied extensively. Approaches are often characterized as *centralized* (a better term would be *coupled*) or *decoupled*: A coupled planner computes a path in a combined configuration space, which essentially treats the robots as a single combined robot. A decoupled planner may compute a path for each robot independently, then use a coordination diagram to plan collision-free trajectories for each robot along its path. Or it may plan a trajectory for each robot in order of priority and avoid the positions of previously planned robots, which are considered as moving obstacles.

Decoupled planners are generally faster, usually because fewer degrees of freedom are considered at one time. Unfortunately, they are usually not *complete* – some coupling may be necessary, as when two robots each have their goal as the other's start position, so a decoupled planner may not find a solution, even when one exists. Centralized or coupled planners, on the other hand, may be complete in theory, but they may need to work in configuration spaces of impractically high dimensions, regardless of how challenging the actual instance of the planning problem is – two robots in separate rooms would still be considered as a system with double the degrees of freedom, even though their tasks can be carried out independently without interference.

In this paper, we demonstrate an algorithm for multiple robot planning problems that decomposes any instance of multi-robot planning into a sequence of sub-problems with the minimum degree of coupled control. Informally, the control of two robots must be directly coupled if they must move at the same time to achieve their goals. The transitive closure of the direct coupling relationship is an equivalence relation that partitions the robots into classes that must be planned together as a composite. The degree of a composite robot is the sum of the number of degrees of freedom of the individual robots that are coupled.

We partition the robots into an ordered sequence of composite robots so that each composite can move from start to goal in turn, and *minimize* the maximum degree of all composite robots. If the problem instance has degree of coupling α, our algorithm is complete if we have access to an external general-purpose path planner that is complete for up to α degrees of freedom. Although the number of robots may appear exponentially in the combinatorial parts of the algorithm (though our experiments show that this worst case may be avoided), their degrees of freedom do not blow up the complexity of planning. Thus, our implementation is able to solve challenging scenarios which cannot be solved by traditional multi-robot planners. It is applicable to robots of any kind with any number of degrees of freedom, provided that our external planner is too.

After a brief review of related work in Section II, we define in Section III the notions of composite robots and execution sequences and constraints upon them, which we assume are returned from our external planner. In Section IV we present our algorithm and prove its properties, and in Section VI we discuss our experimental results. We conclude the paper in Section VII.

II. RELATED WORK

Path planning for multiple robots has been extensively studied for decades. For the general background and theory of motion planning and coordination, we refer readers to [10, 13]. In this section, we review prior work that addresses similar problems as ours.

As mentioned earlier, prior work for multiple robots are often classified into *coupled* and *decoupled* planners. The coupled approaches aggregate all the individual robots into one large composite system and apply single-robot motion planning algorithms. Much of classical motion planning techniques

[1]This research is supported in part by ARO, NSF, RDECOM, and Intel.

for exact motion planning, randomized motion planning and their variants would apply directly [10, 9, 11, 13].

Decoupled approaches plan for each robot individually and then perform a velocity tuning step in order to avoid collisions along these paths [29, 23, 18, 20, 22]. Alternatively, other schemes such as coordination graphs [15], or incremental planning [21] can help to ensure that no collisions occur along the paths. Prioritized approaches plan a trajectory for each robot in order of priority and avoid the positions of previously planned robots, which are considered as moving obstacles [7]. The choice of the priorities can have a large impact on the performance of the algorithm [27]. Some planners also search through a space of prioritizations [3, 4].

In addition, hybrid methods combine aspects of both coupled and decoupled approaches to create approaches that are more reliable or offer completeness but also scale better than coupled approaches [1, 2, 14, 26, 19, 5].

Geometric assembly problems initially seem closely related, especially when they speak of the number of "hands" or distinct motions needed to assemble and configuration of objects [16, 24, 17]. The various blocking graphs described in [28, 8] inspired the constraint graphs that we use. Differences can be seen on closer inspection: Assembly problems often restrict motions to simple translations or screws, and the aim is to create subassemblies that will then move together. (Our individual robots would need coordinated planning to move together as a subassembly.) Start positions for objects are usually uncomplicated, but assembly plans may need to be carried by some manipulator that must be able to reach or grasp the objects. In some sense, our algorithm combines these ideas: it captures constraints and then uses them to reduce the complexity of the motions that are planned.

III. DEFINITIONS AND PRELIMINARIES

Our multi-robot planning problem is formally defined as follows. The input consists of n robots, $r_1, \ldots, r_n$, and a common workspace in which these robots move (imagine a two- or three-dimensional scene with obstacles). The configuration space of robot r_i, denoted $\mathcal{C}(r_i)$, has dimension $\dim(r_i)$ equal to the number of degrees of freedom of robot r_i. Each robot r_i has a start configuration $s_i \in \mathcal{C}(r_i)$ and a goal configuration $g_i \in \mathcal{C}(r_i)$.

The task is to compute a path $\pi : [0,1] \in \mathcal{C}(r_1) \times \ldots \times \mathcal{C}(r_n)$, such that initially $\pi(0) = (s_1, \ldots, s_n)$, finally $\pi(1) = (g_1, \ldots, g_n)$, and at each intermediate time $t \in [0,1]$, with robots positioned at $\pi(t)$, no robot collides with an obstacle in the workspace or with another robot. At times we will refer to trajectories, which are the projections of a path into the space of a subset of robots.

The rest of this section gives precise definitions and properties for notions that we use to construct our algorithm for computing plans for composite robots with minimum degree of coupling.

A. The Coupled Relation from a Solution Path

Consider a path π that solves a multi-robot planning problem, as defined above. For each robot r_i, we define an *active*

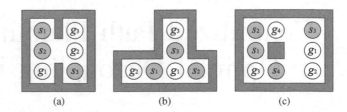

Fig. 1. Three simple example problems for circular planar robots, with start (gray) and goal (white) configurations shown. The dark gray delineates the static obstacles. Solution sequences for (a) = $[r_3, r_2, r_1]$ and (b) = $[r_3, r_1 r_2]$. Instance (c) has four solution sequences, $[r_2, r_1, r_3, r_4], [r_2, r_1, r_4, r_3], [r_2, r_4, r_1, r_3]$, and $[r_4, r_1, r_2, r_3]$.

interval $\tau_i \subset [0,1]$ as the open interval from the first time r_i leaves its start position to the last time r_i reaches its goal position.

Definition 1 (Coupled relation) Two robots r_i, r_j are *directly coupled* if their active intervals intersect, $\tau_i \cap \tau_j \neq \emptyset$. The transitive closure of this relation is an equivalence relation, and we say that robots in the same equivalence class are *coupled*.

On the interval $[0,1]$, the equivalence classes of the coupled relation determine the connected subsets of the union of all active intervals. If we list the equivalence classes in order of increasing time, we get a sequence of coupled or composite robots that can be abstracted from a specific path as an *execution sequence*, as we define in the next subsection.

The *degree of coupling of a solution path* is the maximum of the sum of the degrees of freedom in any connected set of active intervals; these are the maximum number of degrees of freedom that need be considered simultaneously by any coupled planner to construct or validate this sequential plan. The *degree of coupling of a planning instance* is the minimum degree of coupling of any solution path. Our algorithm finds a solution path that achieves the minimum degree of coupling α for the given instance, by using an external coupled planner on problems with at most α degrees of freedom. While hard, puzzle-like problem instances may still require a full-dimensional configuration space, we argue that many "realistic" problems, even those involving many robots, can be solved with much lower dimensional planning.

B. Execution Sequences

In our algorithm, we will decompose the multi-robot motion planning problem into lower-dimensional sub-problems that can be *sequentially* executed. In these sub-problems, some individual robots will be coupled and considered as composite robots by the lower-dimensional planner.

Definition 2 (Composite Robot) We define a *composite robot* $R \subseteq \{r_1, \ldots, r_n\}$ as a subset of the n robots that is treated as one coupled or composite body. R's configuration space, $\mathcal{C}(R)$, is the Cartesian product of the configuration spaces of the robots in R, its dimension is the sum of the degrees of freedom of the robots in R: $\dim(R) = \sum_{r_i \in R} \dim(r_i)$, and its active interval is the smallest interval that contains the union of the active intervals of all $r_i \in R$.

138

Since individual robots can also be thought of as composite, we will tend to omit the word "composite" and just say "robot." When we want to emphasize the composite nature of a robot, we concatenate the individual robots it consists of. For example $\{r_1, r_2\} = r_1 r_2$.

Definition 3 (Execution Sequence) We define an *execution sequence* S as an ordered partition of the n robots into a sequence $S = (R_1, \ldots, R_k)$ of composite robots, such that $R_1 \cup \cdots \cup R_k = \{r_1, \ldots, r_n\}$ and $R_i \cap R_j = \emptyset$ for $i \neq j$.

An execution sequence is *valid* if it is the sequence of equivalence classes of the *coupled* relation for some solution path (with no collisions).

We call a valid execution sequence a *solution sequence*.

Our algorithm will find a solution sequence with the minimum number of degrees of freedom, so we define the dimension $\dim(S)$ of a solution sequence $S = (R_1, \ldots, R_k)$ as the dimension of the largest composite robot in the solution sequence: $\dim(S) = \max_{R_i \in S} \dim(R_i)$. An *optimal solution sequence* is a solution sequence S^* with minimal dimension among all solution sequences: $S^* = \arg\min_S \dim(S)$.

Fig. 1 illustrates three example problems for 2-D circular robots in the plane, each robot r_i must find a path from the gray start postion s_i to the corresponding white goal position g_i without collisions. A valid solution path for (a) is to first move r_3, then r_2 and then r_1 – i.e., execution sequence $S = (r_3, r_2, r_1)$ solves the problem. Hence, the 6-D configuration space of problem (a) can be decomposed into three sequential 2-D sub-problems. For problem (b) there is no solution by moving individual robots. But a valid solution path could first move r_3, and then move r_1 and r_2 simultaneously as a composite robot in a coordinated effort to reach their goal – i.e., execution sequence $S = (r_3, r_1 r_2)$ solves the problem. Hence, problem (b) can be decomposed into one 2-D sub-problem and one 4-D sub-problem. Problem (c) has four possible execution sequences, all of which have either r_4 or both r_1 and r_2 moving before r_3.

C. Order Constraints from a Robot R

Generalizing from these examples, we can observe that valid execution sequences depend only on the start or goal positions of inactive robots.

Observation 4 Execution sequence S is valid if, for all $i \in [1, k]$, robot $R_i \in S$ can move from its start to its goal without collisions, even when the *goal* configurations of robots in $\{R_1, \ldots, R_{i-1}\}$ and the *start* configurations of robots in $\{R_{i+1}, \ldots, R_k\}$ have been added to the obstacles.

By a thought experiment, let's develop notation for constraints on the ordering of robots in solution sequences for specific trajectories before giving the formal definition. Suppose a specific trajectory for (individual or composite) robot R has a collision with the goal configuration of robot r_j. We then write $R \prec r_j$ to indicate that R must either complete its motion before r_j begins, or R and r_j will be coupled since their active intervals overlap. Similarly, if the trajectory

collides with the start configuration of robot r_k, we may write $r_k \prec R$. We collect all collisions with a single trajectory into a conjunction, then write a disjunction of the conjunctions for all possible trajectories, and simplify the resulting expression in disjunctive normal form (DNF), which we denote $\mathcal{P}(R)$. For example, if some trajectory for R collides with no start or goal positions, then $\mathcal{P}(R) = \top$ ('true'), and In general, we need to keep only the *minterms* – those conjunctions that do not contain another as a subset.

For example, in Fig. 1(c) robot r_4 can reach its goal without going through any query configurations of other robots, so $\mathcal{P}(r_4) = \top$ ('true'). If, due to static obstacles, some robot has no path to the goal at all, we could write its expression as $\bot$ ('false'). Robot r_1 can either move through the goal configuration of r_3 and the start configuration of r_4 to reach its goal, *or* it can move through the start configuration of r_2 and the goal configuration of r_4. Hence:

$$\mathcal{P}(r_1) = (r_2 \prec r_1 \land r_1 \prec r_4) \lor (r_1 \prec r_3 \land r_4 \prec r_1).$$

In Fig. 1(b), robot r_2 has to move through both the start and the goal configuration of robot r_1. This gives the constraint $\mathcal{P}(r_2) = r_2 \prec r_1 \land r_1 \prec r_2$, which we may abbreviate as $r_1 \sim r_2$. This means that r_1 and r_2 need to be active simultaneously, so their motion must be planned as a composite robot $r_1 r_2$, for which $\mathcal{P}(r_1 r_2) = (r_3 \prec r_1 r_2)$.

The tilde abbreviation, and the fact that our ordering relation is transitive, allows us to rewrite each conjunction of $\mathcal{P}(R)$ in the following *atomic form*: we arbitrarily choose a representative $r \in R$, replace each capital R with the representative r, and AND the expression $\bigwedge_{r' \in R} r \sim r'$. Thus, for Fig. 1(b), we have the atomic form $\mathcal{P}(r_1 r_2) = r_3 \prec r_1 \land r_1 \sim r_2$.

Each conjunction has a natural interpretation as a directed graph in which robots are vertices and directed edges indicate $\prec$ relations. By transitivity, any directed path from r_i to r_j indicates a relation $r_i \prec r_j$ and any pair of vertices in the same strongly connected component are related by $\sim$. We use such constraint graphs in our implementation in Section IV-A.

Two properties of these constraint expressions are easy to observe. First, by construction,

Property 5 For each atomic constraint $r \prec r'$ in the constraint expression $\mathcal{P}(R)$ of (composite) robot R, either $r \in R$ or $r' \in R$.

Second, because any trajectory for a larger composite robot includes trajectories for subsets, we can observe that the constraints on the larger robot imply those on the smaller.

Property 6 If $R' \supseteq R$, then $\mathcal{P}(R') \Rightarrow \mathcal{P}(R)$.

This means, for instance, that if a robot r_i needs to move before a robot r_j, then any composite robot R involving r_i needs to move before r_j. This is an important property, as it allows us to obtain the constraints on the execution sequence *iteratively*, starting with the robots of smallest dimension (fewest degrees of freedom).

D. Constraints from an Execution Sequence

If we AND the constraints for each robot in an execution sequence, we get the expression that must be satisfied for it to be a solution sequence – a sequence in which each composite robot has a valid trajectory.

Lemma 7 An execution sequence $S = (R_1, \ldots, R_k)$ is a *solution sequence* if and only if S satisfies the constraint expression $\mathcal{P}(R_1) \wedge \cdots \wedge \mathcal{P}(R_k)$.

Proof: If S satisfies $\mathcal{P}(R_i)$, (composite) robot R_i can reach its goal configuration without moving through any of the *goal* configurations of robots in $\{R_1, \ldots, R_{i-1}\}$, and without moving through any of the *start* configurations of robots in $\{R_{i+1}, \ldots, R_k\}$. At the moment of execution of R_i, all robots in $\{R_1, \ldots, R_{i-1}\}$ reside at their goal configuration, as they have already been executed, and all robots in $\{R_{i+1}, \ldots, R_k\}$ reside at their start configuration, as they have not yet been executed. Hence, R_i can validly move to its goal configuration. If S does not satisfy $\mathcal{P}(R_i)$, (composite) robot R_i either has to move through any of the *goal* configurations of robots in $\{R_1, \ldots, R_{i-1}\}$, or through any of the *start* configurations of robots in $\{R_{i+1}, \ldots, R_k\}$ in order to reach its goal. As all robots in $\{R_1, \ldots, R_{i-1}\}$ reside at their goal configuration and all robots in $\{R_{i+1}, \ldots, R_k\}$ reside at their start configuration at the moment of execution of R_i, it is not possible for R_i to validly reach its goal. $\square$

Notice that because we AND the DNF expressions for the robots of an execution sequence, we can convert the resulting expression back to DNF by distributing across the new ANDs. The conjunctions of the resulting disjunction simply take one conjunction from the DNF for each robot in the execution sequence, so any sequence that satisfies the expression can directly seen to satisfy the DNF expressions for each constituent robot.

E. The CR Planner for Low-dimensional Sub-problems

Now, let us postulate a *CR planner* (coupled or composite robot planner), that given the workspace, with start and goal configurations for robots $\{r_1, r_2, \ldots, r_n\}$, and a small subset R of these robots, returns the DNF expression $\mathcal{P}(R)$ with each clause in atomic form.

As a proof of existence, we can construct a CR planner using as a black box any complete planner that can determine feasibility for the composite robot R in a fixed workspace. Simply apply the black box to at most $4^{n-|R|}$ instances, with each other robot $r' \notin R$ added as an obstacle in its start position ($r \prec r'$), its goal position ($r' \prec r$), both ($\top$), or neither ($r \sim r'$). Each feasible path found adds a conjunction of all its constraints to the DNF expression, which is simplified and put in atomic form. Our actual planner, described in Section V, will be less costly.

In the next section, we present an algorithm that efficiently finds an optimal solution sequence S^* to solve the multi-robot planning problem. Our algorithm is able to do this by planning

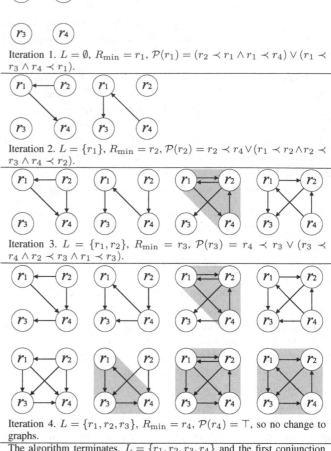

Iteration 1. $L = \emptyset$, $R_{\min} = r_1$, $\mathcal{P}(r_1) = (r_2 \prec r_1 \wedge r_1 \prec r_4) \vee (r_1 \prec r_3 \wedge r_4 \prec r_1)$.

Iteration 2. $L = \{r_1\}$, $R_{\min} = r_2$, $\mathcal{P}(r_2) = r_2 \prec r_4 \vee (r_1 \prec r_2 \wedge r_2 \prec r_3 \wedge r_4 \prec r_2)$.

Iteration 3. $L = \{r_1, r_2\}$, $R_{\min} = r_3$, $\mathcal{P}(r_3) = r_4 \prec r_3 \vee (r_3 \prec r_4 \wedge r_2 \prec r_3 \wedge r_1 \prec r_3)$.

Iteration 4. $L = \{r_1, r_2, r_3\}$, $R_{\min} = r_4$, $\mathcal{P}(r_4) = \top$, so no change to graphs.

The algorithm terminates. $L = \{r_1, r_2, r_3, r_4\}$ and the first conjunction in E has all of its strongly connected components in L. The solution sequence returned is $S = (r_2, r_1, r_4, r_3)$.

Fig. 2. An illustration of the steps of our algorithm on the problem of Fig. 1(c). In each iteration the constraints $\mathcal{P}(R)$ of a new robot are incorporated into E. We show the conjunctions in E as a set of constraint graphs. Strongly connected components in the constraint graphs are indicated by a gray background. Initially $E = \{\top\}$, which corresponds to one empty constraint graph.

only in configuration spaces whose dimension is less than or equal to $\dim(S^*)$.

IV. INCREMENTAL DISCOVERY OF COUPLING

In this section we explain how to use the CR planner, which produces constraints on execution sequences induced by small subsets of robots, to find the lowest degree of coupling that will solve the given instance of a multi-robot planning problem. Our algorithm incrementally calls the planner on higher and higher degree sub-problems, using the discovered constraints to determine what robots must be coupled. First, we describe the constraint graph, a data structure for the constraint expressions we have collected.

A. Constraint Graphs

In our algorithm, we maintain the constraints we have obtained so far in a constraint expression E. We represent E in *disjunctive normal form*, i.e. as a disjunction $E = J_1 \vee J_2 \vee \cdots$

140

of conjunctions J_i. Each conjunction J can be represented by a *graph* $G(J)$, which we call a *constraint graph*. A constraint graph has n nodes, one for each robot r_i, and a set of *directed* edges that indicate constraints on the order of execution of the robots. That is, for each atomic constraint $r_i \prec r_j$ in J, there is an edge from the node of r_i to the node of r_j in $G(J)$ (see, for example, Fig. 2). If $J = \top$, the corresponding constraint graph $G(\top)$ does not contain any edges.

If a constraint graph contains a cycle, there is a contradiction among the constraints. This means that the involved robots need to be coordinated as a composite robot in order to find a solution. To be more precise, the set of nodes (robots) in a graph is partitioned into a set of *strongly connected components*. A strongly connected component is a maximal set of nodes that are *strongly connected* to each other; two nodes r_i and r_j are strongly connected if there is a path in the graph both from r_i to r_j and from r_j to r_i. By definition, each node is strongly connected to itself.

Let $G^{SCC}(J)$ denote the *component graph* of $G(J)$, which contains a node for each strongly connected component of $G(J)$ and a directed edge from node R to node R' if there is an edge in $G(J)$ from any $r \in R$ to any $r' \in R'$. Note that G^{SCC} is a directed *acyclic* graph. Each node in $G^{SCC}(J)$ corresponds to a (composite) robot consisting of the robots involved in the strongly connected component. Topologically sorting $G^{SCC}(J)$ gives an execution sequence $S(J)$ of composite robots. Trivially, the following holds:

Corollary 8 If $G(J)$ is a constraint graph corresponding to conjunction J, then $S(J)$ is an execution sequence that satisfies J.

B. Incrementally Building the Execution Sequence

To build the expression sequence for an instance of multi-robot planning, our algorithm primarily maintains a DNF constraint expression E in the form of constraint graphs for its conjunctions J_i (if there are no conjunctons, $E = \bot$). Our algorithm also maintains a list L of the (composite) robots that have been passed to the CR planner and whose constraints $\mathcal{P}(R)$ have been incorporated into E.

Initially, $E = \{\top\}$, as we begin with no constraints. Now, iteratively, we select the (composite) robot $R_{\min}$ that has the smallest dimension among all (composite) robots for which we have not yet planned in the execution sequences $S(J)$ of all conjunctions $J \in E$:

$$R_{\min} = \underset{R \in \bigcup_{J \in E} S(J) \setminus L}{\arg \min} \dim(R).$$

Next, the CR planner is invoked on $R_{\min}$; it returns the set of constraints $\mathcal{P}(R_{\min})$. For *each* conjunction J in E for which $R_{\min} \in S(J)$, we do the following:

- Let $F = J \wedge \mathcal{P}(R_{\min})$, and transform F into disjunctive normal form. (Note that for each conjunction J' in F the following holds: $J' \Rightarrow J$ and $J' \Rightarrow \mathcal{P}(R_{\min})$.)
- Remove J from E and add the conjunctions of F to E (we replace J in E by the conjunctions of F).

The constraints of $\mathcal{P}(R_{\min})$ have now been incorporated into E, so we add $R_{\min}$ to the set L.

This procedure repeats until either $E = \emptyset$, in which case there is no solution to the multi-robot planning problem, *or* there exists a conjunction $J_{sol} \in E$ for which all composite robots $R \in S(J_{sol})$ have been planned for and are in L. In this case $S(J_{sol})$ is an *optimal solution sequence*, which we will prove below. In Fig. 2, we show the working of our algorithm on the example of Fig. 1(c).

C. Analysis

Here we prove that the above algorithm gives an optimal solution sequence:

Lemma 9 In each iteration of the algorithm, the constraints $\mathcal{P}(R_{\min})$ of composite robot $R_{\min}$ are incorporated into E. Right after the iteration, the following holds for all conjunctions $J \in E$: if $R_{\min} \in S(J)$ then $J \Rightarrow \mathcal{P}(R_{\min})$.

Proof: When we incorporate $\mathcal{P}(R_{\min})$ into E, all $J \in E$ for which $R_{\min} \in S(J)$ are replaced in E by $F = J \wedge \mathcal{P}(R_{\min})$. Now, all conjunctions $J' \in E$ for which $R_{\min} \in S(J')$ must be in F. Hence $J' \Rightarrow F$ and as a result $J' \Rightarrow \mathcal{P}(R_{\min})$. $\square$

Lemma 10 In each iteration of the algorithm, the constraints $\mathcal{P}(R_{\min})$ of composite robot $R_{\min}$ are incorporated into E. Its dimension is *greater than or equal to* the dimensions of all composite robots in L whose constraints were incorporated before: $\dim(R_{\min}) \geq \max_{R \in L} \dim(R)$.

Proof: Assume the converse is true: let R be the composite robot whose constraints were incorporated in the previous iteration, and let $\dim(R_{\min}) < \dim(R)$. Let $J_{\min}$ be the conjunction in E for which $R_{\min} \in S(J_{\min})$, Then, in the iteration R was selected, $J_{\min}$ did not exist yet in E, otherwise $R_{\min}$ would have been selected (our algorithm always selects the lowest-dimensional composite robot). This means that right before $\mathcal{P}(R)$ was incorporated, there was a $J \in E$ for which $R \in S(J)$ and $R_{\min} \notin S(J)$ that was replaced by $F = J \wedge \mathcal{P}(R)$ such that $J_{\min} \in F$. This means that one or more edges were added to $G(J)$ which caused the robots in $R_{\min}$ to form a strongly connected component in $G(J_{\min})$. Hence, these edges must have been between nodes corresponding to robots in $R_{\min}$. As these edges can only have come from $\mathcal{P}(R)$, this means, by Property 5, that $R \cap R_{\min} \neq \emptyset$. However, R forms a strongly connected component in $G(J)$ as $R \in S(J)$, so then $R \cup R_{\min}$ must also be a strongly connected component in $G(J_{\min})$. $R \cup R_{\min}$ and $R_{\min}$ can only be strongly connected components at the same time if $R \subset R_{\min}$. However, this means that $\dim(R) < \dim(R_{\min})$, so we have reached a contradiction. $\square$

Lemma 11 After each iteration of the algorithm, the following holds for all composite robots $R \in L$ whose constraints have been incorporated into E: for all conjunctions $J \in E$, if $R \in S(J)$ then $J \Rightarrow \mathcal{P}(R)$.

Proof: Lemma 9 proves that right after $\mathcal{P}(R)$ is incorporated, that for all conjunctions $J \in E$, if $R \in S(J)$ then $J \Rightarrow$

$\mathcal{P}(R)$. Now, we show that after a next iteration in which the constraints $\mathcal{P}(R')$ of another (composite) robot R' are incorporated, this still holds for R. Let J be a conjunction in E after $\mathcal{P}(R')$ is incorporated for which $R \in S(J)$. Now, either J already existed after the previous iteration, in which case $J \Rightarrow \mathcal{P}(R)$, or there existed a $J' \in E$ after the previous iteration that was replaced by $F = J' \wedge \mathcal{P}(R')$ such that $J \in F$. In the latter case, either $J' \Rightarrow \mathcal{P}(R)$, in which case also $J \Rightarrow \mathcal{P}(R)$, or $J' \not\Rightarrow \mathcal{P}(R)$ and $R \notin S(J')$. In the latter case, one or more edges must have been added to $G(J')$ by $\mathcal{P}(R')$ that caused the robots in R to form a strongly connected component in $G(J)$. Along similar lines as in the proof of Lemma 10, this means that $\dim(R') < \dim(R)$. However, by Lemma 10 this is not possible, as $\mathcal{P}(R')$ was incorporated later than $\mathcal{P}(R)$. The above argument can inductively be applied to all (composite) robots $R \in L$. $\square$

Theorem 12 (Correctness) The execution sequence S returned by the above algorithm is a solution sequence.

Proof: Let J_{sol} be the conjunction whose execution sequence $S(J_{\text{sol}}) = (R_1, \ldots, R_k)$ is returned by the algorithm. Then, all composite robots $R_i \in S(J_{\text{min}})$ are also in L. By Lemma 11, $J_{\text{sol}} \Rightarrow \mathcal{P}(R_1) \wedge \cdots \wedge \mathcal{P}(R_k)$. Hence, by Corollary 8 and Lemma 7, $S(J_{\text{min}})$ is a solution sequence. $\square$

Theorem 13 (Optimality) The execution sequence S returned by the above algorithm is an *optimal* solution sequence.

Proof: Sketch: Consider any other solution path π', and its execution sequence S' that comes from the equivalence classes of the coupled relation defined in Section III. This execution sequence satisfies the contraints induced by its robots, so the only way it could not have been discovered is for a different sequence of composite robots to be formed. Since the degree grows monotonically, and at all times the robot of lowest dimesion/degree is added to the plan, the execution sequence that is found by the algorithm cannot have larger degree. $\square$

Corollary 14 (Efficiency) The dimension of the highest-dimensional configuration space our algorithm plans in is equal to the dimension of the highest-dimensional (composite) robot in an *optimal* solution sequence.

V. IMPLEMENTATION AND OPTIMIZATION

In this section we describe some details of the implementation of our algorithm. We first describe how we maintain the set of constraints E in our main algorithm. We then describe how we implemented the CR planner that gives the constraints $\mathcal{P}(R)$ for a given composite robot R.

A. Main Algorithm

The implementation of our algorithm largely follows the algorithm as we have described in Section IV-B. We maintain the logical expression E in disjunctive normal form $J_1 \vee J_2 \vee \cdots$ as a set of constraint graphs $\{G(J_1), G(J_2), \ldots\}$. Each graph $G(J)$ is stored as an $n \times n$ *boolean matrix*, where a 1 at position (i, j) corresponds to an edge between r_i and r_j in $G(J)$. A boolean matrix can efficiently be stored

in memory, as each of its entries only require one bit. An operation $J \wedge J'$ of conjunctions J and J' can efficiently be performed by computing the bitwise-*or* of its corresponding boolean matrices. Also, checking whether $J \Rightarrow J'$ is easy; it is the case when the bitwise-or of the boolean matrices of $G(J)$ and $G(J')$ is equal to the boolean matrix of $G(J)$.

All graphs $G(J)$ are stored in *transitively closed* form. The transitive closure of a boolean matrix can efficiently be computed using the Floyd-Warshall algorithm [6]. Given a boolean matrix of a transitively closed graph $G(J)$, it is easy to infer the strongly connected components of $G(J)$.

B. Composite Robot Planner

The constraints $\mathcal{P}(R)$ for a (composite) robot R are obtained by path planning between the start configuration of R and the goal configuration of R in its configuration space $\mathcal{C}(R)$, and by considering through which query configurations of other robots R need to move. For our implementation, we sacrifice some completeness for practicality, by discretizing the configuration space $\mathcal{C}(R)$ into a *roadmap* $RM(R)$. Our discretization is similar to the one used for the planner presented in [26]. We sketch our implementation here.

Prior to running our algorithm, we construct for each *individual* robot r_i a roadmap $RM(r_i)$ that covers its configuration space $\mathcal{C}(r_i)$ well. Let us assume that the start configuration s_i and the goal configuration g_i of robot r_i are present as vertices in $RM(r_i)$. Further, each vertex and each edge in $RM(r_i)$ should be collision-free with respect to the static obstacles in the workspace. The roadmaps $RM(r_i)$ can be constructed by, for instance, a Probabilistic Roadmap Planner [9]. They are reused any time a call to the CR planner is made.

The (composite) roadmap $RM(R)$ of a composite robot $R = \{r_1, \ldots, r_k\}$ is defined as follows. There is a *vertex* $(x_1, \ldots, x_k)$ in $RM(R)$ if for all $i \in [1, k]$ x_i is a node in $RM(r_i)$, and for all $i, j \in [1, k]$ (with $i \neq j$) robots r_i and r_j configured at their vertices x_i and x_j, respectively, do not collide. There is an *edge* in $RM(R)$ between vertices $(x_1, \ldots, x_k)$ and $(y_1, \ldots, y_k)$ if for exactly one $i \in [1, k]$ $x_i \neq y_i$ and x_i is connected to y_i by an edge in $RM(r_i)$ that is not blocked by any robot $r_j \in R$ (with $i \neq j$) configured at x_j. This composite roadmap is not explicitly constructed, but explored implicitly while planning.

Now, to infer the constraints $\mathcal{P}(R)$ for a (composite) robot R, we plan in its (composite) roadmap $RM(R)$ using an algorithm very similar to Dijkstra's algorithm. However, instead of computing distances from the start configuration for each vertex, we compute a constraint expression $\mathcal{P}(x)$ for each vertex x. The logical implication "$\Rightarrow$" takes the role of "$\geq$". The algorithm is given in Algorithm 1.

Note that unlike Dijkstra's algorithm, this algorithm can visit vertices multiple times. In the worst case, this algorithm has a running time exponential in the total number of robots n, as the definition of $\mathcal{P}(R)$ in Section III-E suggests. However, in line 16 of Algorithm 1 many paths are pruned from further exploration, which makes the algorithm tractable for most practical cases. The planner is sped up by ordering the vertices

Algorithm 1 $\mathcal{P}(R)$

1: Let $s, g \in RM(R)$ be the start and goal configuration of R.
2: **for all** vertices x in $RM(R)$ **do**
3:     $\mathcal{P}(x) \leftarrow \bot$
4: $\mathcal{P}(s) \leftarrow \bigwedge_{r_i, r_j \in R} r_i \sim r_j$
5: $Q \leftarrow \{s\}$
6: **while not** (priority) queue Q is empty **do**
7:     Pop front vertex x from Q.
8:     **if not** $\mathcal{P}(x) \Rightarrow \mathcal{P}(g)$ **then**
9:         **for all** edges (x, x') in $RM(R)$ **do**
10:            $C \leftarrow \mathcal{P}(x)$
11:            **for all** robots $r_i \notin R$ **do**
12:                **if** robot r_i configured at s_i "blocks" edge (x, x') **then**
13:                   $C \leftarrow C \wedge r_i \prec R$
14:                **if** robot r_i configured at g_i "blocks" edge (x, x') **then**
15:                   $C \leftarrow C \wedge R \prec r_i$
16:            **if not** $C \Rightarrow \mathcal{P}(x')$ **then**
17:                $\mathcal{P}(x') \leftarrow \mathcal{P}(x') \vee C$
18:                **if** $x' \notin Q$ **then**
19:                   $Q \leftarrow Q \cup \{x'\}$
20: **return** $\mathcal{P}(g)$

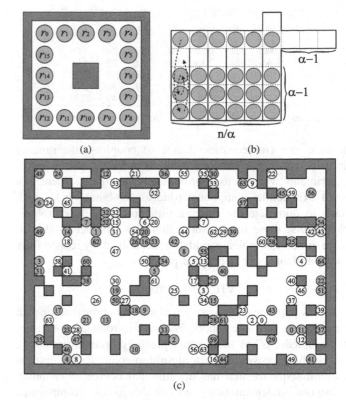

Fig. 3. (a) An environment with sixteen robots that are shown in their start configuration. Each robot r_i has to exchange positions with robot $r_{((i+8) \bmod 16)}$, i.e. the robot on the opposite side of the room. (b) An environment with a variable number of robots n and a variable degree of coupling α. The robots in each column need to reverse their order as shown for the left column. (c) A randomly generated environment with 65 robots. The start and goal configurations are shown in light gray and white, respectively.

in the queue Q according to the partial ordering defined by the implication relation "$\Leftarrow$" on their constraint expressions, such that vertices whose constraint expressions are implied by other constraint expressions are explored first.

VI. RESULTS

We report results of our implementation on the three scenarios shown in Fig. 3. These are challenging scenarios, each for their own reason. Scenario (a) has relatively few robots, but a potentially high degree of coupling among them. Scenario (c) has many robots, but with a low degree of coupling. Problems of the first type are traditionally the domain of coupled planners; decoupled planners are likely to fail in these cases. Problems of the second type are the domain of decoupled methods; for a coupled planner there would be too many robots to be able to compute a solution. In this section, we show that our method copes well with both. In scenario (b), we vary both the number of robots n and the degree of coupling α, and study quantitatively how our algorithm performs in these varying circumstances.

Scenario (a) has sixteen robots $r_0, \ldots, r_{15}$. Each robot r_i has to exchange positions with robot $r_{((i+8) \bmod 16)}$, i.e. the robot on the opposite side of the room. There is very limited space to maneuver. Traditional multi-robot planners are unlikely to solve this problem. A decoupled two-stage planner will only succeed if the paths in the first stage are chosen exactly correct, which is not likely. A prioritized approach will only work if the correct prioritization is chosen. Also coupled planners are probably not able to solve this problem, because planning in a 32-dimensional composite configuration space is generally not feasible. Our method, on the other hand, solved this problem in 2.57 seconds on an Intel Core2Duo P7350 2GHz with 4GByte of memory. It returned the optimal solution sequence $(r_0 r_7 r_8 r_{15}, r_1 r_9, r_2 r_{10}, r_3 r_{11}, r_4 r_5 r_{12} r_{13}, r_6 r_{14}, r_7 r_{15})$. Our algorithm planned eight times for one robot, four times for

two robots and twice for four robots in order to achieve this result.

Scenario (c) involves a randomly generated environment containing as many as 65 robots, giving a full composite configuration space of 130 dimensions. Each robot was assigned a random start and goal configuration. Our method returned a solution sequence solely containing individual robots. That means that our algorithm found a solution by only planning in the 65 two-dimensional configuration spaces of each of the robots. Even though the number of robots is high, the degree of coupling in this example is very low. Our algorithm exploits this; it solved this example in only 73 seconds. After the last iteration of the algorithm, the constraint expression E contained 4104 conjunctions, and the conjunction that provided the solution sequence contained 27 atomic constraints.

Experiments on scenario (b) show how our algorithm performs for a varying number of robots and degree of coupling. The scenario is designed such that it cannot be solved by decoupled planners. We report results in Fig. 4. In the first column, we set the degree of coupling equal to the number of robots, i.e. the problem is fully coupled. In this case, the planner fails for seven or more robots, despite the small size of the workspace. When the degree of coupling is lower, our approach is able to solve problems for a much higher number of robots in reasonable running times. Analysis of the results

$\alpha = n$		$\alpha = 2$		$\alpha = 3$		$\alpha = 4$	
n	time	n	time	n	time	n	time
5	1.39	20	0.30	27	18.8	20	41.7
6	44.0	22	0.69	30	38.8	24	167
7	n/a	24	2.15	33	75.6	28	542
		26	7.69	36	146	32	1254
		28	42.4	39	287	36	3356
		30	261	42	672	40	7244

Fig. 4. Performance results on scenario (b). Running times are in seconds.

indicate that (for constant degree of coupling) the running time increases polynomially with the number of robots for experiments with up to approximately 30 robots. There is a small exponential component (due to combinatorics), which starts to dominate the running time for 30 or more robots. Nontheless, these results show that our approach is able to solve problems which could not be solved by either fully coupled planners or decoupled planners.

VII. CONCLUSION

In this paper, we have presented a novel algorithm for path planning for multiple robots. We have introduced a measure of "coupledness" of multi-robot planning problem instances, and we have proven that our algorithm computes the optimal solution sequence that has the minimal degree of coupling. Using our implementation, we were able to solve complicated problems, ranging from problems with a relatively few robots and high degree of coupling to problems with many robots and a low degree of coupling.

The *quality* of our solutions, usually defined in terms of arrival times of the robots [12], is not optimal as the computed robot plans are executed sequentially. An idea to improve this is to use a traditional *prioritized* planner in a post-processing step, and have it plan trajectories for the (composite) robots in order of the solution sequence our algorithm computes.

Another idea for future work is to exploit *parallelism* to increase the performance of our algorithm. Our current algorithm is iterative, but it seems possible that planning for different (composite) robots can be carried out in parallel.

A limitation of our current approach is that it computes a sequence of complete robot plans only (i.e. plans that go from start to goal). This causes more coupling than what would seem necessary for some problem instances. For example if three robots block the centers of a long hall that a fourth robot is trying to traverse, then all four would be coupled even though the three could each move out of the way in turn. Even though the three have no direct coupling, the are coupled by transitive closure through the fourth. If we add *intermediate goals* for the fourth, however, and use our ordering constraints to specify that these intermediate configurations must be visited in order, then it is easy to reduce the degree of coupling. Thus, the selection and placement of intermediate goals seems a fruitful topic for further study.

REFERENCES

[1] R. Alami, F. Robert, F. Ingrand, S. Suzuki. Multi-robot cooperation through incremental plan-merging. *Proc. IEEE Int. Conference on Robotics and Automation*, pp. 2573–2579, 1995.

[2] B. Aronov, M. de Berg, F. van der Stappen, P. Švestka, J. Vleugels. Motion planning for multiple robots. *Discrete and Computational Geometry* 22(4), pp. 505–525, 1999.

[3] M. Bennewitz, W. Burgard, S. Thrun. Finding and optimizing solvable priority schemes for decoupled path planning techniques for teams of mobile robots. *Robotics and Autonomous Systems* 41(2), pp. 89–99, 2002.

[4] S. Buckley. Fast motion planning for multiple moving robots. *Proc. IEEE Int. Conference on Robotics and Automation*, pp. 322–326, 1989.

[5] C. Clark, S. Rock, J.-C. Latombe. Motion planning for multiple robot systems using dynamic networks. *Proc. IEEE Int. Conference on Robotics and Automation*, pp. 4222–4227, 2003.

[6] T. Cormen, C. Leiserson, R. Rivest. *Introduction to Algorithms*. The MIT Press, 1998.

[7] M. Erdmann, T. Lozano-Perez. On multiple moving objects. *Proc. IEEE Int. Conference on Robotics and Automation*, pp. 1419–1424, 1986.

[8] D. Halperin, J.-C. Latombe, R. Wilson. A general framework for assembly planning: The motion space approach. *Algorithmica* 26(3-4), pp. 577–601, 2000.

[9] L. Kavraki, P. Švestka, J.-C. Latombe, M. Overmars. Probabilistic roadmaps for path planning in high-dimensional configuration spaces. *IEEE Trans. Robotics and Automation* 12(4), pp. 566–580, 1996.

[10] J.-C. Latombe. *Robot Motion Planning*. Kluwer Academic Publishers, 1991.

[11] S. LaValle, J. Kuffner. Rapidly-exploring random trees: Progress and prospects. *Workshop on the Algorithmic Foundations of Robotics*, 2000.

[12] S. LaValle, S. Hutchinson. Optimal motion planning for multiple robots having independent goals. *IEEE Trans. on Robotics and Automation* 14(6), pp. 912–925, 1998.

[13] S. LaValle. *Planning Algorithms*. Cambridge University Press, 2006.

[14] T-Y. Li, H-C. Chou. Motion planning for a crowd of robots. *Proc. IEEE Int. Conference on Robotics and Automation*, pp. 4215–4221, 2003.

[15] Y. Li, K. Gupta, S. Payandeh. Motion planning of multiple agents in virtual environments using coordination graphs. *Proc. IEEE Int. Conference on Robotics and Automation*, pp. 378–383, 2005.

[16] B. Natarajan. On Planning Assemblies. *Proc. Symposium on Computational Geometry*, pp. 299–308, 1988.

[17] B. Nnaji. *Theory of Automatic Robot Assembly and Programming*. Chapman & Hall, 1992.

[18] P. O'Donnell, T. Lozano-Perez. Deadlock-free and collision-free coordination of two robot manipulators. *Proc. IEEE Int. Conference on Robotics and Automation*, pp. 484–489, 1989.

[19] M. Peasgood, C. Clark, J. McPhee. A Complete and scalable strategy for coordinating multiple robots within roadmaps. *IEEE Trans. on Robotics* 24(2), pp. 283–292, 2008

[20] J. Peng, S. Akella. Coordinating multiple robots with kinodynamic constraints along specified paths. *Int. Journal of Robotics Research* 24(4), pp. 295–310, 2005.

[21] M. Saha, P. Isto. Multi-robot motion planning by incremental coordination. *Proc. IEEE/RSJ Int. Conference on Intelligent Robots and Systems*, pp. 5960–5963, 2006.

[22] G. Sanchez, J.-C. Latombe. Using a PRM planner to compare centralized and decoupled planning for multi-robot systems. *Proc. IEEE Int. Conference on Robotics and Automation*, pp. 2112–2119, 2002.

[23] T. Siméon, S. Leroy, J.-P. Laumond. Path coordination for multiple mobile robots: a resolution complete algorithm. *IEEE Trans. on Robotics and Automation* 18(1), pp. 42–49, 2002.

[24] J. Snoeyink, J. Stolfi. Objects that cannot be taken apart with two hands. *Discrete and Computational Geometry* 12, pp. 367–384, 1994.

[25] S. Sundaram, I. Remmler, N. Amato. Disassembly sequencing using a motion planning approach. *Proc. IEEE Int. Conference on Robotics and Automation*, pp. 1475–1480, 2001.

[26] P. Švestka, M. Overmars. Coordinated path planning for multiple robots. *Robotics and Autonomous Systems* 23(3), pp. 125–152, 1998.

[27] J. van den Berg, M. Overmars. Prioritized motion planning for multiple robots. *Proc. IEEE/RSJ Int. Conf. on Intelligent Robots and Systems*, pp. 2217–2222, 2005.

[28] R. Wilson, J.-C. Latombe. Geometric Reasoning About Assembly. *Artificial Intelligence* 71(2), 1994.

[29] K. Kant, S. Zucker. Towards efficient trajectory planning: The path-velocity decomposition. *Int. Journal of Robotics Research* 5(3), pp. 72–89, 1986.

Accurate Rough Terrain Estimation with Space-Carving Kernels

Raia Hadsell, J. Andrew Bagnell, Daniel Huber, Martial Hebert
The Robotics Institute
Carnegie Mellon University
Pittsburgh, Pennsylvania 15213

Abstract—Accurate terrain estimation is critical for autonomous offroad navigation. Reconstruction of a 3D surface allows rough and hilly ground to be represented, yielding faster driving and better planning and control. However, data from a 3D sensor samples the terrain unevenly, quickly becoming sparse at longer ranges and containing large voids because of occlusions and inclines. The proposed approach uses online kernel-based learning to estimate a continuous surface over the area of interest while providing upper and lower bounds on that surface. Unlike other approaches, visibility information is exploited to constrain the terrain surface and increase precision, and an efficient gradient-based optimization allows for realtime implementation.

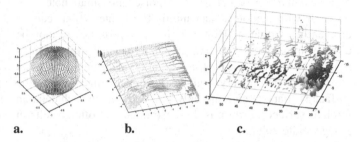

Fig. 1. Evenly sampled 3D objects (left) and laser scans on smooth roadways (center) do not offer the same reconstruction challenges as scans of rough terrain (right), which often have complex structure and very variable resolution that decays rapidly with distance.

I. INTRODUCTION

Terrain estimation is a critical component of mobile robot navigation, but accurate reconstruction of rough, hilly, and cluttered terrain is very difficult. The distribution of data points from a 3D sensor on a mobile robot decays rapidly away from the scanner, and there may be large ground regions that return no points at all. This variable resolution is inevitable as it is due to discrete sampling, use of static scanning patterns, and application to terrain whose geometry is unknown *a priori*. Indeed, if the sampling density is dense enough, such as in scanned 3D objects (Figure 1a) or regular enough, such as on smooth roads (Figure 1b), then 3D reconstruction offers less challenge. In rough outdoor terrain, however, complex natural geometry, uneven ground, and inclines all exacerbate the problem and make accurate terrain estimation difficult (Figure 1c).

Variable distributions make terrain estimation very challenging, and many autonomous systems truncate their terrain model to relatively short ranges or make do with a flat, 2D cost map for this reason [8, 7]. Our approach exploits the *visibility* aspect of laser scanning to improve the terrain estimate even in sparse regions. Data points from a ladar sensor must be visible to the sensor; i.e., the rays connecting sensor source to data points must lie *above the terrain surface*. Thus, the elevation function can be constrained by both the ladar points, which must lie *on* the surface, and the ladar rays, which must lie *above* the surface. This can be thought of as a *space carving* approach, since it uses visibility information. The new visibility constraints are incorporated in an reproducing kernel Hilbert space (RKHS) framework rather than a voxel-based approach, yielding a continuous surface estimate with high accuracy that smooths noisy data.

Many approaches simplify the problem considerably by representing the terrain as a flat cost map, but this is insufficient for modeling offroad terrain because hills and rough ground are not accurately represented, forcing the vehicle to drive at very low speeds and make conservative decisions. However, an explicit 3D model of the world, based on points and triangulated meshes, is infeasible: this sort of representation is very expensive, and mesh interpolation over complex terrain is non-trivial. Elevation maps provide a simplification of the full 3D model, but cannot represent overhanging structures and are limited to a fixed resolution by a discretized grid. If the terrain is represented by a continuous elevation *function*, however, then the effect of the ground on the vehicle can be more precisely predicted, allowing for faster autonomous driving and longer range planning. Modeling the terrain as a 3D surface is difficult because of the sparse, uneven distribution of 3D sensor data. Current methods use interpolation to create a continuous mesh surface, but this can be very difficult if the terrain is complex and the data is sparse. In our approach, the 3D surface is modeled as a elevation function over a 2D domain. This surface is estimated using kernel functions, which allow non-linear, complex solutions.

In order to learn the elevation function, we propose a kernel-based approach that models the surface as a hypothesis in a *reproducing kernel Hilbert space* (RKHS). Using a kernel formulation provides a principled means of optimizing a surface function that can produce a highly nonlinear solution. In order to pose the problem as an RKHS regression constrained by visibility information as well as surface points, we incorporate the space-carving constraint into the mathematical framework and give a rule for functional gradient descent optimization, yielding an efficient realtime program. The proposed method is evaluated using LADAR datasets of rough offroad terrain.

II. RELATED WORK

Kernel-based surface estimation has been adopted by the graphics community in recent years for modeling 3D scanned objects [19, 25, 23]. These approaches fit radial basis functions to scanned surface points, yielding an embedding function f. The zero-set $f^{-1}(0)$ implicitly defines the surface. The advantages of using radial basis functions to model the surface of a scanned 3D object are that noise and small holes can be dealt with smoothly, and multi-object interactions can be efficiently computed. However, these approaches cannot be directly applied to terrain data gathered from laser rangefinders mounted on a mobile robot. Such data is dense and precise at close range, but quickly degrades at longer ranges, where the surface is sparsely and unevenly sampled. Given such data, an implicit surface function is ill-constrained and often results in a degenerate solution.

Explicit elevation maps are a standard approach for modeling rough terrain for mobile robotics. There are many strategies for building these maps, from mesh algorithms to interpolation to statistical methods [5, 2, 14].

Burgard et al., following on the research done by Paciorek and Schervish and Higdon et al. [13, 6], have successfully applied Gaussian process regression to the problem of rough terrain modeling, although their approach is computationally expensive and has not been applied to large datasets [16, 15, 12]. Burgard's research adapts Gaussian process regression for the task of mobile robot terrain estimation by considering issues such as computational constraints, iterative adaptation, and accurate modeling of local discontinuity. Our approach uses a kernel-based methodology as well, but we propose an iterative algorithm that exploits both ray and point information to fit basis functions to solve a system of constraints.

Using ray constraints, or visibility information, to improve a 3D surface model has rarely been proposed in mobile robotics. Space carving algorithms, originally suggested by Kutulakos and Seitz [10], use visibility information to produce a voxel model from calibrated images of a scene (a survey of space-carving approach can be found in [21]), but this strategy has not been adopted by the ladar community. Another approach that exploited the visibility constraints was the *locus* method of Kweon et al. [11]. These approaches produced discrete maps, rather than continuous elevation functions, and relied on unwieldy heuristics to ensure that the map had desirable properties such as a watertight surface. In contrast, the approach we propose exploits visibility information while learning a continuous, bounded surface.

In addition, recent publications from several different research groups have advanced the field of 2 and 3D map construction in significant ways. In particular, Yguel et al. proposed the use of sparse wavelets to model a 3D environment from range data [24], and Fournier et al. use an octree representation to efficiently represent a 3D world model [3].

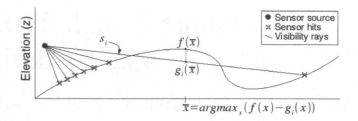

Fig. 2. Visualization in 2D of the point- and ray-based constraints on a terrain elevation function. The estimated surface must intersect the data points ("sensor hits") and lie below the visibility rays from the sensor. If there is a violation of the ray constraint, a support kernel function is added at the most violating point $\overline{x}$.

III. KERNEL-BASED REGRESSION FOR TERRAIN ESTIMATION

Given a set of 3D points from a sensor mounted on a mobile robot, we seek to estimate a continuous elevation function $f(x, y) = z$ that both intersects the data points and does not exceed the height of the rays connecting sensor and data points. A 2D example, in which the elevation map intersects the data points but violates the ray constraint, is shown in Figure 2. The dataset $\mathcal{S}$ consists of n tuples, each with a 3D point (x_i, y_i, z_i) and a corresponding 3D sensor location (sx_i, sy_i, sz_i), which are summarized as a point $\mathbf{x}_i = [x_i \ y_i]$, a height z_i, and a line segment, or ray, connecting source and point, which we denote by $\mathbf{s}_i$. The projection of $\mathbf{s}_i$ on the XY plane is denoted $\hat{\mathbf{s}}_i$, and the function $g_i(\cdot)$ is used to denote the height of $\mathbf{s}_i$ at a given point ($g_i = \infty$ at every location that does not intersect $\hat{\mathbf{s}}_i$). Given this data, we learn a function that meets both a *point*-based constraint (3) and a *ray*-based constraint (4):

$$\text{Given} \quad \mathcal{S} = \{(\mathbf{x}_1, z_1, \mathbf{s}_1), (\mathbf{x}_2, z_2, \mathbf{s}_2), ..., (\mathbf{x}_n, z_n, \mathbf{s}_n)\} \quad (1)$$

$$\text{find} \quad f : \mathbb{R}^2 \rightarrow \mathbb{R} \quad (2)$$

$$\text{s.t.} \quad f(\mathbf{x}_i) = z_i \ \forall \ \mathbf{x}_i, \quad (3)$$

$$f(\mathbf{x}) \leq g_i(\mathbf{x}) \ \forall \ \mathbf{s}_i, \mathbf{x}. \quad (4)$$

In order to solve this problem for complex surfaces and datasets, we use a kernel formulation whereby distances between points can be computed in a highly non-linear, high-dimensional feature space without actually computing the coordinates of the data points in that feature space, since any continuous, symmetric, positive semi-definite kernel function $k(\mathbf{x}_i, \mathbf{x}_j)$ can be expressed as a dot product in a higher dimensional space [1]. Thus the height function $f(x, y)$ is a hypothesis in RKHS and can be expressed by a kernel expansion:

$$f(x, y) = f(\mathbf{x}) = \sum_{i=1}^{n} \alpha_i k(\mathbf{x}, \mathbf{x_i}), \quad (5)$$

where $k(\cdot, \cdot)$ is a radial basis function and α are learned coefficients. For efficiency, the kernel function $k(\cdot, \cdot)$ is a compactly

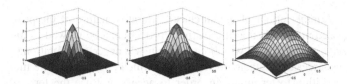

Fig. 3.   The Wu $\phi_{2,1}$ function (6), shown with σ of 0.5, 1, and 2.

supported kernel suggested by Wu [18] (see Figure 3):

$$k(\mathbf{x}_i, \mathbf{x}_j) = k(\rho(\mathbf{x}_i, \mathbf{x}_j)) = (1 - \rho)_+^4 (4 + 16\rho + 12\rho^2 + 3\rho^3), \quad (6)$$

where

$$\rho(\mathbf{x}_i, \mathbf{x}_j) = \frac{\|\mathbf{x}_i - \mathbf{x}_j\|}{\sigma}, \quad (7)$$

and σ is the lengthscale of the basis function. This kernel function was chosen over other possible radial basis functions because of the recommendations of Schaback [18] and the empirical success of this kernel for surface estimation in [25].

A. Optimization by Subgradient Descent

To find $f \in \mathcal{H}$ that meets both point-based constraints and ray-based constraints, we form the following convex optimization problem:

$$\mathcal{L}(f, \mathcal{S}) = \lambda \|f\|_{\mathcal{H}}^2 + \frac{1}{N} \sum_{i=1}^{N} r(\mathbf{x}_i, z_i, g_i, f), \quad (8)$$

where $r(f, g_i, \mathbf{x}_i, z_i)$ penalizes both point and ray constraints. Instead of optimizing an infinite set of linear constraints, i.e., the constraint that each ray must lie above the surface, we optimize a single non-linear constraint and use a max function to find the most offending point in the projection of each ray. The cost function $r(f, g_i, \mathbf{x}_i, z_i)$ is thus:

$$r(\cdot) = \frac{1}{2} \max(0, \max_{\mathbf{x}} (f(\mathbf{x}) - g_i(\mathbf{x})))^2 + \frac{1}{2} (f(\mathbf{x}_i) - z_i)^2. \quad (9)$$

To solve the optimization problem above, which contains linear and non-linear constraints (because of the max operation), we use a functional gradient descent approach that relies on subgradients to tackle non-differentiable problems. Using a stochastic gradient method allows fast online performance which is critical for real-world application. Online subgradient kernel methods have been successfully applied to several problems [17].

The subgradient method iteratively computes a gradient-like vector which is defined using a tangent to the lower bound at a particular point of the non-differentiable function. There is a continuum of subgradients at each point of non-differentiability, but only one subgradient, which is the gradient, at differentiable points. A detailed examination of the subgradient method can be found in the original work by Shor [20]. Online gradient descent with regret bounds for subgradient methods was developed by Zinkevich [26].

To learn using functional gradient descent, the function is updated with the negative of the (sub)gradient stepped by a learning rate η:

$$f_{t+1} = f_t - \eta_t \frac{\partial \mathcal{L}(f_t, g_t, \mathbf{x}_t, z_t)}{\partial f}. \quad (10)$$

Since f is a hypothesis in RKHS,

$$\frac{\partial \mathcal{L}(f(\mathbf{x}_t), g_t, \mathbf{x}_t, z_t)}{\partial f} = r'(f(\mathbf{x}_t), g_t, \mathbf{x}_t, z_t) k(\mathbf{x}_t, \mathbf{x}) + \lambda f. \quad (11)$$

We rely on the kernel expansion of f (Eq. 5) to derive an efficient stochastic update, following the example of [9]. Following this stochastic approach, basis functions are added iteratively, and at the same time the weights of previously added basis functions are decayed. The number and location of the basis functions is not identical to the training points, and the final number of basis functions may be greater or fewer than the sample size. Thus, at time t, a new basis function may be added at location $\mathbf{x}_t$ with coefficient

$$\alpha_t = -\eta_t r'(f(\mathbf{x}_t), g_t, \mathbf{x}_t, z_t), \quad (12)$$

and the existing coefficients are decayed:

$$\alpha_i = (1 - \eta_t \lambda)\alpha_i \text{ for } i < t. \quad (13)$$

For our optimization problem, the gradient of r has 2 components, for the different constraints in the loss function, so up to 2 basis functions are added with different coefficients. If $f(\mathbf{x}_i) \neq z_i$, then the added basis function is centered at $\mathbf{x}_i$, with coefficient $\alpha_t = -\eta_t(f(\mathbf{x}_i - z_i))$. The second basis function is added at the most violating location along ray s_i, if one exists. We compute $\overline{x} = \text{argmax}_{\mathbf{x}}(f(\mathbf{x}) - g_i(\mathbf{x}))$ by line search on ray s_i, and if $\overline{x} > 0$ then a basis function is added at $\overline{x}$ with a coefficient $\alpha_{t+1} = -\eta_{t+1}(f(\overline{\mathbf{x}}) - g_i(\overline{\mathbf{x}}))$. The algorithm is also described in Alg. 1.

The method that has been described in this section uses a fixed lengthscale, σ. A fixed lengthscale with a compactly supported kernel function does not give global support. Many solutions have been proposed for this problem in the graphics community, where radial basis functions are used to estimate 3D surfaces and compactly supported kernels are chosen for efficiency [19, 25, 21]. Most commonly, *backfitting*, a general algorithm that can fit any additive model, is used to decrease the lengthscale incrementally [4]. The training data is partitioned, and each subset is sequentially fit toward the residual of the previous partition using a decaying lengthscale [25]. The same strategy can be employed for our space carving terrain reconstruction algorithm. In fact, there is no need for the training data to be partitioned in our approach: the lengthscale may be decayed on each epoch of gradient-based learning over the full dataset.

B. Uncertainty Bounds

Uncertainty attribution is very valuable for a mobile robot in rough terrain. A terrain estimate coupled with an uncertainty bound is much more powerful than a terrain estimate alone, because the planning and behavior of the robot will be effected. Velocity may be modified, exploration behaviors changed, and safety measures enacted based on the uncertainty

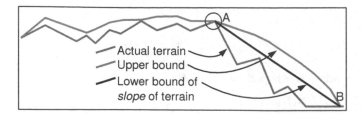

Fig. 4. An upper bound on the terrain gives a lower bound on the slope of the terrain under certain circumstances. For an interior chord to the upper bound $(\overline{AB})$ whose maximum point is coincident with the maximum point on the terrain surface under the chord (at A in the illustration), the slope of the chord is a tight lower bound on the actual slope of the underlying terrain.

attribute. In some circumstances, the upper bound of a region induces a tight lower bound on the slope of the actual terrain, which can be used to identify non-traversable steep slopes, *even if there are no data points in that region*. In other words, the slope of an interior chord of the upper bound is a lower bound on the slope of the actual terrain, *if* the highest elevation of the chord is equal to the local maxima of the terrain under the chord. Thus lethal regions may be located entirely on basis of the upper bound, as long as the upper bound is pinned to the actual surface by at least one data point, an insight described by [8] and provable by application of the mean value theorem. This scenario is depicted in Figure 4: the slope of chord $\overline{AB}$ is a lower bound on the maximum slope of the underlying terrain, because at point A the chord and the terrain are both at local maxima.

Upper and lower bounds can be learned using the kernel-based terrain estimation method that has been proposed. At time 0, the surface estimate is initialized by a positive or negative offset corresponding to the globally maximum or globally minimum expected elevation (in our experiments, +5 meters and -5 meters were the "priors" used to initialize the upper and lower bounds). The subsequent learning is identical to the original algorithm: ray and point constraints are applied to fit the surface by training a kernel function with stochastic gradient-based learning. The outcome is an upper bound surface and a lower bound surface. The three learning processes can be run in parallel on a distributed system.

C. Online Algorithm and Implementation

The algorithm considers each data point on each epoch, although basis functions may not be added if there are no point or ray violations (see Algorithm 1). The learning rate η is reduced at each update step: $\eta_t \propto t^{-\frac{1}{2}}$, and regularization is applied after each learning step. The learning process is halted if the average gradient at time t is below a threshold or if the error rate on a validation set increases, resulting in convergence after a small number of epochs. Since the algorithm is stochastic and iterative in nature, data can be continually added to the training set and the learning will smoothly adapt to the additional information. This is a valuable characteristic for a autonomous navigation system to have, since new data is continuously arriving from sensors.

Another effect of the stochastic, iterative nature of the proposed method is that the learning process can be viewed as an *anytime algorithm* which can be run whenever computational resources and data are available and which does not require the algorithm to terminate in order to access and use the terrain estimate. The upper and lower bounds can be computed in parallel with the surface estimate and thus fit into the *anytime algorithm* formulation. This quality is particularly beneficial for realtime navigation systems that constantly modulate their velocity and trajectory based not only on an estimate of the terrain ahead but also on the uncertainty of that estimate.

Input: Training set $\mathcal{S}$ (points (Nx3) and sources (Nx3))
Output: Surface $f : \mathbb{R}^2 \to \mathbb{R}$
while *Not Converged* **do**
    foreach *Point* $\mathbf{x}_i$ **do**
        Calculate $f(\mathbf{x}_i)$;
        if $f(\mathbf{x}_i) \neq z_i$ **then**
            Add basis $\mathbf{x}_t = \mathbf{x}_i$;
            With weight $\alpha_t = -\eta_t(f(\mathbf{x}_i - z_i)$;
            incr t;
        end
        Calculate $\overline{x} = \text{argmax}_{\mathbf{x}}(f(\mathbf{x}) - g_i(\mathbf{x}))$;
        if $\overline{\mathbf{x}} > 0$ **then**
            Add basis $\mathbf{x}_t = \overline{\mathbf{x}}$;
            With weight $\alpha_t = -\eta_t(f(\overline{\mathbf{x}}) - g_i(\overline{\mathbf{x}}))$;
            incr t;
        end
        foreach *Basis* $x_j, j < t$ **do**
            Decay weight $\alpha_j = (1 - \lambda_t \eta_t)\alpha_j$;
        end
    end
end

Algorithm 1: Online algorithm using subgradients. The first test ($f(\mathbf{x}_i) \neq z_i$) tests for violation of the point constraint, and the second test detects violation of the ray constraint.

One of the benefits of the proposed approach is that it can deliver an accurate, continuous, bounded terrain reconstruction in realtime. The choice of a gradient-based optimization strategy, plus a compact kernel function that allows fast nearest neighbor searching (through use of a kd-tree) to perform the kernel expansion, permit this level of efficiency. The experiments described in this paper were conducted on a 2.2 GHz computer using a non-optimized C++ implementation. The timing results are promising; datasets with 10,000 points converge within a second, and datasets with over 1 million points converge in a few seconds. From this preliminary evaluation, we conclude that deployment on a full realtime navigation system would be feasible.

IV. EVALUATION AND ANALYSIS

Tests of the proposed method were conducted in order to evaluate the effectiveness of the approach on both artificial and natural datasets, and to demonstrate the convergence properties. Mean squared error is calculated by comparing the elevation of a set of test points with the predicted surface elevation:

$$MSE(\mathcal{S}_{\text{test}}) = E_{\text{test}} = \frac{1}{p} \sum_{i=1..p} (f(\mathbf{x}_i) - z_i)^2, \quad (14)$$

where $\mathcal{S}_{\text{test}}$ is a set of p test points $\mathbf{x}_i$ with known elevations z_i.

A. Evaluation: Sinusoid Data (2D)

A synthetic dataset was used to develop and test the proposed method. Data points were sampled irregularly over a sine function, using an exponentially decaying point density and eliminating points that were not visible from a source point, yielding 400 samples (see Figure 5a). The samples were divided into training and test sets, and testing showed that the algorithm converged and that the space-carving kernels improved the surface estimate. Figure 5b shows the surface and bounds after the first learning epoch and 5c shows the surface after convergence. In addition, a single point was added to the dataset during learning to demonstrate the online capability of this method (Figure 5d).

B. Evaluation: Offroad Natural Terrain Data

The approach described in the previous section has been implemented in Matlab and tested on an outdoor scene using data from the 360° HDL-64 ladar scanner, manufactured by Velodyne and installed on Boss, the Carnegie Mellon entry in the 2008 DARPA Urban Challenge [22]. The left and right images in Figure 6 show a slightly narrowed view of one dataset. Note that the data is very dense within a few meters of the sensor, but quickly degenerates to a sparse, uneven distribution. The terrain in this dataset is extremely rugged, composed of uneven piles of rubble.

The results of reconstructing the terrain are shown in Figure 7, and 2D cross sections of the bounded estimate are shown in Figure 8. The surface was estimated using a training set of 10,000 points (shown as a point cloud in Figure 7d), and another set of 5000 points was partitioned for testing. The algorithm was also tested with over one million points to demonstrate computational efficiency. The stochastic algorithm converged after 8 epochs, and a total of 21,392 basis functions were used. To compute the upper and lower bounds, the same training data was learned with an initial surface height of +5 and -5 meters at time 0. The terrain estimate is shown in Figure 7a, the lower bound is shown in 7b, and the upper bound is shown in 7c. The surfaces are similar where data is dense, but the upper and lower bound pull away as the data becomes sparse at the edges of the region shown, and also in the front center of the region, where the robot was occluding the ground.

Looking at the results in 2D sections, as in Figure 8, allows a graphical depiction of the upper and lower bounds, the estimate, and the training points. In this figure, four cross sections of the surfaces shown in Figure 7 are shown, corresponding to the bisections overlaid on the training points in Figure 9. The upper regions in the 2D plots show the uncertainty between the estimated surface and the upper bound on the surface, and

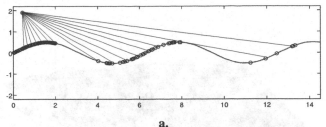

a.

Cross Section of Terrain with Uncertainty Bounds (Sine Wave Dataset)

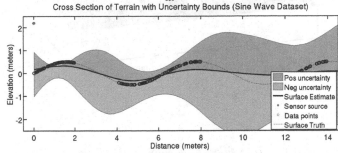

b.

Cross Section of Terrain Surface with Uncertainty Bounds (Sine Dataset)

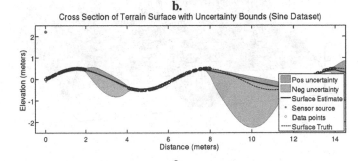

c.

Cross Section of Terrain Surface with Uncertainty Bounds (Sine Dataset)

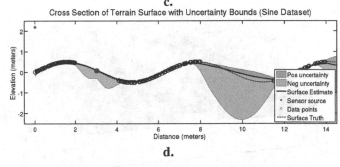

d.

Fig. 5. **a:** An synthetic dataset containing irregularly sampled points on a 2D sine function. The density of points decreases exponentially with distance from the sensor. Points not visible to a sensor were removed. **b:** The estimated surface, plus low and high bounds, after 2 learning epochs. **c:** The estimated surface, plus low and high bounds, after convergence (12 epochs). Note the effect of ray constraints on the upper bound. **d:** A single point (shown in figure) was added interactively to demonstrate the ability of the algorithm to adapt to new data online.

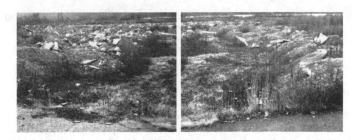

Fig. 6. Left and right views of one of 5 rough terrain datasets used for evaluation of the algorithm.

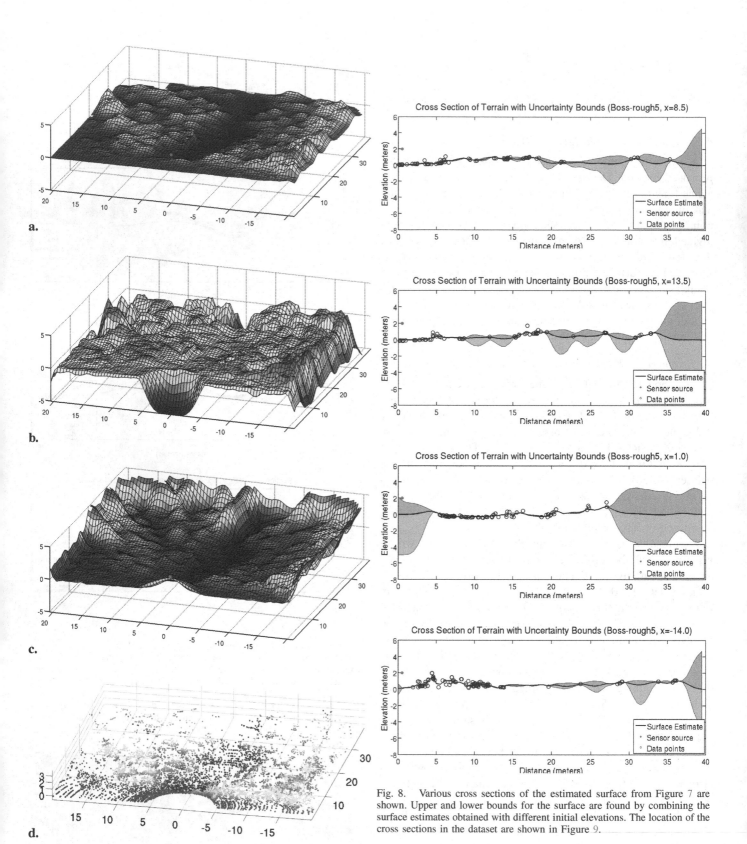

a.

b.

c.

d.

Cross Section of Terrain with Uncertainty Bounds (Boss-rough5, x=8.5)

Cross Section of Terrain with Uncertainty Bounds (Boss-rough5, x=13.5)

Cross Section of Terrain with Uncertainty Bounds (Boss-rough5, x=1.0)

Cross Section of Terrain with Uncertainty Bounds (Boss-rough5, x=-14.0)

Fig. 8. Various cross sections of the estimated surface from Figure 7 are shown. Upper and lower bounds for the surface are found by combining the surface estimates obtained with different initial elevations. The location of the cross sections in the dataset are shown in Figure 9.

Fig. 7. 3 estimated surfaces, using elevation priors of 0 meters (**a.**), 5 meters (**b.**), and -5 meters (**c.**). Using high and low elevation priors gives uncertainty bounds on the estimated surface. The training set point cloud is shown in **d.**

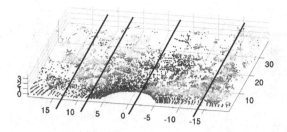

Fig. 9. The locations of the cross sections shown in Figure 8.

TABLE I

ERROR RATES ON FIVE OFFROAD DATASETS.

Dataset	No Visibility Info	With Visibility Info
1	0.059	**0.014**
2	0.045	**0.025**
3	0.120	**0.035**
4	0.098	**0.096**
5	**0.061**	0.065

the lower regions show the negative uncertainty. The upper bound is often much tighter than the lower bound because of the ray constraints that give additional constraint on the upper bound, but do not impact the lower bound. In areas where no data at all is given over an area larger than the maximum lengthscale of the kernel (5 meters, in this case), the surface estimate will be at 0 and the upper bound and lower bound will be at 5 and -5 meters respectively, since 0, 5, and -5 are the elevation priors set for the surface and the bounds.

C. Contribution of Visibility Information

To evaluate the contribution of the ray constraints gained from the laser visibility information, five offroad datasets are used. For each dataset, a surface is estimated with and without ray constraints and the mean squared error of a withheld test set is calculated. Ideally, the test set would be uniformly sampled on a grid across the terrain. Instead the test samples are from the same distribution as the training set, which effectively masks the benefit of the ray constraints by not testing in the sparsest areas where the ray constraints have the greatest effect.

Despite this bias, the results given in Table I and Figure 10 confirm the benefit of using visibility information. Using the ray constraints helps substantially on all datasets but one, in which the margin between the two results is very slim. The datasets that are not aided as much by the ray constraints are both flatter terrain, where the potential gain from the visibility information is less because the data is evenly distributed. The convergence of the online algorithms for each data set is plotted in Figure 11.

Inclines are terrain types that are smooth yet problematic for LADAR systems, typically producing very sparse returns because of the high angle of incidence. However, due to that same grazing angle, the ray information is extremely helpful. To elucidate the point, an examination of the surfaces learned for a downhill slope with and without visibility constraints is

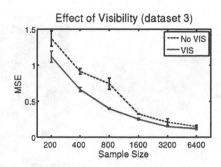

Fig. 10. The performance of the method on a rough terrain dataset. The mean squared error, computed on the test set after convergence of the algorithm, is recorded for increasing sample sizes. The use of ray constraints decreases the error rate at every sample size. Final error rates are given for other datasets (see Table I). Unfortunately, the test data is from the same distribution as the training set, so the benefit of using visibility information is somewhat masked.

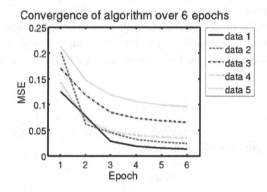

Fig. 11. This graph shows the stable convergence of each natural terrain dataset over 6 epochs.

given. In Figure 12**a.**, the slope is shown (the incline is 30 meters long with a downward slope of roughly 5°). In 12**b.**, the distribution of training data over the test terrain is shown. The data density is very high at the top of the hill, directly in front of the vehicle, and on the embankment at the side of the incline, but almost non-existent toward the bottom of the hill. 12**c.** and 12**d.** show the terrain estimation with and without visibility information. The incline surface is poorly estimated unless the visibility information is used.

V. CONCLUSION

Reconstruction of rough terrain using 3D sensor data is a tough problem because of the highly variable distribution of points in the region of interest. Our proposed approach uses visibility information to carve the surface and produces not only a terrain estimate but also uncertainty bounds. We formulate the learning problem as an RKHS optimization and derive a subgradient-based stochastic solution, which gives computational efficiency, allows data to be added online, and makes the approach an *anytime* algorithm. The evaluation, on both synthetic and natural data, clearly demonstrates the effectiveness of the approach and the utility of the space carving visibility information.

The next phase of this project will incorporate color imagery into the terrain estimation algorithm. The data can be fused

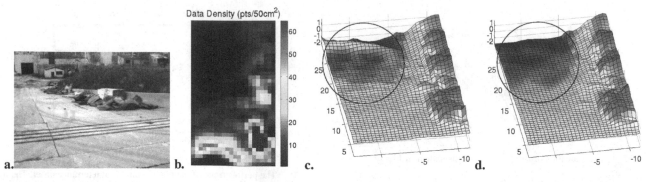

Fig. 12. The laser hits become very sparse if the terrain inclines downward, as for this sloped driveway (**a, b**). If the visibility information is not used, the surface is incorrectly estimated (**c**). When the visibility information is used to constrain the surface, the estimate is far better on the sloping drive (**d**).

using calibration, and the higher resolution, longer range color data can be used to increase both the precision and the range of the terrain model. We also look forward to developing a realtime implementation of the algorithm and expanding the evaluation of the method to include extensive data input over multiple timesteps.

ACKNOWLEDGMENT

This work was supported by the Agency for Defense Development, Republic of Korea.

REFERENCES

[1] M. A. Aizerman, E. M. Braverman, and L. I. Rozoner. Theoretical foundations of the potential function method in pattern recognition learning. *Automation and Remote Control*, 25:821–837, 1964.
[2] J. Bares, M. Hebert, T. Kanade, E. Krotkov, T. Mitchell, R. Simmons, and W. R. L. Whittaker. Ambler: An autonomous rover for planetary exploration. *IEEE Computer*, 22(6):18–26, June 1989.
[3] J. Fournier, B. Ricard, and D. Laurendeau. Mapping and exploration of complex environments using persistent 3d model. In *Conference on Computer and Robot Vision*, pages 403–410, 2007.
[4] J. H. Friedman and W. Stuetzle. Projection pursuit regression. *Journal of the American Statistical Association*, (76):817–823, 1981.
[5] M. Hebert, C. Caillas, E. Krotkov, I. S. Kweon, and T. Kanade. Terrain mapping for a roving planetary explorer. In *Proc. of International Conference on Robotics and Automation (ICRA)*, volume 2, pages 997–1002, May 1989.
[6] D. Higdon, J. Swall, and J. Kern. Non-stationary spatial modeling. *Bayesian Statistics*, 6, 1998.
[7] L. D. Jackel, E. Krotkov, M. Perschbacher, J. Pippine, and C. Sullivan. The DARPA LAGR program: Goals, challenges, methodology, and phase I results. *Journal of Field Robotics*, 23(11-12):945–973, 2006.
[8] A. Kelly, A. T. Stentz, O. Amidi, M. W. Bode, D. Bradley, A. Diaz-Calderon, M. Happold, H. Herman, R. Mandelbaum, T. Pilarski, P. Rander, S. Thayer, N. M. Vallidis, , and R. Warner. *Toward Reliable Off Road Autonomous Vehicles Operating in Challenging Environments*, 25(1):449–483, May 2006.
[9] J. Kivinen, A. J. Smola, and R. C. Williamson. Learning with kernels. In *Advances in Neural Information Processing Systems (NIPS)*. MIT Press, 2002.
[10] K. N. Kutulakos and S. M. Seitz. A theory of shape by space carving. *International Journal of Computer Vision*, 38(3):199–218, 2000.
[11] I. S. Kweon and T. Kanade. High-resolution terrain map from multiple sensor data. *IEEE Trans. Pattern Analysis and Machine Intelligence*, 14(2):278–292, 1992.
[12] T. Lang, C. Plagemann, and W. Burgard. Adaptive non-stationary kernel regression for terrain modeling. In *Proc. of Robotics: Science and Systems (RSS)*. MIT Press, 2007.
[13] C. J. Paciorek and M. J. Schervish. Nonstationary covariance functions for Gaussian process regression. In *Advances in Neural Information Processing Systems (NIPS)*. MIT Press, 2004.
[14] P. Pfaff, R. Triebel, and W. Burgard. An efficient extension to elevation maps for outdoor terrain mapping and loop closing. *Int. Journal of Robotics Research*, 26(2):217–230, 2007.
[15] C. Plagemann, K. Kersting, and W. Burgard. Nonstationary Gaussian process regression using point estimates of local smoothness. In *Proc. of the European Conference on Machine Learning (ECML)*, Antwerp, Belgium, 2008.
[16] C. Plagemann, S. Mischke, S. Prentice, K. Kersting, N. Roy, and W. Burgard. Learning predictive terrain models for legged robot locomotion. In *Proc. of International Conference on Intelligent Robots and Systems (IROS)*. IEEE, 2008.
[17] N. Ratliff, J. D. Bagnell, and M. Zinkevich. (online) subgradient methods for structured prediction. In *Proc. of Conference on Artificial Intelligence and Statistics (AI-STATS)*, 2007.
[18] R. Schaback. Creating surfaces from scattered data using radial basis functions. *Mathematical methods for curves and surfaces*, page 477496, 1995.
[19] B. Scholkopf, J. Giesen, and S. Spalinger. Kernel methods for implicit surface modeling. In *Advances in Neural Information Processing Systems (NIPS)*. MIT Press, 2005.
[20] N. Z. Shor, K. C. Kiwiel, and A. Ruszcayñski. *Minimization methods for non-differentiable functions*. Springer-Verlag New York, Inc., New York, NY, USA, 1985.
[21] G. Slabaugh, R. Schafer, and M. Hans. Multi-resolution space carving using level set methods. In *Proc. of International Conference on Image Processing*, 2002.
[22] C. Urmson, J. Anhalt, D. Bagnell, C. Baker, R. Bittner, M. N. Clark, J. Dolan, D. Duggins, T. Galatali, C. Geyer, M. Gittleman, S. Harbaugh, M. Hebert, T. M. Howard, S. Kolski, A. Kelly, M. Likhachev, M. McNaughton, N. Miller, K. Peterson, B. Pilnick, R. Rajkumar, P. Rybski, B. Salesky, Y.-W. Seo, S. Singh, J. Snider, A. Stentz, W. R. Whittaker, Z. Wolkowicki, J. Ziglar, H. Bae, T. Brown, D. Demitrish, B. Litkouhi, J. Nickolaou, V. Sadekar, W. Zhang, J. Struble, M. Taylor, M. Darms, and D. Ferguson. Autonomous driving in urban environments: Boss and the Urban Challenge. *J. Field Robotics*, 25(8):425–466, 2008.
[23] C. Walder, B. Scholkopf, and O. Chapelle. Implicit surface modelling with a globally regularised basis of compact support. *Eurographics*, 25(3), 2006.
[24] M. Yguel, C. T. M. Keat, C. Braillon, C. Laugier, and O. Aycard. Dense mapping for range sensors: Efficient algorithms and sparse representations. In *Robotics: Science and Systems*, 2007.
[25] J. Zhu, S. Hoi, and M. Lyu. A multi-scale Tikhonov regularization scheme for implicit surface modelling. In *Proc. of Conference on Computer Vision and Pattern Recognition (CVPR)*. IEEE, 2007.
[26] M. Zinkevich. Online convex programming and generalized infinitesimal gradient ascent. In *ICML*, pages 928–936, 2003.

View-based Maps

Kurt Konolige, James Bowman, JD Chen, Patrick Mihelich
Willow Garage
Menlo Park, CA 94025
Email: konolige@willowgarage.com

Michael Calonder, Vincent Lepetit, Pascal Fua
EPFL
Lausanne, Switzerland
Email: michael.calonder@epfl.ch

Abstract— Robotic systems that can create and use visual maps in realtime have obvious advantages in many applications, from automatic driving to mobile manipulation in the home. In this paper we describe a mapping system based on retaining stereo views of the environment that are collected as the robot moves. Connections among the views are formed by consistent geometric matching of their features. Out-of-sequence matching is the key problem: how to find connections from the current view to other corresponding views in the map. Our approach uses a vocabulary tree to propose candidate views, and a strong geometric filter to eliminate false positives – essentially, the robot continually re-recognizes where it is. We present experiments showing the utility of the approach on video data, including map building in large indoor and outdoor environments, map building without localization, and re-localization when lost.

I. INTRODUCTION

Fast, precise, robust visual mapping is a desirable goal for many robotic systems, from transportation to in-home navigation and manipulation. Vision systems, with their large and detailed data streams, should be ideal for recovering 3D structure and guiding tasks such as manipulation of everyday objects, navigating in cluttered environments, and tracking and reacting to people. But the large amount of data, and its associated perspective geometry, also create challenging problems in organizing the data in an efficient and useful manner.

One useful idea for maintaining the spatial structure of visual data is to organize it into a set of representative views, along with spatial constraints among the views, called a *skeleton*. Figure 1 gives an example of a skeleton constructed in an indoor environment. Typically views are matched in sequence as the camera is moved around, so the skeleton mimics the camera trajectory (red trajectory). In loop closure, the camera enters an area already visited, and can re-connect with older views. The overall map is generated by nonlinear optimization of the system [2, 19, 33]. View-based maps have the advantage of *scalability*: using incremental techniques, new views can be added and the skeleton optimized online.

One problem is how to efficiently perform loop closure. Previous approaches used exhaustive search of the current view against all skeleton views that could possibly be in the area, given the relative uncertainty of views. This approach does not scale well to larger skeletons, and involves constant calculation of relative covariance. Instead, to limit the number of views that must be considered for loop closure, we employ a vocabulary tree [27] to suggest candidate views, a type

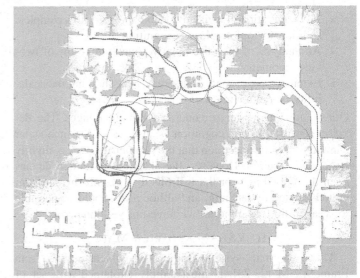

Fig. 1. Top: Skeleton map constructed online from just stereo images, registered against a laser map for reference. Visual odometry shown against the map, and corrected by loop closure from visual place recognition. Tick marks at sides of map are 10m intervals. Bottom shows typical views, with blurring, clutter, people, and blank walls.

of *place recognition* (PR). The vocabulary tree allows us to efficiently filter thousands of skeleton views to find possible matches, as well as add new views online. We call this online PR *re-recognition*: the robot recognizes its position relative to the stored view map on every cycle, without any a priori knowledge of its position (unlike localization, which requires a position hypothesis).

The addition of vocabulary tree PR to view-based maps is a happy alignment of technologies that expands the utility of visual mapping in interesting ways. For example, even without sequence information, it is often possible to quickly reconstruct a skeleton map from a set of views (Figure 9 and Section VI-D). In the Experiments section, we highlight some other applications that show the scalability of view-based maps. Loop closure over large distances is possible: we show indoor maps with 800m trajectories (Figure 1), and outdoor rough-terrain maps with 5km trajectories. On a smaller scale, view matching with large numbers of points is inherently accurate, showing a few centimeter accuracy over a desktop workspace. Additional capabilities include automatic recovery from localization failures (e.g., occlusion and motion blur) and incremental construction of maps.

The main contributions of this paper are

- The construction of a realtime system for robust, accurate visual map making over large and small spaces.
- The use of views (images), view matching, and geometric relations between views as a uniform approach to short-term tracking and longer-term metric mapping and loop closure.
- The integration of a visual vocabulary tree into a complete solution for online place recognition.
- An analysis of the false positive rejection ability of two-view geometry.
- Extensive experiments with real data, showing the scalability of the technique.

Our solution uses stereo cameras for input images. The development of place recognition is also valid for monocular cameras, with the exception that the geometric check is slightly stronger for stereo. However, the skeleton system so far has been developed just for the full 6DOF pose information generated by stereo matching, and although it should be possible to weaken this assumption, we have not yet done so.

II. VSLAM AND VIEW MAPS

The view map system (Figure 2), which derives from FrameSLAM [2, 23], is most simply explained as a set of nonlinear constraints among camera views, represented as nodes and edges (see Figure 5 for a sample graph). Constraints are input to the graph from two processes, visual odometry (VO) and place recognition (PR). Both rely on geometric matching of views to find relative pose relationships; they differ only in their search method. VO continuously matches the current frame of the video stream against the last keyframe, until a given distance has transpired or the match becomes too weak. This produces a stream of keyframes at a spaced distance, which become the backbone of the constraint graph, or *skeleton*. PR functions opportunistically, trying to find any other views that match the current keyframe. This is much more difficult, especially in systems with large loops. Finally, an optimization process finds the best placement of the nodes in the skeleton.

It is interesting to note that current methods in visual SLAM divide in the same way as in laser-based SLAM, namely, those that keep track of landmarks using an EKF filter (monoSLAM [9, 10] and variations [29, 32]), and those that, like ours, maintain a constraint graph of views, similar to the original Lu and Milios method [25]. The main limitation of the landmark methods is the filter size, which is only tractable in small (room-size) environments. An exception is [29], which uses a submap technique, although realtime performance has not yet been demonstrated. Landmark systems also tend to be less accurate, because they typically track only a few tens of landmarks per frame. In contrast, our visual odometry technique tracks 300 points per frame, and we construct

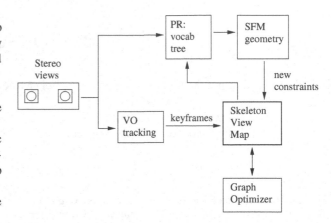

Fig. 2.   System overview.

maps containing several thousand views (and thus hundreds of thousands of points).

In a similar vein, the recent Parallel Tracking and Mapping (PTAM) system [20, 21] also uses 3D landmarks, but employs standard SfM bundle adjustment to build a map from many views. Many more points can be handled in the decoupled tracking phase, leading to accurate and robust performance under many conditions. Still, it is limited to small environments (around 150 keyframes) by the number of points and by bundle adjustment. It is also subject to tracking failures on self-similar textures (e.g., bushes), object motion, and scene changes (e.g., removal of an object). In contrast, view-based maps use consistent view geometry to robustly estimate poses even in the presence of distractors.

The skeleton system deployed here comes directly from the frameSLAM work in [2, 23]. Several other systems employ constraint graphs as the basic map structure. Fraundorfer et al. [13] have a monocular system that represents only direction information between views, and produce only a topological map. Eade and Drummond [11] employ a hybrid approach, using EKF landmarks within a local area called a *node*, then connecting the nodes via similarity relations. An interesting point of their graph optimization is the use of cycles to constrain relative scale between nodes. Other robotics work that employs similar ideas about constructing view-based constraints is in [33, 34]. These systems also keep a constraint network of relative pose information between frames, based on stereo visual odometry, and solve it using nonlinear least square methods. The main difference with our system is that frameSLAM represents the relationships as nonlinear constraints, which are more accurate over angular deformations, and can reduce the size of the skeleton graph to deal with larger areas as required.

III. RELATED PLACE RECOGNITION WORK

Visual place recognition is an image classification problem; new views are classified against a set of previously seen views. For use in VSLAM, the classifier must support efficient online learning of new reference views. Image matching techniques based on bag-of-words matching are ideally suited to this purpose. For fast lookup of similar places, we rely on the hierarchical vocabulary trees proposed by Nistér and Stewénius [27], which has the advantage of fast online learning of new places. Other methods include alternative approximate nearest neighbor algorithms [30, 26] and various refinements for improving the response or efficiency of the tree [8, 17, 18].

Cummins and Newman [8] show how to use visual features for navigation and loop closure over very large trajectories. They use pairwise feature statistics and sequences of views to address the perceptual aliasing problem, especially notable in man-made environments containing repeated structure. Jegou et al. [17] incorporate Hamming embedding and weak geometric consistency constraints into the inverted file to improve performance. In this work, we rely instead on a strong geometric consistency check on single views.

Jegou et al. [18] note that even using inverted files, query time is linear in the number of reference images; they propose a two-level inverted file scheme to improve the complexity. Our experiments do show linearly increasing query/update time, but with a very small constant factor (Figure 6). For our scale of application (in the thousands of images), the query time of the vocabulary tree is nearly constant, and such sophistication is unnecessary.

In application to graph-based VSLAM, Callmer et al. [4] propose a loop closure procedure that uses a vocabulary tree in a manner similar to ours, along with a weak geometric check to weed out some false positives. Eade and Drummond [12] have extended their node approach with a PR method based on bag of words, in which they learn the words online. They give few statistics on the performance of PR, so it isn't possible to compare directly – they have the advantage of learning based on the observed features, but have far fewer words (3000 vs. 100,000 in our case). They have independently introduced some of the same applications of PR as given here: recovery from localization error and stitching together trajectories when common views are found. Finally, Williams et al. [35] also recover from localization errors in a landmark-based VSLAM framework, by training a classifier to recognize landmarks online; so far their system has been limited to 80 landmarks, mostly because of EKF processing.

There is an interesting convergence between our work and recent photo stitching in the vision community [31]. They employ a similar skeletonization technique to limit the extent of bundle adjustment calculations, but run in batch mode, with no attempt at realtime behavior. Klopschitz et al. [22] use a vocabulary tree to identify possible matches in video stream, and then followed by a dynamic programming technique to verify a sequence of view matches. They are similar to our work in emphasizing online operation.

IV. FRAMESLAM BACKGROUND

The skeleton view map encodes 6DOF relations bewteen views. For two views c_i and c_j with a known relative pose, the constraint between them is

$$\Delta z_{ij} = c_i \ominus c_j, \text{ with covariance } \Lambda^{-1} \qquad (1)$$

where $\ominus$ is the inverse motion composition operator – in other words, c_j's position in c_i's frame. The covariance expresses the strength of the constraint, and arises from the geometric matching step that generates the constraint, explained below.

Given a constraint graph, the optimal position of the nodes is a nonlinear optimization problem of minimizing $\sum_{ij} \Delta z_{ij}^\top \Lambda \Delta z_{ij}$; a standard solution is to use preconditioned conjugate gradient [2, 16]. For realtime operation, it is more convenient to run an incremental relaxation step, and the recent work of Grisetti et al. [15] on SGD provides an efficient method of this kind, called Toro, which we use for the experiments. This method selects a subgraph to optimize, and runs only a few iterations on each addition to the graph. Other relaxation methods for nonlinear constraint systems include [14, 28].

A. Geometric Consistency Check and Pose Estimation

Constraints arise from the perspective view geometry between two stereo camera views. The process can be summarized by the following steps:

1) Match features in the left image of one view with features in the left image of the other view ($N \times N$ matching).
2) (RANSAC steps) From the set of matches, pick three candidates, and generate a relative motion hypothesis between the views. Stereo information is essential here for giving the 3D coordinates of the points.
3) Project the 3D points from one view onto the other based on the motion hypothesis, and count the number of inliers.
4) Repeat 2 and 3, keeping the hypothesis with the best number of inliers.
5) Polish the result by doing nonlinear estimation of the relative pose from all the inliers.

The last step iteratively solves a linear equation of the form

$$J^\top J \delta x = -J^\top \Delta z, \qquad (2)$$

where Δz is the error in the projected points, δx is a change in the relative pose of the cameras, and J is the Jacobian of z with respect to x. The inverse covariance derives from $J^\top J$, which approximates the curvature at the solution point. As a practical matter, Toro accepts only diagonal covariances, so instead of using $J^\top J$, we scale a simple diagonal covariance based on the inlier response.

In cases where there are too few inliers, the match is rejected; this issue is explored in detail in Section V-C. The important result is that geometric matching provides an almost foolproof method for rejecting bad view matches.

B. Visual Odometry and Re-detection

Our overriding concern is to make the whole system robust. In outdoor rough terrain, geometric view matching for VO has proven to be extremely stable even under very large image motion [24], because points are re-detected and matched over large areas of the image for each frame. For this paper's experiments, we use a recently-developed scale-space detector called STAR (similar to the CenSurE detector [1]) outdoors, and the FAST detector indoors. There is no motion assumption to drive keypoint match prediction – all keypoints are redetected at each frame. For each keypoint in the current left image, we search a corresponding area of size 128x64 pixels for keypoints in the reference keyframe image for a match using SAD correlation of a 16x16 patch. Robust geometric matching then determines the best pose estimate. Keyframes are switched when the match inlier count goes below 100, or the camera has moved 0.3m or 10 degrees.

In a 400 m circuit of our labs, with almost blank walls, moving people, and blurred images on fast turns, there was not a single VO frame match failure (see Figure 5 for sample frames). The PTAM methods of [20], which employ hundreds of points per frame, can also have good performance, with pyramid techniques to determine large motions. However, they are prone to fail when there is significant object motion, since they do not explore the space of geometrically consistent data associations

C. Skeleton Graph Construction

The VO module provides a constant stream of keyframes to be integrated into the skeleton graph. To control the size of the graph for large environments, only a subset of the keyframes need to be kept in the graph. For example, in the 5km outdoor runs, a typical distance between skeleton views is 5m.

As each keyframe is generated by VO, it is kept in a small sequential buffer until enough distance has accumulated to integrate it into the skeleton. At this point, all the views in the buffer are reduced to a single constraint between the first and last views in the buffer. The reduction process is detailed in [2]; for a linear sequence of constraints, it amounts to compounding the pose differences $\Delta z_{01} \oplus \Delta z_{12} \oplus \cdots \oplus \Delta z_{n,n-1}$.

One can imagine many other schemes for skeleton construction that try to balance the density of the graph, but this simple one worked quite well. In the case of lingering in the same area for long periods of time, it would be necessary to stop adding new views to the graph, which otherwise would grow without limit. The frameSLAM graph reduction supports online node deletion, and we are starting to explore strategies for controlling the density of views in an area.

After incorporating a new view into the skeleton, the Toro optimizer is run for a few iterations to optimize the graph. The optimization can be amortized over time, allowing online operation for fairly large graphs, up to several thousand views (see the timings in Figure 6).

V. MATCHING VIEWS

In this section we describe our approach to achieving efficient view matching against thousands of frames. We develop a filtering technique for matching a new image against a dataset of reference images (PR), using a vocabulary tree to suggest candidate views from large datasets. From a small set of the top candidates, we apply the geometric consistency check, using Randomized Tree signatures [5] as an efficient

viewpoint-invariant descriptor for keypoint matching. Finally, we develop statistics to verify the rejection capability of this check.

A. Compact Randomized Tree Signatures

We use Randomized Tree (RT) signatures [5] as descriptors for keypoints. An RT classifier is trained offline to recognize a number of keypoints extracted from an image database, and all other keypoints are characterized in terms of their response to these classification trees. Remarkably, a fairly limited number of base keypoints—500 in our experiments—is sufficient. However, a limitation of this approach is that storing a pre-trained Randomized Tree takes a considerable amount of memory. A recent extension [6] compacts signatures into much denser and smaller vectors resulting in both a large decrease in storage requirement and substantially faster matching, at essentially the same recognition rates as RT signatures and other competing descriptors; Table I compares creation and matching times. The performance of compact signatures means that the $N \times N$ keypoint match of the geometric consistency check is not a bottleneck in the view matching process.

	Descriptor Creation (512 kpts)	N×N Matching (512×512 kpts)
Sparse RTs (CPU)	31.3 ms	27.7 ms
Compact RTs (CPU)	**7.9 ms**	**6.3 ms**
U-SURF64 (CPU)	150 ms	120 ms
		73 ms (ANN)
U-SURF64 (GPU)		6.8 ms

TABLE I

TIMINGS FOR DESCRIPTOR CREATION AND MATCHING.

B. A Prefilter for Place Recognition

We have implemented a place recognition scheme based on the vocabulary trees of Nistér and Stewénius [27] which has good performance for both inserting and retrieving images based on the compact RT descriptors. We call this step a *prefilter* because it just suggests candidates that could match the current view, which must then be subject to the geometric consistency check for confirmation and pose estimation. VO and PR both use the geometric check, but PR has the harder task of finding matches against all views in the skeleton, while VO only has to match against the reference keyframe. The prefilter is a bag-of-words technique that works with monocular views (the left image of the stereo pairs).

The vocabulary tree is a hierarchical structure that simultaneously defines both the visual words and a search procedure for finding the closest word to any given keypoint. The tree is constructed offline by hierarchical k-means clustering on a large training set of keypoint descriptors. The set of training descriptors is clustered into k centers. Each center then becomes a new branch of the tree, and the subset of training descriptors closest to it are clustered again. The process repeats until the desired number of levels is reached.

The discriminative ability of the vocabulary tree increases with the number of words, at a cost of greater quantization error [3] and increased memory requirements. Nistér and Stewénius have shown that performance improves with the number of words, up to very large (>1M) vocabularies. In our experiments, we use about 1M training keypoints from 500 images in the Holidays dataset [17], with $k = 10$, and create a tree of depth 5, resulting in 100K visual words. The Holidays dataset consists of mostly outdoor images, so the vocabulary tree is trained on data visually dissimilar to the indoor environments of most of our experiments.

The vocabulary tree is populated with the reference images by dropping each of their keypoint descriptors to a leaf and recording the image in a list, or *inverted file*, at the leaf. To query the tree, the keypoint descriptors of the query image are similarly dropped to leaf nodes, and potentially similar reference images retrieved from the union of the inverted files. In either case, the vocabulary tree describes the image as a vector of word frequencies determined by the paths taken by the descriptors through the tree. Each reference image is scored for relevance to the query image by computing the L1 distance between their frequency vectors. The score is entropy-weighted to discount very common words using the Term Frequency Inverse Document Frequency (TF-IDF) approach described in [27, 30].

To evaluate the vocabulary tree as a prefilter, we constructed a small test set of some 180 keyframes over a 20m trajectory, and determined ground truth matches by performing geometric matching across all 180×180 possibilities. In this dataset, each keyframe averages 11.8 ground truth matches. We inserted these keyframes, along with another 553 non-matching distractor keyframes, into the vocabulary tree. Querying the vocabulary tree with each of the 180 test keyframes in turn, we obtained their similarity scores against all the reference images. The sensitivity of the vocabulary tree matching is shown by the ROC curve (Figure 3, left) obtained by varying a threshold on the similarity score.

Since we can only afford to put a limited number of candidates through the geometric consistency check, the critical performance criterion is whether the correct matches appear among the most likely candidates. Varying N, we counted the percentage of the ground truth matches appearing in the top-N results from the vocabulary tree. For robustness, we want to be very likely to successfully relocalize from the current keyframe, so we also count the percentage of test keyframes with at least one or at least two ground truth matches in the top-N results (Figure 3, right).

In our experiments, we take as match candidates the top $N = 15$ responses from place recognition. We expect to find at least one good match for 97% of the keyframes and two good matches for 90% of the keyframes. For any given keyframe, we expect almost 60% of the correct matches to appear in the top 15 results.

C. Geometric Consistency Check

We can predict the ability of the geometric consistency check (Section IV-A) to reject false matches by making a few assumptions about the statistics of matched points, and

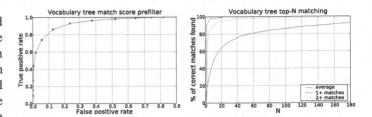

Fig. 3. Left: ROC curve for the vocabulary tree prefilter on the test dataset. Right: "Average" curve shows percentage of the correct matches among the top N results from the vocabulary tree; other curves are the percentage of views with at least 1 or 2 matches in the top N.

estimating the probability that two unrelated views I_0 and I_1 will share at least M matches, given a relative pose estimate. Based on perspective geometry, any point match will be an inlier if the projection in I_1 lies on the epipolar line of the point in I_0. In our case, with 640×480 images, an inlier radius of 3 pixels, the probability of being an inlier is:

$$A_{track}/A_{image} = (6 * 640)/(640 * 480) = .0125 \quad (3)$$

This is for monocular images; for stereo images, the two image disparity checks (assuming disparity search of 128 pixels) yield a further factor of $(6/128)*(6/128)$. In the more common case with dominant planes, one of the image disparity checks can be ignored, and the factor is just $(6/128)$. If the matches are random and independent (i.e., no common objects between images), then counting arguments can be applied. The distribution of inliers over N trials with probability p of being an inlier is $B_{p,N}$, the binomial distribution. We take the maximum inliers over K RANSAC trials, so the probability of having less than x inliers is $(1 - B_{p,N}(x))^K$. The probability of exactly x inliers over all trials is

$$(1 - B_{p,N}(x))^K - (1 - B_{p,N}(x - 1))^K \quad (4)$$

Figure 4 shows the probabilities for the planar stereo case, based on Equation 4. The graph peaks sharply at 2 inliers (out of 250 matches), showing the theoretic rejection ability of the geometric check. However, the real world has structure, and some keypoints form clusters: these factors violate the independent match assumption. Figure 4 compares actual rejections from the three datasets in the Experiments section, with two different types of keypoints, FAST and STAR. These show longer tails, especially FAST, which has very strong clustering at corners. Note that repetitive structure, which causes false positives for bag-of-words matching, as noted in [8], is rejected by the geometric check – for example, the windows in Figure 7. Even with the long tail, probabilities are very low for larger numbers of inliers, and the rejection filter can be set appropriately.

VI. EXPERIMENTS

As explained in Section II, the view-based system consists of a robust VO detector that estimates incremental poses of a stereo video stream, and a view integrator that finds and adds non-sequential links to the skeleton graph, and optimizes the graph. We carried out a series of tests on stereo data from three different environments:

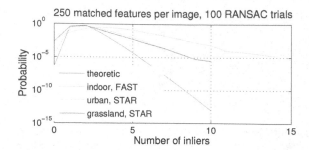

Fig. 4. The probability of getting x inliers from a random unrelated view match. Theoretic probability (see text) compared to three different datasets. Note log scale for probabilities.

Type	length	image res	image rate	stereo base	skeleton views
Office	0.8 km	640x480	30 Hz	9 cm	4.2k
Urban	0.4 km	768x568	25 Hz	100 cm	0.5k
Terrain	10 km	512x384	15 Hz	50 cm	14.6k

Rectification is not counted in timings; for the indoor sequence it is done in the stereo hardware. VO consumes 11 ms per video frame, leaving 22 ms for view integration, 2/3 of the available time at the fastest frame rate. As in PTAM [20], view integration can be run in parallel with VO, so on a dual-core machine view matching and optimization could consume a whole processor. Given its efficiency, we publish results here for a single processor only. In all experiments, we restrict the number of features per image to ~300, and use 100 RANSAC iterations for geometric matching.

A. Large Office Loop

The first experiment is a large office loop of about 800m in length. The trajectory was done by joysticking a robot at around 1m/sec. Figure 1 shows some images: there is substantial blurring during fast turns, sections with almost blank walls, cluttered repetitive texture, and moving people. There are a total of 24K images in the trajectory, with 10k keyframes, 4235 skeleton views, and 21830 edges (Figure 1 shows the first 400m). Most of the edges are added from neighboring nodes along the same trajectory, but a good portion come from loop closures and parallel trajectories (Figure 5).

View matching has clearly captured the major structural aspects of the trajectory, relative to open-loop VO. It closed the large loop from the beginning of the trajectory to the end, as well as two smaller loops in between. We also measured the planarity of the trajectory: for the view-based system, RMS error was 22 cm; for open-loop VO, it was 50 cm.

Note that the vocabulary tree prefilter makes no distinction between reference views that are temporally near or far from the current view: all reference views are treated as places to be recognized. By exploiting the power of geometric consistency, there is no need to compute complex covariance gating information for data association, as is typically done for EKF-based systems [9, 10, 29, 32].

The time spent in view integration is broken down by

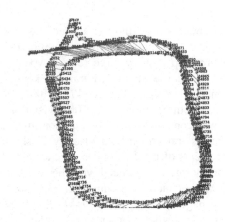

Fig. 5. A closeup from the office dataset showing the matched views on a small loop. The optimizer has been turned off to show the links more clearly.

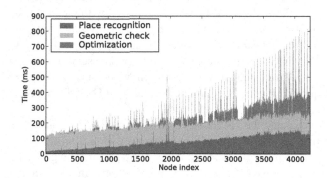

Fig. 6. Timing for view integration per view during the office loop trajectory.

category in Figure 6. The vocab tree prefilter grows linearly, to about 100 ms at the end; the geometry check is constant at 65 ms. Toro does almost no work at the beginning of the trajectory, then grows to average 120 ms at the end, with maximum time of 500 ms. VO can run at frame rates, while simultaneously adding and optimizing skeleton frames at 2 Hz.

B. Versailles Rond

We ran viewmap on an outdoor urban sequence from a car in Versailles, a trajectory of about 400m (Figure 7). The skeleton map contained 140 views, and PR found 12 matches after looping around, even when the car moved into an adjacent lane. The Versailles images have a lot of self-similarity in the windows, but the geometric check rejects false positives. This sequence easily runs online.

C. Rough-Terrain Loops

Large off-road trajectories present the hardest challenge for VSLAM. Grass, trees and other natural terrain have self-similar texture and few distinguishing landmarks. The dataset we used was taken by a very aggressive offroad autonomous vehicle, with typical motion of 0.5 m between frames, and sometimes abrupt roll, pitch, and vertical movement. VO fails on about 2% of the frames, mostly because of complete occlusion of one camera; we fill in with IMU data. There are two 5 km trajectories of 30K frames that overlap occasionally. To test the system, we set the skeleton view distance to only

Fig. 7. Versailles Rond sequence of 700 video frames taken from a moving vehicle, 1m baseline, narrow FOV. (Dataset courtesy of Andrew Comport [7]) Top: matched loop closure frames. Bottom: top-down view of trajectory superimposed on satellite image.

Fig. 8. Matched loop closure frames from the rough-terrain dataset. The match was made between two separate autonomous 5km runs, several hours apart: note the low cloud in the left image.

1m. The resultant graph has 14649 nodes and 69545 edges, of which 189 are cross-links between the two trajectories. The trajectories are largely corrected via the crosslinks – the error at the end of the loop changes from over 100m with raw VO to less than 10m. Note that there are no loop closures within each trajectory, only between them. Figure 8 shows such a match. The PR system has the sensitivity to detect close possibilities, and the geometric check eliminates false positives – in Section V-C we tested 400K random non-matching image pairs from this dataset, and found none with over 10 inliers (Figure 4).

D. TrajectorySynth

To showcase the capability of view integration, we performed a reconstruction experiment without any temporal information provided by video sequencing or VO, relying just on view integration. We take a small portion of the office loop,

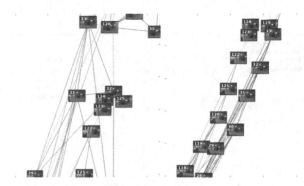

Fig. 9. Trajectory synthesis with no sequencing information: view constraints from PR at left; final optimized map at right.

extract 180 keyframes, and push them into the vocabulary tree. We then choose one keyframe as the seed, and use view integration to add all valid view matches to the view skeleton. The seed is marked as used, and one of the keyframes added to the skeleton is chosen as the next seed. The process repeats until all keyframes are marked as used.

The resultant graph is shown in Figure 9, left. The nodes are placed according to the first constraint found; some of these constraints are long-range and weak, and so the graph is distorted. Optimizing using Toro produces the consistent graph on the right. The time per keyframe is 150 ms, so that the whole trajectory is reconstructed in 37 seconds, about 2 times faster than realtime. The connection to view stitching [31] is obvious, to the point where we both use the same term "skeleton" for a subset of the views. However, their method is a batch process that uses full bundle adjustment over a reduced set of views, whereas our approximate method retains just pairwise constraints between views.

E. Relocalization

Under many conditions, VO can lose its connection to the previous keyframe. If this condition persists (say the camera is covered for a time), then it may move an arbitrary distance before it resumes. The scenario is sometimes referred to as the "kidnapped robot" problem. View-based maps solve this problem with no additional machinery. To illustrate, we took the small loop sequence from the TrajectorySynth experiment, and cut out enough frames to give a 5m jump in the actual position of the robot. Then we started the VO process again, using a very weak link to the previous node so that we could continue using the same skeleton graph. After a few keyframes, the view integration process finds the correct match, and the new trajectory is inserted in the correct place in the growing map (Figure 10). This example clearly indicates the power of constant re-recognition.

F. Accuracy of View-based Maps

To verify the accuracy of the view-based map, we acquired a sequence of video frames that are individually tagged by "ground truth" 3D locations recorded by the IMPULSE Motion Capture System from PhaseSpace Inc. The trajectory is about 23 m in total length, consisting of 4 horizontal loops

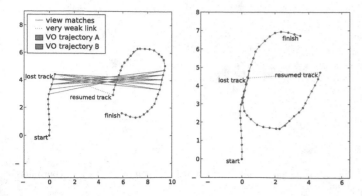

Fig. 10. Kidnapped robot problem. There is a cut in the VO process at the last frame in the left trajectory, and the robot is transported 5m. After continuing a short time, a correct view match inserts the new trajectory into the map.

with diameters of roughly 1.5 m and elevations from 0 to 1m. There are total of 6K stereo images in the trajectory, with 224 graph nodes, and 360 edges. The RMS error of the nodes was 3.2 cm for the view-based system, which is comparable to the observed error for the mocap system. By contrast, open-loop VO had an error of 14 cm.

VII. CONCLUSION

We have presented a complete system for online generation of view-based maps. The use of re-recognition, where the robot's position is re-localized at each cycle with no prior information, leads to robust performance, including automatic relocalization and map stitching.

There are some issues that emerged in performing this research that bear further scrutiny. First, SGD optimization takes too long on very large graphs, since its convergence is sublinear. A better strategy is to use a few iterations of SGD, followed by Gauss-Seidel iterations to converge quickly. Second, we would like to investigate the monocular case, where full 6DOF constraints are not present in the skeleton graph.

REFERENCES

[1] M. Agrawal and K. Konolige. CenSurE: Center surround extremas for realtime feature detection and matching. In *ECCV*, 2008.
[2] M. Agrawal and K. Konolige. FrameSLAM: From bundle adjustment to real-time visual mapping. *IEEE Transactions on Robotics*, 24(5), October 2008.
[3] O. Boiman, E. Shechtman, and M. Irani. In defense of nearest-neighbor based image classification. In *Proceedings of IEEE Conference on Computer Vision and Pattern Recognition*. IEEE, 2008.
[4] J. Callmer, K. Granström, J. Nieto, and F. Ramos. Tree of words for visual loop closure detection in urban slam. In *Proceedings of the 2008 Australasian Conference on Robotics and Automation*, page 8, 2008.
[5] M. Calonder, V. Lepetit, and P. Fua. Keypoint signatures for fast learning and recognition. In *ECCV*, 2008.
[6] M. Calonder, V. Lepetit, K. Konolige, P. Mihelich, and P. Fua. High-speed keypoint description and matching using dense signatures. In *Under review*, 2009.
[7] A. Comport, E. Malis, and P. Rives. Accurate quadrifocal tracking for robust 3d visual odometry. In *ICRA*, 2007.
[8] M. Cummins and P. M. Newman. Probabilistic appearance based navigation and loop closing. In *ICRA*, 2007.
[9] A. Davison. Real-time simultaneaous localisation and mapping with a single camera. In *ICCV*, pages 1403–1410, 2003.
[10] A. J. Davison, I. D. Reid, N. D. Molton, and O. Stasse. Monoslam: Real-time single camera slam. *IEEE PAMI*, 29(6), 2007.
[11] E. Eade and T. Drummond. Monocular SLAM as a graph of coalesced observations. In *Proc. ICCV*, 2007.
[12] E. Eade and T. Drummond. Unified loop closing and recovery for real time monocular slam. In *BMVC*, 2008.
[13] F. Fraundorfer, C. Engels, and D. Nistér. Topological mapping, localization and navigation using image collections. In *IROS*, pages 3872–3877, 2007.
[14] U. Frese, P. Larsson, and T. Duckett. A multilevel relaxation algorithm for simultaneous localisation and mapping. *IEEE Transactions on Robotics*, 21(2):1–12, 2005.
[15] G. Grisetti, D. L. Rizzini, C. Stachniss, E. Olson, and W. Burgard. Online constraint network optimization for efficient maximum likelihood mapping. In *ICRA*, 2008.
[16] J. Gutmann and K. Konolige. Incremental mapping of large cyclic environments. In *Proc. IEEE International Symposium on Computational Intelligence in Robotics and Automation (CIRA)*, pages 318–325, Monterey, California, November 1999.
[17] H. Jegou, M. Douze, and C. Schmid. Hamming embedding and weak geometric consistency for large scale image search. In *ECCV*, 2008.
[18] H. Jegou, H. Harzallah, and C. Schmid. A contextual dissimilarity measure for accurate and efficient image search. *Computer Vision and Pattern Recognition, IEEE Computer Society Conference on*, 0:1–8, 2007.
[19] A. Kelly and R. Unnikrishnan. Efficient construction of globally consistent ladar maps using pose network topology and nonlinear programming. In *Proceedings 11th International Symposium of Robotics Research*, 2003.
[20] G. Klein and D. Murray. Parallel tracking and mapping for small AR workspaces. In *Proc. Sixth IEEE and ACM International Symposium on Mixed and Augmented Reality (ISMAR'07)*, Nara, Japan, November 2007.
[21] G. Klein and D. Murray. Improving the agility of keyframe-based slam. In *ECCV*, 2008.
[22] M. Klopschitz, C. Zach, A. Irschara, and D. Schmalstieg. Generalized detection and merging of loop closures for video sequences. In *3DPVT*, 2008.
[23] K. Konolige and M. Agrawal. Frame-frame matching for realtime consistent visual mapping. In *Proc. International Conference on Robotics and Automation (ICRA)*, 2007.
[24] K. Konolige, M. Agrawal, and J. Solà. Large scale visual odometry for rough terrain. In *Proc. International Symposium on Research in Robotics (ISRR)*, November 2007.
[25] F. Lu and E. Milios. Globally consistent range scan alignment for environment mapping. *Autonomous Robots*, 4:333–349, 1997.
[26] M. Muja and D. Lowe. Fast approximate nearest neighbors with automatic algorithm configuration. In *VISAPP*, 2009.
[27] D. Nistér and H. Stewénius. Scalable recognition with a vocabulary tree. In *CVPR*, 2006.
[28] E. Olson, J. Leonard, and S. Teller. Fast iterative alignment of pose graphs with poor estimates. In *In ICRA*, 2006.
[29] L. Paz, J. Tardós, and J. Neira. Divide and conquer: EKF SLAM in O(n). *IEEE Transactions on Robotics*, 24(5), October 2008.
[30] J. Sivic and A. Zisserman. Video google: A text retrieval approach to object matching in videos. *Computer Vision, IEEE International Conference on*, 2:1470, 2003.
[31] N. Snavely, S. M. Seitz, and R. Szeliski. Skeletal sets for efficient structure from motion. In *Proc. Computer Vision and Pattern Recognition*, 2008.
[32] J. Solà, M. Devy, A. Monin, and T. Lemaire. Undelayed initialization in bearing only slam. In *ICRA*, 2005.
[33] B. Steder, G. Grisetti, C. Stachniss, S. Grzonka, A. Rottmann, and W. Burgard. Learning maps in 3d using attitude and noisy vision sensors. In *IEEE International Conference on Intelligent Robots and Systems (IROS)*, 2007.
[34] R. Unnikrishnan and A. Kelly. A constrained optimization approach to globally consistent mapping. In *Proceedings International Conference on Robotics and Systems (IROS)*, 2002.
[35] B. Williams, G. Klein, and I. Reid. Real-time slam relocalisation. In *ICCV*, 2007.

Generalized-ICP

Aleksandr V. Segal
Stanford University
Email: avsegal@cs.stanford.edu

Dirk Haehnel
Stanford University
Email: haehnel@stanford.edu

Sebastian Thrun
Stanford University
Email: thrun@stanford.edu

Abstract—In this paper we combine the Iterative Closest Point ('ICP') and 'point-to-plane ICP' algorithms into a single probabilistic framework. We then use this framework to model locally planar surface structure from both scans instead of just the "model" scan as is typically done with the point-to-plane method. This can be thought of as 'plane-to-plane.' The new approach is tested with both simulated and real-world data and is shown to outperform both standard ICP and point-to-plane. Furthermore, the new approach is shown to be more robust to incorrect correspondences, and thus makes it easier to tune the maximum match distance parameter present in most variants of ICP. In addition to the demonstrated performance improvement, the proposed model allows for more expressive probabilistic models to be incorporated into the ICP framework. While maintaining the speed and simplicity of ICP, the Generalized-ICP could also allow for the addition of outlier terms, measurement noise, and other probabilistic techniques to increase robustness.

I. INTRODUCTION

Over the last decade, range images have grown in popularity and found increasing applications in fields including medical imaging, object modeling, and robotics. Because of occlusion and limited sensor range, most of these applications require accurate methods of combining multiple range images into a single model. Particularly in mobile robotics, the availability of range sensors capable of quickly capturing an entire 3D scene has drastically improved the state of the art. A striking illustration of this is the fact that virtually all competitors in the DARPA Grand Challenge relied on fast-scanning laser range finders as the primary input method for obstacle avoidance, motion planning, and mapping. Although GPS and IMUs are often used to calculate approximate displacements, they are not accurate enough to reliably produce precise positioning. In addition, there are many situation (tunnels, parking garages, tall buildings) which obstruct GPS reception and further decrease accuracy. To deal with this shortcoming, most applications rely on scan-matching of range data to refine the localization. Despite such wide usage, the typical approach to solving the scan-matching problem has remained largely unchanged since its introduction.

II. SCANMATCHING

Originally applied to scan-matching in the early 90s, the ICP technique has had many variations proposed over the past decade and a half. Three papers published around the same time period outline what is still considered the state of the art solution for scan-matching. The most often cited analysis of the algorithm comes from Besl and McKay[1]. [1] directly addresses registration of 3D shapes described either geometrically or with point clouds. Chen and Medioni[7] considered the more specific problem of aligning range data for object modeling. Their approach takes advantage of the tendency of most range data to be locally planar and introduces the "point-to-plane" variant of ICP. Zhang[5] almost simultaneously describes ICP, but adds a robust method of outlier rejection in the correspondence selection phase of the algorithm.

Two more modern alternatives are Iterative Dual Correspondence [15] and Metric-Based ICP [16]. IDC improves the point-matching process by maintaining two sets of correspondences. MbICP is designed to improve convergence with large initial orientation errors by explicitly putting a measure of rotational error as part of the distance metric to be minimized.

The primary advantages of most ICP based methods are simplicity and relatively quick performance when implemented with kd-trees for closest-point look up. The drawbacks include the implicit assumption of full overlap of the shapes being matched and the theoretical requirement that the points are taken from a known geometric surface rather than measured [1]. The first assumption is violated by partially overlapped scans (taken from different locations). The second causes problems because different discretizations of the physical surface make it impossible to get exact overlap of the individual points even after convergence. Point-to-plane, as suggested in [7], solves the discretization problem by not penalizing offsets along a surface. The full overlap assumption is usually handled by setting a maximum distance threshold in the correspondence.

Aside from point-to-plane, most ICP variations use a closed form solution to iteratively compute the alignment from the correspondences. This is typically done with [10] or similar techniques based on cross-correlation of the two data sets. Recently, there has been interest in the use of generic non-linear optimization techniques instead of the more specific closed form approaches [9]. These techniques are advantageous in that they allow for more generic minimization functions rather then just the sum of euclidean distances. [9] uses non-linear optimization with robust statistics to show a wider basin of convergence.

We argue that among these, the probabilistic techniques are some of the best motivated due to the large amount of theoretical work already in place to support them. [2] applies a probabilistic model by assuming the second scan is generated from the first through a random process. [4] Applies ray tracing techniques to maximize the probability of alignment.

[8] builds a set of compatible correspondences, and then maximizes probability of alignment over this distribution. [17] introduces a fully probabilistic framework which takes into account a motion model and allows estimates of registration uncertainty. An interesting aspect of the approach is that a sampled analog of the Generalized Hough Transform is used to compute alignment without explicit correspondences, taking both surface normals into account for 2D data sets.

There is also a large amount of literature devoted to solving the global alignment problem with multiple scans ([18] and many others). Many approaches to this ([18] in particular) use a pair-wise matching algorithm as a basic component. This makes improvements in pairwise matching applicable to the global alignment problem as well.

Our approach falls somewhere between standard IPC and the fully probabilistic models. It is based on using MLE as the non-linear optimization step, and computing discrete correspondences using kd-trees. It is unique in that it provides symmetry and incorporates the structural assumptions of [7]. Because closest point look up is done with euclidean distance, however, kd-trees can be used to achieve fast performance on large pointclouds. This is typically not possible with fully probabilistic methods as these require computing a MAP estimate over assignments. In contrast to [8], we argue that the data should be assumed to be locally planar since most environments sampled for range data are piecewise smooth surfaces. By giving the minimization processes a probabilistic interpretation, we show that is easy to extend the technique to include structural information from both scans, rather then just one as is typically done in "point-to-plane" ICP. We show that introducing this symmetry improves accuracy and decreases dependence on parameters.

Unlike the IDC [15] and MbICP [16] algorithms, our approach is designed to deal with large 3D pointclouds. Even more fundamentally both of these approaches are somewhat orthogonal to our technique. Although MbICP suggests an alternative distance metric (as do we), our metric aims to take into account structure rather then orientation. Since our technique does not rely on any particular type (or number) of correspondences, it would likely be improved by incorporating a secondary set of correspondences as in IDC.

A key difference between our approach and [17] is the computational complexity involved. [17] is designed to deal with planar scan data – the Generalized Hough Transform suggested requires comparing every point in one scan with every point in the other (or a proportional number of comparisons in the case of sampling). Our approach works with kd-trees for closest point look up and thus requires $O(n \log(n))$ explicit point comparisons. It is not clear how to efficiently generalize the approach in [17] to the datasets considered in this paper. Furthermore, there are philosophical differences in the models.

This paper proceeds by summarizing the ICP and point-to-plane algorithms, and then introducing Generalized-ICP as a natural extension of these two standard approaches. Experimental results are then presented which highlight the advantages of Generalized-ICP.

A. ICP

The key concept of the standard ICP algorithm can be summarized in two steps:

1) compute correspondences between the two scans.
2) compute a transformation which minimizes distance between corresponding points.

Iteratively repeating these two steps typically results in convergence to the desired transformation. Because we are violating the assumption of full overlap, we are forced to add a maximum matching threshold d_{max}. This threshold accounts for the fact that some points will not have any correspondence in the second scan (e.g. points which are outside the boundary of scan A). In most implementations of ICP, the choice of d_{max} represents a trade off between convergence and accuracy. A low value results in bad convergence (the algorithm becomes "short sighted"); a large value causes incorrect correspondences to pull the final alignment away from the correct value. Standard ICP is listed as Alg. 1.

---

input : Two pointclouds: $A = \{a_i\}, B = \{b_i\}$
        An initial transformation: T_0
output: The correct transformation, T, which aligns A
        and B

1   $T \leftarrow T_0$;
2   **while** *not converged* **do**
3   **for** $i \leftarrow 1$ **to** N **do**
4       $m_i \leftarrow \texttt{FindClosestPointInA}(T \cdot b_i)$;
5   **if** $\|m_i - T \cdot b_i\| \leq d_{max}$ **then**
6         $w_i \leftarrow 1$;
7   **else**
8         $w_i \leftarrow 0$;
9   **end**
10   **end**
11    $T \leftarrow \underset{T}{\mathrm{argmin}} \{\sum_i w_i \|T \cdot b_i - m_i\|^2\}$;
12 **end**

Algorithm 1: Standard ICP

B. Point-to-plane

The point-to-plane variant of ICP improves performance by taking advantage of surface normal information. Originally introduced by Chen and Medioni[7], the technique has come into widespread use as a more robust and accurate variant of standard ICP when presented with 2.5D range data. Instead of minimizing $\Sigma \|T \cdot b_i - m_i\|^2$, the point-to-plane algorithm minimizes error along the surface normal (i.e. the projection of $(T \cdot b_i - m_i)$ onto the sub-space spanned by the surface normal). This improvement is implemented by changing line 11 of Alg. 1 as follows:

$$T \leftarrow \underset{T}{\mathrm{argmin}} \{\sum_i w_i \|\eta_i \cdot (T \cdot b_i - m_i)\|^2\}$$

where η_i is the surface normal at m_i.

III. GENERALIZED-ICP

A. Derivation

Generalized-ICP is based on attaching a probabilistic model to the minimization step on line 11 of Alg. 1. The technique keeps the rest of the algorithm unchanged so as to reduce complexity and maintain speed. Notably, correspondences are still computed with the standard Euclidean distance rather then a probabilistic measure. This is done to allow for the use of kd-trees in the look up of closest points and hence maintain the principle advantages of ICP over other fully probabilistic techniques – speed and simplicity.

Since only line 11 is relevant, we limit the scope of the derivation to this context. To simplify notation, we assume that the closest point look up has already been performed and that the two point clouds, $A = \{a_i\}_{i=1,...,N}$ and $B = \{b_i\}_{i=1,...,N}$, are indexed according to their correspondences (i.e. a_i corresponds with b_i). For the purpose of this section, we also assume all correspondences with $||m_i - T \cdot b_i|| > d_{max}$ have been removed from the data.

In the probabilistic model we assume the existence of an underlying set of points, $\hat{A} = \{\hat{a}_i\}$ and $\hat{B} = \{\hat{b}_i\}$, which generate A and B according to $a_i \sim \mathcal{N}(\hat{a}_i, C_i^A)$ and $b_i \sim \mathcal{N}(\hat{b}_i, C_i^B)$. In this case, $\{C_i^A\}$ and $\{C_i^B\}$ are covariance matrices associated with the measured points. If we assume perfect correspondences (geometrically consistent with no errors due to occlusion or sampling), and the correct transformation, T^*, we know that

$$\hat{b}_i = T^* \hat{a}_i \qquad (1)$$

For an arbitrary rigid transformation, T, we define $d_i^{(T)} = b_i - T a_i$, and consider the distribution from which $d_i^{(T^*)}$ is drawn. Since a_i and b_i are assumed to be drawn from independent Gaussians,

$$d_i^{(T^*)} \sim \mathcal{N}(\hat{b}_i - (T^*)\hat{a}_i, C_i^B + (T^*)C_i^A(T^*)^T)$$
$$= \mathcal{N}(0, C_i^B + (T^*)C_i^A(T^*)^T)$$

by applying Eq. (1).

Now we use MLE to iteratively compute T by setting

$$T = \underset{T}{\operatorname{argmax}} \prod_i p(d_i^{(T)}) = \underset{T}{\operatorname{argmax}} \sum_i \log(p(d_i^{(T)}))$$

The above can be simplified to

$$T = \underset{T}{\operatorname{argmin}} \sum_i {d_i^{(T)}}^T (C_i^B + T C_i^A T^T)^{-1} d_i^{(T)} \qquad (2)$$

This defines the key step of the Generalized-ICP algorithm.

The standard ICP algorithm can be seen as a special case by setting

$$C_i^B = I$$
$$C_i^A = 0$$

In this case, (2) becomes

$$T = \underset{T}{\operatorname{argmin}} \sum_i {d_i^{(T)}}^T d_i^{(T)}$$
$$= \underset{T}{\operatorname{argmin}} \sum_i ||d_i^{(T)}||^2 \qquad (3)$$

which is exactly the standard ICP update formula.

With the Generalized-IPC framework in place, however, we have more freedom in modeling the situation; we are free to pick any set of covariances for $\{C_i^A\}$ and $\{C_i^B\}$. As a motivating example, we note that the point-to-plane algorithm can also be thought of probabilistically.

The update step in point-to-plane ICP is performed as:

$$T = \underset{T}{\operatorname{argmin}} \left\{ \sum_i ||P_i \cdot d_i||^2 \right\} \qquad (4)$$

where P_i is the projection onto the span of the surface normal at b_i. This minimizes the distance of $T \cdot a_i$ from the plane defined by b_i and its surface normal. Since P_i is an orthogonal projection matrix, $P_i = P_i^2 = P_i^T$. This means $||P_i \cdot d_i||^2$ can be reformulated as a quadratic form:

$$||P_i \cdot d_i||^2 = (P_i \cdot d_i)^T \cdot (P_i \cdot d_i)$$
$$= d_i^T \cdot P_i \cdot d_i$$

Looking at (4) in this format, we get:

$$T = \underset{T}{\operatorname{argmin}} \left\{ \sum_i d_i^T \cdot P_i \cdot d_i \right\} \qquad (5)$$

Observing the similarity between the above and (2), it can be shown that point-to-plane ICP is a limiting case of Generalized-ICP. In this case

$$C_i^B = P_i^{-1} \qquad (6)$$
$$C_i^A = 0 \qquad (7)$$

Strictly speaking P_i is non-invertible since it is rank deficient. However, if we approximate P_i with an invertible Q_i, Generalized-ICP approaches point-to-plane as $Q_i \to P_i$. We can intuitively interpret this limiting behavior as b_i being constrained along the plane normal vector with nothing known about its location inside the plane itself.

B. Application: plane-to-plane

In order to improve performance relative to point-to-plane and increase the symmetry of the model, Generalized-ICP can be used to take into account surface information from both scans. The most natural way to incorporate this additional structure is to include information about the local surface of the second scan into (7). This captures the intuitive nature of the situation, but is not mathematically feasible since the matrices involved are singular. Instead, we use the intuition of point-to-plane to motivate a probabilistic model.

The insight of the point-to-plane algorithm is that our point cloud has more structure then an arbitrary set of points in 3-space; it is actually a collection of surfaces sampled by a range-measuring sensor. This means we are dealing with

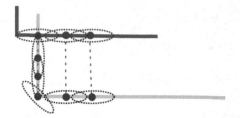

Fig. 1. illustration of plane-to-plane

a sampled 2-manifold in 3-space. Since real-world surfaces are at least piece-wise differentiable, we can assume that our dataset is locally planar. Furthermore, since we are sampling the manifold from two different perspectives, we will not in general sample the exact same point (i.e. the correspondence will never be exact). In essence, every measured point only provides a constraint along its surface normal. To model this structure, we consider each sampled point to be distributed with high covariance along its local plane, and very low covariance in the surface normal direction. In the case of a point with e_1 as its surface normal, the covariance matrix becomes

$$\begin{pmatrix} \epsilon & 0 & 0 \\ 0 & 1 & 0 \\ 0 & 0 & 1 \end{pmatrix}$$

where ϵ is a small constant representing covariance along the normal. This corresponds to knowing the position along the normal with very high confidence, but being unsure about its location in the plane. We model both a_i and b_i as being drawn from this sort of distribution.

Explicitly, given μ_i and ν_i – the respective normal vectors at b_i and a_i – C_i^B and C_i^A are computed by rotating the above covariance matrix so that the ϵ term represents uncertainty along the surface normal. Letting $\mathbf{R}_x$ denote one of the rotations which transform the basis vector $e_1 \rightarrow x$, set

$$C_i^B = \mathbf{R}_{\mu_i} \cdot \begin{pmatrix} \epsilon & 0 & 0 \\ 0 & 1 & 0 \\ 0 & 0 & 1 \end{pmatrix} \cdot \mathbf{R}_{\mu_i}^T$$

$$C_i^A = \mathbf{R}_{\nu_i} \cdot \begin{pmatrix} \epsilon & 0 & 0 \\ 0 & 1 & 0 \\ 0 & 0 & 1 \end{pmatrix} \cdot \mathbf{R}_{\nu_i}^T$$

The transformation, $\mathbf{T}$, is then computed via (2).

Fig. 1 provides an illustration of the effect of the algorithm in an extreme situation. In this case all of the points along the vertical section of the light gray scan are incorrectly associated with a single point in the dark gray scan. Because the surface orientations are inconsistent, plane-to-plane will automatically discount these matches: the final summed covariance matrix of each correspondence will be isotropic and will form a very small contribution to the objective function relative to the thin and sharply defined correspondence covariance matrices. An alternative view of this behavior is as a soft constraint for each correspondence. The inconsistent matches allow the dark gray scan-point to move along the x-axis while the light gray scan-points are free to move along the y-axis. The incorrect correspondences thus form very weak and uninformative constraints for the overall alignment.

Computing the surface covariance matrices requires a surface normal associated with every point in both scans. There are many techniques for recovering surface normals from point clouds, and the accuracy of the normals naturally plays an important role in the performance of the algorithm. In our implementation, we used PCA on the covariance matrix of the 20 closest points to each scan point. In this case the eigenvector associated with the smallest eigenvalue corresponds with the surface normal. This method is used to compute the normals for both point-to-plane and Generalized-ICP. For Generalized-ICP, the rotation matrices are constructed so that the ϵ component of the variance lines up with the surface normal.[1]

IV. RESULTS

We compare all three algorithms to test performance of the proposed technique. Although efficient closed form solutions exist for $\mathbf{T}$ in standard ICP, we implemented the minimization with conjugate gradients to simplify comparison. Performance is analyzed in terms of convergence to the correct solution after a known offset is introduced between the two scans. We limit our tests to a maximum of 250 iterations for standard ICP, and 50 iterations for the other two algorithms since convergence was typically achieved before this point (if at all).

Both simulated (Fig. 3) and real (Fig. 4) data was used in order to demonstrate both theoretical and practical performance. The simulated data set also allowed tests to be performed on a wider range of environments with absolutely known ground truth. The outdoor simulated environment differs from the collected data primarily in the amount of occlusion presented, and in the more hilly features of the ground plane. The real-world outdoor tests also demonstrate performance with more detailed features and more representative measurement noise.

Simulated data was generated by ray-tracing a SICK scanner mounted on a rotating joint. Two 3D environments were created to test performance against absolute ground truth both in the indoor (Fig. 2(a)) and an outdoor (Fig. 2(b)) scenario. The indoor environment was based on an office hallway, while the outdoor setting reflects a typical landscape around a building. In both cases, we simulated a laser-scanner equipped robot traveling along a trajectory and taking measurements at fixed points along the path. Gaussian noise was added to make the tests more realistic.

Tests were also performed on real data from the logs of an instrumented car. The logs included data recorded by a roof-mounted Velodyne range finder as the car made a loop through a suburban environment and were annotated with GPS and IMU data. This made it possible to apply a pairwise constraint-based SLAM technique to generate ground truth

---

[1]In our implementation we compute these transformations by considering the eigen decomposition of the empirical covariance of the 20 closest points, $\hat{\Sigma} = \mathbf{U}\mathbf{D}\mathbf{U}^T$. We then use $\mathbf{U}$ in place of the rotation matrix (in effect replacing $\mathbf{D}$ with $\mathbf{diag}(\epsilon, 1, 1)$ to get the final surface-aligned matrix).

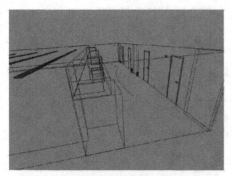

(a) indoor scene

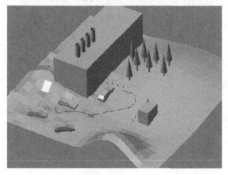

(b) outdoor scene

Fig. 2.   simulated 3D environments

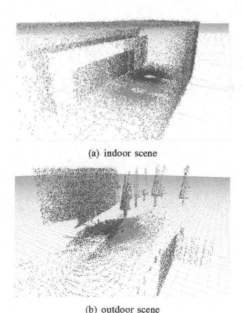

(a) indoor scene

(b) outdoor scene

Fig. 3.   ray-traced scans – scan A is shown in light gray, scan B in dark gray

Fig. 4.   Velodyne scans – scan A is shown in light gray, scan B in dark gray

positioning. Although (standard) ICP itself was used in the pairwise matching to generate the ground truth, the spacing of scans used for the SLAM approach was an order of magnitude smaller. In contrast, the scan pairs used for testing were extracted with much higher spacing (15-20+ meters) in order to pose a much more challenging problem. This is not a perfect method to generate ground truth, but we believe it provides a reasonable baseline to make comparisons between the algorithms.

To measure performance, all algorithms were run on pairs of scans from each of the three data sets. For each scan pair, the initial offset was set to the true offset with a uniformly generated error term added. The error term was set within ±1.5m and $\pm15°$ along all axes. Performance was measured by averaging positioning error over all scan pairs for a particular algorithm. In all cases tested, rotational error was negligible.

As mentioned before, selection of d_{max} plays an important role in the convergence of ICP. Fig. 5 shows the average error for different values of d_{max}; the plot shows average performance across all scan pairs. Fig. 9 shows the averages for individual scan pairs based on ideal values of d_{max}; it demonstrates the distribution of error across the range of scan pairs. In contrast to Fig. 5, the large number off random initial offsets averaged into each data point of Fig. 9 serves to sample the space of possible offsets. For Fig. 5, the algorithms were run on each scan pair with 10 randomly generated starting positions. For the plots in Fig. 9, each data point was generated with 50 random initial poses using best-case values for d_{max}. In all cases, error bars were computed as $\frac{\sigma}{\sqrt{N}}$.

The plots in Fig. 5 show that the proposed algorithm is more robust to choice of the matching threshold and demonstrates better performance in general. This is to be expected since it more completely models the environment and will automatically discount many incorrect matches based on the structure of the scene. In particular, Fig. 5 shows that in the simulated environments, the accuracy of the algorithm is not sensitive to overestimated values of d_{max}. For the real data, Generalized-ICP is still shown to be less sensitive due to the smaller slope of average error as $d_{max} \to \infty$. The discrepancy between simulated and real data can be explained by the difference in their respective frequency profiles. Whereas the simulated environments only have high-level features modeled

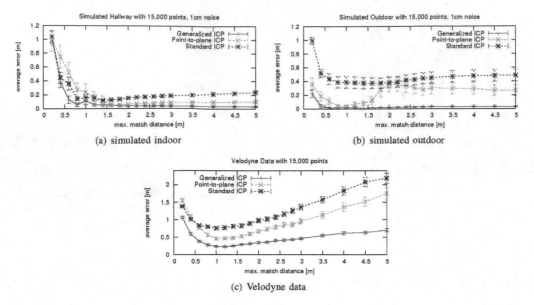

(a) simulated indoor   (b) simulated outdoor

(c) Velodyne data

Fig. 5.   average error as a function of d_{max}

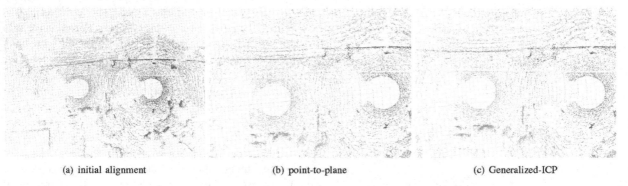

(a) initial alignment   (b) point-to-plane   (c) Generalized-ICP

Fig. 6.   example of results for Velodyne scan pair #31

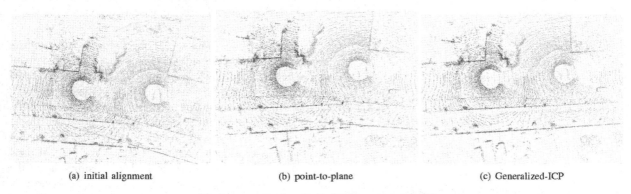

(a) initial alignment   (b) point-to-plane   (c) Generalized-ICP

Fig. 7.   example of results for Velodyne scan pair #45

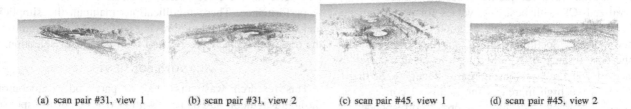

(a) scan pair #31, view 1   (b) scan pair #31, view 2   (c) scan pair #45, view 1   (d) scan pair #45, view 2

Fig. 8. Velodyne scan pairs #31 and #45 shown in perspective to illustrate scene complexity

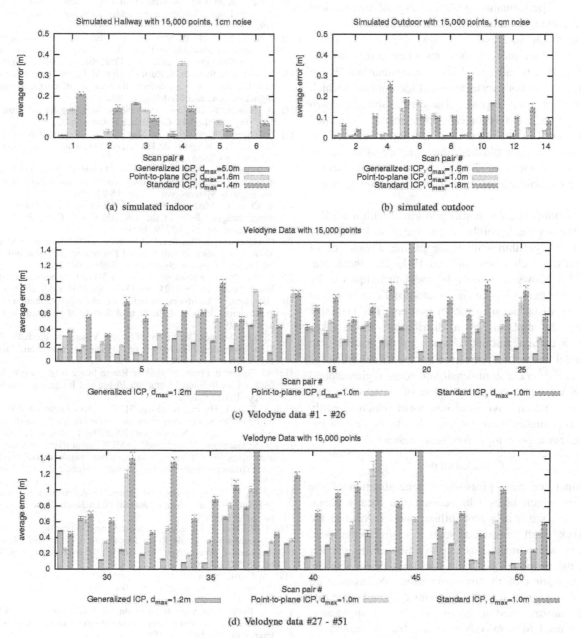

Fig. 9. average error with ideal values of d_{max} which minimize Fig. 5

by hand, the real world data contains much more detailed, high-frequency data. This increases the chances of incorrect correspondences which share a common surface orientation – a situation which is not taken into account by our algorithm. Nonetheless, even when comparing worst-cast values of d_{max} for Generalized-ICP with best-case values for point-to-plane, Generalized-ICP performs roughly as good.

As mentioned in Section II, the d_{max} plays an important role in the performance of ICP. Setting a low value decreases the chance of convergence, but increases accuracy. Setting a value which is too high increases the radius of convergence, but decreases accuracy since more incorrect correspondences are made. The algorithm proposed in this paper heavily reduces the penalty of picking a large value of d_{max} by discounting the effect of incorrect correspondences. This makes it easier to get good performance in a wide range of environment without hand-picking a value of d_{max} for each one.

In addition to the increased accuracy, the new algorithm gives equal consideration to both scans when computing the transformation. Fig. 6 and Fig. 7 show two situations where using the structure of both scans removed local minima which were present with point-to-plane. These represent top-down views of Velodyne scans recorded approximately 30 meters apart and aligned. Fig. 8 shows some additional views of the same scan pairs to better illustrate the structure of the scene. The scans cover a range of 70-100 meters from the sensor in an outdoor environment as seen from a car driving on the road.

Because this minimization is still performed within the ICP framework, the approach combines the speed and simplicity of the standard algorithm with some of the advantages of fully probabilistic techniques such as EM. The theoretical framework also allows standard robustness techniques to be incorporated. For example, the Gaussian kernel can be mixed with a uniform distribution to model outliers. The Gaussian RVs can also be replaced by a distribution which takes into account a certain amount of slack in the matching to explicitly model the inexact correspondences (by assigning the distribution of $d_i^{(\mathbf{T})}$ a constant density on some region around 0). Although we have considered some of these variations, none of them have an obvious closed form which is easily minimized. This makes them too complex to include in the current work, but a good topic for future research.

V. Conclusion

In this paper we have proposed a generalization of the ICP algorithm which takes into account the locally planar structure of both scans in a probabilistic model. Most of the ICP framework is left unmodified so as to maintain the speed and simplicity which make this class of algorithms popular in practice; the proposed generalization only deals with the iterative computation of the transformation. We assume all measured points are drawn from Gaussians centered at the true points which are assumed to be in perfect correspondence. MLE is then used to iteratively estimate transformation for aligning the scans. In a range of both simulated and real-world experiments, Generalized-ICP was shown to increase accuracy. At the same time, the use of structural information from both scans decreased the influence of incorrect correspondences. Consequently the choice of maximum matching distance as a parameter for the correspondence phase becomes less critical to performance. These modifications maintain the simplicity and speed of ICP, while improving performance and removing the trade off typically associated with parameter selection.

Acknowledgment

This research was supported in part under subcontract through Raytheon Sarcos LLC with DARPA as prime sponsor, contract HR0011-04-C-0147.

References

[1] P. Besl, N. McKay. "A Method for Registration of 3-D Shapes," IEEE Trans. on Pattern Analysis and Machine Intel., vol. 14, no. 2, pp. 239-256, 1992.

[2] P. Biber, S. Fleck, W. Strasser. "A Probabilistic Framework for Robust and Accurate Matching of Point Clouds," Pattern Recognition, Lecture Notes in Computer Science, vol. 3175/2004, pp. 280-487, 2004.

[3] N. Gelfan, L. Ikemoto, S. Rusinkiewicz, M. Levoy. "Geometrically Stable Sampling for the ICP Algorithm," Fourth International Conference on 3-D Digital Imaging and Modeling, p. 260, 2003.

[4] D. Haehnel, W. Burgard. "Probabilistic Matching for 3D Scan Registration," Proc. of the VDI-Conference Robotik, 2002.

[5] Z. Zhang. "Iterative Point Matching for Registration of Free-Form Curves," IRA Rapports de Recherche, Programme 4: Robotique, Image et Vision, no. 1658, 1992.

[6] D. Hahnel, W. Burgard, S. Thrun. "Learning compact 3D models of indoor and outdoor environments with a mobile robot," Robotics and Autonomous Systems, vol. 44, pp. 15-27, 2003.

[7] Y. Chen, G. Medioni. "Object Modeling by Registration of Multiple Range Images," Proc. of the 1992 IEEE Intl. Conf. on Robotics and Automation, pp. 2724-2729, 1991.

[8] L. Montesano, J. Minguez, L. Montano. "Probabilistic Scan Matching for Motion Estimation in Unstructured Environments," IEEE Intl. Conf. on. Intelligent Robots and Systems, pp. 3499-3504, 2005.

[9] A. Fitzgibbon. "Robust registration of 3D and 3D point sets," Image and Vision Computing, vol. 21, no. 13-14, pp. 1145-1153, 2003.

[10] B. Horn. "Closed-form solution of absolute orientation using unit quaternions," Journal of the Optical Society of America A, vol. 4, pp. 629-642, 1987.

[11] S. Rusinkiewicz, M. Levoy. "Efficient Variants of the ICP Algorithm," Third International Conference on 3-D Digital Imaging and Modeling, p. 145, 2001.

[12] G. Dalley, P. Flynn. "Pair-Wise Range Image Registration: A Study in Outlier Classification," Computer Vision and Image Understanding, vol. 87, pp. 104-115, 2002.

[13] S. Kim, Y. Hwang, H. Hong, M. Choi. "An Improved ICP Algorithm Based on the Sensor Projection for Automatic 3D Registration," Lecture Notes in Computer Science, vol. 2972/2004 pp. 642-651, 2004.

[14] J.-S. Gutmann, C. Schlegel, "AMOS: comparison of scan matching approaches for self-localization in indoor environments," eurobot, p.61, 1st Euromicro Workshop on Advanced Mobile Robots (EUROBOT), 1996.

[15] F. Lu, E. Milos. "Robot Pose Estimation in Unknown Environments by Matching 2D Range Scans," Journal of Intelligent Robotics Systems 18: pp. 249-275, 1997.

[16] J. Minguez, F. Lamiraux, L. Montesano. "Metric-Based Scan Matching Algorithms for Mobile Robot Displacement Estimation," Robotics and Automation, Proceedings of the 2005 IEEE International Conference on, pp. 3557-3563, 2005.

[17] A. Censi, "Scan matching in a probabilistic framework," Robotics and Automation, Proceedings of the 2006 IEEE International Conference on, pp. 2291-2296, 2006.

[18] K. Pulli, "Mutliview Registration for Large Data Sets," 3-D Digital Imaging and Modeling, 1999. Proceedings. Second International Conference on, pp. 160-168, 1999.

3D Laser Scan Classification Using Web Data and Domain Adaptation

Kevin Lai   Dieter Fox

University of Washington, Department of Computer Science & Engineering, Seattle, WA

Abstract— Over the last years, object recognition has become a more and more active field of research in robotics. An important problem in object recognition is the need for sufficient labeled training data to learn good classifiers. In this paper we show how to significantly reduce the need for manually labeled training data by leveraging data sets available on the World Wide Web. Specifically, we show how to use objects from Google's 3D Warehouse to train classifiers for 3D laser scans collected by a robot navigating through urban environments. In order to deal with the different characteristics of the web data and the real robot data, we additionally use a small set of labeled 3D laser scans and perform *domain adaptation*. Our experiments demonstrate that additional data taken from the 3D Warehouse along with our domain adaptation greatly improves the classification accuracy on real laser scans.

I. INTRODUCTION

In order to navigate safely and efficiently through populated urban environments, autonomous robots must be able to distinguish between objects such as cars, people, buildings, trees, and traffic lights. The ability to identify and reason about objects in their environment is extremely useful for autonomous cars driving on urban streets as well as robots navigating through pedestrian areas or operating in indoor environments. Over the last years, several robotics research groups have developed techniques for classification tasks based on visual and laser range information [22, 1, 7, 21, 15, 17]. A key problem in this context is the availability of sufficient labeled training data to learn classifiers. Typically, this is done by manually labeling data collected by the robot, eventually followed by a procedure to increase the diversity of that data set [17]. However, data labeling is error prone and extremely tedious. We thus conjecture that relying solely on manually labeled data does not scale to the complex environments robots will be deployed in.

The goal of this research is to develop learning techniques that significantly reduce the need for labeled training data for classification tasks in robotics by leveraging data available on the World Wide Web. The computer vision community has recently demonstrated how web-based data sets can be used for various computer vision tasks such as object and scene recognition [16, 14, 20] and scene completion [10]. These techniques take a radically different approach to the computer vision problem; they tackle the complexity of the visual world by using millions of weakly labeled images along with non-parametric techniques instead of parametric, model-based approaches. In robotics, Saxena and colleagues [18] recently used synthetically generated images of objects to learn

Fig. 1.   (Upper row) Part of a 3D laser scan taken in an urban environment (ground plane points shown in dark gray). The scan contains multiple cars, a person, and trees and buildings in the background. (lower row) Example models from Google's 3D Warehouse.

grasp points for manipulation. Their system learned good grasp points solely based on synthetic training data.

Based on these successes, it seems promising to investigate how external data sets can be leveraged to help sensor-based classification tasks in robotics. Unfortunately, this is not as straightforward as it seems. A key problem is the fact that the data available on the World Wide Web is often very different from that collected by a mobile robot. For instance, a robot navigating through an urban environment will often observe cars and people from very close range and angles different from those typically available in data sets such as LabelMe [16]. Furthermore, weather and lighting conditions might differ significantly from web-based images.

The difference between web-based data and real data collected by a robot is even more obvious in the context of classifying 3D laser scan data. Here, we want to use objects from Google's 3D Warehouse to help classification of 3D laser scans collected by a mobile robot navigating through urban terrain (see Fig. 1). The 3D Warehouse dataset [9] contains thousands of 3D models of user-contributed objects such as furniture, cars, buildings, people, vegetation, and street signs. On the one hand, we would like to leverage such an extremely rich source of freely available and labeled training data. On the other hand, virtually all objects in this dataset are generated manually and thus do not accurately reflect the data observed by a 3D laser scanner.

The problem of leveraging large data sets that have different characteristics than the target application is prominent in natural language processing (NLP). Here, text sources from very different topic domains are often combined to help classification. Several relevant techniques have been developed for transfer learning [4] and, more recently, domain adaptation [11, 6, 5]. These techniques use large sets of labeled text from one domain along with a smaller set of labeled text from the target domain to learn a classifier that works well on the target domain.

In this paper we show how domain adaptation can be applied to the problem of 3D laser scan classification. Specifically, the task is to recognize objects in data collected with a 3D Velodyne laser range scanner mounted on a car navigating through an urban environment. The key idea of our approach is to learn a classifier based on objects from Google's 3D Warehouse along with a small set of labeled laser scans. Our classification technique builds on an exemplar-based approach developed for visual object recognition [14]. Instead of labeling individual laser points, our system labels a soup of segments [13] extracted from a laser scan. Each segment is classified based on the labels of exemplars that are "close" to it. Closeness is measured via a learned distance function for spin-image signatures [12, 2] and other shape features. We show how the learning technique can be extended to enable domain adaptation. In the experiments we demonstrate that additional data taken from the 3D Warehouse along with our domain adaptation greatly improves the classification accuracy on real laser scans.

This paper is organized as follows. In the next section, we provide background on exemplar-based learning and on the laser scan segmentation used in our system. Then, in Section III, we show how the exemplar-based technique can be extended to the domain adaptation setting. Section IV introduces a method for probabilistic classification. Experimental results are presented in Section V, followed by a discussion.

II. LEARNING EXEMPLAR-BASED DISTANCE FUNCTIONS FOR 3D LASER SCANS

In this section we review the exemplar-based recognition technique introduced by Malisiewicz and Efros [14]. While the approach was developed for vision-based recognition tasks, we will see that there is a rather natural connection to object recognition in laser scans. In a nutshell, the approach takes a set of labeled segments and learns a distance function for each segment, where the distance function is a linear combination of feature differences. The weights of this function are learned such that the decision boundary maximizes the margin between the associated subset of segments belonging to the same class and segments belonging in other classes. We describe the details of the approach in the context of our 3D laser classification task.

A. Laser Scan Segmentation and Feature Extraction

Fig. 2. (left) Laser points of a car extracted from a 3D scan. (right) Segmentation via mean-shift. The soup of segments additionally contains a merged version of these segments.

Given a 3D laser scan point cloud of a scene, we first segment out points belonging to the ground from points belonging to potential objects of interest. This is done by fitting a ground plane to the scene. To do this, we first bin the points into grid cells of size $25 \times 25 \times 25 cm^3$, and run RANSAC plane fitting on each cell to find the surface orientations of each grid cell. We take only the points belonging to grid cells whose orientations are less than 30 degrees with the horizontal and run RANSAC plane fitting again on all of these points to obtain the final ground plane estimation. The assumption here is that the ground has a slope of less than 30 degrees, which is usually the case and certainly for our urban data set. Laser points close to the ground plane are labeled as ground and not considered in the remainder of our approach. Fig. 1 displays a scan with the automatically extracted ground plane points shown in dark gray.

Since the extent of each object is unknown, we perform segmentation to obtain individual object hypotheses. We experimented with the Mean-Shift [3] and Normalized Cuts [19] algorithms at various parameter settings and found that the former provided better segmentation. In the context of vision-based recognition, Malisiewicz and Efros recently showed that it is beneficial to generate multiple possible segmentations of a scene, rather than relying on a single, possibly faulty segmentation [13]. Similar to their technique, we generate a "soup of segments" using mean-shift clustering and considering merges between clusters of up to 3 neighboring segments. An example segmentation of a car automatically extracted from a complete scan is shown in Fig. 2. The soup also contains a segment resulting from merging the two segments.

We next extract a set of features capturing the shape of a segment. For each laser point, we compute spin image features [12], which are 16×16 matrices describing the local shape around that point. Following the technique introduced by Assfalg and colleagues [2] in the context of object retrieval, we compute for each laser point a spin image signature, which compresses information from its spin image down to an 18-dimensional vector. Representing a segment using the spin image signatures of all its points would be impractical, so the final representation of a segment is composed of a smaller set of spin image signatures. In [2], this final set of signatures is computed by clustering all spin image signatures describing an object. The resulting representation is rotation-invariant,

Fig. 3. (left) Tree model from the 3D Warehouse and (right) point cloud extracted via ray casting.

which is beneficial for object retrieval. However, in our case the objects of concern usually appear in a constrained range of orientations. Cars and trees are unlikely to appear upside down, for example. The orientation of a segment is actually an important distinguishing feature and so unlike in [2], we partition the laser points into a $3 \times 3 \times 3$ grid and perform k-means clustering on the spin image signatures within each grid cell, with a fixed $k = 3$. Thus, we obtain for each segment $3 \cdot 3 \cdot 3 = 27$ shape descriptors of length $3 \cdot 18 = 54$ each. We also include as features the width, depth and height of the segment's bounding box, as well as the segment's minimum height above the ground. This gives us a total of 31 descriptors.

In order to make segments extracted from a 3D laser scan comparable to objects in the 3D-Warehouse, we perform segmentation on a point cloud generated via ray casting on the object (see Fig. 3).

B. Learning the Distance Function

Assume we have a set of n labeled laser segments, $\mathcal{E} = \{e_1, e_2, \ldots, e_n\}$. We refer to these segments as *exemplars*, e, since they serve as examples for the appearance of segments belonging to a certain class. Let $\mathbf{f}_e$ denote the features describing an exemplar e, and let $\mathbf{f}_z$ denote the features of an arbitrary segment z, which could also be an exemplar. $\mathbf{d}_{ez}$ is the vector containing component-wise, L_2 distances between individual features describing e and z: $\mathbf{d}_{ez}[i] = ||\mathbf{f}_e[i] - \mathbf{f}_z[i]||$. In our case, features $\mathbf{f}_e$ and $\mathbf{f}_z$ are the 31 descriptors describing segment e and segment z, respectively. $\mathbf{d}_{ez}$ is a $31 + 1$ dimensional distance vector where each component, i, is the L_2 distance between feature i of segments e and z, with an additional bias term as described in [14]. Distance functions between two segments are linear functions of their distance vector. Each exemplar has its own distance function, D_e, specified by the weight vector $\mathbf{w}_e$:

$$D_e(z) = \mathbf{w}_e \cdot \mathbf{d}_{ez} \qquad (1)$$

To learn the weights of this distance function, it is useful to define a binary vector $\boldsymbol{\alpha}_e$, the length of which is given by the number of exemplars with the same label as e. During learning, $\boldsymbol{\alpha}_e$ is non-zero for those exemplars that are in e's class and that should be similar to e, and zero for those that are considered irrelevant for exemplar e. The key idea behind these vectors is that even within a class, different segments can have very different feature appearance. This could depend, for example, on the angle from which an object is observed.

The values of $\boldsymbol{\alpha}_e$ and $\mathbf{w}_e$ are determined for each exemplar separately by the following optimization:

$$\{\mathbf{w}_e^*, \boldsymbol{\alpha}_e^*\} = \operatorname*{argmin}_{\mathbf{w}_e, \boldsymbol{\alpha}_e} \sum_{i \in \mathcal{C}_e} \alpha_{ei} L(-\mathbf{w}_e \cdot \mathbf{d}_{ei}) + \sum_{i \notin \mathcal{C}_e} L(\mathbf{w}_e \cdot \mathbf{d}_{ei})$$

$$\text{subject to } \mathbf{w}_e \geq 0; \ \alpha_{ei} \in \{0, 1\}; \ \sum_i \alpha_{ei} = K \qquad (2)$$

Here, $\mathcal{C}_e$ is the set of examplars that belong to the same class as e, α_{ei} is the i-th component of $\boldsymbol{\alpha}_e$, and L is an arbitrary positive loss function. The constraints ensure that K values of $\boldsymbol{\alpha}_e$ are non-zero. Intuitively, this ensures that the optimization aims at maximizing the margin of a decision boundary that has K segments from e's class on one side, while keeping exemplars from other classes on the other side. The optimization procedure alternates between two steps. The $\boldsymbol{\alpha}_e$ vector in the k-th iteration is chosen such that it minimizes the first sum in (2):

$$\boldsymbol{\alpha}_e^k = \operatorname*{argmin}_{\boldsymbol{\alpha}_e} \sum_{i \in \mathcal{C}_e} \alpha_{ei} L(-\mathbf{w}_e^k \cdot \mathbf{d}_{ei}) \qquad (3)$$

This is done by simply setting α_e^k to 1 for the K smallest values of $L(-\mathbf{w}_e \cdot \mathbf{d}_{ei})$, and setting it to zero otherwise. The next step fixes $\boldsymbol{\alpha}_e$ to $\boldsymbol{\alpha}_e^k$ and optimizes (2) to yield the new $\mathbf{w}_e^{k+1}$:

$$\mathbf{w}_e^{k+1} = \operatorname*{argmin}_{\mathbf{w}_e} \sum_{i: \in \mathcal{C}_e} \alpha_{ei}^k L(-\mathbf{w}_e \cdot \mathbf{d}_{ei}) + \sum_{i \notin \mathcal{C}_e} L(\mathbf{w}_e \cdot \mathbf{d}_{ei}) \qquad (4)$$

When choosing the loss function L to be the square hinge-loss function, this optimization yields standard Support Vector Machine learning. The iterative procedure converges when $\boldsymbol{\alpha}_e^k = \boldsymbol{\alpha}_e^{k+1}$.

Malisiewicz and Efros showed that the learned distance functions provide excellent recognition results for image segments [14].

III. DOMAIN ADAPTATION

So far, the approach assumes that the exemplars in the training set $\mathcal{E}$ are drawn from the same distribution as the segments on which the approach will be applied. While this worked well for Malisiewicz and Efros, it does not perform well when training and test domain are significantly different. In our scenario, for example, the classification is applied to segments extracted from 3D laser scans, while most of the training data is extracted from the 3D-Warehouse data set. As we will show in the experimental results, combining training data from both domains can improve classification over just using data from either domain, but this performance gain cannot be achieved by simply combining data from the two domains into a single training set.

In general, we distinguish between two domains. The first one, the *target domain*, is the domain on which the classifier will be applied after training. The second domain, the *source domain*, differs from the target domain but provides additional data that can help to learn a good classifier for the target domain. In our context, the training data now consists of exemplars chosen from these two domains: $\mathcal{E} = \mathcal{E}^t \cup \mathcal{E}^s$.

Here, $\mathcal{E}^t$ contains exemplars from the target domain, that is, labeled segments extracted from the real laser data. $\mathcal{E}^s$ contains segments extracted from the 3D-Warehouse. As typical in domain adaptation, we assume that we have substantially more labeled data from the source domain than from the target domain: $|\mathcal{E}^s| \gg |\mathcal{E}^t|$. We now describe two methods of domain adaptation in the context of the exemplar-based learning technique.

A. Domain Adaptation via Feature Augmentation

Daume introduced feature augmentation as a general approach to domain adaptation [5]. It is extremely easy to implement and has been shown to outperform various other domain adaptation techniques and to perform as well as the thus far most successful approach to domain adaptation [6]. The approach performs adaptation by generating a stacked feature vector from the original features used by the underlying learning technique. Specifically, let $\mathbf{f}_e$ be the feature vector describing exemplar e. Daume's approach generates a stacked vector $\mathbf{f}_e^*$ as follows:

$$\mathbf{f}_e^* = \begin{pmatrix} \mathbf{f}_e \\ \mathbf{f}_e^s \\ \mathbf{f}_e^t \end{pmatrix} \qquad (5)$$

Here, $\mathbf{f}_e^s = \mathbf{f}_e$ if e belongs to the source domain, and $\mathbf{f}_e^s = \mathbf{0}$ if it belongs to the target domain. Similarly, $\mathbf{f}_e^t = \mathbf{f}_e$ if e belongs to the target domain, and $\mathbf{f}_e^t = \mathbf{0}$ otherwise. Using the stacked feature vector, it becomes clear that exemplars from the same domain are automatically closer to each other in feature space than exemplars from different domains. Daume argued that this approach works well since data points from the target domain have more influence than source domain points when making predictions about test data.

B. Domain Adaption for Exemplar-based Learning

We now present a method for domain adaptation specifically designed for the exemplar-based learning approach. The key difference between our domain adaptation technique and the single domain approach described in Section II lies in the specification of the binary vector $\boldsymbol{\alpha}_e$. Instead of treating all exemplars in the class of e the same way, we distinguish between exemplars in the source and the target domain. Specifically, we use the binary vectors $\boldsymbol{\alpha}_e^s$ and $\boldsymbol{\alpha}_e^t$ for the exemplars in these two domains. The domain adaptation objective becomes

$$\{\mathbf{w}_e^*, \boldsymbol{\alpha}_e^{s*}, \boldsymbol{\alpha}_e^{t*}\} = \operatorname*{argmin}_{\mathbf{w}_e, \boldsymbol{\alpha}_e^s, \boldsymbol{\alpha}_e^t}$$
$$\sum_{i \in \mathcal{C}_e^s} \alpha_{ei}^s L(-\mathbf{w}_e \cdot \mathbf{d}_{ei}) + \sum_{i \in \mathcal{C}_e^t} \alpha_{ei}^t L(-\mathbf{w}_e \cdot \mathbf{d}_{ei}) +$$
$$\sum_{i \notin \mathcal{C}_e} L(\mathbf{w}_e \cdot \mathbf{d}_{ei}), \qquad (6)$$

where $\mathcal{C}_e^s$ and $\mathcal{C}_e^t$ are the source and target domain exemplars with the same label as e. The constraints are virtually identical to those for the single domain objective (2), with the constraints on the vectors becoming $\sum_i \alpha_{ei}^s = K^s$ and $\sum_i \alpha_{ei}^t = K^t$. The values for K^s and K^t give the number of source and target exemplars that must be considered during the optimization.

The subtle difference between (6) and (2) has a substantial effect on the learned distance function. To see this, imagine the case where we train the distance function of an exemplar from the source domain. Naturally, this exemplar will be closer to source domain exemplars from the same class than to target domain exemplars from that class. In the extreme case, the vectors determined via (3) will contain 1s only for source domain exemplars, while they are zero for all target domain exemplars. The single domain training algorithm will thus not take target domain exemplars into account and learn distance functions for source domain exemplars that are good in classifying source domain data. There is no incentive to make them classify target domain exemplars well. By keeping two different α-vectors, we can force the algorithm to optimize for classification on the target domain as well. The values for K^s and K^t allow us to trade off the impact of target and source domain data. They are determined via grid search using cross-validation, where the values that maximize the area under the precision-recall curve are chosen.

The learning algorithm is extremely similar to the single domain algorithm. In the k-th iteration, optimization of the α-vectors is done by setting $\alpha_e^{s\ k}$ and $\alpha_e^{t\ k}$ to 1 for the exemplars yielding the K^s and K^t smallest loss values, respectively. Then, the weights $\mathbf{w}_e^{k+1}$ are determined via convex SVM optimization using the most recent α-vectors within (6).

IV. PROBABILISTIC CLASSIFICATION

To determine the class of a new segment, z, Malisiewicz and Efros determine all exemplars e for which $\mathbf{d}_{ez} \leq 1$ and then choose the majority among the classes of these exemplars. However, this approach does not model the reliability of individual exemplars and does not lend itself naturally to a probabilistic interpretation. Furthermore, it does not take into account that the target domain is different from the source domain.

To overcome these limitations, we choose the following naïve Bayes model over exemplars. For each exemplar e and each segment class c we compute the probability p_{ec} that the distance $\mathbf{d}_{ez}$ between the exemplar and a segment from that class is less than 1:

$$p_{ec} := p(\mathbf{d}_{ez} \leq 1 \mid C(z) = c) \qquad (7)$$

Here, $C(z)$ is the class of segment z. Since the ultimate goal is to label segments from the target domain only, we estimate this probability solely based on the labeled segments from the target domain. Specifically, p_{ec} is determined by counting all segments z in $\mathcal{E}^t$ that belong to class c *and* that are close to e, that is, for which $\mathbf{d}_{ez} \leq 1$. Normalization with the total number of target domain segments in class c gives the desired probability.

Assuming independence among the distances to all exemplars given the class of a segment z, the probability distribution over z's class can now be computed as

$$p(C(z) = c \mid \mathcal{E}) \propto p(C(z) = c) \prod_{e \in \mathcal{E}, \mathbf{d}_{ez} \leq 1} p_{ec} \prod_{e \in \mathcal{E}, \mathbf{d}_{ez} > 1} (1 - p_{ec})$$

$$(8)$$

where $p(C(z) = c)$ is estimated via class frequencies in the target domain data. We found experimentally that using eq. 8 as described, where it includes influence from both associated ($d_{ez} \leq 1$) and unassociated ($d_{ez} > 1$) exemplars, led to worse results than including just the associated exemplars. This is because there are many more unassociated exemplars than associated ones, and so they have undue influence over the probability. We instead compute the *positive support* using just the associated exemplars.

We can apply the results of segment classification to individual laser points. As described in Section II-A, we extract a soup of segments from a 3D laser scan. Thus, each laser point might be associated to multiple segments. Using the probability distributions over the classes of these segments (with *positive support* only), the distribution over the class of a single laser point l is given by

$$p(C(l) = c \mid \mathcal{E}) \propto p(C(z) = c) \prod_{z \in Z_l} \prod_{e \in \mathcal{E}, \mathbf{d}_{ez} \leq 1} p_{ec} \quad (9)$$

where Z_l is the set of segments associated with point l. In our setup, a test segment is assigned to the class with the highest probability.

V. EXPERIMENTAL RESULTS

We evaluate different approaches to 3D laser scan classification based on real laser scans and objects collected from the Google 3D Warehouse. The task is to classify laser points into the following seven classes: cars, people, trees, street signs, fences, buildings, and background. Our experiments demonstrate that both domain adaptation methods lead to improvements over approaches without domain adaptation and alternatives including LogitBoost. In particular, our exemplar-based domain adaptation approach obtains the best performance.

A. Data Set

We evaluated our approach using models from Google 3D Warehouse as our source domain set, $\mathcal{E}^s$, and ten labeled scans of real street scenes as our target domain set, $\mathcal{E}^t$. The ten real scans, collected by a vehicle navigating through Boston, were chosen such that they did not overlap spatially. Labeling of these scans was done by inspecting camera data collected along with the laser data. We automatically downloaded the first 100 models of each of cars, people, trees, street signs, fences and buildings from Google 3D Warehouse and manually pruned out low quality models, leaving around 50 models for each class. We also included a number of models to serve as the background class, consisting of various other objects that commonly appear in street scenes, such as garbage cans, traffic barrels and fire hydrants. We generated 10 simulated laser scans from different viewpoints around each of the downloaded models, giving us a total of around 3200

exemplars in the source domain set. The ten labeled scans totaled to around 400 exemplars in the six actual object classes. We generate a "soup of segments" from these exemplars, using the data points in real scans not belonging to the six actual classes as candidates for additional background class exemplars. After this process, we obtain a total of 4,900 source domain segments and 2,400 target domain segments.

B. Comparison with Alternative Approaches

We compare the classification performance of our exemplar-based domain adaptation approach to several approaches, including training the single domain exemplar-based technique only on Warehouse exemplars, training it only on the real scans, and training it on a mix of Warehouse objects and labeled scans. The last combination can be viewed as a naïve form of domain adaptation. We also tested Daume's feature augmentation approach to domain adaptation. Our software is based on the implementation provided by Malisiewicz.

The optimal K values (length of the α vectors) for each approach were determined separately using grid search and cross validation. Where training involves using real scans, we repeated each experiment 10 times using random train/test splits of the 10 total available scans. Each labeled scan contains around 240 segments on average.

The results are summarized in Fig. 4. Here the probabilistic classification described in Section IV was used and the precision-recall curves are generated by varying the probabilistic classification threshold between $[0.5, 1]$. The precision and recall values are calculated on a per-laser-point basis. Each curve corresponds to a different experimental setup. The left plot shows the approaches trained on five real laser scans, while the right plot shows the approaches trained on three real laser scans. All approaches are tested on real laser scans only. 3DW stands for exemplars from the 3D-Warehouse, and Real stands for exemplars extracted from real laser scans. Note that since the first setup (3DW) does not use real laser scans, the curves for this approach on the two plots are identical. Where exemplars from both the 3D-Warehouse and real scans are used, we also specify the domain adaptation technique used. By Simple we denote the naïve adaptation of only mixing real and Warehouse data. Stacked refers to Daume's stacked feature approach, applied to the single domain exemplar technique. Finally, Alpha is our technique.

It comes as no surprise that training on Warehouse exemplars only performs worst. This result confirms the fact that the two domains actually have rather different characteristics. For instance, the windshields of cars are invisible to the real laser scanner, thereby causing a large hole in the object segment. In Warehouse cars, however, the windshields are considered solid, causing a locally very different point cloud. Also, Warehouse models, created largely by casual hobbyists, tend to be composed of simple geometric primitives, while the shape of objects from real laser scans can be both more complex and more noisy.

Somewhat surprisingly, the naïve approach of training on a mix of both Warehouse and real scans leads to worse

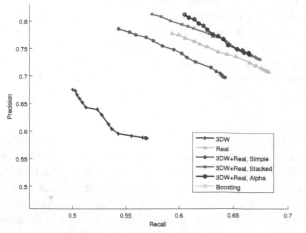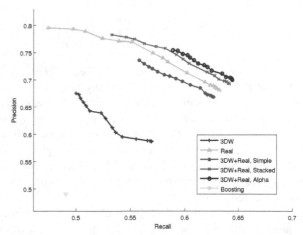

Fig. 4. Precision-recall curves comparing. (left) Performance of the various approaches, trained five real scans where applicable. (right) Performance of the various approaches, trained on three real scans where applicable.

performance than just training on real scans alone. This shows that domain adaptation is indeed necessary when incorporating training data from multiple domains. Both domain adaptation approaches outperform the approaches without domain adaptation. Our exemplar-based approach outperforms Daume's feature augmentation approach when target domain training data is very scarce (when trained with only 3 real scans).

To gauge the overall difficulty of the classification task, we also trained a LogitBoost [8] classifier on the mix of Warehouse and real scans. LogitBoost achieved a maximum F-score of 0.48 when trained on five scans, and a maximum F-score of 0.49 when trained on three scans (see Fig. 4). The F-score is the harmonic mean between precision and recall: $F = 2 \cdot Precision \cdot Recall/(Precision + Recall)$. As a comparison, our approach achieves an F-score of 0.70 when trained on five scans and 0.67 when trained on three scans. The inferior results achieved by LogitBoost demonstrate that this is not a trivial classification problem and that the exemplar-based approach is an extremely promising technique for 3D laser scan classification. Our approach has an overall accuracy of 0.57 for cars, 0.31 for people 0.55 for trees, 0.35 for street signs, 0.32 for fences and 0.73 for buildings.

C. Feature Selection and Thresholding Comparisons

To verify that all of the selected features contribute to the success of our approach, we also compared the performance of our approach using three different sets of features. We looked at using just bounding box dimensions and the minimum height off the ground (dimensions only), adding in the original, rotation-invariant Spin Image Signatures as described in [2] (Original Spin Signatures + dimensions), and adding in our $3 \times 3 \times 3$ grid of Spin Image Signatures (Grid Spin Signatures + dimensions). When trained on 3 scans using dimensions features only, our approach achieves a maximum F-score of 0.63. Using Original Spin Signatures + dimensions, we achieved an F-score of 0.64. Finally, using Grid Spin Signatures and dimensions achieved an F-score of 0.67. Due to noise and occlusions in the scans, as well as imperfect segmentation,

the classes are not easily separable just based on dimensions. Also, our Grid Spin Image Signature features perform better than the original, rotation-invariant, Spin Image Signatures, justifying our modification to remove their rotation-invariance.

We also compared our probabilistic classification approach to the recognition confidence scoring method described by Malisiewicz in [14] and found that the precision-recall curves generated by probabilistic classification attain recalls between $30 - 50$ percentage points above recognition confidence scoring for corresponding precision values.

D. Examples

Fig. 5 provides examples of exemplars matched to the three laser segments shown in the panels in the left column. The top row gives ordered matches for the car segment on the left, the middle and bottom row show matches for a person and tree segment, respectively. As can be seen, the segments extracted from the real scans are successfully matched against segments from both domains, real and Warehouse. The person is mis-matched with one object from the background class "other" (second row, third column). Part of a laser scan and its ground truth labeling is shown in Fig. 6, along with the labeling achieved by our approach.

VI. CONCLUSION

The computer vision community has recently shown that using large sets of weakly labeled image data can help tremendously to deal with the complexity of the visual world. When trying to leverage large data sets to help classification tasks in robotics, one main obstacle is that data collected by a mobile robot typically has very different characteristics from data available on the World Wide Web, for example. For instance, our experiments show that simply adding Google 3D Warehouse objects when training 3D laser scan classifiers can *decrease* the accuracy of the resulting classifier.

In this paper we presented a domain adaptation approach that overcomes this problem. Our technique is based on an exemplar learning approach developed in the context of image-based classification [14]. We showed how this approach can be

Fig. 5. Exemplar matches. The leftmost column shows example segments extracted from 3D laser scans: car, person, tree (top to bottom). Second to last columns show exemplars with distance below threshold, closer exemplars are further to the left.

applied to 3D laser scan data and be extended to the domain adaptation setting. For each laser scan, we generate a "soup of segments" in order to generate multiple possible segmentations of the scan. The experimental results show that our domain adaptation improves the classification accuracy of the original exemplar-based approach. Furthermore, our approach clearly outperformed a boosting technique trained on the same data.

There are several areas that warrant further research. First, we classified laser data solely based on shape. While adding other sensor modalities is conceptually straightforward, we believe that the accuracy of our approach can be greatly improved by adding visual information. Here, we might also be able to leverage additional data bases on the Web. We only distinguish between six main object classes and treat all other segments as belonging to a background class. Obviously, a realistic application requires us to add more classes, for example distinguishing different kinds of street signs. So far, we only used small sets of objects extracted from the 3D Warehouse. A key question will be how to incorporate many thousands of objects for both outdoor and indoor object classification. Finally, our current implementation is far from being real time. In particular, the scan segmentation and spin image feature generation take up large amounts of time. An efficient implementation and the choice of more efficient features will be a key part of future research. Despite all these shortcomings, however, we believe that this work is a promising first step toward robust many-class object recognition for mobile robots.

ACKNOWLEDGMENTS

We would like to thank Michael Beetz for the initial idea to use 3D-Warehouse data for object classification, and Albert Huang for providing us with the urban driving data set. This work was supported in part by a ONR MURI grant number N00014-07-1-0749, by the National Science Foundation under Grant No. 0812671, and by a postgraduate scholarship from the Natural Sciences and Engineering Research Council of Canada. Any opinions, findings, and conclusions or recommendations expressed in this material are those of the authors and do not necessarily reflect the views of the National Science Foundation.

REFERENCES

[1] D. Anguelov, B. Taskar, V. Chatalbashev, D. Koller, D Gupta, G. Heitz, and A. Ng. Discriminative learning of Markov random fields for segmentation of 3D scan data. In *Proc. of the IEEE Computer Society Conference on Computer Vision and Pattern Recognition (CVPR)*, 2005.

[2] J. Assfalg, M. Bertini, A. Del Bimbo, and P. Pala. Content-based retrieval of 3-D objects using spin image signatures. *IEEE Transactions on Multimedia*, 9(3), 2007.

[3] D. Comaniciu and P. Meer. Mean shift: A robust approach toward feature space analysis. *IEEE Transactions on Pattern Analysis and Machine Intelligence (PAMI)*, 24(5), 2002.

[4] W. Dai, Q. Yang, G. Xue, and Y. Yu. Boosting for transfer learning. In *Proc. of the International Conference on Machine Learning (ICML)*, 2007.

[5] H. Daumé. Frustratingly easy domain adaptation. In *Proc. of the Annual Meeting of the Association for Computational Linguistics (ACL)*, 2007.

[6] H. Daumé and D. Marcu. Domain adaptation for statistical classifiers. *Journal of Artificial Intelligence Research (JAIR)*, 26, 2006.

[7] B. Douillard, D. Fox, and F. Ramos. Laser and vision based outdoor object mapping. In *Proc. of Robotics: Science and Systems (RSS)*, 2008.

[8] J. Friedman, T. Hastie, and R. Tibshirani. Additive logistic regression: A statistical view of boosting. *The Annals of Statistics*, 28(2), 2000.

[9] Google. 3d warehouse. http://sketchup.google.com/3dwarehouse/.

[10] J. Hays and A. Efros. Scene completion using millions of photographs. *ACM Transactions on Graphics (Proc. of SIGGRAPH)*, 26(3), 2007.

[11] J. Jiang and C. Zhai. Instance weighting for domain adaptation in NLP. In *Proc. of the Annual Meeting of the Association for Computational Linguistics (ACL)*, 2007.

[12] A. Johnson and M. Hebert. Using spin images for efficient object recognition in cluttered 3D scenes. *IEEE Transactions on Pattern Analysis and Machine Intelligence (PAMI)*, 21(5), 1999.

[13] T. Malisiewicz and A. Efros. Improving spatial support for objects via multiple segmentations. In *Proc. of the British Machine Vision Conference*, 2007.

[14] T. Malisiewicz and A. Efros. Recognition by association via learning per-examplar distances. In *Proc. of the IEEE Computer Society Conference on Computer Vision and Pattern Recognition (CVPR)*, 2008.

Fig. 6. (top) Ground truth classification for part of a 3D laser scan. Shades indicate ground plane and object types (from dark to light): ground, person, building, car, tree, street sign, other, and unclassified. (bottom) Classification achieved by our approach. As can be seen, most of the objects are classified correctly. The street signs in the back and the car near the center are not labeled since they are not close enough to any exemplar.

[15] I. Posner, M. Cummins, and P. Newman. Fast probabilistic labeling of city maps. In *Proc. of Robotics: Science and Systems (RSS)*, 2008.

[16] B. Russell, K. Torralba, A. Murphy, and W. Freeman. Labelme: a database and web-based tool for image annotation. *International Journal of Computer Vision*, 77(1-3), 2008.

[17] B. Sapp, A. Saxena, and A. Ng. A fast data collection and augmentation procedure for object recognition. In *Proc. of the National Conference on Artificial Intelligence (AAAI)*, 2008.

[18] A. Saxena, J. Driemeyer, and A. Ng. Robotic grasping of novel objects using vision. *International Journal of Robotics Research*, 27(2), 2008.

[19] J. Shi and J. Malik. Normalized cuts and image segmentation. *IEEE Transactions on Pattern Analysis and Machine Intelligence (PAMI)*,

22(8), 2000.

[20] A. Torralba, R. Fergus, and W. Freeman. 80 million tiny images: a large dataset for non-parametric object and scene recognition. *IEEE Transactions on Pattern Analysis and Machine Intelligence (PAMI)*, 30(11), 2008.

[21] R. Triebel, R. Schmidt, O. Martinez Mozos, and W. Burgard. Instance-based amn classification for improved object recognition in 2d and 3d laser range data. In *Proc. of the International Joint Conference on Artificial Intelligence (IJCAI)*, 2007.

[22] C. Wellington, A. Courville, and T. Stentz. Interacting Markov random fields for simultaneous terrain modeling and obstacle detection. In *Proc. of Robotics: Science and Systems (RSS)*, 2005.

Adaptive Relative Bundle Adjustment

Gabe Sibley, Christopher Mei, Ian Reid, Paul Newman
Department of Engineering Science
University of Oxford, OX1 3PJ, Oxford, UK
{gsibley,cmei,ian,pnewman}@robots.ox.ac.uk

Abstract—It is well known that bundle adjustment is the optimal non-linear least-squares formulation of the simultaneous localization and mapping problem, in that its maximum likelihood form matches the definition of the Cramer Rao Lower Bound. Unfortunately, computing the ML solution is often prohibitively expensive – this is especially true during loop closures, which often necessitate adjusting all parameters in a loop. In this paper we note that it is precisely the choice of a single privileged coordinate frame that makes bundle adjustment costly, and that this expense can be avoided by adopting a completely relative approach. We derive a new relative bundle adjustment, which instead of optimizing in a single Euclidean space, works in a metric-space defined by a manifold. Using an adaptive optimization strategy, we show experimentally that it is possible to solve for the full ML solution incrementally in constant time – even at loop closure. Our system also operates online in real-time using stereo data, with fast appearance-based loop closure detection. We show results for sequences of 23k frames over 1.08km that indicate the accuracy of the approach.

I. INTRODUCTION

Bundle adjustment is the optimal solution to the so-called "full" simultaneous localization and mapping problem, in that it solves for the maximum likelihood solution given all measurements over all time. The goal in bundle adjustment is to minimize error between observed and predicted image-measurements of n 3D landmarks sensed from m 6D sensor poses (or frames). Measurements and parameter estimates are usually considered to be normally distributed, and the problem is typically tackled with non-linear least-squares optimization routines like Levenberg–Marquardt or the Gauss-Newton method. The linearized system matrix that appears in this process matches the form of the Fisher Information matrix, which in turn defines the Cramer Rao Lower Bound that is used to assess estimator consistency and optimality. It is not surprising therefore that bundle adjustment is the optimal non-linear least-squares simultaneous localization and mapping algorithm.

The cost of optimizing the bundle adjustment objective-function is cubic in complexity (in either m or n). For large and growing problems, this can quickly become prohibitive. This is especially true during loop-closure, when often all parameters in the loop must be adjusted. In a single coordinate frame, the farther the robot travels from the origin, the larger position uncertainty becomes. Errors at loop closure can therefore become arbitrarily large, which in turn makes it impossible to compute the *full* maximum likelihood solution in constant time (here the "full" solution is the one that finds the optimal estimates for all parameters).

It is not clear that it is necessary to estimate everything in a single coordinate frame – for instance most problems of autonomous navigation, such as path planning, obstacle avoidance or object manipulation, can be addressed within the confines of a metric manifold. Taking this route, we structure the problem as a graph of relative poses with landmarks specified in relation to these poses. In 3D this graph defines a connected Riemannian manifold with a distance metric based on shortest paths. Notice that this is not a sub-mapping approach, as there are no distinct overlapping estimates, and there is only one objective function with a *minimal* parameter vector; similarly, this is not a pose-graph relaxation approach, as it solves for landmark structure as well.

Together with an adaptive optimization scheme that only ever solves for a small sub-portion of the state vector, we find evidence that the full maximum likelihood solution in the manifold can be found using an incrementally constant time algorithm. Crucially, this appears true *even at loop closure*. We stress at the outset that the relative solution is not equivalent to the normal Euclidean-space solution and it does not produce an estimate that can be easily embedded in a single Euclidean frame. Converting from the relative manifold into a single Euclidean space is a difficult problem that we argue is best handled by external resources that do not have constant run-time requirements - e.g. by operator computers, not on the robot.

In the next section we describe the related literature. In Section III we derive the new relative objective function. Results from simulation and initial results on real sequences are presented in Section IV. We conclude with a discussion of the pros and cons of the relative approach.

II. RELATED WORK

There has been much interest in Gaussian non-linear least-squares solutions to SLAM based on "full-SLAM" or bundle adjustment [29][31][8][12][19], though the problem is an old one [3][22]. The full SLAM problem tries to optimize the joint vehicle trajectory and map structure simultaneously given all measurements ever made. There are approximate incremental solutions that only optimize a small local subset of the map [7], and there are methods that approximate the full solution with various forms of marginalization [19][27], or by ignoring small dependency information [30][21]. Recently some have successfully employed techniques from the linear algebra and numerical optimization communities to greatly reduce the cost of finding the full solution [17]. Many use key-frames to reduce complexity, though at the expense of accuracy

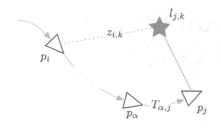

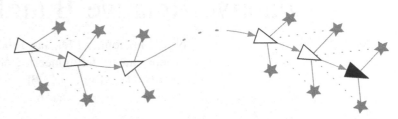

(a) Notation.   (b) Relative graph representation. Landmarks relative to "base-frames", each frame is stored relative to it's "parent".

Figure 1. (a) Notation for a simple trajectory: poses are indicated with triangles, landmarks with stars. Landmark base-frames are indicated with solid lines — e.g. here landmark k is stored relative to frame j. Each inter-pose edge in the graph includes an error-state transform defined about $t_j = 0$ — that is, $T_{\alpha,j}=\hat{T}_{\alpha,j}T_{(t_j)}$, where $\hat{T}_{\alpha,j}$ is the current estimate of the relative transform between frame α and frame j. Notice that predicting the measurement $z_{i,k}$ of landmark k in frame i will rely on all parameters in the kinematic chain from p_j to p_i. Figure (b) shows how landmarks are stored relative to the poses; clearly there is no reference to a privileged global coordinate frame.

[10][23][18]. All these techniques suffer from computational complexity issues during loop closures.

In the context of long-term autonomy, roboticists recognize the need for online, real-time, navigation and mapping algorithms. This means that localization and mapping algorithms must operate incrementally within a constant-time budget. Driven by this need, many authors have recognized the benefit of relative representations [2][9][19][1][15][4][13][20]. The most common solution is probably sub-mapping [2][25][6][9], which breaks the estimation into many smaller mapping regions, computes individual solutions for each region, and then estimates the relationships between these sub-maps. Many difficult issues arise in sub-mapping, including map overlap, data duplication, map fusion and breaking, map alignment, optimal sub-map size, and consistent global estimation in a single Euclidean frame. The relative bundle adjustment we propose can be seen as a *continuous* sub-mapping approach that avoids these complications.

To solve large SLAM problems with many loops, the most successful methods currently are the pose-graph optimization algorithms. Instead of solving the full SLAM problem, these methods optimize a set of relative pose constraints [24][14]. This is attractive because using forward substitution it is possible to transform full SLAM into a generally sparse set of pose constraints [11][29], and even to make the resulting system of equations relative [19]. Note that, given the assumed Gaussian problem structure, this kind of forward substitution to a pose-graph is algebraically equivalent to marginalization; methods that marginalize landmark parameters onto pose parameters so as to define a pose-graph are executing the forward substitution phase of sparse bundle adjustment. In this light, pose-graph relaxation, which solves for the optimal path estimate, can be seen as one-half of one iteration of full SLAM, because full SLAM also back-substitutes for the map parameters, and iterates the procedure to convergence. Like other methods, pose-graph solvers have worst-case complexity at loop closure that is dependent on the length of the loop.

The work most similar to relative bundle adjustment is the relative formulations given by Eade [9] and Konolige [19]. The former is akin to sub-mapping methods with constraints to enforce global Euclidean consistency at loop closure; the latter formulates the cost function relative to a single Euclidean frame and then makes a series of approximations to produce a sparse relative pose-graph. Neither method derives the purely relative objective function (incrementally, both rely on some form of single-reference frame), neither formulates the objective function completely without privileged frames, and both methods carry the burden of finding a globally consistent estimate in a single Euclidean frame. Our approach is substantially different because of the completely relative underlying objective function that we derive.

Finally, a number of adaptive region approaches have been explored within the privileged Euclidean frame paradigm [28][26]. These techniques, together with all of the methods presented in this section, are not constant time at loop closure, and all but one [2] solve for a solution in a single Euclidean space. We find that using adaptive region estimation in conjunction with the relative formulation is the key that enables constant time operation.

III. METHODS

Instead of optimizing an objective function parameterized in a single privileged coordinate frame, we now derive a completely relative formulation.

A. Problem Formulation

Bundle adjustment seeks to minimize error between the observed and predicted measurements of n landmarks sensed from m sensor poses (or frames). Likewise we minimize the difference between predicted and measured values. Let $l_{j,k}$, $k \in 1,...,n$, $j \in 1,...,m$ be a set of n 3D landmarks each parameterized relative to some *base-frame* j. Let t_j, $j \in 1,....,m$ be a set of m 6D relative pose relationships associated with edges in an undirected graph of frames. The graph is built incrementally as the vehicle moves through the environment, and extra edges are added during loop closure. The graph defines a connected Riemannian manifold that is by definition everywhere locally Euclidean, though globally it is not embedded in a single Euclidean space. The relationship between parent-frame α and child-frame j is defined by a 4×4 homogeneous transform matrix, $T_{\alpha,j}=\hat{T}_{\alpha,j}T_{(t_j)}$, where

$\hat{T}_{\alpha,j}$ is the current estimate and $T_{(t_j)}$ is the 4×4 homogeneous matrix defined by t_j. An example trajectory and graph with this notation is shown in Figure I.

Each t_j parameterizes an infinitesimal delta transform applied to the relationship from its parent frame in the graph (i.e. an error-state formulation). The kinematic chain from frame j to frame i is defined by a sequence of 4×4 homogeneous transforms

$$T_{ji} = \hat{T}_{j,j+1}T_{(t_{j+1})}\hat{T}_{j+1,j+2}T_{(t_{j+2})}, ..., \hat{T}_{i-1,i}T_{(t_i)};$$

the sensor model for a single measurement is

$$h_{i,k}(l_{j,k}, t_i, ...t_j) = Proj\left(T_{j,i}^{-1}l_{j,k}\right)$$
$$= Proj\left(g_{i,k}(l_{j,k}, t_{j+1}, ...t_i)\right)$$

where $g_{i,k} : \mathbb{R}^{\dim(x)} \to \mathbb{R}^4$, $x \mapsto T_{j,i}^{-1}l_{j,k}$ transforms the homogeneous point $l_{j,k}$ from base-frame j to the observation frame i. This describes how landmark k, stored relative to base-frame j, is transformed into sensor frame i and then projected into the sensor. We make the usual assumption that measurements $z_{i,k}$ are normally distributed: $z_{i,k} \sim N(h_{i,k}, R_{i,k})$. The cost function we associate with this formulation is

$$J = \sum_{k \in 1}^{n} \sum_{i \in 1}^{m_k} (z_{i,k} - h_{i,k}(x))^T R_{i,k}^{-1} (z_{i,k} - h_{i,k}(x)) \quad (1)$$
$$= \|z - h(x)\|_{R^{-1}}, \quad (2)$$

which depends on the landmark estimate, $l_{j,k}$ and *all the transform estimates* $t_{j+1}, ...t_i$ *on the kinematic chain from the base-frame j to the measurement-frame i*. This problem is solved using iterative non-linear least-squares Gauss-Newton minimization for the values of x that minimize re-projection error — this yields the *maximum likelihood* estimate (subject to local minima). Projecting via kinematic chains like this is novel, but it changes the sparsity patterns in the system Jacobian. Compared to normal bundle adjustment, this new pattern increases the cost of solving the sparse normal equations for updates δx to the state vector x — though as we will see, the ultimate computational complexity is the same.

B. Sparse Solution

The *normal equations* associated with the iterative non-linear least squares Gauss-Newton solution to equation (1) are

$$H^T R^{-1} H \delta x = H^T R^{-1}(z - h(x)). \quad (3)$$

where $H = \frac{\partial h}{\partial x}$ is the Jacobian of the sensor model, R is the block diagonal covariance matrix describing the uncertainty of the collective observation vector z (the stacked vector of all measurements). Referring to the example in Figure 3 we see that $H^T = \begin{bmatrix} H_l^T H_t^T \end{bmatrix}$ and $\delta x = [\delta l; \delta t]$, which exposes a well known 2×2 block structure for equation (3),

$$\begin{bmatrix} V & W \\ W^T & U \end{bmatrix} \begin{bmatrix} \delta l \\ \delta t \end{bmatrix} = \begin{bmatrix} r_l \\ r_t \end{bmatrix},$$

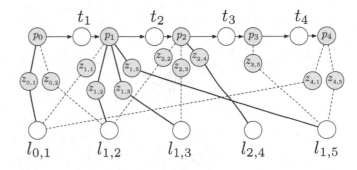

Figure 2. Graphical example for the sequence of 12 observations, $\{z_{0,1}, z_{1,1}, z_{4,1}\}$, $\{z_{0,2}, z_{1,2}, z_{2,2}\}$, $\{z_{1,3}, z_{2,3}\}$, $\{z_{2,4}\}$, $\{z_{1,5}, z_{3,5}, z_{4,5}\}$. There are five poses, $p_{0,...,4}$, four edge estimates $t_{1,...,4}$, and five landmarks $l_{0,1}$, $l_{1,2}$, $l_{1,3}$, $l_{2,4}$ and $l_{1,5}$. This example has the Jacobian $H = \frac{\partial h}{\partial x}$ that is depicted in Figure 3. Bold lines from poses indicate which frames are base-frames.

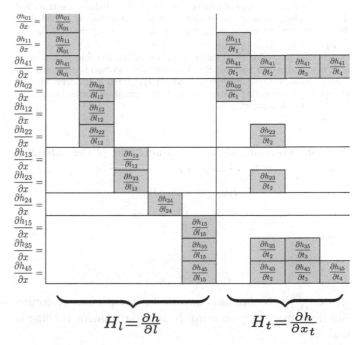

Figure 3. Example relative bundle adjustment Jacobian structure for the sequence of 12 observations in Figure 2. Grey indicates non-zero entries. The horizontal stripes in the right hand H_t term above correspond to projections that rely on transforming state estimates along kinematic chains from frame j to frame i. These stripes are the only difference in sparsity pattern between the relative formulation and traditional bundle adjustment.

where δl and δt are state updates for the map and edge transforms that we are solving for; $r_l = H_l^T R^{-1}(z - h(x))$, $r_t = H_t^T R^{-1}(z - h(x))$, $V = H_l^T R^{-1} H_l$, $W = H_l^T R^{-1} H_t$, and $U = H_t^T R^{-1} H_t$. Building this linear system is the dominant cost in solving each iteration, which makes it important to compute the sparse Jacobian of h efficiently.

C. Relative Jacobians

Due to the functional dependence of the projection model on the kinematic chain of relative poses, the Jacobians in the relative formulation are very different from their Euclidean counterpart. With reference to Figure 4, focus for a moment on a single infinitesimal transform $T_{(t_c)}$ that is somewhere

179

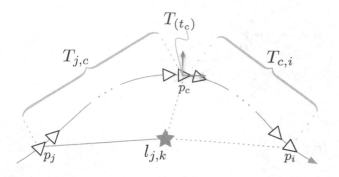

$$g_{i,k}(x) = (T_{j,c}T_{(t_c)}T_{c,i})^{-1}l_{j,k}$$

Figure 4. This diagram shows the sensor following a path through p_j and p_i while making measurements of landmark $l_{j,k}$ (indicated with dashed lines). Landmark k is stored relative to frame j (indicated by a solid line). To compute the projection of landmark k in frame i, we evaluate $h_{i,k} = Proj(g_{i,k}(x))$, where $g_{i,k}(x) = T_{j,i}^{-1}l_{j,k} = (T_{j,c}T_{(c)}T_{c,i})^{-1}l_{j,k}$, which encapsulates projection along the kinematic chain between frame j and frame i. To help understand how the relative formulation Jacobian is computed, this diagram focuses on the error-state transform $T_{(t_c)}$. The state-vector terms of interest when computing derivatives are 1) the transform parameters t_c, and 2) the landmark parameters $l_{j,k}$.

along the kinematic chain from frame i to j. The individual derivatives shown in Figure 3 are

$$\frac{\partial h_{i,k}}{\partial l_{j,k}} = \frac{\partial k}{\partial g_{i,k}}\frac{\partial g_{i,k}}{\partial l_{j,k}},$$

and

$$\frac{\partial h_{i,k}}{\partial t_c} = \frac{\partial k}{\partial g_{i,k}}\frac{\partial g_{i,k}}{\partial t_c}$$

where $\frac{\partial k}{\partial g_{i,k}}$ is the Jacobian of the perspective projection function (using the standard K intrinsic camera calibration matrix).

The Jacobian of $g_{i,k}$ with respect to the 3D point $l_{j,k}$ is

$$\frac{\partial g_{i,k}}{\partial l_{j,k}} = \begin{bmatrix} R_{i,j} \\ 0 \end{bmatrix}.$$

The Jacobian of $g_{i,k}$ with respect to t_c has three cases that depend on the direction of the transform $T_{(t_c)}$ on the path from frame i to j

$$\frac{\partial g_{i,k}}{\partial t_c} = \begin{cases} T_{i,c}\frac{\partial T_{(t_c)}}{\partial t_c}T_{c,j}l_{j,k} & \text{if } T_{(t_c)}\text{points towards } j \\ T_{i,c}\frac{\partial T_{(-t_c)}}{\partial t_c}T_{c,j}l_{j,k} & \text{if } T_{(t_c)}\text{points towards } i \\ 0 & \text{if } i = j \end{cases}$$

and $\frac{\partial T_{(t_c)}}{\partial t_c}$ is the canonical generators of SE(3) (a $4 \times 4 \times 6$ tensor). We now address the cost of solving each update.

D. Complexity of Computing the Relative Sparse Solution

Similar to sparse bundle adjustment, the following steps are used to exploit the structure of H to compute the *normal equations* and state-updates efficiently:

1) *Build linear system*, computing the terms U, V, W, r_t, and r_l. Complexity is $O(m^2n)$ using key-frames.
2) *Forward substitute*, computing $A = U - W^TV^{-1}W$, and $b = r_t - W^TV^{-1}r_l$. Complexity is $O(m^2n)$.
3) *Solve reduced system* of equations, $A\delta t = b$ for the update δt. Complexity is $O(m^3)$.
4) *Back substitute* to solve for the map update, $\delta l = V^{-1}(r_l - W\delta t)$. Complexity is $O(mn)$.

The first step is completely different in the relative framework so we describe it in more detail in algorithm 1. The overall complexity is $O(m^3)$, which matches traditional sparse bundle adjustment. Note that it is easy to convert algorithm 1 into a robust m-estimator by replacing the weights, $w_{i,k}$, with robust weight kernels, $w_{i,k} = R_{i,k}^{-1}\mathcal{W}(e_{i,k})$ — for example we use the Huber kernel [16]. Section IV gives results of applying this sparse optimization routine to large simulated and real sequences.

---

algorithm 1 Build linear system. Computes U, V, W, r_t, and r_l in $O(m^2n)$

Clear U, V, W, r_t, and r_l
for all landmarks k **do**
  for all key-frames i with a measurement of landmark k **do**
    Compute $\frac{\partial h_{i,k}}{\partial l_{j,k}}$
    $e_{i,k} = z_{i,k} - h_{i,k}(x)$
    $w_{i,k} = R_{i,k}^{-1}$
    $V_k = V_k + \frac{\partial h_{i,k}}{\partial l_{j,k}}^T w_{i,k}^{-1} \frac{\partial h_{i,k}}{\partial l_{j,k}}$
    $r_{l_k} = r_{l_k} + \frac{\partial h_{i,k}}{\partial l_{j,k}}^T w_{i,k}^{-1} e_{i,k}$
    for all $p \in Path(i,j)$ **do**
      Compute $\frac{\partial h_{i,k}}{\partial t_p}$
      $r_{t_P} = r_{t_P} + \frac{\partial h_{i,k}}{\partial t_p}^T w_{i,k} e_{i,k}$
      $W_{k,p} = W_{k,p} + \frac{\partial h_{i,k}}{\partial l_{j,k}}^T w_{i,k} \frac{\partial h_{i,k}}{\partial t_p}$
      for all $q \in Path(p,j)$ **do**
        Compute $\frac{\partial h_{i,k}}{\partial t_q}$
        $U_{p,q} = U_{p,q} + \frac{\partial h_{i,k}}{\partial t_p}^T w_{i,k} \frac{\partial h_{i,k}}{\partial t_q}$
        $U_{q,p} = U_{q,p} + \frac{\partial h_{i,k}}{\partial t_q}^T w_{i,k} \frac{\partial h_{i,k}}{\partial t_p}$
      end for
    end for
  end for
end for

Finally, notice that if feature tracks are contiguous over numerous frames (which they typically are), then the sparsity pattern in W will be the same in the relative-formulation as it is in the traditional one – hence the relative-formulation cost of forward-substitution, solving the reduced system, and back-substitution (steps 2-4) should be approximately equivalent.

E. Adaptive Updates

To reduce computation, it is important to optimize only those parameters that might change in light of new information [26][28]. Below we outline one approach to limit the parameters that are actively optimized.

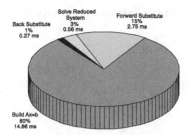

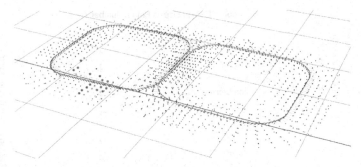

Figure 5. Average run-times for the main steps of relative bundle adjustment on an Intel Core 2 Duo 2.8GHz processor. The average adaptive region from the Monte Carlo simulation was 4.6 frames. Note that it is the cost of building the linear system of equations that dominates the cubic complexity of solving for the adaptive region of poses.

Figure 7. Figure-of-eight sequence used in Monte Carlo simulation. This sequence has 288 frames, 3,215 landmarks and 12,591 measurements with 1 pixel standard deviation Gaussian measurement noise added.

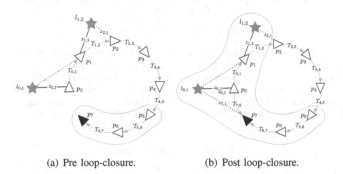

(a) Pre loop-closure. (b) Post loop-closure.

Figure 6. Discovery of local active region. In (a), re-projection errors have changed by more than $\Delta \epsilon$ in the local frames p_5, p_6, and p_7. In (b) a new edge $T_{7,0}$ is added during loop-closure, and the graph search leads to a larger active region with frames p_0, p_1, p_5, p_6, and p_7.

A breadth-first-search from the most recent frame is used to discover local parameters that might require adjustment. During the search, all frames in which the average re-projection error changes by more than a threshold, $\Delta \epsilon$, are added to an *active region* that will be optimized. The search stops when no frame being explored has a change in re-projection error greater than $\Delta \epsilon$. Landmarks visible from active frames are activated, and all non-active frames that have measurements of these landmarks are added to a list of static frames, which forms a slightly larger set we call the static region. Measurements made from static frames are included in the optimization, but the associated relative pose-error parameters are not solved for. Example active regions are shown in Figure 6.

IV. RESULTS

The iterative nonlinear least-squares solution that exploits the sparse relative structure and the four steps in section III-D results in the run-time break-down shown in Figure 5. This illustrates that building the sparse system of equations is the dominant cost.

A. Simulation Results

To determine the performance of the relative framework, a batch of Monte Carlo simulations were run. The sequence contains a realistic trajectory, landmark distribution, and a 1 pixel standard deviation Gaussian measurement noise (see Figure 7).

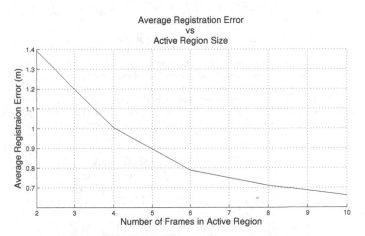

Figure 8. Average Registration Error vs. Number of Frames being updated. In the relative formulation, as the local region grows the average RMS error drops quickly toward the same as when computed with all frames active. This motivates the use of an adaptive region that allows parameters to vary only if it has an effect on the cost function.

We compute errors in the following way: for each pose in the trajectory, we register that pose to its ground truth counterpart, and then *localize* the rest of the relative trajectory in that frame. Note that "localizing" the relative trajectory is done with a breadth-first-search that computes each frame's pose in the coordinate system of the root frame; this process projects from the relative manifold into a single Euclidean frame, and may cause "rips" to appear at distant loop closures. Finally, the total trajectory registration error is computed as the average Euclidean distance between ground truth and the localized frames. The average of all frames and all registrations is the error plotted. Not surprisingly, initial results in Figure 8 indicate that error reduces towards the full solution (in the relative space) as the local region increases in size.

The results here use an adaptive region threshold of $\Delta \epsilon = 0.05$ pixels. With this threshold we find that the discovery of new frames to include in the active region quickly drops to between 4 and 5 poses, except at loop closure where it jumps to accommodate the larger region of poses found by the breadth-first-search. Figure 9 shows the adaptive region size discovered for two different loop closures, one 50m long and another 100m long. The point to note is that the discovered adaptive region is independent of loop size, and that errors do not propagate around the loop even though loop closure error is ~75cm on average for the 500 frame sequence. Using

181

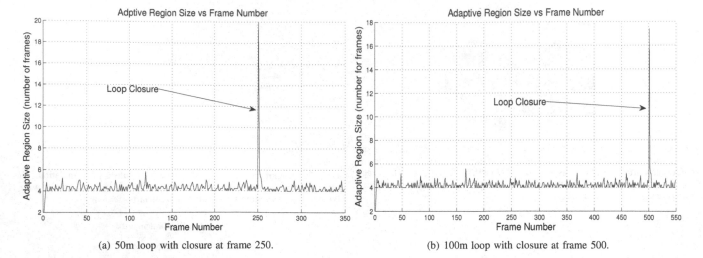

(a) 50m loop with closure at frame 250.

(b) 100m loop with closure at frame 500.

Figure 9. This figure shows how the number of frames in the adaptive region fluctuates over time and during loop closure. During loop closure the size of the adaptive region jumps to accommodate all the local frames that have been added to the active region, as well as any neighboring frames that will be affected. Notice that errors do not propagate all the way around the loop, and only a fraction of the state vector needs to be updated. Loop closure at 250 and 500 frames induces updates in approximately the same number of parameters, which strongly indicates that optimization at loop closure will remain constant time, independent of loop size. Before loop closure, the average metric position error is over 75cm for the 500 frame loop. Using the same adaptive region criteria, Euclidean bundle adjustment would require adjusting *all* parameters in the loop - whereas the adaptive relative approach only adjusts 20 poses.

	Science Park		
	Avg.	Min.	Max.
Distance Traveled (km)	—	—	1.08
Frames Processed	—	—	23,268
Velocity (m/s)	0.93	0.0	1.47
Angular Velocity (deg/s)	9.49	0.0	75.22
Frames Per Second	22.2	10.6	31.4
Features per Frame	93	44	143
Feature Track Length	13.42	2	701
Re-projection Error	0.17	0.001	0.55

Table I
TYPICAL PERFORMANCE OF OUR ONLINE SYSTEM FOR THE BEGBROKE SCIENCE PARK DATA SET PROCESSED ON AN INTEL CORE 2 DUO 2.8GHz.

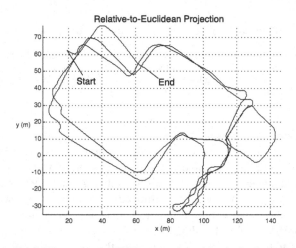

Figure 10. 1.08km path over 23K frames estimated for the Begbroke Science Park sequence. Table I shows typical performance results.

the same adaptive region criteria, Euclidean bundle adjustment would require adjusting *all* parameters in the loop - whereas the adaptive relative approach adjusts just 20 poses.

Our adaptive strategy for discovering the active region is designed to have a rippling effect: when parameter estimates change, it effects the re-projection error in nearby frames, which, if greater than $\Delta\epsilon$, will add those parameters to the active region, potentially causing them to change... etc. A key result of the relative formulation is that these errors *stop propagating* and balance out with distance from the new information - that is, the network of parameters is critically damped.

B. Real Data

The system operates online at 20-40Hz, this includes all image processing, feature tracking, robust initialization routines, and calls to FABMAP [5] to detect loop closures. We have run it successfully on sequences with up to 110K frames over tens of kilometers. Figure 10 shows a 1.08 kilometer trajectory computed from 23K frames. Table I gives an indication of typical system performance.

V. DISCUSSION

The privileged-frame approach and the relative formulations are very different; their objective functions are different and they solve for different quantities. The former embeds the trajectory in a single Euclidean space; the latter in a connected Riemannian manifold. At first reading it may appear that the lack of a simple Euclidean distance metric between two points, and the fact that we cannot render the solution very easily, is a disadvantage of the relative formulation. Note however that the manifold is a metric space, and distance between two points can be computed from shortest paths in the graph. With this in mind, the relative representation should *still* be amenable to planning algorithms which are commonly defined over graphs in the first place. Furthermore, because the manifold is (by definition) locally Euclidean, algorithms that require precise *local* metric estimates, such as obstacle avoidance or object manipulation, can operate without impediment.

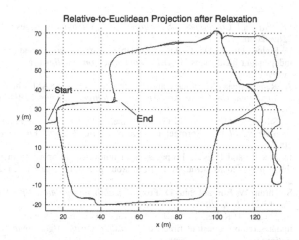

Figure 11. This figure shows a globally consistent relaxed view of the Beg-broke Science Park sequence. To view relative estimates in a consistent fashion (single global frame) we have to transform from the relative representation to a single Euclidean coordinate system. The sequence here has 23K poses over 1.08 kilometers which makes the conversion computationally expensive. This relative-to-global transformation process is designed to run on the user interface, *not* on the robot.

We posit that a *topometric* relative formulation is suffi-cient for many mobile robot navigation tasks, and that a single global Euclidean representation is rarely necessary. Certainly the benefits afforded by incrementally constant-time performance are tremendous, and in the light of that, some inconvenience may be acceptable. If a unified global Euclidean picture is deemed essential by a particular external application or technique, our choice would be to push responsibility for generating the single Euclidean embedding into that process - for example undertaking fast approximate pose-graph relax-ation in order to render consistent results in a user interface [24], [14].

A. Rendering in a Single Euclidean Space

As an example, Figure 11 shows the result of transforming a large relative state estimate into a single Euclidean frame using pose-graph relaxation. Note that even this state-of-the art global Euclidean estimate fails to discover the true rectilinear structure. Arguably the best way to improve the map would be to *schedule* new measurements across the diagonal of the map, thereby considerably constraining the solution. While this interventionist approach is used extensively in surveying, we are not comfortable with placing such a requirement on a mobile platform — ideally navigation and mapping should be a quiet background task producing estimates for consumption by any interested client process. With this example in mind, perhaps accurate global Euclidean state estimates are the wrong goal to aim for — what matters is relative metric accuracy and topological consistency — all of which can be attained with a relative manifold approach.

VI. CONCLUSION

The fact that the variables in bundle adjustment are defined relative to a single coordinate frame has a large impact on the algorithm's iterative convergence rate. This is especially true at loop closure, when large errors must propagate around the entire loop to correct for global errors that have accumulated along the path. As an alternative, we have presented an adap-tive relative formulation that can be viewed as a *continuous* sub-mapping approach – in many ways our relative treatment is an intuitive simplification of previous sub-mapping methods. Furthermore by solving all parameters within an adaptive region, the proposed method attempts to match the full max-imum likelihood solution within the metric space defined by the manifold. In stark contrast to traditional bundle adjustment, our evaluations and results indicate that state updates in the relative approach are constant time, and crucially, remain so even during loop closure events.

ACKNOWLEDGMENTS

The work reported in this paper undertaken by the Mobile Robotics Group was funded by the Systems Engineering for Autonomous Systems (SEAS) Defence Technology Centre established by the UK Ministry of Defence, Guidance Ltd, and by the UK EPSRC (CNA and Platform Grant EP/D037077/1). The work reported in this paper undertaken by the Ac-tive Vision Lab acknowledges the support of EPSRC grant GR/T24685/01.

REFERENCES

[1] J. Blanco, J. Fernandez-Madrigal, and J. Gonzalez. Toward a unified bayesian approach to hybrid metric–topological SLAM. *IEEE Transac-tions on Robotics and Automation*, 24(2):259–270, 2008.

[2] M. C. Bosse, P. M. Newman, J. J. Leonard, and S. Teller. SLAM in large-scale cyclic environments using the atlas framework. *International Journal of Robotics Research*, 23(12):1113–1139, December 2004.

[3] D.C. Brown. A solution to the general problem of multiple station analytical stereotriangulation. Technical report, RCP-MTP Data Reduc-tion Technical Report No. 43, Patrick Air Force Base, Florida (also designated as AFMTC 58-8), 1958.

[4] L. A. Clemente, A. J. Davison, I. Reid, J. Neira, and J. D. Tardos. Mapping large loops with a single hand-held camera. In *Robotics: Science and Systems*, 2007.

[5] M. Cummins and P. Newman. Probabilistic appearance based navigation and loop closing. In *IEEE Conference on Robotics and Automation*, 2007.

[6] A. Davison, I. Reid, N. Molton, and O. Stasse. MonoSLAM: Realtime single camera SLAM. *IEEE Transactions Pattern Analysis and Machine Intelligence*, 29(6):1113–1139, 2007.

[7] M. C. Deans. *Bearings-Only Localization and Mapping*. PhD thesis, School of Computer Science, Carnegie Mellon University, 2005.

[8] F. Dellaert. Square root SAM. In *Proceedings of Robotics: Science and Systems*, pages 1181–1203, 2005.

[9] E. Eade and T. Drummond. Unified loop closing and recovery for real time monocular SLAM. In *Proceedings British Machine Vision Conference*, September 2008.

[10] C. Engels, H. Stewenius, and D. Nister. Bundle adjustment rules. In *Photogrammetric Computer Vision*, 2006.

[11] R. Eustice, H. Singh, J. Leonard, M. Walter, and R. Ballard. Visually navigating the RMS Titanic with SLAM information filters. In *Robotics: Science and Systems*, pages 57–64, 2005.

[12] A. W. Fitzgibbon and A. Zisserman. *Automatic Camera Recovery for Closed or Open Image Sequences*. Springer, 2004.

[13] U. Frese and T. Duckett. A multigrid approach for accelerating relaxation-based SLAM. In *Proceedings IJCAI Workshop on Reasoning with Uncertainty in Robotics (RUR 2003)*, pages 39–46, Acapulco, Mexico, 2003.

[14] G. Grisetti, C. Stachniss, S. Grzonka, and W. Burgard. A tree param-eterization for efficiently computing maximum likelihood maps using gradient descent. In *Proceedings Robotics: Science and Systems*, 2007.

[15] J.E. Guivant and E.M. Nebot. Optimization of the simultaneous localization and map-building algorithm for real-time implementation. *IEEE Transactions on Robotics and Automation*, 17(3):242–257, June 2001.

[16] P. J. Huber. Robust estimation of a location parameter. *The Annals of Mathematical Statistics*, 35(2):73–101, 1964.

[17] M. Kaess. *Incremental Smoothing and Mapping*. PhD thesis, Georgia Institute of Technology, 2008.

[18] G. Klein and D. Murray. Improving the agility of keyframe-based SLAM. In *European Conference on Computer Vision*, 2008.

[19] K. Konolige and M. Agrawal. FrameSLAM: from bundle adjustment to realtime visual mapping. *IEEE Transactions on Robotics and Automation, IEEE Journal of Robotics and Automation, International Journal of Robotics Research*, 24(5):1066–1077, 2008.

[20] A. Martinelli, V. Nguyen, N. Tomatis, and R. Siegwart. A relative map approach to SLAM based on shift and rotation invariants. *Robotics and Autonomous Systems*, 55(1):50–61, 2007.

[21] P. F. McLauchlan. The variable state dimension filter applied to surface-based structure from motion. Technical report, University of Surrey, 1999.

[22] E. M. Mikhail. *Observations and Least Squares*. Rowman & Littlefield, 1983.

[23] E. Mouragnon, M. Lhuillier, M. Dhome, F. Dekeyse, and P. Sayd. Real time localization and 3d reconstruction. In *Proceedings of Computer Vision and Pattern Recognition*, 2006.

[24] E. Olson, J. Leonard, and S. Teller. Fast iterative alignment of pose graphs with poor initial estimates. In *Proceedings of the IEEE International Conference on Robotics and Automation*, pages 2262–2269, 2006.

[25] P. Pinies and J. D. Tardos. Scalable slam building conditionally independent local maps. In *IEEE conference on Intelligent Robots and Systems*, 2007.

[26] A. Ranganathan, M. Kaess, and F. Dellaert. Loopy SAM. In *International Joint Conferences on Artificial Intelligence*, pages 2191–2196, 2007.

[27] G. Sibley, L. Matthies, and G. Sukhatme. *A Sliding Window Filter for Incremental SLAM*. Springer Lecture Notes in Electrical Engineering, 2007.

[28] D. Steedly and I. Essa. Propagation of innovative information in non-linear least-squares structure from motion. In *ICCV01*, pages 223–229, 2001.

[29] S. Thrun, W. Burgard, and D. Fox. *Probabilistic Robotics*. MIT Press, Cambridge, MA, 2005.

[30] S. Thrun, D. Koller, Z. Ghahmarani, and H. Durrant-Whyte. SLAM updates require constant time. In *Workshop on the Algorithmic Foundations of Robotics*, December 2002.

[31] B. Triggs, P. McLauchlan, R. Hartley, and A. Fitzgibbon. Bundle adjustment – A modern synthesis. In W. Triggs, A. Zisserman, and R. Szeliski, editors, *Vision Algorithms: Theory and Practice*, LNCS, pages 298–375. Springer Verlag, 2000.

Underwater Human-Robot Interaction via Biological Motion Identification

Junaed Sattar and Gregory Dudek
Center for Intelligent Machines
McGill University
Montréal, Québec, Canada H3A 2A7.
Email: {junaed, dudek}@cim.mcgill.ca

Abstract— We present an algorithm for underwater robots to visually detect and track human motion. Our objective is to enable human-robot interaction by allowing a robot to follow behind a human moving in (up to) six degrees of freedom. In particular, we have developed a system to allow a robot to detect, track and follow a scuba diver by using frequency-domain detection of biological motion patterns. The motion of biological entities is characterized by combinations of periodic motions which are inherently distinctive. This is especially true of human swimmers. By using the frequency-space response of spatial signals over a number of video frames, we attempt to identify signatures pertaining to biological motion. This technique is applied to track scuba divers in underwater domains, typically with the robot swimming behind the diver. The algorithm is able to detect a range of motions, which includes motion directly away from or towards the camera. The motion of the diver relative to the vehicle is then tracked using an Unscented Kalman Filter (UKF), an approach for non-linear estimation. The efficiency of our approach makes it attractive for real-time applications on-board our underwater vehicle, and in future applications we intend to track scuba divers in real-time with the robot. The paper presents an algorithmic overview of our approach, together with experimental evaluation based on underwater video footage.

Fig. 1. An underwater robot servoing off a target carried by a diver.

I. INTRODUCTION

Motion cues have been shown to be powerful indicators of human activity and have been used in the identification of their position, behavior and identity. In this paper we exploit motion signatures to facilitate visual servoing, as part of a larger human-robot interaction framework. From the perspective of visual control of an autonomous robot, the ability to distinguish between mobile and static objects in a scene is vital for safe and successful navigation. For the vision-based tracking of human targets, motion patterns are an important signature, since they can provide a distinctive cue to disambiguate between people and other non-biological objects, including moving objects, in the scene. We look at both of these features in the current work.

Our work exploits motion-based tracking as one input cue to facilitate human-robot interaction. While the entire framework is outside the scope of this paper, an important sub-task for our robot, like many others, is for it to follow a human operator (as can be seen in Fig.1). We facilitate the detection and tracking of the human operator using the spatio-temporal signature of human motion. In practice, this detection and servo-control behavior is just one of a suite of vision-based interaction mechanisms. In the context of servo-control, we need to detect a human, estimate his image coordinates (and possible image velocity), and exploit this in a control loop. We use the periodicity inherently present in biological motion, and swimming in particular, to detect human scuba divers. Divers normally swim with a distinctive kicking gait which, like walking, is periodic, but also somewhat individuated. In many practical situations, the preferred applications of UAV technologies call for close interactions with humans. The underwater environment poses new challenges and pitfalls that invalidate assumptions required for many established algorithms in autonomous mobile robotics. While truly autonomous underwater navigation remains an important goal, having the ability to guide an underwater robot using sensory inputs also has important benefits; for example, to train the robot to perform a repetitive observation or inspection task, it might very well be convenient for a scuba diver to perform the task as the robot follows and learns the trajectory. For future missions, the robot can use the information collected by following the diver to carry out the inspection. This approach also has the added advantage of not requiring a second person teleoperating the robot, which simplifies the operational loop and reduces the associated overhead of robot deployment.

Our approach to track scuba divers in underwater video footage and real-time streaming video arises thus from the need for such semi-autonomous behaviors and visual human-robot interaction in arbitrary environments. The approach is computationally efficient for deployment on-board an autonomous underwater robot. Visual tracking is performed in the spatio-temporal domain in the image space; that is, spatial frequency variations are detected in the image space in different motion directions across successive frames. The frequencies associated with a diver's gaits (flipper motions) are identified and tracked. Coupled with a visual servoing mechanism, this feature enables an underwater vehicle to follow a diver without any external operator assistance, in environments similar to that shown in Fig. 2.

The ability to track spatio-temporal intensity variations using the frequency domain is not only useful for tracking scuba divers, but also can be useful to detect motion of particular species of marine life or surface swimmers [6]. It is also associated with terrestrial motion like walking or running, and our approach seems appropriate for certain terrestrial applications as well. It appears that most biological motion underwater as well as on land is associated with periodic motion, but in this paper we concentrate our attention to tracking human scuba divers and servoing off their position. Our robot is being developed with marine ecosystem inspection as a key application area. Recent initiatives taken for protection of coral reefs call for long-term monitoring of such reefs and species that depend on reefs for habitat and food supply. We envision our vehicle to have the ability to follow scuba divers around such reefs and assist in monitoring and mapping of distributions of different species of coral.

The paper is organized in the following sections: in Sec. II we look at related work in the domains of tracking, oriented filters and spatio-temporal pattern analysis in image sequences, Kalman filtering and underwater vision for autonomous vehicles. Our Fourier energy-based tracking algorithm is presented in Sec. III. Experimental results of running the algorithm on video sequences are shown in Sec. IV. We draw conclusions and discuss some possible future directions of this work in Sec. V.

Fig. 2. Typical visual scene encountered by an AUV while tracking scuba divers.

II. RELATED WORK

The work presented in this paper combines previous research in different domains, and its novelty is in the use of frequency signatures in visual target recognition and tracking, combined with the Unscented Kalman Filter for tracking 6-DOF human motion. In this context, 6-DOF refers to the number of degrees of freedom of just the body center, as opposed to the full configuration space. In the following paragraphs we consider some of the extensive prior work on tracking of humans in video, underwater visual tracking and visual servoing in general.

A key aspect of our work is a filter-based characterization of the motion field in an image sequence. This has been a problem of longstanding relevance and activity, and were it not for the need for a real-time low-overhead solution, we would be using a full family of steerable filters, or a related filtering mechanism [2, 3]. In fact, since our system needs to be deployed in a hard real-time context on an embedded system, we have opted to use a sparse set of filters combined with a robust tracker. This depends, in part, on the fact that we can consistently detect the motion of our target human from a potentially complex motion field. Tracking humans using their motion on land, in two degrees of freedom, was examined by Niyogi and Adelson [8]. They look at the positions of head and ankles, respectively, and detect the presence of a human walking pattern by looking at a "braided pattern" at the ankles and a straight-line translational pattern at the position of the head. In their work, however, the person has to walk across the image plane roughly orthogonal to the viewing axis for the detection scheme to work.

There is evidence that people can be discriminated from other objects, as well as from one another, based on motion cues alone (although the precision of this discrimination may be limited). In the seminal work using "moving light displays", Rashid observed [10] that humans are exquisitely sensitive to human-like motions using even very limited cues. There has also been work, particularly in the context of biometric person identification, based on the automated analysis of human motion or walking gaits [16, 7, 15]. In a similar vein, several research groups have explored the detection of humans on land from either static visual cues or motion cues. Such methods typically assume an overhead, lateral or other view that allows various body parts to be detected, or facial features to be seen. Notably, many traditional methods have difficulty if the person is walking directly away from the camera. In contrast, the present paper proposes a technique that functions without requiring a view of the face, arms or hands (either of which may be obscured in the case of scuba divers). In addition, in our particular tracking scenario the diver can point directly away from the robot that is following him, as well as move in an arbitrary direction during the course of the tracking process.

While tracking underwater swimmers visually has not been explored in great depth in the past, some prior work has been done in the field of underwater visual tracking and visual servoing for autonomous underwater vehicles. Naturally, this is closely related to generic servo-control. In that context, on-line real-time performance is crucial. On-line tracking systems, in conjunction with a robust control scheme, provide underwater robots the ability to visually follow targets underwater [14]. Previous work on spatio-temporal detection and tracking of biological motion underwater has been shown to work well [12], but only when the motion of the diver is directly towards or away from the camera. Our current work looks at motion in a variety of directions over the spatio-temporal domain, incorporates a variation of the Kalman filter and also estimates diver distance and is thus a significant improvement over that particular technique.

In terms of the tracking process itself, the Kalman filter is, of course, the preeminent classical methodology for real-time tracking. It depends, however, on a linear model of system dynamics. Many real systems, including our model of human swimmers, are non-linear and the linearization needed to implement a Kalman filter needs to be carefully managed to avoid poor performance or divergence. The Unscented Kalman Filter [5] we deploy was developed to facilitate non-linear control and tracking, and can be regarded as a compromise between Kalman Filtering and fully non-parametric Condensation [4].

III. METHODOLOGY

To track scuba divers in the video sequences, we exploit the periodicity and motion invariance properties that characterize biological motion. To fuse the responses of the multiple frequency detectors, we combine their output with an Unscented Kalman Filter. The core of our approach is to use periodic motion as the signature of biological propulsion and specifically for person-tracking, to detect the kicking gait of a person swimming underwater. While different divers have distinct kicking gaits, the periodicity of swimming (and walking) is universal. Our approach, thus, is to examine the amplitude spectrum of rectangular slices through the video sequence along the temporal axis. We do this by computing a windowed Fourier transform on the image to search for regions that have substantial band-pass energy at a suitable frequency. The flippers of a scuba diver normally oscillate at frequencies between 1 and 2 Hz. Any region of the image that exhibits high energy responses in those frequencies is a potential location of a diver. The essence of our technique is therefore to convert a video sequence into a sampled frequency-domain representation in which we accomplish detection, and then use these responses for tracking. To do this, we need to sample the video sequence in both the spatial and temporal domain and compute local amplitude spectra. This could be accomplished via an explicit filtering mechanism such as steerable filters which might directly yield the required bandpass signals. Instead, we employ windowed Fourier transforms on the selected space-time region which are, in essence, 3-dimensional blocks of data from the video sequence (a 2-dimensional region of the image extended in time). In principle, one could directly employ color information at this stage as well, but due to the need to limit computational cost and the low mutual information content between color channels (especially underwater), we perform the frequency analysis on luminance signals only.

We look at the method of *Fourier Tracking* in Sec. III-A. In Sec. III-B, we describe the multi-directional version of the Fourier tracker and motion detection algorithm in the XYT domain. The application of the Unscented Kalman Filter for position tracking is discussed in Sec. III-C.

A. Fourier Tracking

The core concept of the tracking algorithm presented here is to take a time-varying spatial signal (from the robot) and use the well-known discrete-time Fourier transform to convert the signal from the spatial to the frequency domain. Since the target of interest will typically occupy only a region of the image at any time, we naturally need to perform spatial and temporal windowing. The standard equations relating the spatial and frequency domain are as follows.

$$x[n] = \frac{1}{2\pi} \int_{2\pi} X(e^{j\omega}) e^{j\omega} d\omega \tag{1}$$

$$X(e^{j\omega}) = \sum_{n=-\infty}^{+\infty} x[n] e^{-j\omega n} \tag{2}$$

where $x[n]$ is a discrete aperiodic function, and $X(e^{j\omega})$ is periodic with length 2π and frequency ω. Equation 1 is referred to as the *synthesis* equation, and Eq. 2 is the *analysis* equation where $X(e^{j\omega})$ is often called the *spectrum* of $x[n]$ [9]. The coefficients of the converted signal correspond to the amplitude and phase of complex exponentials of harmonically-related frequencies present in the spatial domain.

For our application, we do not consider phase information, but look only at the absolute amplitudes of the coefficients of the above-mentioned frequencies. The phase information might be useful in determining relative positions of the undulating flippers, for example. It might also be used to provide a discriminator between specific individuals. Moreover, by not differentiating between the individual flippers during tracking, we achieve a speed-up in the detection of high-energy responses, at the expense of sacrificing relative phase information.

Spatial sampling is accomplished using a Gaussian windowing function at regular intervals and in multiple directions over the image sequence. The Gaussian is appropriate since it is well known to simultaneously optimize localization in both space and frequency space. It is also a separable filter, making it computationally efficient. Note, as an aside, that some authors have considered tracking using a box filter for sampling, but these produce undesirable ringing in the frequency domain, which can lead to unstable tracking. The Gaussian filter has good frequency domain properties and it can be computed recursively making it exceedingly efficient.

B. Multi-directional Motion Detection

To detect motion in multiple directions, we use a predefined set of vectors, each of which is composed of a set of small

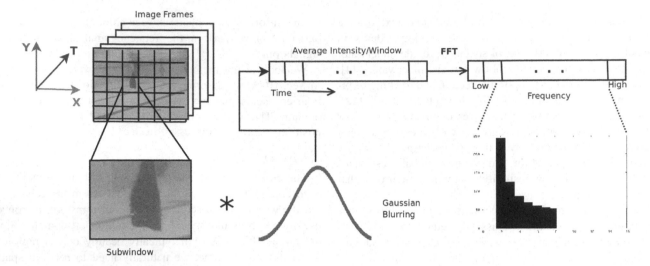

Fig. 3. Outline of the Directional Fourier motion detection and tracking process. The Gaussian-filtered temporal image is split into subwindows, and the average intensity of each subwindow is calculated for every timeframe. For the length of the filter, a one-dimensional intensity vector is formed, which is then passed through an FFT operator. The resulting amplitude plot can be seen, with the symmetric half removed.

rectangular subwindows in the spatio-temporal space. The trajectories of each of these subwindows are governed by a corresponding starting and ending point in the image. In any given time T, this rectangular window resides in a particular position along this trajectory and represents a Gaussian-weighted grayscale intensity value of that particular region in the image. Over the entire trajectory, these windows generate a vector of intensity values along a certain direction in the image, producing a purely temporal signal for amplitude computation. We weight these *velocity vectors* with an exponential filter, such that intensity weights of a more recent location of the subwindow have a higher weight than another at that same location in the past. This weighting helps to maintain the causal nature of the frequency filter applied to this velocity vector. In the current work, we extract 17 such velocity vectors (as seen in Fig. 4) and apply the Fourier transform to them (17 is the optimum number of vectors we can process in quasi-real time in our robot hardware). The space formed by the velocity vectors is a conic in the XYT space, as depicted in Fig. 5. Each such signal provides an amplitude spectrum that can be matched to a profile of a typical human gait. A statistical classifier trained on a large collection of human gait signals would be ideal for matching these amplitude spectra to human gaits. However, these human-associated signals appear to be easy to identify, and as such, an automated classifier is not currently used. Currently, we use two different approaches to select candidate spectra. In the first, we choose the particular direction that exhibits significantly higher energy amplitudes in the low-frequency bands, when compared to higher frequency bands. In the second approach, we precompute by hand an amplitude spectrum from video footage of a swimming diver, and use this amplitude spectrum as a true reference. To find possible matches, we use the Bhattacharyya measure [1] to find similar amplitude spectra, and choose those as possible candidates.

C. Position Tracking Using an Unscented Kalman Filter

Each of the directional Fourier motion operators outputs an amplitude spectrum of different frequencies present in each associated direction. As described in Sec. III-B, we look at the amplitudes of the low-frequency components of these directional operators, the ones that exhibit high responses are chosen as possible positions of the diver, and thus the position of the diver can be tracked across successive frames.

To further enhance the tracking performance, we run the output of the motion detection operators through an Unscented Kalman Filter (UKF). The UKF is a highly effective filter for state estimation problems, and is suitable for systems with a non-linear process model. The track trajectory and the motion perturbation are highly non-linear, owing to the undulating propulsion resulting from flipper motion and underwater currents and surges. We chose the UKF as an appropriate filtering mechanism because of this inherent non-linearity, and also its computational efficiency.

According to the UKF model, an N-dimensional random variable $\mathbf{x}$ with mean $\hat{x}$ and covariance P_{xx} is approximated by $2N+1$ points known as the *sigma points*. The sigma points at iteration $k-1$, denoted by $\chi^i_{k-1|k-1}$, are derived using the following set of equations:

$$\chi^0_{k-1|k-1} = \mathbf{x}^a_{k-1|k-1}$$
$$\chi^i_{k-1|k-1} = \mathbf{x}^a_{k-1|k-1} + (\sqrt{(N+\lambda)(P)^a_{k-1|k-1}})_i$$
$$i = 1 \dots N$$
$$\chi^i_{k-1|k-1} = \mathbf{x}^a_{k-1|k-1} + (\sqrt{(N+\lambda)(P)^a_{k-1|k-1}})_{i-N}$$
$$i = N+1 \dots 2N$$

where $(\sqrt{(N+\lambda)(P)^a_{k-1|k-1}})_i$ is the i-th column of the matrix square-root of $((N+\lambda)(P)^a_{k-1|k-1})$, and λ is a predefined constant.

188

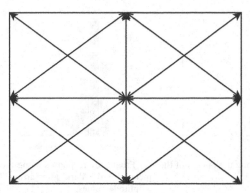 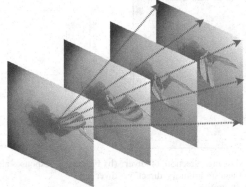

(a) Motion directions covered by the various directional Fourier operators, depicted in a 2D spatial arrangement.

(b) Image slices along the time axis showing 5 out of 17 possible track directions while tracking a scuba diver.

Fig. 4. Directions of motion for Fourier tracking, also depicted in 3D in a diver swimming sequence.

For the diver's location, the estimated position **x** is a two-dimensional random variable, and thus the filter requires 5 sigma points. The sigma points are generated around the mean position estimate by projecting the mean along the X and Y axes, and are propagated through a non-linear motion model (*i.e.*, the transition model) f, and the estimated mean (*i.e.*, diver's estimated location), $\hat{\mathbf{x}}$, is calculated as a weighted average of the transformed points:

$$\chi^i_{k|k-1} = f(\chi^i_{k-1|k-1}) \quad i = 0 \dots 2N$$

$$\hat{\mathbf{x}}_{k|k-1} = \sum_{i=0}^{2N} W^i \chi^i_{k|k-1}$$

where W^i are the constant weights for the state (*position*) estimator.

As an initial position estimate of the diver's location for the UKF, we choose the center point of the vector producing the highest low-frequency amplitude response. Ideally, the non-linear motion model for the scuba diver can be learned from training using video data, but for this application we use a hand-crafted model created from manually observing such footage. The non-linear motion model we employ predicts forward motion of the diver with a higher probability than up and down motion, which in turn is favored over sideways motion. For our application, a small number of iterations (approximately between 5 and 7) of the UKF is sufficient for convergence.

IV. EXPERIMENTAL RESULTS

The proposed algorithm has been experimentally validated on video footage recorded of divers swimming in open- and closed-water environments (*i.e*, pool and open ocean, respectively). Both types of video sequences pose significant challenges due to the unconstrained motion of the robot and the diver, and the poor imaging conditions, particularly observed in the open-water footage due to suspended particles, water salinity and varying lighting conditions. The algorithm outputs

a direction corresponding to the most dominant biological motion present in the sequence, and a location of the most likely position of the entity generating the motion response. Since the Fourier tracker looks backward in time every N frames to find the new direction and location of the diver, the output of the computed locations are only available after a "bootstrap phase" of N frames. We present the experimental setup below in Sec. IV-A findings and the results in Sec. IV-B.

A. Experimental Setup

As mentioned, we conduct experiments offline on video sequences recorded from the cameras of an underwater robot. The video sequences contain footage of one or more divers swimming in different directions across the image frame, which make them suitable for validating our approach. We run our algorithm on a total of 2530 frames of a diver swimming in a pool, and 2680 frames of a diver swimming in the open-ocean, collected from open ocean field trials of the robot. In total, the frames amounted to over 10 minutes video footage of both environments. The *Xvid*-compressed video frames have dimensions of 768×576 pixels, the detector operated at a rate

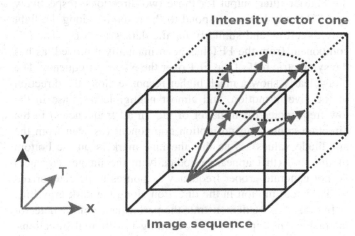

Fig. 5. Conic space covered by the directional Fourier operators.

 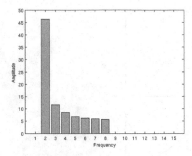 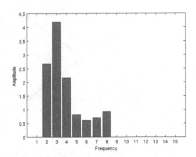

(a) Snapshot image showing direction of diver motion (in light gray) and an arbitrary direction without a diver (in dark gray).

(b) Frequency responses along the motion of the diver.

(c) Frequency responses along the direction depicted by the dark gray arrow. Note the low amplitude values.

Fig. 6. Contrasting frequency responses for directions with and without diver motion in a given image sequence.

of approximately 10 frames *per* second, and the time window for the Fourier tracker for this experiment is 15 frames, corresponding to approximately 1.5 seconds of footage. Each rectangular subwindow is 40×30 pixels in size (one-fourth in each dimension). The subwindows do not overlap each other on the trajectory along a given direction.

For visually servoing off the responses from the frequency operators, we couple the motion tracker with a simple Proportional-Integral-Derivative (PID) controller. The PID controller accepts image space coordinates as input and provides as output motor commands for the robot such that the error between the desired position of the tracked diver and the current position is minimized. While essential for following divers, the servoing technique is not an integral part of the motion detection algorithm, and thus runs independently of any specific visual tracking algorithm.

B. Results

Figure 6(a) shows a diver swimming along a diagonal direction away from the camera, as depicted by the light gray arrow. No part of the diver falls on the direction shown by the dark gray arrow, and as such there is no component of motion present in that direction. Figure 6(b) and 6(c) show the Fourier filter output for those two directions, respectively (the light gray bars correspond to the response along the light gray direction, and similarly for the dark gray bars). The DC component from the FFT has been manually removed, as has the symmetric half of the FFT over the Nyquist frequency. The plots clearly show a much higher response along the direction of the diver's motion, and almost negligible response in the low frequencies (as a matter of fact in all frequencies) in the direction containing no motion component (as seen from the amplitude values). Note that the lane markers on the bottom of the pool (that appear periodically in the image sequence) do not generate proper frequency responses to be categorized as biological motion in the direction along the dark gray line.

In Fig. 7(a), we demonstrate the performance of the detector in tracking multiple divers swimming in different directions. The sequence shows a diver swimming in a direction away from the robot, while another diver is swimming in front of

Direction	Lowest-Frequency Amplitude response
Left-to-right	205.03
Right-to-left	209.40
Top-to-bottom	242.26
Up-from-center	251.61
Bottom-to-top	281.22

TABLE I

LOW-FREQUENCY AMPLITUDE RESPONSES FOR MULTIPLE MOTION DIRECTIONS.

her across the image frame in an orthogonal direction. The amplitude responses obtained from the Fourier operators along the directions of the motion for the fundamental frequency are listed in ascending order in Tab. I. The first two rows correspond to the direction of motion of the diver going across the image, while the bottom three rows represent the diver swimming away from the robot. As expected, the diver closer and unobstructed to the camera produces the highest responses, but motion of the other diver also produces significant low-frequency responses. The other 12 directions exhibit negligible amplitude responses in the proper frequencies compared to the directions presented in the table. The FFT plots for motion in the bottom-to-top and left-to-right direction are seen in Figs. 7(b) and 7(c), respectively. As before, the FFT plot has the DC component and the symmetric half removed for presentation clarity.

An interesting side-effect of the Fourier tracker is the effect of the diver's distance from the robot (and hence the camera) on the low-frequency amplitude. Figure 8 shows two sequences of scuba divers swimming away from the robot, with the second diver closer to the camera. The amplitude responses have similar patterns, exhibiting high energy at the low-frequency regions. The spectrum on top, however, has more energy in the low-frequency bands than the one on the bottom, where the diver is closer to the camera. The close proximity to the camera results in a lower variation of the intensity amplitude, and thus the resulting Fourier amplitude spectra shows lower energy in the low-frequency bands.

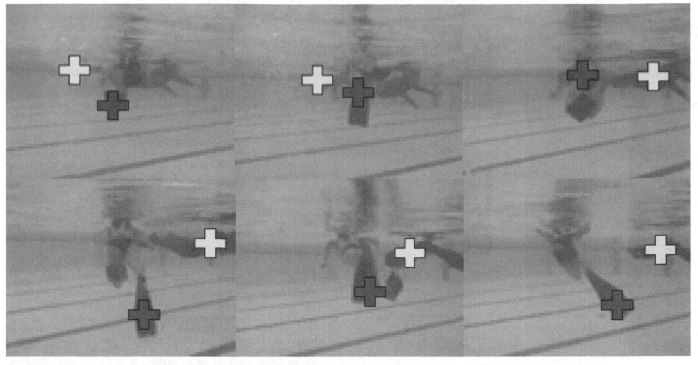

(a) An image sequence capturing two divers swimming in orthogonal directions.

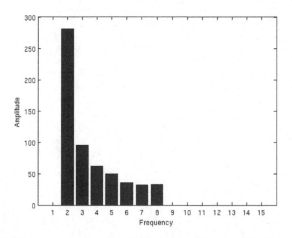

(b) Frequency responses for the diver swimming away from the robot (dark gray cross) in Fig. 7(a).

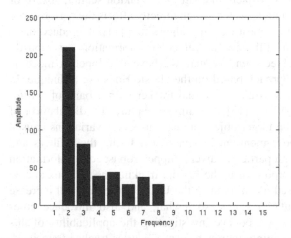

(c) Frequency responses for the diver swimming across the robot (light gray cross) in Fig. 7(a).

Fig. 7. Frequency responses for two different directions of diver motion in a given image sequence.

V. CONCLUSIONS AND FUTURE WORK

In this paper, we present a technique for robust detection and tracking of biological motion underwater, specifically to track human scuba divers. We consider the ability to visually detect biological motion an important feature for any mobile robot, and especially for underwater environments to interact with a human operator. In a larger scale of visual human-robot interaction, such a feature forms an essential component of the communication paradigm, using which an autonomous vehicle can effectively recognize and accompany its human controller. The algorithm presented here is conceptually simple and easy to implement. Significantly, this algorithm is optimized for real-time use on-board an underwater robot. In the very near future, we aim to focus our experiments on our platform, and measure performance statistics of the algorithm implemented on real robotic hardware. While we apply a heuristic for modeling the motion of the scuba diver to feed into the UKF for position tracking, we strongly believe that with the proper training data, a more descriptive and accurate model can be learned. Incorporating such a model promises to increase the performance of the motion tracker.

While color information can be valuable as a tracking cue, we do not look at color in this work. Hues are affected by the optics of the underwater medium, which changes

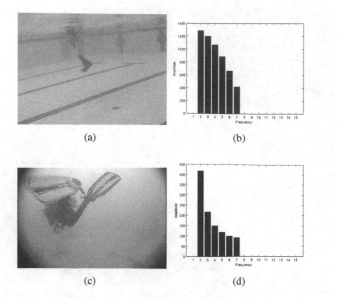

(a)   (b)

(c)   (d)

Fig. 8. Effect of diver's distance from camera on the amplitude spectra. Being farther away from the camera produces higher energy responses (Fig. 8(b)) in the low-frequency bands, compared to divers swimming closer (Fig. 8(d)).

object appearances drastically. Lighting variations, suspended particles and artifacts like silt and plankton scatter, absorb or refract light underwater, which directly affects the performance of otherwise-robust tracking algorithms [11]. To reduce these effects and still have useful color information for robustly tracking objects underwater, we have developed a machine learning approach based on the classic Boosting technique. In that work, we train our visual tracker with a bank of *spatio-chromatic* filters [13] that aim to capture the distribution of color on the target object, along with color variations caused by the above-mentioned phenomena. Using these filters and training for a particular diver's flipper, robust color information can be incorporated in the Fourier tracking mechanism, and be directly used as an input to the UKF. While this will increase the computational cost somewhat, and also introduce color dependency, we believe investigating the applicability of this machine learning approach in our Fourier tracker framework is a promising avenue for future research.

REFERENCES

[1] A. Bhattcharyya. On a measure of divergence between two statistical populations defined by their probability distributions. *Bulletin Calcutta Math Society*, 35:99–110, 1943.

[2] D. J. Fleet and A. D. Jepson. Computation of component velocity from local phase information. *Internation Journal of Computer Vision*, 5(1):77–104, August 1990.

[3] W. T. Freeman and E. H. Adelson. The design and use of steerable filters. *IEEE Transactions on Pattern Analysis and Machine Intelligence*, 13(9):891–906, 1991.

[4] M. Isard and A. Blake. Condensation – conditional density propagation for visual tracking. *International Journal of Computer Vision*, 29(1):5–28, 1998.

[5] S. Julier and J. Uhlmann. A new extension of the kalman filter to nonlinear systems. In *International Symposium on Aerospace/Defense Sensing, Simulation and Controls*, Orlando, FL, USA, 1997.

[6] M. F. Land. Optics of the eyes of marine animals. In P. J. H. A. K. C. M. W. L. maddock, editor, *Light and life in the sea*, pages 149–166. Cambridge University Press, Cambridge, UK, 1990.

[7] M. S. Nixon, T. N. Tan, and R. Chellappa. *Human Identification Based on Gait*. The Kluwer International Series on Biometrics. Springer-Verlag New York, Inc. Secaucus, NJ, USA, 2005.

[8] S. A. Niyogi and E. H. Adelson. Analyzing and recognizing walking figures in xyt. In *Proceedings of the IEEE Computer Society Conference on Computer Vision and Pattern Recognition*, pages 469–474, 1994.

[9] A. V. Oppenheim, A. S. Willsky, and S. H. Nawab. *Signals & systems (2nd ed.)*. Prentice-Hall, Inc., Upper Saddle River, NJ, USA, 1996.

[10] R. Rashid. Toward a system for the interpretation of moving light display. *IEEE Transactions on Pattern Analysis and Machine Intelligence*, 2(6):574–581, November 1980.

[11] J. Sattar and G. Dudek. On the performance of color tracking algorithms for underwater robots under varying lighting and visibility. In *Proceedings of the IEEE International Conference on Robotics and Automation ICRA2006*, pages 3550–3555, Orlando, Florida, May 2006.

[12] J. Sattar and G. Dudek. Where is your dive buddy: Tracking humans underwater using spatio-temporal features. In *Proceedings of the IEEE/RSJ International Conference on Intelligent Robots and Systems IROS2007*, pages 3654–3659, San Diego, California, October 2007.

[13] J. Sattar and G. Dudek. Robust servo-control for underwater robots using banks of visual filters. In *Proceedings of the IEEE International Conference on Robotics and Automation, ICRA2009*, pages 3583–3588, Kobe, Japan, May 2009.

[14] J. Sattar, P. Giguere, G. Dudek, and C. Prahacs. A visual servoing system for an aquatic swimming robot. In *Proceedings of the IEEE/RSJ International Conference on Intelligent Robots and Systems, IROS2005*, pages 1483–1488, Edmonton, Alberta, Canada, August 2005.

[15] H. Sidenbladh and M. J. Black. Learning the statistics of people in images and video. *Int. J. Comput. Vision*, 54(1-3):181–207, 2003.

[16] H. Sidenbladh, M. J. Black, and D. J. Fleet. Stochastic tracking of 3d human figures using 2d image motion. In *ECCV (2)*, pages 702–718, 2000.

Robustness of the Unscented Kalman Filter for State and Parameter Estimation in an Elastic Transmission

Edvard Naerum[1,2], H. Hawkeye King[3] and Blake Hannaford[3]
[1]The Interventional Centre, Rikshospitalet University Hospital, Oslo, Norway
[2]Faculty of Medicine, University of Oslo, Norway
[3]Department of Electrical Engineering, University of Washington, Seattle, WA, USA
E-mail: edvard.narum@medisin.uio.no, (hawkeye1, blake)@u.washington.edu

Abstract— The Unscented Kalman Filter (UKF) was applied to state and parameter estimation of a one degree of freedom robot link with an elastic, cable-driven transmission. Only motor encoder and command torque data were used as input to the filter. The UKF was used offline for joint state and model-parameter estimation, and online for state estimation. This paper presents an analysis of the robustness of the UKF to unknown/unmodeled variation in inertia, cable tension and contact forces, using experimental data collected with the robot.

Using model parameters found offline the UKF successfully estimated motor and link angles and velocities online. Although the transmission was very stiff, and hence the motor and link states almost equal, information about the individual states was obtained. Irrespective of variation from nominal conditions the UKF link angle estimate was better than using motor position as an approximation (i.e. inelastic transmission assumption). The angle estimates were particularly robust to variation in operating conditions, velocity estimates less so. A near-linear relationship between contact forces and estimation errors suggested that contact forces might be estimated using this error information.

I. INTRODUCTION

A. Background

State and parameter knowledge are of prime importance in robot manipulation. Sensors for state measurement are often noisy so filtering may be necessary. Filters, however, can introduce a phase lag and degrade performance. Model based estimation may be a good alternative for estimating states and parameters from noisy sensors, or a reduced set of sensors.

The BioRobotics Lab at the University of Washington has developed the RAVEN robot, a new six degree of freedom (DoF) surgical robot prototype [1]. The RAVEN is cable-driven, allowing the motors to be removed from the arms and attached to the base. This decreases the arms' inertia, but the cables introduce elasticity into the system. That, in turn, creates additional states since the link is elastically coupled to the motor. The RAVEN has encoders mounted on the motors but not on the link side, since encoders on the links themselves would require additional, complex wiring. Furthermore, the encoder signal is highly quantized making noise-free velocity estimation difficult to compute. In addition, several of the model parameters are not directly observable, namely the cable stiffness and damping, and friction. This paper focuses on a single-DoF RAVEN-like testbed to study the problems of state estimation and parameter estimation in an elastic transmission.

B. Related Work

Most published studies on elastic transmissions have focused on elastic-joint robots. For a cable-driven robot the analysis is normally more complicated due to coupling between cable runs for multiple joints. However, there is no coupling in a 1-DoF cable-driven robot, and the modeling coincides with that of a 1-DoF elastic-joint robot. When extrapolating the results of this paper back to a multi-DoF robot such as the RAVEN the coupling, which is a kinematic relationship, must be taken into account.

Research on state and parameter estimation in elastic-joint robots gained momentum in the 1980's, together with the development of controller designs that required knowledge of the robot's state and model. Nicosia and Tornambé [2] developed a method to estimate the parameters of an elastic-joint robot using the output injection technique used in observers for nonlinear systems. The state of the robot was assumed known. In [3] Nicosia et al. designed approximate state observers for elastic-joint robots, and derived conditions under which exact observers exist. Their work required some measurement of the state at both sides of the transmission. Jankovic [4] proposed a reduced order high-gain state observer for elastic-joint robots that only required measurement of the motor side state, i.e. motor angle and velocity. A variable structure observer was designed by Léchevin and Sicard in [5] for elastic-joint manipulators. Their observer used link angle measurements. Abdollahi et al. [6] built a general state observer for nonlinear systems where a neural network was used to learn the system's dynamics. The observer's potential was demonstrated with an application to elastic-joint robots with measurement of motor angle and velocity.

C. Present Work

The task at hand in this paper is simultaneous state and parameter estimation of an elastic transmission with only motor angle measurements available. The aforementioned studies either require the knowledge of more states, or they only allow the estimation of certain parameters, such as the transmission elasticity. We would like to estimate several parameters, including friction. Therefore, we approach the problem through straight-forward application of the Unscented Kalman Filter (UKF) [7, 8, 9]. The primary difference between the UKF and the Extended Kalman Filter (EKF) is that the

UKF is capable of performing nonlinear estimation without linearization, whereas the EKF linearizes the nonlinear system and applies a regular Kalman filter algorithm. The UKF was chosen as it has been argued that it should replace the EKF in estimation applications for nonlinear systems because of its simpler implementation and better performance [7, 9]. Araki et al. [10] estimated the parameters of a robot with rigid joints using a UKF, and experiments were done with a two-DoF robot. Other works not related to robotics that employ the UKF for state and/or parameter estimation include Zhao et al. [11], Gove and Hollinger [12] and Zhan and Wan [13].

We present a study of simultaneous state and parameter estimation for offline system identification, and an online state estimator for smooth, noise-free state measurement with no phase distortion. The UKF is used in both cases. *The main contribution of this work is an experimental examination of robustness of the UKF to unknown variations in inertia, cable tension and force disturbances.* We are primarily interested in applications to teleoperation, and the motivating factors behind the robustness experiments are

1) to investigate the possibility of estimating contact forces with an elastic-joint robot with partial state measurement in a bilateral teleoperation setting, and

2) to get an indication of the performance of a simple joint-level UKF for state estimation and feedback control in a multi-DoF robot.

A more general, but important question that also needs to be answered is whether, and with what accuracy we can estimate the link state using motor angle measurements. In other words, can we infer better information about the actual link state by the application of an advanced filter like the UKF, or will using the motor state as an approximation to the link state be just as useful, i.e. a rigid robot assumption?

II. SYSTEM & MODELING

A. Hardware

To test the UKF robustness in a real world setting, a simple 1-DoF test platform was employed (see Fig. 1). Termed the "pulleyboard" it uses all the same hardware as the RAVEN surgical robot, and is intended as a testbed for that device.

The pulleyboard uses a single Maxon EC40 brushless DC motor and a DES 70/10 motor controller (Maxon Motors). The end effector is a rotational link with nominal moment of inertia $3.63 \cdot 10^{-4}$ kgm^2 about the axis of rotation. The moment of inertia can be varied by additional weight rigidly affixed to the link. The link is cable driven with a cable run from the motor to the link and back for a total cable length of 1.58 m. The stainless steel cable goes through a total of four idler pulleys with a diameter of 15 mm and eight idler pulleys with a diameter of 7.5 mm, in addition to the motor and link shafts with a diameter of 14 mm. This cable route is similar to the second axis of the surgical robot. Also, a force gauge with an additional 15 mm pulley is included in one part of the cable run to measure cable tension.

On this simplified testbed it is easy to have two shaft encoders, one colocated at the motor and another noncolocated

Fig. 1. The pulleyboard with a force gauge in the lower right corner.

at the link. These sensors directly measure motor angle q_m and link angle q_l (see Fig. 2). So that the results will be applicable to the RAVEN robot, the noncolocated sensor was used only for validation, and not for control or parameter estimation.

Also, identical to the RAVEN the pulleyboard includes an RTAI Linux PC. The PC is a 32-bit AMD Athlon XP with 1GB RAM. I/O with the the pulleyboard uses the BRL USB board developed in the BioRobotics Lab at the University of Washington [14]. The robot control software is an RTAI-Linux kernel module, and the control loop runs at 1kHz.

B. Modeling

The main modeling assumption is that, although the pulleyboard has several idler pulleys, it can be modeled as one motor side and one link side connected by one cable run. Idler pulley inertias and transmission friction can be lumped into either one of the two sides without any significant loss of accuracy.

Fig. 2 shows a schematic drawing of the pulleyboard the way it is modeled. The pulleyboard model comprises a motor inertia, J_m, and a link inertia, J_l, connected by two cables each with a longitudinal stiffness k_e and damping b_e. For the purpose of modeling, the two cables are considered as one torsional spring/damper connecting the two inertias, where the total torsional effect is the combination of the longitudinal effect of the two cables. The spring is modeled as an exponential spring, i.e. for a generic displacement d we have

$$F_{spring} = k_e(e^{\alpha d} - 1).$$

Initial estimation tests were carried out where both k_e and α were considered unknown. However, we were unable to

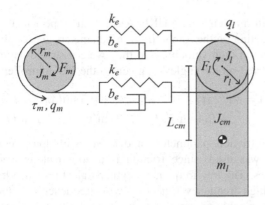

Fig. 2. The pulleyboard model.

make the UKF estimate for α converge. Possibly, this is because the relative displacements of the stiff transmission are very small, and within these displacements a wide range of values for α yield approximately the same force. Instead, $\alpha = 1$ throughout this paper. The use of an exponential spring model is nevertheless worthwhile, as the increased degree of nonlinearity will put the UKF further to the test. The damper is modeled as a linear damper. The mass of the cables is neglected. It is assumed that the cables are pretensioned equally, and to a sufficiently high tension that neither cable goes slack under normal operation. Friction in the transmission is lumped into a motor side term F_m and a link side term F_l.

Let $r_m = r_l$ be the motor and link capstan radii, respectively. Since they are equal the pulleyboard's gear ratio is 1:1. Let τ_m be the torque applied to the motor, q_m the angle of the motor and q_l the angle of the link. The motor and link angles are zero when the link is in a vertical position, the tensions in the two cables are equal and the system is not moving. The dynamic equation is then written as

$$J\ddot{q} + T + F + N = \tau \tag{1}$$

where $q = [q_m, q_l]^T$, $\tau = [\tau_m, 0]^T$, and

$$J = \begin{bmatrix} J_m & 0 \\ 0 & J_l \end{bmatrix}, \qquad T = \begin{bmatrix} r_m\gamma \\ -r_l\gamma \end{bmatrix},$$

$$F = \begin{bmatrix} F_m \\ F_l \end{bmatrix} \quad \text{and} \quad N = \begin{bmatrix} 0 \\ m_l L_{cm} g \sin(q_l) \end{bmatrix}.$$

The motor inertia J_m can be found by consulting the motor's data sheet, and the link inertia J_l can be calculated using the parallel axis theorem. The variable γ is used to denote the total longitudinal force of the spring/damper effect of both cables. It is calculated as

$$\gamma = k_e(e^{q_m r_m - q_l r_l} - e^{q_l r_l - q_m r_m}) + 2b_e(\dot{q}_m r_m - \dot{q}_l r_l).$$

The lumped transmission friction F is modeled by the simple Coulomb and viscous combination:

$$F_i = F_{c,i}\text{sign}(\dot{q}_i) + F_{v,i}\dot{q}_i, \quad i = m, l$$

where $F_{c,i}$ and $F_{v,i}$ are the Coulomb and viscous friction constants, respectively. Motor and link shaft friction is also

included in these terms. Finally, N contains the gravitational torque acting on the link, with m_l being the mass of the link, L_{cm} is the distance from the axis of rotation to the center of mass, and g is the acceleration of gravity.

Some of the parameters in the dynamic equation are known, while some are unknown. In particular, the inertias J_m and J_l are known, and so are m_l and L_{cm}, so that the gravitational term N can be computed. The capstan radii r_m and r_l are also known. On the other hand, the stiffness constant k_e and the damping constant b_e are unknown. So are the four friction constants $F_{c,i}$ and $F_{v,i}$, for a total of six unknown parameters. These parameters have to be estimated.

III. METHODS

The Unscented Kalman Filter (UKF) was used in two ways: offline for simultaneous state and parameter estimation and online for state estimation. System parameters identified offline were used online for state-only estimation. Although the pulleyboard has motor and link angle available, only motor angle was used so as to emulate a system with colocated sensors and actuators.

A PD controller was implemented for all data collection, both during the offline estimation and the following robustness experiments. The pulleyboard performed sinusoidal trajectory following tasks, and except for the force disturbance tests the pulleyboard was operated in free-space motion. As the topic of our study was state estimation performance, and not control performance, the state estimate was *not* used for feedback control. Instead, the motor shaft encoder information was used directly for PD control. A low-pass filter was applied to avoid the resonant high-frequency modes inherent in the elastic cable transmission, and velocity was calculated by a first-difference of the low-pass filtered position samples.

A. Offline State and Parameter Estimation

The key to parameter estimation with the UKF is to regard the unknown parameters as part of the state. That way the basic UKF algorithm does not have to be modified. The pulleyboard's dynamic equation (1) is written in state-space form, discretized, and the state vector is augmented with the unknown parameters.

To write the dynamic equation in state-space form we first define the state vector of the continuous-time system $x(t)$ as

$$x := \begin{bmatrix} q_m & q_l & \dot{q}_m & \dot{q}_l \end{bmatrix}^T.$$

We also define the generic input signal $u := \tau$. The pulleyboard's dynamics in state-space form are then given as

$$\dot{x} = f(x, u), \tag{2}$$
$$y = Cx \tag{3}$$

where

$$f := \begin{bmatrix} \dot{q} \\ J^{-1}(\tau - N - F - T) \end{bmatrix},$$
$$C := \begin{bmatrix} 1 & 0 & 0 & 0 \end{bmatrix}.$$

The measurement y is equal to motor angle $q_m = x_1$.

195

Since f is nonlinear we cannot compute the exact discrete-time equivalent of the continuous-time system. Instead we numerically integrate the state-space equations using a fourth order Runga-Kutta method. Denoting the Runga-Kutta 4 operator by F_{RK4} we get

$$x(k+1) = F_{RK4}(f(x(k), u(k))) =: F(x(k), u(k)), \quad (4)$$
$$y(k) = Cx(k) \quad (5)$$

where k is the time index. The measurement equations (3) and (5) for the continuous and discrete-time systems are the same.

The final step before applying the UKF algorithm is to construct the augmented dynamic equation of the pulleyboard. We define the augmented state vector as

$$x^a := \begin{bmatrix} x^T & k_e & b_e & F_{c,m} & F_{c,l} & F_{v,m} & F_{v,l} \end{bmatrix}^T.$$

The augmented system is then described by

$$x^a(k+1) = F^a(x^a(k), u(k)), \quad (6)$$
$$y(k) = C^a x^a(k). \quad (7)$$

where

$$F^a(x^a(k), u(k)) := \begin{bmatrix} F(x(k), u(k)) \\ k_e(k) \\ b_e(k) \\ F_{c,m}(k) \\ F_{c,l}(k) \\ F_{v,m}(k) \\ F_{v,l}(k) \end{bmatrix},$$
$$C^a := \begin{bmatrix} C & 0_{1\times 6} \end{bmatrix}.$$

With the system (6), (7) we are ready for direct application of the UKF algorithm. Essentially, the UKF uses a deterministic sampling approach to calculate the state estimate $\hat{x}^a$ and covariance. A set of samples are chosen that completely capture the true mean and covariance of the state x^a. These samples are called *sigma points* and they are propagated through the nonlinearity F^a instead of the state itself. The state estimate and covariance are then found by weighted average computation. For the offline state and parameter estimation a square-root implementation of the UKF algorithm is used to shorten the execution time [9]. Although process noise $v(k)$ and measurement noise $n(k)$ are not included in (6) and (7) the knowledge of the 10×10 process noise covariance matrix R_v and the measurement noise covariance R_n is required. In this paper these are used solely to control the convergence properties of the algorithm. For further details on the derivation and use of the UKF, see e.g. [7, 8, 9].

Before data collection, the six unknown parameters were divided into two groups; one group consisted of the cable parameters k_e and b_e, while the other group consisted of the friction parameters $F_{c,m}$, $F_{c,l}$, $F_{v,m}$ and $F_{v,l}$. A separate data set was collected for each group, both with a duration of 140 seconds. For the cable parameters a data set with high frequency motion was used, because it maximizes the relative motion between motor and link side, and thereby provides the

most information to the UKF. Also, with respect to the total torque acting on the joints, friction parameters are usually less dominant at high velocities, so they were assumed negligible. The desired motor angle q_m^d given to the PD controller was

$$q_m^d(t) = \sin(2\pi t) - 0.7\sin(3\pi t) + 0.5\sin(4\pi t)$$
$$+ 0.4\sin(5\pi t) - 0.2\sin(6\pi t) + 0.1\sin(7\pi t).$$

For the friction parameters a data set with low frequency motion was used, since friction is a dominant term at low velocities. The low frequency angular trajectory was identical to the high frequency trajectory with frequencies divided by three. The cable parameters found with the first data set were used during this second round of estimation.

In the end all six parameters were estimated. For the online estimation and robustness tests these parameters were considered known and fixed.

B. Online State Estimation Under Nominal Conditions

Online state estimation was initially tested under nominal conditions to check the baseline state estimation performance, and to validate the parameters found offline. For state-only estimation there is no need to define the augmented system (6), (7), the original system (4), (5) can be used as is. Therefore state-only estimation is easier to implement than combined state and parameter estimation, and the smaller system dimension makes execution faster. Thus, a square-root implementation of the online state estimator was not deemed necessary. As before, the 4×4 process noise covariance matrix R_v and the measurement noise covariance R_n were only set to control the convergence properties of the UKF, and did not reflect the real noise in the system. The trajectory used for validation was

$$q_m^d(t) = 0.75\sin(1.6\pi t) - 0.63\sin(2.6\pi t)$$
$$+ 0.225\sin(3.66\pi t) + 0.15\sin(4.4\pi t)$$
$$- 0.12\sin(6.6\pi t) + 0.075\sin(6.9\pi t). \quad (8)$$

Data were collected for 60 seconds.

The main reason not to adaptively estimate the unknown parameters online is that contact force estimates would be adversely affected. If the end effector were in contact with the environment an adaptive UKF would alter the system parameters to reflect the coupled system's (pulleyboard+environment) behavior. Hence, wrong parameter estimates would be used for contact force estimation.

C. Robustness Experiments

The main purpose of the present work is to study the robustness of the Unscented Kalman Filter to changes in system parameters. Robustness was measured in terms of the ability of the UKF to maintain satisfactory real-time state estimation using system parameters identified offline *even as the system undergoes dynamic changes*. The robustness of the UKF was studied when

- the link inertia was increased from its nominal value,
- the cable tension varied around its nominal value,

- a contact force on the link increased from its nominal value of zero, and
- the motion of the link was physically constrained.

For each test the motor and link angles were recorded together with the estimated state and input motor torque. Actual velocities were computed during post processing using a zero-phase low-pass filter on the first differences of the encoder signals. The relative root-mean-square (RMS) error and maximum error between actual and estimated state was computed for each test.

1) Link Inertia: When a robot picks up an object the effective inertia changes, which impacts the system dynamics. Also, the inertia matrix of a multi-DoF robot is not constant, as is the case for the 1-DOF pulleyboard, but varies with the robot's configuration. To simplify the system dimension and reduce computing time, it may be desirable to use a 1-DoF UKF for each joint of a multi-DoF robot. For these reasons the UKF must be robust to changes in inertia.

For this experiment the link inertia was changed by attaching weights to the link at several distances from the axis of rotation. Inertia was increased from the nominal link inertia of $3.63 \cdot 10^{-4}$ kgm^2 up to $1.33 \cdot 10^{-3}$ kgm^2, an increase by a factor of 3.65. For practical reasons it was impossible to decrease the inertia. At each inertia the same sum-of-sinusoids angular trajectory was used as for the validation in section III-B.

2) Cable Tension: The amount of pretension in the cable indirectly determines the cable's spring and damping parameters, and also the lumped friction parameters. Since the cable tension will change over time, so will the parameters that depend on it. By varying the tension around the nominal value used for parameter estimation we measured the robustness of the UKF to changes in estimated parameters.

Tension was varied in steps of approximately 260g from the maximum tension of 3.3kg, via the nominal tension of 2.2kg, down to 1.17kg. The tensions tested ranged from about 0.5 to 1.5 times the nominal tension. Again, the same sinusoid was used as for the validation in section III-B.

3) Contact Force: Contact force estimation is of interest in teleoperation as a potential replacement for force sensors. The contact force can be computed with complete knowledge of the input torque, system state and dynamics. However, it is often the case that the complete state is not known, and it is therefore desirable to see how the state estimates are affected as a result of the applied contact force.

Accurate contact forces were simulated by attaching a cable perpendicular to the link and pulling with weights in a hanging basket. Meanwhile the pulleyboard followed the desired trajectory:

$$q_m^d(t) = 0.5 \sin(\pi t).$$

A single 0.5Hz sine wave was used instead of (8) to keep the basket from jumping around. The quasi-static conditions ensured that the additional inertia was not affecting the dynamics, and that the weight was simply a constant force. At first the contact force corresponded to the weight of the basket only (12.5 g), and then weights of 12.5g were

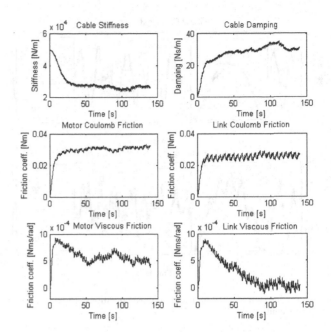

Fig. 3.   Offline convergence of the six estimated parameters.

added every 20 seconds. Weights were added 7 times for a total of 8 data points. The maximum contact force was $8 \times 12.5\text{g} \times 9.81\text{m/s}^2 = 0.981\text{N}$. Since there was a small transient when new weights were added only the data in the middle 10 seconds of every 20-second interval were used.

4) Constrained Motion: When the link of the pulleyboard is physically constrained, a worst case scenario is created wherein the UKF assumes the robot is moving when it is not. Due to the constraint, the link will not move at all, but the motor will because of the elastic transmission. Since the cable transmission is very stiff ($k_e = 3 \cdot 10^4 \pm 1 \cdot 10^4$ N/m), the difference between motor angle and link angle will always be small (typically < 0.15 rad). Hence, the UKF state estimator must be very accurate in order to provide any additional information about the actual link state. The alternative would be to assume that the motor and link states are equal. We can get a clear picture of whether the implementation of the UKF is worthwhile if the UKF link state estimate is better than using the colocated sensor to approximate the link state.

To check this, the pulleyboard was operated with the link physically constrained for 20 seconds. The link was rigidly fixed to the base using a C-clamp, and a sinusoidal trajectory of 0.5 Hz was commanded to the PD controller.

IV. RESULTS

A. Offline State and Parameter Estimation

The results of the offline parameter estimation are shown in Fig. 3. The top two graphs show the convergence of the spring and damper constants, which were found with the high frequency data set. The bottom four graphs show the friction terms, which were found with the low frequency data set. All six of the estimated parameters showed convergence. For this data set, the Coulomb friction tended to converge quickly in

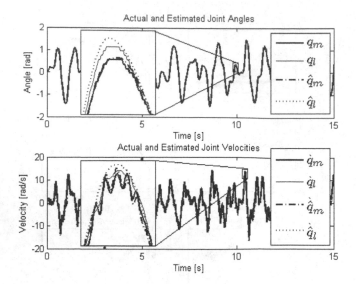

Fig. 4. Online state estimation under nominal conditions: 15-second excerpt and zoom detail.

10-20 seconds, while the stiffness and damping parameters showed slower convergence. The viscous friction took even longer to converge; the link viscous friction in particular took nearly 100 seconds to reach steady state.

B. Online State Estimation Under Nominal Conditions

Fig. 4 shows a 15-second excerpt from the 60-second data set, plus a detail at high zoom level. The state estimates are denoted by $\hat{q}_m$, $\hat{q}_l$, $\hat{\dot{q}}_m$ and $\hat{\dot{q}}_l$. At high magnification the difference between motor and link angles becomes evident, whereas the difference in velocity is less pronounced. The zoom window in the lower plot shows oscillations in the actual motor velocity, and the state estimator tracking those oscillations *in phase*. The link velocity is less oscillatory, and the link velocity estimate reflects that.

C. Robustness Experiments

Table I contains relative RMS estimation errors in percent for nominal and off-nominal experiments, calculated as

$$\text{RMS}\% = 100 * \text{RMS(estimation error)}/\text{RMS(actual value)}.$$

Absolute errors are used for constrained motion. The errors are $\tilde{q}_m := q_m - \hat{q}_m$, $\tilde{q}_l := q_l - \hat{q}_l$, $\dot{\tilde{q}}_m := \dot{q}_m - \hat{\dot{q}}_m$ and $\dot{\tilde{q}}_l := \dot{q}_l - \hat{\dot{q}}_l$.

1) Link Inertia: Fig. 5 shows the relative RMS error and the maximum error of the state estimation as link inertia is increased. All four graphs show the same phenomenon; the error stays almost constant until the inertia reaches 2.5 times its nominal inertia. It is then much higher for an inertia of 3.5 times nominal, and especially pronounced for the link angle.

2) Cable Tension: Fig. 6 shows the relative RMS and maximum values of the state estimation error as the cable tension varies around its nominal value. The motor angle estimation error stays relatively constant. Link angle error, however, is lower than nominal for lower tension, and gets worse with higher tensions. The RMS error of the velocity estimates stays flat up to the nominal tension, and then

TABLE I

RELATIVE RMS ERRORS (%) FOR ROBUSTNESS EXPERIMENTS. FOR COMPARISON, THE RELATIVE RMS ERROR OF MOTOR ANGLE TO LINK ANGLE IS INCLUDED. †THE LAST ROW CONTAINS ABSOLUTE RMS ERROR VALUES IN RAD OR RAD/S.

		$\tilde{q}_m$	$\tilde{q}_l$	$\dot{\tilde{q}}_m$	$\dot{\tilde{q}}_l$	$q_m - q_l$
Nominal		0.29	0.98	17	13	1.5
Relative Inertia	1.08	0.33	1.2	19	15	1.6
	1.14	0.37	1.0	21	17	1.7
	1.2	0.40	1.1	23	19	1.8
	1.71	0.44	1.1	26	23	2.3
	2.52	0.58	1.2	34	32	3.5
	3.65	1.5	4.4	93	82	4.8
Relative Tension	0.54	0.29	0.77	16	13	1.7
	0.65	0.30	0.79	17	13	1.6
	0.77	0.29	0.90	16	13	1.6
	0.88	0.28	1.0	16	13	1.5
	1.1	0.30	1.3	18	16	1.5
	1.2	0.32	1.3	20	17	1.6
	1.4	0.33	1.5	21	18	1.6
	1.5	0.35	1.4	23	20	1.7
Force [N]	0.12	0.77	2.8	126	116	3.4
	0.25	1.1	3.1	179	164	3.9
	0.37	1.4	3.2	234	218	4.9
	0.49	1.6	3.1	288	267	5.9
	0.61	1.9	3.3	346	321	7.0
	0.74	2.1	3.3	402	374	7.9
	0.86	2.3	3.2	458	432	8.8
	0.98	2.4	3.0	516	495	9.7
Constrained Motion†		0.01	0.01	6.6	6.0	0.05

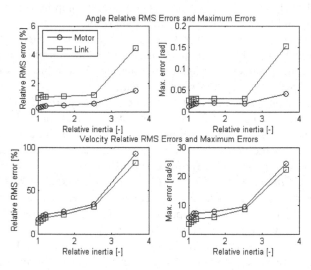

Fig. 5. State estimation performance expressed in terms of relative RMS error and maximum error for increasing link inertia.

increases almost linearly with higher tension. It is hard to find a consistent pattern in the maximum errors. However, the link angle error does seem to go up as tension increases.

3) Contact Force: Fig. 7 shows the effect of constant applied force on estimation error. Motor angle and velocity, and link velocity all have errors with near-linear characteristics, whereas the link angle estimate has a somewhat more random shape and shows less variation. It should be noted that velocity estimation performance is very poor.

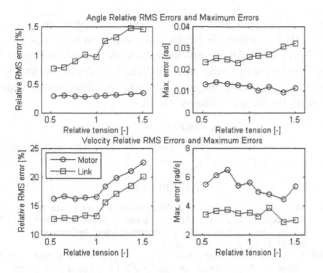

Fig. 6. State estimation performance expressed in terms of relative RMS error and maximum error for changing cable tension.

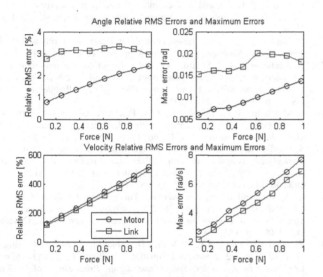

Fig. 7. State estimation performance expressed in terms of relative RMS error and maximum error for increasing contact force.

4) Constrained Motion: Fig. 8 shows the system response to an input of 0.5 radians at 0.5 Hz under constrained motion. The motor angle estimate is still close to the actual angle, and the link angle estimate is smaller than the motor angle, closer to the real link angle. The link angle appears to be out of phase with the motor trajectory tracking. Both velocity estimates are far away from the actual values, at or near zero.

V. DISCUSSION

Offline parameter estimation yielded reasonable values for all six of the desired parameters. Dividing the parameters into two groups and estimating them separately was necessary to have all parameters converge to good estimates. Also, the choice of initial value for the cable stiffness was important for the convergence outcome. Setting the initial value of k_e to zero did not lead to convergence. In our case the initial stiffness guess was set at $5 \cdot 10^4$ N/m. It is interesting to note that the

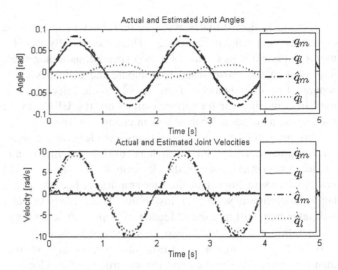

Fig. 8. State estimation when link is physically constrained.

results showed Coulomb friction to be greater than viscous friction by about two orders of magnitude. This agrees with findings in [15], where it was found that pulley friction was primarily Coulombic. The assumption made in section III-A that friction is less dominant at higher velocities also hinged on this observation.

In the nominal online tests motor and link states were accurately estimated using the system parameters found in the offline estimation step. Velocity estimation was not as good as angle estimation. However, the velocity estimate was still significantly improved from the regular low-pass filtered first difference and, in particular, there was no phase lag in the UKF velocity estimate. Phase lag is a common factor in controller instability and poor controller performance. With the improved estimates we will be able to increase the controller gains and improve tracking performance.

The right-most column in Table I contains the relative RMS values of $q_m - q_l$. This is the error in using motor angle as a substitute for link angle. The RMS error of the motor and link angle estimation $\tilde{q}_m$ and $\tilde{q}_l$ were consistently smaller showing that performance of the UKF was satisfactory and derived more useful information. This result is also visible in Fig. 4. We therefore conclude that the UKF is worthwhile for estimating link state from colocated sensors. Note that the RMS error shot up when the inertia of the link was increased from a factor of 2.5 to 3.6. This suggests a "safe range" of inertial changes, or weights that the robot can hold without affecting state estimation too much. Also, care should be taken in using the velocity estimate during environmental contact.

Fig. 7 reveals a near-linear relationship between contact force and estimation error for all states except link angle. The reverse relationship would suggest using error to estimate contact force, especially from the motor angle estimate, since it is measurable in real time. A similar technique was also proposed in [16]. However, we would only feel confident doing this under quasi-static conditions, because that is how the experiments were carried out.

Overall, angle estimates were more robust to changes in dynamics than velocity estimates. This is especially true for motor angle estimates. That is to be expected, as motor angle is closest to the sensor used to compute the estimates (only separated by quantization). Thus, if only angle information is used in controllers or for further estimation, the UKF may be used to achieve desired robustness in changes to the operating conditions. Although velocity estimates were less robust, phase information was kept intact, at least during the inertia and tension experiments. In feedback control the phase of the velocity is more important than its amplitude, because phase lag is a common source of instability. Hence, our velocity estimates may still be used in feedback loops over some range of operating conditions, at least in free-space motion.

In all Figs. 5-7 motor angle estimation errors were smaller than link errors, while the opposite was true for the velocities. We believe that the reason why the motor angle estimates were better than their link counterparts is simply that the encoder was located on the motor side (remember that the link side encoder was only used for validation). Furthermore, we believe that the link velocity estimation errors were smaller because the motor velocity signal contained more high frequency oscillations making it harder to predict.

Observing the RMS error versus changes in inertia, tension or contact force, we expect the minimum error at the nominal conditions. For the inertia and contact force tests this was the case, but not for the tension tests. Fig. 6 shows that instead it seems that RMS values decreased slightly for smaller tensions, while increasing almost linearly with tensions above nominal. This suggests that when cable tension is reduced due to cable stretch or other factors, state estimate will not be adversely affected. Overall, change in error due to unmodeled changes in tension were relatively small, meaning UKF performance is quite robust in that respect. This is important, since cable tension is bound to change with time in a cable-driven robot.

VI. CONCLUSION & FUTURE WORK

The Unscented Kalman Filter has been tested for state and parameter estimation in an elastic transmission using only motor angle measurements. The performance and robustness of the UKF to changes in system dynamics has been studied.

During normal operation the UKF was able to distinguish between motor and link angles although the transmission was very stiff, and it provided additional information about the actual link state that could not be measured directly with the motor encoder. This is useful in the context of state feedback in control loops.

Robustness experiments showed that angle estimates (motor and link) were robust to changes in operating conditions, such as contact with the surroundings or variation in dynamic parameters. Velocity estimates were less robust.

An almost linear relationship between increasing contact force and estimation errors suggested using the errors for estimating the contact force.

Future work will mainly focus on implementing the UKF for state and parameter estimation on the RAVEN robot. While our results show that under a range of conditions a joint level UKF can provide better state estimation than is possible without such a filter, implementation on a multi-DoF system will undoubtedly bring new challenges and discoveries. We also intend to study the importance of nonlinear versus linear models for the transmission elasticity.

ACKNOWLEDGEMENT

This work was partially funded by The Research Council of Norway project 167529/V30.

REFERENCES

[1] M. J. H. Lum, D. Friedman, J. Rosen, G. Sankaranarayanan, H. H. King, K. Fodero, R. Leuschke, M. Sinanan, and B. Hannaford, "The RAVEN - design and validation of a telesurgery system," *International Journal of Robotics Research*, May 2009.

[2] S. Nicosia and A. Tornambé, "A new method for the parameter estimation of elastic robots," in *Proc. IEEE International Conference on Systems, Man and Cybernetics*, vol. 1, Beijing, China, Aug. 1988, pp. 357–360.

[3] S. Nicosia, P. Tomei, and A. Tornambé, "A nonlinear observer for elastic robots," *IEEE Trans. Robot. Automat.*, vol. 4, no. 1, pp. 45–52, Feb. 1988.

[4] M. Jankovic, "Observer based control for elastic joint robots," *IEEE Trans. Robot. Automat.*, vol. 11, no. 4, pp. 618–623, Aug. 1995.

[5] N. Léchevin and P. Sicard, "Observer design for flexible joint manipulators with parameter uncertainties," in *Proc. IEEE International Conference on Robotics and Automation*, vol. 3, Albuquerque, NM, USA, Apr. 1997, pp. 2547–2552.

[6] F. Abdollahi, H. A. Talebi, and R. V. Patel, "A stable neural network-based observer with application to flexible-joint manipulators," *IEEE Trans. Neural Networks*, vol. 17, no. 1, pp. 118–129, Jan. 2006.

[7] S. J. Julier and J. K. Uhlmann, "A new extension of the Kalman filter to nonlinear systems," in *Proc. SPIE 11th International Symposium on Aerospace/Defense Sensing, Simulation and Controls*, vol. 3068, Orlando, FL, USA, Apr. 1997, pp. 182–193.

[8] E. A. Wan and R. var der Merwe, "The unscented Kalman filter for nonlinear estimation," in *Proc. IEEE Symposium on Adaptive Systems for Signal Processing, Communications and Control*, Lake Louise, Canada, Oct. 2000, pp. 153–158.

[9] R. van der Merwe and E. A. Wan, "The square-root unscented Kalman filter for state and parameter estimation," in *Proc. IEEE International Conference on Acoustics, Speech and Signal Processing*, vol. 6, Salt Lake City, UT, USA, May 2001, pp. 3461–3464.

[10] N. Araki, M. Okada, and Y. Konishi, "Parameter identification and swing-up control of an acrobot system," in *Proc. IEEE International Conference on Industrial Technology*, Hong Kong, China, Dec. 2005, pp. 1040–1045.

[11] X. Zhao, J. Lu, W. P. A. Putranto, and T. Yahagi, "Nonlinear time series prediction using wavelet networks with Kalman filter based algorithm," in *Proc. IEEE International Conference on Industrial Technology*, Hong Kong, China, Dec. 2005, pp. 1226–1230.

[12] J. H. Gove and D. Y. Hollinger, "Application of a dual unscented Kalman filter for simultaneous state and parameter estimation in problems of surface-atmosphere exchange," *Journal of Geophysical Research*, vol. 111, Apr. 2006.

[13] R. Zhan and J. Wan, "Neural network-aided adaptive unscented Kalman filter for nonlinear state estimation," *IEEE Signal Processing Letters*, vol. 13, no. 7, pp. 445–448, July 2006.

[14] K. Fodero, H. H. King, M. J. H. Lum, C. Bland, J. Rosen, M. Sinanan, and B. Hannaford, "Control system architecture for a minimally invasive surgical robot," *Medicine Meets Virtual Reality 14: Accelerating Change in Healthcare: Next Medical Toolkit*, 2006.

[15] C. R. Johnstun and C. C. Smith, "Modeling and design of a mechanical tendon actuation system," *Journal of Dynamic Systems, Measurement and Control*, vol. 114, no. 2, pp. 253–261, June 1992.

[16] P. J. Hacksel and S. E. Salcudean, "Estimation of environment forces and rigid-body velocities using observers," in *Proc. IEEE International Conference on Robotics and Automation*, San Diego, CA, USA, May 1994, pp. 931–936.

POMDPs for Robotic Tasks with Mixed Observability

Sylvie C.W. Ong   Shao Wei Png   David Hsu   Wee Sun Lee
Department of Computer Science, National University of Singapore
Singapore 117590, Singapore

Abstract— **Partially observable Markov decision processes (POMDPs) provide a principled mathematical framework for motion planning of autonomous robots in uncertain and dynamic environments. They have been successfully applied to various robotic tasks, but a major challenge is to scale up POMDP algorithms for more complex robotic systems. Robotic systems often have *mixed observability*: even when a robot's state is not fully observable, some components of the state may still be fully observable. Exploiting this, we use a factored model to represent separately the fully and partially observable components of a robot's state and derive a compact lower-dimensional representation of its belief space. We then use this factored representation in conjunction with a point-based algorithm to compute approximate POMDP solutions. Separating fully and partially observable state components using a factored model opens up several opportunities to improve the efficiency of point-based POMDP algorithms. Experiments show that on standard test problems, our new algorithm is many times faster than a leading point-based POMDP algorithm.**

I. INTRODUCTION

Planning under uncertainty is a critical ability for autonomous robots operating in uncontrolled environments, such as homes or offices. In robot motion planning, uncertainty arises from two main sources: a robot's action and its perception. If the effect of a robot's action is uncertain, but its state is fully observable, then Markov decision processes (MDPs) provide an adequate model for planning. MDPs with a large number of states can often be solved efficiently (see, *e.g.*, [2]). When a robot's state is not fully observable, maybe due to noisy sensors, partially observable Markov decision processes (POMDPs) become necessary, and solving POMDPs is much more difficult. Despite the impressive progress of point-based POMDP algorithms in recent years [9], [11], [16], [19], [20], solving POMDPs with a large number of states remains a challenge. It is, however, important to note that robotic systems often have *mixed observability*: even when a robot's state is not fully observable, some components of the state may still be fully observable. For example, consider a mobile robot equipped with an accurate compass, but not a geographic positioning system (GPS). Its orientation is fully observable, though its position may only be partially observable. We refer to such problems as *mixed observability MDPs* (MOMDPs), a special class of POMDPs. In this work, we separate the fully and partially observable components of the state through a factored model and show that the new representation drastically improves the speed of POMDP planning, leading to a much faster algorithm for robotic systems with mixed observability.

In a POMDP, a robot's state is not fully observable. Thus we model it as a *belief*, which is a probability distribution over all possible robot states. The set of all beliefs form the *belief space* $\mathcal{B}$. The concept of belief space is similar to that of configuration space, except that each point in $\mathcal{B}$ represents a probability distribution over robot states rather than a single robot configuration or state. Intuitively, the difficulty of solving POMDPs is due to the "curse of dimensionality": in a POMDP with discrete states, the belief space $\mathcal{B}$ has dimensionality equal to $|\mathcal{S}|$, the number of robot states. The size of $\mathcal{B}$ thus grows exponentially with $|\mathcal{S}|$. Consider, for example, the navigation problem for an autonomous underwater vehicle (AUV). The state of the robot vehicle consists of its 3-D position and orientation. Suppose that after discretization, the robot may assume any of 100 possible positions on a 10×10 grid in the horizontal plane, 5 depth levels, and 24 orientations. The resulting belief space is 12,000-dimensional!

Now if the robot has an accurate pressure sensor and a gyroscope, we may reasonably assume that the depth level and the orientation are known exactly and fully observable, and only maintain a belief on the robot's uncertain horizontal position. In this case, the belief space becomes a union of 120 disjoint 100-dimensional subspaces. Each subspace corresponds to an exact depth level, an exact orientation, and beliefs on the uncertain horizontal positions. These 100-dimensional subspaces are still large, but a substantial reduction from the original 12,000-dimensional space.

The main idea of our approach is to exploit full observability whenever possible to gain computational efficiency. We separate the fully and partially observable state components using a factored model and represent explicitly the disjoint belief subspaces in a MOMDP so that all operations can be performed in these lower-dimensional subspaces.

The observability of a robot's state is closely related to sensor limitations. Two common types of sensor limitations are addressed in this work. In the first case, some components of a robot's state are sensed accurately and considered fully observable. Our approach handles this by separating the state components with high sensing precision from the rest. The second case is more subtle. Some sensors have *bounded* errors, but are not accurate enough to allow any assumption of fully observable state components *a priori*. Nevertheless we show that a robot with such sensors can be modeled as a MOMDP by (re)parameterizing the robot's state space. The reparameterization technique enables a much broader class of planning problems to benefit from our approach.

We tested our algorithm on three distinct robotic tasks with large state spaces. The results show that it significantly outperforms a leading point-based POMDP algorithm.

II. BACKGROUND

A. POMDPs

A POMDP models an agent taking a sequence of actions under uncertainty to achieve a goal. Formally a discrete POMDP with an infinite horizon is specified as a tuple $(\mathcal{S}, \mathcal{A}, \mathcal{O}, T, Z, R, \gamma)$, where $\mathcal{S}$ is a set of states, $\mathcal{A}$ is a set of actions, and $\mathcal{O}$ is a set of observations.

In each time step, the agent lies in some state $s \in \mathcal{S}$. It takes some action $a \in \mathcal{A}$ and moves from s to a new state s'. Due to the uncertainty in action, the end state s' is modeled as a conditional probability function $T(s, a, s') = p(s'|s, a)$, which gives the probability that the agent lies in s', after taking action a in state s. The agent then makes an observation to gather information on its own state. Due to the uncertainty in observation, the observation result $o \in \mathcal{O}$ is again modeled as a conditional probability function $Z(s, a, o) = p(o|s, a)$. See Fig. 1 for an illustration.

To elicit desirable agent behavior, we define a suitable reward function $R(s, a)$. In each time step, the agent receives a reward $R(s, a)$ if it takes action a in state s. The agent's goal is to maximize its expected total reward by choosing a suitable sequence of actions. For infinite-horizon POMDPs, the sequence of actions has infinite length. We specify a discount factor $\gamma \in [0, 1)$ so that the total reward is finite and the problem is well defined. In this case, the expected total reward is $\mathrm{E}\left[\sum_{t=0}^{\infty} \gamma^t R(s_t, a_t)\right]$, where s_t and a_t denote the agent's state and action at time t, respectively.

For a POMDP, planning means computing an *optimal policy* that maximizes the expected total reward. Normally, a policy is a mapping from the agent's state to an action. It tells the agent what action to take in each state. However, in a POMDP, the agent's state is partially observable and not known exactly. We rely on the concept of a belief, which is a probability distribution over $\mathcal{S}$. A POMDP policy $\pi: \mathcal{B} \to \mathcal{A}$ maps a belief $b \in \mathcal{B}$ to a prescribed action $a \in \mathcal{A}$.

A policy π induces a value function $V(b)$ that specifies the expected total reward of executing policy π starting from b. It is known that V^*, the value function for the optimal policy π^*, can be approximated arbitrarily closely by a convex, piecewise-linear function

$$V(b) = \max_{\alpha \in \Gamma}(\alpha \cdot b), \quad (1)$$

where Γ is a finite set of vectors called α-vectors, b is the discrete vector representation of a belief, and $\alpha \cdot b$ denotes the inner product of an α-vector and b. Each α-vector is associated with an action. The policy can be executed by evaluating (1) to find the action corresponding to the best α-vector at the current belief. So a policy can be represented by a value function $V(b)$ consisting of a set Γ of α-vectors. Policy computation, which, in this case, involves the construction of Γ, is usually performed offline.

Given a policy, represented as a value function $V(b)$, the control of the agent's actions, also called policy execution, is performed online in real time. It consists of two steps executed repeatedly. The first step is action selection. If the agent's current belief is b, it finds the action a that maximizes $V(b)$ by evaluating (1). The second step is belief estimation. After the agent takes an action a and receives an observation o, its new belief b' is given by

$$b'(s') = \tau(b, a, o) = \eta Z(s', a, o) \sum_{s \in \mathcal{S}} T(s, a, s')b(s), \quad (2)$$

where η is a normalizing constant. The process then repeats.

B. Related Work

POMDPs are a powerful framework for planning under uncertainty [8], [17]. It has a solid mathematical foundation and wide applicability. Its main disadvantage is high computational complexity [10]. As mentioned in Section I, the belief space $\mathcal{B}$ used for POMDP policy computation grows exponentially with the number of states that an agent has. The resulting curse of dimensionality is one major obstacle to efficient solution of POMDPs. There are several approaches to overcome the difficulty, including sampling $\mathcal{B}$, building factored models of $\mathcal{B}$ [1], or building lower-dimensional approximations of $\mathcal{B}$ [13], [14]. In recent years, point-based algorithms, which are based on the idea of sampling, have made impressive progress in computing approximate solutions to large POMDPs [9], [11], [16], [19], [20]. They have been successfully applied to a variety of non-trivial robotic tasks, including coastal navigation, grasping, target tracking, and exploration [5], [11], [12], [19]. In some cases, POMDPs with hundreds of states have been solved in a matter of seconds (see, *e.g.*, [7], [16], [19]).

Our work aims at overcoming the difficulty of high-dimensional belief spaces and further scaling up point-based POMDP algorithms for realistic robotic tasks. The main idea is to use a factored model to represent separately the fully and partially observable components of an agent's state and derive a compact lower-dimensional representation of $\mathcal{B}$. We then use this new representation in conjunction with a point-based algorithm to compute approximate POMDP solutions. A related idea has been used in medical therapy planning [4].

III. MIXED OBSERVABILITY MDPs

A. The MOMDP Model

In the standard POMDP model, the state lumps together multiple components. Consider again our AUV navigation example from Section I. The robot's state consists of its horizontal position, depth, and orientation. In contrast, a factored POMDP model separates the multiple state components and represents each as a distinct state variable. If three variables p, d, and θ represent the AUV's horizontal position, depth, and orientation, respectively, then the state space is a cross product of three subspaces: $\mathcal{S} = \mathcal{S}_p \times \mathcal{S}_d \times \mathcal{S}_\theta$. This allows for a more structured and compact representation of transition, observation, and reward functions in a POMDP.

We propose to represent a robotic system with mixed observability as a factored POMDP, specifically, a factored POMDP model with mixed state variables. We call our

202

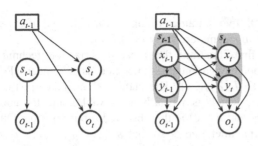

Fig. 1. The standard POMDP model (left) and the MOMDP model (right). A MOMDP state s is factored into two variables: $s = (x, y)$, where x is fully observable and y is partially observable.

model a MOMDP. In a MOMDP, the fully observable state components are represented as a single state variable x, while the partially observable components are represented as another state variable y. Thus (x, y) specifies the complete system state, and the state space is factored as $\mathcal{S} = \mathcal{X} \times \mathcal{Y}$, where $\mathcal{X}$ is the space of all possible values for x and $\mathcal{Y}$ is the space of all possible values for y. In our AUV example, x represents the depth and the orientation (d, θ), and y represents the horizontal position p.

Formally a MOMDP model is specified as a tuple $(\mathcal{X}, \mathcal{Y}, \mathcal{A}, \mathcal{O}, T_{\mathcal{X}}, T_{\mathcal{Y}}, Z, R, \gamma)$. The conditional probability function $T_{\mathcal{X}}(x, y, a, x') = p(x'|x, y, a)$ gives the probability that the fully observable state variable has value x' if the robot takes action a in state (x, y), and $T_{\mathcal{Y}}(x, y, a, x', y') = p(y'|x, y, a, x')$ gives the probability that the partially observable state variable has value y' if the robot takes action a in state (x, y) and the fully observable state variable has value x'. Compared with the standard POMDP model, the MOMDP model uses a factored state-space representation $\mathcal{X} \times \mathcal{Y}$, with the corresponding probabilistic state-transition functions $T_{\mathcal{X}}$ and $T_{\mathcal{Y}}$. All other aspects remain the same. See Fig. 1 for a comparison.[1]

So far, the changes introduced by the MOMDP model seem mostly notational. The computational advantages become apparent when we consider the belief space $\mathcal{B}$. Since the state variable x is fully observable and known exactly, we only need to maintain a belief $b_{\mathcal{Y}}$, a probability distribution on the state variable y. Any belief $b \in \mathcal{B}$ on the complete system state $s = (x, y)$ is then represented as $(x, b_{\mathcal{Y}})$. Let $\mathcal{B}_{\mathcal{Y}}$ denote the space of all beliefs on y. We now associate with each value x of the fully observable state variable a belief space for y: $\mathcal{B}_{\mathcal{Y}}(x) = \{(x, b_{\mathcal{Y}}) \mid b_{\mathcal{Y}} \in \mathcal{B}_{\mathcal{Y}}\}$. $\mathcal{B}_{\mathcal{Y}}(x)$ is a subspace in $\mathcal{B}$, and $\mathcal{B}$ is a union of these subspaces: $\mathcal{B} = \bigcup_{x \in \mathcal{X}} \mathcal{B}_{\mathcal{Y}}(x)$. Observe that while $\mathcal{B}$ has $|\mathcal{X}||\mathcal{Y}|$ dimensions, where $|\mathcal{X}|$ and $|\mathcal{Y}|$ are the number of states in $\mathcal{X}$ and $\mathcal{Y}$, each $\mathcal{B}_{\mathcal{Y}}(x)$ has only $|\mathcal{Y}|$ dimensions. Effectively we represent the high-dimensional space $\mathcal{B}$ as a union of lower-dimensional subspaces. When the uncertainty in a system is small, specifically, when $|\mathcal{Y}|$ is small, the MOMDP model leads to dramatic improvement in computational efficiency, due to the reduced dimensionality of the space.

Now consider how we would represent and execute a

MOMDP policy. As mentioned in Section II-A, a POMDP policy can be represented as a value function $V(b) = \max_{\alpha \in \Gamma}(\alpha \cdot b)$, where Γ is a set of α-vectors. In a MOMDP, a belief is given by $(x, b_{\mathcal{Y}})$, and the belief space $\mathcal{B}$ is union of subspaces $\mathcal{B}_{\mathcal{Y}}(x)$ for $x \in \mathcal{X}$. Correspondingly, a MOMDP value function $V(x, b_{\mathcal{Y}})$ is represented as a collection of α-vector sets: $\{\Gamma_{\mathcal{Y}}(x) \mid x \in \mathcal{X}\}$, where for each x, $\Gamma_{\mathcal{Y}}(x)$ is a set of α-vectors defined over $\mathcal{B}_{\mathcal{Y}}(x)$. To evaluate $V(x, b_{\mathcal{Y}})$, we first find the right α-vector set $\Gamma_{\mathcal{Y}}(x)$ using the x value and then find the maximum α-vector from the set:

$$V(x, b_{\mathcal{Y}}) = \max_{\alpha \in \Gamma_{\mathcal{Y}}(x)}(\alpha \cdot b_{\mathcal{Y}}). \qquad (3)$$

In general, any value function $V(b) = \max_{\alpha \in \Gamma}(\alpha \cdot b)$ can be represented in this new form, as stated in the theorem below.

Theorem 1: Let $\mathcal{B} = \bigcup_{x \in \mathcal{X}} \mathcal{B}_{\mathcal{Y}}(x)$ be the belief space of a MOMDP with state space $\mathcal{X} \times \mathcal{Y}$. If $V(b) = \max_{\alpha \in \Gamma}(\alpha \cdot b)$ is any value function over $\mathcal{B}$ in the standard POMDP form, then $V(b)$ is equivalent to a MOMDP value function $V'(x, b_{\mathcal{Y}}) = \max_{\alpha \in \Gamma_{\mathcal{Y}}(x)}(\alpha \cdot b_{\mathcal{Y}})$ such that for any $b = (x, b_{\mathcal{Y}})$ with $b \in \mathcal{B}$, $x \in \mathcal{X}$, and $b_{\mathcal{Y}} \in \mathcal{B}_{\mathcal{Y}}(x)$, $V(b) = V'(x, b_{\mathcal{Y}})$.[2]

Geometrically, each α-vector set $\Gamma_{\mathcal{Y}}(x)$ represents a restriction of the POMDP value function $V(b)$ to the subspace $\mathcal{B}_{\mathcal{Y}}(x)$: $V_x(b_{\mathcal{Y}}) = \max_{\alpha \in \Gamma_{\mathcal{Y}}(x)}(\alpha \cdot b_{\mathcal{Y}})$. In a MOMDP, we compute only these lower-dimensional restrictions $\{V_x(b_{\mathcal{Y}}) \mid x \in \mathcal{X}\}$, because $\mathcal{B}$ is simply a union of subspaces $\mathcal{B}_{\mathcal{Y}}(x)$ for $x \in \mathcal{X}$.

A comparison of (1) and (3) also indicates that (3) often results in faster policy execution, because action selection can be performed more efficiently. First, each α-vector in $\Gamma_{\mathcal{Y}}(x)$ has length $|\mathcal{Y}|$, while each α-vector in Γ has length $|\mathcal{X}||\mathcal{Y}|$. Furthermore, in a MOMDP value function, the α-vectors are partitioned into groups according to the value of x. We only need to calculate the maximum over $\Gamma_{\mathcal{Y}}(x)$, which is potentially much smaller than Γ in size.

In summary, by factoring out the fully and partially observable state variables, a MOMDP model reveals the internal structure of the belief space as a union of lower-dimensional subspaces. We want to exploit this structure and perform all operations on beliefs and value functions in these lower-dimensional subspaces rather than the original belief space. Before we describe the details of our algorithm, let us first look at how MOMDPs can be used to handle uncertainty commonly encountered in robotic systems.

B. Modeling Robotic Tasks with MOMDPs

Sensor limitations are a major source of uncertainty in robotic systems and are closely related to observability. If a robot's state consists of several components, some are fully observable, possibly due to accurate sensing, but others are not. This is a natural case for modeling with MOMDPs. All fully observable components are grouped together and modeled by the variable x. The other components are modeled by the variable y.

---

[1]A MOMDP can be regarded as an instance of dynamic Bayesian network (DBN). Following the DBN methodology, we could factor x or y further, but this may lead to difficulty in value function representation.

[2]Due to space limitations, the proofs of all the theorems are provided in the full version of the paper, which will be available on-line at http://motion.comp.nus.edu.sg/papers/rss09.pdf

Sometimes, however, a system does not appear to have mixed observability: none of the sensed state components is fully observable. Is it still possible to model it as a MOMDP? The answer is yes under certain conditions, despite the absence of obvious fully observable state components.

1) Pseudo Full Observability: All sensors are ultimately limited in resolution. It is task-dependent to decide whether the sensor resolution is accurate enough to make the sensed state component fully observable. For example, a robot searches for an unseen target and has small uncertainty on its own position. It is reasonable to assume that the robot position is fully observable, as the uncertainty on the robot position is small compared to that on the target position and the robot's behavior depends mostly on the latter. By treating the robot position as fully observable, we can take advantage of the MOMDP model for faster policy computation and execution. Note, however, that treating the unseen target's position as fully observable is not reasonable and unlikely to lead to a useful policy.

For increased robustness, we can actually execute a computed MOMDP policy on the corresponding POMDP model, which does *not* assume any fully observable state variables. The POMDP treats both state variables x and y as partially observable and maintains a belief b over them. To account for the additional uncertainty on x in the POMDP, our idea is to define a new value function $V_P(b)$ by averaging over the value function $V(x, b_y)$, which represents the computed MOMDP policy. For this, we first calculate a belief b_x on x by marginalizing out y: $b_x(x) = \sum_{y \in \mathcal{Y}} b(x, y)$. We then calculate the belief $b_{y|x}$ on y, conditioned on the x value: $b_{y|x}(y) = b(x, y)/b_x(x)$. Now the new value function

$$V_P(b) = \sum_{x \in \mathcal{X}} b_x(x)V(x, b_{y|x})$$

can be used to generate a policy for the POMDP through one-step look-ahead search.

Let $V^*(b)$ and $V^*(x, b_y)$ be respectively the optimal value functions for a POMDP and the corresponding MOMDP under the assumption of fully observable state variables. It can be shown that the value function $V_P(b)$ constructed from $V^*(x, b_y)$ is an upper bound on $V^*(b)$. In this sense, the MOMDP model is an approximation to the POMDP model. The MOMDP model is less accurate due to the additional assumption, but has substantial computational advantages.

2) Reparameterized Full Observability: Sometimes sensors are not accurate enough to justify an assumption of full observability for any sensed component. However, we can still model such a system as a MOMDP by reparameterizing the state space $\mathcal{S}$. For example, in a navigation task, the position sensor is noisy, but accurate enough to localize the robot to a small region around the actual position. Define $h(o)$, the *preimage* of an observation o, to be the set of states that have non-zero probability of emitting o. We say that a system has *bounded uncertainty* if the preimage of its observation is always a small subset of $\mathcal{S}$: $\max_{o \in \mathcal{O}}(|h(o)|/|\mathcal{S}|) < c$ for some constant $c \ll 1$. We show below that any system modeled as a POMDP, can be equivalently constructed as an

MOMDP by reparameterizing $\mathcal{S}$, even if none of the sensed state components are fully observable.

We first illustrate the reparameterization technique on the robot navigation task, modeled as a standard POMDP $(\mathcal{S}, \mathcal{A}, \mathcal{O}, T, Z, R, \gamma)$. The state $s \in \mathcal{S}$ indicates the robot position, which is partially observable, due to inaccurate sensing, and $o \in \mathcal{O}$ is the observed robot position. Our goal is to reparameterize $\mathcal{S}$ so that $s = (x, y)$, where x is fully observable and y is partially observable. We choose x to be o, the observed robot position, and define y as the offset of the actual position from the observed position. Using this parameterization, we can construct the new state-transition functions $T_\mathcal{X}$ and $T_\mathcal{Y}$ from the old state-transition function T and observation function Z:

$$T_\mathcal{X}(x, y, a, x') = \sum_{\sigma \in \mathcal{S}} T(s, a, \sigma)Z(\sigma, a, x'), \quad (4)$$

$$T_\mathcal{Y}(x, y, a, x', y') = T(s, a, s')Z(s', a, x')/T_\mathcal{X}(x, y, a, x'), (5)$$

where $s = (x, y)$ and $s' = (x', y')$. The correctness of this construction can be verified by applying the definitions and the basic probability rules. Similarly, we construct the new observation function Z_M and the new reward function R_M: $Z_M(x, y, a, o) = 1$ if and only if $x = o$, and $R_M(x, y, a) = R(s, a)$, where $s = (x, y)$.

In the general case, we use x to index the preimage of each observation $o \in \mathcal{O}$ and use y to indicate the exact state within the preimage. The resulting reparameterized MOMDP has the following property:

Theorem 2: The POMDP $(\mathcal{S}, \mathcal{A}, \mathcal{O}, T, Z, R, \gamma)$ and the reparameterized MOMDP $(\mathcal{X}, \mathcal{Y}, \mathcal{A}, \mathcal{O}, T_\mathcal{X}, T_\mathcal{Y}, Z_M, R_M, \gamma)$ with $\mathcal{X} = \mathcal{O}$ are equivalent: Let b and b' be the beliefs reached after an arbitrary sequence of actions and observations, $a_1, o_1, \cdots, a_t, o_t$, by the POMDP and MOMDP models respectively. Then b and b' represent the same belief in the original POMDP state space. The probability of observation given the history, $p(o_t|a_1, o_1, \cdots, a_t)$, and the expected total reward for any sequence of actions and observations, is the same in both models. Consequently, any policy has the same expected reward for both models.

The reparameterization can be performed on any POMDP. The resulting MOMDP brings computational advantages if $|\mathcal{Y}|$ is small. This happens when a system has bounded uncertainty with a small constant c, meaning that the observations are informative. Furthermore, we can relax the condition of bounded uncertainty and require that the preimages are bounded for some, rather than all state components. The reparameterization is beneficial in this general case as well.

IV. COMPUTING MOMDP POLICIES

A. Overview

For policy computation, we combine the MOMDP model with a point-based POMDP algorithm. Point-based algorithms have been highly successful in computing approximate solutions to large POMDPs. Their key idea is to sample a set of points from $\mathcal{B}$ and use it as an approximate representation of $\mathcal{B}$, rather than represent $\mathcal{B}$ exactly. They

Algorithm 1 Point-based MOMDP policy computation.

1: Initialize the α-vectors, $\Gamma = \{\Gamma_y(x) \mid x \in \mathcal{X}\}$, representing the lower bound $\underline{V}$ on the optimal value function V^*. Initialize the upper bound $\overline{V}$ on V^*.

2: Insert the initial belief point (x_0, b_{y0}) as the root of the tree $T_{\mathcal{R}}$.

3: **repeat**

4:     SAMPLE($T_{\mathcal{R}}$, Γ).

5:     Choose a subset of nodes from $T_{\mathcal{R}}$. For each chosen node (x, b_y), BACKUP($T_{\mathcal{R}}, \Gamma, (x, b_y)$).

6:     PRUNE($T_{\mathcal{R}}$, Γ).

7: **until** termination conditions are satisfied.

8: **return** Γ.

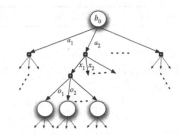

Fig. 2. The belief search tree rooted at $b_0 = (x_0, b_{y0})$. Each circle indicates a node representing a MOMDP belief state (x, b_y).

also maintain a set of α-vectors as an approximation to the optimal value function. The various point-based algorithms differ mainly in their strategies for sampling $\mathcal{B}$ and constructing α-vectors.

The MOMDP model enables us to treat a high-dimensional belief space as a union of lower-dimensional subspaces. We represent explicitly these subspaces and compute only the restriction of the value function to these subspaces. The idea is general and is independent of the strategies for belief point sampling or α-vector construction. Hence the MOMDP model can be used in conjunction with any of the existing point-based algorithms. However, to make the presentation concrete, we describe our approach based on the SARSOP algorithm [9], one of the leading point-based POMDP solvers.

Our point-based MOMDP algorithm is based on *value iteration* [15]. Exploiting the fact that the optimal value function V^* must satisfy the Bellman equation, value iteration starts with an initial approximation to V^* and performs backup operations on the approximation by iterating on the Bellman equation until the iteration converges.

In our algorithm, we sample incrementally a set of points from $\mathcal{B}$ and maintain a set of α-vectors, which represents a piecewise-linear lower-bound approximation $\underline{V}$ to V^*. To improve the approximation $\underline{V}$, we perform backup operations on the α-vectors at the sampled points. A backup operation is essentially an iteration of dynamic programming, which improves the approximation by looking ahead one step further. With suitable initialization, $\underline{V}$ is always a lower bound on V^*, and converges to V^* under suitable conditions [6], [11], [19]. The MOMDP model enables the primitive operations in the algorithm, such as α-vector backup and pruning, to be performed more efficiently.

B. The Algorithm

After initialization, our algorithm iterates over three main functions, SAMPLE, BACKUP, and PRUNE (Algorithm 1).

1) SAMPLE: Let $\mathcal{R} \subset \mathcal{B}$ be the set of points reachable from a given initial belief point $b_0 = (x_0, b_{y0})$ under arbitrary sequences of actions. Most of the recent point-based POMDP algorithms sample from $\mathcal{R}$ instead of $\mathcal{B}$ for computational efficiency. The SARSOP algorithm aims to

be even more efficient by focusing the sampling near $\mathcal{R}^*$, the subset of points reachable from (x_0, b_{y0}) under *optimal* sequences of actions, usually a much smaller space than $\mathcal{R}$.

To sample near $\mathcal{R}^*$, we maintain both a lower bound $\underline{V}$ and an upper bound $\overline{V}$ on the optimal value function V^*. The lower bound $\underline{V}$ is represented by a collection of α-vector sets, $\{\Gamma_y(x) \mid x \in \mathcal{X}\}$, and initialized using fixed-action policies [3]. The upper bound $\overline{V}$ is represented by a collection of sets of belief-value pairs: $\{\Upsilon_y(x) \mid x \in \mathcal{X}\}$, where $\Upsilon_y(x) = \{(b_y, \overline{v}) \mid b_y \in \mathcal{B}_y(x)\}$. A belief-value pair $(b_y, \overline{v}) \in \Upsilon_y(x)$ gives an upper bound $\overline{v}$ on the value function at (x, b_y). They are used to perform sawtooth approximations [3] in each of the subspaces $\mathcal{B}_y(x)$. The upper bound can be initialized in various ways, using the MDP or the Fast Informed Bound technique [3].

The sampled points form a tree $T_{\mathcal{R}}$ (Fig. 2). Each node of $T_{\mathcal{R}}$ represents a sampled point in $\mathcal{R}$. In the following, we use the notation (x, b_y) to denote both a sampled point and its corresponding node in $T_{\mathcal{R}}$. The root of $T_{\mathcal{R}}$ is the initial belief point (x_0, b_{y0}).

To sample new belief points, we start from the root of $T_{\mathcal{R}}$ and traverse a single path down. At each node along the path, we choose a with the highest upper bound and choose x' and o that make the largest contribution to the gap ϵ between the upper and the lower bounds at the root of $T_{\mathcal{R}}$. New tree nodes are created if necessary. To do so, if at a node (x, b_y), we choose a, x', and o, a new belief on y is computed:

$$b'_y(y') = \tau(x, b_y, a, x', o)$$
$$= \eta Z(x', y', a, o)$$
$$\times \sum_{y \in \mathcal{Y}} T_x(x, y, a, x') T_y(x, y, a, x', y') b_y(y), \quad (6)$$

where η is a normalization constant. A new node (x', b_y') is then inserted into $T_{\mathcal{R}}$ as a child of (x, b_y). Clearly, every point sampled this way is reachable from (x_0, b_{y0}). By carefully choosing a, x', and o based on the upper and lower bounds, we can keep the sampled belief points near $\mathcal{R}^*$.

When the sampling path ends under a suitable set of conditions, we go up back to the root of $T_{\mathcal{R}}$ along the same path and perform backup at each node along the way.

2) BACKUP: A backup operation at a node (x, b_y) collates the value function information in the children of (x, b_y) and propagates it back to (x, b_y). The operations are performed on both the lower and the upper bounds. For the lower bound, we perform α-vector backup (Algorithm 2). A new α-vector resulting from the backup operation at (x, b_y) is inserted into

Algorithm 2 α-vector backup at a node (x, b_y) of $T_{\mathcal{R}}$.

BACKUP$(T_{\mathcal{R}}, \Gamma, (x, b_y))$

1: For all $a \in \mathcal{A}$, $x' \in \mathcal{X}$, $o \in \mathcal{O}$,
$$\alpha_{a,x',o} \leftarrow \operatorname{argmax}_{\alpha \in \Gamma_y(x')}(\alpha \cdot \tau(x, b_y, a, x', o)).$$
2: For all $a \in \mathcal{A}$, $y \in \mathcal{Y}$,
$$\alpha_a(y) \leftarrow R(x, y, a) + \gamma \sum_{x',o,y'} (T_{\mathcal{X}}(x, y, a, x')$$
$$\times T_{\mathcal{Y}}(x, y, a, x', y') Z(x', y', a, o)\alpha_{a,x',o}(y')).$$
3: $\alpha' \leftarrow \operatorname{argmax}_{a \in \mathcal{A}}(\alpha_a \cdot b_y)$
4: Insert α' into $\Gamma_y(x)$.

$\Gamma_y(x)$, the set of α-vectors associated with observed state value x. For the upper bound backup at (x, b_y), we perform the standard Bellman update to get a new belief-value pair and insert it into $\Upsilon_y(x)$.

3) PRUNE: Invocation of SAMPLE and BACKUP generates new sampled points and α-vectors. However, not all of them are useful for constructing an optimal policy and are pruned to improve computational efficiency. We iterate through the α-vector sets in the collection $\{\Gamma_y(x) \mid x \in \mathcal{X}\}$ and prune any α-vector in $\Gamma_y(x)$ that does not dominate the rest at some sampled point (x, b_y), where $b_y \in \mathcal{B}_y(x)$.

Our description of the algorithm is quite brief due to space limitations. More details and justifications for our particular choices of the sampling, backup, and pruning strategies can be found in [9].

We would like to point out that to solve a MOMDP, the main modifications required in SARSOP are the belief update operation (Eq. (6)) and the backup operation (Algorithm 2). They are common to most point-based POMDP algorithms, such as PBVI [11], Perseus [20], HSVI [18], and FSVI [16]. So using Eq. (6) and Algorithm 2 to replace the corresponding parts in these algorithm would allow them to benefit from the MOMDP approach as well.

C. Correctness and Computational Efficiency

Modifying the belief update and backup operations does not affect the convergence property of SARSOP. The algorithm above provides the same theoretical guarantee as the original SASRSOP algorithm. The formal statement and the proof are given in the full version of the paper.

MOMDPs allow the belief space $\mathcal{B}$ to be represented as a union of low-dimensional subspaces $\mathcal{B}_y(x)$ for $x \in \mathcal{X}$. This brings substantial computational advantages. Specifically, the efficiency gain of our algorithm comes mainly from BACKUP and PRUNE, where α-vectors are processed. In a MOMDP, α-vectors have length $|\mathcal{Y}|$, while in the corresponding POMDP, α-vectors have length $|\mathcal{X}||\mathcal{Y}|$. As a result, all operations on α-vectors in BACKUP and PRUNE are faster by a factor of $|\mathcal{X}|$ in our algorithm. Furthermore, in a MOMDP, α-vectors are divided into disjoint sets $\Gamma_y(x)$ for $x \in \mathcal{X}$. When we need to find the best α-vector, *e.g.*, in line 1 of Algorithm 2, we only do so within $\Gamma_y(x)$ for some fixed x rather than over all α-vectors, as in a POMDP.

The efficiency gain in BACKUP sometimes comes at a cost: although α-vectors in a MOMDP are shorter, they also contain less information, compared with POMDP α-vectors. In Algorithm 2, performing backup at (x, b_y) generates an

α-vector that spans only the subspace $\mathcal{B}_y(x)$. The backup does not generate any information in any other subspace $\mathcal{B}_y(x')$ with $x' \neq x$. In contrast, POMDP algorithms do more computation while performing backup and generate α-vectors that span the entire space $\mathcal{B}$. If a problem has many similar observable states in the sense that the α-vectors in one belief subspace $\mathcal{B}_y$ are useful in other subspaces as well, then POMDP algorithms may obtain more useful information in each backup operation and perform better than our algorithm, despite the higher cost of each backup operation. This, however, requires a special property which may not hold in general for complex systems.

V. EXPERIMENTS

We used MOMDPs to model several distinct robotic tasks, all having large state spaces, and tested our algorithm on them. In this section, we describe the experimental setup and the results.

A. Robotic Tasks

1) Tag: The Tag problem first appeared in the work on PBVI [11], one of the first point-based POMDP algorithms. In Tag, the robot's goal is to follow a target that intentionally moves away. The robot and the target operate in an environment modeled as a grid. They can start in any grid positions, and in one step, they can either stay or move to one of four adjacent positions (above, below, left, and right). The robot knows its own position exactly, but can observe the target position only if they are in the same position. The robot pays a cost for each move and receives a reward when it arrives in the same position as that of the target.

In the MOMDP for this task, the robot position, which is known exactly, is modeled by the fully observable state variable x. The x variable can also take one extra value that indicates that the robot and the target are in the same position. The target position is modeled by the partially observable state variable y, as the robot does not see the target in general. Experiments were performed on environment maps with different resolutions. Tag(M) denotes an experiment on a map with M positions. Here $|\mathcal{X}| = M + 1$ and $|\mathcal{Y}| = M$, while in the standard POMDP model, the state space has $|\mathcal{S}| = (M + 1)M$ dimensions.

2) Rock Sample: The Rock Sample problem [18] has frequently been used to test the scalability of new POMDP algorithms. In this problem, a rover explores an area modeled as a grid and searches for rocks with scientific value. The rover always knows its own position exactly, as well as those of the rocks. However, it does not know which rocks are valuable. The rover can take noisy long-range sensor readings to gather information on the rocks. The accuracy of the readings depends on the distance between the rover and the rocks. The rover can also sample a rock in the immediate vicinity. It receives a reward or a penalty, depending on whether the sampled rock is valuable.

Here, the x variable in the MOMDP represents the robot position, and the y variable is a binary vector in which each entry indicates whether a rock is valuable or not. Experiments

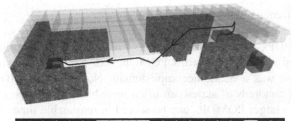

Fig. 3. AUV navigation. Top: The 3-D environment and an AUV simulation trajectory generated from the computed policy. The AUV can localize its horizontal position only at the surface level (lightly shaded in the picture). It rises to the surface level to localize, navigates through the rocks, and then dives to reach the goal. Bottom: The grid map for the deepest level. "S" marks the AUV starting positions, which are all located at this level. The AUV is equally likely to start in any of them. "E" marks the end positions, also located at this level only. "R" marks the rocks.

were performed on maps of different sizes and with different number of rocks. RockSample(M, R) denotes a map size of $M \times M$ and R rocks.

3) AUV Navigation: An AUV navigates in an oceanic environment modeled as a 3-D grid with 4 levels and 7×20 positions at each level (Fig. 3). It needs to navigate from the right boundary of the deepest level to some goal locations near the left boundary and must avoid rock formations, which are present in all levels except the surface. In each step, the AUV may either stay in the current position or move to any adjacent position along its current orientation. Whether the action is stay or move, the AUV may drift to a neighboring horizontal position due to control uncertainty or ocean currents. The AUV does not know its exact starting position. It knows its horizontal position only at the surface level, where GPS signals are available. However, surfacing causes heavy fuel consumption and must be avoided if possible. Using its pressure sensor and gyroscope, the AUV can acquire accurate information on the depth and the orientation, which is discretized into 24 values.

In the MOMDP model, the x variable represents the AUV's depth and orientation, and the y variable represents the AUV's horizontal position. Although in our specific setup, the belief over the AUV's horizontal position is always unimodal, the MOMDP approach is general and can be applied without change if the belief is multimodal.

B. Results

We applied the MOMDP algorithm to the three tasks above. For each task, we first performed long preliminary runs to determine approximately the reward level for the optimal policies and the amount of time needed to reach it. We then ran the algorithm for a maximum of half an hour to reach this level. To estimate the expected total reward of the resulting policy, we performed sufficiently large number of simulation runs until the variance in the estimated value was small. For comparison, we also ran SARSOP [9], a leading POMDP algorithm, on the same tasks modeled as standard

TABLE I

PERFORMANCE COMPARISON ON TASKS WITH MIXED OBSERVABILITY.

			Reward	Time (s)
Tag(29)				
$\|\mathcal{X}\|=30, \|\mathcal{Y}\|=29$		MOMDP	-6.03 ± 0.04	4.7
$\|\mathcal{S}\|=870, \|\mathcal{A}\|=5, \|\mathcal{O}\|=30$		SARSOP	-6.03 ± 0.12	16.5
Tag(55)				
$\|\mathcal{X}\|=56, \|\mathcal{Y}\|=55$		MOMDP	-9.90 ± 0.11	19
$\|\mathcal{S}\|=3,080, \|\mathcal{A}\|=5, \|\mathcal{O}\|=56$		SARSOP	-9.90 ± 0.12	736
RockSample(7,8)				
$\|\mathcal{X}\|=50, \|\mathcal{Y}\|=256$		MOMDP	21.47 ± 0.04	160
$\|\mathcal{S}\|=12,545,^3 \|\mathcal{A}\|=13, \|\mathcal{O}\|=2$		SARSOP	21.39 ± 0.01	810
RockSample(10,10)				
$\|\mathcal{X}\|=101, \|\mathcal{Y}\|=1,024$		MOMDP	21.47 ± 0.04	318
$\|\mathcal{S}\|=102,401, \|\mathcal{A}\|=15, \|\mathcal{O}\|=2$		SARSOP	21.47 ± 0.11	1589
RockSample(11,11)				
$\|\mathcal{X}\|=122, \|\mathcal{Y}\|=2,048$		MOMDP	21.80 ± 0.04	188
$\|\mathcal{S}\|=247,809, \|\mathcal{A}\|=16, \|\mathcal{O}\|=2$		SARSOP	21.56 ± 0.11	1369
AUV Navigation				
$\|\mathcal{X}\|=96, \|\mathcal{Y}\|=141$		MOMDP	1020.0 ± 8.5	124
$\|\mathcal{S}\|=13,536, \|\mathcal{A}\|=6, \|\mathcal{O}\|=144$		SARSOP	1019.8 ± 9.7	409

POMDPs. Both algorithms are implemented in C++, and the experiments were performed on a PC with a 2.66GHz Intel processor and 2GB memory.

1) Mixed Observability Tasks: The results are shown in Table I. Column 3 of the table lists the estimated expected total rewards for the computed policies and the 95% confidence intervals. Column 4 lists the running times.

For all tasks, the MOMDP algorithm obtained good approximate solutions well within the time limit and outperformed SARSOP by many times. We ran multiple experiments for Rock Sample with different map sizes and numbers of rocks. As the problem size increases, both $|\mathcal{X}|$ and $|\mathcal{Y}|$ increase. The performance gap between our algorithm, which uses the MOMDP model, and SARSOP, which uses the standard POMDP model, tends to increase as well. In particular, for the largest problem, RockSample(11,11), SARSOP never reached the same reward level attained by the MOMDP algorithm within the time limit. This is not surprising as the computational efficiency gain achievable from the MOMDP model depends on $|\mathcal{X}|$ (Section IV-C). The experiments on Tag show a similar trend.

2) Pseudo Full Observability: We also tested our algorithm on robotic tasks in which all sensed state components are partially observable and obtained promising results.

The original Tag problem is modified to create a noisy version, in which the robot has $p\%$ chance of observing its own position correctly and $(1/8)(100 - p)\%$ chance of observing each of the 8 surrounding positions. In this case, both the robot position and the target position are partially observable. However, applying the technique in Section III-B.1, we modeled the robot position by the fully observable state variable x in the MOMDP and solved the MOMDP with our algorithm. The value function computed by the algorithm was used to generate a policy to control the robot without assuming any fully observable state variables.

We ran experiments with different sensor accuracy p for the Noisy Tag problem. The results are shown in Table II.

---

^{3}Small optimizations can be performed to reduce the number of states in the POMDP model. So $|\mathcal{S}|$ may not be equal to $|\mathcal{X}||\mathcal{Y}|$ for the corresponding MOMDP. However, this is not significant enough to affect $|\mathcal{S}|=O(|\mathcal{X}||\mathcal{Y}|)$.

TABLE II
PERFORMANCE COMPARISON ON TASKS USING PSEUDO OR REPARAMETERIZED FULL OBSERVABILITY TECHNIQUES.

		Reward	Time (s)
NoisyTag(29,90%)			
$|\mathcal{X}|=30, |\mathcal{Y}|=29$	MOMDP	-11.12 ± 0.14	4.5
$|\mathcal{S}|=870, |\mathcal{A}|=5, |\mathcal{O}|=30$	SARSOP	-11.12 ± 0.14	228.0
NoisyTag(29,50%)			
$|\mathcal{X}|=30, |\mathcal{Y}|=29$	MOMDP	-12.14 ± 0.14	1.5
$|\mathcal{S}|=870, |\mathcal{A}|=5, |\mathcal{O}|=30$	SARSOP	-12.15 ± 0.14	11.6
NoisyTag(29,10%)			
$|\mathcal{X}|=30, |\mathcal{Y}|=29$	MOMDP	-12.53 ± 0.14	1.5
$|\mathcal{S}|=870, |\mathcal{A}|=5, |\mathcal{O}|=30$	SARSOP	-12.59 ± 0.14	176.4
NoisyTag(55, 3×3)			
$|\mathcal{X}|=56, |\mathcal{Y}|=495$	MOMDP	-10.62 ± 0.10	32
$|\mathcal{S}|=3,080, |\mathcal{A}|=5, |\mathcal{O}|=56$	SARSOP	-10.61 ± 0.08	927

NoisyTag($M, p\%$) denotes a modified Tag problem with a map of M positions and sensor accuracy $p\%$. The MOMDP algorithm drastically outperformed SARSOP, even when p was as low as 10%. This is a little surprising. The MOMDP model brings computational advantages, but is less accurate than the POMDP model, due to the assumption of fully observable state variables. One would expect that the MOMDP algorithm may reach a reasonable reward level faster, but lose to SARSOP in the long run. However, in this case we did not observe any significant performance loss for the MOMDP algorithm. To confirm the results, we ran SARSOP for two additional hours, but its reward level did not improve much beyond those reported in the table.

3) Reparameterized Full Observability: The Tag problem is again modified so that the robot never observes its own position exactly, but only the 3×3 region around it in the grid. Both the robot and the target positions are partially observable. However, the preimage of any observation on the robot position is always bounded. We can apply the approach described in Section III-B.2 and reparameterize the robot position as (o_r, δ_r), where o_r indicates a 3×3 region in the grid and δ_r indicates the actual position of the robot within the region. We then model the reparameterized problem as a MOMDP: the x variable represents o_r, and the y variable represents δ_r and the target position.

Again the MOMDP algorithm significantly outformed SARSOP (Table II). This suggests that extracting fully observable state variables through reparameterization is a promising idea and deserves further investigation.

VI. CONCLUSION

POMDPs have been successfully used for motion planning under uncertainty in various robotic tasks [5], [11], [12], [19]. A major challenge remaining is to scale up POMDP algorithms for complex robotic systems. Exploiting the fact that many robotic systems have mixed observability, our MOMDP approach uses a factored model to separate the fully and partially observable components of a robot's state. We show that the factored representation drastically improves the speed of POMDP planning, when combined with a point-based POMDP algorithm. We further show that even when a robot does not have obvious fully observable state components, it still can be modeled as a MOMDP by reparameterizing the robot's state space.

Ten years ago, the best POMDP algorithm could solve POMDPs with a dozen states. Five years ago, a point-based algorithm solved a POMDP with almost 900 states, and it was a major accomplishment. Nowadays, POMDPs with hundreds of states can often be solved in seconds, and much larger POMDPs can be solved in reasonable time. We hope that our work is a step further in scaling up POMDP algorithms and ultimately making them practical for robot motion planning in uncertain and dynamic environments.

Acknowledgments. We thank Yanzhu Du for helping with the software implementation. We also thank Tomás Lozano-Pérez and Leslie Kaelbling from MIT for many insightful discussions. This work is supported in part by AcRF grant R-252-000-327-112 from the Ministry of Education of Singapore.

REFERENCES

[1] C. Guestrin, D. Koller, and R. Parr. Solving factored POMDPs with linear value functions. In *Int. Jnt. Conf. on Artificial Intelligence Workshop on Planning under Uncertainty & Incomplete Information*, pp. 67–75, 2001.

[2] C. Guestrin, D. Koller, R. Parr, and S. Venkataraman. Efficient solution algorithms for factored MDPs. *J. Artificial Intelligence Research*, 19:399–468, 2003.

[3] M. Hauskrecht. Value-function approximations for partially observable Markov decision processes. *J. Artificial Intelligence Research*, 13:33–94, 2000.

[4] M. Hauskrecht and H. Fraser. Planning medical therapy using partially observable Markov decision processes. In *Proc. Int. Workshop on Principles of Diagnosis*, pp. 182–189, 1998.

[5] K. Hsiao, L. Kaelbling, and T. Lozano-Pérez. Grasping POMDPs. In *Proc. IEEE Int. Conf. on Robotics & Automation*, pp. 4485–4692, 2007.

[6] D. Hsu, W. Lee, and N. Rong. What makes some POMDP problems easy to approximate? In *Advances in Neural Information Processing Systems (NIPS)*, 2007.

[7] ——. A point-based POMDP planner for target tracking. In *Proc. IEEE Int. Conf. on Robotics & Automation*, pp. 2644–2650, 2008.

[8] L. Kaelbling, M. Littman, and A. Cassandra. Planning and acting in partially observable stochastic domains. *Artificial Intelligence*, 101(1–2):99–134, 1998.

[9] H. Kurniawati, D. Hsu, and W. Lee. SARSOP: Efficient point-based POMDP planning by approximating optimally reachable belief spaces. In *Proc. Robotics: Science and Systems*, 2008.

[10] C. Papadimitriou and J. Tsisiklis. The complexity of Markov decision processes. *Mathematics of Operations Research*, 12(3):441–450, 1987.

[11] J. Pineau, G. Gordon, and S. Thrun. Point-based value iteration: An anytime algorithm for POMDPs. In *Proc. Int. Jnt. Conf. on Artificial Intelligence*, pp. 477–484, 2003.

[12] J. Pineau, M. Montemerlo, M. Pollack, N. Roy, and S. Thrun. Towards robotic assistants in nursing homes: Challenges and results. *Robotics & Autonomous Systems*, 42(3–4):271–281, 2003.

[13] P. Poupart and C. Boutilier. Value-directed compression of POMDPs. In *Advances in Neural Information Processing Systems (NIPS)*, 15:1547–1554. The MIT Press, 2003.

[14] N. Roy, G. Gordon, and S. Thrun. Finding aproximate POMDP solutions through belief compression. *J. Artificial Intelligence Research*, 23:1–40, 2005.

[15] S. Russell and P. Norvig. *Artificial Intelligence: A Modern Approach*. Prentice Hall, 2003.

[16] G. Shani, R. Brafman, and S. Shimony. Forward search value iteration for POMDPs. In *Proc. Int. Jnt. Conf. on Artificial Intelligence*, 2007.

[17] R. Smallwood and E. Sondik. The optimal control of partially observable Markov processes over a finite horizon. *Operations Research*, 21:1071–1088, 1973.

[18] T.Smith and R.Simmons. Heuristic search value iteration for POMDPs. In *Proc. Uncertainty in Artificial Intelligence*, pp. 520–527, 2004.

[19] ——. Point-based POMDP algorithms: Improved analysis and implementation. In *Proc. Uncertainty in Artificial Intelligence*, 2005.

[20] M. Spaan and N. Vlassis. A point-based POMDP algorithm for robot planning. In *Proc. IEEE Int. Conf. on Robotics & Automation*, 2004.

Policy Search via the Signed Derivative

J. Zico Kolter and Andrew Y. Ng
Computer Science Department, Stanford University
{kolter,ang}@cs.stanford.edu

Abstract—**We consider policy search for reinforcement learning: learning policy parameters, for some fixed policy class, that optimize performance of a system. In this paper, we propose a novel policy gradient method based on an approximation we call the *Signed Derivative*; the approximation is based on the intuition that it is often very easy to guess the *direction* in which control inputs affect future state variables, even if we do not have an accurate model of the system. The resulting algorithm is very simple, requires no model of the environment, and we show that it can outperform standard stochastic estimators of the gradient; indeed we show that Signed Derivative algorithm can in fact perform as well as the *true* (model-based) policy gradient, but without knowledge of the model. We evaluate the algorithm's performance on both a simulated task and two real-world tasks — driving an RC car along a specified trajectory, and jumping onto obstacles with an quadruped robot — and in all cases achieve good performance after very little training.**

I. INTRODUCTION

In this paper we consider policy search for reinforcement learning. In this setting, one considers a parametrized control policy and then, by interacting with the environment, modifies the parameters to optimize some cost function. For example, if our control task was to drive a car along a desired trajectory, the cost function could penalize deviations from the trajectory, and the control policy could determine the steering and throttle as a simple (say, linear) function of current state features; in this domain, the policy search task would involve learning the coefficients on the state features to obtain a low cost (i.e., follow the trajectory well). We focus in particular on the policy gradient approach, where we optimize the cost function using gradient descent with respect to the policy parameters.

While there exist many different methods for approximating the policy gradients, in this paper we propose a new algorithm that makes use of what we call the *Signed Derivative* approximation. This method allows us to compute an approximation to the policy gradient *without* a model of the system. Our algorithm is based on the following simple insight: the only term in the policy gradient that depends on the dynamics model is the *derivative* of future state elements with respect to the control inputs. However, while these true derivatives are difficult to compute, we claim that often it is very easy to estimate the *sign* of many of these derivatives; that is, we only want to know the general *direction* of how control adjustments will affect the state. Consider again the example of the car mentioned earlier, where for instance one of the state variables is lateral deviation from the desired trajectory, and one of the controls is the steering angle. While it may be very difficult to know the true derivative of future lateral deviations with respect to the steering angle, the sign of the derivative is in

fact quite obvious: turning more to the left typically results in a lateral deviation that is also more to the left. While such "obvious" derivative signs clearly don't apply to all control tasks, we demonstrate in this paper that they do apply in many interesting domains; in such situations, we show that we can drastically improve the performance of policy gradient methods by using these signed derivatives. Indeed, we show that in many cases, the Signed Derivative method not only outperforms standard stochastic policy gradient algorithms (such as the REINFORCE [15] family of algorithm), but actually performs as well as the true (model-based) policy gradient algorithm, but without any knowledge of the model, only the ability to simulate a single trace.

The remainder of this paper is organized as follows. In Section II we present preliminary material and describe the general Signed Derivative algorithm more formally. In Section III we present theoretical results. In Section IV we present empirical results for the algorithm on a number of different domains, both simulated and real-world. Finally, in Section V we discuss related work, and conclude the paper in Section VI.

II. THE SIGNED DERIVATIVE ALGORITHM

A. Preliminaries and Notation

We consider control in a Markov Decision Process (MDP), which is a tuple $M = (S, A, T, H, C)$, where S is a set of states, A is a set of actions, T is the (unknown, but temporarily assumed to be deterministic) system dynamics $T : S \times A \to S$, H is a time horizon and C is a known (one-step) cost function $C : S \times A \to \mathbb{R}$. Since we are concerned with general, continuous state and action domains, we let $S \subseteq \mathbb{R}^n$ and $A \subseteq \mathbb{R}^m$. We can capture time-varying dynamics and costs by including time as a state variable, though for the remainder of this paper we will make any time dependence explicit. Finally, although our algorithm is extendable to general cost functions, for the sake of concreteness we will here assume a common quadratic form of the reward function

$$C_t(s_t, u_t) = (s_t - s_t^\star)^T Q_t (s_t - s_t^\star) + u_t^T R_t u_t$$

where $s_t^\star$ denotes the desired state of system at time t, and Q_t and R_t are diagonal positive semidefinite matrices that penalize state deviation and control respectively.

A (time-dependent) policy $\pi : S \times \mathbb{R} \to A$ is a mapping from states and times to actions. As we are focused on the policy-search setting in this paper, here we consider policies parametrized by some set of parameters θ — we use the notation $u = \pi(s, t; \theta)$ to denote the policy π, parametrized by θ, evaluated at state s and time t. For example, a common class of policies that we will consider in this paper is policies that are linear in the state features

$$u = \pi(s; \theta) = \theta^T \phi(s, t)$$

where $\phi : S \times \mathbb{R} \rightarrow \mathbb{R}^k$ is a mapping from states and times to features and $\theta \in \mathbb{R}^{m \times k}$ is a set of parameters that linearly map these features into controls.

Given a policy, we define the *multi-step cost* function (also called the value function, or just the cost function, as opposed to the one-step cost function defined above above) as the sum of all one-step costs over the horizon H,

$$J(s_0, \theta) = \sum_{t=1}^{H} C_t(s_t, u_{t-1})$$

where $u_t = \theta^T \phi(s_t, t)$ and where $s_{t+1} = T(s_t, u_t)$. We can now more formally define the policy gradient algorithm as a gradient descent method that repeatedly updates the parameters according to

$$\theta \leftarrow \theta - \alpha \nabla_\theta J(s_0, \theta)$$

where α is a step size and $\nabla_\theta J(s_0, \theta)$ is the gradient of the cost function with respect to the policy parameters. Although, computing this gradient term can be quite complicated without a model of the system, in the next section we describe a simple approximation method.

B. The Signed Derivative Approximation

In this section we derive a simple approximation to the policy gradient, using an approximation we called the *signed derivative*. We want to emphasize that the final form of the algorithm, shown in Algorithm 1, is quite simple, even though the derivation is somewhat involved.

To motivate the signed derivative method, we first consider the basic question of *why* ones needs a model of the system to compute the policy gradient $\nabla_\theta J(s, \theta)$. We will derive this result shortly, but it turns out that the policy gradient depends on the model only through terms of the form

$$\left(\frac{\partial s_t}{\partial u_{t'}} \right)$$

for $t > t'$. These terms are the *Jacobians* of future states with respect to previous inputs. They provide the critical motivation for the signed derivative approximation, so it is worth looking at them more closely. These Jacobians are matrices $\left(\frac{\partial s_t}{\partial u_{t'}} \right) \in \mathbb{R}^{n \times m}$ where the i, j element of the $\left(\frac{\partial s_t}{\partial u_{t'}} \right)$ denotes the derivative of the ith element of the state s_t with respect to the jth element of $u_{t'}$, i.e.,

$$\left(\frac{\partial s_t}{\partial u_{t'}} \right)_{ij} \equiv \frac{\partial (s_t)_i}{\partial (u_{t'})_j}.$$

In other words $(\frac{\partial s_t}{\partial u_{t'}})_{ij}$ indicates how the ith element of s_t would change if we made a small adjustment to the j element of the control at a previous time t' (and assuming we are following the policy θ).

In general, the elements of these Jacobians are quite difficult to compute, as they depend on the true dynamics model of the environment and the policy parameters θ. However, the signed derivative approximation is based on the insight that often times it is fairly easy to guess the *signs* of the dominant entries of these matrices: this only requires knowing the general *direction* of how previous control inputs will affect future states. Returning to the example of driving a car, it may be very difficult to determine the derivative of a future state with respect to the steering wheel, but the *direction* of the gradient seems fairly obvious: turning the wheel more to the left will likely result in future states also more to the left.

Furthermore, there is another property of these Jacobians that allows us to come up with a reasonable approximation: in many control settings, each state is primarily affected by only *one* control input. For example, if we are driving our car along a straight line, one state variable (for example, the distance traveled along the line), would be primarily affected by only one control (in this case, the gas pedal). Indeed, many control tasks seem to be expressly designed such that this is the case. For example, imagine trying to drive a car where both the steering wheel and gas pedal controlled some different combinations of both the wheel angle and the throttle; while such a control system is technically "equivalent" to a standard car, it would take much more work to learn. This suggests, at least anecdotally, that humans also exploit these orthogonal control effects, and so we can expect many control tasks to be designed in this way. In other words, we can expect one element in each row of the Jacobians to be larger than the others, corresponding to the "dominant" control element, and these are precisely those elements where we can guess their sign.

Given this discussion, the signed derivative approximation is quite straightforward. We approximate *all* the Jacobian terms with a single matrix $S \in \mathbb{R}^{n \times m}$, called the signed derivative, where entries in S correspond to the signs of the dominant entries in the Jacobians (which, by our discussion above, means that S has only one non-zero entry per row). Consider one last time the driving example, and suppose that the car is facing primarily along the x axis. If we represent the state of the car as its position and orientation (x, y, θ), and let u_1 and u_2 be the throttle and steering angle respectively. Then a reasonable estimate for the signed derivative would be

$$S = \begin{bmatrix} 1 & 0 \\ 0 & 1 \\ 0 & 1 \end{bmatrix}.$$

For instance, $S_{11} = 1$ means that the first state variable (x) is primarily controlled by the first control input (throttle). This makes sense, since the car is mostly aligned with the x axis, so throttle will primarily affect this state. Similarly, y and θ are primarily affected by the second control (steering), which also makes sense, because the steering wheel can cause the car to both turn and veer to the side.

Finally recall, from the beginning of this section, that the Jacobians were the only terms in the policy gradient that

Algorithm 1 Policy Gradient w/ Signed Derivative (PGSD)

Input:

  $S \in \mathbb{R}^{m \times n}$: signed derivative matrix
  $H \in \mathbb{Z}_+$: horizon
  $Q_t \in \mathbb{R}^{n \times n}, R_t \in \mathbb{R}^{m \times m}$: diagonal cost function matrices
  $\alpha \in \mathbb{R}_+$: learning rate
  $\phi : \mathbb{R}^n \times \mathbb{R} \to \mathbb{R}^k$: feature vector function
  $\theta_0 \in \mathbb{R}^{k \times m}$: initial policy parameters

Repeat:

1. Execute policy for H steps to obtain
   $u_0, s_1, \dots, u_{H-1}, u_H$.
2. Compute approximate gradients w.r.t. controls:
$$\tilde{\nabla}_{u_t} J(s_0, \Theta) \leftarrow \sum_{t'=t+1}^{H} S^T Q_{t'}(s_{t'} - s_{t'}^\star) + R_t u_t$$
3. Update parameters:
$$\theta \leftarrow \theta - \frac{\alpha}{H} \sum_{t=0}^{H-1} \phi(s_t, t)(\tilde{\nabla}_{u_t} J(s_0, \Theta))^T$$

---

required a model. Therefore, after making such the signed derivative approximation we can now perform (approximate) policy gradient *without* the need for any model of the system. This is precisely the method that we show in Algorithm 1. The precise form of the gradient updates is derived in the next section, but the basic idea of the algorithm is simple: we are just performing policy gradient, replacing the Jacobian terms with the signed derivative approximation S. It may seem surprising that the method would perform well, given that the signed derivative is a very crude approximation to the true Jacobians; but, we will show, from both a theoretical and empirical perspective, that we can expect the algorithm to perform well in many situations.

C. Formal Derivation of the Policy Gradient

Here we prove the claim made in the previous section, that the policy gradient depends only on the dynamics model through the Jacobian terms, and we derive the precise form of the gradient given in Algorithm 1. The derivation is slightly technical, but the algorithm itself can be understood just from the discussion above.

To avoid certain dependencies, we have to initially consider the gradient of the cost function with respect to H different sets of policy parameters for each time, $\Theta = \{\theta_0, \dots, \theta_{H-1}\}$. We will then take a gradient step in terms of these parameters, projected back into the space where they are all equal. The gradients are given by

$$\nabla_{\theta_t} J(s, \Theta) = \left(\frac{\partial u_t}{\partial \theta_t}\right)^T \nabla_{u_t} J(s, \Theta)$$
$$= \phi(s_t, t) \left(\nabla_{u_t} J(s, \Theta)\right)^T.$$

using the fact that $u_t = \theta_t^T \phi(s_t, t)$ and that s_t doesn't depend

on θ_t. Furthermore, using the definition of J,

$$\nabla_{u_t} J(s, \Theta) = \nabla_{u_t} \sum_{t'=1}^{H} (s_{t'} - s_{t'}^\star)^T Q_{t'}(s_{t'} - s_{t'}^\star) + u_t^T R_t u_t$$
$$= \sum_{t'=t+1}^{H} \left(\frac{\partial s_{t'}}{\partial u_t}\right)^T Q_{t'}(s_{t'} - s_{t'}^\star) + R_t u_t$$

This gives a gradient with respect to each θ_i (where, as stated, the only model-dependent terms are the Jacobians $\left(\frac{\partial s_{t'}}{\partial u_t}\right)^T$). Therefore, the gradient of the cost with respect to a *single* θ is equivalent to taking a step in the direction of all these θ_i's then projecting onto the space where $\theta_0 = \theta_1 = \dots = \theta_{H-1}$. This is accomplished by updating each θ_t according to

$$\theta_t \leftarrow \theta_t - \frac{\alpha}{H} \sum_{t'=0}^{H-1} \nabla_{\theta_{t'}} J(s, \Theta),$$

i.e., the policy gradient with respect to a single parameter θ is also given by

$$\nabla_\theta J(s, \theta) = \frac{1}{H} \sum_{t=0}^{H-1} \nabla_{\theta_t} J(s, \Theta).$$

As stated, the only terms that depend on the model in this sum are the Jacobians. This derivation should make it apparent that Algorithm 1 is simply approximating the policy gradient by substituting the signed derivative for all the Jacobian terms.

III. THEORETICAL RESULTS

Given that the signed derivative is admittedly a rather crude approximation to the true Jacobians, there remains some question as to why we might expect such an approach to work. While the ultimate test of the algorithm's usefulness is, of course, its empirical performance, the results we present here can give insight and intuition into why we obtain the positive results shown in latter sections. We first describe the basic intuition behind the analysis.

There are many possible sources of error for any policy gradient algorithm, but here we analyze the two types of error introduced by the signed gradient approximation itself. First, the signed gradient allows only one control variable to influence a given state variable; even if a state element is *primarily* affected by one control, there most likely exist smaller influences from the other controls as well. Second, the signed gradient makes no attempt to capture the relative magnitudes (or the magnitudes of any kind) of the entries in the Jacobian. Formally, these approximations are represented as

$$\frac{\partial s_t}{\partial u_{t'}} = D(S + E_{t,t'}) \tag{1}$$

for the signed derivative S and matrices $E_{t,t'} \in \mathbb{R}^{m \times n}$ and positive diagonal $D \in \mathbb{R}^{n \times n}$. The D matrix scales the entries in the signed derivative, accounting for the second type of error mentioned above. As we will see more formally below, this type of error isn't overly costly, since it has the effect of simple scaling the entries of the cost functions. Especially in the extreme case where policy gradient finds a solution that obtains near-zero cost, the actual entries of Q_t become unimportant.

The $E_{t,t'}$ terms capture other errors: they add arbitrary constants to the entries in the signed derivative, accounting for the effects of additional control inputs on the states and for time-dependent variation in the relative scaling of the Jacobian. In the worst case, there is little that can be done about such errors: if the entries of $E_{t,t'}$ are large, then the gradient approximation using the signed derivative can be very far from the true gradient. However, there is a great deal of reason to believe that, in many situations $E_{t,t'}$ won't be too large: time-varying scaling should be relatively small over short horizons, and from the discussion in the previous section, we expect cross-terms in the Jacobian to be relatively small in magnitude. And as formalized below, if the $E_{t,t'}$ are small, then we expect the signed derivative to perform well.

Theorem 1: Using the notation from (1), suppose $\|E_{t,t'}\|_2 \leq \epsilon$ for all t, t'.[1] Define the modified cost function $\tilde{Q}_t = DQ_t$. Then, given additional technical assumptions (described fully in the appendix), PGSD will converge with probability one to some solution $\tilde{\theta}$ that is "close" to a local minimum of the cost function $J_{\tilde{Q}}(\theta)$, the cost function that uses $\tilde{Q}_t$ (but the same R_t) as the cost matrices:

$$\|\nabla_\theta J_{\tilde{Q}}(\tilde{\theta})\| \leq O(\epsilon).$$

Furthermore, if performing gradient descent with respect to the true gradient (of the actual cost function) results in η-optimal policy parameters — i.e., $J_Q(\theta^\star) \leq \eta$ — then PGSD also obtains an order η-optimal solution[2]

$$J_Q(\tilde{\theta}) \leq \kappa(D)\eta + O(\epsilon).$$

Proof: (sketch) The full proof is given in the appendix,[3] but we provide a very brief sketch. The proof proceeds in three steps. First, we show that the gradient approximated using the signed derivative is equivalent to the true gradient using the $\tilde{Q}$ costs, plus a bounded error term

$$\widetilde{\nabla}_\theta J_Q(\theta) = \nabla_\theta J_{\tilde{Q}}(\theta) + \tilde{E}$$

for $\|\tilde{E}\| \leq O(\epsilon)$. Second, we show that following this approximate gradient using a stochastic gradient method will converge, with probability one, to a point that is close to a minimum of $J_{\tilde{Q}}(\theta)$. Finally, we show that given suitable assumptions about the region of convergence, a policy that is close to locally optimal for $J_{\tilde{Q}}(\theta)$ will also be close to locally optimal for $J_Q(\theta)$. ∎

IV. EXPERIMENTAL RESULTS

A. Simulated Two-Link Arm

While we will present experiments on real systems shortly, we begin by presenting an evaluation of our proposed method

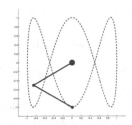

Fig. 1.   Two-link pendulum trajectory following task.

on a simulated two-link arm, in order to rigorously compare to previous policy gradient approaches, and to provide a readily available implementation of our approach. Code for the all the results in this section is available at `http://cs.stanford.edu/~kolter/rss09sd`. We emphasize that the purpose of this section is to specifically compare PGSD with other policy gradient approaches. The control task itself is fairly straightforward, and many other approaches such as adaptive control or iterative learning control could also be applied, though this is beyond the scope of this paper; we will discuss these related works more in Section V.

The two-link pendulum is a well-known control task in robotics and control. The system, shown in Figure 1 consists of two planar links; the state consists of the joint angles and velocities of both joints and the control specifies a torque at each of the joints. The equations of motion can be easily derived from Lagrangian dynamics, and we introduce stochasticity to the system by adding Gaussian noise to the torques before integrating the equations of motion. The task we consider here, also shown in the figure, is to move the end effector along some desired trajectory. When the model of the system is known, it is fairly easy to apply classical control methodologies such as inverse dynamics or LQR to find an optimal controller, but of course we don't provide this model to PGSD or other comparable algorithms. We feel that this is a particularly demonstrative example for the Signed Derivative algorithm, since it is well-known that there *are* cross terms that cause all joints to be affected by all the control inputs — for instance, a common (more challenging) task is to swing the pendulum upright and balance by applying torques only to the elbow — yet we claim that the Signed Derivative approximation is still reasonable, since joints are *primarily* affected by their own control.

The cost function for this domain penalizes deviations from the desired joint angles (we first computed the trajectory in joint space), and we use a time horizon of $H = 5$. Note that this doesn't mean that the controller only needs to follow the trajectory for 5 steps, but rather that at each time the controller should ideally act optimally with respect to a receding horizon of $H = 5$; since the cost function itself "guides" the arm along the trajectory, such a horizon is suitable. We use a linear control policy $u_t = \theta^T \phi(s_t, t)$ where ϕ contains 1) deviations from desired joint angles, 2) deviations from desired joint velocities, 3) desired joint accelerations, and 4) $\sin(2\pi t/t_{\text{total}})$

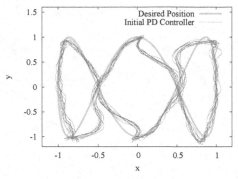

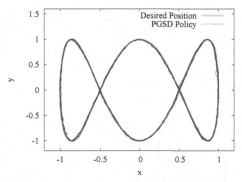

Fig. 2. (top) Trajectory from initial PD controller. (bottom) Trajectories from controller learned using PGSD.

Fig. 3. (top) Trajectory from initial PD controller. (bottom) Trajectories from controller learned using PGSD.

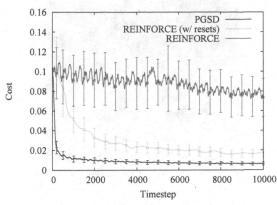

Fig. 4. Average cost versus time for different policy gradient methods. Costs are averaged over 20 runs, and shown with 95% confidence intervals.

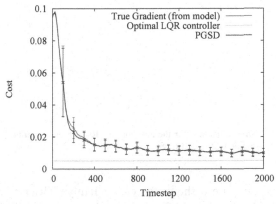

Fig. 5. Average cost versus time for PGSD versus model-based methods. Costs are averaged over 20 runs, and shown with 95% confidence intervals.

where t_{total} is total time for the complete trajectory (this last term was added to account for a visible periodic pattern in the controls). This leads to s total of 14 parameters for the policy. For algorithms that require a stochastic policy, we added Gaussian noise to the parameters: $u_t = (\theta + \epsilon_t)^T \phi(s_t, t)$, $(\epsilon_t)_{ij} \sim \mathcal{N}(0, \sigma)$.

Figure 4 compares the performance versus time of PGSD, and a well-known policy gradient RL algorithm, the REINFORCE algorithm.[4] All free parameters of the learning algorithms (gradient step sizes, policy noise, number of episodes) were hand-optimized to give that fastest convergence that didn't cause any divergence issues. As the figure shows, PGSD drastically outperform the other methods, converging much faster to a low-cost policy. This improvement is especially notable given that the REINFORCE algorithm is actually given an advantage: since the task we're considering is not episodic (at least not at the time-scale of the horizon), episodic algorithms don't immediately apply, and so we instead allow the algorithm the ability to reset to previous states observed along the trajectory. The REINFORCE without resets in the fiture does not have such an advantage, but also performs much worse. Figures 3 and 2 show the resulting controller learned

by the PGSD algorithm after 2000 time steps (4 times through the trajectory), along with the trajectory achieved by the initial PD controller (used to initialize all the learning algorithms).

We also compare, in Figure 5, the performance of the PGSD algorithm, policy gradient using the true gradient from the model, and an optimal LQR controller. Not surprisingly, the LQR controller performs best: this controller is built by linearizing around the (known) dynamics at each operating point, then computing a series of non-stationary policies for each point (in total, the LQR controller has 9000 parameters). However, using only 14 parameters, the true policy gradient and PGSD algorithm are able to obtain a controller that performs relatively close to this full LQR controller. Furthermore, the most important result is that the learning curve for PGSD is virtually *indistinguishable* from the true policy gradient learning curve; despite the rather crude approximation made by the signed derivative, this resulting algorithm performs *just as well* on this task, and requires no model of the system (and therefore also less computation time, since there is no need for time-consuming finite difference computations).

B. Autonomous RC Driving

In this section we apply the PGSD algorithm to the task of learning to drive an autonomous RC car along a desired

---

[4]We intentionally scaled the parameters of this control task to be the same order of magnitude, so more advanced techniques such as natural gradients[6, 11] didn't improve performance significantly. In preliminary experiments we also evaluated a variety of finite difference and weight perturbation methods, but didn't notice a substantial improvement over REINFORCE for this task.

Fig. 6. RC car used for the driving experiments.

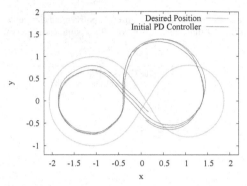

Fig. 7. Desired trajectory for the autonomous RC driving experiments, with trajectory for initial PD controller.

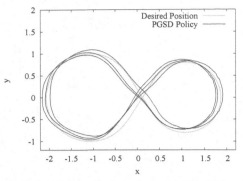

Fig. 8. Desired trajectory for the autonomous RC driving experiments, with typical trajectory learned using PGSD after approximately 20 seconds of learning.

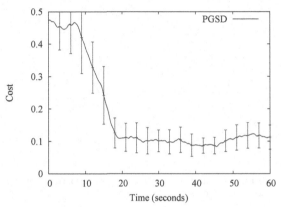

Fig. 9. Average cost versus time for the PGSD algorithm on the RC car task. Costs are averaged over 10 runs, and shown with 95% confidence intervals.

trajectory. Figure 6 shows the car, a Tamiya TRF415, which is about 40cm long and 20cm wide. A pattern of LED lights is attached to the car, and tracked by an external PhaseSpace motion capture system for pose estimation. All processing is done on a workstation PC, with controls transmitted to the car at 50hz.

The simplest representation of the car's state is as six dimensional vector representing the 2D position x, y, the orientation θ, and the time derivatives $\dot{x}, \dot{y}, \dot{\theta}$. However, a more natural representation for the signed derivative approach is to represent the car's state relative to some desired trajectory — here the trajectory is specified as a continuous spline that gives the desired state as a function of time. In this alternate representation, the state consists of the longitudinal, lateral, and angular deviation (and their derivatives) from the desired trajectory. The control is two dimensional, consisting of a commanded throttle and steering angle.

We use the same form of linear controller as in the previous sections, but where $\phi(s, t)$ now contains 1) the full state (represented as the deviation terms), 3) the desired velocities, relative to the car frame, 3) the deviations for a target state 0.5 seconds and 4) a constant term. Some of the θ parameters are forced to be zero (so that, for instance, the throttle doesn't depend on the lateral deviation), for a total of 16 parameters in the policy. The cost function penalizes the longitudinal, lateral, and angular deviation, any control outside a specified valid range, and control that changes more that some amount between two time steps (to minimize oscillations). We used a time horizon of $H = 25$.

Figures 8 and 7 show the control task we consider: driving the car in an irregular figure-eight pattern at varying speeds (2.0 m/s along the larger loop, 1.5 m/s along the smaller loop). The figure also shows the trajectory followed by an initial PD controller: while the PD controller follows the overall pattern of the trajectory, it clearly does not perform very well. Figure 9 shows the learning curve of the PGSD algorithm. As the figure shows, PGSD is able to very quickly — within an average of 20 seconds, about 3 times around the trajectory — obtain a policy that performs far better than the initial PD controller. We show a typical trajectory from one of these learned controllers in Figure 8. The learned policies do not perform flawlessly — the car still sometimes veers off the desired path — but we feel this is largely due to the limited policy class itself; to perform better, one might need more complex, time-varying policies, to capture the fact that the car needs to behave differently at different points along the path. Nonetheless, PGSD converges to a very reasonable policy — in fact, better than any we were able to hand tune in the same policy class — in just 20 seconds of learning.

C. LittleDog Jumping

In this section we present results on applying PGSD to the task of "jumping" the front legs of a quadruped robot to quickly climb up large steps. The "LittleDog" robot that we

Fig. 10.   The LittleDog robot.

Fig. 11.   The desired task for the LittleDog: climb over three large steps.

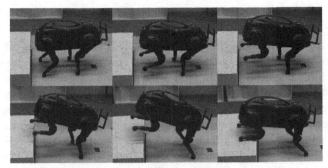

Fig. 12.   A properly executed jump.

use for this task, shown in Figure 10, is designed and built by Boston Dynamics, Inc. The task we are concerned with here is shown in Figure 11: we want to quickly climb up three steps, whose height is approximately equal to the robot's ground clearance. Because the steps are so large, the most efficient motion to climb up is to jump the two front legs on a step, then pull the rest of the body up.

However, jumping the front legs on the LittleDog robot is not a trivial task. The LittleDog's legs are not powerful enough to force its body off the ground, so the only means of jumping is to lean the body backwards until the virtually all the mass rests on the hind legs, then quickly raise the front legs and push forward before the robot falls over. Figure 12 shows a properly executed front leg jump. However, if the weight is not shifted properly, the robot will either plant it's feet into the step, or flip over backward. It's very difficult to correct such failures, because usually by the time it is apparent that the robot has failed to jump properly, the robot does not have the power to correct itself. Therefore, jumping is a "one-off" maneuver: we guess an amount to shift backward, then apply an open-loop sequence of joint commands, hoping to jump successfully. The situation is made complicated because the "correct"amount to shift the weight depends, for example, on the state of the robot, namely the current position of the COG relative to the back feet, and the forward velocity of the robot. Because we want a policy that can jump regardless of the initial conditions, we applied the PGSD algorithm to learn a jumping policy that predicts the correct amount to shift given features of the current state.

Although full state space for the LittleDog is 36 dimensional (12 joints and twelve joint velocities plus a 6D pose and 6D pose velocities), we don't need to take into account the complete state. Rather, the only state element that is particularly crucial for the jumping maneuver is the pitch of the body: if the pitch is too small, the dog won't clear the step, but if it is too large, the dog will flip backward. Therefore, the cost function can depend only on the pitch of the dog. The control is a number that indicates how far back to shift the weight before pushing forward; we determine the control as a linear function of three features: 1) the current shift of the center of mass, 2) the forward velocity of the dog and 3) a constant term.

There is one straightforward generalization of the PGSD algorithm, as presented so far, that we make for this task. Although the cost function depends only on the pitch of the robot, it is difficult to know the "optimal" pitch — unlike previous tasks where the optimal state value was clearly defined. Instead, the readily observable quantity is simply whether the jump succeeded, or whether the robot either didn't clear the step or flipped over. Therefore, if we define the one-step cost as the ℓ_1 error between the pitch and optimal pitch, then the gradient is just the sign of the direction we should move our control in. When $H = 1$, the PGSD update then takes on a very simple form:

$$\theta \leftarrow \begin{cases} \theta - \alpha\phi(s) & \text{robot didn't clear step} \\ \theta & \text{jump succeeded} \\ \theta + \alpha\phi(s) & \text{robot flipped backwards} \end{cases}$$

Despite the simplicity of this update rule, it works well in practice. We evaluated this PGSD variant on the LittleDog robot, attempting to climb the three steps as shown in Figure 11. After 28 failures (either flipping backwards or failing to clear the step), the robot successfully jumped all three steps for the first time. After 59 failures, the learning process had converged on a stable controller: the robot succeeded in climbing all three steps for 13 out of the next 20 trials. This is far better than any policy we had been able to code by hand.[5] A video of the learning process on the dog is available at the website mentioned previously.

[5]While it is possible to increase the reliability of the system by adding extra steps to ensure that the robot always enters a similar configuration before each jump, in these experiments we wanted to test precisely how well a controller could perform under many different circumstances.

V. Related Work

As mentioned in the introduction, there is a great deal of work on policy gradient methods for reinforcement learning. If a model of the system is known, then we can compute the gradient using simple finite difference methods — this holds even in stochastic domains if we are allowed to fix the random seeds which lead to this stochasticity, an approach known as the PEGASUS algorithm [10]. These model-based methods have been applied to many robotics domains. However, such a model might not always be available, or might be difficult to learn from data. Additionally, as we have shown, our PGSD method can sometimes perform as well as the model-based methods without any model other than the signed derivative approximation.

In situations where we have no model, we can still apply finite difference methods or weight perturbation, so long as the step sizes are large enough to overcome noise. Such an approach was successfully applied to the task of learning a quadruped trotting gait in [8]. Recently, [13] investigated the effect of sampling distributions on the signal-to-noise ratio of these and similar gradient updates.

A related but different approach uses a likelihood ratio trick to obtain an estimate of the gradient using a number of episodes run under the system and policy of interest: the REINFORCE [15] algorithm was the first of such methods, but many extensions and generalization have been proposed [4, 5, 11, 7]. There has also been work on estimating and using the natural gradient, a gradient that is invariant to reparameterizations of the policy [6, 3, 12]. However, most of these algorithms require running multiple episodes in order to obtain a reasonable estimate of the gradient (or natural gradient), which is difficult for non-episodic tasks such as those we consider. In these domains, PGSD has the strong advantage of only requiring a single episode to obtain an estimate of the gradient.

Our work also shares a strong connection to [1]. This paper proposes a method for using inaccurate simulation models by using only the local gradient information implied by these models. This is quite similar in spirit to our approach, except we discard any need for even an inaccurate simulator, and encode all necessary information directly in the signed gradient: the approximation may be rougher, but unlike this past approach PGSD does not require performing any local policy search in a simulator.

Finally, we want to note the connection between the algorithm we propose here and the field of adaptive control [14, 2] — in particular the subtopics of Model Reference Adaptive Control (MRAC) and Self-Tuning Regulators — and Iterative Learning Control (ILC) [9]. The general philosophy of these approaches is similar to PGSD: they use an error signal (i.e., between the actual and desired state) to directly adapt the parameters. However, typical formulations of MRAC or ILC use hand-crafted update rules to modify the controller, with the focus on analyzing stability properties of the resulting controllers. In contrast, PGSD uses a general update rule that derives completely from the Reinforcement Learning setting of long time horizons and general cost functions, plus the Signed Derivative approximation of the model derivatives. derivatives. Generally speaking however, PGSD could be viewed somewhat as an instance of MRAC or ILC, with a very particular form for the update rule.

VI. Conclusion

In this paper, we proposed the Signed Derivative method, a method for approximating policy gradients, using the insight that often times it is very easy to guess the direction in which control inputs will affect future states. We show that this algorithm, Policy Gradient with the Signed Derivative (PGSD) can perform very well compared to stochastic gradient estimators, and in fact can perform *as well* as the true gradient, even though it has no knowledge of the true environment's model. We further evaluated our algorithm on two real-world control tasks — driving an RC car and jumping with a quadruped robot — and demonstrated very good performance on both domains. While we stress that the PGSD approach is not suitable for all situations (for instance, if the effects of controls on the system is entirely unknown), we feel that in many situations the approach applies quite easily, and offers very substantial performance benefits.

References

[1] Pieter Abbeel, Morgan Quigley, and Andrew Y. Ng. Using innaccurate models in reinforcement learning. In *Proceedings of the International Conference on Machine Learning*, 2006.

[2] Karl Johan Astrom and Bjorn Wittenmark. *Adaptive Control*. Prentice Hall, 1994.

[3] J. Andrew Bagnell and Jeff Schneider. Covariant policy search. In *Proceedings of the International Joint Conference on Artificial Intelligence*, 2003.

[4] Jonathan Baxter and Peter L. Bartlett. Infinite-horizon gradient-based policy search. *Journal of Artificial Intelligence Research*, 15:319–350, 2001.

[5] Evan Greensmith, Peter L. Bartlett, and Jonathan Baxter. Variance reduction techniques for gradient estimates in reinforcement learning. *Journal of Machine Learning Research*, 5:1471–1530, 2004.

[6] Sham Kakade. A natural policy gradient. In *Neural Information Processing Systems 14*, 2001.

[7] Jens Kober and Jan Peters. Policy search for motor primitives in robotics. In *Neural Information Processing Systems 21*, 2009.

[8] Nate Kohl and Peter Stone. Machine learning for fast quadrupedal locomotion. In *Proceedings of the AAAI*, pages 611–616, July 2004.

[9] Kevin L. Moore. Iterative learning control: an expository overview. *Applied and Computational Controls, Signal Processing, and Circuits*, 1(1):151–214, 1999.

[10] Andrew Y. Ng and Michael Jordan. Pegasus: A policy search method for large mdps and pomdps. In *Proceedings of the Conference on Uncertainty in Artificial Intelligence*, 2000.

[11] Jan Peters and Stefan Schaal. Policy gradient methods for robotics. In *Proceedings of the IEEE Conference on Intelligent Robotics Systems*, 2006.

[12] Jan Peters, Sethu Vijayakumar, and Stefan Schaal. Natural actor-critic. In *Proceedings of the European Conference on Machine Learning*, 2005.

[13] John W. Roberts and Russ Tedrake. Signal-to-noise ratio analysis of policy gradient algorithms. In *Neural Information Processing Systems 21*, 2009.

[14] Shankar Sastry and Marc Bodson. *Adaptive Control: Stability, Convergence, and Robustness*. Prentice-Hall, 1994.

[15] Ronald J. Williams. Simple statistical gradient-following algorithms for connectionist reinforcement learning. *Machine Learning*, 8:229–256, 1992.

Non-parametric Learning to Aid Path Planning Over Slopes

Sisir Karumanchi, Thomas Allen, Tim Bailey and Steve Scheding
ARC Centre of Excellence For Autonomous Systems (CAS),
Australian Centre For Field Robotics (ACFR),
The University of Sydney,
NSW. 2006, Australia.
Email: s.karumanchi/t.allen/t.bailey/s.scheding@cas.edu.au

Abstract— This paper addresses the problem of closing the loop from perception to action selection for unmanned ground vehicles, with a focus on navigating slopes. A new non-parametric learning technique is presented to generate a mobility representation where maximum feasible speed is used as a criterion to classify the world. The inputs to the algorithm are terrain gradients derived from an elevation map and past observations of wheel slip. It is argued that such a representation can aid in path planning with improved selection of vehicle heading and operating velocity in off-road slopes. Results of mobility map generation and its benefits to path planning are shown.

I. INTRODUCTION

Learning techniques that close the loop from perception to action selection are of particular interest for off-road robotics. This loop closure refers to the need for an intermediate module that processes sensed exteroceptive[1] information into a representation which can directly aid in decision making (such as path planning). In ground vehicle robotics the focus is usually on identifying hard hazards such as obstacles or classifying predefined environmental states into different degrees of traversibility [1]–[4]. Assumptions such as terrain homogeneity, or perpetual existence of a road are often made to simplify the problem. In the absence of such assumptions, theoretical techniques that use sensed information to aid decision making need to be investigated. In this paper an intermediate scene interpretation module is proposed in Section II to close the loop from perception to action.

Autonomous navigation in unstructured conditions such as non-homogeneous uneven terrain is a challenging problem to solve. In such environments two main issues need to be addressed. First, explicit assumptions about the terrain should be avoided. Second, in addition to hard hazards (such as obstacles), soft hazards (situations where behaviour needs to be adapted) need to be identified and dealt with. For example, terrain slopes are soft hazards and to successfully negotiate them, vehicle behaviour such as velocity, operating gear and vehicle heading needs to be adjusted.

Due to recent developments in Bayesian non-parametric techniques, *learning from experience* architectures offer promise. In such architectures, no assumptions need to be made about the environment. The environment representation is only limited by the available sensor suite and the variables used to define the exteroceptive state. Therefore, experience-based learning techniques are viable to address the problem of closing the loop from perception to action.

Existing 'learning from experience' techniques include Reinforcement Learning [5] and model predictive techniques [6]. However, they make assumptions such as the existence of a reward function in Reinforcement Learning, or the existence of accurate models for model predictive techniques. It is difficult to quantify such reward functions or develop accurate models in unstructured environments.

Imitation Learning [7] and Inverse Reinforcement Learning [8] are relatively new concepts and have been applied to the problem of learning a reward function from example behaviour. However, they rely heavily on expert input. Controller limitations are usually ignored when systems rely heavily on human input. For example, tuning involved in a PID controller limits the mobility of the platform as it was tuned for a few selected conditions. Such limitations can be dealt with implicitly when the vehicle explores its behavioural capabilities on its own terms.

A *learning from proprioception*[2] approach is demonstrated in [9] for a Mars-Rover platform where the authors represent the environment in proprioception space in terms of expected slip. This approach ignores the influence of operating velocity on wheel slip. For a Mars-Rover, proprioceptive measures (such as wheel slip) are mainly dependent on environment conditions and behavioral influences (such as velocity) could be ignored because the platform moves slowly. However, this assumption cannot be made for larger platforms where there is a distribution of slip values for a given condition pertaining to all possible behaviours. Also, such an approach is limited to the case when proprioception is a scalar or a weighted average of scalars. The latter usually involves manual tuning of weights which is not an intuitive process. A single scalar cost cannot capture all the objectives in unstructured conditions. Instead it is beneficial to use a collage of proprioceptive stimuli to judge actions.

---

[1]Exteroception: perception of external factors that are not under agent control

[2]Proprioception: perception of internal factors that are affected by environment and one's own behaviour.

217

This paper addresses the problem in question with a specific focus on negotiating two dimensional slopes given range sensor measurements. Section II formally introduces the scene interpretation problem as conditional density estimation and a non-parametric solution using Gaussian Processes is proposed. In Section IV, results are shown on an elevation map derived from laser scans. Mobility maps are derived from the given elevation map by analysing terrain gradients with past observations of vehicle slip collected from a 8x8 skid-steered vehicle (see Figure 1). Finally, in Section V results compare path planning over mobility maps with planning over heuristic costs.

Fig. 1.   Argo 8x8 Unmanned Ground Vehicle

II. PROPRIOCEPTIVE SCENE INTERPRETATION

A. Motivation

Current terrain perception modules in unmanned ground vehicles (UGVs) are focused on creating an accurate internal representation of the environment. Exteroceptive parameters such as terrain colour and terrain slope have little value if the vehicle cannot associate them with a value of cost/utility of movement. This task of interpreting exteroceptive data by associating a scalar value of cost or utility is referred to as *Scene Interpretation*. Better representations in the scene interpretation problem can aid the purpose of bridging the gap between perception and action selection. One cost/utility representation of the world that is of interest is a mobility map [10] (see Figure 2). Here the maximum feasible speed of the vehicle between two points is used as a criterion for continuous classification as such capturing the net utility of an environment condition with a single value. Such a mobility map explicitly represents traversibility of occupied, admissible and unexplored regions and can be used as an objective map for high level trajectory planning algorithms such as A^*.

For a path planning application, negotiating slopes demands two key requirements in cost representations i) Orientation sensitivity, as navigating down hill and up hill need to be judged differently ii) Ability to encapsulate platform and controller limitations as performance on slopes is very sensitive to controller tuning in practice. Tuning a controller to a certain condition (such as flat terrain) can limit its performance in other conditions (such as non-flat terrain).

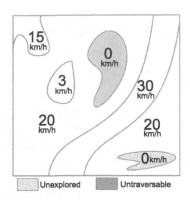

Fig. 2.   Sample Mobility Map Indicating Maximum Feasible Speed (Shaded areas indicate immobile and unexplored areas)

A vectorised representation of a mobility map can offer orientation sensitivity by making the mobility values dependent on the direction of pitch and roll slope definitions. An additional benefit of mobility representation over the traditional heuristic cost spaces is that environment utility is defined in behaviour space. In behaviour space, scene interpretation can be treated as a learning problem where agents learn about behavioural limits by physically interacting with the environment. The observed extent of behaviour limitation provides information about environment utility/risk. Both hard and soft hazards are encapsulated in a continuum as different degrees of behavioural limits. Such an interpretation also captures both platform and controller limitations implicitly.

For experience-based scene interpretation, practicality of collecting training data is critical. For example, the learning task of determining behavioural limits can be achieved either in an unsupervised or supervised manner. The former involves optimisation to judge behaviour. Data needs to be collected under all circumstances (including worst case scenarios) to determine optimal behavioural limits. For safety reasons, worst case exploration is not practical on large platforms. Instead, a supervised approach can be developed with an aim to provide an upper bound on feasible actions given access to proprioceptive feedback (wheel slip). Unlike the unsupervised case, data collection is practical as the vehicle only explores what it can negotiate comfortably. Unexplorable behaviour contributes to scene interpretation by indirectly providing information about behaviour limitation.

The need for proprioceptive feedback is to relate exteroceptive states ($\tilde{e}$) such as terrain slopes with action states ($\tilde{a}$) such as vehicle velocity. If a relation can be drawn, a bound on operating velocity can be determined for a given environment condition to create mobility maps. Given no additional information, the two states are independent ($\tilde{e} \perp \tilde{a}|\emptyset$). However, when the right proprioceptive feedback ($\tilde{j}$) is observed, the two become indirectly related ($\tilde{e} \not\perp \tilde{a}|\tilde{j}$) . This is because of the causal dependencies as both environment and vehicle behaviour affect proprioception and this causes the two sources to be related when the right proprioception is observed.

The process of analysing and selecting useful proprioceptive measures is a separate problem of its own, and is not dealt with in this paper. For the skid-steered vehicle of interest, slip estimates are chosen as proprioceptive feedback. The slip values cannot be measured directly, so they are estimated with an Unscented Kalman Filter [11] (UKF) using the two-track process model mentioned in [12]. The reason for using a Kalman filter is to efficiently deal with sensor noise. The test platform has an onboard Inertial Navigation System (INS) and its used to sense vehicle actions such as velocity with good accuracy. In addition, pitch and roll information from the INS are used to sense the current terrain slope (exteroceptive conditions).[3]

In the next subsections, the notation is summarised in one place to provide easy reference to all the variables and then the theory is introduced.

B. Nomenclature

$\tilde{x}$ - tilde is used to indicate that a particular variable is a vector.

$\tilde{a}$ - Action vector (operating velocity)

$\tilde{e}$ - Exteroceptive stimuli (terrain slopes)

$\tilde{j}$ - Proprioceptive stimuli (wheel slips): A vector of measures that indicate dependence of performance on environmental conditions and vehicle behaviour.

$\mathbf{H}$ - Experience set (training set)
$$\begin{pmatrix} \tilde{e}_1 & \tilde{e}_2 & \cdots & \tilde{e}_N \\ \tilde{j}_1 & \tilde{j}_2 & \cdots & \tilde{j}_N \\ \tilde{a}_1 & \tilde{a}_2 & \cdots & \tilde{a}_N \end{pmatrix}$$

$\mathbf{J}^*$ - Set of proprioceptive stimuli observed in ideal conditions $-\{\tilde{j}_1^*, \tilde{j}_2^*, \cdots, \tilde{j}_M^*\}$.

$\mathbf{J}^*$ is the set of samples derived from a constrained *region* in proprioception space that is indicative of feasible conditions.

$\mathbf{E}_{test}$ - Set of test conditions which need to be interpreted (Test set) - $\{\tilde{e}_{test1}, \tilde{e}_{test2}, \cdots, \tilde{e}_{testT}\}$.

for example, the set of both horizontal and vertical gradients for each grid cell form the set of test conditions to interpret terrain slopes from an elevation map.

C. Problem Definition

Gathering experience corresponds to collecting co-occurrent observations of $\tilde{e}$, $\tilde{a}$ and $\tilde{j}$ in as many varied conditions as possible.[4] This experience set ($\mathbf{H}$) serves as a training set for learning. The exploration philosophy for collecting training data is to explore the natural feasibility of vehicle behaviour in

as many varied conditions as possible either under manual or autonomous control. The latter has the advantage of exploring controller limitations.

Before velocity limits can be derived from experience data, an intermediate goal is to infer the *feasible behaviour distribution* for any test condition given the set of all past observations ($\mathbf{H}$) and the comfortable proprioception set ($\mathbf{J}^*$) which is chosen by the user to be observations in ideal/nominal conditions. For interpreting slopes, observations from flat terrain conditions are labelled as ideal and used as a reference. This process can be intuitively understood as training the robot what to look for (in proprioception) when exploring feasibility of actions in unknown conditions.

Once feasible behaviour distribution is inferred, an upper bound using the commulative density function (CDF) can determine velocity limits for use in mobility maps. The process of deriving a mobility map given a set of test conditions is outlined below.

Scene Interpretation Process For An Elevation Map
Input: Elevation Map (A set of elevation values)
- Apply the Sobel operator [13] to determine gradient maps in pitch and roll directions (Test set$-\mathbf{E}_{test}$).
foreach $\tilde{e}_{test}$ in $\mathbf{E}_{test}$ **do**
| - Infer feasible behaviour distribution from past experience ($\mathbf{H}$ & $\mathbf{J}^*$)
| - Determine operational limit (Maximum Feasible Speed)
end
Output: Mobility Map (Set of all associated mobility values ordered according to their respective test condition in $\mathbf{E}_{test}$)

Determining the feasible behaviour distribution is a conditional density estimation problem. The feasible behaviour distribution for a selected environment condition is $p(\tilde{a}|\tilde{e}_{test}, i = 1)$[5] where i is an indicator variable to represent the feasibility constraint $\tilde{j} \in \mathbf{J}^*$.

$$i = \begin{cases} 1 & \tilde{j} \in \mathbf{J}^* \\ 0 & \tilde{j} \notin \mathbf{J}^* \end{cases} \quad (1)$$

$p(\tilde{a}|\tilde{e}_{test}, i = 1)$ is a *measure of confidence* in taking an action $\tilde{a}$ given past experience ($\mathbf{H}$). Confidence for an action is based on how often proprioception observed under that action was within the set of proprioceptive stimuli observed in ideal conditions ($\mathbf{J}^*$) i.e. actions that generated stimuli in the region of accustomed proprioception $\mathbf{J}^*$ are preferred.

D. A Non-parametric Approximation

In this section a hierarchical non-parametric[6] approach is presented to approximate the global conditional density

---

[3]The test platform used in this work does not have any suspension, so pitch and roll information from the INS reflects the terrain slope accurately. For other platforms, terrain slopes must be derived from exteroceptive sensors.

[4]Existence of a stationary joint distribution $p(\tilde{e}, \tilde{j}, \tilde{a})$ is assumed. Therefore the experience/training set is a collection of i.i.d samples from the joint.

[5]Equivalent to $p(\tilde{a}|\tilde{e}_{test}, i = 1, \mathbf{H})$- dependence on the training set $\mathbf{H}$ is not shown for conciseness.

[6]Non-parametric techniques are preferred for learning from experience (memory-based learning) problems as they make the least assumptions about the global form of the distribution.

$p(\tilde{a}|\tilde{e}_{test}, i = 1)$. The local module approximates the function $\tilde{a} = f(\tilde{e}, \tilde{j})$ within a Bayesian non-parametric framework using Gaussian Processes. While the regression module infers local conditional distributions, the global conditional distribution is treated as a kernel density estimation problem where the number of kernels grow as the number of elements in the $\mathbf{J}^*$ set grows. Together, the density $p(\tilde{a}|\tilde{e}_{test}, i = 1, \mathbf{H})$ can be adapted online as the sets $\mathbf{J}^*$ and $\mathbf{H}$ grow.

The whole process is captured in the following equation where the desired distribution is derived by marginalising $p(\tilde{a}, \tilde{j}|\tilde{e}, i)$ over $\tilde{j}$.

$$p(\tilde{a}|\tilde{e}, i = 1) = \int p(\tilde{a}, \tilde{j}|\tilde{e}, i = 1)d\tilde{j} \qquad (2)$$

$$= \int p(\tilde{a}|\tilde{e}, \tilde{j})p(\tilde{j}|i = 1)d\tilde{j} \qquad (3)$$

$p(\tilde{j}|i = 1)$ corresponds to the distribution of desired proprioception. In this application, since one has access to M samples of $\tilde{j}$ in the $\mathbf{J}^*$ region (i.i.d. samples from $p(\tilde{j}|i = 1)$), the above equation can be approximated as a weighted sum of conditional distributions at the observed $\tilde{j}$ locations.

$$p(\tilde{a}|\tilde{e}, i = 1) \approx \sum_{\tilde{j}_i \in \mathbf{J}^*} \pi_{\tilde{j}_i} p(\tilde{a}|\tilde{e}, \tilde{j} = \tilde{j}_i) \qquad (4)$$

where $\pi_{\tilde{j}_i}$ are mixing components; $\sum \pi_{\tilde{j}_i} = 1$

If all samples in the $\mathbf{J}^*$ set are given equal importance.

$$p(\tilde{a}|\tilde{e}, i = 1) \approx \frac{1}{M} \sum_{\tilde{j}_i \in \mathbf{J}^*} p(\tilde{a}|\tilde{e}, \tilde{j} = \tilde{j}_i) \qquad (5)$$

Equation 5 is in the form of kernel density estimation, but with variable kernels, as $p(\tilde{a}|\tilde{e} = \tilde{e}_{test}, \tilde{j} = \tilde{j}_i)$ is inferred from data. This can be viewed as an infinite mixture of conditional densities as the number of components grows when the $\mathbf{J}^*$ set is allowed to grow. If the local conditionals are approximated to be Gaussian then the global approximation turns out to be a Gaussian mixture.

1) Gaussian Process Regression: Inferring the local conditional distribution $p(\tilde{a}|\tilde{e}, \tilde{j} = \tilde{j}_i)$ from observed data at each $\tilde{j}_i$ location can be treated as a Bayesian Regression problem ($f : \{\tilde{e}, \tilde{j}\} \to a$) [14], [15], where $\tilde{e}, \tilde{j}$ are augmented together to form the input vector $\tilde{x}$ and a is the output y.

$$a_i = f(\tilde{e}_i, \tilde{j}_i) + \varepsilon \qquad (6)$$

where

ε - Noise $- N(0, \beta^{-1})$

$\beta -$ Noise precision

A Gaussian Process (GP) is completely specified by its covariance function $K(x, x')$ and its choice defines the space of functions (latent variables - f) that can be generated [14]. Further, the output is assumed to be zero mean. This is reflected in the prior over the latent variables $p(f)$. Because of this zero mean assumption in GP's , predictions are biased towards null behaviour region (zero) if no data is observed in the test conditions. In the scene interpretation problem, this translates to being cautious in unexplored or underexplored environments which is desired.

$$P(f) = N(0, K) \text{ - Prior On Functions} \qquad (7)$$

where

$f -$ latent variables

$K(x, x') -$ covariance function

$x = \{\tilde{e}, \tilde{j}\} -$ input values

The predictive distribution is a Gaussian with the following form:

$$P(a|\tilde{e}_{test}, \tilde{j}_{test}, \mathbf{H}) = N(\mu(x), \text{var}(x)) \qquad (8)$$

where

$\mu(x) = K(x_{test}, X)[K(X, X) + \beta^{-1}I_N]^{-1}\mathbf{y}$

$\text{var}(x) = K_{test,test} + \beta^{-1} - K_{test}^T[K + \beta^{-1}I_N]^{-1}K_{test}$

$\mathbf{H} - \{a_{train}, \tilde{e}_{train}, \tilde{j}_{train}\}_{1...N} -$ Training data

$\mathbf{y} - \{a_{train}\}_{1...N} -$ Training outputs

$X - \{\tilde{e}_{train}, \tilde{j}_{train}\}_{1...N} -$ Training inputs

$x_{test} - \{\tilde{e}_{test}, \tilde{j}_{test}\} -$ Test input

$K_{test} = K(X, x_{test})$

$K_{test,test} = K(x_{test}, x_{test})$

For the scene interpretation problem, the commonly used squared exponential covariance function is chosen. This choice has the stationarity property of associating observations within a local neighbourhood which is desired.

$$K(x, x') = \sigma_f^2 \exp\left(-\frac{1}{2l^2}(x - x')^2\right) \qquad (9)$$

The hyper-parameters of the GP $\theta = \{l, \sigma_f, \beta^{-1}\}$ are learned by maximising the log likelihood of the training data ($\mathbf{H}$) using a numerical optimisation technique.

GP regression is a discriminative approach, additional exteroceptive or proprioceptive states could be augmented into the input vector. This allows for incorporating additional sensors or proprioceptive measures into the scene interpretation process. However, inversion of an $N \times N$ matrix ($[K(X, X) + \beta^{-1}I_N]^{-1}$) is its main limitation which is an $O(N^3)$ operation (N is the size of the dataset). In this work, the inversion was done off-line after the experience data was collected. For online viability, further work needs to be done to investigate techniques that limit the size of the dataset on the fly by either selecting an informative subset within the dataset or dividing the input space with a gating network in a mixture of experts architecture as mentioned in [16].

III. TEST PLATFORM, TESTING ENVIRONMENT AND DATA COLLECTION

The testing platform is a skid-steered vehicle (Figure 1). The platform is equipped with sensors to measure wheel speed, engine RPM, gearbox RPM and brake pressures. It also has an onboard Inertial Navigation System (INS) with access to raw accelerometer and gyro readings from the onboard IMU. The testing environment has access to DGPS (Differential GPS) corrections for the navigation module. The INS system along with GPS/DGPS observations delivers very good localisation (5cm accuracy) and vehicle actions such as velocity are available with good accuracy. Pitch and roll information from the INS are used to sense terrain slope (exteroceptive conditions) so that an elevation map can be interpreted from terrain gradients.

Training data was collected while executing 30 second exploration maneuvers in various terrain conditions. The exploration maneuvers included an acceleration phase, a coasting phase, a turning phase (both left and right turning) and a braking phase to ensure sufficient proprioceptive excitation. The different terrain conditions include flat terrain, uphill, downhill, positive and negative side slope conditions on grass, and a few runs over flat tarmac and a flat gravel road. The exploration runs were repeated for three distinct behaviours (*slow*: < 1m/s, *normal*: $1 - 2$m/s and *fast*: $2 - 3$m/s) on each of the terrain conditions, so as to achieve sufficient exploration in behaviour space. In total, 20 minutes of data was collected at 20Hz.

IV. SCENE INTERPRETATION RESULTS

Given training data, and the set of test conditions, the extent of observed movement limitation for each of the test condition needs to be derived. In this section, laser data collected over 100x100m off-road terrain is used to derive an elevation map shown in Figure 3 (top). The Sobel operator [13] was applied to the elevation map image to derive pitch and roll gradients, that together form the set of test conditions ($\mathbf{E}_{test}$). Each grid cell has its corresponding exteroceptive state ($\tilde{e}_{test} - \{slope_{Pitch}, slope_{Roll}\}$) value which needs to be associated with a corresponding velocity limit.

$$\mathbf{E}_{test} - \begin{pmatrix} slope_{Pitch1} & slope_{Pitch2} & \cdots & slope_{PitchT} \\ slope_{Roll1} & slope_{Roll2} & \cdots & slope_{RollT} \end{pmatrix}$$

The slip estimate is two dimensional as observations from the UKF using the two-track process model mentioned in [12] consist of slips observations for both left and right tracks ($\tilde{j} - \{slip_{Left}, slip_{Right}\}$).

The set of all slip observations obtained from flat terrain conditions are chosen to be the nominal proprioception set $\mathbf{J}^*$. Given the desired proprioception set and the experience data from the training runs, a Gaussian Process with a squared exponential covariance function was optimised and the proprioceptive scene interpretation process mentioned in

Section II-C was implemented for the set of slope queries ($\mathbf{E}_{test}$) derived from the gradient maps (see Figure 3).

Each slope condition query results in a Gaussian mixture (see Equation 5). By selecting an upper bound on each of such conditional distribution the maximum feasible speed is determined. The upper bound can be determined from the cumulative density function. Also, a caching data structure is used to prevent interpretation of the same condition twice. This significantly improves the speed of the interpretation process.

The end result of such queries on an elevation map is a mobility map shown in Figure 3. The mobility map interprets the obstacles in the scene (trees) as untraversable with a velocity limit of zero, and the rest of the traversable regions on a continuous scale between 0-7kmph. Brighter the pixel intensity easier it is to traverse.

Mobility is defined in vehicle frame, the direction of movement affects pitch and roll slopes which in turn affects mobility values. In this paper, A^* path planning in performed on a grid based representation, hence eight possible directions for slope are considered. Mobility for a given grid cell is a vector of values pertaining to eight possible orientations. Only one such mobility map is shown in Figure 3. All the eight mobility maps are shown in Figure 4. Particularly of interest are maps in Subfigures 4(e) and 4(a), the values for going downhill (4(e)) are significantly smaller than going uphill (4(a)), indicating the need for increased caution.

V. PATH PLANNING FOR SLOPES

In Figure 5, path planning over a vector of mobility maps is compared with planning over a scalar cost map. This scalar cost is the maximum gradient of all eight orientations, and the corresponding 'traversibility' map is shown in Figure 4(i).

The key benefit of these mobility maps with respect to planning is that the cost is orientation sensitive. To leverage this benefit in the A^* algorithm, the arc cost of a connection between two nodes was given as a function of the particular mobility map associated with the direction of this arc. Figure 5(b) demonstrates the desired sensitivity to platform configuration, whereby the path $\overrightarrow{AB}$, and the reverse path $\overrightarrow{BA}$ take different routes, since the path taken to go downhill is treated differently from going uphill. In the scalar cost map case, shown in Figure 5(a), the paths $\overrightarrow{AB}$ and $\overrightarrow{BA}$ are the same.

The A^* paths in Figure 5 only offer heading commands and no information about velocity, so the heuristic cost path in Figure 5(a) needs to be operated with a constant speed preselected for cautious navigation (usually about 1m/s or 3.6 km/hr for the platform in question). For the second case, information from the eight mobility maps can be used to regulate or bound velocities. Average 'maximum feasible speeds' for the paths in Figure 5(b) are 6.1121 km/hr in the forward path (white) and 5.9079 km/hr in the backward path (green). These values are an improvement from that of the cautious case as the mobility values adjust to situations of caution by slowing down and situations of confidence by speeding up.

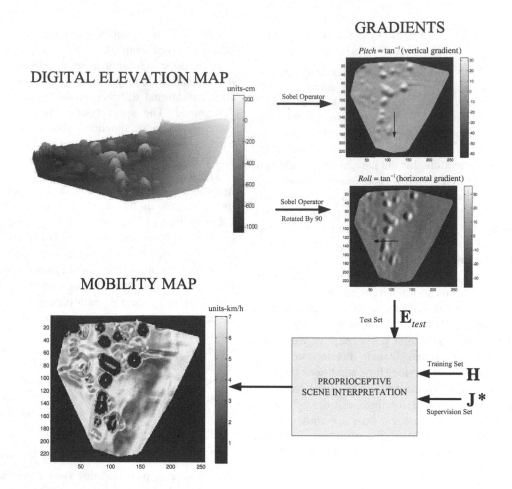

DIGITAL ELEVATION MAP

units-cm

GRADIENTS

Pitch = tan^{-1}(vertical gradient)

Sobel Operator

Roll = tan^{-1}(horizontal gradient)

Sobel Operator
Rotated By 90

MOBILITY MAP

units-km/h

Test Set $\mathbf{E}_{test}$

PROPRIOCEPTIVE
SCENE INTERPRETATION

Training Set $\mathbf{H}$

$\mathbf{J*}$

Supervision Set

Fig. 3. Scene interpretation of an elevation map ($\sim$ 100x120m at 0.5m grid resolution) derived from laser data (units of mobility = km/hr). The mobility map shown assumes that the vehicle intends to travels in the downwards direction shown in the pitch gradient map.

In a separate experiment, controller performance is compared for paths planned on a 5° hill (see Figure 6(a)) with and without mobility paths. For a given set of four waypoints, $A*$ was used to plan across the heuristic scalar cost representation and across mobility maps. The path given by the scalar frame work is executed for constant values of velocity starting from 1m/s to 3m/s (3.6 - 10.8km/h). In contrast the second path planned over mobility maps uses values from the maps to regulate velocity. Figures 6(b) and 6(c) show the input waypoints, the planned paths and the executed paths.

For this experiment, training data were collected under autonomous control where the tuned controller is part of the system. The control system was tuned on flat grass conditions, hence it does not perform well in non-flat conditions (especially downhill where the system becomes significantly underdamped for higher velocities). By performing the same scene interpretation process as before with this new training data the controller limitations are captured into the mobility representation. In Table I the paths are compared for average speed and tracking performance. Tracking performance is judged by standard control theoretic metrics such as root mean squared error (RMSE), L_∞ norm (max error value) and the

L_2 norm. It can be seen that the path over mobility maps shows improved controller performance for similar average and maximum speeds as the path with the highest constant velocity of $3m/s$. This improvement is due to the vehicle slowing down in downhill conditions and speeding up on flat terrain.

While the results in Figure 5 illustrate the desired orientation sensitivity in paths, results in Table I show that controller limitations are captured in the mobility representation and help in regulating velocity to improve controllability. These results illustrate that accounting for directional mobility are of benefit for UGV planning. Use of mobility information leads to a safer choice of paths with a reduced risk of excessive slippage and improved controllability as the system automatically decides on the feasible velocity of operation.

VI. CONCLUSION AND DISCUSSION

This paper presents a non-parametric learning technique to interpret terrain slopes from wheel-slip observations, and describes the benefits of defining costs in behaviour space (as operational limits). First, this enables experience-based learning and encapsulates platform and controller limitations in a common representation. Second, behaviour dependent

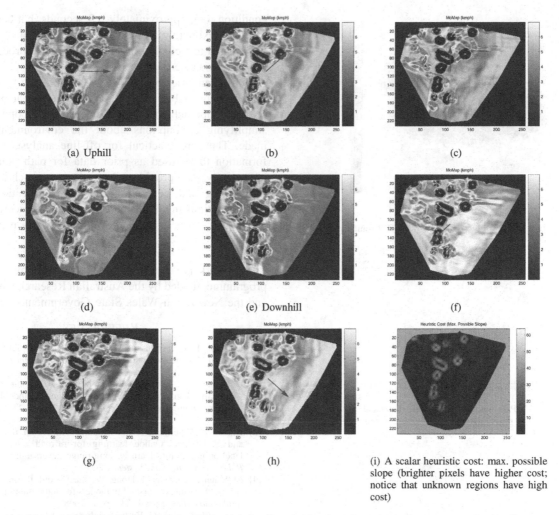

(a) Uphill

(b)

(c)

(d)

(e) Downhill

(f)

(g)

(h)

(i) A scalar heuristic cost: max. possible slope (brighter pixels have higher cost; notice that unknown regions have high cost)

Fig. 4.   Mobility Maps for eight possible vehicle headings (a-h) and an alternative scalar cost representation (i)

(a) A^* path planning on heuristic cost map (path distance: 101m)   (b) A^* path planning on directional mobility maps (path distance: 121m forward and 109m backward)

Fig. 5.   Path planning on heuristic cost maps vs. directional mobility maps. [start waypoint - ●, goal waypoint - *, forward path - solid line, backward path - dotted line]

TABLE I

TRAJECTORY FOLLOWING RESULTS: AVERAGE SPEED AND TRACKING PERFORMANCE

	Speed performance			Cross track error performance		
	time(s)	Mean Vel(m/s)	Max. Vel(m/s)	RMSE	L_∞ norm	L_2 norm
$A^* + 1m/s$	130	0.95	1.70	1.15	2.34	158.65
$A^* + 2m/s$	73	1.71	2.62	1.61	3.66	254.56
$A^* + 3m/s$	65	1.99	3.41	2.14	5.69	536.04
A^*+Mobility Maps	64	2.07	3.29	1.98	4.54	360.03

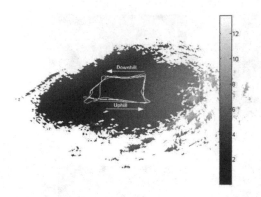

(a) Trajectory test environment - A 5° Hill (units: m)

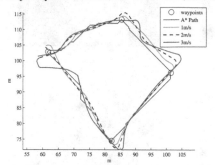

(b) A* Path on scalar heuristic cost and the executed trajectories with different constant velocities

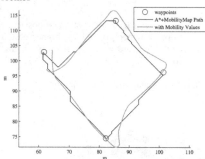

(c) A* Path on mobility maps and the executed trajectory with mobility values

Fig. 6. Trajectory following experiment

costs can be created for aiding decision making, such as orientation sensitive costs for UGV path planning. Finally, proprioceptive feedback such as wheel slip can be incorporated and used to learn effectively in complex environments.

The current process of creating the mobility map is slow, as it involves inverting an $N \times N$ matrix (where N is the size of the dataset). In this work, the inversion was done offline, and the size of dataset was of the order of 10000 data points. The mobility maps shown took about 15 minutes to generate on a 2GHz PC in a Matlab implementation. After the first interpretation, the inverted matrix and the interpreted values were cached and the subsequent queries were very quick in comparison. A new query for an unknown test

condition took approximately 0.3 seconds, but cached queries took approximately 0.03 seconds. Depending on the number of repeated test conditions, the speed of the process will vary for a given scene.

Non-parametric techniques and experience based learning, although not viable as a full online process at present, are a sensible approach in off-road unstructured conditions where simplifying assumptions about the environment cannot be made. They are practical for off-line analysis of sensor information to be used as prior data for path planning. More importantly, they provide a theoretical approach to the process of defining and generating costs. Sparsification and local approximations in GP's is an active area of research, and these techniques can be used in the future to make it more viable.

ACKNOWLEDGMENT

This work is supported by the ARC Centre of Excellence programme, funded by the Australian Research Council (ARC) and the New South Wales State Government.

REFERENCES

[1] L. Ojeda, J. Borenstein, G. Witus, and R. Karlsen, "Terrain characterization and classification with a mobile robot," *Journal Of Field Robotics*, 2006.
[2] L. D. Jackel, E. Krotkov, M. Perschbacher, J. Pippine, and C. Sullivan, "The darpa lagr program: Goals, challenges, methodology, and phase i results," *Journal of Field Robotics*, vol. 23, no. 11-12, pp. 945–973, 2006.
[3] R. Hadsell, P. Sermanet, A. N. Erkan, J. Ben, J. Han, B. Flepp, U. Muller, and Y. LeCun, "Online learning for offroad robots: Using spatial label propagation to learn by long-range traversability," *Proceedings of Robotics: Science and Systems*, 2007.
[4] M. Shneier, T. Chang, T. Hong, W. Shackleford, R. Bostelman, and J. S. Albus, "Learning traversability models for autonomous mobile vehicles," *Autonomous Robots*, vol. 24, no. 1, 2008.
[5] R. S. Sutton and A. G. Barto, *Reinforcement Learning: An Introduction*. The MIT Press, 1998.
[6] A. R. Green and D. Rye, "Sensible planning for vehicles operating over difficult unstructured terrains," in *IEEE Aerospace Conference*, 2007, pp. 1–8.
[7] D. Silver, J. Bagnell, and A. Stentz, "High performance outdoor navigation from overhead data using imitation learning," in *Robotics: Science and Systems IV*, Zurich, Switzerland, 2008.
[8] P. Abbeel and A. Y. Ng, "Apprenticeship learning via inverse reinforcement learning," in *21st International Conference on Machine Learning*, Banff, Canada, 2004.
[9] A. Angelova, L. Matthies, D. Helmick, and P. Perona, "Learning and prediction of slip using visual information," *Journal of Field Robotics*, 2007.
[10] J. Y. Wong, *Theory Of Ground Vehicles*, 3rd ed. New York: Wiley, 2001.
[11] S. J. Julier, J. K. Uhlmann, and H. F. Durrant-Whyte, "A new method for the nonlinear transformation of means and covariances in filters and estimators," in *IEEE Transactions on Automatic Control*, vol. 45, 2000.
[12] A. T. Le, D. Rye, and H. Durrant-Whyte, "Estimation of track-soil interactions for autonomous tracked vehicles," *IEEE International Conference on Robotics and Automation*, vol. 2, pp. 1388–1393 vol.2, Apr 1997.
[13] R. C. Gonzalez and R. E. Woods, *Digital Image Processing*, 3rd ed. Prentice Hall, 2008.
[14] C. E. Rasmussen and C. K. I. Williams, *Gaussian Processes for Machine Learning*. MIT Press, 2006.
[15] C. M. Bishop, *Pattern Recognition And Machine Learning*. Springer, 2006.
[16] C. E. Rasmussen and Z. Ghahramani, "Infinite mixtures of gaussian process experts," in *In Advances in Neural Information Processing Systems 14*. MIT Press, 2002, pp. 881–888.

Learning GP-BayesFilters via Gaussian Process Latent Variable Models

Jonathan Ko   Dieter Fox

University of Washington, Department of Computer Science & Engineering, Seattle, WA

Abstract— **GP-BayesFilters are a general framework for integrating Gaussian process prediction and observation models into Bayesian filtering techniques, including particle filters and extended and unscented Kalman filters. GP-BayesFilters learn nonparametric filter models from training data containing sequences of control inputs, observations, and ground truth states. The need for ground truth states limits the applicability of GP-BayesFilters to systems for which the ground truth can be estimated without prohibitive overhead. In this paper we introduce GPBF-LEARN, a framework for training GP-BayesFilters without any ground truth states. Our approach extends Gaussian Process Latent Variable Models to the setting of dynamical robotics systems. We show how weak labels for the ground truth states can be incorporated into the GPBF-LEARN framework. The approach is evaluated using a difficult tracking task, namely tracking a slotcar based on IMU measurements only.**

I. INTRODUCTION

Over the last years, Gaussian processes (GPs) have been applied with great success to robotics tasks such as reinforcement learning [3] and learning of prediction and observation models [5, 16, 8]. GPs learn probabilistic regression models from training data consisting of input-output examples [17]. GPs combine extreme modeling flexibility with consistent uncertainty estimates, which makes them an ideal tool for learning of probabilistic estimation models in robotics. The fact that GP regression models provide Gaussian uncertainty estimates for their predictions allows them to be seamlessly incorporated into filtering techniques, most easily into particle filters [5, 16].

GP-BayesFilters are a general framework for integrating Gaussian process prediction and observation models into Bayesian filtering techniques, including particle filters and extended and unscented Kalman filters [9, 7]. GP-BayesFilters learn GP filter models from training data containing sequences of control inputs, observations, and ground truth states. In the context of tracking a micro-blimp, GP-BayesFilters have been shown to provide excellent performance, significantly outperforming their parametric Bayes filter counterparts. Furthermore, GP-BayesFilters can be combined with parametric models to improve data efficiency and thereby reduce computational complexity [7]. However, the need for ground truth training data requires substantial labeling effort or special equipment such as a motion capture system in order to determine the true state of the system during training [8]. This requirement limits the applicability of GP-BayesFilters to systems for which such ground truth states are readily available.

The need for ground truth states in GP-BayesFilter training stems from the fact that standard GPs only model noise in the output data, input training points are assumed to be noise-free [17]. To overcome this limitation, Lawrence [11] recently introduced Gaussian Process Latent Variable Models (GPLVM) for probabilistic, non-linear principal component analysis. In contrast to the standard GP training setup, GPLVMs only require output training examples; they determine the corresponding inputs via optimization. Just like other dimensionality reduction techniques such as principal component analysis, GPLVMs learn an embedding of the output examples into a low-dimensional latent (input) space. In contrast to PCA, however, the mapping from latent space to output space is not a linear function but a Gaussian process. While GPLVMs were originally developed in the context of visualization of high-dimensional data, recent extensions enabled their application to dynamic systems [4, 21, 19, 12].

In this paper we introduce GPBF-LEARN, a framework for learning GP-BayesFilters from partially or fully unlabeled training data. The input to GPBF-LEARN are temporal sequences of observations and control inputs along with partial information about the underlying state of the system. GPBF-LEARN proceeds by first determining a state sequence that best matches the control inputs, observations, and partial labels. These states are then used along with the control and observations to learn a GP-BayesFilter, just as in [7]. Partial information ranges from noisy ground truth states, to sparse labels in which only a subset of the states are labeled, to completely label-free data. To determine the optimal state sequence, GPBF-LEARN extends recent advances in GPLVMs to incorporate robot control information and probabilistic priors over the hidden states.

We demonstrate the capabilities of GPBF-LEARN using the autonomous slotcar testbed shown in Fig. 1. The car moves along a slot on a race track while being controlled remotely. Position estimation is performed based on an inertial measurement unit (IMU) placed on the car. Note that tracking solely based on the IMU is difficult, since the IMU provides only turn information. Using this testbed, we demonstrate that GPBF-LEARN outperforms alternative approaches to learning GP-BayesFilters. We furthermore show that GPBF-LEARN can be used to automatically align multiple demonstration traces and learn a filter from completely unlabeled data.

This paper is organized as follows. After discussing related work, we provide background on Gaussian process regression, Gaussian process latent variable models, and GP-BayesFilters.

Then, in Section IV, we introduce the GPBF-LEARN framework. Experimental results are given in Section V, followed by a discussion.

II. RELATED WORK

Lawrence [11] introduced Gaussian Process Latent Variable Models (GPLVMs) for visualization of high-dimensional data. Original GPLVMs impose no smoothness constraints on the latent space. They are thus not able to take advantage of the temporal nature of dynamical systems. One way to overcome this limitation is the introduction of so-called back-constraints [13], which have been applied successfully in the context of WiFi-SLAM, where the goal is to learn an observation model for wireless signal strength data without relying on ground truth location data [4].

Wang and colleagues [21] introduced Gaussian Process Dynamic Models (GPDM), which are an extension of GPLVMs specifically aimed at modeling dynamical systems. GPDMs have been applied successfully to computer animation [21] and visual tracking [19] problems. However, these models do not aim at tracking the hidden state of a physical system, but rather at generating good observation sequences for animation. They are thus not able to incorporate control input or information about the desired structure of the latent space. Furthermore, the tracking application introduced by Urtasun and colleagues [19] is not designed for real-time or near real-time performance, nor does is provide uncertainty estimates as GP-BayesFilters. Other alternatives for non-linear embedding in the context of dynamical systems are hierarchical GPLVMs [12] and action respecting embeddings (ARE) [1]. None of these techniques are able to incorporate control information or impose prior knowledge on the structure of the latent space. We consider both capabilities to be extremely important for robotics applications.

The system identification community has developed various subspace identification techniques [14, 20]. The goal of these techniques is the same as that of GPBF-LEARN, namely to learn a model for a dynamical system from sequences of control inputs and observations. The model underlying N4SID [20] is a linear Kalman filter. Due to its flexibility and robustness, N4SID is extremely popular. It has been applied successfully for human motion animation [6]. In our experiments, we demonstrate that GPBF-LEARN provides superior performance due to its ability to model non-linear systems. We also show that N4SID provides excellent initialization for GPLVMs for dynamical systems.

III. PRELIMINARIES

This section provides background on Gaussian Processes (GPs) for regression, their extension to latent variable models (GPLVMs), and GP-BayesFilters, which use GP regression to learn observation and prediction models for Bayesian filtering.

A. Gaussian Process Regression

Gaussian processes (GP) are non-parametric techniques for learning regression functions from sample data [17]. Assume we have n d-dimensional input vectors: $\mathbf{X} = [\mathbf{x}_1, \mathbf{x}_2, ..., \mathbf{x}_n]$. A GP defines a zero-mean, Gaussian prior distribution over the outputs $\mathbf{y} = [y_1, y_2, ..., y_n]$ at these values:[1]

$$p(\mathbf{y} \mid \mathbf{X}) = \mathcal{N}(\mathbf{y}; 0, \mathbf{K}_y + \sigma_n^2 \mathbf{I}), \tag{1}$$

The covariance of this Gaussian distribution is defined via a kernel matrix, $\mathbf{K}_y$, and a diagonal matrix with elements σ_n^2 that represent zero-mean, white output noise. The elements of the $n \times n$ kernel matrix $\mathbf{K}_y$ are specified by a kernel function over the input values: $\mathbf{K}_y[i, j] = k(\mathbf{x}_i, \mathbf{x}_j)$. By interpreting the kernel function as a distance measure, we see that if points $\mathbf{x}_i$ and $\mathbf{x}_j$ are close in the input space, their output values y_i and y_j are highly correlated.

The specific choice of the kernel function k depends on the application, the most widely used being the squared exponential, or Gaussian, kernel:

$$k(\mathbf{x}, \mathbf{x}') = \sigma_f^2 \, e^{-\frac{1}{2}(\mathbf{x}-\mathbf{x}')W(\mathbf{x}-\mathbf{x}')^T} \tag{2}$$

The kernel function is parameterized by W and σ_f. The diagonal matrix W defines the length scales of the process, which reflect the relative smoothness of the process along the different input dimensions. σ_f^2 is the signal variance.

Given training data $D = \langle \mathbf{X}, \mathbf{y} \rangle$ of n input-output pairs, a key task for a GP is to generate an output prediction at a test input $\mathbf{x}_*$. It can be shown that conditioning (1) on the training data and $\mathbf{x}_*$ results in a Gaussian predictive distribution over the corresponding output y_*

$$p(y_* \mid \mathbf{x}_*, D) = \mathcal{N}\left(y_*, \mathrm{GP}_\mu\left(\mathbf{x}_*, D\right), \mathrm{GP}_\Sigma\left(\mathbf{x}_*, D\right)\right) \tag{3}$$

with mean

$$\mathrm{GP}_\mu\left(\mathbf{x}_*, D\right) = \mathbf{k}_*^T [K + \sigma_n^2 I]^{-1} \mathbf{y} \tag{4}$$

and variance

$$\mathrm{GP}_\Sigma\left(\mathbf{x}_*, D\right) = k(\mathbf{x}_*, \mathbf{x}_*) - \mathbf{k}_*^T \left[K + \sigma_n^2 I\right]^{-1} \mathbf{k}_*. \tag{5}$$

Here, $\mathbf{k}_*$ is a vector of kernel values between $\mathbf{x}_*$ and the training inputs $\mathbf{X}$: $\mathbf{k}_*[i] = k(\mathbf{x}_*, \mathbf{x}_i)$. Note that the prediction uncertainty, captured by the variance GP_Σ, depends on both the process noise and the correlation between the test input and the training inputs.

The hyperparameters $\boldsymbol{\theta}_y$ of the GP are given by the parameters of the kernel function and the output noise: $\boldsymbol{\theta}_y = \langle \sigma_n, W, \sigma_f \rangle$. They are typically determined by maximizing the log likelihood of the training outputs [17]. Making the dependency on hyperparameters explicit, we get

$$\boldsymbol{\theta}_y^* = \underset{\boldsymbol{\theta}_y}{\mathrm{argmax}} \ \log \ p(\mathbf{y} \mid \mathbf{X}, \boldsymbol{\theta}_y). \tag{6}$$

The GPs described thus far depend on the availability of fully labeled training data, that is, data containing ground truth input values $\mathbf{X}$ and possibly noisy output values $\mathbf{y}$.

---

[1]For ease of exposition, we will only describe GPs for one-dimensional outputs, multi-dimensional outputs are handled by assuming independence between the output dimensions.

B. Gaussian Process Latent Variable Models

GPLVMs were introduced in the context of visualization of high-dimensional data [10]. GPLVMs perform nonlinear dimensionality reduction in the context of Gaussian processes. The underlying probabilistic model is still a GP regression model as defined in (1). However, the input values $\mathbf{X}$ are not given and become latent variables that need to be determined during learning. In the GPLVM, this is done by optimizing over both the latent space $\mathbf{X}$ and the hyperparameters:

$$\langle \mathbf{X}^*, \boldsymbol{\theta}_y^* \rangle = \operatorname*{argmax}_{\mathbf{X}, \boldsymbol{\theta}_y} \log \, p(\mathbf{Y} \mid \mathbf{X}, \boldsymbol{\theta}_y) \qquad (7)$$

This optimization can be performed using scaled conjugate gradient descent. In practice, the approach requires a good initialization to avoid local maxima. Typically, such initializations are done via PCA or Isomap [11, 21].

The standard GPLVM approach does not impose any constraints on the latent space. It is thus not able to take advantage of the specific structure underlying dynamical systems. Recent extensions of GPLVMs, namely Gaussian Process Dynamical Models [21] and hierarchical GPLVMs [12], can model dynamic systems by introducing a prior over the latent space $\mathbf{X}$, which results in the following joint distribution over the observed space, the latent space, and the hyperparameters:

$$p(\mathbf{Y}, \mathbf{X}, \boldsymbol{\theta}_y, \boldsymbol{\theta}_x) = p(\mathbf{Y} \mid \mathbf{X}, \boldsymbol{\theta}_y)p(\mathbf{X} \mid \boldsymbol{\theta}_x)p(\boldsymbol{\theta}_y)p(\boldsymbol{\theta}_x) \quad (8)$$

Here, $p(\mathbf{Y} \mid \mathbf{X}, \boldsymbol{\theta}_y)$ is the standard GPLVM term, $p(\mathbf{X} \mid \boldsymbol{\theta}_x)$ is the prior modeling the dynamics in the latent space, and $p(\boldsymbol{\theta}_y)$ and $p(\boldsymbol{\theta}_x)$ are priors over the hyperparameters. The dynamics prior is again modeled as a Gaussian process

$$p(\mathbf{X} \mid \boldsymbol{\theta}_x) = \mathcal{N}(\mathbf{X}; 0, \mathbf{K}_x + \sigma_m^2 \mathbf{I}), \qquad (9)$$

where $\mathbf{K}_x$ is an appropriate kernel matrix. In Section IV, we will discuss different dynamics kernels in the context of learning GP-BayesFilters. The unknown values for this model are again determined via maximizing the log posterior of (8):

$$\langle \mathbf{X}^*, \boldsymbol{\theta}_y^*, \boldsymbol{\theta}_x^* \rangle = \operatorname*{argmax}_{\mathbf{X}, \boldsymbol{\theta}_y, \boldsymbol{\theta}_x} \Big(\log p(\mathbf{Y} \mid \mathbf{X}, \boldsymbol{\theta}_y) +$$

$$\log p(\mathbf{X} \mid \boldsymbol{\theta}_x) + \log p(\boldsymbol{\theta}_y) + \log p(\boldsymbol{\theta}_x) \Big) (10)$$

Such extensions to GPLVMs have been used successfully to model temporal data such as motion capture sequences [21, 12] and visual tracking data [19].

C. GP-BayesFilters

GP-BayesFilters are Bayes filters that use GP regression to learn prediction and observation models from training data. Bayes filters recursively estimate posterior distributions over the state $\mathbf{x}_t$ of a dynamical system at time t conditioned on sensor data $\mathbf{z}_{1:t}$ and control information $\mathbf{u}_{1:t-1}$. Key components of every Bayes filter are the prediction model, $p(\mathbf{x}_t \mid \mathbf{x}_{t-1}, \mathbf{u}_{t-1})$, and the observation model, $p(\mathbf{z}_t \mid \mathbf{x}_t)$. The prediction model describes how the state $\mathbf{x}_t$ changes based on time and control input $\mathbf{u}_{t-1}$, and the observation model describes the likelihood of making an observation $\mathbf{z}_t$ given

the state $\mathbf{x}_t$. In robotics, these models are typically parametric descriptions of the underlying processes, see [18] for several examples.

GP-BayesFilters use Gaussian process regression models for both prediction and observation models. Such models can be incorporated into different versions of Bayes filters and have been shown to outperform parametric models [7]. Learning the models of GP-BayesFilters requires ground truth sequences of a dynamical system containing for each time step a control command, $\mathbf{u}_{t-1}$, an observation, $\mathbf{z}_t$, and the corresponding ground truth state, $\mathbf{x}_t$. GP prediction and observation models can then be learned based on training data

$$D_p = \langle (\mathbf{X}, \mathbf{U}), \mathbf{X}' \rangle$$
$$D_o = \langle \mathbf{X}, \mathbf{Z} \rangle,$$

where $\mathbf{X}$ is a matrix containing the sequence of ground truth states, $\mathbf{X} = [\mathbf{x}_1, \mathbf{x}_2, \ldots, \mathbf{x}_T]$, $\mathbf{X}'$ is a matrix containing the state changes, $\mathbf{X}' = [\mathbf{x}_2 - \mathbf{x}_1, \mathbf{x}_3 - \mathbf{x}_2, \ldots, \mathbf{x}_T - \mathbf{x}_{T-1}]$, and $\mathbf{U}$ and $\mathbf{Z}$ contain the sequences of controls and observations, respectively. By plugging these training sets into (4) and (5), one gets GP prediction and observation models mapping from a state, $\mathbf{x}_{t-1}$, and a control, $\mathbf{u}_{t-1}$, to change in state, $\mathbf{x}_t - \mathbf{x}_{t-1}$, and from a state, $\mathbf{x}_t$, to an observation, $\mathbf{z}_t$, respectively. These probabilistic models can be readily incorporated into Bayes filters such as particle filters and unscented Kalman filters. An additional derivative of (4) provides the Taylor expansion needed for extended Kalman filters [7].

The need for ground truth training data is a key limitation of GP-BayesFilters and other applications of GP regression models in robotics. While it might be possible to collect ground truth data using accurate sensors [7, 15, 16] or manual labeling [5], the ability to learn GP models based on weakly labeled or unlabeled data significantly extends the range of problems to which such models can be applied.

IV. GPBF-LEARN

In this section we show how GP-BayesFilters can be learned from weakly labeled data. While the extensions of GPLVMs described in Section III-B are designed to model dynamical systems, they lack important abilities needed to make them fully useful for robotics applications. First, they do not consider control information, which is extremely important for learning accurate prediction models in robotics. Second, they optimize the values of the latent variables (states) solely based on the output samples (observations) and GP dynamics in the latent space. However, in state estimation scenarios, one might want to impose stronger constraints on the latent space $\mathbf{X}$. For example, it is often desirable that latent states $\mathbf{x}_t$ correspond to physical entities such as the location of a robot. To enforce such a relationship between latent space and physical robot locations, it would be advantageous if one could label a subset of latent points with their physical counterparts and then constrain the latent space optimization to consider these labels.

We now introduce GPBF-LEARN, which overcomes limitations of existing techniques. The training data for GPBF-LEARN, $D = [\mathbf{Z}, \mathbf{U}, \widehat{\mathbf{X}}]$, consists of time stamped sequences containing observations, $\mathbf{Z}$, controls, $\mathbf{U}$, and weak labels, $\widehat{\mathbf{X}}$, for the latent states. In the context discussed here, the labels provide noisy information about subsets of the latent states. Given training data D, the posterior over the sequence of hidden states and hyperparameters is as follows:

$$p(\mathbf{X}, \boldsymbol{\theta}_x, \boldsymbol{\theta}_z \mid \mathbf{Z}, \mathbf{U}, \widehat{\mathbf{X}}) \propto$$
$$p(\mathbf{Z} \mid \mathbf{X}, \boldsymbol{\theta}_z)\, p(\mathbf{X} \mid \mathbf{U}, \boldsymbol{\theta}_x)\, p(\mathbf{X} \mid \widehat{\mathbf{X}})\, p(\boldsymbol{\theta}_z) p(\boldsymbol{\theta}_x) \quad (11)$$

In GPBF-LEARN, both the observation model, $p(\mathbf{Z} \mid \mathbf{X}, \boldsymbol{\theta}_z)$, and the prediction model, $p(\mathbf{X} \mid \mathbf{U}, \boldsymbol{\theta}_x)$, are Gaussian processes, and $\boldsymbol{\theta}_x$ and $\boldsymbol{\theta}_z$ are the hyperparameters of these GPs. While the observation model in (11) is the same as in the GPLVM for dynamical systems (8), the prediction GP now includes control information. Furthermore, the GPBF-LEARN posterior contains an additional term for labels, $p(\mathbf{X} \mid \widehat{\mathbf{X}})$, which we describe next.

A. Weak Labels

The labels $\widehat{\mathbf{X}}$ represent prior knowledge about individual latent states $\mathbf{X}$. For instance, it might not be possible to generate highly accurate ground truth states for every data point in the training set. Instead, one might only be able to provide accurate labels for a small subset of states, or noisy estimates for the states. At the same time, such labels might still be extremely valuable since they guide the latent variable model to determine a latent space that is similar to the desired, physical space. While the form of prior knowledge can take on various forms, we here consider labels that represent independent Gaussian priors over latent states:

$$p(\mathbf{X} \mid \widehat{\mathbf{X}}) = \prod_{\hat{\mathbf{x}}_t \in \widehat{\mathbf{X}}} \mathcal{N}(\mathbf{x}_t; \hat{\mathbf{x}}_t, \sigma^2_{\hat{\mathbf{x}}_t}) \quad (12)$$

Here, $\sigma^2_{\hat{\mathbf{x}}_t}$ is the uncertainty in label $\hat{\mathbf{x}}_t$. As note above, $\widehat{\mathbf{X}}$ can impose priors on all or any subset of latent states. As we will show in the experiments, these additional terms generate more consistent tracking results on test data.

B. GP Dynamics Models

GP dynamics priors, $p(\mathbf{X} \mid \mathbf{U}, \boldsymbol{\theta}_x)$, do not constrain individual states but model prior information of how the system evolves over time. They provide substantial flexibility for modeling different aspects of a dynamical system. Intuitively, these priors encourage latent states $\mathbf{X}$ that correspond to smooth mappings from past states and controls to future states. Even though the dynamics GP is an integral part of the posterior model (11), for exposure reason it is easier to treat it as if it was a separate GP.

Different dynamics models are achieved by changing the specific values for the input and output data used for this dynamics GP. We denote by $\mathbf{X}^{\text{in}}$ and $\mathbf{X}^{\text{out}}$ the input and output data for the dynamics GP, where $\mathbf{X}^{\text{in}}$ is typically derived from states at specific points in time, and $\mathbf{X}^{\text{out}}$ is derived from states at the next time step. To more strongly emphasize the sequential aspect of the dynamics model we will use time t to index data points. Using the GP dynamics model we get

$$p(\mathbf{X} \mid \mathbf{U}, \boldsymbol{\theta}_x) = \mathcal{N}(\mathbf{X}^{\text{out}}; 0, \mathbf{K}_{\text{x}} + \sigma^2_x \mathbf{I}), \quad (13)$$

where σ^2_x is the noise of the prediction model, and the kernel matrix $\mathbf{K}_{\text{x}}$ is defined via the kernel function on input data to the dynamics GP: $\mathbf{K}_{\text{x}}[t, t'] = k\left(\mathbf{x}^{\text{in}}_t, \mathbf{x}^{\text{in}}_{t'}\right)$, where $\mathbf{x}^{\text{in}}_t$ and $\mathbf{x}^{\text{in}}_{t'}$ are input vectors for time steps t and t', respectively.

The specification of $\mathbf{X}^{\text{in}}$ and $\mathbf{X}^{\text{out}}$ determines the dynamics prior. To see, consider the most basic dynamics GP, which solely models a mapping from the state at time $t - 1$, $\mathbf{x}_{t-1}$, to the state at time t, $\mathbf{x}_t$. In this case we get the following specification:

$$\mathbf{x}^{\text{in}}_t = \mathbf{x}_{t-1} \quad (14)$$
$$\mathbf{x}^{\text{out}}_t = \mathbf{x}_t \quad (15)$$

Optimization with such a dynamics model encourages smooth state sequences $\mathbf{X}$. Generating smooth *velocities* can be achieved by setting $\mathbf{x}^{\text{in}}_t$ to $\dot{\mathbf{x}}_{t-1}$ and $\mathbf{x}^{\text{out}}_t$ to $\dot{\mathbf{x}}_t$, where $\dot{\mathbf{x}}_t$ represents the velocity $[\mathbf{x}_t - \mathbf{x}_{t-1}]$ at time t [21]. It should be noted that such a velocity model can be incorporated without adding a velocity dimension to the latent space. A more complex, localized dynamics model that takes control and velocity into account can be achieved by the following settings:

$$\mathbf{x}^{\text{in}}_t = [\mathbf{x}_{t-1}, \dot{\mathbf{x}}_{t-1}, \mathbf{u}_{t-1}]^T \quad (16)$$
$$\mathbf{x}^{\text{out}}_t = \dot{\mathbf{x}}_t \quad (17)$$

This model encourages smooth changes in velocity depending on control input. By adding $\mathbf{x}_{t-1}$ to $\mathbf{x}^{\text{in}}_t$, the dynamics model becomes *localized*, that is, the impact of control on velocity can be different for different states. While one could also model higher order dependencies, we here stick to the one given in (17), which corresponds to a relatively standard prediction model for Bayes filters.

C. Optimization

Just as regular GPLVM models, GPBF-LEARN determines the unknown values of the latent states $\mathbf{X}$ by optimizing the log of posterior over the latent state sequence and the hyperparameters. The log of (11) is given by

$$\log p(\mathbf{X}, \boldsymbol{\theta}_x, \boldsymbol{\theta}_z \mid D) =$$
$$\log p(\mathbf{Z} \mid \mathbf{X}, \boldsymbol{\theta}_z) + \log p(\mathbf{X} \mid \mathbf{U}, \boldsymbol{\theta}_x) +$$
$$\log p(\mathbf{X} \mid \widehat{\mathbf{X}}) + \log p(\boldsymbol{\theta}_z) + \log p(\boldsymbol{\theta}_x) + \text{const}, \quad (18)$$

where D represents the training data $[\mathbf{Z}, \mathbf{U}, \widehat{\mathbf{X}}]$. We perform this optimization using scaled conjugate gradient descent [21]. The gradients of the log are given by:

$$\frac{\partial \log p(\mathbf{X}, \boldsymbol{\theta}_x, \boldsymbol{\theta}_z \mid \mathbf{Z}, \mathbf{U})}{\partial \mathbf{X}} =$$
$$\frac{\partial \log p(\mathbf{Z} \mid \mathbf{X}, \boldsymbol{\theta}_z)}{\partial \mathbf{X}} + \frac{\partial \log p(\mathbf{X} \mid \mathbf{U}, \boldsymbol{\theta}_x)}{\partial \mathbf{X}} + \frac{\partial \log p(\mathbf{X} \mid \widehat{\mathbf{X}})}{\partial \mathbf{X}} (19)$$
$$\frac{\partial \log p(\mathbf{X}, \boldsymbol{\theta}_x, \boldsymbol{\theta}_z \mid D)}{\partial \boldsymbol{\theta}_x} = \frac{\partial \log p(\mathbf{X} \mid \mathbf{U}, \boldsymbol{\theta}_x)}{\partial \boldsymbol{\theta}_x} + \frac{\partial \log p(\boldsymbol{\theta}_x)}{\partial \boldsymbol{\theta}_x} \quad (20)$$
$$\frac{\partial \log p(\mathbf{X}, \boldsymbol{\theta}_x, \boldsymbol{\theta}_z \mid D)}{\partial \boldsymbol{\theta}_z} = \frac{\partial \log p(\mathbf{Z} \mid \mathbf{X}, \boldsymbol{\theta}_z)}{\partial \boldsymbol{\theta}_z} + \frac{\partial \log p(\boldsymbol{\theta}_z)}{\partial \boldsymbol{\theta}_z} . (21)$$

TABLE I

THE GPBF-LEARN ALGORITHM.

The individual derivatives follow as

$$\frac{\partial \log p(\mathbf{Z} \mid \mathbf{X}, \boldsymbol{\theta}_z)}{\partial \mathbf{X}} = \frac{1}{2} trace \left(\mathbf{K}_Z^{-1} \mathbf{Z}\mathbf{Z}^t \mathbf{K}_Z^{-1} - \mathbf{K}_Z^{-1} \right) \frac{\partial \mathbf{K}_Z}{\partial \mathbf{X}}$$

$$\frac{\partial \log p(\mathbf{Z} \mid \mathbf{X}, \boldsymbol{\theta}_z)}{\partial \boldsymbol{\theta}_Z} = \frac{1}{2} trace \left(\mathbf{K}_Z^{-1} \mathbf{Z}\mathbf{Z}^t \mathbf{K}_Z^{-1} - \mathbf{K}_Z^{-1} \right) \frac{\partial \mathbf{K}_Z}{\partial \boldsymbol{\theta}_z}$$

$$\frac{\partial \log p(\mathbf{X} \mid \boldsymbol{\theta}_x, \mathbf{U})}{\partial \mathbf{X}} = \frac{1}{2} trace \left(\mathbf{K}_X^{-1} \mathbf{X}_{out} \mathbf{X}_{out}^T \mathbf{K}_X^{-1} - \mathbf{K}_X^{-1} \right) \frac{\partial \mathbf{K}_X}{\partial \mathbf{X}}$$
$$- \mathbf{K}_X^{-1} \mathbf{X}_{out} \frac{\partial \mathbf{X}_{out}}{\partial \mathbf{X}}$$

$$\frac{\partial \log p(\mathbf{X} \mid \boldsymbol{\theta}_x, \mathbf{U})}{\partial \boldsymbol{\theta}_x} = \frac{1}{2} trace \left(\mathbf{K}_X^{-1} \mathbf{X}_{out} \mathbf{X}_{out}^T \mathbf{K}_X^{-1} - \mathbf{K}_X^{-1} \right) \frac{\partial \mathbf{K}_X}{\partial \boldsymbol{\theta}_x}$$

$$\frac{\partial \log p(\mathbf{X} \mid \widehat{\mathbf{X}})}{\partial \mathbf{X}[i,j]} = -(\mathbf{X}[i,j] - \widehat{\mathbf{X}}[i,j])/\sigma_{\widehat{\mathbf{x}}_t}^2,$$

where $\frac{\partial \mathbf{K}}{\partial \mathbf{X}}$ and $\frac{\partial \mathbf{K}}{\partial \theta}$ are the matrix derivatives. They are formed by taking the partial derivative of the individual elements of the Gram matrix with respect to $\mathbf{X}$ or the hyperparameters, respectively.

D. GPBF-LEARN Algorithm

A high level overview of the GPBF-LEARN algorithm is given in Table I. The input to GPBF-LEARN consists of training data containing a sequence of observations, $\mathbf{Z}$, control inputs, $\mathbf{U}$, and weak labels, $\widehat{\mathbf{X}}$. In the first step, the unknown latent states $\mathbf{X}$ are initialized using the information provided by the weak labels. This is done by setting every latent state to the estimate provided by $\widehat{\mathbf{X}}$. In the sparse labeling case, the states without labels are initialized by linear interpolation between those for which a label is given. In the fully unsupervised case, where $\widehat{\mathbf{X}}$ is empty, we use N4SID to initialize the latent states [20]. In our experiments, N4SID provides initialization that is far superior to the standard PCA initialization used by [11, 21]. Then, in Step 2, scaled conjugate gradient (SCG) descent determines the latent states and hyperparameters via optimization of the log posterior (18). This iterative procedure computes the gradients (19) – (21) during each iteration using the dynamics model and the weak labels. Finally, the resulting latent states $\mathbf{X}^*$, along with the observations and controls are used to learn a GP-BayesFilter, as described in Section III-C.

In essence, the final step of the algorithm "compiles" the complex latent variable model into an efficient, online GP-BayesFilter. The key difference between the filter model and the latent variable model is due to the fact that the filter model makes a first order Markov assumption. The latent variable model, on the other hand, optimizes all latent points jointly and these points are all correlated via the GP kernel matrix. To reflect the difference between these models, we learn new hyperparameters for the GP-BayesFilter.

V. EXPERIMENTS

In these experiments we evaluate different properties of GPBF-LEARN using the computer controlled slotcar platform shown in Fig. 1. Specifically, we demonstrate the ability of GPBF-LEARN to incorporate prior knowledge over the latent states, to learn robust GP-BayesFilters from noisy and sparse labeled data, and to perform system identification without any ground truth states.

In an additional experiment not reported here, we compared the two dynamics models described in Section IV-B. Using 10-step ahead prediction as evaluation criteria, we found that our control based model (17) significantly outperforms the simpler model (15) that is typically used for GPLVMs. In fact, our model reduces the prediction error by almost 50%, from 29.2 to 16.1 cm.

A. Slotcar Evaluation Platform

The experimental setup consists of a track and a miniature car which is guided along the track by a groove, or slot, cut into the track. The left panel in Fig. 1 shows the track, which contains elevation changes as well as banked curves with a total length of about 14m. An overhead camera tracks the car and is used as ground truth data for evaluation of the algorithms. The car is a standard 1:32 scale model manufactured by Carrera International and augmented with a Microstrain 3DM-GX1 inertial measurement unit (IMU), as shown in the next panel in Fig. 1. The IMU tracks the relative orientation of the car. These measurements are sent off-board in real-time via a WiFi interface. Control signals to the car are supplied by an offboard computer. These controls signals are directly proportional to the amperage supplied the the car motor.

The data, $\langle \mathbf{Z}, \mathbf{U} \rangle$, is collected at 15 frames per second. From the IMU data, we extract the 3D orientation of the car in Euler angles. During the evaluations, we take the *difference* between consecutive angle readings as the observation data $\mathbf{Z}$. The raw angles could potentially be used as observations as well. However, even though this might make tracking and system identification easier over short periods of time, we observed substantial IMU angle drift and thus decided to use a more realistic scenario that does not depend on a global heading sensor. As can be seen in the third panel in Fig. 1, the resulting turning rate data is very noisy and includes substantial amounts of aliasing, in which the same angle measurements occur at many different locations on the track. For instance, all angle differences are close to zero whenever the car moves through a straight section of the track. This kind of aliasing makes learning the latent space particularly challenging since it does not provide a unique mapping from the observation sequence $\mathbf{Z}$ to the latent space $\mathbf{X}$.

  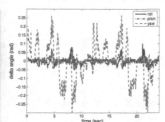 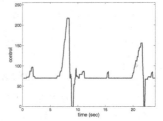

Fig. 1. (left) The slotcar track used during the experiments. An overhead camera is used to ground truth evaluation. (left middle) The test vehicle moves along a slot in the track, velocity control is provided remotely by a desktop PC. The state of the vehicle is estimated based on an on-board IMU. (right middle) IMU turning rate in roll, pitch, and yaw. Shown is data collected over two rounds around the track. (right) Control inputs for the same run.

In all experiments we use a GP-UKF to generate tracking results [9]. In the first set of experiments we demonstrate that GPBF-LEARN can learn a latent (state) space $\mathbf{X}$ that is consistent with a desired latent space specified via weak labels $\widehat{\mathbf{X}}$. Here, the desired latent space is the 1D position of the car along the track. In this scenario, we assume that the training data contains noisy or sparse labels $\widehat{\mathbf{X}}$, as below.

B. Incorporating Noisy Labels

Here we consider the scenario in which one is not able to provide extremely accurate ground truth states for the training data. Instead, one can only provide noisy labels $\widehat{\mathbf{X}}$ for the states. We evaluate four possible approaches to learning a GP-BayesFilter from such data. The first, called INIT, simply ignores the fact that the labels are noisy and learns a GP-BayesFilter using the initial data $\widehat{\mathbf{X}}$. The next two use the noisy labels to initialize the latent variables $\mathbf{X}$, but performs optimization *without* the weak label terms described in Section IV-A. We call this approach GPDM, since it results from applying the model of Wang *et.al.* [21] to this setting. We do this with and without the use of control data $\mathbf{U}$ in order to distinguish the contributions of the various components. Finally, GPBFL denotes our GPBF-LEARN approach that considers the noisy labels during optimization.

The system state in this scenario is the 1D position of the car along the track, that is, the approach must learn to project the 3D IMU observations $\mathbf{Z}$ along with the control information $\mathbf{U}$ into a 1D latent space $\mathbf{X}$. Training data consist of 5 manually controlled cycles of the car around the track. We perform cross-validation by applying the different approaches to four loops and testing tracking performance on the remaining loop. The overhead camera provides fairly accurate 1D track position via background subtraction and simple blob tracking, followed by snapping xy pixel locations to an aligned model of the track. To simulate noisy labels, we added different levels of Gaussian noise to the camera based 1D track locations and used these as $\widehat{\mathbf{X}}$. For each noise level applied to the labels we perform a total of 10 training and test runs. For each run, we extract GP-BayesFilters using the resulting optimized latent states $\mathbf{X}^*$ along with the controls and IMU observations. Currently, learning is done with each loop treated as a separate episode as we do not handle the jump between the beginning and end of the loop for the 1D latent space. This could likely be handled by a periodic kernel as future work. The quality

of the resulting models is tested by checking how close $\mathbf{X}^*$ is to the ground truth states provided by the camera, and by tracking with a GP-UKF on previously unseen test data.

The left panel in Fig. 2 shows a plot of the differences between the learned hidden states, $\mathbf{X}^*$, and the ground truth for different values of noise applied to the labels $\widehat{\mathbf{X}}$. As can be seen, GPBFL is able to recover the correct 1D latent space even for high levels of noise. GPDM which only considers the labels by initializing the latent states generates a high error. This is due to the fact that the optimization performed GPDM lets these latent states "drift" from the desired values. The optimization performed by GPDM without control is even higher than that with control. GPDM without control ends up overly smooth since it does not have controls to constrain the latent states. Not surprisingly, the error of INIT increases linearly in the noise of the labels, since INIT uses these labels as the latent states without any optimization.

The middle panel in Fig. 2 shows the RMS error when running a GP-BayesFilter that was extracted based on the learned hidden states using the different approaches. For clarity, we only show the averages over those runs that did not produce a tracking error. A run is considered a failure if the RMS error is greater than 70 cm. Out of its 80 runs, INIT produced 18 tracking failures, GPDM without controls 11, GPDM with controls 7, while our approach GPBFL produced only one failure. Note that a tracking failure can occur due to both mis-alignment between the learned latent space and high noise in the observations.

As can be seen in the figure, GPBFL is able to learn a GP-BayesFilter that maintains a low tracking RMS error even when the labels $\widehat{\mathbf{X}}$ are very noisy. On the other hand, simply ignoring noise in labels results in increasingly bad tracking performance, as shown by the graph for INIT. In addition, GPDM generates significantly poorer tracking performance than our approach.

C. Incorporating Sparse Labels

In some settings it might not be possible to provide even noisy labels for all training points. Here we evaluate this scenario by randomly removing noisy labels from the training data. For the approach INIT we generated full labels by linearly interpolating between the sparse labels. The right panel in Fig. 2 shows the errors between ground truth 1D latent space and the learned latent space, $\mathbf{X}^*$, for different

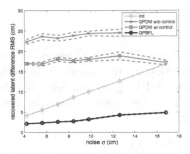 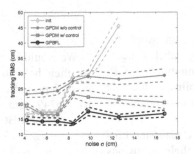 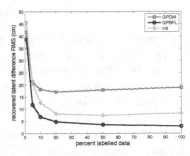

Fig. 2. Evaluation of INIT, GPDM, and GPBFL on noisy and sparse labels. Dashed lines provide 95% confidence intervals. (left) Difference between the learned latent states $\mathbf{X}^*$ and ground truth as a function of noise level in the labels $\hat{\mathbf{X}}$. (middle) Tracking errors for different noise levels. (right) Difference between the learned latent states and ground truth as a function of label sparsity.

levels of label sparsity. Again, our approach, GPBFL, learns a more consistent latent space as GPDM, which uses the labels only for initialization. The linear interpolation approach, INIT, outperforms GPDM since it does not learn anything and thereby avoids drifting from the provided labels.

D. GPBF-LEARN *for Subspace Identification without Labels*

The final experiment demonstrates that GPBF-LEARN can learn a model without any labeled data. Here, the training input consists solely of turning rate observations $\mathbf{Z}$ and control inputs $\mathbf{U}$. No weak labels $\hat{\mathbf{X}}$ are provided and no information about the structure of the latent space is given. To encode less knowledge about the underlying race track, we make GPBF-LEARN learn a 2D latent space. Overall, this is an extremely challenging task for latent variable models. To see, we initialized the latent state of GPBF-LEARN using PCA, as is typically done for GPLVMs [21, 11, 19]. In this case, GPBF-LEARN was not able to learn a smooth model of the latent space. This is because PCA does not take the dynamics in latent space into account.

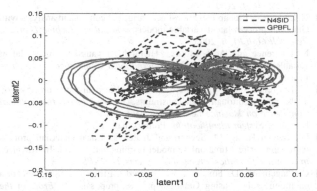

Fig. 3. 2D latent space learned by N4SID and GPBF-LEARN with N4SID initialization.

A different approach for initialization is N4SID, which is a well known, linear model for system identification of dynamical systems [20]. N4SID provides an estimate of the hidden state which does take into account the system dynamics. The latent space recovered by N4SID is given by the dashed line in Fig. 3. N4SID can only generate an extremely un-smooth latent space that does not reflect the smooth structure of the underlying track. When running GPBF-LEARN on the data, initialized with N4SID, we get the solid line shown in the

same figure. Obviously, GPBF-LEARN takes advantage of its underlying non-linear GP model to recover a smooth latent space that nicely reflects the cyclic structure of the race track. Note that we would not expect all cycles through the track to be mapped exactly on top of each other, since the slotcar has very different observations depending on its velocity.

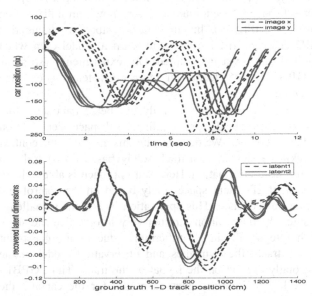

Fig. 4. Plots showing misalignment of the input data (top), and alignment in the latent space learned by GPBF-LEARN (bottom).

An important aspect of the optimization problem is to learn a proper alignment of the data, that is, each portion of the real track should correspond to similar latent states. The top of Fig. 4 shows the true x and y positions of the car in physical space for the different cycles through the track. As can be seen, the car moves through the track with different velocities, resulting in the mis-aligned graphs. To visualize that GPBF-LEARN was able to recover an aligned latent space from the unaligned input data (note that we used IMU, xy is only used for visualization), we plot the two latent dimensions vs. the position on the 1D track model. This plot is shown in the bottom of Fig. 4. As can be seen, the latent states are well aligned with the 1D model over the different cycles through the track. This result is extremely encouraging, since it shows that we might be able to learn an imitation control model based on such demonstrations, as done by Coates and colleagues [2].

VI. CONCLUSION

This paper introduced GPBF-LEARN, a framework for learning GP-BayesFilters from only weakly labeled training data. We thereby overcome a key limitation of GP-BayesFilters, which so far required the availability of accurate ground truth states for learning Gaussian process prediction and observation models [7].

GPBF-LEARN builds on recently introduced Gaussian Process Latent Variable Models (GPLVMs) and their extensions to dynamical systems [11, 21]. GPBF-LEARN improves on existing GPLVM systems in various ways. First, it can incorporate weak labels on the latent states. It is thereby able to learn a latent space that is consistent with a desired physical space, as demonstrated in the context of our slotcar track. Second, GPBF-LEARN can incorporate control information into the dynamics model used for the latent space. Obviously, this ability to use control information is extremely important for complex dynamical systems. Third, we introduce N4SID, a linear subspace ID technique, as a very powerful initialization method for GPLVMs. In our slotcar testbed we found that N4SID enabled GPBF-LEARN to learn a model even when the initialization via PCA failed. Our experiments also show that GPBF-LEARN learns far more consistent models than N4SID alone.

Additional experiments on fully unlabeled data show that GPBF-LEARN can perform nonlinear subspace identification and data alignment. We demonstrate this ability in the context of tracking a slotcar on a track solely based on control and IMU turn rate information. Here, our approach is able to learn a consistent 2D latent space solely based on the control and observation sequence. This application is extremely challenging, since the observations are not very informative and show a high rate of aliasing. Furthermore, due to the constraint onto the track, the dynamics and observation model of the car strongly depend on the layout of the track. Thus, GPBF-LEARN has to jointly recover a model for the car and the track.

We have also obtained some preliminary results of tracking the slotcar in the 3D latent space. For this task, the automatically learned GP hyperparameters turned out to be insufficient for tracking, requiring additional manual tuning. To overcome this problem, we intend to explore the use of discriminative learning to optimize the hyperparameters for filtering.

In future work, GPBF-LEARN could be applied to imitation learning, similar to the approach introduced by Coates and colleagues for helicopter control [2]. In this context we would take advantage of the automatic alignment of different demonstrations given by GPBF-LEARN . An integration of GP-BayesFilter with model predictive control techniques is an interesting question in this context. Other possible extensions include the incorporation of parametric models to improve learning and generalization. Finally, the latent model underlying GPBF-LEARN is by no means restricted to GP-BayesFilters. It can be applied to improve learning quality whenever there is no accurate ground truth data available for training Gaussian processes.

ACKNOWLEDGMENTS

We would like to thank Michael Chung and Deepak Verma for their help in running the slotcar experiments. This work was supported in part by ONR MURI grant number N00014-07-1-0749 and by the NSF under contract numbers IIS-0812671 and BCS-0508002.

REFERENCES

[1] M. Bowling, D. Wilkinson, A. Ghodsi, and A. Milstein. Subjective localization with action respecting embedding. In *Proc. of the International Symposium of Robotics Research (ISRR)*, 2005.

[2] A. Coates, P. Abbeel, and A. Ng. Learning for control from multiple demonstrations. In *Proc. of the International Conference on Machine Learning (ICML)*, 2008.

[3] Y. Engel, P. Szabo, and D. Volkinshtein. Learning to control an octopus arm with Gaussian process temporal difference methods. In *Advances in Neural Information Processing Systems 18 (NIPS)*, 2006.

[4] B. Ferris, D. Fox, and N. Lawrence. WiFi-SLAM using Gaussian process latent variable models. In *Proc. of the International Joint Conference on Artificial Intelligence (IJCAI)*, 2007.

[5] B. Ferris, D. Hähnel, and D. Fox. Gaussian processes for signal strength-based location estimation. In *Proc. of Robotics: Science and Systems (RSS)*, 2006.

[6] E. Hsu, K. Pulli, and J. Popović. Style translation for human motion. *ACM Transactions on Graphics (Proc. of SIGGRAPH)*, 2005.

[7] J. Ko and D. Fox. GP-BayesFilters: Bayesian filtering using Gaussian process prediction and observation models. In *Proc. of the IEEE/RSJ International Conference on Intelligent Robots and Systems (IROS)*, 2008.

[8] J. Ko, D. Klein, D. Fox, and D. Hähnel. Gaussian processes and reinforcement learning for identification and control of an autonomous blimp. In *Proc. of the IEEE International Conference on Robotics & Automation (ICRA)*, 2007.

[9] J. Ko, D. Klein, D. Fox, and D. Hähnel. GP-UKF: Unscented Kalman filters with Gaussian process prediction and observation models. In *Proc. of the IEEE/RSJ International Conference on Intelligent Robots and Systems (IROS)*, 2007.

[10] N. Lawrence. Gaussian process latent variable models for visualization of high dimensional data. In *Advances in Neural Information Processing Systems (NIPS)*, 2003.

[11] N. Lawrence. Probabilistic non-linear principal component analysis with Gaussian process latent variable models. *Journal of Machine Learning Research (JMLR)*, 6, 2005.

[12] N. Lawrence and A. J. Moore. Hierarchical gaussian process latent variable models. In *Proc. of the International Conference on Machine Learning (ICML)*, 2007.

[13] N. Lawrence and J. Quiñonero Candela. Local distance preservation in the gp-lvm through back constraints. In *Proc. of the International Conference on Machine Learning (ICML)*, 2006.

[14] L. Ljung. *System Identification*. Prentice hall, 1987.

[15] D. Nguyen-Tuong, M. Seeger, and J. Peters. Local Gaussian process regression for real time online model learning and control. In *Advances in Neural Information Processing Systems 22 (NIPS)*, 2008.

[16] C. Plagemann, D. Fox, and W. Burgard. Efficient failure detection on mobile robots using Gaussian process proposals. In *Proc. of the International Joint Conference on Artificial Intelligence (IJCAI)*, 2007.

[17] C.E. Rasmussen and C.K.I. Williams. *Gaussian processes for machine learning*. The MIT Press, 2005.

[18] S. Thrun, W. Burgard, and D. Fox. *Probabilistic Robotics*. MIT Press, Cambridge, MA, September 2005. ISBN 0-262-20162-3.

[19] R. Urtasun, D. Fleet, and P. Fua. Gaussian process dynamical models for 3D people tracking. In *Proc. of the IEEE Computer Society Conference on Computer Vision and Pattern Recognition (CVPR)*, 2006.

[20] P. Van Overschee and B. De Moor. *Subspace Identification for Linear Systems: Theory, Implementation, Applications*. Kluwer Academic Publishers, 1996.

[21] J. Wang, D. Fleet, and A. Hertzmann. Gaussian process dynamical models for human motion. In *IEEE Transactions on Pattern Analysis and Machine Intelligence (PAMI)*, 2008.

3D Relative Pose Estimation from Six Distances

Nikolas Trawny, Xun S. Zhou, and Stergios I. Roumeliotis
Department of Computer Science and Engineering, University of Minnesota, Minneapolis, MN 55455
E-mail: {trawny,zhou,stergios}@cs.umn.edu

Abstract—In this paper, we present three fast, hybrid numeric-algebraic methods to solve polynomial systems in floating point representation, based on the eigendecomposition of a so-called multiplication matrix. In particular, these methods run using standard double precision, use only linear algebra packages, and are easy to implement. We provide the proof that these methods do indeed produce valid multiplication matrices, and show their relationship. As a specific application, we use our algorithms to compute the 3D relative translation and orientation between two robots, based on known egomotion and six robot-to-robot distance measurements. Equivalently, the same system of equations arises when solving the forward kinematics of the general Stewart-Gough mechanism. Our methods can find all 40 solutions, trading off speed (0.08s to 1.5s, depending on the choice of method) for accuracy.

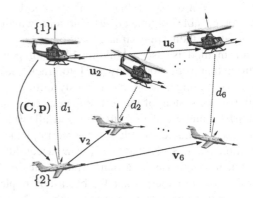

Fig. 1. 3DD problem geometry.

I. INTRODUCTION

For the successful operation of multi-robot systems, accurate knowledge of the 3D robot-to-robot position and orientation (pose) is essential. Common methods to determine this relative transformation, such as direct manual measurements, absolute measurements with respect to a common frame (e.g., using GPS), or indirect correlation of sensor measurements (e.g., map or image matching) either provide insufficient accuracy, or can be infeasible in GPS-denied or featureless areas. In recent work [1], we have addressed this problem by using robot-to-robot distance measurements from different vantage points. Our previous method uses 10 distance measurements plus known robot egomotion to linearly compute the robot-to-robot relative pose. In order to increase robustness to outliers and noise (e.g., using RANSAC), the aim of this paper is to present a solution algorithm for the *minimal* problem that requires only 6 distance measurements to arrive at a discrete set of solutions for the six unknowns.

The forward kinematics of the general Stewart-Gough mechanism [2] represent the mechanical analogue to this minimal problem. The general Stewart-Gough mechanism consists of two platforms connected by six legs, whose lengths determine the pose of the end platform relative to the base platform. The forward-kinematics problem is then to determine the relative pose of both platforms given the leg lengths and their attachment coordinates. For conciseness, we will from now on refer to both this, as well as the 3D robot relative pose problem as the 3DD (3D-distance) problem. It has been shown that the 3DD problem requires solving a system of polynomial equations that has 40 (generally complex) solutions [3], [4], all of which can be real [5]. Unfortunately, solving this polynomial system has proven quite difficult. The currently available techniques capable of finding all 40 solutions are either slow

and not suitable for real-time application (run-time can reach minutes, even hours) and/or they are highly non-trivial to implement (they require non-standard data types, specialized libraries, and a large amount of symbolic calculations).

The contribution of this paper is twofold: (i) We present the theoretical proofs for the correctness of two polynomial system solvers. In particular, these proofs cover a generalization of [6], [7], as well as the method of [8], and establish their relationship. Both approaches employ the paradigm of solving polynomial systems using the eigendecomposition of a so-called multiplication matrix [9], also referred to as action matrix. These solvers are generally applicable to a wide range of problems arising in robotics and computer vision. (ii) We apply these solvers to the 3DD problem, and provide solution algorithms that are very fast (0.08s - 1.5s), use standard double precision data types, and can be very easily implemented using only linear algebra libraries.

Following the description of related work (Section II), we present the 3DD problem in Section III, motivating the need for fast polynomial solvers. We provide a brief background on the theory of polynomial system solving in Section IV and outline the solution algorithms in Section V. We end the paper with simulation results (Section VI), and conclusions and an outlook on future work in Section VII.

II. RELATED WORK

Although the two incarnations of the 3DD problem - determining the relative pose of two robots from distance measurements, or solving the forward kinematics of the general Stewart-Gough mechanism - are mathematically equivalent, in that they both lead to the same system of multivariate polynomials, research has mostly concentrated on the latter. Current algorithms for solving the forward kinematics problem of the general Stewart-Gough mechanism can be broadly divided into numeric and algebraic methods.

Newton's method is a widely used numeric approach to obtain solutions to the forward kinematics of the Stewart mechanism [10] due to its high speed and ease of implementation. It works extremely well when initialized close to a true solution, as is the case when tracking a slowly-moving manipulator at high update rates. However, it may converge very slowly or even diverge when initialized poorly, and will generally not be able to find all 40 solutions without prior information about their approximate locations.

Newton's method is also a building block of polynomial continuation, that uses homotopy [11] to track the known solutions of a starting system continuously until they match those of the actual system of interest. Polynomial continuation was the earliest numerical tool used to find all 40 solutions of the forward kinematics problem [3]. Free implementations of this method, such as PHCpack [12], can successfully solve this problem, and were used as ground-truth in our simulations. Unfortunately, in our experiments, the black-box implementation of PHCpack required on average more than 120 s to solve one instance of the problem and is hence too slow for real-time application (although the forward kinematics have reportedly been solved using continuation methods in 14 s [4]). Moreover, continuation methods may sometimes miss solutions [13].

One algebraic method for solving a polynomial system is to compute its Gröbner basis [14] with respect to lexicographical monomial ordering. Such a Gröbner basis is essentially an equivalent polynomial system from which the solution can be easily obtained. In particular, for this problem it will contain one univariate polynomial of degree 40, whose roots form the solutions of the original system for that variable. The values of the remaining variables can be obtained from the other polynomials of the Gröbner basis via back substitution. Despite recent advances in algorithms to compute Gröbner bases, computing one symbolically for this problem has proven intractable, and even for specific numerical instances using rational numbers still requires several seconds up to minutes [13].

A second algebraic method is to use dialytic elimination or resultants to obtain the 40-degree univariate polynomial. This has first been achieved by Husty [15], and since then refined [16], [17]. To our knowledge, Lee and Shim's algorithm [17] is currently the fastest solution method for the forward kinematics (the authors report timings of 0.02 s). Unfortunately, all the elimination algorithms mentioned above are quite challenging to implement, they require non-standard, high-precision data-types (e.g., with 30 digits precision [17]) to cope with numerical error accumulation, and a large amount of symbolic computations.

The algorithms presented in this paper solve a system of polynomials by an eigendecomposition of an associated multiplication matrix, a general hybrid algebraic-numeric approach pioneered by Auzinger and Stetter [9] and described in detail in [18]. This concept has recently been successfully applied to solve minimal problems in computer vision [19], [20]. More precisely, our methods are inspired by the general techniques to solve polynomial systems in floating point representation presented recently in [19] and [8]. Two modifications to [19] have been introduced with the aim of improving numerical stability [6], [7], however, without proving that these methods will indeed yield a valid multiplication matrix, and without providing conditions under which they will work. In this paper, we prove that a generalized version of [6], [7], as well as the approach in [8] do indeed yield valid multiplication matrices for a specific class of problems in which certain rank conditions are fulfilled (which are tacitly implied in [6], [7]). Our methods make extensive use of linear algebra, in particular QR factorization, and require only standard double precision. Their high speed and accessibility make them therefore attractive for real-time applications in industrial settings, e.g., in flight simulator control.

III. PROBLEM FORMULATION

To motivate the need for fast polynomial solvers, in this section we will outline the 3DD problem and show how it results in a system of polynomials. In the next two sections we will then discuss algorithms to solve this system.

Assume that two robots move in 3D and take six robot-to-robot distance measurements d_i, $i = 1, \ldots, 6$. We assume without loss of generality that the global frames of each robot, $\{1\}$ and $\{2\}$, are attached to the points where the first mutual measurement takes place (cf. Fig. 1). Further, we assume that each robot knows the coordinates, $\mathbf{u}_i := {}^1\mathbf{u}_i$ and $\mathbf{v}_i := {}^2\mathbf{v}_i$, of its location at the time of the remaining five measurements with respect to its own global frame of reference. The objective is to find the 6 degree-of-freedom transformation, i.e., the translation $\mathbf{p} := {}^1\mathbf{p}_2$ and rotation $\mathbf{C} := {}^1_2\mathbf{C}$ of the second frame with respect to the first.

The distance measurements can be expressed as the length of vector $\mathbf{w}_i$, $i = 1, \ldots, 6$, connecting the two robots at the time of measurement.

$$d_i = \|\mathbf{w}_i\|_2 = \sqrt{\mathbf{w}_i^T \mathbf{w}_i}, \quad \mathbf{w}_i := \mathbf{p} + \mathbf{C}\mathbf{v}_i - \mathbf{u}_i$$

Squaring each distance, and noting that $\mathbf{u}_1 = \mathbf{v}_1 = \mathbf{0}$, we obtain the following six polynomial constraints

$$\mathbf{p}^T \mathbf{p} - d_1^2 = 0 \tag{1}$$

$$-\mathbf{u}_i^T \mathbf{p} + \mathbf{v}_i^T \mathbf{C}^T \mathbf{p} + f_i = 0 \quad i = 2, \ldots, 6 \tag{2}$$

$$\text{where } f_i = \frac{1}{2}(d_1^2 + \mathbf{v}_i^T \mathbf{v}_i + \mathbf{u}_i^T \mathbf{u}_i - d_i^2) - \mathbf{u}_i^T \mathbf{C}\mathbf{v}_i$$

Due to its lack of singularities, we choose the (unit) quaternion to represent orientation. It is defined as $\bar{q} = \begin{bmatrix} q_1 & q_2 & q_3 & q_4 \end{bmatrix}^T = \begin{bmatrix} \mathbf{q}^T & q_4 \end{bmatrix}^T$, and related to the rotational matrix by

$$\mathbf{C}(\bar{q}) = \mathbf{I}_3 - 2q_4\lfloor \mathbf{q} \times \rfloor + 2\lfloor \mathbf{q} \times \rfloor^2 \tag{3}$$

where $\lfloor \mathbf{a} \times \rfloor$ is the skew-symmetric cross-product matrix of a vector $\mathbf{a}$. Since the quaternions $\bar{q}$ and $-\bar{q}$ both represent the same rotation, the number of solutions is doubled. One can easily eliminate the spurious solutions by discarding those with $q_4 < 0$. When using the unit quaternion representation, we need to add the unit-norm constraint

$$\bar{q}^T \bar{q} - 1 = 0 \qquad (4)$$

to the polynomial system of (1) and (2).

A. Normalization of **p**

In order to guarantee a bounded norm of the system's solution, in addition to using the unit-norm quaternion, we introduce the normalized translation

$$\bar{\mathbf{p}} = \mathbf{p}/\|\mathbf{p}\|_2 = \mathbf{p}/d_1 \qquad (5)$$

This normalization can easily be realized by normalizing the position coordinates $\bar{\mathbf{u}}_i := \mathbf{u}_i/d_1$, $\bar{\mathbf{v}}_i := \mathbf{v}_i/d_1$, and distances, $\bar{d}_i := d_i/d_1$. Then, the system (1) and (2) becomes

$$\bar{\mathbf{p}}^T \bar{\mathbf{p}} - 1 = 0 \qquad (6)$$

$$-\bar{\mathbf{u}}_i^T \bar{\mathbf{p}} + \bar{\mathbf{v}}_i^T \mathbf{C}^T \bar{\mathbf{p}} + \bar{f}_i = 0 \quad i = 2,\ldots,6 \qquad (7)$$

$$\text{with } \bar{f}_i = \frac{1}{2}(1 + \bar{\mathbf{v}}_i^T \bar{\mathbf{v}}_i + \bar{\mathbf{u}}_i^T \bar{\mathbf{u}}_i - \bar{d}_i^2) - \bar{\mathbf{u}}_i^T \mathbf{C} \bar{\mathbf{v}}_i \qquad (8)$$

From the solution of this normalized system, we can easily recover the actual translation from $\mathbf{p} = d_1 \bar{\mathbf{p}}$.

B. Increasing speed by prior elimination of $\bar{\mathbf{p}}$

Since the complexity and hence the speed of polynomial solvers depend heavily on the number of unknowns, we reduce the number of variables by eliminating the translation, $\bar{\mathbf{p}}$. As a result, we obtain a new system of polynomials only in the elements of the quaternion $\bar{q}$. To this end, we first define the rotated normalized translation vector,

$$\bar{\mathbf{r}} := \mathbf{C}^T \bar{\mathbf{p}} \qquad (9)$$

which allows us to write (7) as

$$\begin{bmatrix} \bar{\mathbf{v}}_i^T & -\bar{\mathbf{u}}_i^T & \bar{f}_i \end{bmatrix} \begin{bmatrix} \bar{\mathbf{r}} \\ \bar{\mathbf{p}} \\ 1 \end{bmatrix} = 0, \quad i = 2,\ldots,6 \qquad (10)$$

We also obtain the following constraints from (9)

$$(q_4 \mathbf{I}_3 - \lfloor \mathbf{q} \times \rfloor)\bar{\mathbf{r}} - (q_4 \mathbf{I}_3 + \lfloor \mathbf{q} \times \rfloor)\bar{\mathbf{p}} = \mathbf{0} \qquad (11)$$

$$\mathbf{q}^T \bar{\mathbf{r}} - \mathbf{q}^T \bar{\mathbf{p}} = 0 \qquad (12)$$

where (11) results from substituting (3) in (9) and multiplying by $(q_4 \mathbf{I}_3 - \lfloor \mathbf{q} \times \rfloor)$ (using $\lfloor \mathbf{q} \times \rfloor^3 = -\mathbf{q}^T \mathbf{q} \lfloor \mathbf{q} \times \rfloor$) while we arrive at (12) by noting that $\mathbf{q}$ is the unit eigenvector of $\mathbf{C}$ with corresponding eigenvalue 1 and pre-multiplying both sides of (9) with $\mathbf{q}^T$.

By stacking these equations, we obtain

$$\underbrace{\begin{bmatrix} \bar{\mathbf{v}}_2^T & -\bar{\mathbf{u}}_2^T & \bar{f}_2 \\ \vdots & \vdots & \vdots \\ \bar{\mathbf{v}}_6^T & -\bar{\mathbf{u}}_6^T & \bar{f}_6 \\ (q_4 \mathbf{I}_3 - \lfloor \mathbf{q} \times \rfloor) & -(q_4 \mathbf{I}_3 + \lfloor \mathbf{q} \times \rfloor) & \mathbf{0} \\ \mathbf{q}^T & -\mathbf{q}^T & 0 \end{bmatrix}}_{\Xi} \begin{bmatrix} \bar{\mathbf{r}} \\ \bar{\mathbf{p}} \\ 1 \end{bmatrix} = \mathbf{0} \qquad (13)$$

In order for (13) to have a solution, the matrix Ξ must be rank deficient, or the determinants of its (7×7)-submatrices must vanish. These determinants are polynomials in the elements of $\bar{q}$ only. Out of the 36 possibilities, we define the following five 4-th order polynomials (using Matlab notation)

$$\theta_1(\bar{q}) = \det\left(\Xi\left(\begin{bmatrix} 1:5 & 6 & 7 \end{bmatrix}, :\right)\right) \qquad (14)$$

$$\theta_2(\bar{q}) = \det\left(\Xi\left(\begin{bmatrix} 1:5 & 6 & 8 \end{bmatrix}, :\right)\right) \qquad (15)$$

$$\theta_3(\bar{q}) = \det\left(\Xi\left(\begin{bmatrix} 1:5 & 7 & 8 \end{bmatrix}, :\right)\right) \qquad (16)$$

$$\theta_4(\bar{q}) = \det\left(\Xi\left(\begin{bmatrix} 1:5 & 6 & 9 \end{bmatrix}, :\right)\right) \qquad (17)$$

$$\theta_5(\bar{q}) = \det\left(\Xi\left(\begin{bmatrix} 1:5 & 7 & 9 \end{bmatrix}, :\right)\right) \qquad (18)$$

Notice that the $\theta_i(\bar{q})$ have not yet used the information in (6). To this end, we solve (13) for $\bar{\mathbf{p}}$ symbolically, using Cramer's rule. From the submatrix $\widetilde{\Xi}_1 = \Xi\left(\begin{bmatrix} 1:6 \end{bmatrix}, :\right)$ consisting of the first six rows of Ξ, we obtain a first candidate

$$\bar{\mathbf{p}}_1 = -\frac{1}{p_{d,1}} \begin{bmatrix} p_{n_x,1} \\ p_{n_y,1} \\ p_{n_z,1} \end{bmatrix} \qquad (19)$$

where

$$p_{n_x,1} = \det\left(\widetilde{\Xi}_1\left(:, \begin{bmatrix} 1:3 & 7 & 5 & 6 \end{bmatrix}\right)\right) \qquad (20)$$

$$p_{n_y,1} = \det\left(\widetilde{\Xi}_1\left(:, \begin{bmatrix} 1:3 & 4 & 7 & 6 \end{bmatrix}\right)\right) \qquad (21)$$

$$p_{n_z,1} = \det\left(\widetilde{\Xi}_1\left(:, \begin{bmatrix} 1:3 & 4 & 5 & 7 \end{bmatrix}\right)\right) \qquad (22)$$

$$p_{d,1} = \det\left(\widetilde{\Xi}_1\left(:, \begin{bmatrix} 1:6 \end{bmatrix}\right)\right) \qquad (23)$$

Analogously, we obtain three additional translation candidates, $\bar{\mathbf{p}}_2, \bar{\mathbf{p}}_3, \bar{\mathbf{p}}_4$ from the submatrices

$$\widetilde{\Xi}_2 = \Xi\left(\begin{bmatrix} 1:5 & 7 \end{bmatrix}, :\right) \qquad (24)$$

$$\widetilde{\Xi}_3 = \Xi\left(\begin{bmatrix} 1:5 & 8 \end{bmatrix}, :\right) \qquad (25)$$

$$\widetilde{\Xi}_4 = \Xi\left(\begin{bmatrix} 1:5 & 9 \end{bmatrix}, :\right) \qquad (26)$$

Notice that the denominators and the three numerators of each candidate are polynomials in $\bar{q}$.

Direct substitution of each translation candidate $\bar{\mathbf{p}}_i$ in (6) yields four 6-th order polynomials in $\bar{q}$, $i = 1,\ldots,4$

$$\phi_i(\bar{q}) = p_{n_x,i}^2 + p_{n_y,i}^2 + p_{n_z,i}^2 - p_{d,i}^2 = 0 \qquad (27)$$

The new polynomial system resulting from the original equations (6), (7) after eliminating the translation $\bar{\mathbf{p}}$, is

$$\left.\begin{array}{r} \theta_i(\bar{q}) = 0, \ i = 1,\ldots,5 \\ \phi_i(\bar{q}) = 0, \ i = 1,\ldots,4 \\ \bar{q}^T \bar{q} - 1 = 0 \end{array}\right\} \qquad (28)$$

Once we have obtained a solution for $\bar{q}$, we can recover the translation by evaluating (20)-(23) at that solution and back-substituting the resulting values in (19).

IV. BACKGROUND

Before presenting the algorithms to solve the 3DD problem, we briefly introduce necessary notation and sketch the theoretical basis of the multiplication matrix techniques for polynomial system solving. We recommend [18] and [14] for a detailed reference on this subject. Intuitively, multiplication matrices are the generalization of companion matrices (for solving univariate polynomials) to systems of polynomials in multiple variables.

Let a *monomial* in n variables be denoted by $\mathbf{x}^\alpha :=$ $x_1^{\alpha_1} x_2^{\alpha_2} \ldots x_n^{\alpha_n}$, $\alpha_i \in \mathbb{Z}_{\geq 0}$, and a *polynomial* in n variables with complex coefficients correspondingly by $\psi :=$ $\sum_j c_j \mathbf{x}^{\alpha_j}$, $c_j \in \mathbb{C}$. The ring of all polynomials in n variables with complex coefficients will be denoted as $\mathbb{C}\left[x_1, \ldots, x_n\right]$. The *total degree* of a monomial is defined as $\sum_{i=1}^n \alpha_i$, and the total degree of a polynomial is the maximum total degree of all its monomials. In order to represent a polynomial ψ in vector notation, let ℓ denote the total degree of ψ, and stack all monomials up to and including total degree ℓ in a vector $\mathbf{x}_\ell := \begin{bmatrix} x_1^\ell & x_1^{\ell-1}x_2 & \ldots & x_n & 1 \end{bmatrix}^T$. We can then write $\psi = \mathbf{c}^T\mathbf{x}_\ell$ where $\mathbf{c}$ is a potentially very sparse coefficient vector. For a polynomial system, $\psi_1 = 0, \ldots, \psi_u = 0$, we can analogously define ℓ as the maximum total degree of all ψ_i, and by stacking the corresponding coefficient vectors $\mathbf{c}_i$ form the matrix system $\mathbf{C}\mathbf{x}_\ell = \mathbf{0}$. Let $\mathbf{x}_\ell(\mathbf{p})$ denote the vector of monomials evaluated at a point $\mathbf{p} \in \mathbb{C}^n$. Solving the system then means finding all points $\mathbf{p}$ that fulfill $\mathbf{C}\mathbf{x}_\ell(\mathbf{p}) = \mathbf{0}$.

Next, we define the *ideal* $I := \langle \psi_1, \ldots, \psi_u \rangle =$ $\{\sum_i h_i \psi_i, h_i \in \mathbb{C}\left[x_1, \ldots, x_n\right]\}$, i.e., the set of all linear combinations of the original system multiplied with arbitrary polynomials. Each element of the ideal will equal zero when evaluated at a solution of the original system. Solving the system involves finding specific new elements of the ideal, which can be generated by (i) multiplication of an existing polynomial ψ_i with a monomial (which essentially shifts the coefficients inside $\mathbf{c}_i$), and (ii) linear combinations of polynomials, i.e., multiplication from the left of $\mathbf{C}\mathbf{x}_\ell$ by arbitrary matrices of full column rank. Notice in particular, that diagonal scaling does not change the ideal. We therefore scale each row of $\mathbf{C}$ to unit norm, which is a crucial prerequisite for numerical accuracy of our algorithms. New members of the ideal can be added to the system as additional rows in $\mathbf{C}$.

Let the set of polynomials $G = \{\gamma_1, \ldots, \gamma_t\}$ be a *Gröbner basis* of I. We can define *division* by G by writing any polynomial $f \in \mathbb{C}\left[x_1, \ldots, x_n\right]$ as $f = \sum_{k=1}^t h_k\gamma_k + r$, where $h_k, r \in \mathbb{C}\left[x_1, \ldots, x_n\right]$, and no monomial in r is divisible by the leading term of any $\gamma_k \in G$. r is called the *remainder* of f on division by G, or $r = \overline{f}^G$. The special properties of Gröbner bases ensure that (i) $\overline{f}^G$ is unique, and (ii) $\overline{f}^G = 0 \Leftrightarrow f \in I$.

Our solution methods rely on the special structure of the so-called *quotient-ring* $A := \mathbb{C}\left[x_1, \ldots, x_n\right]/I$, defined as the set of all remainders under division by G. In what follows, we assume that the polynomial system has a finite number s

of discrete solutions.[1] Based on the Finiteness Theorem [18, p.39], A is then a finite dimensional vector space with exactly s dimensions over $\mathbb{C}$, and we can choose a monomial basis $B = \{\mathbf{x}^{\beta_1}, \ldots, \mathbf{x}^{\beta_s}\}$, e.g., the standard monomials or *normal set* obtained from a Gröbner basis [18, p.38]. Importantly, this normal set usually remains constant for different numerical instances of the same problem, and therefore needs to be computed only once. Denote $\mathbf{x}_B$ as the vector of basis monomials $\mathbf{x}_B = \begin{bmatrix} \mathbf{x}^{\beta_1} & \ldots & \mathbf{x}^{\beta_s} \end{bmatrix}^T$. Then, any remainder r can be written as $r = \mathbf{c}_r^T\mathbf{x}_B$.

Within the quotient ring we can define a multiplication map, which can be represented by a *multiplication matrix*.

Proposition 1: [18, Prop. 4.1, p. 56] Let $f \in \mathbb{C}\left[x_1, \ldots, x_n\right]$, and $A = \mathbb{C}\left[x_1, \ldots, x_n\right]/I$. Then we define the mapping $\mathbf{m}_f : A \to A$ by the rule: if $\overline{g}^G \in A$, then $\mathbf{m}_f(\overline{g}^G) := \overline{f \cdot g}^G \in A$. The map $\mathbf{m}_f(.)$ is a linear map from A to A. Furthermore, if $r' = \mathbf{m}_f(r)$, and $r = \mathbf{c}_r^T\mathbf{x}_B$, $r' = \mathbf{c}_{r'}^T\mathbf{x}_B$, then $\mathbf{c}_{r'} = \mathbf{M}_f\mathbf{c}_r$, where $\mathbf{M}_f$ is an $s \times s$ matrix, called the *multiplication matrix*.

The following two properties of the multiplication matrix are the key to solving the system of polynomials.

Proposition 2: Let the polynomial system have a finite number s of discrete solutions. Further, $f \in \mathbb{C}\left[x_1, \ldots, x_n\right]$ is chosen such that the values $f(\mathbf{p}_j)$ evaluated at a solution $\mathbf{p}_j$, $j = 1, \ldots, s$ are distinct. Then

1) Eigenvalues [18, Thm 4.5, p.59]: The eigenvalues of the multiplication matrix $\mathbf{M}_f$ are equal to the function f evaluated at each solution. In particular, the eigenvalues of the multiplication matrix $\mathbf{M}_{x_i}$ on A coincide with the x_i-coordinates of the solutions.

2) Eigenvectors [18, Prop. 4.7, p.64]: The left eigenvectors of the matrix $\mathbf{M}_f$ are given by the row vectors $\begin{bmatrix} \mathbf{p}_j^{\beta_1} & \ldots & \mathbf{p}_j^{\beta_s} \end{bmatrix}$ for all solutions $\mathbf{p}_j$, $j = 1, \ldots, s$.

Assuming a normal set $\mathbf{x}_B$ is known, the problem of solving a polynomial system has now been reduced to finding a multiplication matrix $\mathbf{M}_{x_i}$, for which we need to determine the column vectors $\mathbf{M}_{x_i}(:, j)$ as the coefficients of the remainders $\overline{x_i\mathbf{x}^{\beta_j}}^G$, $j = 1, \ldots, s$. The key observation [21] is that

$$x_i\mathbf{x}^{\beta_j} = \sum h_k\gamma_k + \mathbf{M}_{x_i}(:, j)^T\mathbf{x}_B \qquad (29)$$

$$\Leftrightarrow x_i\mathbf{x}^{\beta_j} - \mathbf{M}_{x_i}(:, j)^T\mathbf{x}_B = \sum h_k\gamma_k \in I \qquad (30)$$

In other words, we can read off the columns of $\mathbf{M}_{x_i}$ from certain elements of the ideal. Byröd *et al.* [19] describe one potential method how these elements can be found numerically if the normal set is known, using only linear algebra, without having to compute a Gröbner basis every time. This approach has two main drawbacks: First, it requires a normal set, which has to be computed once from a Gröbner basis (using a suitable computer algebra system, e.g., Macaulay 2). The normal set depends on the monomial order of this Gröbner basis, which can be unnecessarily restrictive [22]. Further, the

---

[1]More precisely, we assume I is zero-dimensional, radical and does not contain 1. We refer to [18] for what to do if these relatively mild assumptions are not met.

matrix inversion involved in computing $\mathbf{M}_{x_i}$ can be poorly conditioned for a particular normal set [6], which reduces numerical accuracy.

To mitigate these issues, [6], [7] and [8] proposed to use a set of (potentially) *polynomial* basis functions resulting from coordinate transformations of the monomials. Not only can an adaptively chosen transformation improve the conditioning, it also avoids having to compute a Gröbner basis to obtain a normal set. Unfortunately, [6], [7], [8] are all lacking a proof that their proposed transformations produce indeed a valid multiplication matrix. In what follows, we provide this proof for a more general version of [6], as well as for two solution algorithms based on [7] and [8] for a specific class of polynomial systems. In Section III we apply these algorithms to the 3DD problem.

The starting point for both algorithms is an expanded version, $\mathbf{C}_e\mathbf{x}_{l+1} = \mathbf{0}$, of the original system of polynomials. The matrix $\mathbf{C}_e$ is obtained as follows: define a vector of monomials of degree up to and including an integer l as $\mathbf{x}_l = \begin{bmatrix} x_1^l & x_1^{l-1}x_2 \dots & x_n & 1 \end{bmatrix}^T$. The length of this vector is $n_l = \binom{n+l}{n}$. Similarly, define the vector of monomials of total degree up to and including $l+1$ as $\mathbf{x}_{l+1}$ with length n_{l+1}. Finally, define the vector of monomials of total degree exactly $l+1$ as $\mathbf{x}_k$ with length $n_k = \binom{n+l}{n-1} = \binom{n+l+1}{n} - \binom{n+l}{n}$. Let ℓ_i be the total degree of ψ_i. Multiply ψ_i by each monomial of total degree up to and including $l + 1 - \ell_i$, and add the resulting polynomials to the system. Repeat this process for all polynomials $\psi_1, \dots, \psi_u$, and write the resulting extended system of m polynomials as

$$\mathbf{C}_e\mathbf{x}_{l+1} = \begin{bmatrix} \mathbf{C}_k & \mathbf{C}_l \end{bmatrix} \begin{bmatrix} \mathbf{x}_k \\ \mathbf{x}_l \end{bmatrix} = \mathbf{0} \qquad (31)$$

so that $\mathbf{C}_k$ and $\mathbf{C}_l$ have dimensions $(m \times n_k)$ and $(m \times n_l)$. Notice that this process does not require any algebraic computations, but merely assigning the coefficients of the original polynomials to specific elements of $\mathbf{C}_e$. This matrix can be very large, but is usually extremely sparse (e.g., 0.5-5% nonzero elements for instances of 3DD, depending on the parameterization).

We will further need the *unreduced multiplication matrix* $\mathbf{M}'_{x_i}$ of dimension $(n_{l+1} \times n_l)$ that maps each monomial in $\mathbf{x}_l$ to $x_i \cdot \mathbf{x}_l$. In particular, $\mathbf{M}'_{x_i}(j, k) = 1$ if $\mathbf{x}_{l+1}(j) = x_i \cdot \mathbf{x}_l(k)$. As a consequence of this construction, we have

$$\mathbf{M}'^T_{x_i}\mathbf{x}_{l+1}(\mathbf{p}) = p_i\mathbf{x}_l(\mathbf{p}) \qquad (32)$$

For illustration, for $n = 2$, $l = 1$, $i = 2$, we have $\mathbf{x}_1 = \begin{bmatrix} x_1 & x_2 & 1 \end{bmatrix}$, $\mathbf{x}_2 = \begin{bmatrix} x_1^2 & x_1x_2 & x_2^2 & x_1 & x_2 & 1 \end{bmatrix}$, and $\mathbf{M}'_{x_2} = \begin{bmatrix} 0 & 1 & 0 & 0 & 0 & 0 \\ 0 & 0 & 1 & 0 & 0 & 0 \\ 0 & 0 & 0 & 0 & 1 & 0 \end{bmatrix}^T$.

V. POLYNOMIAL SYSTEM SOLVERS

In this section, we describe 3 different polynomial system solvers, starting with the normal-set-based method, then introducing coordinate transformations to improve accuracy, and finally using the right nullspace directly to compute the multiplication matrix. We then see how these methods are related.

A. Computing the multiplication matrix using a normal set

As described in [19], if a normal set $\mathbf{x}_B$ is available for the problem at hand (e.g., computed from the Gröbner bases of example systems using a computer algebra program), the multiplication matrix is computed as follows.

First, define the monomials $\mathbf{x}_R$ as the set $x_i\mathbf{x}_B \setminus \mathbf{x}_B$, i.e., the monomials $x_i\mathbf{x}_B$ that do not belong to the normal set. These are the monomials that need to be reduced through division by the ideal. Finally, let $\mathbf{x}_E$ be the monomials in $\mathbf{x}_{l+1}$ that belong neither in $\mathbf{x}_B$ nor $\mathbf{x}_R$. By reordering the monomials and the columns of $\mathbf{C}_e$, we obtain

$$\mathbf{C}_e\mathbf{x}_{l+1} = \begin{bmatrix} \mathbf{C}_E & \mathbf{C}_R & \mathbf{C}_B \end{bmatrix} \begin{bmatrix} \mathbf{x}_E \\ \mathbf{x}_R \\ \mathbf{x}_B \end{bmatrix} = \mathbf{0} \qquad (33)$$

Let $\mathbf{N}^T$ be the left nullspace of $\mathbf{C}_E$, i.e., $\mathbf{N}^T\mathbf{C}_E = \mathbf{0}$, and decompose $\mathbf{N}^T\mathbf{C}_R = \mathbf{Q}\mathbf{R} = \begin{bmatrix} \mathbf{Q}_1 & \mathbf{Q}_2 \end{bmatrix} \begin{bmatrix} \mathbf{R}_1 \\ \mathbf{0} \end{bmatrix}$ using QR factorization. Assuming that $\mathbf{R}_1$ is of full rank, we arrive at

$$\begin{bmatrix} \mathbf{N}^T\mathbf{C}_R & \mathbf{N}^T\mathbf{C}_B \end{bmatrix} \begin{bmatrix} \mathbf{x}_R \\ \mathbf{x}_B \end{bmatrix} = \mathbf{Q} \begin{bmatrix} \mathbf{R}_1 & \mathbf{Q}_1^T\mathbf{N}^T\mathbf{C}_B \\ \mathbf{0} & \mathbf{Q}_2^T\mathbf{N}^T\mathbf{C}_B \end{bmatrix} \begin{bmatrix} \mathbf{x}_R \\ \mathbf{x}_B \end{bmatrix} = \mathbf{0}$$

$$\Rightarrow \mathbf{Q}_1\mathbf{R}_1 \begin{bmatrix} \mathbf{I}_R & \mathbf{T} \end{bmatrix} \begin{bmatrix} \mathbf{x}_R \\ \mathbf{x}_B \end{bmatrix} = \mathbf{0}, \quad \text{where } \mathbf{T} = \mathbf{R}_1^{-1}\mathbf{Q}_1^T\mathbf{N}^T\mathbf{C}_B$$

In a slight deviation from our previous notation, let the unreduced multiplication matrix $\mathbf{M}'_{x_i}$ correspond to the map $\mathbf{M}'_{x_i} : \mathbf{x}_B \to \begin{bmatrix} \mathbf{x}_R \\ \mathbf{x}_B \end{bmatrix}$ induced by multiplication with x_i. Then, the multiplication matrix becomes [19]

$$\mathbf{M}_{x_i} = \begin{bmatrix} -\mathbf{T}^T & \mathbf{I}_s \end{bmatrix} \mathbf{M}'_{x_i} \qquad (34)$$

This method has two drawbacks: first, it requires determining the normal set, and second, the inverse in the expression for $\mathbf{T}$ can be very ill-conditioned.

B. Using a general coordinate transformation

In [6], a specific coordinate transformation was proposed to improve conditioning of this inverse. In what follows, we prove that a very general form of coordinate transformation indeed yields a valid multiplication matrix, of which the methods presented in [6], [7], [8] are special cases.

The involutive test described in [8] provides a means to determine the degree to which the system needs to be expanded or "prolonged" (cf. (31)). Passing this test also eliminates the need to determine a normal set. The 3DD polynomial system (28) passed the involutive test after prolongation or expansion to total degree $l + 1 = 9$, and no projection. Another condition implied by this specific outcome of the involutive test is that $\mathbf{C}_k$ is of full rank, and that $\mathbf{C}_e$ is of rank $n_{l+1} - s$, i.e., it has an s-dimensional nullspace. These important conditions ensure that the rank conditions needed in the derivations below are fulfilled.

To start, we first decompose $\mathbf{C}_k$ (cf. (31)), such that

$$\mathbf{C}_k = \mathbf{Q}_1\mathbf{R}_1\mathbf{E}_1^T = \mathbf{Q}_1 \begin{bmatrix} \mathbf{R}_{11} \\ \mathbf{0} \end{bmatrix} \mathbf{E}_1^T \qquad (35)$$

where $\mathbf{E}_1$ is an $(n_k \times n_k)$ column permutation matrix, i.e., $(\mathbf{E}_1)^{-1} = \mathbf{E}_1^T$, and $\mathbf{R}_{11}$ is upper triangular.

We now split $\mathbf{Q}_1^T \mathbf{C}_l = \begin{bmatrix} \mathbf{D}_{l1}^T & \mathbf{D}_{l2}^T \end{bmatrix}^T$, so that $\mathbf{D}_{l1}$ is of dimension $(n_k \times n_l)$ and we can write

$$\mathbf{C}_e = \mathbf{Q}_1 \begin{bmatrix} \mathbf{R}_{11} & \mathbf{D}_{l1} \\ \mathbf{0} & \mathbf{D}_{l2} \end{bmatrix} \begin{bmatrix} \mathbf{E}_1^T & \mathbf{0} \\ \mathbf{0} & \mathbf{I}_{n_l} \end{bmatrix} \quad (36)$$

A second QR-decomposition yields

$$\mathbf{D}_{l2} = \mathbf{Q}_2 \mathbf{R}_2 \mathbf{E}_2^T = \mathbf{Q}_2 \begin{bmatrix} \mathbf{R}_{21} & \mathbf{R}_{22} \\ \mathbf{0} & \mathbf{0} \end{bmatrix} \mathbf{E}_2^T \quad (37)$$

where $\mathbf{R}_{21}$ is upper-triangular, $\mathbf{R}_{22}$ is of dimension $(n_l - s) \times s$, and $\mathbf{E}_2$ is an $(n_l \times n_l)$ permutation matrix (which is required, since $\mathbf{D}_{l2}$ is rank-deficient). We can further split $\mathbf{E}_2 = \begin{bmatrix} \mathbf{E}_{21} & \mathbf{E}_{22} \end{bmatrix}$, such that $\mathbf{E}_{22}$ is of dimension $(n_l \times s)$.

We now have

$$\mathbf{C}_e = \mathbf{Q}_1 \begin{bmatrix} \mathbf{I}_{n_k} & \mathbf{0} \\ \mathbf{0} & \mathbf{Q}_2 \end{bmatrix} \begin{bmatrix} \mathbf{R}_{11} & \mathbf{D}_{l1}\mathbf{E}_{21} & \mathbf{D}_{l1}\mathbf{E}_{22} \\ \mathbf{0} & \mathbf{R}_{21} & \mathbf{R}_{22} \\ \mathbf{0} & \mathbf{0} & \mathbf{0} \end{bmatrix} \begin{bmatrix} \mathbf{E}_1^T & \mathbf{0} \\ \mathbf{0} & \mathbf{E}_2^T \end{bmatrix}$$

$$= \begin{bmatrix} \mathbf{Q}_1' & \mathbf{Q}_2' \end{bmatrix} \begin{bmatrix} \mathbf{R}' \\ \mathbf{0} \end{bmatrix} \mathbf{E}^T \quad (38)$$

where $\mathbf{R}'$ is of dimension $(n_{l+1} - s \times n_{l+1})$ and upper triangular, and $\mathbf{E} = \begin{bmatrix} \mathbf{E}_1 & \mathbf{0} \\ \mathbf{0} & \mathbf{E}_2 \end{bmatrix}$. Further, partition $\mathbf{R}' = \begin{bmatrix} \mathbf{R}_k' & \mathbf{R}_l' \end{bmatrix}$ so that $\mathbf{R}_k'$ is of dimension $(n_{l+1} - s \times n_k)$, and $\mathbf{R}_l'$ is of dimension $(n_{l+1} - s \times n_l)$.

We can now introduce a general, invertible coordinate transformation matrix $\mathbf{V}_e$ $(n_{l+1} \times n_{l+1})$,

$$\mathbf{V}_e = \begin{bmatrix} \mathbf{V}_{11} & \mathbf{V}_{12} \\ \mathbf{0} & \mathbf{V}_{22} \end{bmatrix} \quad (39)$$

where $\mathbf{V}_{22}$ is of dimension $(s \times n_l)$, and also define its inverse as $\mathbf{V}_e^{-1} = \begin{bmatrix} \boldsymbol{\Lambda}_1 & \boldsymbol{\Lambda}_2 \end{bmatrix}$, where $\boldsymbol{\Lambda}_1$ is of dimension $(n_{l+1} \times n_{l+1} - s)$, and $\boldsymbol{\Lambda}_2$ is of dimension $(n_{l+1} \times s)$.

Using this coordinate transformation and (38) in (31) yields

$$\mathbf{Q}_1' \mathbf{R}' \mathbf{V}_e^{-1} \mathbf{V}_e \mathbf{E}^T \mathbf{x}_{l+1} = \mathbf{0} \quad (40)$$

$$\mathbf{Q}_1' \mathbf{R}' \boldsymbol{\Lambda}_1 \begin{bmatrix} \mathbf{I}_{n_{l+1}-s} & \mathbf{T} \end{bmatrix} \begin{bmatrix} \mathbf{V}_{11}\mathbf{E}_1^T & \mathbf{V}_{12}\mathbf{E}_2^T \\ \mathbf{0} & \mathbf{V}_{22}\mathbf{E}_2^T \end{bmatrix} \mathbf{x}_{l+1} = \mathbf{0} \quad (41)$$

$$\Leftrightarrow \begin{bmatrix} \mathbf{V}_{11}\mathbf{E}_1^T & \mathbf{V}_{12}\mathbf{E}_2^T \end{bmatrix} \mathbf{x}_{l+1} = -\mathbf{T} \begin{bmatrix} \mathbf{0} & \mathbf{V}_{22}\mathbf{E}_2^T \end{bmatrix} \mathbf{x}_{l+1} \quad (42)$$

where $\mathbf{T} = (\mathbf{R}'\boldsymbol{\Lambda}_1)^{-1}\mathbf{R}'\boldsymbol{\Lambda}_2$. Notice that the introduction of the coordinate transformation has given us leverage to improve the conditioning of the inversion of $(\mathbf{R}'\boldsymbol{\Lambda}_k)$ when computing $\mathbf{T}$ through judicious choice of $\boldsymbol{\Lambda}_k$. Equation (42) means that we can now express every polynomial in $\begin{bmatrix} \mathbf{V}_{11}\mathbf{E}_1^T & \mathbf{V}_{12}\mathbf{E}_2^T \end{bmatrix} \mathbf{x}_{l+1}$ as a linear combination of monomials in $\mathbf{x}_B = \mathbf{V}_{22}\mathbf{E}_2^T \mathbf{x}_l$, which we will choose as a basis of the quotient ring.

For this basis, we will now prove that the multiplication matrix can be formed as

$$\mathbf{M}_{x_i} = \begin{bmatrix} -\mathbf{T}^T & \mathbf{I}_s \end{bmatrix} \mathbf{V}_e^{-T} \mathbf{E}^T \mathbf{M}_{x_i}' \mathbf{E}_2 \mathbf{V}_{22}^T \quad (43)$$

using the unreduced multiplication matrix $\mathbf{M}_{x_i}'$ defined in Section IV.

Proposition 3: If we choose the basis as $\mathbf{x}_B = \mathbf{V}_{22}\mathbf{E}_2^T \mathbf{x}_l$, and compute the multiplication matrix as in (43), then $\mathbf{x}_B(\mathbf{p}_j)$ is a right eigenvector of $\mathbf{M}_{x_i}^T$.

Proof:

$$\mathbf{M}_{x_i}^T \mathbf{x}_B(\mathbf{p}_j) \overset{(43)}{=} \mathbf{V}_{22}\mathbf{E}_2^T \mathbf{M}_{x_i}'^T \mathbf{E} \mathbf{V}_e^{-1} \begin{bmatrix} -\mathbf{T} \\ \mathbf{I}_s \end{bmatrix} \mathbf{V}_{22}\mathbf{E}_2^T \mathbf{x}_l(\mathbf{p}_j)$$

$$= \mathbf{V}_{22}\mathbf{E}_2^T \mathbf{M}_{x_i}'^T \mathbf{E} \mathbf{V}_e^{-1} \begin{bmatrix} -\mathbf{T} \begin{bmatrix} \mathbf{0} & \mathbf{V}_{22}\mathbf{E}_2^T \end{bmatrix} \\ \begin{bmatrix} \mathbf{0} & \mathbf{V}_{22}\mathbf{E}_2^T \end{bmatrix} \end{bmatrix} \mathbf{x}_{l+1}(\mathbf{p}_j) \quad (44)$$

$$\overset{(42)}{=} \mathbf{V}_{22}\mathbf{E}_2^T \mathbf{M}_{x_i}'^T \mathbf{E} \mathbf{V}_e^{-1} \mathbf{V}_e \mathbf{E}^T \mathbf{x}_{l+1}(\mathbf{p}_j) \quad (45)$$

$$\overset{(32)}{=} \mathbf{V}_{22}\mathbf{E}_2^T p_i \mathbf{x}_l(\mathbf{p}_j) = p_i \mathbf{x}_B(\mathbf{p}_j) \quad (46)$$

■

Finally, we need to compute the solutions $\mathbf{p}_j$ from the eigenvectors $\mathbf{v}_j := c\mathbf{V}_{22}\mathbf{E}_2^T \mathbf{x}_l(\mathbf{p}_j)$ where c is some unknown scaling coefficient (resulting, e.g., from restricting the eigenvectors to unit norm). We define $\mathbf{v}_j' := \mathbf{E}\mathbf{V}_e^{-1} \begin{bmatrix} -\mathbf{T} \\ \mathbf{I}_s \end{bmatrix} \mathbf{v}_j$, knowing from the proof that $\mathbf{v}_j' = c\mathbf{x}_{l+1}(\mathbf{p}_j)$. Noting that the last $n+1$ elements of $\mathbf{x}_{l+1}$ are given by $\begin{bmatrix} x_1 & \cdots & x_n & 1 \end{bmatrix}$, we undo the scaling and recover the solution as

$$\mathbf{p}_j = \mathbf{v}_j'(n_{l+1} - n : n_{l+1} - 1)/\mathbf{v}_j'(n_{l+1}) \quad (47)$$

Obviously, only the last $n+1$ elements of the vector $\mathbf{v}_j'$ will actually need to be computed.

C. Using QR with column pivoting

A trivial choice for the coordinate transformation is $\mathbf{V}_e = \mathbf{V}_e^{-1} = \mathbf{I}$, which is precisely the case presented in [7]. In this case, the basis becomes $\mathbf{x}_B = \begin{bmatrix} \mathbf{0} & \mathbf{I}_s \end{bmatrix} \mathbf{E}_2^T \mathbf{x}_l = \mathbf{E}_{22}^T \mathbf{x}_l$. Notice that this algorithm only requires two QR factorizations, and computing $\mathbf{T}$ reduces to solving an upper triangular system.

Contrary to [6], [7], this paper provides more flexibility in the choice of $\mathbf{V}_e$, the proof that it actually yields a valid multiplication matrix, as well as guidance for the degree to which to expand (via the involutive test [8]). The appeal of this method is further that it requires no normal set and can be implemented purely using standard linear algebra tools.

D. Using the right nullspace

The following method [8] proposed by Reid and Zhi (which we will call R.Z. method for short), uses a different paradigm to find the multiplication matrix. The key observation is that if all solutions are distinct, and the dimension of the right nullspace of $\mathbf{C}_e$ equals the number of solutions, s, then the vectors $\mathbf{x}_{l+1}(\mathbf{p}_j)$ evaluated at the solutions span the right nullspace. In other words, if $\mathbf{B}_{l+1}$ is any matrix whose columns span the nullspace of $\mathbf{C}_e$, then $\mathbf{x}_{l+1}(\mathbf{p}_j) = \mathbf{B}_{l+1}\mathbf{c}_j$ for some coefficient vector $\mathbf{c}_j$.

Let $\mathbf{B}_{l+1} = \begin{bmatrix} \mathbf{B}_k \\ \mathbf{B}_l \end{bmatrix}$ be the right nullspace of $\mathbf{C}_e$ so that $\mathbf{B}_l$ is of dimension $(n_l \times s)$. Notice that $\mathbf{x}_{l+1}(\mathbf{p}_j) = \mathbf{B}_{l+1}\mathbf{c}_j$ and $\mathbf{x}_l(\mathbf{p}_j) = \begin{bmatrix} \mathbf{0} & \mathbf{I} \end{bmatrix} \mathbf{x}_{l+1}(\mathbf{p}_j) = \mathbf{B}_l\mathbf{c}_j$.

Form the SVD of $\mathbf{B}_l$

$$\mathbf{B}_l = \begin{bmatrix} \mathbf{U}_1 & \mathbf{U}_2 \end{bmatrix} \begin{bmatrix} \mathbf{S} \\ \mathbf{0} \end{bmatrix} \mathbf{V}^T \quad (48)$$

such that $\mathbf{U}_1$ is of dimension $(n_l \times s)$, and $\mathbf{S}$ is square. Moreover, based on the involutive test, $\mathbf{B}_l$ is of rank s, i.e., $\mathbf{S}$ is invertible.

We choose $\mathbf{x}_B = \mathbf{U}_1^T \mathbf{x}_l$ as basis for the quotient ring, and form the multiplication matrix as [8]

$$\mathbf{M}_{x_i} = \mathbf{S}^{-1} \mathbf{V}^T \mathbf{B}_{l+1}^T \mathbf{M}'_{x_i} \mathbf{U}_1 \qquad (49)$$

Since $\mathbf{S}$ is diagonal, the computation of its inverse is trivial.

Proposition 4: If we choose the basis of the quotient ring as $\mathbf{x}_B = \mathbf{U}_1^T \mathbf{x}_l$, and compute the multiplication matrix as in (49), then $\mathbf{x}_B(\mathbf{p}_j)$ is a right eigenvector of $\mathbf{M}_{x_i}^T$.

Proof:

$$
\begin{aligned}
\mathbf{M}_{x_i}^T \mathbf{x}_B(\mathbf{p}_j) &\overset{(49)}{=} \mathbf{U}_1^T \mathbf{M}'^T_{x_i} \mathbf{B}_{l+1} \mathbf{V} \mathbf{S}^{-1} \mathbf{U}_1^T \mathbf{B}_l \mathbf{c}_j \\
&\overset{(48)}{=} \mathbf{U}_1^T \mathbf{M}'^T_{x_i} \mathbf{B}_{l+1} \mathbf{c}_j \\
&= \mathbf{U}_1^T \mathbf{M}'^T_{x_i} \mathbf{x}_{l+1}(\mathbf{p}_j) \\
&\overset{(32)}{=} p_i \mathbf{U}_1^T \mathbf{x}_l(\mathbf{p}_j)
\end{aligned}
$$

∎

We can recover the solution $\mathbf{p}_j$ from an eigenvector $\mathbf{v}_j = c\mathbf{U}_1^T \mathbf{x}_l(\mathbf{p}_j)$ of $\mathbf{M}_{x_i}^T$ by writing

$$
\begin{aligned}
\mathbf{v}'_j &= \mathbf{U}_1 \mathbf{v}_j = c\mathbf{U}_1 \mathbf{U}_1^T \mathbf{x}_l(\mathbf{p}_j) \\
&= c\mathbf{U}_1 \mathbf{U}_1^T \mathbf{B}_l \mathbf{c}_j = c\mathbf{U}_1 \mathbf{U}_1^T \mathbf{U}_1 \mathbf{S} \mathbf{V}^T \mathbf{c}_j \\
&= c\mathbf{B}_l \mathbf{c}_j = c\mathbf{x}_l(\mathbf{p}_j)
\end{aligned}
$$

and by undoing the scaling as before.

Interestingly, as we will show, the R.Z. method can be cast as a special instance of the general transformation described in Section V-B. For this we proceed as follows (cf. (40)):

$$\mathbf{Q}'_1 \mathbf{R}' \mathbf{V}_e^{-1} \mathbf{V}_e \mathbf{E}^T \mathbf{x}_{l+1} = 0 \qquad (50)$$

$$\mathbf{Q}'_1 \begin{bmatrix} \mathbf{I}_{n_{l+1}-s} & \mathbf{T} \end{bmatrix} \begin{bmatrix} \mathbf{V}_{11}\mathbf{E}_1^T & \mathbf{V}_{12}\mathbf{E}_2^T \\ 0 & \mathbf{V}_{22}\mathbf{E}_2^T \end{bmatrix} \mathbf{x}_{l+1} = 0 \qquad (51)$$

where we have (arbitrarily) imposed the structure

$$\begin{bmatrix} \mathbf{R}'_k & \mathbf{R}'_l \end{bmatrix} = \begin{bmatrix} \mathbf{I}_{n_{l+1}-s} & \mathbf{T} \end{bmatrix} \mathbf{V}_e = \begin{bmatrix} \mathbf{V}_{11} & \mathbf{V}_{12} + \mathbf{T}\mathbf{V}_{22} \end{bmatrix} \qquad (52)$$

To make the correspondence to R.Z., we require that $\mathbf{V}_{22}\mathbf{E}_2^T = \mathbf{U}_1^T$. Also, we choose $\mathbf{V}_{12}\mathbf{E}_2^T\mathbf{U}_1 = 0$. We obtain

$$\mathbf{V}_{11} = \mathbf{R}'_k \qquad (53)$$

$$\mathbf{V}_{22} = \mathbf{U}_1^T \mathbf{E}_2 \qquad (54)$$

$$\mathbf{R}'_l = \mathbf{V}_{12} + \mathbf{T}\mathbf{U}_1^T\mathbf{E}_2 \qquad | \cdot \mathbf{E}_2^T\mathbf{U}_1 \qquad (55)$$

$$\mathbf{T} = \mathbf{R}'_l\mathbf{E}_2^T\mathbf{U}_1 \qquad (56)$$

Backsubstitution of $\mathbf{T}$ yields

$$\mathbf{R}'_l = \mathbf{V}_{12} + \mathbf{R}'_l\mathbf{E}_2^T\mathbf{U}_1\mathbf{U}_1^T\mathbf{E}_2 \qquad (57)$$

$$\mathbf{V}_{12} = \mathbf{R}'_l(\mathbf{I}_{n_l} - \mathbf{E}_2^T\mathbf{U}_1\mathbf{U}_1^T\mathbf{E}_2) \qquad (58)$$

$$= \mathbf{R}'_l\mathbf{E}_2^T(\mathbf{U}_1\mathbf{U}_1^T + \mathbf{U}_2\mathbf{U}_2^T - \mathbf{U}_1\mathbf{U}_1^T)\mathbf{E}_2 \qquad (59)$$

$$= \mathbf{R}'_l\mathbf{E}_2^T\mathbf{U}_2\mathbf{U}_2^T\mathbf{E}_2 \qquad (60)$$

In summary, the (not necessarily unique) choice of $\mathbf{V}_e = \begin{bmatrix} \mathbf{R}'_k & \mathbf{R}'_l\mathbf{E}_2^T\mathbf{U}_2\mathbf{U}_2^T\mathbf{E}_2 \\ 0 & \mathbf{U}_1^T\mathbf{E}_2 \end{bmatrix}$ will yield exactly the same multiplication matrix as the R.Z. method.

VI. SIMULATION RESULTS

In this section, we evaluate the accuracy and efficiency of solving the 3DD problem using QR decomposition (QR) (cf. Section V-C), the R.Z. method (cf. Section V-D), and the normal set-based method (Trad).

The simulation has been conducted on 1000 robot trajectories which are randomly generated with the two robots being 1 m - 2 m apart, and moving an average 3 m - 6 m between taking distance measurements. We computed the solutions for these 1000 samples using the three proposed techniques, and using homotopy continuation (PHC) [12] as ground truth.

Our code was implemented in C++ on a Pentium T9400, 2.53 GHz Laptop, using inverse iterations with sparse LU factorization [23] to compute the right nullspace of $\mathbf{C}_e$ for R.Z. We obtained $\mathbf{C}_e$ of size (1100×715), with 41340 non-zero elements, of total degree $l+1 = 9$ according to the involutive test [8]. The QR method and R.Z. were run in 1.5 s and 0.37 s, respectively, while the normal set-based method runs significantly faster (0.08 s), at the cost of having to determine the normal set and potentially worse numerical conditioning. This runtime is competitive with the fastest reported time of any numerical solver for the 3DD problem.

To evaluate the accuracy of all the 40 solutions produced by the presented three algorithms, we compared our solutions to the PHC solutions. We first matched the 40 solutions of QR, R.Z., and traditional normal set method to the closest PHC solutions, and computed the average 2-norm of their errors over all 40 solutions. The error distributions of these three methods are shown in Fig. 2. The results show that the accuracy of the QR and R.Z. methods is about two orders of magnitude higher than the traditional normal set method, particularly for the orientation. In case the accuracy obtained is insufficient, the solutions of any of the three methods can easily be refined using Newton steps.

We also tried to compare the accuracy of our algorithms to that of Lee and Shim [17], which is currently the fastest reported solution to the 3DD problem, but which requires a special data type (30 significant digits) for accuracy. We implemented their method in Matlab using the symbolic toolbox to represent the coefficients symbolically for the required precision. Unfortunately, at the time of publication we were still unable to reproduce results of useable accuracy other than for problems with integer coefficients. Improving our implementation of their method is work in progress.

VII. CONCLUSION

In this paper, we have presented three fast algorithms to compute the forward kinematics of the general Stewart-Gough mechanism, or, equivalently, to compute the relative pose between two robots based on distance measurements. The algorithms use three different ways to compute the multiplication matrix, based on [19], [7], and [8], from whose eigenvectors we can extract the solutions. We have proven that these methods indeed produce valid multiplication matrices, and shown their relationship.

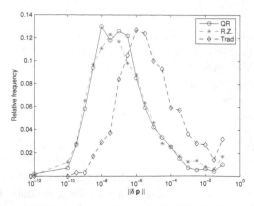

(a) Position error with respect to PHC solutions

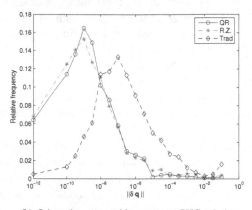

(b) Orientation error with respect to PHC solutions

Fig. 2. Relative pose error distribution generated from the error histogram of 1000 samples. The markers on each curve indicate the centroids of each bin in the histogram, and the y-axis shows the relative frequency of samples in the bins. The QR and R.Z. methods shows significantly improved accuracy over the traditional normal set method, at the cost of increased run time.

Our simulation results have shown that we can compute the solution in less than 1.5 s, and most solutions achieve an accuracy in the order of 10^{-8}. Our methods' major advantage over currently available algorithms is their ease of implementation and the fact that they only use standard double data types and linear algebra libraries.

The hybrid algebraic-numeric techniques for solving polynomial systems used in this paper have only very recently appeared in the robotics and computer vision. Keeping in mind that many geometric problems in robotics can be converted to a system of polynomials, we believe that fast and principled polynomial system solvers such as the ones presented in this paper can be the key to solving interesting long-standing as well as new research problems.

ACKNOWLEDGEMENTS

This work was supported by the University of Minnesota (DTC), and the National Science Foundation (IIS-0643680, IIS-0811946, IIS-0835637). The authors gratefully acknowledge Ryan Elmquist for helping with code implementation.

REFERENCES

[1] N. Trawny, X. S. Zhou, K. X. Zhou, and S. I. Roumeliotis, "3D relative pose estimation from distance-only measurements," in *Proc. of the IEEE/RSJ International Conference on Intelligent Robots and Systems*, San Diego, CA, Oct. 29 – Nov. 2, 2007, pp. 1071–1078.

[2] D. Stewart, "A platform with six degrees of freedom," in *Proc. of the Inst. of Mechanical Engineers*, vol. 180, no. 15, 1965, pp. 371–376.

[3] M. Raghavan, "The Stewart platform of general geometry has 40 configurations," *Journal of Mechanical Design*, vol. 115, pp. 277–282, 1993.

[4] C. W. Wampler, "Forward displacement analysis of general six-in-parallel SPS (Stewart) platform manipulators using soma coordinates," *Mechanism and Machine Theory*, vol. 31, no. 3, pp. 331–337, 1996.

[5] P. Dietmaier, "The Stewart-Gough platform of general geometry can have 40 real postures," in *Advances in Robot Kinematics: Analysis and Control*, J. Lenarcic and M. L. Husty, Eds. Kluwer Academic Publishers, 1998, pp. 7–16.

[6] M. Byröd, K. Josephson, and K. Åström, "Improving numerical accuracy of Gröbner basis polynomial equation solvers," in *Proc. of the IEEE International Conference on Computer Vision*, Rio de Janeiro, Oct. 14–20, 2007, pp. 1–8.

[7] ——, "A column-pivoting based strategy for monomial ordering in numerical Gröbner basis calculations," in *Proc. of the 10th European Conference on Computer Vision*, Marseille, France, Oct. 12–18, 2008.

[8] G. Reid and L. Zhi, "Solving polynomial systems via symbolic-numeric reduction to geometric involutive form," *Journal of Symbolic Computation*, vol. 44, no. 3, pp. 280–291, Mar. 2009.

[9] W. Auzinger and H. J. Stetter, "An elimination algorithm for the computation of all zeros of a system of multivariate polynomial equations," in *Numerical Mathematics (Singapore 1988), Int. Ser. Numer. Math.* Basel: Birkhäuser, 1988, vol. 86, pp. 11–30.

[10] P. R. McAree and R. W. Daniel, "A fast, robust solution to the Stewart platform forward kinematics," *Journal of Robotic Systems*, vol. 13, no. 7, pp. 407–427, 1996.

[11] C. W. Wampler, A. P. Morgan, and A. J. Sommese, "Numerical continuation methods for solving polynomial systems arising in kinematics," *Journal of Mechanical Design*, vol. 112, no. 1, pp. 59–68, 1990.

[12] J. Verschelde, "Algorithm 795: PHCpack: A general-purpose solver for polynomial systems by homotopy continuation," *ACM Transactions on Mathematical Software*, vol. 25, no. 2, pp. 251–276, 1999.

[13] L. Rolland, "Synthesis of the forward kinematics problem algebraic modeling for the general parallel manipulator: displacement-based equations," *Advanced Robotics*, vol. 21, no. 9, pp. 1071–1092, Sep. 2007.

[14] D. Cox, J. Little, and D. O'Shea, *Ideals, Varieties, and Algorithms*, 3rd ed. Springer, 2008.

[15] M. L. Husty, "An algorithm for solving the direct kinematics of general Stewart-Gough platforms," *Mechanism and Machine Theory*, vol. 31, no. 4, pp. 365–380, 1996.

[16] C. Innocenti, "Forward kinematics in polynomial form of the general Stewart platform," *Journal of Mechanical Design*, vol. 123, no. 2, pp. 254–260, 2001.

[17] T.-Y. Lee and J.-K. Shim, "Improved dialytic elimination algorithm for the forward kinematics of the general Stewart-Gough platform," *Mechanism and Machine Theory*, vol. 38, pp. 563–577, Jun. 2003.

[18] D. Cox, J. Little, and D. O'Shea, *Using Algebraic Geometry*, 2nd ed. Springer, 2005.

[19] M. Byröd, Z. Kukelova, K. Josephson, T. Pajdla, and K. Åström, "Fast and robust numerical solutions to minimal problems for cameras with radial distortion," in *Proc. of the IEEE Conference on Computer Vision and Pattern Recognition*, Anchorage, AK, Jun. 23–28, 2008, pp. 1–8.

[20] H. Stewénius, C. Engels, and D. Nistér, "Recent developments on direct relative orientation," *International Journal of Photogrammetry and Remote Sensing*, vol. 60, no. 4, pp. 284–294, Jun. 2006.

[21] Z. Kukelova and T. Pajdla, "A minimal solution to the autocalibration of radial distortion," in *Proc. of the IEEE Conference on Computer Vision and Pattern Recognition*, Minneapolis, MN, Jun. 18–23, 2007, pp. 1–7.

[22] H. J. Stetter, *Numerical Polynomial Algebra*. SIAM: Society for Industrial and Applied Mathematics, 2004.

[23] C. Gotsman and S. Toledo, "On the computation of null spaces of sparse rectangular matrices," *SIAM Journal on Matrix Analysis and Applications*, vol. 30, no. 2, pp. 445–463, May 2008.

An Ab-initio Tree-based Exploration to Enhance Sampling of Low-energy Protein Conformations

Amarda Shehu

Department of Computer Science, George Mason University, Fairfax, Virginia 22030

Email: {amarda}@cs.gmu.edu

Abstract— This paper proposes a robotics-inspired method to enhance sampling of native-like protein conformations when employing only amino-acid sequence. Computing such conformations, essential to associate structural and functional information with gene sequences, is challenging due to the high-dimensionality and the rugged energy surface of the protein conformational space. The contribution of this work is a novel two-layered method to enhance the sampling of geometrically-distinct low-energy conformations at a coarse-grained level of detail. The method grows a tree in conformational space reconciling two goals: (i) guiding the tree towards lower energies and (ii) not over-sampling geometrically-similar conformations. Discretizations of the energy surface and a low-dimensional projection space are employed to select more often for expansion low-energy conformations in under-explored regions of the conformational space. The tree is expanded with low-energy conformations through a Metropolis Monte Carlo framework that uses a move set of physical fragment configurations. Testing on sequences of seven small-to-medium structurally-diverse proteins shows that the method rapidly samples native-like conformations in a few hours on a single CPU. Analysis shows that computed conformations are good candidates for further detailed energetic refinements by larger studies in protein engineering and design.

I. INTRODUCTION

A globular protein molecule repeatedly populates its functional (native) state at room temperature after denaturation [1]. Despite this discovery in 1973 by Anfinsen, the problem of computing the conformations that comprise the protein native state from knowledge of amino-acid sequence alone continues to challenge structural biology [2]. Computing native conformations, however, is essential in associating structural and functional information with newly discovered gene sequences, engineering novel proteins, predicting protein stability, and modeling protein-ligand or protein-protein interactions [3]–[5].

Sampling native conformations is inherently difficult due to the vast high-dimensional conformational space available to a protein chain. The high-dimensionality challenge has drawn robotics researchers to adapt and apply algorithms that plan motions for articulated mechanisms with many degrees of freedom (dofs) to the study of protein conformations [6]–[11]. Though these methods often have to be adapted to deal with hundreds of dofs in protein chains (from dozens of dofs in articulated mechanisms), the motion-planning framework has allowed addressing the problem of computing paths from a given initial to a given goal protein conformation [6], [7].

The problem addressed in this work is the discovery of native conformations from knowledge of amino-acid sequence. No information is available on the goal conformations besides the energy landscape view that associates native conformations with lowest energies [12]. Energetic considerations complicate the search for native conformations. Interactions among atoms in a protein, which scale quadratically with the number of modeled atoms, give rise to an energy surface that can currently be probed only with empirical energy functions. A fundamental challenge in probing the native state is efficiently computing native-like conformations associated with the true global minimum (or competing minima) in an approximated energy surface constelled with local minima [12].

Faced with a vast space rich in local minima, some methods narrow the search space relevant for the native state by employing additional information about this state, often obtained from experiment [13]–[15]. Such information, however, is not available for millions of newly discovered protein-encoding gene sequences, nor is it easily obtained for novel sequences proposed in silico [16]. In such cases, ab-initio methods that employ only amino-acid sequence become very important.

In order to explore a vast conformational space, ab-initio methods often proceed in two stages: they first obtain a broad view of the conformational space, usually at a coarse-grained level of detail, to reveal candidate conformations that then undergo further refinement at a second stage [3], [17]. Coarse-grained representations of a protein chain are employed to reduce the number of dofs (from thousands to hundreds on small-to-medium proteins), and as a result, the dimensionality of the ensuing conformational space. The current generation of physically-realistic coarse-grained energy functions allows employing coarse-grained representations of protein chains as a practical alternative to all-atom representations without sacrificing predictive power [18].

Obtaining a broad view of the conformational space is crucial when postponing the computationally demanding all-atom detail and energetic refinement on coarse-grained conformations deemed to be native-like. Sampling a large number of low-energy conformations often entails enhanced sampling methods that build on Molecular Dynamics (MD) or Monte Carlo (MC) (cf. [19]). While MD-based techniques systematically search the conformational space, MC-based techniques often exhibit higher sampling efficiency [19]. Most notable among MC-based ab-initio methods are those that employ fragment assembly. These methods simplify the search space by using a limited move set of physical fragment configurations to assemble conformations [3], [14], [17].

Enhanced sampling on a simplified search space, coupled

with realistic energy functions, have allowed fragment assembly methods to achieve high prediction accuracy of the native state, albeit at a time cost. It takes weeks on multiple CPUs to obtain a large number of low-energy conformations potentially relevant for the native state. It also remains difficult to ensure that computed conformations are geometrically-distinct and not representative of only a few regions of the conformational space [17]. Part of the difficulty lies in the inability to define a few conformational (reaction) coordinates on which to define distance measures. Popular measures like least Root-Mean-Squared-Deviation (lRMSD) and radius of gyration (Rg) can mask away differences among conformations [17].

The contribution of this work is a novel two-layered method to enhance the sampling of geometrically-distinct low-energy conformations. The method uses no information about the native state of a protein beyond amino-acid sequence. Employing a coarse-grained representation of a protein chain, the method focuses on efficiently obtaining diverse native-like conformations. The goal of the method is to serve as a filtering step that, in a matter of a few hours on a single CPU, reveals native-like conformations that can be further refined through detailed studies. From now on the method is referred to as FeLTr for Fragment Monte CarLo Tree Exploration.

As in fragment assembly, FeLTr assembles a conformation with configurations of short fragments compiled over a nonredundant database of protein structures. Employing fragments from the database (rather than random) improves the likelihood of assembling a physical conformation. A physically-realistic coarse-grained energy function estimates the energy of assembled conformations. The fragment assembly is implemented in a Metropolis MC framework, where the chain of a current conformation is scanned and new fragment configurations are proposed to replace old ones. Replacements that meet the Metropolis criterion are accepted, resulting in a new conformation. Using a limited move set of fragment configurations and a coarse-grained representation, FeLTr is able to rapidly obtain low-energy native-like conformations. Most importantly, the method obtains diverse native-like conformations through a novel tree-based exploration of the conformational space.

Inspired by tree-based methods in robotics motion planning, FeLTr grows a tree in conformational space, reconciling two goals: (i) expanding the tree towards conformations with lower energies while (ii) not oversampling geometrically-similar conformations. The first goal is warranted due to the fact that the desired native-like conformations are associated with the lowest energies in the energy surface sculpted by an amino-acid sequence. The second goal attempts to obtain a broad view of the conformational space near the native state by not oversampling geometrically-similar conformations.

To achieve the first goal, FeLTr partitions energies of computed conformations into levels through a discretized one-dimensional (1d) grid. The grid is used to select conformations associated with lower energy levels more often for expansion. The second goal is achieved by keeping track of computed conformations in a three-dimensional (3d) projection space using the recently proposed ultrafast shape recognition (USR)

features [20]. The projection space is discretized in order to select for expansion low-energy conformations that fall in under-explored regions of the conformational space. After a conformation is selected for expansion, a short Metropolis MC trajectory summarized above is employed to expand the tree with a new low-energy conformation. This two-layered exploration, detailed in section II, is illustrated in Fig. 1.

The employment of the projection space in FeLTr is inspired by recent sampling-based motion-planning methods that make use of decompositions, subdivisions, and projections of the configurational space or the workspace of a robot to balance the exploration between coverage and progress toward the goal [21]–[25]. It is worth mentioning that, while the focus of this work is on protein chains, the projections employed in FeLTr are not tied to protein conformations. Since the projections rely only on geometry, they can be used for any articulated mechanism (manipulators, humanoid, modular robots) (see section IV for more on this point).

FeLTr is tested on sequences of seven proteins of different lengths (20-76 amino acids) – amounting to 40-152 dofs – and native topologies (β, α/β, α). Results show that FeLTr obtains compact low-energy conformations on all proteins. Clustering the lowest-energy conformations reveals that the native state is captured among the computed conformations. These conformations are good candidates for further refinement with detailed all-atom energy functions. The proposed method can serve as an efficient initial filtering step in larger detailed studies focused on extracting structural and functional properties of uncharacterized protein sequences [17], [26].

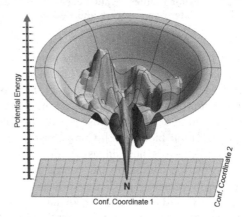

Fig. 1. According to the landscape view of protein folding, the native state, labeled N, is associated with the global minimum of a funnel-like energy surface [12]. The surface shown here is adapted from [12]. The proposed FeLTr method cross-sects this surface by discretizing potential energy values in a 1d grid, illustrated here with the z-axis. The axis changes color from light to dark gray to denote energy values that reach lower values with the native state. The grid on the xy-plane discretizes the projection of the conformational space onto a few conformational coordinates. The illustration here shows two coordinates for visualization purposes. FeLTr uses three coordinates to describe projected conformations, as detailed in section II.

The following summary of related work further places the proposed FeLTr in context. The method is detailed in section II. Analysis of the conformations computed on the seven sequences chosen for testing is presented in section III. Discussion follows in section IV.

Related Work

Computational methods search conformational space systematically or at random [19]. Systematic searches rely on MD to sample conformations by numerically solving Newton's equations of motion. Since the solution accuracy dictates a small timestep (femtoseconds) between consecutive conformations in an MD trajectory, a broad view of the space demands multiple long trajectories. MC-based searches conduct biased probabilistic walks in conformational space [19].

Since both MD and MC are prone to converge to local minima in the energy surface, many methods build on them to enhance sampling. Such methods include simulated annealing, importance and umbrella sampling, replica exchange (also known as parallel tempering), local elevation, activation relaxation, local energy flattening, jump walking, multicanonical ensemble, conformational flooding, Markov state models, discrete timestep MD, and many more (cf. [19]).

Some of these methods narrow the search space by employing Nuclear Magnetic Resonance (NMR) data such as order parameters, residual couplings, and other NMR data [15], [27]. Such data are often incorporated in a new term in the energy function to construct a pseudo-energy function that biases the search towards conformations that reproduce the experimental data. Other methods employ an experimental structure representing the native state as a semi-rigid template and conduct a geometrically-constrained search around it for additional native-like conformations [13], [28].

Ab-initio methods employ only amino-acid sequence for a protein at hand. Some rely on sophisticated energy functions to guide MD trajectories to native-like conformations [29]. The most successful methods employ fragment assembly [3]. While the fragment length varies among methods, the basic process is to assemble conformations with physical fragment configurations extracted from a nonredundant database of protein structures [3], [14], [17]. Deciding on a suitable fragment length depends on the richness of the database to provide a comprehensive picture of fragment configurations. The current diversity of the Protein Data Bank (PDB) supports a minimum length of three amino acids [17].

While a limited move set of fragment configurations speeds up an MC-based exploration, there is no guarantee that assembled conformations are distinct or capture the native state. A large number of low-energy coarse-grained conformations are often computed to increase the probability that a few are sufficiently close to the native state that they will reach this state upon detailed energetic refinements [3], [17]. Since these refinements, often in all-atom detail, are computationally expensive, they can be conducted only on few conformations. Hence, it is important that the exploration reveal diverse coarse-grained conformations near the native state. It is currently difficult to find reaction coordinates on which to measure diversity [30]. Popular but lacking measures like lRMSD and Rg are not integrated in the exploration but largely confined to analysis over computed conformations [17].

FeLTr integrates coordinates proposed in [20] in its tree-based search. The tree is expanded in conformational space through an MC framework, keeping track of computed conformations in the low-dimensional projection space. MC has been used in sampling-based motion planing to escape local minima [31] and enhance sampling for closed chains with spherical joints [32]. The employment of the projection space in FeLTr is also inspired by motion-planning methods that decompose, subdivide, and project a robot's configurational space or workspace to balance the exploration between coverage and progress toward the goal [21]–[25].

II. FeLTr

In the usual MC framework, the probabilistic walk in the conformational space resumes from the last conformation computed. FeLTr enhances this framework by conducting a tree-based exploration of the conformational space. Given that only amino-acid sequence information is available, the root of the tree is an extended coarse-grained conformation. FeLTr then explores the conformational space iteratively, at each iteration selecting a conformation and then expanding it. While every expansion involves a short Metropolis MC trajectory, the important decision about which conformation to select for expansion depends on (i) the energy levels populated and (ii) the projection space covered by computed conformations. Pseudocode is given in Algo. 1. Sections describing the main steps in FeLTr are referenced at the end of each line.

Algo. 1 FeLTr: Fragment Monte CarLo Tree Exploration

Input: α, amino-acid sequence
Output: ensemble Ω_α of conformations

1: $C_{\text{init}} \leftarrow$ extended coarse-grained conf from α ▷II-A
2: $G_E \leftarrow$ explicit 1d energy grid   ▷II-B.1
3: **for** $\ell \in G_E$ **do**
4:    $\ell.G_{\text{USR}} \leftarrow$ implicit 3d geom projection grid   ▷II-B.2
5: ADDCONF(C_{init}, G_E, G_{USR})
6: **while** TIME AND $|\Omega_\alpha|$ do not exceed limits **do**
7:    $\ell \leftarrow$ SELECTENERGYLEVEL(G_E)   ▷II-B.1
8:    cell $\leftarrow$ SELECTGEOMCELL($\ell.G_{\text{USR}}$.cells)   ▷II-B.2
9:    $C \leftarrow$ SELECTCONF(cell.confs)   ▷II-B.2
10:    $C_{\text{new}} \leftarrow$ MC_EXPANDCONF(C)   ▷II-C
11:    **if** $C_{\text{new}} \neq$ NIL **then**   ▷MC succeeded
12:      ADDCONF(C_{new}, G_E, G_{USR})
13:      $\Omega_\alpha \leftarrow \Omega_\alpha \cup \{C_{\text{new}}\}$ ▷add conf to ensemble

ADDCONF(C, G_E, G_{USR})
▷add C to appropriate energy level in G_E
1: $E(C) \leftarrow$ COARSEGRAINEDENERGY(C)   ▷II-C.2
2: $\ell \leftarrow$ level in G_E where $E(C)$ falls into
3: ℓ.confs $\leftarrow \ell$.confs $\cup \{C\}$
▷add C to appropriate cell in geom grid associated with ℓ
4: $P(C) \leftarrow$ USRGEOMPROJ(C)   ▷II-B.2
5: cell $\leftarrow$ cell in $\ell.G_{\text{USR}}$ where $P(C)$ falls into
6: **if** cell = NIL **then**   ▷cell had not yet been created
7:    cell $\leftarrow$ new geom projection cell
8:    $\ell.G_{\text{USR}}$.cells $\leftarrow \ell.G_{\text{USR}}$.cells $\cup \{$cell$\}$
9: cell.confs $\leftarrow$ cell.confs $\cup \{C\}$

A. Coarse-grained Representation of a Protein Chain

Only backbone heavy atoms and side-chain C_β atoms are explicitly represented. Employing the idealized geometry model, which fixes bond lengths and angles to idealized (native) values, positions of backbone atoms are computed from ϕ, ψ angles. These angles are set to $120°, -120°$ in an extended conformation (Algo. 1:1). Positions of C_β atoms are determined from the computed backbone as in [33].

B. Selection: Combination of Energy Layers and Low-dimensional Geometric Projections

1) Energy Layers: A 1d grid, G_E (Algo. 1:2), is defined on the segment $[E_{\min}, 0]$. $E_{\min}$ refers to the lowest expected energy on computed conformations (set to -200 kcal/mol on the tested proteins), and 0 refers to the highest energy. Since the Metropolis MC expansion quickly obtains negative-energy conformations (and conformations with nonnegative energies are not relevant because they are highly infeasible), it is not necessary to maintain a grid over nonnegative energies. Energy levels are generated every δE units. δE is set to a small value (2 kcal/mol), so that the average energy $E_{\text{avg}}(\ell)$ over conformations populating a specific energy level $\ell \in G_E$ captures well the distribution of energies in ℓ.

This discretization is used to bias the selection towards conformations in the lower energy levels. The weight function $w(\ell)$ associated with an energy level $\ell \in G_E$ is set to $w(\ell) = E_{\text{avg}}(\ell) \cdot E_{\text{avg}}(\ell) + \epsilon$, where $\epsilon = 1.0/2^{22}$ ensures that conformations with higher energies have a nonzero probability of selection. An energy level ℓ is then selected with probability $w(\ell)/\sum_{\ell' \in G_E} w(\ell')$ (Algo. 1:7). This quadratic weight function biases the selection towards conformations with lower energies while allowing for some variation. Allowing higher-energy conformations to be selected provides FeLTr with the ability to jump over barriers in the energy surface. The Metropolis MC expansion also allows jumping over barriers.

2) Low-dimensional Geometric Projections: Conformations in a chosen energy level are then projected onto a low-dimensional space. Borrowing from the USR features proposed in [20], three projection coordinates are defined on each computed conformation: the mean atomic distance μ_{ctd}^1 from the centroid (ctd), the mean atomic distance μ_{fct}^1 from the atom farthest from the centroid (fct), and the mean atomic distance μ_{ftf}^1 from the atom farthest from fct (ftf). The ctd, fct, and ftf atoms capture well-separated extremes of a conformation. So, the distribution of atomic distances from each extreme point (approximated with the first moment) is likely to yield new geometric information on a conformation.

The three projection coordinates capture overall topologic differences among conformations. As discussed in section IV, the coordinates are not specific to protein conformations but can be applied to any articulated mechanism. The coordinates allow introducing a second layer of discretization in FeLTr. The reason for the second layer is that conformations with similar energies may be geometrically different (a notion captured by entropy in statistical mechanics), and FeLTr aims to compute geometrically-distinct low-energy conformations.

Conformations in an energy level are partitioned in cells of the projection space to employ coverage in this space as a second criterion for selection. An implicit 3d grid, G_{USR}, is associated with each energy level (Algo.1:3-4), based on a uniform discretization of the projection coordinates. The selection is then biased towards cells with fewer conformations through the weight function $1.0/[(1.0 + \text{nsel}) \cdot \text{nconfs}]$, where nsel records how often a cell is selected, and nconfs is the number of conformations that project to the cell (Algo. 1:8). Similar selection schemes have been advocated in motion-planning literature as a way to increase geometric coverage during exploration [21], [24]. Once a cell is chosen, the actual conformation selected for expansion is obtained at random over those in the cell (Algo. 1:9), since conformations in the same cell have similar energies (within δE).

C. Expansion: Metropolis MC with Fragment Assembly

After a conformation is selected for expansion, its chain of N amino acids is scanned, defining N-2 consecutive fragments of three amino acids referred to as trimers. A conformation can now be updated by replacing configurations of its trimers. A trimer configuration consists of 6 ϕ, ψ angles defined over its backbone. The expansion procedure (Algo. 1:10) iterates N-2 times, at each iteration choosing a trimer at random over the chain. Upon choosing a trimer, a database of trimer configurations, whose construction is detailed in section II-C.1, is then queried with the amino-acid sequence of the trimer.

Of all configurations available for the trimer in the database, one obtained at random is proposed to replace the configuration in the current conformation. The reason for the random rather than consecutive iteration over the trimers in a chain is that an iterative scanning of the chain may result in local minima; that is, configurations are proposed but not accepted. The decision on whether to accept a trimer configuration (Algo. 1:11) is done under the Metropolis criterion, with the coarse-grained energy function defined in section II-C.2.

1) Database of Fragment Configurations: A PDB subset of nonredundant protein structures (as of November 2008) is extracted through the PISCES server [34] to contain proteins that have $\leq 40\%$ sequence similarity, ≤ 2.5 Å resolution and R-factor ≤ 0.2. Proteins studied in this work are removed from the database. The 40% similarity cutoff ensures that topologies that are over-populated by similar protein sequences in the PDB are not over-represented in the database. Around $6,000$ obtained protein chains are split into all possible overlapping trimers. The database maintains a list of configurations populated by each trimer over all extracted chains - a total of more than ten million configurations. No less than 10 configurations are populated for any trimer of each tested sequence.

2) Coarse-grained Energy Function: The function that evaluates the energy of a coarse-grained conformation, recently proposed in [17], is a linear combination of non-local terms (local terms are excluded since conformations are assembled with physical trimer configurations): $E = E_{\text{Lennard-Jones}} + E_{\text{H-Bond}} + E_{\text{contact}} + E_{\text{burial}} + E_{\text{water}} + E_{\text{Rg}}$. The $E_{\text{Lennard-Jones}}$ term is implemented after the 12-6

Lennard-Jones potential in AMBER9 [35], with a modification that allows a soft penetration of van der Waals spheres. The E_{H-Bond} term allows formation of local and non-local hydrogen bonds. The terms $E_{contact}$, E_{burial}, and E_{water}, implemented as in [29], allow formation of non-local contacts, a hydrophobic core, and water-mediated interactions.

The E_{Rg} term in this work penalizes a conformation by $(Rg - Rg_{PDB})^2$ if the conformation's Rg value is above the Rg_{PDB} value predicted for a chain of same length from proteins in the PDB. The predicted value fits well to the line $2.83 \cdot N^{0.34}$ [14], which is used to compute Rg_{PDB} for each sequence of N amino acids. The E_{Rg} term penalizes non-compact conformations, since native-like conformations are compact and with a well-packed hydrophobic core. Moreover, Rg_{PDB} and the Rg value of an extended conformation are used to define the boundaries of the projection space.

3) Metropolis Criterion: After replacing a trimer configuration in a selected conformation, the resulting energy is evaluated with the above energy function. The proposed replacement is accepted if it results in a lower energy (Algo. 1:10-11). Otherwise, it is accepted with probability $e^{-\beta \cdot \Delta E}$, where ΔE is the difference in energy after the replacement, and β is a temperature scaling factor. In this work, β is chosen to allow an energy increase of 10 kcal/mol with probability 0.1 so the tree is expanded with conformations that cross energy barriers.

D. Analysis of Computed Conformations

As shown in Algo. 1, conformations computed by FeLTr are gathered in the ensemble Ω_α. The distribution of energies of conformations in Ω_α is analyzed to obtain the average energy $\langle E \rangle$ and standard deviation σE. Let Ω_α^* denote the subensemble of conformations with energies no higher than $\langle E \rangle - \sigma E$. Ω_α^* is clustered with a simple leader-like algorithm [36], using a conservative cluster radius of 2.0 Å. The lowest-energy conformations of each cluster are offered by FeLTr as candidates for further detailed refinement. The results below show that this analysis reveals distinct clusters of native-like conformations that capture the native state.

The purpose of the analysis is to reveal possibly more than one energy minimum. Since exact quantum mechanics calculations cannot be afforded on long chains, empirical energy functions are used instead. These functions (like the one employed in this work) need to rank lower in energy those computed conformations that are more native-like. The lowest energy value reported, however, may not correspond to the most native-like conformation.

III. Experiments and Results

Since the conformational space available to a protein chain is high-dimensional, the ability of a method to reproduce conformations that populate the protein native state provides an important benchmark [26]. FeLTr is applied to seven structurally-diverse protein sequences of varying lengths. Comparing computed conformations with experimentally-available native structures of each protein reveals that FeLTr captures the native state. The results below, which benchmark

FeLTr against a Metropolis MC simulation, show that FeLTr consistently obtains lower energies than the MC simulation.

The low-energy conformations obtained by FeLTr are analyzed not only for the presence of native-like conformations, but also for their geometric diversity. The results presented below show that FeLTr populates diverse energy minima significantly better than the MC simulation. While the native state is usually present among the highest-populated minima, other obtained minima contain compact low-energy conformations that differ on content of secondary structure segments or overall 3d arrangement of these segments.

A. Chosen Systems

The seven proteins chosen to test FeLTr, listed in Table III-A, include tryptophan cage (Trp-cage), Pin1 Trp-Trp ww domain (wwD), villin headpiece (hp36), engrailed homeodomain (eHD), bacterial ribosomal protein (L20), immunoglobulin binding domain of streptococcal protein G (GB1), and calbindin D_{9k}. These proteins are chosen because they vary in size (number of amino acids) and ultimately number of dofs, native fold (3d global arrangement of local secondary structure segments), and are actively studied both in silico and in the wet lab due to the importance of their biological functions.

Protein	Trp-cage	wwD	hp36	eHD	L20	GB1	Calbindin D_{9k}
Fold	α	β	α	α	α	α/β	α
Size	20	26	36	54	60	60	76
Dofs	40	52	72	108	120	120	152

FeLTr is implemented in C++, run on an Intel Core2 Duo machine with 4GB RAM and 2.66GHz CPU, and allowed to compute no more than $50,000$ conformations in no more than 3 hours. Limiting the number of conformations ensures that the exploration tree and ensemble Ω_α fit in memory. On small proteins like Trp-cage, wwD, and hp36, this number of conformations is reached in 40 minutes, 1 hour, and 2 hours, respectively. (Time grows quadratically with chain length due to the Lennard-Jones energy term; e.g., computing $50,000$ conformations takes ~ 36 hours on 200 amino-acid chains.)

B. Analyzing the Efficiency of FeLTr

The efficiency of FeLTr is estimated in comparison with a Metropolis MC simulation that is limited by the same time and number of conformations like FeLTr. To keep all other conditions similar, the MC simulation also employs the same fragment assembly to compute conformations in its trajectory and the same coarse-grained representation and energy function to calculate the energy of a computed conformation. This comparison allows to directly probe the effect of the FeLTr tree-based exploration guided by the novel two-layer discretization in sampling native-like conformations.

Fig. 2(a1) plots energies versus Rg values of computed conformations. Significantly lower energies are obtained with FeLTr (shown in dark gray) than the MC simulation (shown in light gray). This result, also revealed when plotting energies versus lRMSDs of computed conformations from an

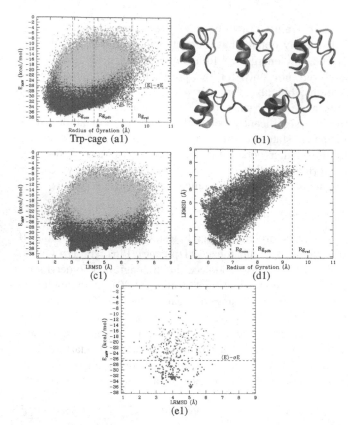

Fig. 2. Plots superimpose Trp-cage MC-computed data in light gray over FeLTr-computed data in dark gray. (a1) Energies of conformations are plotted vs. Rg values. The horizontal line marks the energetic cutoff. The three vertical lines show different Rg thresholds. (b1) FeLTr-computed conformations that meet the energetic criterion are clustered. Lowest-energy conformations of five main clusters (dark gray) are superimposed over the native structure (transparent light gray). (c1) Energies are plotted vs. lRMSDs from native. (d1) For conformations that meet the energetic criterion, lRMSDs from native are plotted vs. Rg values. (e1) For conformations that map to same cell of the projection space as the native, energies are plotted vs. lRMSDs from native.

experimentally-available native structure (Fig. 2(c1) for Trp-cage and 3(b2-7) for the other proteins), is not surprising. FeLTr guides the tree towards lower energies, whereas the MC simulation resumes from the last conformation generated.

The MC simulation employed for comparison is a powerful probabilistic walk that uses fragment assembly to make large hops in conformational space; that is, the lRMSD between two consecutive conformations can be significant. Figs. 2(c1) and 3(b2-7) show that the MC simulation comes close (in lRMSD) to the native structure of most tested proteins (FeLTr comes closer in Figs 2(c1) and 3(b3, b6)). However, FeLTr populates more energy minima, revealed when plotting energies of computed conformations versus their Rg values and versus lRMSDs from the native structure. This indicates the geometric projection layer helps to explore different conformational topologies with which to populate low energy levels. It is worth noting that, though Rg and lRMSD are not general conformational coordinates, they are useful to visualize 2d projections of the probed energy surface.

The inability of an MC trajectory to populate diverse energy minima is addressed in recent work by executing numer-

ous trajectories. The trajectories are initiated from different carefully selected conformations in a simulated annealing framework [17]. This exploration, while offering a broad view of the conformational space near the native state, requires 1-3 weeks of computation on 50 CPUs [17]. The selection of conformations from which to initiate MC trajectories is seamlessly integrated in FeLTr through the tree-based exploration.

Native structures of many of the proteins presented here have been computed by detailed MD simulations that employ discrete timesteps [26]. Running times of such simulations, while resulting in high-quality native-like conformations, are limited by the dynamics of the proteins considered. While fast folders require ns-long MD simulations, slow folders may require longer than μs-long simulations (more than one week on one CPU). Other work that computes coarse-grained native-like conformations with MC trajectories (of similar length, $\geq 50,000$ consecutive conformations) employs long angle-based (rather than sequence-based) fragments [14].

C. Extracting Native-like Conformations with FeLTr

The goal of FeLTr is not to obtain single structures with high accuracy but to compute coarse-grained native-like conformations whose accuracy can be later improved through further refinements. Determining what makes a FeLTr-computed conformation native-like depends only on energetic considerations. The $\langle E \rangle - \sigma E$ cutoff defines a subensemble Ω_α^* of conformations that can be considered for further refinement. Employing other measures such as Rg or lRMSD from the native structures employs information that is not available from knowledge of amino-acid sequence only. Focusing on Ω_α^* reduces the number of conformations by more than 50%.

Fig. 2(a1) shows that conformations with energies no higher than $\langle E \rangle - \sigma E$ have diverse Rg values. The vertical lines in Fig. 2(a1) mark three Rg thresholds: Rg_{rel} - the Rg value of an extended conformation, Rg_{PDB} - the Rg value predicted for a chain of same length from the PDB, and Rg_{con} - a smaller Rg value proposed in [14] for more compact conformations. Specifically, $Rg_{con} = 2.5 \cdot N^{0.34}$, where N is the number of amino acids in a protein sequence. The three vertical lines show that (i) almost all computed conformations are more compact than an extended conformation; (ii) the number of conformations proposed for refinement can be further reduced (down to $0.25|\Omega_\alpha|$) by discarding those with $Rg > Rg_{PDB}$; and (iii) the method obtains more compact conformations than rewarded by the coarse-grained energy function.

Clustering Ω_α^* allows offering only the lowest-energy conformations of the top-populated clusters for further refinement. These conformations are superimposed in opaque dark gray over the transparent light gray native structure of each considered protein - see Fig. 2(b1) and 3(a2-7). The native structures are obtained from the PDB: PDB id 1l2y for Trp-cage, 1i6c for wwD, 1vii for hp36, 1enh for eHD, 1gyz for L20, 1gb1 for GB1, 4icb for calbindin. With the exception of GB1, the native structure is captured among the top clusters. As Fig. 2(b1) shows, the top two clusters are very similar to the Trp-cage native structure (2-3 Å lRMSD), with some variability in the loop. Plotting Rg values of conformations in Ω_α^* versus their lRMSDs from the native in Fig. 2(d1) shows a well-separated cluster of conformations. The cluster is around 2.0 Å in lRMSD from the native and around an Rg value of 6.5 Å, which is similar to the Rg value of 6.93 Å of the native structure.

D. The Projection Space Layer Helps Obtain Geometrically Distinct Conformations

Conformations that map to the same cell in the 3d grid over the projection space can still be significantly different (in terms of lRMSD) from one another. Fig. 2(e1), which plots energies versus lRSMDs from the native structure, shows that `FeLTr` obtains lower energies even for conformations that map to the same cell as the native structure. In addition, these conformations have diverse lRMSDs, up to 7 Å.

Projection coordinates capture overall topology, with fine structural details handled by the energy function. For example, the GB1 conformation representative of the top cluster projects to the same cell as the native structure. The native topology is captured, but the β-sheets are not fully formed. Since β-sheets arise from non-local interactions, they cannot be captured at the fragment level but through an energy function. Improvements in energy functions to capture non-local backbone pairings are the subject of much research [3], [26].

IV. DISCUSSION

The coarse graining in `FeLTr` is based on the backbone-based theory of protein folding [37]. Other work that also employs coarse graining shows that geometry presculpts the protein energy surface [38]. `FeLTr` leverages the role of geometry in shaping the protein energy surface by employing both geometry and energy to guide its tree-based exploration.

Since time grows quadratically with the number of atoms (due to the Lennard-Jones term), coarse graining (which reduces the number of atoms) and the focus on computing diverse low-energy conformations make `FeLTr` particularly effective to handle high-dof chains. Coarser protein representations like C_α traces may extend applicability to longer chains.

Coarse graining can also benefit methods that search high-dimensional spaces of articulated robots. The importance of coarse graining is indeed starting to emerge in sampling-based motion planning. Together with work in [21], which shows benefits in using different layers of granularity (from geometric to kinematic to dynamic), `FeLTr` also supports the use of reduced models to address high dimensionality.

The projection coordinates employed here are not proposed as general reaction coordinates on which to project the energy surface. Finding such coordinates remains the subject of much research in computational biology [30]. Rather, these coordinates are a first attempt towards integrating a projection space in the exploration of conformational space. Directions for future research include the design of novel projections and parallel implementations to further enhance the exploration.

Since the USR projections rely only on geometry, they are not tied to protein chains but can apply to any articulated

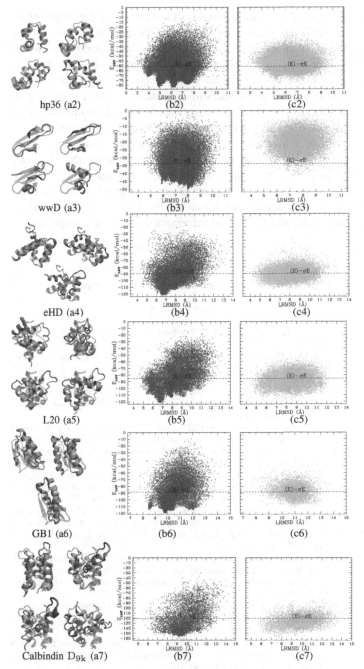

Fig. 3. (a2-7) `FeLTr`-computed conformations that meet the energetic criterion are clustered. The lowest-energy conformations of the most populated clusters are superimposed in dark gray over the native structure drawn in transparent light gray. Energies of conformations are plotted vs. their lRMSDs from the native structure in dark gray for `FeLTr`-computed conformations in (b2-7) and light gray for MC-computed conformations in (c2-7).

mechanism. In particular, sampling-based motion planners like DSLX [21], PDST [24] and [39] that rely on low-dimensional projections can potentially benefit from using USR projections to effectively explore high-dimensional spaces.

`FeLTr` also offers an interesting insight on how to generate valid samples for articulated mechanisms. For instance, since random sampling of dofs in manipulation planning often results in self-colliding configurations, the equivalent of a fragment database can be employed to extract good configurations for different fragments. Work in [40] has also proposed the usage of chain fragments in sampling valid configurations.

The Rg_{PDB} employed to obtain compact conformations does not capture proteins with diverse functional states. Different values of Rg thresholds, obtained from experiment or defined systematically over a range as in [17], can be employed in future work to extend applications on such proteins.

FeLTr makes a first step towards rapidly computing coarse-grained native-like conformations from sequence. Analysis shows the native structure is among computed conformations. The lowest-energy conformations are good candidates for further refinement in all-atom detail in order to associate structural and functional information with novel sequences.

ACKNOWLEDGMENT

The author thanks Erion Plaku, Srinivas Akella, and the anonymous reviewers for comments on this paper.

REFERENCES

[1] C. B. Anfinsen, "Principles that govern the folding of protein chains," *Science*, vol. 181, no. 4096, pp. 223–230, 1973.

[2] R. H. Lathrop, "The protein threading problem with sequence amino acid interaction preferences is NP-complete," *Protein Eng*, vol. 7, no. 9, pp. 1059–1068, 1994.

[3] P. Bradley, K. M. S. Misura, and D. Baker, "Toward high-resolution de novo structure prediction for small proteins," *Science*, vol. 309, no. 5742, pp. 1868–1871, 2005.

[4] S. Yin, F. Ding, and N. V. Dokholyan, "Eris: an automated estimator of protein stability," *Nat Methods*, vol. 4, no. 6, pp. 466–467, 2007.

[5] T. Kortemme and D. Baker, "Computational design of protein-protein interactions," *Curr. Opinion Struct. Biol.*, vol. 8, no. 1, pp. 91–97, 2004.

[6] N. M. Amato, K. A. Dill, and G. Song, "Using motion planning to map protein folding landscapes and analyze folding kinetics of known native structures," *J. Comp. Biol.*, vol. 10, no. 3-4, pp. 239–255, 2002.

[7] M. S. Apaydin, D. L. Brutlag, C. Guestrin, D. Hsu, and J.-C. Latombe, "Stochastic roadmap simulation: an efficient representation and algorithm for analyzing molecular motion," *J. Comp. Biol.*, vol. 10, no. 3-4, pp. 257–281, 2003.

[8] J. Cortes, T. Simeon, R. de Angulo, D. Guieysse, M. Remaud-Simeon, and V. Tran, "A path planning approach for computing large-amplitude motions of flexible molecules," *Bioinformatics*, vol. 21, no. S1, pp. 116–125, 2005.

[9] A. Lee, I. Streinu, and O. Brock, "A methodology for efficiently sampling the conformation space of molecular structures," *J. Phys. Biol.*, vol. 2, no. 4, pp. 108–S115, 2005.

[10] K. M. Kim, R. L. Jernigan, and G. S. Chirikjian, "Efficient generation of feasible pathways for protein conformationa transitions," *Biophys. J.*, vol. 83, no. 3, pp. 1620–1630, 2002.

[11] I. Georgiev and B. R. Donald, "Dead-end elimination with backbone flexibility," *Bioinformatics*, vol. 23, no. 13, pp. 185–194, 2007.

[12] K. A. Dill and H. S. Chan, "From levinthal to pathways to funnels," *Nat. Struct. Biol.*, vol. 4, no. 1, pp. 10–19, 1997.

[13] A. Shehu, C. Clementi, and L. E. Kavraki, "Modeling protein conformational ensembles: From missing loops to equilibrium fluctuations," *Proteins: Struct. Funct. Bioinf.*, vol. 65, no. 1, pp. 164–179, 2006.

[14] H. Gong, P. J. Fleming, and G. D. Rose, "Building native protein conformations from highly approximate backbone torsion angles," *Proc. Natl. Acad. Sci. USA*, vol. 102, no. 45, pp. 16 227–16 232, 2005.

[15] K. Lindorff-Larsen, R. B. Best, M. A. DePristo, C. M. Dobson, and M. Vendruscolo, "Simultaneous determination of protein structure and dynamics," *Nature*, vol. 433, no. 7022, pp. 128–132, 2005.

[16] D. Lee, O. Redfern, and C. Orengo, "Predicting protein function from sequence and structure," *Nat. Rev. Mol. Cell Biol.*, vol. 8, no. 12, pp. 995–1005, 2007.

[17] A. Shehu, L. E. Kavraki, and C. Clementi, "Multiscale characterization of protein conformational ensembles," *Proteins: Struct. Funct. Bioinf.*, 2009, in press.

[18] C. Clementi, "Coarse-grained models of protein folding: Toy-models or predictive tools?" *Curr. Opinion Struct. Biol.*, vol. 18, pp. 10–15, 2008.

[19] W. F. van Gunsteren and et al., "Biomolecular modeling: Goals, problems, perspectives," *Angew. Chem. Int. Ed. Engl.*, vol. 45, no. 25, pp. 4064–4092, 2006.

[20] P. J. Ballester and G. Richards, "Ultrafast shape recognition to search compound databases for similar molecular shapes," *J. Comput. Chem.*, vol. 28, no. 10, pp. 1711–1723, 2007.

[21] E. Plaku, L. Kavraki, and M. Vardi, "Discrete search leading continuous exploration for kinodynamic motion planning," in *Robotics: Sci. and Syst.*, Atlanta, GA, USA, 2007.

[22] G. Sánchez and J.-C. Latombe, "On delaying collision checking in PRM planning: Application to multi-robot coordination," *Int. J. Robot. Res.*, vol. 21, no. 1, pp. 5–26, 2002.

[23] Y. Yang and O. Brock, "Efficient motion planning based on disassembly," in *Robotics: Sci. and Syst.*, Cambridge, MA, 2005, pp. 97–104.

[24] A. M. Ladd and L. E. Kavraki, "Motion planning in the presence of drift, underactuation and discrete system changes," in *Robotics: Sci. and Syst.*, Boston, MA, 2005, pp. 233–241.

[25] H. Choset and et al., *Principles of Robot Motion: Theory, Algorithms, and Implementations*, 1st ed. Cambridge, MA: MIT Press, 2005.

[26] F. Ding, D. Tsao, H. Nie, and N. V. Dokholyan, "Ab initio folding of proteins with all-atom discrete molecular dynamics," *Structure*, vol. 16, no. 7, pp. 1010–1018, 2008.

[27] M. Vendruscolo, E. Pacci, C. Dobson, and M. Karplus, "Rare fluctuations of native proteins sampled by equilibrium hydrogen exchange," *J. Am. Chem. Soc.*, vol. 125, no. 51, pp. 15 686–15 687, 2003.

[28] S. Wells, S. Menor, B. Hespenheide, and M. F. Thorpe, "Constrained geometric simulation of diffusive motion in proteins," *J. Phys. Biol.*, vol. 2, no. 4, pp. 127–136, 2005.

[29] G. A. Papoian, J. Ulander, M. P. Eastwood, Z. Luthey-Schulten, and P. G. Wolynes, "Water in protein structure prediction," *Proc. Natl. Acad. Sci. USA*, vol. 101, no. 10, pp. 3352–3357, 2004.

[30] P. Das, M. Moll, H. Stamati, L. E. Kavraki, and C. Clementi, "Low-dimensional free energy landscapes of protein folding reactions by nonlinear dimensionality reduction," *Proc. Natl. Acad. Sci. USA*, vol. 103, no. 26, pp. 9885–9890, 2006.

[31] J. Barraquand and J.-C. Latombe, "Robot motion planning: A distributed representation approach," *Int. J. Robot. Res.*, vol. 10, pp. 628–649, 1991.

[32] L. Han, "Hybrid probabilistic roadMap-Monte Carlo motion planning for closed chain systems with spherical joints," in *ICRA*, New Orleans, LA, 1994, pp. 920–926.

[33] M. Milik, A. Kolinski, and J. Skolnick, "Algorithm for rapid reconstruction of protein backbone from alpha carbon coordinates," *J. Comput. Chem.*, vol. 18, no. 1, pp. 80–85, 1997.

[34] G. Wang and R. L. Dunbrack, "Pisces: a protein sequence culling server," *Bioinformatics*, vol. 19, no. 12, pp. 1589–1591, 2003.

[35] D. A. Case and et al., "Amber 9," University of California, San Francisco, 2006.

[36] A. K. Jain, R. C. Dubes, and C. C. Chen, "Bootstrap techniques for error estimation," *IEEE Trans. Pattern Analysis and Machine Intelligence*, vol. 9, no. 55, pp. 628–633, 1987.

[37] G. D. Rose, P. J. Fleming, J. R. Banavar, and A. Maritan, "A backbone-based theory of protein folding," *Proc. Natl. Acad. Sci. USA*, vol. 103, no. 45, pp. 16 623–16 633, 2006.

[38] T. H. Hoang, A. Trovato, F. Seno, J. R. Banavar, and A. Maritan, "Geometry and symmetry presculpt the free-energy landscape of proteins," *Proc. Natl. Acad. Sci. USA*, vol. 101, no. 21, pp. 7960–7964, 2007.

[39] H. Kurniawati and D. Hsu, "Workspace-based connectivity oracle: An adaptive sampling strategy for PRM planning," in *WAFR*, ser. Springer Tracts in Advanced Robotics, New York, NY, 2006, vol. 47, pp. 35–51.

[40] L. Han and N. M. Amato, "A kinematics-based probabilistic roadmap method for closed chain systems," in *Algorithmic and Computational Robotics: New Directions*, B. R. Donald, K. M. Lynch, and D. Rus, Eds. MA: AK Peters, 2001, pp. 233–246.

Cellular Muscle Actuators with Variable Resonant Frequencies

Thomas W. Secord and H. Harry Asada
d'Arbeloff Laboratory for Information Systems and Technology
Department of Mechanical Engineering
Massachusetts Institute of Technology
Cambridge, MA 02139, USA
Email: {secord, asada}@mit.edu

Abstract—This paper presents the design and analysis of a novel variable stiffness and variable resonance actuator based on a cellular arrangement of piezoelectric devices. The cellular muscle actuator design concept is presented followed by a general dynamic model for establishing the theoretical bounds on achievable resonant frequencies. A model that is specific to the proposed design is then formulated to include the effects of parasitic dynamics. The resonance characteristics of a three cell prototype system are identified experimentally. The theoretical model and experimental results agree over a large frequency range and illustrate the variable resonance concept.

I. INTRODUCTION

Variable stiffness actuators have several unique properties and have been an active research area in robotics during the last two decades. For mobile robots, tunable joint compliance effectively absorbs impulsive reaction forces from the ground and allows a robot to better negotiate with rough terrain. Compliance is also essential in manipulating an object and physically interacting with environments. The compliance in physical contact tasks must often be altered to meet task requirements. Although feedback control is an easy way of varying "servo" stiffness, the bandwidth of a control loop is usually much lower than the desired speed of the physical interaction. Therefore, variable mechanical compliance must be inherent to the actuator's physical construction to meet bandwidth requirements.

Several groups have developed variable stiffness actuators for robotic applications. However, many of these designs are variations on traditional electromechanical motors [1]. Furthermore, much work in this field focuses mainly on legged locomotion (e.g. [2],[3]) and human-robot interaction (e.g. [4]).

In addition to variable stiffness, resonance has also been a key property in robotic systems and can be traced to the first hopping robots of Raibert [5]. The early work of Pratt also demonstrated that traditionally stiff actuators can resonate with a load using series elastic stiffness in the drive train [6]. Resonance has also seen recent use in flapping flight robots [7],[8]. With respect to research focusing on actuators, the work in [9] and [10] examines resonance amplitudes for piezoelectric materials and electrostrictive materials, respectively. Resonance extends into the biomechanics realm as well. Animals, especially humans, prefer gait frequencies that closely correspond to the natural frequencies of their limb pendular motion because it is energetically favorable and it appears to simplify the neural control of movement [11].

The goal of the present work is to combine the beneficial characteristics of variable stiffness and resonance into a single actuator. This paper describes the design of an actuator based on cellular architecture where both variable stiffness and variable resonance properties are independently tunable by changing both stiffness and mass distribution within the actuator. Actuator performance can be improved in many ways if the stiffness and mass properties can be modulated depending on task requirements. Specifically, for a class of periodic motions including flapping, swimming, walking, and running, power efficiency and amplitude of motion can be improved significantly if the frequency of the periodic motion matches the resonance frequency of the combined load and actuator. Tuning resonant frequencies to minimize energy expenditure is especially important for mobile robot applications. The resonance tuning capability is also crucial for energy harvesting applications [12].

The design concept of an actuator with tunable stiffness and mass distribution will be described for cellular muscle actuators consisting of many discrete actuator units. The concept of cellular actuators has been described in previous work of the authors' group. For example, the application of cellular architecture is discussed for shape memory alloy (SMA) systems in [13] and for piezoelectric materials (specifically PZT) in [14]. In this work, existing design strategies for PZT are extended to incorporate variable stiffness.

The paper will first review the cellular architecture concept for PZT and introduce the variable stiffness design. Next, the variable stiffness concept is extended to include variable resonant frequencies and theoretical bounds on the resonant frequencies are established using a simplified model. In the final sections, the model is extended and compared to experimental data. Both the model and experiments show the versatility of the actuator design strategy.

II. CELLULAR ARCHITECTURE AND VARIABLE STIFFNESS DESIGN

This section summarizes basic principles of variable stiffness cellular actuators (VSCA). A particular PZT cellular actuator design will be described to clarify the concepts, yet the basic principles described in this section are applicable to a class of cellular actuators using various smart materials.

A. PZT Cellular Actuators with Large Strain Amplification

Fig. 1 shows the design concept for a nested-flexure PZT cellular actuator. Each cellular unit consists of a PZT stack and double-layer nested flexures that amplify displacement by a combined factor of 20 (Fig. 1 (a)). This large amplification gain is necessary because PZT has very small inherent strain (0.1%). Fig. 1 (b) shows a prototype of serially connected PZT cellular units. The six cellular units yield a 65 mm net body length and can produce a 7 mm total free displacement and a 5 N blocking force.

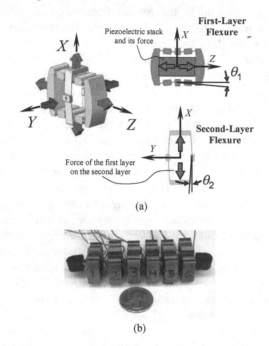

(a)

(b)

Fig. 1. (a) Motion of nested amplification flexures. (b) PZT cellular actuator having 6 serially connected cells.

From Fig. 1 (a), observe that as a voltage is applied to the PZT stacks, the first layer flexure is pushed outward along the Z direction, which results in an outward amplified displacement in the X direction provided that $\theta_1 \ll 1$. The second layer flexure is also pushed outward in the X direction, which results in a further amplified displacement in the inward Y direction with a small θ_2. Therefore, a contraction force is generated along the Y-axis output as the PZT stacks are activated. This contractile double-layer flexure design provides a muscle-like motion and allows the connection of multiple units in series without buckling of the flexures.

B. Variable Stiffness Mechanism

Fig. 2 shows a modification to the above design to achieve variable stiffness. Outside the second layer flexure is a rigid structure that limits the stroke of the output displacement in the Y direction. When the PZT is not activated, the output node of the second layer flexure rests on the stroke limiting beam. The output node movement is also limited when an excessive tensile force acts on the output node. As the applied PZT voltage increases, the output node is pulled inward and is detached from the stroke limiting beam.

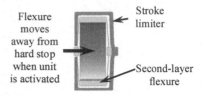

Fig. 2. Design of a variable stiffness PZT-based cell. The system consists of two strain amplification layers. The second layer flexure incorporates a stroke limiting beam.

Fig. 3 is a schematic diagram depicting the mechanism of the cellular unit with a stroke limiter. The outer thin line indicates the stroke limiter, while the vertical thick line shows the output node of the flexure that rests on the stroke limiter when the PZT is inactive (Fig 3 (a)). As the PZT force f_p increases, the thick vertical line detaches from the limiter (Fig 3 (b)). In the ON state, the equivalent compliance seen at the output nodes is $1/k$.

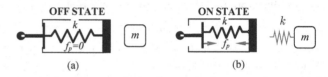

(a)   (b)

Fig. 3. Schematic model of a variable stiffness, PZT-actuated cell. (a) OFF-strain limited rigid state. (b) ON - compliant state.

The compliance characteristics associated with the cellular unit are shown in Fig. 4. The compliance of a single cell is variable in the sense that it can be either zero or a finite value $c = 1/k$. To demonstrate the approach, it is assumed that the compliance follows the ideal curve denoted by the solid line in Fig. 4. In reality, the hard stop imposed by the flexure still affords some finite compliance c_f. Furthermore, the compliance will actually increase as the cell becomes contracted due to the geometric nonlinearity of the flexure; this is shown as the dashed line in the figure. The nonlinearity is assumed to be negligible within the operation range $0 < y < y_l$ as shown in the figure.

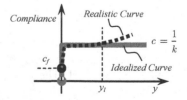

Fig. 4. Compliance characteristics of a variable stiffness PZT-actuated cell.

C. Serial, Parallel, and Antagonistic Connections

The above cellular units can be connected in series, parallel, or antagonistic configurations, creating diverse stiffness characteristics as a collective sum. The most fundamental VSCA is a serial connection of N units, as shown in Fig. 1 (b). This arrangement is referred to as a strand. For simplicity, assume that the stiffness of an OFF state unit is infinitely large, while stiffness in the ON state is a constant k. If n units are ON and the others are OFF, then the resultant stiffness of an N-unit strand reduces to

$$k_S = \frac{k}{n} \quad 1 \le n \le N. \tag{1}$$

Next, consider a parallel arrangement of strands. As shown in Fig. 5 (a), N_p strands of N serially connected units may be arranged in parallel. If the ith strand of serially connected units has stiffness $k_{S,i}$, the resultant stiffness of the entire system is given by

$$k_P = \sum_{i=1}^{N_p} k_{S,i}. \tag{2}$$

To accommodate the stiffness to a desired value, one can determine the number of parallel strands, N_p, and the number of ON units n_i in each strand of serial units.

The N_p strands can be divided into two sets of strands to form an agonist-antagonist arrangement. As shown in Fig. 5 (b), both the agonist and antagonist strands contribute to an output stiffness as a direct sum: $k_{out} = k_{P,ag} + k_{P,an}$. The antagonistic arrangement allows for varying stiffness independent of position, as in the case of natural skeletal muscles [15]. However, unlike skeletal muscles, the stiffness seen at the output of a VSCA will decrease as more cells are activated.

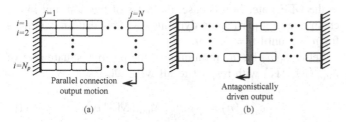

Fig. 5. (a) Parallel arrangement of cellular units. Note that one end of the actuator is shown connected to ground only for clarity, but myriad other boundary conditions are possible. (b) Antagonistic pairing of actuator strands.

Any VSCA implementation must possess two fundamental characteristics: 1) each cellular unit has discrete stiffness states and can be selectively switched ON or OFF and 2) the units are connected to form an aggregate output. Within this framework many implementations are possible. The next section presents the general dynamic properties of VSCAs and shows how these properties lead to variable resonant frequencies.

III. VARIABLE RESONANT FREQUENCIES

A. Principle

The resonant frequencies of a collection of cellular units can be varied by exploiting the cellular architecture. The key behavior is that turning on a specific number of units to accommodate the static stiffness and displacement still allows for a multitude of ON-OFF unit combinations, each of which exhibit different vibration modes with distinct resonant frequencies. Although the total number of ON state units remains the same, the resonant frequencies may be different depending upon which units are in the ON state.

To illustrate the basic concept, recall the simplified dynamic model shown in Fig. 3. Here, the distributed mass of the entire single-unit structure is lumped together and modeled as a single mass, m, while the compliance is lumped into a single compliance element with spring constant k. Consider three of these units connected serially as shown in Fig. 6. Suppose that two out of the three units are turned on so that the static stiffness of the serial connection is $k_S = k/2$. There are three different ways of selecting two units to turn on and they are shown in the upper portion of Fig. 6. Depending upon which units are on, the assembly dynamic behavior is different.

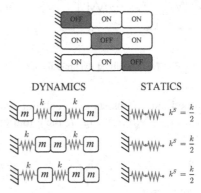

Fig. 6. Diverse ON-OFF state distributions of three serially connected units attached to ground.

Since the system in Fig. 6 has two degrees of freedom, there are two distinct vibration modes and resonant frequencies. Table I shows the resonant frequencies for each of the three ON-OFF states with $n = 2$. The first configuration, OFF-ON-ON, provides the highest resonant frequency overall, while the lowest occurs for the third configuration, ON-ON-OFF. Furthermore, for the first configuration, the grounded mass does not participate dynamically at all, whereas in the second and third cases the OFF unit participates as a translating mass in different parts of the strand.

Since all three cases have the same static stiffness k_S but different resonance properties, this example demonstrates that stiffness and resonant frequencies can be changed independently. Although the resonant frequencies do not vary continuously, multiple choices are available for different task requirements. As the number of cellular units increases, the number of possible stiffness levels and resonant frequencies increases. The next subsection describes a simple model for the general case of N serially connected units.

B. Basic Model of Serial Connection Dynamics

Consider a single serially connected strand of N cellular units. Suppose that n units are in ON state so that n springs are detached from the stroke limiters. This creates an n degree-of-freedom system with each d.o.f. corresponding to a single lumped mass that moves independently. This means

Configuration	$\frac{\omega_1}{\sqrt{k/m}}$	$\frac{\omega_2}{\sqrt{k/m}}$
1) OFF-ON-ON	0.62	1.62
2) ON-OFF-ON	0.54	1.31
3) ON-ON-OFF	0.47	1.51

TABLE I
FIRST AND SECOND NATURAL FREQUENCIES OF EACH OF THE THREE
CONFIGURATIONS

that $N - n$ OFF-state units are rigidly connected to other units as illustrated in Fig 7. Let x_i be the position of the ith combined masses and $\mathbf{x}$ be a vector collectively representing the n lumped mass displacements: $\mathbf{x} = [x_1, x_2, \ldots, x_n]^T$.

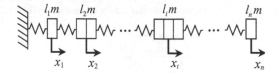

Fig. 7. Simplified dynamic model.

The ith lumped mass consists of l_i cellular units with total mass of $l_i m$ with $l_i \in \mathbb{N}$. From Fig. 7, the equations of motion are obtained as

$$m_i \ddot{x}_i = -2kx_i + kx_{i-1} + kx_{i+1}, \quad 2 \leq i \leq n-1, \quad (3)$$

and for $i = 1$ and $i = n$,

$$m_1 \ddot{x}_1 = -2kx_1 + kx_2, \quad m_n \ddot{x}_n = -kx_n + kx_{n-1}. \quad (4)$$

These equations can be arranged into vector-matrix form:

$$\mathbf{M}\ddot{\mathbf{x}} + \mathbf{K}\mathbf{x} = \mathbf{0}, \quad (5)$$

where $\mathbf{M} = m \cdot diag(l_1, l_2, \ldots, l_n) = m\mathbf{L}$ and

$$\mathbf{K} = k \begin{pmatrix} 2 & -1 & & 0 \\ -1 & 2 & \ddots & \\ & \ddots & 2 & -1 \\ 0 & & -1 & 1 \end{pmatrix}$$

$$= k \cdot \mathbf{S}. \quad (6)$$

C. Dynamic Range of Serially Connected Cells

It is now of interest to examine how widely the resonant frequencies can be varied for a given static stiffness. The maximum and minimum resonant frequencies of serially connected cellular units are obtained using the model from the previous subsection. A more complete dynamic model for a specific implementation will be obtained in IV to include the effects of distributed mass and damping.

The squared resonant frequencies of the n d.o.f. mass-spring system are given by the eigenvalues of matrix product $\mathbf{M}^{-1}\mathbf{K}$:

$$\omega_j^2 = \lambda_j(\mathbf{M}^{-1}\mathbf{K}) = \frac{k}{m}\lambda_j(\mathbf{L}^{-1}\mathbf{S}). \quad (7)$$

The n eigenvalues of the matrix are ordered from the minimum to the maximum and written as

$$\lambda_{\min}(\mathbf{M}^{-1}\mathbf{K}) \leq \lambda_2(\mathbf{M}^{-1}\mathbf{K}) \leq \cdots \leq \lambda_{\max}(\mathbf{M}^{-1}\mathbf{K}). \quad (8)$$

Note that for a fixed number of ON-state units, n, the stiffness matrix, $\mathbf{K}$, remains the same regardless of the arrangement of ON-OFF units. The mass matrix, $\mathbf{M}$, on the other hand, varies depending on the ON-OFF arrangement within the strand. Therefore, the above $\lambda_{min}(\lambda_{max})$ can be further decreased (increased) by changing the ON-OFF unit configurations and thereby changing the mass matrix.

Given a specific static stiffness k_S, which will uniquely determine n, consider the set of all mass matrices associated with the possible distributions of N cells into n clusters of masses:

$$\mathfrak{M}_n = \left\{ \mathbf{M} = m \cdot diag(l_1, \ldots, l_n) \,\Big|\, \sum_{i=1}^n l_i = N - p, \right.$$

$$\left. p = 0, \ldots, N - n; \; l_i \geq 1 \right\}. \quad (9)$$

The number of possible arrangements of n ON units within a strand is equivalent to the cardinality of $\mathfrak{M}_n$ and will be denoted by P_n where $P_n = \binom{N}{n}$.

Notice that the subset of $\mathfrak{M}_n$ defined by $p = 0$ in (9) is the set of all possible mass matrices assuming that the first unit in the strand is in the ON state. For these cases, all of the cell masses will participate dynamically. The subset of $\mathfrak{M}_n$ defined by $p = 1$ corresponds to the unit connected to ground in the OFF state. In this case, the mass of the first unit in the chain does not participate dynamically. This pattern continues for $p > 1$ until $p = N - n$.

Let $\sigma_{\min}(n)$ be the minimum of the positive square root of $\lambda_{\min}(\mathbf{M}^{-1}\mathbf{K})$ with respect to all $\mathbf{M} \in \mathfrak{M}_n$:

$$\sigma_{\min}^2(n) = \min_{\mathbf{M} \in \mathfrak{M}_n} \lambda_{\min}(\mathbf{M}^{-1}\mathbf{K}). \quad (10)$$

$\sigma_{\min}(n)$ provides the lowest resonant frequency among all the ON-OFF unit distributions. Similarly, let

$$\sigma_{\max}^2(n) = \max_{\mathbf{M} \in \mathfrak{M}_n} \lambda_{\max}(\mathbf{M}^{-1}\mathbf{K}). \quad (11)$$

$\sigma_{\max}(n)$ gives the highest possible resonant frequency.

Recall that minimum and maximum eigenvalues of $\mathbf{M}^{-1}\mathbf{K}$ are given, respectively, by the minimum and maximum of the Rayleigh quotient:

$$\lambda_{\min}(\mathbf{M}^{-1}\mathbf{K}) = \min_{\mathbf{x}} \frac{\mathbf{x}^T\mathbf{K}\mathbf{x}}{\mathbf{x}^T\mathbf{M}\mathbf{x}}, \quad (12)$$

and

$$\lambda_{\max}(\mathbf{M}^{-1}\mathbf{K}) = \max_{\mathbf{x}} \frac{\mathbf{x}^T\mathbf{K}\mathbf{x}}{\mathbf{x}^T\mathbf{M}\mathbf{x}}, \quad (13)$$

where min and max are taken assuming $\mathbf{x} \in \mathbb{R}^n$.

Using (12) and (13), one can prove the following proposition concerning the highest resonant frequency $\sigma_{\max}$ and the ON-OFF unit distribution that provides $\sigma_{\max}$.

Proposition 1:

Let $\mathbf{M} = m\mathbf{L}$ and $\mathbf{K} = k\mathbf{S}$ be, respectively, the $n \times n$ mass and stiffness matrices of N serially-connected cellular units among which n units are active (turned ON). The highest resonant frequency occurs when the first $N - n$ units are turned OFF. This maximum frequency is given by

$$\sigma_{\max} = \sqrt{\frac{k}{m}\lambda_{\max}(\mathbf{S})}. \tag{14}$$

Proof:

Since $\mathbf{M} \geq m\mathbf{I}$, $\forall \mathbf{M} \in \mathfrak{M}_n$,

$$\begin{aligned}
\lambda_{\max}(\mathbf{M}^{-1}\mathbf{K}) &= \max_{\mathbf{x}} \frac{\mathbf{x}^T\mathbf{K}\mathbf{x}}{\mathbf{x}^T\mathbf{M}\mathbf{x}} \\
&\leq \max_{\mathbf{x}} \frac{k\mathbf{x}^T\mathbf{S}\mathbf{x}}{m\,\mathbf{x}^T\mathbf{I}\mathbf{x}} \\
&= \frac{k}{m}\lambda_{\max}(\mathbf{S}).
\end{aligned} \tag{15}$$

Therefore,

$$\operatorname*{arg\,max}_{\mathbf{M} \in \mathfrak{M}_n} \lambda_{\max}(\mathbf{M}^{-1}\mathbf{K}) = m\mathbf{I}. \tag{16}$$

Hence, the arrangement where the first $N - n$ units are turned OFF yields $\sigma_{\max}$. $\square$

For the lowest possible resonant frequency, $\sigma_{\min}$, the following proposition applies.

Proposition 2:

The arrangement having all the inactive units placed at the unconstrained end of the serial connection gives the lowest un-damped resonant frequency $\sigma_{\min}$.

Proof:

The n d.o.f. undamped mass-spring system with fixed-free boundary conditions and a mass distribution $l_1 m, l_2 m, \ldots, l_n m$ has a fundamental vibration mode shape Φ_1 with monotonically increasing mass displacement amplitudes $\bar{x}_i$ from the fixed end to the unconstrained end ($i = 1, 2, \ldots, n$). Without loss of generality, the mode shape has been taken as positive. This fact can be shown using the necessary conditions on the fix-free mode shape as described in [16]. Thus, the displacement amplitudes in the first mode shape satisfy

$$0 \leq \bar{x}_1 \leq \bar{x}_2 \leq \ldots \leq \bar{x}_n. \tag{17}$$

The lowest resonant frequency $\sigma_{\min}$ is the frequency of the fundamental vibration mode that has a monotonically increasing modal vector from the fixed end to the unconstrained end, according to (17). Therefore, $\sigma_{\min}$ is given by minimizing the Rayleigh quotient for all $0 \leq x_1 \leq x_2 \leq \cdots \leq x_n$,

$$\sigma_{\min}^2(n) = \min_{\substack{\mathbf{M}\in\mathfrak{M}_n \\ \mathbf{x}\in\mathbb{R}^n}} \frac{k\mathbf{x}^T\mathbf{S}\mathbf{x}}{\mathbf{x}^T\mathbf{M}\mathbf{x}} = \min_{\substack{\mathbf{M}\in\mathfrak{M}_n \\ 0\leq x_1\leq\cdots\leq x_n}} \frac{k\mathbf{x}^T\mathbf{S}\mathbf{x}}{\mathbf{x}^T\mathbf{M}\mathbf{x}}. \tag{18}$$

Without loss of generality, one can assume $x_i \geq 0$ $i = 1, 2, \ldots, n$ in minimizing the Rayleigh quotient. Among all the mass distributions, the one having the most masses at the unconstrained end maximizes the denominator of the quotient for all x_i satisfying the monotonically increasing modal vector condition in (17). Therefore,

$$\min_{\substack{\mathbf{M}\in\mathfrak{M}_n \\ 0\leq x_1\leq\cdots\leq x_n}} \frac{k\mathbf{x}^T\mathbf{S}\mathbf{x}}{\mathbf{x}^T\mathbf{M}\mathbf{x}} = \min_{0\leq x_1\leq\cdots\leq x_n} \frac{k\mathbf{x}^T\mathbf{S}\mathbf{x}}{\mathbf{x}^T\mathbf{M}_0\mathbf{x}}, \tag{19}$$

where $\mathbf{M}_0 \triangleq m \cdot diag(1, 1, \ldots, N - n + 1)$. This represents the case where all of the inactive units are placed at the unconstrained end. $\square$

IV. DETAILED DYNAMIC MODEL FOR PZT-BASED VSCA

The analysis in the previous section provides useful insights as to the mechanism of variable resonant frequencies and their variable range. This section will consider a specific implementation of the VSCA and extend the previous analysis to account for distributed mass, damping, and compliance of the stroke limiter for the PZT cell design. The model of this section can be used for any set of meaningful boundary conditions and can be applied to arbitrary parallel or antagonistic connections.

A. Single Cell Dynamic Model

A detailed lumped parameter model for the PZT-based design is shown in Fig. 8. This model includes the mass m of the hard stop beam as well as its stiffness K. The lower case symbol for the hard stop mass indicates that $m < M$. Likewise, the symbol K for the stroke limiting beam stiffness indicates that K is much larger than a single flexure's stiffness k. The stiffness k does vary slightly with displacement, but this variation is assumed to be small for the purpose of model development. The viscous damping elements are taken as constant with a value b. The damping elements represent losses in the flexures and are interposed between the output nodes of the cell. The damping of the flexures as well as any damping in the load will shift the resonant frequencies away from their undamped counterparts. However, this shift is small insofar as b is relatively small. The PZT-generated force at the output of the second layer is again denoted by f_p. Each cell in an actuator is subject to two tensile forces T_m and T_M. The tension T_M is the tension acting at the output node. Note that it will be useful to express the output node tension as the sum of a preload component $T_{M,PL}$ and a time varying component $\tilde{T}_M$: $T_M = T_{M,PL} + \tilde{T}_M$.

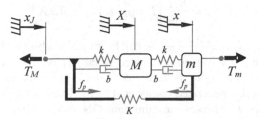

Fig. 8. Single cell model that includes mass distribution, parasitic damping b, and parasitic stiffness K of the stroke limiting beam.

Each cell requires three generalized coordinates to describe its configuration with respect to an inertial reference frame. The first coordinate, x, determines the position of the hard stop mass. The second coordinate, X, determines the position of the second layer unit suspended mass. The third coordinate, x_J, determines the location of the output node. Note that if a second unit is connected to the output node, then x_J is equal to the inertial displacement of the second unit's hard stop mass.

B. Piezoelectric Induced Force

The generation of the self-equilibrating force f_p is achieved through PZT stack expansion and application of the custom-designed nested flexures. For simplicity of analysis, the assembly of the piezoelectric stack and flexures is treated as a two port element, having one port in the electrical domain and the other port in the mechanical domain at the point of application in the lumped model of Fig. 8. The constitutive law for the generation of the force may be written in a structural stiffness form as

$$\left\{ \begin{array}{c} V_{pzt} \\ -f_p \end{array} \right\} = \left(\begin{array}{cc} \kappa_{11} & -\kappa_{12} \\ -\kappa_{12} & \kappa_{22} \end{array} \right) \left\{ \begin{array}{c} Q_{pzt} \\ \Delta X \end{array} \right\}, \qquad (20)$$

where V_{pzt} is the voltage applied to the piezoelectric stacks, Q_{pzt} is the charge accumulated on the stacks (neglecting hysteresis between applied voltage and charge), $\Delta X = x - x_J$, and the elements of the matrix κ_{11}, κ_{12}, and κ_{22} represent generalized stiffness constants. The minus sign preceding f_p accounts for the model sign convention. Each of the stiffness constants has a physical meaning that is important in the design of the actuator cell. However, for brevity, the κ_{ij} constants will not be discussed explicitly. Solving for f_p from (20) yields

$$f_p = \left(\frac{\kappa_{12}}{\kappa_{11}} \right) V_{pzt} - \left(\frac{\kappa_{11}\kappa_{22} - \kappa_{12}^2}{\kappa_{11}} \right) \Delta X. \qquad (21)$$

C. Cell Recruitment

The condition for turning a cell ON may be stated in terms of the applied voltage as

$$V_{min} < V_{pzt} < V_{max}, \qquad (22)$$

where V_{max} is the maximum voltage as specified by the material manufacturer and V_{min} is the minimum voltage required to move the detachable flexure beams away from the hard stop (Fig 2). Based on the model in IV-B and a force analysis of the output node, the value of V_{min} is found to be

$$V_{min} = \frac{\kappa_{11}}{\kappa_{12}} \left[T_{M,PL} + \tilde{T}_M + \beta \Delta X \right], \qquad (23)$$

where $\beta = (\frac{\kappa_{11}\kappa_{22} - \kappa_{12}^2}{\kappa_{11}} - k)$. From (23), observe that the minimum ON voltage is reduced when κ_{11} is reduced, κ_{12} is increased, or $T_{M,PL}$ is decreased. Eq. (23) also shows that the dynamically generated component of the force acting between actuators $\tilde{T}_M$ makes the minimum ON voltage time varying. Therefore, it is crucial that operating conditions are selected for the actuator so that inadvertent switching from ON to OFF

or OFF to ON is avoided because this would result in structural chattering and unwanted internal vibrations.

D. Assembled System Equations of Motion

The equations of motion for an assembled system may now be obtained by isolating three cells within a rectangular array of dimensions $N_p \times N$ as previously discussed in II-C. The equations of motion are formulated for the ij^{th} cell, with i as the strand index and j as the index of a cell within a strand.

Three separate cases need to be considered because the forces acting on the mass m of the ij^{th} cell and the $ij + 1^{th}$ cell will be different depending upon the states of these two adjacent cells. For the ON-ON case, the equations of motion are

$$\left\{ \begin{array}{l} f_{p,ij-1} - f_{p,ij} + k(X_{ij-1} - x_{ij}) + b(\dot{X}_{ij-1} - \dot{x}_{ij}) \\ \quad - k(x_{ij} - X_{ij}) - b(\dot{x}_{ij} - \dot{X}_{ij}) = m\ddot{x}_{ij} \\ k(x_{ij} - X_{ij}) + b(\dot{x}_{ij} - \dot{X}_{ij}) \\ \quad - k(X_{ij} - x_{ij+1}) - b(\dot{X}_{ij} - \dot{x}_{ij+1}) = M\ddot{X}_{ij}. \end{array} \right. \qquad (24)$$

For the OFF-ON case, the equations become

$$\left\{ \begin{array}{l} f_{p,ij-1} - f_{p,ij} + k(X_{ij-1} - x_{ij}) + b(\dot{X}_{ij-1} - \dot{x}_{ij}) \\ \quad + K(x_{ij-1} - x_{ij}) - k(x_{ij} - X_{ij}) \\ \quad - b(\dot{x}_{ij} - \dot{X}_{ij}) = m\ddot{x}_{ij} \\ k(x_{ij} - X_{ij}) + b(\dot{x}_{ij} - \dot{X}_{ij}) \\ \quad - k(X_{ij} - x_{ij+1}) - b(\dot{X}_{ij} - \dot{x}_{ij+1}) = M\ddot{X}_{ij}. \end{array} \right. \qquad (25)$$

And for the OFF-OFF case, the equations are

$$\left\{ \begin{array}{l} f_{p,ij-1} - f_{p,ij} + k(X_{ij-1} - x_{ij}) + b(\dot{X}_{ij-1} - \dot{x}_{ij}) \\ \quad + K(x_{ij-1} - x_{ij}) - k(x_{ij} - X_{ij}) \\ \quad - b(\dot{x}_{ij} - \dot{X}_{ij}) - K(x_{ij} - x_{ij+1}) = m\ddot{x}_{ij} \\ k(x_{ij} - X_{ij}) + b(\dot{x}_{ij} - \dot{X}_{ij}) \\ \quad - k(X_{ij} - x_{ij+1}) - b(\dot{X}_{ij} - \dot{x}_{ij+1}) = M\ddot{X}_{ij}. \end{array} \right. \qquad (26)$$

For any set of linear boundary conditions, the system equations of motion may be written in the standard vibratory form as

$$\mathbf{M}\ddot{\mathbf{q}} + \mathbf{B}\dot{\mathbf{q}} + \mathbf{K}(t)\mathbf{q} = \mathbf{Q}(t), \qquad (27)$$

where $\mathbf{q}$ is the generalized coordinate vector. For simplicity, the second term in (21) is assumed to be small and therefore

$$f_{p,ij} \approx \left(\frac{\kappa_{12}}{\kappa_{11}} \right) V_{pzt,ij}. \qquad (28)$$

Thus, the vector $\mathbf{Q}$ is the vector containing each cell's voltage inputs scaled by the ratio of the stiffness constants κ_{12} and κ_{11}. Both the input $\mathbf{Q}(t)$ and the stiffness matrix $\mathbf{K}(t)$ depend on the ON-OFF arrangement within each strand, thereby making the dynamics equations a switched linear system.

Rearrangement of (24) to (26) leads to a row-wise construction of the mass, stiffness, and damping matrices in (27). Notice that the equations of motion for the masses M

do not change even when a cell is OFF. Furthermore, the equations corresponding to the large masses M are unforced. Hence, a convenient block diagonal form of the stiffness and damping matrices may be constructed. The model structure can be readily extended to include any load behaving as a linear mechanical impedance and this extension is considered explicitly in the next section.

V. EXPERIMENTAL RESULTS

A. Equipment and Procedure

The cellular PZT design described in II-B was used to construct a three unit VSCA. The identical units were connected in series to give a net length of 41 mm. The actuator was then connected to a plastic cantilever, which provided the tension necessary for operation and acted as a linear spring-mass-damper load.

The complete experimental apparatus is shown in Fig. 9. As shown in the figure, vertical deflections of the cantilever, denoted by x_L, were measured using a MicroEpsilon laser sensor with a resolution of 1 μm. Voltages were applied to the piezoelectric stacks using a Cedrat CA-45 amplifier.

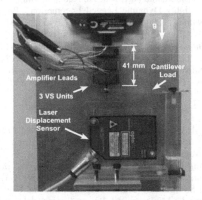

Fig. 9. Experimental apparatus for measuring dynamic properties of a 3-cell system.

A chirp voltage input was used to obtain the frequency response characteristics of the actuator. The chirp signal ranged from 5 Hz to 150 Hz and contained a DC offset of 100 V to assure that the conditions in (22) were satisfied. The maximum voltage used was 150 V. The experimental results for system frequency response were obtained using the ratio of windowed power spectra from the input voltage and output displacement. All measured signals were sampled at 1 kHz with a National Instruments data acquisition board.

Two representative test cases were selected to illustrate the tunability of resonant frequencies. For the first case, the top unit (the uppermost unit in Fig. 9) was turned ON and the lower two units were kept OFF. For the second case, the middle unit was turned ON while the top and lower unit were kept OFF. Units were maintained in the OFF state by disconnecting their leads from the amplifier. The fundamental resonant frequency is compared for both cases in the next subsection.

B. Experimental and Theoretical Model Comparison

Before considering the experimental results, the model from IV-D was used to generate a theoretical response for comparison with the data. The cantilever beam shown in Fig. 9 was modeled as a mass, m_L, connected to ground through a parallel arrangement of a spring and dashpot having values of k_L and b_L, respectively. The resulting model structure for the entire system is given by the following matrix equation:

$$\mathbf{M}\ddot{\mathbf{q}} + b \begin{pmatrix} 2\mathbf{I}_{3\times3} & \mathbf{A} \\ \mathbf{A}^T & 2\mathbf{I}_{3\times3} \end{pmatrix} \dot{\mathbf{q}} + k \begin{pmatrix} 2\mathbf{I}_{3\times3} & \mathbf{A} \\ \mathbf{A}^T & \mathbf{S}_{22} \end{pmatrix} \mathbf{q}$$
$$= \mathbf{Q}' \left(\frac{\kappa_{12}}{\kappa_{11}} V_{pzt}(t) \right), \quad (29)$$

where $\mathbf{q}$ is the vector of displacements of the large masses, small masses, and the cantilever tip: $[X_1, X_2, X_3, x_2, x_3, x_L]^T$; $\mathbf{M} = diag(M, M, M, m, m, m_L)$; $\mathbf{A}$ is a lower diagonal Toeplitz matrix of -1's; $\mathbf{S}_{22}$ is a matrix of stiffness ratios involving K/k and k_L/k; and $\mathbf{Q}'$ is the vector determining the force input distribution. Because only one unit was ON at a time, the input $V_{pzt}(t)$ is a scalar. The matrix block $\mathbf{S}_{22}$ and the vector $\mathbf{Q}'$ are the only quantities that change with the ON-OFF configuration. For frequency response prediction, the model structure in (29) is readily converted to a transfer function from V_{pzt} to x_L using the state $[\mathbf{q}\ \dot{\mathbf{q}}]^T$ and the output equation $x_L = [\mathbf{0}_{1\times5}, 1, \mathbf{0}_{1\times6}]\mathbf{q}$.

The parameters of the experimental system used in the theoretical model are given in Table II. The stiffness parameters were measured directly using a Transducer Techniques load cell and the laser displacement sensor described previously. The damping b_L and mass m_L were identified from an initial condition response of the cantilever, and the ratio $\frac{\kappa_{12}}{\kappa_{11}}$ was obtained from the DC response of the assembled system. As is standard in vibration models the damping is used as a free parameter. Therefore, the damping constant b was tuned to match the experimental data as closely as possible.

Parameter	Value	Units
k	4.1	$\frac{N}{mm}$
K	42	$\frac{N}{mm}$
k_L	2.9	$\frac{N}{mm}$
b	0.58	$\frac{N \cdot s}{mm}$
b_L	2.8	$\frac{N \cdot s}{mm}$
m	5.9	g
M	35	g
m_L	1.8	g
$\frac{\kappa_{12}}{\kappa_{11}}$	8.5×10^{-3}	$\frac{C}{m}$

TABLE II
THEORETICAL MODEL PARAMETERS

The theoretical model and experimental model are compared in Fig. 10 for both the ON-OFF-OFF case and the OFF-ON-OFF case. The experimentally obtained gain values and dominant resonant peak locations are well predicted by the linear theoretical model. For the ON-OFF-OFF case, the first resonant peak predicted by the model occurred at $f_{n1} = 33$ Hz, while the measured peak occurs at approximately 32 Hz, resulting in a difference of 3%. For the OFF-ON-OFF case, notice that the resonant peak shifted upwards by approximately 8 Hz since both the theoretical and the experimental results show a fundamental frequency of approximately 41 Hz. In the magnitude curve for the ON-OFF-OFF case, the minor resonant peaks at 26 Hz and 38 Hz are likely unmodeled vibration modes in the cantilever. These resonant peaks are well damped and not viewed as a significant source of error in the model order.

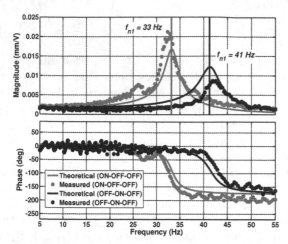

Fig. 10. Comparison of theoretical model and experimental measurement for the ON-OFF-OFF and OFF-ON-OFF configuration.

One important feature of these results is the displacement amplification effect at resonance. The amplification of displacement is an important feature particularly for piezoelectric actuators because their inherent strain is extremely small (0.1 %). The amplification flexures (Fig. 1) amplify the piezoelectric stack strain to approximately 10 % and tuning the resonant frequency to a desired value allows the actuator to obtain an additional 5 to 10 times larger output displacement. For the first experimental case, the measured ratio of the resonance magnitude to the DC magnitude was 11.1. For the second case, this ratio was 5.7.

The most important feature of these results is the change in fundamental frequency. Specifically, the increase in fundamental frequency from the first case (ON-OFF-OFF) to the second case (OFF-ON-OFF) illustrates that the resonance properties of the system are indeed tunable and that the tuning may be achieved by simply changing the configuration of the ON-OFF units within the strand.

VI. CONCLUSION

The two primary objectives of this work were to quantify the basic properties of VSCAs and describe a novel PZT-based design that achieves variable resonant frequencies through variable stiffness. The change in stiffness and resonant frequencies is achieved by selectively turning variable stiffness units ON or OFF within a serial strand. The strands may also exist within a greater parallel structure, which is one subject of future research. Experiments and models both illustrate the shift in resonant frequency based on a simple change in ON-OFF configuration.

The PZT-based actuators considered in this paper show particular promise for VSCAs because they achieve static strain that is commensurate with skeletal muscle (10-20%) and their strain is further amplified under resonance conditions. Furthermore, VSCA designs based on PZT actuator cells are durable, low power, high bandwidth devices. One particular advantage of PZT is its near-zero power consumption when holding a load in DC. We are presently exploring application areas for our VSCA technology include deep-sea robotics, aircraft manufacturing, and nuclear reactor inspection robots. Further extensions of the technology include using cells with differing stiffnesses to achieve more uniform modal density across the tuning range and using the cellular system as both an actuator and an energy harvesting device for battery recharge in mobile robots.

REFERENCES

[1] J. Choi, S. Park, W. Lee, and S. Kang, "Design of a Robot Joint with Variable Stiffness," *Proc. of IEEE Int. Conf. on Robotics and Automation*, pp. 1760-1765, 2008.
[2] J. Hurst, J. Chestnutt, and A. Rizzi, "An Actuator with Physically Variable Stiffness for Highly Dynamic Legged Locomotion," *Proc. of IEEE Int. Conf. on Robotics and Automation*, pp. 4662-4667, 2004.
[3] R. Blickhan, "The Spring-Mass Model for Running and Hopping," *Journal of Biomechanics*, Vol. 22, No. 11/12, pp. 1217-1227, 1989.
[4] R. Schiavi, G. Grioli, S. Sen, and A. Bicchi, "VSA-II: A Novel Prototype of Variable Stiffness Actuator for Safe and Performing Robots Interacting with Humans," *Proc. of IEEE Int. Conf. on Robotics and Automation*, pp. 1760-1765, 2008.
[5] M. Raibert, "Legged Robots That Balance," Cambridge, MA: MIT Press, 1986.
[6] G. Pratt and M. Williamson, "Series Elastic Actuators," *Proc. of IEEE/RSJ Int. Conf. on Intelligent Robots and Systems*, Vol. 1, pp. 399-406, 1995.
[7] J. Yan, R. Wood, S. Avandhanula, M. Sitti, and R. Fearing, "Towards Flapping Wing Control for a Micromechanical Flying Insect," *Proc. of IEEE Int. Conf. on Robotics and Automation*, pp. 3901-3908, 2001.
[8] K. Issac and S. Agrawal, "An Investigation Into the Use of Springs and Wing Motions to Minimize the Power Expended by a Pigeon- Sized Mechanical Bird for Steady Flight," *Journal of Mech. Design*, Vol. 129, pp. 381-389, 2007.
[9] N. Lobontiu, M. Goldfarb, and E. Garcia, "Maximizing the Resonant Displacement of Piezoelectric Beams," *Proc. of SPIE*, Vol. 3668, No. 154, 1999.
[10] R. Kornbluh, R. Pelrine, J. Eckerle, and J. Joseph, "Electrostrictive Polymer Artificial Muscle Actuators," *Proc. of IEEE Int. Conf. on Robotics and Automation*, Vol. 3, pp. 2147-2154, 1998.
[11] L. Goodman, M. Riley, and S. Mitra, and M. Turvey, "Advantages of Rhythmic Movements at Resonance: Minimal Active Degrees of Freedom, Minimal Noise, and Maximal Predictability," *Journal of Motor Behavior*, Vol. 32, No. 1, pp. 3-8, 2000.
[12] V. Challa, et.al, "A Vibration Energy Harvesting Device with Bidirectional Resonance Frequency Tunability," *Smart Materials and Structures*, Vol. 17, No. 015035, 2008.
[13] K. Cho, J. Rosmarin, and H. Asada, "Design of Vast DOF Artificial Muscle Actuators with a Cellular Array Structure and its Application to a Five-fingered Robotic Hand," *Proc. of IEEE Int. Conf. on Robotics and Automation*, pp. 2214-2219, 2006.
[14] J. Ueda, T. Secord, and H. Asada, "Design of PZT Cellular Actuators with Power-law Strain Amplification," *Proc. of IEEE/RSJ Int. Conf. on Robots and Systems*, pp. 1160-1165, 2007.
[15] N. Hogan, "Adaptive Control of Mechanical Impedance by Coactivation of Antagonist Muscles," *IEEE Trans. on Auto Cont.*, Vol. 29, pp. 681-690, 1984.
[16] G.M.L Gladwell, *Inverse Problems in Vibration*, Martinus Nijhoff, Dordrecht, 1986.

Positioning Unmanned Aerial Vehicles as Communication Relays for Surveillance Tasks

Oleg Burdakov⋆, Patrick Doherty†, Kaj Holmberg⋆, Jonas Kvarnström†, Per-Magnus Olsson†,‡
⋆ Dept. of Mathematics. E-mail: {olbur, kahol}@mai.liu.se
† Dept. of Computer and Information Science. E-mail: {pdy, jonkv, perol}@ida.liu.se
Linköping University, SE-581 83 Linköping, Sweden
‡ Corresponding author

Abstract— When unmanned aerial vehicles (UAVs) are used to survey distant targets, it is important to transmit sensor information back to a base station. As this communication often requires high uninterrupted bandwidth, the surveying UAV often needs a free line-of-sight to the base station, which can be problematic in urban or mountainous areas. Communication ranges may also be limited, especially for smaller UAVs. Though both problems can be solved through the use of relay chains consisting of one or more intermediate relay UAVs, this leads to a new problem: Where should relays be placed for optimum performance? We present two new algorithms capable of generating such relay chains, one being a dual ascent algorithm and the other a modification of the Bellman-Ford algorithm. As the priorities between the number of hops in the relay chain and the cost of the chain may vary, we calculate chains of different lengths and costs and let the ground operator choose between them. Several different formulations for edge costs are presented. In our test cases, both algorithms are substantially faster than an optimized version of the original Bellman-Ford algorithm, which is used for comparison.

I. INTRODUCTION

A wide variety of applications for unmanned aerial vehicles (UAVs) include the need for surveillance of distant targets, including search and rescue operations, traffic surveillance and forest fire monitoring as well as military applications. In most cases the information gathered must be transmitted continuously from a *survey UAV* to a base station where the current operation is being coordinated. As this information may include urgent high-volume sensor data such as live video, one often requires considerable uninterrupted communications bandwidth. To minimize quality degradation, UAV applications therefore tend to require line-of-sight (LOS) communications, which is problematic in urban or mountainous areas. The maximum communication range is typically also limited, especially when smaller and lower-cost UAVs are used as relays.

Though the problem of achieving line-of-sight can be ameliorated by increasing altitude, this would require greater communication ranges and is not always permitted by aviation regulations. UAVs may also be unable to fly at sufficient altitude to avoid detection in the case of military or police surveillance, preferring instead to hide behind obstacles. Instead, both intervening obstacles and limited range can be handled using a chain of one or more intermediate *relay UAVs* [15] passing on information to the base station (Figure 1).

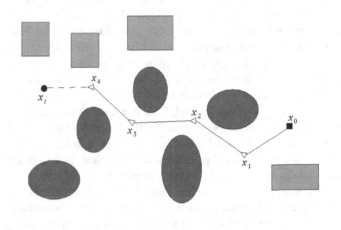

Fig. 1. UAVs at x_1, x_2 and x_3 are acting as relays, connecting the base station at x_0 with the survey UAV at x_4, which is surveying the target at x_t. This is a relay chain of length 4.

Rather than limiting ourselves to the use of a single relay UAV, we are interested in the general relay positioning problem for an arbitrary number of relays. A solution to this problem should be sufficiently scalable to enable the use of a large number of comparatively inexpensive miniature UAVs with highly limited communication range. While the number of relays used is an obvious quality measure of a relay chain, one can also benefit greatly from taking other scenario-dependent measures into consideration, even at the cost of using additional UAVs. For example, position flexibility and safety margins may be important in case of wind drift or moving targets, and in some cases one may want to minimize the risk of detection by others. As the desired trade-off may vary from case to case and can be difficult to formalize, we prefer to generate multiple relay chains using different numbers of UAVs and leave the final choice of solution to a ground operator.

In this paper we formally define the relay positioning problem in a continuous setting. As this problem is intractable, we continue by generating a corresponding discrete graph, where nodes correspond to physical locations where relays may be placed, edges connect nodes between which communication is possible, and edge costs are used to model quality measures related to UAV placement and communication paths. The desired set of relay chains can then be generated by solving the *all hops optimal path* (AHOP) graph search problem [7]. We present two distinct algorithms for solving the AHOP problem considerably more efficiently than previous algorithms, the first being a dual ascent algorithm and the second being a modification of the Bellman-Ford [6] algorithm. Though examples in this article focus on the use of UAVs, both algorithms can equally well be used for relay placement for unmanned ground vehicles (UGVs).

II. PREVIOUS WORK

Control behavior for teams of unmanned ground vehicles involving line-of-sight was investigated in Sweeney et al. [17]. In an indoor setting, a lead UGV advances from the base station towards the goal position and incrementally determines where to place relay UGVs along the way in order to maintain communication with the base station. Various strategies are evaluated in terms of time and energy usage. A small survey of positioning algorithms for UGVs is available in Nguyen et al. [13]. The algorithms presented in these articles have several commonalities. No quality or cost measure is used and it is not certain that the goal position will be reached, as no a priori calculation or evaluation of paths is performed.

Arkin and Diaz [2] used a behavior-based architecture to allow teams of ground robots with line-of-sight communication to explore buildings and to find stationary objects, using only limited knowledge about the area in which those objects may be placed. This problem differs significantly from ours, where the goal position is known in advance.

An algorithm for maintaining LOS between groups of planetary rovers exploring an area is presented by Anderson et al. [1]. The algorithm is based on several heuristics and although testing indicates that the algorithm performs well, it does not guarantee a solution even if one exists. Similarly, no quality measure exists.

The concept of using a UAV as a communication relay, including intended platforms and communications equipment, is discussed in Pinkney et al. [15], but no algorithms are presented. In limited cases, where a single relay UAV is sufficient, the survey UAV could plan and fly a trajectory to the goal while the relay UAV continuously attempts to maintain line-of-sight to both the survey UAV and the base station [16]. The benefit of using a single relay UAV in an urban environment has also been simulated [5]. Here the UAV works as a relay between two entities on the ground. The focus of the work is to determine the percentage of an urban area with acceptable coverage for UAVs positioned at different heights. Only a single relay is used and no algorithm for determining the positioning of the UAV is provided.

Problems that are superficially very similar to the multiple relay positioning problem are encountered in ad-hoc networks, where messages are to be delivered in a network where there is no control of the network topology. Routing algorithms for such networks must be able to handle addition and removal of

nodes at runtime [10, 12]. Using a swarm of UAVs to improve the range and reliability of an ad-hoc network has also been investigated [14]. Good results are achieved, mainly through significantly increasing the range using the same transmission power, compared to using a direct ground link. However, significant differences exist between the problem investigated here and ad-hoc networks in that we have full control over node (UAV) positioning and in that we are only interested in transferring information between the survey UAV and the base station.

III. PROBLEM FORMULATION

We formally define the *relay positioning problem* as follows.

Let $L \geq 1$ UAVs with identical communication capabilities be available, one of which is the survey UAV.

Assume as given the free space $X \subseteq R^3$ where relay and survey UAVs may safely be placed, together with the position $x_0 \in R^3 \setminus X$ of a base station and the position $x_t \in R^3 \setminus X$ of a survey target.

Assume also as given a boolean *communication reachability function* $g(x, x')$ that holds iff communication is possible between points $x, x' \in R^3$ and a *communication cost function* $f(x, x')$ denoting the non-negative cost of establishing such a communication link. Finally, assume as given a *survey reachability function* $s(x, x')$ that holds iff a survey UAV at $x \in X$ is able to survey a target at $x' \in R^3 \setminus X$, and a *survey cost function* $t(x, x')$ denoting the non-negative cost of such a survey.

A *relay chain* of length l between x_0 and x_t is a tuple of free UAV positions $\langle x_1, \ldots, x_l \rangle$, where $\{x_1, \ldots, x_l\} \subseteq X$, such that $\forall i, \ 0 \leq i < l \rightarrow g(x_i, x_{i+1})$ and $s(x_l, x_t)$. The *cost* of a relay chain $\langle x_1, \ldots, x_l \rangle$, denoted by $cost(\langle x_1, \ldots, x_l \rangle)$, is defined as $(\sum_{i=0}^{l-1} f(x_i, x_{i+1})) + t(x_l, x_t)$.

A relay chain c of length l between x_0 and x_t is *relevant* iff there exists no relay chain c' of length $l' < l$ between x_0 and x_t such that $cost(c') \leq cost(c)$. This reflects the fact that one would not use c if there is an alternative chain that uses fewer relays and is not more expensive.

The relay positioning problem consists of finding, for each $1 \leq l \leq L$, a relevant minimum-cost relay chain of length l between x_0 and x_t, or determining that no such chain exists. Note that any problem instance may admit multiple (optimal) solutions, since there may be multiple minimum-cost relay chains of any given length.

A. Reachability Functions

The communication reachability function g determines whether communication should be considered feasible between two points, regardless of the quality of the resulting link. For our purposes, g is usually defined by a limited radius r and a requirement of free line-of-sight between positions. More formally, let $O \subseteq R^3$ be the set of coordinates that are free from obstacles. Note that this may differ from the positions in X, where it should also be possible to place UAVs safely and minimum distances to obstacles may have to be taken into account. Then, $g(x_i, x_j)$ holds iff $\|x_j - x_i\| \leq r$ and

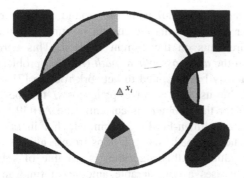

Fig. 2. Example of a node with UAV communication radius r. The obstructed volume is the sum of the black obstacles and the shaded area representing the non-visible volume inside the sphere.

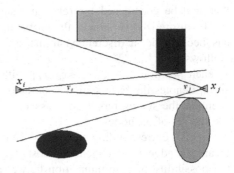

Fig. 3. Minimum free angle used as edge cost.

$$[x_i, x_j] = \{x \in R^3 : \alpha x_i + (1 - \alpha)x_j, \alpha \in [0,1]\} \subseteq O.$$
However, the algorithms to be defined below work equally well with other definitions of g, with or without free LOS.

The survey reachability function s is usually defined in a similar manner, though often with a different maximum distance between the survey UAV and its target.

Note that reachability functions may be asymmetric. For example, a UAV may only be able to survey targets below it.

B. Cost Functions

In the definition of communication and survey cost functions, the term "cost" is used in a very general sense, where both functions can be used to model arbitrary position-related quality measures determining the overall quality of a relay chain. For example, a high survey cost $t(x, x')$ may be used to indicate that surveying a target at x' from a UAV positioned at x would yield information of comparatively low quality. Similarly, given a suitable model of the environment and the communications hardware being used, distance- or position-dependent transmission power requirements as well as many forms of communication performance degradation can be modeled as communication costs and taken into account by positioning algorithms. Finally, communication costs can be used to model arbitrary penalties for specific relay positions.

We will now provide two specific examples of communication cost functions $f(x, x')$ related to the existence of obstacles that may hinder line-of-sight communications. Note that one can easily use a weighted sum of such measures together with costs related to continuous communication degradation such as path loss. Survey cost functions can be defined analogously.

Obstructed volume. Given a maximum communication range r, a UAV positioned at $x \in X$ can communicate with other UAVs in a sphere of volume $4\pi r^3/3$. Given line of sight requirements, parts of this volume may be obstructed by obstacles such as buildings or hills. Though we normally operate in three dimensions, a 2D example is shown in Figure 2 for simplicity. The greater the obstructed volume is, the less flexible the given position is in terms of future UAV movements. Thus, one can directly define $f(x, x')$ as the obstructed volume within a sphere of radius r centered

at x'.

Minimum free angle. While the obstructed volume measure is useful if one anticipates attempting to maintain communication during future movements to entirely new UAV positions, other measures are more relevant for maintaining flexibility around current relay positions. This can be important for several reasons, such as the need to minimize susceptibility to disturbances and be robust against temporary wind drift. One such measure of flexibility is the minimum free angle between two successive positions x_i and x_j in a relay chain. Figure 3 shows a two-dimensional example, where the UAV at x_i must not move outside the angle originating in x_j and vice versa. The cost associated with this edge might then be inversely correlated to the minimum of these angles. One might also use the inverse of the square root or logarithm of the angle, which increases the relative penalty of including small-angle edges in a path.

For both of these cost functions, terrains with higher and more variable obstacle densities will be more likely to give rise to a spectrum of alternative relevant relay chains. In this situation, the shortest chains (in terms of the number of edges) are close to obstacles and consequently quite expensive, while the cheapest chains take long detours in order to maintain a greater distance from the obstacles. Conversely, terrains with lower and less diverse obstacle densities will be more likely to give rise to only a few relevant relay chains, as longer chains are also likely to be more expensive.

IV. SOLVING THE RELAY POSITIONING PROBLEM

The relay positioning problem is a multi-extremal problem where the feasible set is typically disjoint, and in many cases the number of disjoint feasible subsets is large, which makes the continuous problem intractable. We therefore suggest to discretize the environment, in order to decrease the computational burden. The solution is thus divided into two distinct steps: First, generate a discrete visibility graph, and second, use this graph to solve the all hops optimal path problem.

A. Discretization and Visibility Graph Generation

The first step of discretizing an instance of the continuous relay positioning problem consists of selecting a finite set of positions $X' \subseteq X$ among the free coordinates. These are the positions that will be considered for relay and survey UAV placement in the discretized version of the problem. Once this selection has been made, a discrete visibility graph can be created as follows.

Associate each position $x \in X' \cup \{x_0, x_t\}$ with a unique node, where n_0 denotes the base station node associated with x_0 and n_t denotes the survey target node associated with x_t, and let N be the set of all nodes.

For each $x \in X'$ corresponding to $n \in N$ and satisfying $g(x_0, x)$, create an edge $e = (n_0, n)$ of cost $f(x_0, x)$ representing the possibility of communication between the base station and position x. For each $x, x' \in X'$ corresponding to $n, n' \in N$ and satisfying $g(x, x')$, create an edge $e = (n, n')$ of cost $f(x, x')$ representing the possibility of communication between positions x and x'. Finally, for each $x \in X'$ corresponding to $n \in N$ and satisfying $s(x, x_t)$, create an edge $e = (n, n_t)$ of cost $t(x, x_t)$ representing the fact that a survey UAV at x would be able to survey the target at x_t. Let E be the set of all edges.

Then, $G(N, E)$ is a directed graph corresponding to the original continuous problem instance. Note in particular that the survey target node has no outgoing edges and that all its incoming edges satisfy the survey reachability function, ensuring that its predecessor in any path from the base station to the survey target must be suitable for a survey UAV. Note also that most parts of this graph only depend on the environment and not on the position of the base station or survey target, and can be precalculated.

A number of different alternatives are possible when determining which free positions to include in the graph. One approach consists of placing a regular three-dimensional grid over the terrain and constructing a graph from the unobstructed grid cells, where the size of the grid cells would depend on the maximum obstacle density. Regular grids are quite suitable for the type of urban and mountainous terrain we are interested in, where we are unlikely to find pathological obstacle distributions that only permit communication if relays are positioned exactly in the right place. Alternative approaches may distribute and place nodes depending on local obstacle density, for example by using structures with variable cell size or by using techniques such as bridge sampling [9] to increase the density of nodes in narrow passages.

B. The All Hops Optimal Path Problem

Given a discrete graph representation, a number of associated problems can be solved using well-known algorithms. For example, Dijkstra's algorithm or A* can be used to find the cheapest relay chain (path), the relay chain with the smallest number of hops, or the cheapest relay chain from those having the smallest number of hops[1] [6, 8].

However, these algorithms only generate a single relay chain, whereas we are interested in generating a spectrum of relevant relay chains ranging from the shortest in terms of

---

[1] The latter can be calculated using compound path costs of the form $\langle l, c \rangle$, where l is the length of the path, c is its cost, and $\langle l_1, c_1 \rangle < \langle l_2, c_2 \rangle$ iff $(l_1 < l_2)$ or $(l_1 = l_2$ and $c_1 < c_2)$.

hops to the cheapest, allowing the ground operator to choose whichever chain best fits the current resource constraints and the requirements on the current mission. This corresponds directly to the *all hops optimal path* (AHOP) problem, which has previously been applied to network routing [7].

Given the use of additive path costs, the best known algorithm for this problem is currently the AHOP version of the standard Bellman-Ford algorithm, which at its kth iteration finds all cheapest paths using k hops from a given source [7]. However, although this algorithm is capable of solving the problem, it uses a considerable amount of time and is not practically useful for larger problem instances involving large maps, fine-grained grids, or long communication ranges. This is especially true in mixed-initiative settings where a ground operator initiating a survey mission should be given a prompt response.

We have therefore developed two distinct algorithms for solving the all hops optimal path problem more efficiently than the standard Bellman-Ford algorithm. The following terminology and variables are common to both algorithms: n_0 is the start node corresponding to the base station, n_t is the target node, and given a node n, n_- is its set of predecessor nodes and n_+ is its set of successor nodes. The cost associated with an edge between nodes n and n' is denoted by $c_{n,n'}$.

We use the term *cheapest path tree* for a tree rooted in n_0, storing the least-cost path to each node in N given compound path costs of the form $\langle c, l \rangle$, where c is the cost of the path, l is its length, and $\langle c_1, l_1 \rangle < \langle c_2, l_2 \rangle$ iff $(c_1 < c_2)$ or $(c_1 = c_2$ and $l_1 < l_2)$. This is in a sense the inverse of the previously discussed compound path cost, in that it prefers the shortest path among those that have the lowest cost. The cheapest path tree can be calculated by any shortest path tree algorithm, for example, Dijkstra's algorithm.

C. Dual Ascent Algorithm

Our first algorithm (Figure 4) is a dual ascent algorithm that repeatedly calculates the cheapest path p from n_0 to n_t using modified edge costs $c'_{n,n'} = c_{n,n'} + \alpha$, where α is increased in every iteration. This is also assumed to yield, for each node n, its depth q_n (the number of hops along the path from the root to n) and its optimal path cost y_n given the current value of α.

Let l be initialized to the maximum number of acceptable hops, that is, one more than the number of available UAVs, where the last hop is between the survey UAV and the target. Give α the initial value α_0. In most cases we use $\alpha_0 = 0$, causing the algorithm to begin by generating the cheapest path possible using original edge costs. Higher values of α_0 can be used to discourage longer paths, with the effect that the algorithm no longer generates the longest relevant relay chains.

If the length of the generated path is at most l, it is feasible and uses fewer relays than any path found so far. In this case, the path corresponds directly to a relevant relay chain and can be yielded as partial output of the algorithm. Also, l can be updated to reflect that we are now only interested in paths strictly shorter than the one we just found.

1 $l \leftarrow L + 1$
2 $\alpha \leftarrow \alpha_0$
3 Calculate cheapest path tree from n_0
      using costs $c'_{n,n'} = c_{n,n'} + \alpha$.
      From the tree, obtain p and y_n, q_n for all n
4 **if** length$(p) \leq l$ **then**
5   Yield p // *Feasible and shorter than previous solutions*
6     $l \leftarrow$ length$(p) - 1$
7 $F \leftarrow \{(n, n') \in E : q_{n'} \geq q_n + 2\}$
8 **if** $F = \emptyset$ **then**
9   **return**   // *Path can not be shortened*
10 Calculate $\epsilon_{n,n'} \leftarrow \frac{c'_{n,n'} + y_n - y_{n'}}{q_{n'} - q_n - 1} \; \forall (n, n') \in F$
11 $\epsilon \leftarrow \min \epsilon_{n,n'}$
12 $\alpha \leftarrow \alpha + \epsilon$
13 **Goto** 3

Fig. 4.   Dual Ascent Algorithm

k (path length)	g_k (cost)	p_k (predecessor)
4	18	n_{20}
6	12	n_{10}
7	10	n_{11}

TABLE I

EXAMPLE OF INFORMATION STORED IN EACH NODE AFTER EXECUTION
OF THE MODIFIED BELLMAN-FORD ALGORITHM.

It is clear that increasing α by ϵ will increase the cost of a path containing k edges by $k\epsilon$, thus penalizing paths in proportion to hop counts. The algorithm therefore calculates such an ϵ in a way that guarantees that some paths are shortened but no relevant paths between n_0 and n_t are missed. This calculation considers all edges (n, n') in the original graph where reaching n' from n_0 in the current cheapest path tree requires at least two hops more than reaching n. The current path to n' could then be shortened by replacing it with the current path to n and the edge between n and n'. Since this has not already been done, we know that we currently have $y_n + c'_{n,n'} > y'_n$, that is, going through n is currently more expensive. Making the path through n *equally* expensive[2] entails increasing the cost of the longer path by $y_n + c'_{n,n'} - y_{n'}$. (To be more exact, the cost of the longer path must be increased that much *more* than the cost of the shorter path.) This additional cost has to be split by $q'_n - (q_n + 1) = q'_n - q_n - 1$ nodes (the length of the old path to n', minus the length of the potential new path through n), yielding the expression for $\epsilon_{n,n'}$ found in the algorithm above. Finally, to ensure no solutions are missed, we increase α by the minimum of all such $\epsilon_{n,n'}$ and iterate. The formal proof that increasing α in this manner yields exactly the intended answer is somewhat lengthy and for details we refer the reader to Burdakov et al. [4].

When no path can be shortened, the algorithm terminates.

It should be noted that this algorithm can be used in an anytime manner, initially generating the cheapest path and then generating incrementally shorter but more expensive paths.

D. Modified Bellman-Ford Algorithm

Our second algorithm (Figure 5) is a modification of the original Bellman-Ford algorithm, or to be precise, a modification of its AHOP version [11]. The algorithm incrementally

[2]Recall that the cost function for the cheapest path tree will prefer the shorter of two equally expensive paths. Thus, making the path through n strictly less expensive is not necessary.

generates and updates a set of *reachability records* for each node. This set can also be represented as one table for each node as shown in Table I. Each reachability record $\langle k, g_k, p_k \rangle$ associated with a node indicates that the node can be reached from the base station n_0 in k hops at a cost of g_k using the predecessor p_k. A complete path from n_0 to n can always be reconstructed (in reverse order) by considering the reachability record of the predecessor p_k for $k - 1$ hops and continuing recursively until n_0 is reached.

After termination, each reachability record is guaranteed to correspond to a relevant path. Conversely, if there is a relevant path of length l between n_0 and n, then n is guaranteed to have a reachability record for $k = l$. Thus, the fact that a target node n_t is associated with a reachability record $\langle k, g_k, p_k \rangle$ corresponds exactly to the existence of a relevant relay chain between n_0 and n_t of length $k - 1$ and cost g_k. In other words, n_t can be reached from the base station using $k - 1$ intermediate UAVs, one of which is the survey UAV. For example, Figure 1 shows a path of length $k = 5$ corresponding to a relay chain of length $k - 1 = 4$, which can be extracted using the reachability record for $k = 5$ for the target node.

The algorithm requires a preprocessing step consisting of calculating the cheapest path tree with n_0 as root. We then retrieve the height k^*_{max} of this tree. Clearly, no relay chain consisting of more than k^*_{max} nodes can be relevant, since all longer paths must also be at least as expensive. This limits the number of relay positions ever required for this graph as well as the depth to which the graph needs to be searched.

N is partitioned into sets N^*_k of nodes occurring at depth k in the cheapest path tree, where $0 \leq k \leq k^*_{max}$. Clearly, any relay chain using more than k hops to reach a node $n \in N^*_k$ cannot be relevant, as it would be possible to shorten the chain (by reaching n in exactly k hops) without increasing its cost (since only k hops were required in the *cheapest* path). In addition to using this fact in the main algorithm as described below, we also use it to create an initial reachability record corresponding to the cheapest relevant relay chain for each node. In Table I, this corresponds to the row where $k = 7$.

In the main algorithm, $g(n)$ denotes the cost of the cheapest path from n_0 to node n found so far, calculated using standard additive edge costs. Initially, $g(n) = \infty$ for all nodes. The algorithm will construct and use a sequence of sets V_k, each of which is characterized by the fact that any relevant relay chain of length k must consist of a path to a node $n \in V_k$ in $k - 1$ hops followed by a single outgoing edge from n.

```
1   for each n ∈ N do g(n) ← +∞
2   g(n_0) ← 0
3   for each n ∈ n_0- do              // Incoming edges...
4       E ← E \ {(n, n_0)}            // ...are removed
5   V_1 ← {n_0}
6   for k = 1, ..., min(L + 1, k*_max - 1) do
7       for each n' ∈ N*_k do
8           for each n ∈ n'_- do      // Incoming edges...
9               E ← E \ {(n, n')}     // ...are removed
10      V_{k+1} ← N*_k
11      for each n ∈ V_k do
12          for each n' ∈ n_+ do
13              c ← g_{k-1}(n) + c_{n,n'}   // To n' through n in k hops
14              if c < g(n') then
15                  g(n') ← c                // Lowest cost so far
16                  g_k(n') ← c              // Lowest cost in k hops
17                  p_k(n') ← n              // Predecessor for k hops
18                  V_{k+1} ← V_{k+1} ∪ {n'}
```

Fig. 5. Modified Bellman-Ford Algorithm

Lines 2–5 take care of the initial node n_0. No reachability record needs to be created for this node, but $g(n_0)$ must be updated to indicate that it can be reached with cost 0 (line 2). Clearly no relevant relay chain ending in n_0 can pass *through* n_0, so all incoming edges to n_0 can be removed (lines 3–4). Finally, line 5 prepares for the first iteration by setting $V_1 = \{n_0\}$, indicating that any relevant relay chain of length 1 must consist of a path to n_0 in 0 hops (the empty path) followed by a single outgoing edge. This handles all paths of length 0.

In lines 6–18, each iteration considers paths of length $k \geq 1$, up to a maximum of $\min(L + 1, k^*_{max} - 1)$. This reflects the fact that (i) we allow at most L UAVs, which yields a total path length of up to $L+1$ when edges to the base station and target are included, (ii) no paths of length greater than k^*_{max} can be relevant, since k^*_{max} reflects the depth of the *cheapest* path tree, and (iii) any relevant path of length exactly k^*_{max} was already generated during preprocessing. Setting $L = \infty$ ensures that the algorithm finds relevant relay chains of all possible lengths.

Recall that any node $n' \in N^*_k$ occurs at depth k in the cheapest path tree. Reachability records for strictly shorter paths to n' were already created in earlier iterations, and a record for a path of length k was created during preprocessing. Records for paths of length strictly greater than k cannot be created, as this would correspond to finding a path which is longer but strictly cheaper than the one of length k, which was by definition a cheapest path. Thus, no new paths ending in any such n' can be relevant, and we can remove all incoming edges to n' without affecting correctness or optimality (lines 7–9). However, we do need to consider longer paths going *through* these nodes (line 10, explained further below).

In lines 11–18, we consider all potentially relevant relay chains of length k. Recall that any such chain must consist of a path to a node $n \in V_k$ followed by a single outgoing

edge from n. We consider each such path in turn, determining its destination n' and calculating its cost c (lines 11–13). If the path is cheaper than the cheapest path found previously (with a cost of $g(n')$) we create a new reachability record with $g_k(n')$ set to the new path cost and $p_k(n')$ set to the new predecessor n (lines 14–17). Note that though reachability records are sparse, we know that any node in V_k does have a reachability record for $k - 1$ due to the construction of V_k.

What remains is to prepare for the next iteration by constructing V_{k+1} according to its definition. That is, we should construct V_{k+1} in such a way that any relevant given relay chain of length $k + 1$ necessarily consists of a path of length k to a node $n \in V_{k+1}$ followed by a single outgoing edge from n. It is clear that no relevant relay chain would use k hops to reach a node n if it could be reached in $k - 1$ hops without incurring additional costs. Thus, it is only necessary to consider nodes whose costs were decreased by allowing paths of length k or that could not be reached at all with paths of length $k-1$. This is achieved in line 10, for nodes in N^*_k where we found a cheapest path of length k during preprocessing, and in line 18, for nodes where we found a cheaper path of length k in this iteration.

This algorithm calculates the path with the fewest hops, as well as the cheapest path regardless of the number of hops and all relevant paths in between. Like the Dual Ascent algorithm, this algorithm can be used in an anytime manner, though in the reverse direction, where the main algorithm generates chains in order of increasing length and decreasing cost.

When the algorithm terminates, a table similar to the one in Table I exists for each node. This table contains all the information necessary to trace a path back to the start node. The first row corresponds to using the minimal number of UAVs and the following rows correspond to using an increasingly larger number of UAVs, giving a path with progressively lower cost, until the least-cost path with the largest number of UAVs, which is found on the last row. For path lengths lower than the smallest k in the table, no path can be found. For example, no path of length $k = 3$ could be found for the node represented by Table I regardless of cost. Other "missing" values of k indicate that even if a path of length k could be found, it would be at least as expensive as some shorter path in the table. For example, the fact that Table I contains no record for $k = 5$ means that a path of length 5 to the given node would have a cost of at least 18, the cost of the closest record for a smaller value of k, and that using such a path would be pointless.

When the operator makes a decision about how many UAVs to allocate to the mission, he chooses whether to allocate the minimum number of UAVs required, or any other number greater than this. This requires more UAVs but yields a better solution in terms of whatever quality measure was provided to the algorithm.

See Burdakov et al. [3] for additional details and proofs.

E. Multiple Survey Targets

The graph creation procedure discussed in Section IV-A can easily be altered to generate separate relay chains to an arbitrary number of survey targets. Instead of generating a single target node, one node t_i is created for every position x_i in a set $T \subseteq R^3$ of survey target positions. Similarly, instead of creating incoming edges to a single target node, incoming edges are created for every $x \in X', t \in T$ satisfying $s(x, t)$.

This provides support for a preprocessing phase where a single call to either of the new algorithms generates relay chains to hundreds or thousands of potential target positions. Relay chains can then quickly be extracted from the given graphs. Especially for the modified Bellman-Ford algorithm, this would take only marginally more time than generating relay chains to a single target position, given that the number of additional nodes and edges is small compared to the number of nodes and edges used for free positions where relays and survey UAVs may be located.

Though this is not a solution to the multiple target surveillance problem, since no attempt will be made to minimize the total number of relays used for a given set of targets, it can still be highly useful in the case where the position of the survey target is not known in advance.

V. TEST RESULTS

The AHOP version of the original Bellman-Ford algorithm has a time complexity of $O(|N||E|) \subseteq O(|N|^3)$. Applying the widely used optimization of terminating as soon as no node cost has been decreased during iteration k, this bound can be tightened to $O(k_{max}^*|E|) \subseteq O(|N||E|) \subseteq O(|N|^3)$.

Though the *worst* case complexity of the modified Bellman-Ford algorithm is identical, using Dijkstra's algorithm for preprocessing and partitioning ensures that one almost always considers far fewer than $|E|$ edges in each iteration. Using Dijkstra in this manner is possible due to the use of non-negative edge costs, and sensible because we want to generate multiple alternative paths, possibly to multiple targets, as opposed to a unique cheapest path to a single target, where Dijkstra alone would have sufficed.

Dual Ascent performs at most $|N|^2$ iterations, each time calling Dijkstra. Thus, its worst case complexity is in $O(|N|^2(|E| + |N| \log |N|)) \subseteq O(|N|^4)$. However, only extremely specific graph structures and cost functions require $|N|^2$ iterations, and the algorithm can still outperform Bellman-Ford in practice.

In our test cases, the obstructed volume was used for both communication cost and survey cost. We used line-of-sight requirements with varying maximum range for communication reachability and survey reachability. The environment graph was constructed using a three-dimensional grid and the grid cell size was varied between 10 and 40 meters, though it was constant within each problem instance. The test cases were taken from both rural and urban environments and the positions of base stations and targets were chosen with the intention to create challenging test cases. The lengths of relevant relay chains varied, using between 3 and 19 UAVs.

All test results were obtained from a C++ implementation on a PC running Windows Vista with an Intel Core 2 Duo 2.4 GHz CPU and 2 GB RAM. Only one core was used in testing. For comparison, we also implemented an AHOP version of the standard Bellman-Ford algorithm using the common optimization discussed above. Dijkstra's algorithm was used to calculate the cheapest path tree for both the dual ascent algorithm and the pre-processing phase of the modified Bellman-Ford algorithm. To ensure that all solutions were found, we used $L = \infty$ and $\alpha_0 = 0$.

Table II shows empirical test results for all three algorithms, averaged from 1000 executions.

The first three columns show the number of nodes and edges of each discretized graph together with the height k_{max}^* of the associated cheapest path tree rooted in the base station node.

Opt. BF shows the time requirements for our optimized implementation of the standard Bellman-Ford algorithm.

As the modified Bellman-Ford algorithm requires the calculation of the cheapest path tree as well as the determination of the N_k^* sets, the times for this algorithm are displayed in four separate columns. The column labelled *CPT* contains the times for calculating the cheapest path tree and the column labelled N_k^* shows the times for determining the N_k^* sets. The time used by the main algorithm are shown in the column marked *Mod. BF*. The time used by the complete algorithm including preprocessing is the sum of the three previous columns and is shown in the column marked *Total*. The times might not add up due to rounding.

The final column shows the complete time requirements for the Dual Ascent algorithm.

In the smaller cases tested here, the modified Bellman-Ford appears to offer the best performance, while the dual ascent algorithm is somewhat faster for the larger test cases. In all cases, both of the new algorithms are considerably faster than the standard Bellman-Ford algorithm, even when using the optimization mentioned above. Thus, the new algorithms enable us to solve considerably larger problem instances than before within reasonable time limits and permit the use of larger maps as well as finer-grained discretizations.

VI. FUTURE WORK

For simplicity, in this paper all UAVs had the same communication range. In the future we plan to extend the algorithms to handle UAVs with different communication ranges.

We also intend to use the algorithms and formulations used here as a basis for solving the relay problem for a moving target. One of the suggested formulations for edge cost was to use the obstructed volume within a certain radius. In general this will place the UAVs at positions with high visibility. If the target starts to move, it is beneficial to be located at such positions as it is more likely that the UAVs can stay in the same place and maintain communication to the other UAVs, as compared to being placed at positions with lower visibility.

VII. CONCLUSION

The use of relay chains is essential to a large variety of UAV and UGV applications where communication range is limited,

$|N|$	$|E|$	k^*_{max}	Opt. BF	Timing in milliseconds Modified Bellman-Ford				Dual Ascent
				CPT	N^*_k	Mod. BF	Total	
1395	14322	35	34.6	0.018	0.001	0.014	0.033	2.4
1654	33340	27	68.6	1.75	0.000	6.75	8.5	16.5
1993	10272	33	45.6	0.036	0.002	0.027	0.065	3.7
3411	122246	19	175	2.3	0.005	15.1	17.4	51.0
5294	1134634	14	1079	17.1	0.013	12.2	29.4	28.1
5824	624386	19	823	10.1	0.024	6.7	16.8	16.1
7117	1625726	15	1680	25.0	0.050	18.1	43.1	40.5
9226	5190976	10	3923	79.7	0.000	221.4	301.5	139.0
9226	23086152	5	12344	324.6	0.083	435.0	759.7	574.3
15105	6311322	14	6422	97.3	1.28	81.1	179.7	156.0
18801	13225274	13	12529	194.0	2.17	163.3	359.5	323.7
38542	28327214	17	36382	429.2	6.53	379.3	815.1	729.6

TABLE II

TEST RESULTS. ALL TIMES ARE IN MILLISECONDS AND ARE AVERAGES OF 1000 EXECUTIONS.

including but not limited to surveillance tasks. We have presented two algorithms for the relay positioning problem that are efficient both in terms of the number of relays used and in terms of the amount of time required to generate a solution. These algorithms build on a discretization of the continuous problem, but differ in the methods used for solving the all hops optimal path problem in the resulting directed graph: While the first algorithm uses a dual ascent technique, the second builds on the Bellman-Ford algorithm. An extension to efficiently generate independent paths to multiple survey targets was also presented. Both algorithms were shown to be significantly faster than the standard Bellman-Ford algorithm on a variety of problems of different sizes, generating solutions in a small fraction of the time previously required.

ACKNOWLEDGEMENTS

This work has been supported by LinkLab (www. linklab.se), the National Aeronautics Research Program NFFP04 S4203, the Swedish Foundation for Strategic Research (SSF) Strategic Research Center MOVIII, the Center for Industrial Information Technology CENIIT (06.09) and the Linnaeus Center for Control, Autonomy, and Decision-making in Complex Systems (CADICS), funded by the Swedish Research Council (VR).

REFERENCES

[1] Stuart O. Anderson, Reid Simmons, and Dani Goldberg. Maintaining Line of Sight Communications Network between Planetary rovers. In *Proceedings of the 2003 IEEE/RSJ International Conference of Intelligent Robots and Systems*. IEEE, 2003.

[2] Ronald C. Arkin and Jonathan Diaz. Line-of-sight constrained exploration for reactive multiagent robotic teams. In *7th International Workshop on Advanced Motion Control*, 2002.

[3] Oleg Burdakov, Patrick Doherty, Kaj Holmberg, and Per-Magnus Olsson. Optimal placement of communications relay nodes. Technical Report LiTH-MAT-R-2009-03, Linköping University, Department of Mathematics, 2009.

[4] Oleg Burdakov, Kaj Holmberg, and Per-Magnus Olsson. A dual-ascent method for the hop-constrained shortest path with application to positioning of unmanned aerial vehicles. Technical Report LiTH-MAT-R-2008-07, Linköping University, Department of Mathematics, 2008.

[5] Carmen Cerasoli. An analysis of unmanned airborne vehicle relay coverage in urban environments. In *Proceedings of MILCOM 2007*. IEEE, 2007.

[6] Thomas Cormen, Charles Leiserson, and Ronald Rivest. *Introduction to Algorithms*. MIT Press, 1990.

[7] Roch Guérin and Ariel Orda. Computing shortest paths for any number of hops. *IEEE/ACM Transactions on Networking*, 10(5), 2002.

[8] Peter E. Hart, Nils J. Nilsson, and Bertram Raphael. A formal basis for the heuristic determination of minimum cost graphs. *IEEE Transactions on Systems Science and Cybernetics*, 4(2), 1968.

[9] David Hsu, Tingting Jiang, John Reif, and Zheng Sun. The bridge test for sampling narrow passages with probabilistic roadmap planners. In *Proceedings of the IEEE International Conference on Robotics and Automation*, 2003.

[10] David B. Johnson and David A. Maltz. *Dynamic Source Routing In Ad Hoc Wireless Networks*. Kluwer Academic Publishers, 1996.

[11] Eugene L. Lawler. *Combinatorial Optimization: Networks and Matroids*. Holt, Rinehart and Winston, 1976.

[12] Qun Li and Daniela Rus. Sending messages to mobile users in disconnected ad-hoc wireless networks. In *Proceedings of the 6th annual international conference on Mobile computing and networking*. ACM, 2000.

[13] Hoa G. Nguyen, Narek Pezeshkian, Michelle Raymond, A. Gupta, and Joseph M. Spector. Autonomous communication relays for tactical robots. In *Proceedings of the 11th International Conference on Advanced Robotics*, 2003.

[14] Ramesh Palat, Annamalai Annamalai, and Jeffrey Reed. Cooperative relaying for ad-hoc ground networks using swarm UAVs. In *Proceedings of MILCOM 2006*. IEEE, 2006.

[15] Frank J. Pinkney, Dan Hampel, and Stef DiPierro. Unmanned aerial vehicle (UAV) communications relay. In *Proceedings of MILCOM 1996*. IEEE, 1996.

[16] Tom Schouwenaars. *Safe Trajectory Planning of Autonomous Vehicles*. PhD thesis, Massachusetts Institute of Technology, February 2006.

[17] John Sweeney, TJ Brunette, Yuandong Yang, and Roderic Grupen. Coordinated teams of reactive mobile platforms. In *Proceedings of the 2002 IEEE International Conference on Robotics and Automation*, 2002.

Efficient, Guaranteed Search
with Multi-agent Teams

Geoffrey Hollinger and Sanjiv Singh
Robotics Institute
Carnegie Mellon University
Pittsburgh, PA 15213
Email: {gholling, ssingh}@ri.cmu.edu

Athanasios Kehagias
Division of Mathematics
Aristotle University of Thessaloniki
Thessaloniki, Greece GR54124
Email: kehagiat@gen.auth.gr

Abstract— Here we present an anytime algorithm for clearing an environment using multiple searchers. Prior methods in the literature treat multi-agent search as either a worst-case problem (i.e., clear an environment of an adversarial evader with potentially infinite speed), or an average-case problem (i.e., minimize average capture time given a model of the target's motion). We introduce an algorithm that combines finite-horizon planning with spanning tree traversal methods to generate plans that clear the environment of a worst-case adversarial target *and* have good average-case performance considering a target motion model. Our algorithm is scalable to large teams of searchers and yields theoretically bounded average-case performance. We have tested our proposed algorithm through a large number of experiments in simulation and with a team of robot and human searchers in an office building. Our combined search algorithm both clears the environment and reduces average capture times by up to 75% when compared to a purely worst-case approach.

I. INTRODUCTION

Imagine you are the leader of a team of agents (humans, robots, and/or virtual agents), and you enter a building looking for a person, moving object, or contaminant. You wish either to locate a target in the environment or authoritatively say that no target exists. Such a scenario may occur in urban search and rescue [1], military operations, network decontamination [2], or even aged care [3]. In some special cases, you may have a perfect model of how the target is moving; however, in most cases you will only have an approximate model or even no model at all. To complicate the situation further, the target may be adversarial and actively avoiding being found.

Known algorithms would force you, the leader, to make a choice in this situation. Do you make the worst-case assumption and choose to treat the target as adversarial? This would allow you to utilize graph search algorithms to guarantee finding the target (if one exists), but it would not allow you to take advantage of any model of the target's motion. As a result your search might take a very long time. Or do you decide to trust your motion model of the target and assume that the target is non-adversarial? This assumption would allow the use of efficient (average-case) search methods from the optimization literature, but it would eliminate any guarantees if the model is inaccurate. In this case, your target may avoid you entirely. It is necessary to make one of these choices because no existing method provides fast search times and also guarantees finding a target if the model is wrong.

Fig. 1. Volunteer firefighter searching with a Pioneer mobile robot. The robot and humans execute a combined schedule that both clears the environment of an adversarial target and optimizes a non-adversarial target motion model. Our decentralized algorithm allows for robots and humans flexibly to fill the different search roles. The agents share their paths and an estimation of the target's position.

In this paper, we bridge the gap between worst-case (or guaranteed) search and average-case (or efficient) search. We propose a novel algorithm that augments a guaranteed clearing schedule with an efficient component based on a non-adversarial target motion model. We extend a guaranteed search spanning tree traversal algorithm to optimize clearing time, and we augment it with a decentralized finite-horizon planning method in which agents implicitly coordinate by sharing plans. We show that the average-case performance of the combined algorithm is bounded, and we demonstrate how our algorithm can be used in an anytime fashion by providing additional search schedules with increasing runtime. This produces a family of clearing schedules that can easily be selected before the search or as new information becomes available during the search. The contribution of this paper is the first algorithm to provide guaranteed solutions to the clearing problem and additionally improve the performance of the search with fortuitous information.

The remainder of this paper is organized as follows. We first discuss related work in both worst-case search and average-case search highlighting the lack of a combined treatment

(Section II). We then define both the worst-case and average-case search problems and show the formal connection between the two (Section III). This leads us to the presentation of our search algorithm, including a description of the finite-horizon and spanning tree traversal components, as well as theoretical analysis of performance bounds (Sections IV and V). We then test our algorithm through simulated trials and through experiments on a heterogeneous human/robot search team (Section VI). Finally, we conclude and discuss avenues for future work (Section VII).

II. RELATED WORK

As mentioned above, literature in autonomous search can be partitioned into average-case and worst-case problems. The initial formulation of the worst-case graph search problem is due to Parsons [4]. He described a scenario where a team of agents searches for an omniscient, adversarial evader with unbounded speed in a cave-like topology represented by a graph. In this formulation, the edges of the graph represent passages in the cave, and the nodes represent intersections. The evader hides in the edges (passages) of the graph, and it can only move through nodes that are not guarded by searchers. This problem was later referred to as *edge search* since the evader hides in the edges of the graph. Parsons defined the edge search number of a graph as the number of searchers needed to guarantee the capture of an evader on that graph. Megiddo et al. later showed that finding the edge search number is NP-hard for arbitrary graphs [5].

Several variations of the edge search problem appear in the literature that place restrictions on the form of the *cleared set*: the nodes on the graph that have been cleared of a potential target. *Connected edge search* requires that the cleared set be a connected subgraph at all times during the search. Barrière et al. argued that connected edge search is important for network decontamination problems because decontaminating agents should not traverse dangerous contaminated parts of the graph. They also formulated a linear-time algorithm that generates search schedules with the minimal number of searchers on trees [2].

Unfortunately, edge search does not apply directly to many robotics problems. The possible paths of an evader in many indoor and outdoor environments in the physical world cannot be accurately represented as the edges in a graph.[1] For this reason, robotics researchers have studied alternative formulations of the guaranteed search problem. LaValle et al. formulated the search problem in polygonal environments and presented a complete algorithm for clearing with a single searcher [6]. However, their algorithm shows poor scalability to large teams and complex environments. Gerkey et al. demonstrate how a stochastic optimization algorithm (PARISH) can be used to coordinate multiple robotic searchers in small indoor environments [7]. Though it is more scalable than complete algorithms, PARISH still requires considerable computation and

communication between searchers, which makes it difficult to apply to complex environments.

In our prior work, we examined the *node search* problem in which the evader hides in the nodes of a graph, and we showed how search problems in the physical world can be represented as node search.[2] We proposed the Guaranteed Search with Spanning Trees (GSST) anytime algorithm that finds connected node search clearing schedules with a near-minimal number of searchers [9]. GSST is the first anytime algorithm for guaranteed search, and it is linearly scalable in the number of nodes in the graph, which makes it scalable to large teams and complex environments. Note that the algorithms described above (including our own prior work) do not optimize the time to clear the environment, and they do not take into account a model of the target's movement beyond a worst-case model.

A different, but closely related, search problem arises if we relax the need to deal with an adversarial target. If the target's motion model is non-adversarial and approximately known to the searchers, the searchers can optimize the average-case performance of their search schedule given this motion model. Assuming a Markovian motion model (i.e., the target's next position is dependent only on the target's current position), the average-case search problem can be expressed as a Partially Observable Markov Decision Process (POMDP). Near-optimal solutions to fairly large POMDPs are possible [10]; however, the size of search problems scales exponentially in the number of searchers. Thus, search problems with even moderately sized teams are outside the scope of general POMDP solvers.

In response to the intractability of optimal solutions, researchers have proposed several approximation algorithms for average-case search. Sarmiento et al. examined the case of a stationary target and presented a scalable heuristic [11]. Singh et al. discussed the related problem of multi-robot informative path planning and showed a scalable constant factor approximation algorithm in that domain [12]. In our prior work, we extended these approximation guarantees to average-case search with our Finite-Horizon Path Enumeration with Sequential Allocation (FHPE+SA) algorithm [13]. Our algorithm provides near-optimal performance in our test domains, but it does not consider the possibility that the model is inaccurate. Thus, the search may last for infinite time if the target is acting in a way that is not properly modeled. For instance, if a target is modeled as moving but is actually stationary, the searchers may never examine parts of the environment because they believe the target would have moved out of them.

To the best of our knowledge, no search algorithm exists that can clear an environment of a worst-case target and improve average-case search performance based on additional

---

[1]One exception would be a road network, which could be modeled as an edge search problem. The application of our combined algorithm to these types of environments is left for future work.

[2]Note that several alternative versions of "node search" appear in the literature (see Alspach [8] for a survey). In one formulation, the evader resides in the edges of the graph, and these edges are cleared by trapping (i.e., two searchers occupy the adjacent nodes). In another, the pursuers have knowledge of the evader's position while attempting to capture the evader by moving onto the same node.

information (e.g., a target motion model or online sensor data). Our algorithm fills this gap.

III. PROBLEM SETUP

In this section, we define the search problem with respect to both a worst-case adversarial target and a non-adversarial target. We also show the formal connection between worst-case search and the guaranteed search problem found in the literature. Assume we are given K searchers and a graph $G(N, E)$ with $|N|$ nodes and $|E|$ edges. The nodes in the graph represent possible locations in the environment, and the edges represent connections between them (see Figure 2 for examples in indoor environments). At all times $t = 1, 2, \ldots, T$, searchers and a target exist on the nodes of this graph and can move through an edge to arrive at another node at time $t + 1$. Given a graph of possible locations and times, we can generate a time-evolving, time-unfolded graph $G'(N', E')$, which gives a time-indexed representation of the nodes in the environment (the formal definition of G' is given in the Appendix).

The searchers' movements are controlled, and they are limited to feasible paths on G'. We will refer to the path for searcher k as $A_k \subset N'$, and their combined paths as $A = A_1 \cup \ldots \cup A_K$. The searchers receive reward by moving onto the same node as the target. This reward is discounted by the time at which this occurs. Given that a target takes path Y, the searchers receive reward $F_Y(A) = \gamma^{t_A}$, where $t_A = \min \{t : (m, t) \in A \cap Y\}$ (i.e., the first time at which path Y intersects path A), with the understanding that $\gamma \in (0, 1)$, $\min \emptyset = \infty$, and $\gamma^\infty = 0$. Thus, if the paths do not intersect, the searchers receive zero reward.

This paper considers two possible assumptions on the target's behavior, which yield the average-case and worst-case search problems. If we make a non-adversarial assumption on the target's behavior, we can utilize a target motion model independent of the locations of the searchers. This yields a probability $P(Y)$ for all possible target paths $Y \in \Psi$. We can now define the optimization problem in Equation 1. Equation 1 maximizes the *average-case* reward given a motion model defined by $P(Y)$. Note that if the target's motion model obeys the Markov property, we can estimate a probability distribution of its location efficiently using matrix algebra.

$$A^* = \operatorname*{argmax}_A \sum_{Y \in \Psi} P(Y) F_Y(A) \qquad (1)$$

The average-case optimization problem in Equation 1 does not consider the possibility that the motion model may be incorrect. An alternative assumption on the target's behavior is to assume that it actively avoids the searchers as best possible. For the search problem, this implies that the target chooses path Y that minimizes $F_Y(A)$. This yields the game theoretic optimization problem in Equation 2. Here, the searchers' goal is to maximize the *worst-case* reward if the target acts as best it can to reduce reward.

$$A^* = \operatorname*{argmax}_A \min_Y F_Y(A) \qquad (2)$$

It is important to note that this formulation assumes that the searchers have a "perfect" sensor that will always detect the target if it resides in the same cell. We can relax this assumption somewhat by modeling a non-unity capture probability into the average-case reward function. For the worst-case reward function, the searchers could run the schedule several times to generate a bound on the worst-case probability of missing the target (i.e., each schedule will be guaranteed to have some nonzero probability of locating the target).

Given the worst-case and average-case assumptions on the target's behavior (either of which could be correct), the searchers' goal is to generate a feasible set of paths A such that the reward of both optimization problems are maximized. One option is to use scalarization to generate a final weighted optimization problem as shown in Equation 3. The variable α is a weighting value that can be tuned depending on how likely the target is to follow the average-case model.

$$A^* = \operatorname*{argmax}_A \quad \alpha \sum_{Y \in \Psi} P(Y) F_Y(A) + (1 - \alpha) \min_Y F_Y(A),$$
$$(3)$$

where $\alpha \in [0, 1]$ is a weighting variable to be tuned based on the application.

While scalarization is a viable approach, we argue that it makes the problem more difficult to solve. Note that the underlying function $F_Y(A)$ is nondecreasing and submodular (see Appendix for definition) as is its expectation in Equation 1. In fact, Equation 1 is the efficient search (MESPP) optimization problem as defined by Hollinger and Singh [13], which can be optimized using the FHPE+SA algorithm. FHPE+SA yields a bounded approximation and has been shown to perform near-optimally in practice. In contrast, $\min_Y F_Y(A)$ is nondecreasing but is *not* submodular. Thus, FHPE+SA is not bounded and, in fact, performs poorly in practice. Furthermore, Krause et al. show that game theoretic sensor placement problems of similar form to Equation 2 do not yield any bounded approximation (unless $P = NP$) [14]. Thus, scalarization combines an easier problem with a more difficult problem, which prevents exploiting the structure of the easier MESPP problem.

We propose treating the combined search problem as a resource allocation problem. More specifically, some searchers make the average-case assumption while others make the worst-case assumption. One case where this approach can improve the search schedule is in the (very common) scenario where a portion of the map must be cleared before progressing. In this scenario, several searchers are often assigned to guard locations while waiting for the other searchers to finish clearing. Some of these guards can be used to explore the uncleared portion of the map. An example of this case is shown in our human-robot experiments in Section VI.

If the searchers are properly allocated to the average-case and worst-case tasks, we can generate search schedules with good performance under both assumptions. The decentralized nature of the multi-robot search task makes this approach feasible by allowing different robots to optimize the two separate components of the combined problem. The question now becomes: how many searchers do we assign to each task? Several observations relating worst-case search to graph theoretic node search can help answer this question.

The graph theoretic node search optimization problem is to find a feasible set of paths A for a minimal number of searchers K_{min} such that A is a clearing path [9, 2]. In other words, find a path that clears the environment of an adversarial evader using the smallest number of searchers. Propositions 1 and 2 show the connection between node clearing and the worst-case search problem defined in Equation 2.[3]

Proposition 1: The value $\min_Y F_Y(A) > 0$ if and only if A is a clearing path (i.e., a set of paths that clear the environment of any target within it).

Proof: Assume that A is not a clearing path and $\min_Y F_Y(A) > 0$. Since A is not a clearing path, this implies that one or more nodes in G are contaminated (i.e., may contain a target) at all times $t = 1, \ldots, T$. W.l.o.g. let Y be a target path that remains within the contaminated set for all t. Such a path is feasible due to the assumptions of the evader and the recontamination rules. The contaminated set at a given time is by definition not observed by the searchers at that time. Thus, $A \cap Y = \emptyset$, which implies that $F_Y(A) = 0$ for these A and Y. If the target chooses this path, we have a contradiction.

Now, assume that A is a clearing path and $\min_Y F_Y(A) = 0$. This assumption implies that $A \cap Y = \emptyset$. By definition of clearing path, the contaminated set of $G = \emptyset$ at some time T. However, this implies that $A \cap Y \neq \emptyset$ or else there would exist a contaminated cell. Again we have a contradiction. ∎

Proposition 2: For a given number of searchers K and graph G, $\max_A \min_Y F_Y(A) = 0$ if and only if $K < s(G)$, where $s(G)$ is the node search number of G.

Proof: The value of $s(G)$ implies that a clearing schedule exists for all $K \geq s(G)$. By Proposition 1, this implies that a schedule A exists such that $\min_Y F_Y(A) > 0$ for all $K \geq s(G)$.

Similarly, the value of $s(G)$ implies that a clearing schedule does not exist with $K < s(G)$. By Proposition 1, this implies that a schedule A does not exist such that $\min_Y F_Y(A) > 0$ for $K < s(G)$. ∎

Proposition 1 shows that any non-zero solution to the optimization problem in Equation 2 will also be a node search solution with K searchers. Proposition 2 shows that $s(G)$, the node search number of G, will affect the optimization problem in Equation 2; if $K < s(G)$, then $\min_Y F_Y(A) = 0$ for all A. Drawing on this analysis, we now know that at least $s(G)$ searchers must make the worst-case assumption to generate a schedule with any nonzero worst-case reward. This motivates the use of guaranteed search algorithms that minimize the attained search number. However, current guaranteed search algorithms in the literature do not minimize clearing time. We show how this limitation can be overcome in the next section.

[3]Note that Propositions 1 and 2 hold for both monotone and non-monotone clearing schedules. In other words, the propositions hold regardless of whether recontamination is allowed in the schedule.

IV. Algorithm Description

Drawing off the observations in the previous section, we can design an algorithm that both clears the environment of an adversarial target and performs well with respect to a target motion model. The first step in developing a combined algorithm is to generate a guaranteed search schedule that optimizes clearing time with a given number of searchers. We extend the Guaranteed Search with Spanning Trees (GSST) algorithm [9] to do just this.

The GSST algorithm is an anytime algorithm that leverages the fact that guaranteed search is a linear-time solvable problem on trees. Barrière et al. show how to generate a recursive labeling of a tree (we will refer to this labeling as B-labeling) in linear-time [2]. Informally, the labeling determines how many searchers must traverse a given edge of the tree in the optimal schedule (see Algorithm 1). If an edge label is positive, it means that one or more searchers must still traverse that edge to clear the tree. From the labeling, a guaranteed search algorithm can be found using the minimal number of searchers on the tree. GSST generates spanning trees of a given graph and then B-labels them. The labeling is then combined with the use of "guards" on edges that create cycles to generate a guaranteed search schedule on arbitrary graphs (see reference [9] for more detail). However, GSST gives no mechanism for utilizing more searchers than the minimal number. Thus, it does not optimize clearing time.

Algorithm 1 B-labeling for Trees

1: Input: Tree $T(N, E)$, Start node $b \in N$
2: $O \leftarrow$ leafs of T
3: **while** $O \neq \emptyset$ **do**
4:     $l \leftarrow$ any node in O, $O \leftarrow O \setminus l$
5:     **if** l is a leaf **then**
6:         $e \leftarrow$ only edge of l, $\lambda(e) = 1$
7:     **else if** l has exactly one unlabeled edge **then**
8:         $e \leftarrow$ unlabeled edge of l
9:         $e_1, \ldots, e_d \leftarrow$ labeled edges of l
10:         $\lambda_m \leftarrow \max\{\lambda(e_1), \ldots, \lambda(e_d)\}$
11:         **if** multiple edges of l have label λ_m **then**
12:             $\lambda(e) \leftarrow \lambda_m + 1$
13:         **else**
14:             $\lambda(e) \leftarrow \lambda_m$
15:         **end if**
16:         **if** $l \neq b$ and $parent(l)$ ready for labeling **then**
17:             $O \leftarrow O \cup parent(l)$
18:         **end if**
19:     **end if**
20: **end while**
21: Output: B-labeling $\lambda(E)$

Algorithm 2 shows how GSST can be modified to improve clearing time with additional searchers. The algorithm uses B-labeling to guide different groups of searchers into subgraphs that are cleared simultaneously. Algorithm 2 does not explicitly use guards on edges creating cycles; the guards are instead

Algorithm 2 Guardless Guaranteed Search with Spanning Trees (G-GSST)

1: Input: Graph G, Spanning tree T with B-labeling, Searchers K_g
2: **while** graph not cleared **do**
3:    **for** all searchers **do**
4:       **if** moving will recontaminate **then**
5:          Do not move
6:       **else if** positive adjacent edge label exists **then**
7:          Travel along edge with smallest positive label
8:          Decrement edge label
9:       **else**
10:          Move towards closest node (on spanning tree) adjacent to positive labeled edge
11:       **end if**
12:    **end for**
13:    **if** no moves possible without recontamination **then**
14:       Return failure
15:    **end if**
16: **end while**
17: Return feasible clearing schedule

Algorithm 3 Combined G-GSST and FHPE+SA search

1: Input: Graph G, Spanning tree T with B-labeling, Total searchers K, Guaranteed searchers $K_g \leq K$
2: **while** graph not cleared **do**
3:    **for** all searchers **do**
4:       **if** guaranteed searcher **then**
5:          Run next step of G-GSST
6:       **else**
7:          Run FHPE+SA step
8:       **end if**
9:    **end for**
10: **end while**

Algorithm 4 Anytime combined search algorithm

1: Input: Graph G, Searchers K
2: **while** time left **do**
3:    Generate spanning tree T of G and B-label it
4:    **for** $K_g = K$ down to $K_g = 1$ **do**
5:       Run Algorithm 3 with K_g guaranteed searchers and $K - K_g$ efficient searchers
6:       **if** clearing feasible **then**
7:          Store strategy
8:       **else**
9:          Break
10:       **end if**
11:    **end for**
12: **end while**
13: **if** strategies stored **then**
14:    Return strategy with maximal $\alpha R_{avg} + (1 - \alpha)R_{worst}$
15: **else**
16:    Run FHPE+SA to maximize R_{avg} (cannot clear)
17: **end if**

---

determined implicitly from the B-labeling. Thus, we refer to it as Guardless GSST or G-GSST. Note that the use of B-labeling allows the searchers to perform the schedule asynchronously. For example, a searcher who arrives at a node does not need to wait for other searchers to finish their movement before clearing the next node (assuming that the move does not cause a recontamination).

Unlike other algorithms in the literature, G-GSST can be combined with model-based search algorithms to generate a combined search schedule. We augment G-GSST with Finite-Horizon Path Enumeration and Sequential Allocation (FHPE+SA) to yield our combined search algorithm. In short, FHPE+SA algorithm has each searcher plan its own path on a finite-horizon and then share that path with other searchers in a sequential fashion (see reference [13] for a formal description). In other words, one searcher plans its finite-horizon path and shares it with the other searchers; then another searcher plans its finite-horizon path and shares it, and so on. After the initial sequential allocation, searchers replan asynchronously as they reach replanning points in the environment.

Algorithm 3 shows how G-GSST can be augmented with FHPE+SA to yield a combined algorithm. An important quality of the combined search is that, depending on the actions of the FHPE+SA searchers, the schedule may be non-monotone (i.e., it may allow for recontamination). However, since the schedule of the G-GSST searchers is monotone, the search will still progress towards clearing the environment of a worst-case target.

It is important to note that the schedule of the G-GSST searchers (i.e., the searchers performing the guaranteed schedule) can be generated in conjunction with the FHPE+SA plans. For instance, we can modify the G-GSST searcher schedule

based on the actions of the FHPE+SA searchers by pruning portions of the map that happen to be cleared by the FHPE+SA searchers.[4] This provides a tighter coupling between the FHPE+SA searchers and the G-GSST searchers. Alternative methods for integrating the search plans are discussed in Section VII.

Finally, we generate many spanning trees (similarly to the GSST algorithm) to develop many search schedules (see Algorithm 4). These schedules range in quality and also tradeoff worst-case and average-case performance. At any time, the user can stop the spanning tree generation and choose to execute the search schedule that best suits his/her needs. This yields an anytime solution to the worst-case/average-case search problems: one that continues to generate solutions with increasing runtime.

---

[4]Note that implementing this extension requires ensuring that ignoring a portion of the map cleared by the FHPE+SA searchers does not lead to later recontamination. This can be determined by dynamically relabeling the spanning tree and taking into account the cleared set. Since labeling is a linear-time operation, this does not significantly affect runtime.

V. THEORETICAL ANALYSIS

A. Average-Case Performance Bounds

We now show that the average-case component of the combined algorithm is a bounded approximation. Let A be the set of nodes visited by the FHPE+SA algorithm. Let O be the set of nodes visited by the optimal average-case path (maximizes Equation 1), and let O_k be the set of nodes visited by searcher k on the optimal path. The average-case reward is bounded as in Theorem 1. This bound is simply the FHPE+SA bound for $K - K_g$ searchers.

Theorem 1:

$$F_{AC}(A) \geq \frac{F_{AC}(O_1 \cup \ldots \cup O_{K-K_g}) - \epsilon}{2}, \quad (4)$$

where K is the total number of searchers, K_g is the number of searchers used for the clearing schedule, and ϵ is the finite-horizon error ($\epsilon = R\gamma^{d+1}$, where R is the reward received for locating the target, γ is the discount factor, and d is the search depth). $F_{AC}(\cdot) = \sum_{Y \in \Psi} P(Y)F_Y(\cdot)$ as described in Section III.

Proof: This bound is immediate from the theoretical bounds on sequential allocation [12], the monotonic submodularity of F_{AC} [13], and the fact that $K - K_g$ searchers are deployed to perform sequential allocation. The addition of G-GSST searchers cannot decrease $F_{AC}(A)$ due to the monotonicity of F_{AC} and the monotonicity of the cleared set in the G-GSST schedule. ∎

We can extend the FHPE+SA component of the combined algorithm to optimize over several different known models. If there are M models being considered, and each has a probability of $\beta_1 \ldots \beta_M$, we can optimize the weighted average-case objective function in Equation 5.

$$F(A) = \beta_1 \sum_{Y \in \Psi} P_1(Y)F_Y(A) + \ldots + \beta_M \sum_{Y \in \Psi} P_M(Y)F_Y(A), \quad (5)$$

where $\beta_1 + \ldots + \beta_M = 1$, and $P_m(Y)$ describes the probability of the target taking path Y if it is following model m.

If all models obey the Markov assumption, this linear combination of models can be estimated using matrices as if it were a single model without an increase in computation. Additionally, monotonic submodularity is closed under non-negative linear combination, so the theoretical bounds hold in this extended case.

B. Computational Complexity

The labeling component of G-GSST requires visiting each node once and is $O(N)$, where N is the number of nodes in the search graph. The G-GSST traversal is $O(NK_g)$, where K_g is the number of guaranteed searchers. Finally, the FHPE+SA component replanning at each step is $O(N(K-K_g)b^d)$, where b is the average branching factor of the search graph, and d is the FHPE search depth. Thus, generating a single plan with the combined algorithm is $O(N + NK_g + N(K-K_g)b^d)$. This is linear in all terms except the FHPE search depth, which

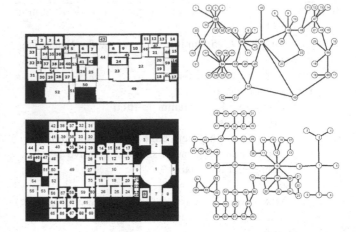

Fig. 2. Example floorplans (left) and graphical representations (right) of environments used for simulated coordinated search. The office (top) required three searchers to clear, and the museum (bottom) required five. Starting cells are denoted with a square.

often can be tuned to a very low number depending on the application.

VI. EXPERIMENTAL RESULTS

A. Simulated Results

We performed simulated testing of our combined search algorithm in two complex environments. The first map has two major cycles representing the hallways in a single floor of the Newell-Simon Hall office building. The second map is that of the U.S. National Gallery museum, which contains many cycles. Increasing the number of cycles in a graph complicates the search problem because it increases the paths by which the target can avoid being found. Figure 2 shows floorplans of these environments as well as the resulting search graphs.

We ran our algorithm on 1000 random spanning trees on both maps, and we compared these results to the schedules found by pure average-case and pure worst-case algorithms. Table I gives a summary of these results. The first row shows the average-case steps (i.e., at each step one or more searchers moves between nodes) to capture a randomly moving target using FHPE+SA with all searchers.[5] This strategy does not have worst-case guarantees if the model is incorrect. The second row shows the best clearing time using GSST on 10,000 random spanning trees. Note that GSST does not optimize for clearing time and only utilizes the minimal number of searchers required to clear. The results show that our combined algorithm yields much lower clearing times than those found with GSST. This is due to the use of additional searchers as in Algorithm 2. In addition, our combined algorithm reduces the average capture time by over 75% in the office when compared to GSST. Furthermore, a worst-case guarantee in the office can be gained by sacrificing only one step of average capture time.

---

[5]The expected capture steps for FHPE+SA were determined over 200 trials with a randomly moving target. The expected capture steps for the combined algorithm were computed in closed form. This is possible because the environment is cleared.

TABLE I

AVERAGE-CASE AND WORST-CASE CAPTURE STEPS COMPARISON IN
MUSEUM AND OFFICE. THE AVERAGE-CASE IS THE EXPECTED STEPS TO
CAPTURE A RANDOMLY MOVING TARGET. THE WORST-CASE IS THE
NUMBER OF STEPS TO CLEAR THE ENVIRONMENT.

	Office ($K = 5$)	Museum ($K = 7$)
FHPE+SA [13]	A.C. 4.8 W.C. ∞	A.C. 9.4 W.C. ∞
GSST [9]	A.C. 24.7 W.C. 72	A.C. 21.8 W.C. 56
Combined ($\alpha = 0.5$)	A.C. 14.0 W.C. 36	A.C. 17.3 W.C. 41
Combined ($\alpha = 0.75$)	A.C. 7.2 W.C. 47	A.C. 13.5 W.C. 45
Combined ($\alpha = 0.99$)	A.C. 5.9 W.C. 77	A.C. 11.6 W.C. 61

The user can determine the weighting of average-case vs. worst-case performance by tuning the value of the α parameter. This explores the Pareto-optimal frontier of solutions. This frontier forms a convex hull of solutions, any of which can be selected after runtime. Figure 3 gives a scatter plot of average-case versus worst-case capture steps for all feasible schedules. The figure also shows the number of G-GSST searchers, K_g, used for each data point. Note that the total number of searchers is fixed throughout the trials, and the remainder of the searchers performed FHPE+SA. All results with lowest average-case capture times are from the lowest K_g. This is because more searchers are used for average-case search and less for clearing. This demonstrates the utility of using FHPE+SA searchers in the schedule. Similarly, the lowest clearing times are with $K_g = K$ (i.e., all searchers are guaranteed searchers). Note that better solutions yield points to the left/bottom of these plots.

B. Human-Robot Teams

To examine the feasibility of our algorithm, we ran several experiments with a human/robot search team. We conducted these experiments on a single floor of an office building as shown in Figure 4. Two humans and a single Pioneer robot share their position information through a wireless network, and the entire team is guided by a decentralized implementation of Algorithm 3. The robot's position is determined by laser AMCL, and it is given waypoints through the Player/Stage software [15]. The humans input their position through the keyboard, and they are given waypoints through a GUI.

In the first experiment, all three searchers (humans and robots) were assigned as G-GSST searchers. This configuration takes 178 seconds to clear the floor of a potential adversary, and it yields an expected capture time of 78 seconds w.r.t. the random model. We also calculated the expected capture time if the model is 50% off (i.e., the target is moving 50% slower than expected). With this modification, the expected capture time increases to 83 seconds.

In the second experiment, the mobile robot was switched to an FHPE searcher, which took 177 seconds to clear and yielded an expected capture time of 73 seconds. The expected capture time with a 50% inaccurate model was 78 seconds. Thus, switching the mobile robot to an FHPE searcher yields a

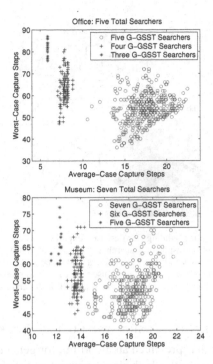

Fig. 3. Scatter plot of search schedules from the combined algorithm. Each data point represents a schedule generated by a different spanning tree input to Algorithm 3. The data point shapes are based on the allocation of searchers to the guaranteed and efficient search roles. The total searchers remains constant throughout the trials.

5 second decrease in expected capture time without sacrificing any worst-case capture time. The 5 second decrease remains even if the model is inaccurate. This shows that the schedules generated in this scenario are fairly robust to changes in the target's motion model.

In both experiments, the schedule clears the left portion of the map first and then continues to clear the right portion. Accompanying videos are available at the following URL that include playback of both simulated and human-robot results.

http://www.andrew.cmu.edu/user/gholling/RSS09/

In this environment, the assignment of the robot as an FHPE searcher does not increase the clearing time. The reason for this is that the robot spends a significant portion of the schedule as a redundant guard. Consequently, the right portion of the map is not searched until very late in the clearing schedule. In contrast, when the robot is used as an FHPE searcher, the right hallway is searched early in the schedule, which would locate a non-adversarial target moving in that area. This confirms our simulated results.

In addition, our results with a human/robot team demonstrate the feasibility of the communication and computational requirements of our algorithm on a small team. The initial plan generation stage is distributed among the searcher network, and once stopped by the user, the best plan is chosen. During execution, there is a broadcast communication requirement as the searchers share their positions on the search graph. This small amount of information was easily handled by a standard wireless network in an academic building.

271

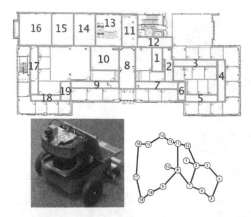

Fig. 4. Map of office building (top) used for experiments with a human/robot search team, and corresponding search graph (bottom right). Pioneer robot searcher (bottom left) with laser rangefinder and camera. The robot used the laser rangefinder for localization, and it used the camera to provide video feed to the humans. The robot and human team started at the entrance in cell 8.

VII. CONCLUSION AND FUTURE WORK

This paper has presented an algorithm for generating multi-agent search paths that both clear an environment of a potential adversary *and* optimize over a non-adversarial target motion model. Our algorithm bridges the gap between guaranteed search and efficient search by providing a combined search algorithm with both worst-case and average-case guarantees. We have shown through simulated experiments that our algorithm performs well when compared to algorithms in both guaranteed and efficient search. In addition, we have demonstrated the feasibility of our algorithm on a heterogeneous human/robot search team in an office environment.

Our current algorithm does not directly use a weighting variable α to incorporate confidence in the model. Instead, we cache many solutions and allow the user to choose one at runtime in an anytime fashion. One method for directly incorporating α would be first to determine the lowest number of searchers capable of clearing and assign the remainder of the searchers proportional to α. An alternative would be to have searchers switch between G-GSST and FHPE+SA during the schedule with some probability related to α. This would allow for dynamic switching but would require careful tuning of the switching function.

Additional future work includes more extensive robustness testing with inaccurate motion models and analysis of the communication requirements with very large search teams. Future field testing includes applications in emergency response, military reconnaissance, and aged care. Combining search guarantees against an adversarial target with efficient performance using a model has the potential to improve autonomous search across this wide range of applications. This paper has contributed the first algorithm and results towards this goal.

ACKNOWLEDGMENT

The authors gratefully acknowledge Joseph Djugash, Ben Grocholsky, and Debadeepta Dey from CMU for their insightful comments and assistance with experiments. We thank volunteer firefighter (and professional roboticist) Seth Koterba for feedback on the system.

REFERENCES

[1] V. Kumar, D. Rus, and S. Singh, "Robot and sensor networks for first responders," *Pervasive Computing*, pp. 24–33, 2004.
[2] L. Barrière, P. Flocchini, P. Fraigniaud, and N. Santoro, "Capture of an intruder by mobile agents," in *Proc. 14th ACM Symp. Parallel Algorithms and Architectures*, 2002, pp. 200–209.
[3] N. Roy, G. Gordon, and S. Thrun, "Planning under uncertainty for reliable health care robotics," in *Proc. Int. Conf. Field and Service Robotics*, 2003.
[4] T. Parsons, "Pursuit-evasion in a graph," in *Theory and Applications of Graphs*, Y. Alavi and D. Lick, Eds. Springer, 1976, pp. 426–441.
[5] N. Megiddo, S. Hakimi, M. Garey, D. Johnson, and C. Papadimitriou, "The complexity of searching a graph," *J. ACM*, vol. 35, pp. 18–44, 1988.
[6] L. Guibas, J. Latombe, S. LaValle, D. Lin, and R. Motwani, "Visibility-based pursuit-evasion in a polygonal environment," *Int. J. Comp. Geometry and Applications*, vol. 9, no. 5, pp. 471–494, 1999.
[7] B. Gerkey, S. Thrun, and G. Gordon, "Parallel stochastic hill-climbing with small teams," in *Proc. 3rd Int. NRL Workshop Multi-Robot Systems*, 2005.
[8] B. Alspach, "Searching and sweeping graphs: A brief survey," *Matematiche*, pp. 5–37, 2006.
[9] G. Hollinger, A. Kehagias, S. Singh, D. Ferguson, and S. Srinivasa, "Anytime guaranteed search using spanning trees," Robotics Institute, Carnegie Mellon Univ., Tech. Rep. CMU-RI-TR-08-36, 2008.
[10] T. Smith, "Probabilistic planning for robotic exploration," Ph.D. dissertation, Robotics Institute, Carnegie Mellon Univ., 2007.
[11] A. Sarmiento, R. Murrieta-Cid, and S. Hutchinson, "A multi-robot strategy for rapidly searching a polygonal environment," in *Proc. 9th Ibero-American Conf. Artificial Intelligence*, 2004.
[12] A. Singh, A. Krause, C. Guestrin, W. Kaiser, and M. Batalin, "Efficient planning of informative paths for multiple robots," in *Proc. Int. Joint Conf. Artificial Intelligence*, 2007.
[13] G. Hollinger and S. Singh, "Proofs and experiments in scalable, near-optimal search with multiple robots," in *Proc. Robotics: Science and Systems Conf.*, 2008.
[14] A. Krause, B. McMahan, C. Guestrin, and A. Gupta, "Selecting observations against adversarial objectives," in *Proc. Neural Information Processing Systems*, 2007.
[15] B. Gerkey, R. Vaughan, and A. Howard, "The player/stage project: Tools for multi-robot and distributed sensor systems," in *Proc. Int. Conf. Advanced Robotics*, 2003, pp. 317–323.

APPENDIX: DEFINITIONS

Definition 1: The time-augmented search graph G' is a *directed graph* and is obtained from G as follows: if u is a node of G then (u, t) is a node of G', where $t = 1, 2, ..., T$ is the *time stamp*; if uv is an edge of G, then $(u, t)(v, t+1)$ and $(v, t)(u, t+1)$ are *directed* edges of G' for every t. There is also a directed edge from (u, t) to $(u, t+1)$ for all u and t. In other words, G' is a "time evolving" version of G and every path in G' is a "time-unfolded" path in G.

Definition 2: A function $F : \mathbf{P}(N') \to \Re_0^+$ is called *nondecreasing* iff for all $A, B \in \mathbf{P}(N')$, we have

$$A \subseteq B \Rightarrow F(A) \leq F(B).$$

Definition 3: A function $F : \mathbf{P}(N') \to \Re_0^+$ is called *submodular* iff for all $A, B \in \mathbf{P}(N')$ and all *singletons* $C = \{(m, t)\} \in \mathbf{P}(N')$, we have

$$A \subseteq B \Rightarrow F(A \cup C) - F(A) \geq F(B \cup C) - F(B).$$

Time-extended Multi-robot Coordination for Domains with Intra-path Constraints

E. Gil Jones   M. Bernardine Dias   Anthony Stentz

Robotics Institute
Carnegie Mellon University
Pittsburgh, PA, USA
Email: egjones,mbdias,axs@cs.cmu.edu

Abstract—**Many applications require teams of robots to co-operatively execute complex tasks. Among these domains are those where successful coordination solutions must respect constraints that occur on the intra-path level. This work focuses on multi-agent coordination for disaster response with intra-path constraints, a compelling application that is not well addressed by current coordination methods. In this domain a group of fire trucks agents attempt to address a number of fires spread throughout a city in the wake of a large-scale disaster. The disaster has also caused many city roads to be blocked by impassable debris, which can be cleared by bulldozer robots. A high-quality coordination solution must determine not only a task allocation but also what routes the fire trucks should take given the intra-path precedence constraints and which bulldozers should be assigned to clear debris along those routes.**

This work presents two methods for generating time-extended coordination solutions – solutions where more than one task is assigned to each agent – for domains with intra-path constraints. While a number of approaches have employed time-extended coordination for domains with independent tasks, few approaches have used time-extended coordination in domains where agents' schedules are interdependent at the path planning level. Our first approach uses tiered auctions and two heuristic techniques, clustering and opportunistic path planning, to perform a bounded search of possible time-extended schedules and allocations. Our second method uses a centralized, non-heuristic, genetic algorithm-based approach that provides higher quality solutions but at substantially greater computational cost. We compare our time-extended approaches with a range of single task allocation approaches in a simulated disaster response domain.

I. INTRODUCTION

Research efforts in robotics have increasingly turned towards using teams of robots to collectively address tasks. A team of robots acting together can often outperform single robot solutions in terms of quality and robustness; however, employing multiple robots requires coordinating robot efforts. Research in multi-robot systems has centered around domains where tasks can be addressed by single robots in isolation – we call these independent task domains. Some approaches to coordination in independent task domains exclusively reason about instantaneous allocation, where each agent is assigned only a single task at a time [4]. Other approaches use time-extended allocation, where agents are assigned a set of tasks to accomplish over a period of time [3] [1]. Reasoning about time-extended allocation can improve performance, as agents can discover synergies and dependencies between tasks. This

increased performance comes at a cost, however, as time-extended coordination introduces substantial challenges for task allocation. For one thing, agents must determine the order in which to perform a number of tasks, a potentially difficult problem that can adversely affect performance if done poorly. Additionally, there are exponentially more potential multi-task assignments than single-task assignments, making it infeasible to consider all possible multi-task allocations and orderings.

The problem of determining high-quality time-extended allocations becomes even more difficult in domains where constraints introduce coupling between agents' schedules; in these domains it can no longer be assumed that agents can independently and accurately determine their fitness for tasks. We are particularly interested in domains where the coupling between agents' schedules occurs at the path planning level. Specifically, we focus on domains where intra-path constraints prevent agents from moving freely within the environment, and where other agents can provide assistance to the constrained agents to permit passage. The presence of intra-path constraints makes it necessary for agents to recruit assistance in order to determine which paths to take through an environment and form accurate fitness estimates for tasks.

One domain where agents' schedules are coupled due to intra-path constraints is precedence-constrained disaster response, a compelling domain that we argue is not well-addressed by existing coordination techniques. In this domain a team of robots capable of extinguishing fires operates in a city that has been ravaged by a disaster of significant proportions. These fire truck agents are tasked with moving to various locations around the city to extinguish fires. In addition to causing fires the disaster has rendered many roads impassable; they are covered with debris and wreckage, creating obstacles around which the bulky fire extinguishing robots cannot navigate. Suppose that there is another group of bulldozer robots in the team that are designed to move freely in rough terrain, and can clear roads of wreckage and debris. We assume that the goal of the team is to maximize a global objective function value that is computed as a sum over time-decreasing rewards offered for extinguishing fires. As in independent task domains, to formulate a high-quality multi-agent plan a coordination solution must allocate fire fighting tasks to fire trucks. Unlike in independent task domains a

273

coordination solution must also allocate debris-clearing tasks to bulldozers so that fire trucks can quickly and easily move to task locations. Determining time-extended assignments of fires and debris piles requires reasoning about the routes that fire trucks will take and the allocation of intra-path constraint satisfaction duties to bulldozers, adding substantial complexity to the already difficult problem of multi-agent task allocation.

In previous work we made some strides in adapting market-based approaches to domains with intra-path constraints [6]. We used a novel method, tiered auctions, and a number of domain-specific bounding methods to efficiently search the space of task allocations, path plans, and intra-path constraint satisfaction duties. One deficiency of this previous work was that we considered only instantaneous assignment for fire trucks, assigning a single fire to each fire truck at a time. Considering only instantaneous fire truck assignment aids in bounding and helps to drastically decrease the size of the search space; however, it may result in an unacceptable loss in solution quality. Time-extended allocation has been shown to substantially improve performance in domains with independent tasks [1]; we believe that reasoning about time-extended allocation can also offer significant performance increases for domains with intra-path constraints.

Efficiently determining high quality time-extended solutions for domains with intra-path constraints is an open research question that is not well addressed by existing approaches to multi-agent coordination. Some previous work has explicitly considered domains where tasks are interrelated by precedence and simultaneity constraints [9] [8]. Other work has explicitly addressed domains where tasks require the joint efforts of multiple agents [15] [14] [12]. While these approaches support reasoning about precedence constraints and interdependent agent schedules, none of these approaches extensively explore time-extended allocation or are well suited for efficiently searching a space of possible routes and associated task assignment problems. Work in genetic algorithms for multi-agent coordination has been widely studied in vehicle routing [11] and for scheduling in both job-shop [10] and multi-processor [2] contexts, but has not been extensively applied to domains that require multi-robot task allocation, scheduling, and path planning. Shima et al. use a genetic algorithm to do time-extended allocation of tasks that may involve precedence constraints to a set of unmanned air vehicles [13]. Their problem domain does not involve precedence constraint consideration in path planning, however, and they demonstrate their work only in scenarios that involve a maximum of 10 tasks, making the scope much smaller than in our scenarios. Koes et al. use an MILP-based formulation for a search and rescue domain with similar elements to our disaster response domain [7]. While this MILP-based technique could give optimal solutions in our domain the algorithm complexity grows exponentially in the number of tasks and intra-path constraints, making it impractical for any but the smallest problems.

The central contributions of this paper are two novel methods for time-extended coordination in domains with intra-path constraints, each of which strikes a different balance between computation time and solution quality. In the first method, we extend tiered auctions to time-extended allocation using two heuristic techniques, clustering and opportunistic path planning, to efficiently search the immense space of multi-task assignments. This approach improves on the solutions achieved by our previous instantaneous assignment approach but takes somewhat more time to determine the coordination solution. Our second method is a centralized genetic algorithm (GA)-based approach which searches in the full space of coordination solutions, including possible routes for agents, using randomized evolutionary methods. This approach can determine higher quality solutions than time-extended tiered auctions if allowed sufficient running time, but takes orders of magnitude more computation time to do so.

In the next section we describe particular characteristics of the disaster response with intra-path constraints domain. We then detail our methods for time-extended coordination and evaluate their performance in a simulated disaster response domain. We finally conclude and suggest future work.

II. DOMAINS ASSUMPTIONS

For the purposes of this work we suppose that the goal of the disaster response domain is to maximize the sum of time-decreasing rewards offered for extinguishing fires at affected buildings. When a fire occurs at a building the entire value of the building is put at risk - the building's value becomes the maximum reward offered for addressing the fire. Fires can be of differing magnitudes, where the magnitude governs the linear rate of decay of the building value. The building value is exhausted when the linear decay causes the value to reach zero - no reward can be obtained from extinguishing the fire once the value reaches zero, though no associated penalties are levied. By these assumptions the reward for extinguishing a particular fire is the affected building value less the linear rate of decay multiplied by the number of cycles that have passed since fire onset. In this work the maximum task rewards and linear decay values will be drawn from independent normal distributions. We assume that other performance metrics, such as resource usage, makespan, or number of tasks addressed are not relevant - only maximizing the sum of individual fire rewards. Note that by these assumptions intra-path constraint satisfaction has no intrinsic value to the objective function, and is only useful in allowing access to fire locations.

We assume that each agent has full knowledge of the domain map including debris pile locations, that all fire reward characteristics, locations, and durations are known precisely, and that agents can reliably communicate. While handling greater uncertainty in domain knowledge is certainly a goal for future work, in this work we focus on a low-uncertainty environment; even in domains with low uncertainty the size of the coordination search space offers substantial challenges.

We model the fire trucks as operating on a graph-based representation of our road network - at any given time fire trucks either occupy nodes or are traveling along road edges between nodes. Those nodes can represent intersections, debris piles, or fires. Bulldozers are not restricted to the road network;

they must avoid buildings but are otherwise can move freely. We model bulldozers as operating in an occupancy grid-based representation of the environment.

III. TIME-EXTENDED COORDINATION USING TIERED AUCTIONS

In this section we first describe our background work in tiered auctions for instantaneous allocation in domains with intra-path constraints. We then discuss how we use clustering and opportunistic path planning to enable tractable and efficient time-extended coordination.

A. Background

Market-based techniques have most frequently been applied to domains with independent tasks. One such approach is our work in a fire-fighting disaster response domain with independent tasks [5]. Our approach used a single tier of auctions; agents determine their task bids in isolation without consulting other agents. To employ a market-based approach in domains with intra-path constraints we needed a market mechanism that would allow agents to assess their fitness for tasks given that schedule quality depends on the schedules of other agents. Efficiency was the primary goal in system design, as the coupling of intra-task constraints and path planning causes a combinatorial explosion in problem complexity.

In order to allow agents to reason about intra-path constraints during the bidding process we added a second tier of auctions to the standard single-tier auction [6]. At the top auction tier fire allocation is determined; in order to bid on fires in those auctions fire trucks search through the space of possible routes, holding second-tier sub-auctions to solicit bulldozer involvement in clearing debris piles to satisfy intra-path constraints. To make the process tractable we structure the auction sequence and route search to permit extensive bounding of the search space.

In our approach during an auction cycle a central Dispatcher/Auctioneer holds auctions periodically to allocate fire fighting tasks. Fire trucks bid on the auctions based on determining the routes to tasks that will allow them to achieve the most reward in terms of the objective function for the task - by our assumptions this is equivalent to searching for the fastest route to the fire. Determining the fastest route to a particular task requires searching the space of possible routes and assigning bulldozers to clear debris piles along those routes. Debris piles are allocated using a series of sequential single item sub-auctions, where bulldozers bid based on adding debris clearing requirements to the end of their schedules. Bulldozers can potentially be assigned clearing duties in many different sub-auctions associated with different fire trucks, different fires, and different routes, though only a single set of assignments corresponding to the fastest route to a single fire will actually be adopted at the end of an auction cycle. Once fire trucks have bid the D/A awards a single fire to the truck that submitted the highest bid for any fire; that truck can then task bulldozers with clearing duties associated with the fastest route. The auction cycle repeats for any idle fire trucks.

Having each fire truck consider each possible route to every unallocated fire during an auction cycle would be prohibitively time consuming. We limit the search by having agents only consider routes that could potentially improve on the best bid submitted thus far. Additionally, we limit the search to routes that need to pass through fewer debris piles than a parameter $NumDebrisCons$; this serves to limit the search to paths which are more likely to yield fast passage and to limit over-exploitation of bulldozer resources.

As a final design consideration, we noted during testing that agents seeking to strictly maximize reward independent of time considerations tended to make somewhat inefficient decisions - going to a fire a long distance away that yielded slightly higher reward, for instance, rather than addressing a much closer fire. While this strategy is effective for maximizing the reward obtained for a single task, it doesn't take into account that the value available for all other tasks is constantly decaying. We determined that agents can obtain more reward over all tasks by trying to concentrate the reward they receive in a given duration and moved to using a model where agents seek to maximize their average reward per time cycle.

B. Methods

While our previous work in tiered auctions for domains with intra-path constraints achieved a substantial performance increase over single-tiered market-based approaches, considering only instantaneous allocations can limit the quality of the produced solutions. At the same time, the space of possible multi-task allocations is exponentially larger than the space of single-task allocations, making maintaining efficiency and tractability difficult. In our approach we attempt to heuristically determine the multi-task assignments that are most likely to yield good performance given the presence of intra-path constraints. We use two central observations in constructing our heuristics for determining likely multi-task assignments. First, we note that addressing a set of intra-path constraints that allow a fire truck to reach one task along a particular route may allow the agent to reach not only the target task but also other tasks near the target task. To exploit this observation we developed a clustering algorithm to group tasks that are located near each other and that are separated by only a few debris; any route from an agent's current location that allows the agent to reach some cluster member will also allow easy access to the other cluster tasks. Second, we note that along many potential routes to a cluster there may be fires that can be addressed with little or no delay to tasks later in the schedule. In fact, it may even result in higher average reward to take a slower path to a task if it allows an agent to address additional fires. We call this heuristic opportunistic path planning, and invoke it during route search. In the rest of this section we describe our two heuristic techniques for time-extended allocation and how they are incorporated into the tiered auction framework.

1) Clustering on Shared Preconditions: Our approach to clustering is to group together tasks that are spatially near each other and that are separated by a small number of debris; these tasks will then be auctioned as a group rather than being auctioned individually. In our implementation clustering occurs at the auctioneer level. When initiating task auctions the DA runs a clustering algorithm over all unassigned tasks. The algorithm iteratively creates a new cluster centered at an unassigned fire task. It then runs a breadth-first search from the selected fire task node. The BFS considers all unassigned fire tasks connected by edges with a total length less than a distance parameter $ClusterDistance$ and that are separated by fewer debris than a parameter $ClusterDebrisMax$; all such fire nodes that have not been assigned to another cluster are included in the new cluster. A clustering of a number of tasks is shown in Figure 1.

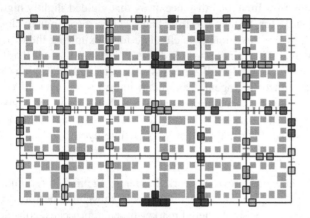

Fig. 1. Output of the clustering algorithm with $ClusterDistance = 2.5$ and $ClusterDebrisMax = 5$ on a disaster response scenario in a five-by-seven road network with 100 fire tasks (squares) and 100 debris (short lines that run perpendicular to the roads). Clusters are grouped by shades.

There are a few important differences between the tiered auction for task clusters and the procedure described in Section III-A. First, the DA offers a cluster of tasks in each task auction instead of just a single task. Second, the fire trucks must now assess the reward that can be obtained from addressing a set of tasks rather than a single task. To do this the agent must determine a route to the cluster, the order in which to address cluster tasks, and the routes to take between cluster tasks. Selecting the optimal ordering of tasks within a cluster so as to maximize average reward is a difficult problem; we use a heuristic based on shortest path distance. When bidding on a task cluster a fire truck first sorts the cluster tasks based on each task's shortest path distance from the truck's current location without considering debris. It then selects the task with the shortest distance as its initial target and run the route searching sub-auction process with this task as the goal, producing an efficient fire truck route and corresponding bulldozer assignments. Next the closest task to the initial target task is selected as the next cluster task to address. The route and bulldozer assignments that were previously determined are passed into the route-searching sub-auction process along with the new target fire. Note that debris may lie between cluster tasks, so the new route search may involve assigning additional debris piles to the ends of the bulldozer schedules associated with previous route searches. The new search is explicitly linked to previous searches to enable bulldozers

to form accurate bids given that they may have assignments associated with previous cluster tasks. The process continues until the fire truck has a route that includes all fires in a given cluster. The fire truck then bids for the cluster based on the average reward accumulated for all cluster fires. The DA assigns all relevant cluster tasks to the agent with the highest bid for any cluster, repeating the auction if any agents remain untasked. The winning fire truck adopts the schedule associated with the winning bid and informs the bulldozers they should begin executing their assigned duties.

2) Opportunistic Path Planning: We next focus on our second heuristic approach to time-extended coordination - opportunistic path planning. In all of our tiered auction work described thus far the goal of route planning has been to determine which route will allow the fire truck to reach and extinguish a target fire as quickly as possible. In opportunistic path planning the goal of route search is instead to determine the route to the target fire that will allow the fire truck to accumulate the greatest average reward, where reward can come both from addressing the target fire and from attending to additional fires along or near the route. In many cases taking a somewhat more indirect route to a target fire may result in higher average reward. We use two different techniques to identify unallocated fires along particular routes that can be opportunistically added to the route. The first technique reasons about adding nodes that lie directly along the node sequence that constitutes the route. The second technique reasons about adding tasks that require slight deviations from the route. When considering whether to add a task assignment it must be determined whether or not the added reward achieved by addressing tasks along the route outweighs the reward lost by delaying the completion of tasks later in the route.

a) On-path tasks: The on-path segment of our opportunistic path planning approach reasons only about tasks that lie directly along the route being evaluated. In this algorithm we start the evaluation at the first node in the route. If this node is an unassigned fire we create a new route schedule that is an exact copy of the current route schedule except that the new schedule will include a stop at the unassigned task for the required duration to address the task. The rewards for the schedules are then compared; whichever schedule has the higher per-cycle average reward is adopted. The search continues until the end of the route schedule.

b) Off-path tasks: The off-path segment of our opportunistic path planning approach reasons about adding tasks that do not lie precisely along the given route but require only a slight deviation from the current route. We note that the primary points where a truck can potentially beneficially deviate from the route lie at the beginning of the route and at each intersection. Thus at each of these points we run a distance-limited breadth-first search; the search will continue until the distance limit parameter $OppMaxDistance$ is reached or debris is encountered. When the search finds an unassigned fire a new route schedule is created with the additional required detour to reach the candidate fire and return to the route. If this new route schedule achieves greater average reward we adopt it; otherwise it is discarded. The process continues until no additions can improve the route.

c) Usage: Opportunistic path planning is used during the route planning portion of the cluster auction. Each route under consideration is passed first to the on-path method and then to the off-path method. The resulting schedule is then passed to the sub-auctioning routine so that a final average reward can be determined. If the route becomes part of a winning bid then the fire truck informs the D/A of all the unassigned fires it is claiming. A plan determined using opportunistic path planning and clustering is shown in Figure 2.

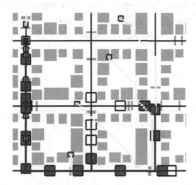

Fig. 2. An example from our simulator of a route determined using clustering and opportunistic path planning. Filled-in squares are fires that are scheduled to be addressed. Small boxes are bulldozers with their associated paths. The square-within-a-square is the fire truck agent.

3) Bounding for Time-extended Allocation: In Section III-A we noted that a primary bounding technique used in tiered auctions with instantaneous allocation was designed to limit search to routes that could potentially improve on the best task bid. This method no longer is applicable in the time-extended approach, as determining a bound on the potential reward associated with a fire is impossible when opportunistic path planning is being used. Instead we use a method we call g-value bounding. The g-value bound for a particular fire is the time that the fastest route thus far discovered has taken to reach the site of a fire including any extra time taken for addressing additional fires or time spent waiting for debris piles to be cleared. The search for routes to a particular fire will terminate when the remaining potential routes are longer than the current minimum g value. The general intuition is that if a relatively fast path to a fire has been found then it is not necessarily effective to explore increasingly indirect paths, even though some high-value indirect paths go unconsidered. A g-value multiplier parameter can be set that will explore paths that are some factor larger than the real g-value bound, but in practice this does not seem to raise overall system performance and can greatly increase running time.

IV. TIME-EXTENDED COORDINATION USING GENETIC ALGORITHMS

Our time-extended tiered auction approach is designed to quickly converge to a desirable coordination solution; this solution will in all likelihood be a local maximum in terms of performance. Our genetic algorithm approach, on the other hand, operates in the full space of possible coordinated plans and can converge to a global objective maximum if given sufficient time. Our GA approach is most appropriate for situations where a centralized solution is possible, where computation time is virtually unlimited, and where maximizing solution quality is paramount. We chose to use genetic algorithms as our randomized non-heuristic algorithm for two main reasons. First, they have been successfully used in both the vehicle routing literature [11], which involves elements of path planning and in job-shop scheduling [10], which involves precedence constraints. Second, some approaches, such as simulated annealing, focus on a single solution hypothesis, but genetic algorithms simultaneously maintain many disparate hypotheses, making searching large coordination spaces easier.

Our GA approach consists of the following algorithm, invoked at time zero:

1) Initialize the population of N genomes.
2) For a preset M generations:
   a) Evaluate genomes in terms of fitness.
   b) Select genomes for seeding the next generation.
   c) Crossover selected genomes.
   d) Apply mutation operators to genomes.
3) Take single highest fitness genome from any generation and use it to set agent schedules.

We will next describe our encoding of a problem solution into a genome as well as each of the above genetic algorithm components in more detail.

A. Encoding

The choice of encoding is a very important part of designing a genetic algorithm. The encoding needs to represent a complete solution to a given instance of a problem. In our case we need to specify the full sequence of actions taken by both the fire trucks and the bulldozers. We can represent fire trucks' actions as an ordered sequence of nodes that the fire truck will visit; this sequence can be turned into a full schedule using the methods we will discuss in Section IV-C. In addition to a node sequence each fire truck entry in a genome has a Boolean work vector, with an entry indicating whether the truck will attempt to address a fire as it passes. Note that we explicitly represent the routes that fire trucks but only implicitly represent the assignment of fire trucks to fires - determining which fires will be addressed by particular trucks only occurs during the evaluation phase. As bulldozers are not restricted to the road network the node sequence for each bulldozer needs only consist of a sequence of debris nodes that each will be assigned to clear.

B. Initialization

We use a random method to initialize genomes, both to seed the first generation and to introduce new individuals into the population in subsequent generations to maintain diversity. In this method for each fire truck in the agent population we select a random fire node and determine a shortest path to that fire node independent of debris considerations. That shortest path node sequence becomes each fire truck's initial entry in the genome. Work vectors are initialized such that trucks will attempt to address all fires on the route. We initialize bulldozers using a simple market-based algorithm that seeks to minimize makespan; the market-based method leads to substantially higher initial population quality than randomly dividing debris clearing responsibilities among agents. This method insures that each debris is assigned to only a single bulldozer and that all debris are assigned to some bulldozer.

C. Fitness Evaluation

The next main component of our GA is the evaluation method - this method must take a genome and determine a fitness value for the genome. In our case the value we want is the total objective function score obtained given that the agents follow the sequences of actions specified by the genome. To determine this value we must simulate the actions of agents in the domain to turn the sequences into schedules, as we cannot determine completion times for tasks without considering how agents' routes or sequences interact. We use an event-driven deterministic simulation method that produces the exact objective score yielded by agents following the sequences indicated in the genome.

D. Selection

We use three methods to select individuals to populate subsequent generations. The first method we use is a binary tournament method. In this method two members of the genome are randomly selected, and the member with the higher fitness is put into the reproduction pool. In this method higher fitness individuals generally are selected but a tournament may involve two low fitness individuals, maintaining diversity. The second method is "Hall-of-fame" (HOF) selection. We maintain a list of a set size of the distinct genomes with the highest fitness that have existed in any generation during the search - this is the "Hall-of-fame." During HOF selection we pick a random individual from the hall and put it in the reproduction pool. The HOF method means that we spend a substantial portion of our time exploring promising hypotheses. Finally, we also can add a randomly generated genome to the pool.

E. Crossover

We do crossover for both fire truck node sequences and bulldozer node sequences. We pair each genome in the selected pool with another node such that each genome has a single partner. Fire truck and bulldozer crossover then occur with some probability. For fire truck crossover we need a method to swap some portion of two agents' node sequences. While work in genetic algorithms for vehicle routing uses somewhat heuristic approaches that attempt to insert sub-routes in beneficial points in agents' schedule [11], we chose to use a simple non-heuristic approach, leaving other approaches for future work. In our method we select a single fire truck in each parent genome for crossover. We select a random crossover

point in each sequence and divide the schedules into front and back portions. For each parent we then create a new sequence consisting of its own front portion and the other parent's back portion, with a shortest path segment connecting the two. Corresponding work vectors are crossed over appropriately.

Bulldozer crossover is more complicated, as we want to preserve the genome property that all remaining debris tasks are assigned to one and only one bulldozer. To this end we used the crossover algorithm discussed in Mesghouni et al. in their work on genetic algorithms for job-shop scheduling [10].

F. Mutation

Each of the mutation operators acts on a single genome in isolation. We use four different mutation vectors that operate on the fire truck node sequences, a mutation operator for the fire truck work vectors, and three mutation operators for bulldozer sequences.

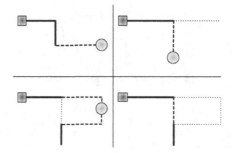

Fig. 3. Examples of the fire truck mutations: Append (upper left), Redirect (upper right), Reduce (lower right), and Lengthen (lower left). The square is the agent's location, the solid line the unaffected part of the original sequence, the smaller dashed line the replaced sub-sequence, the larger dashed line the new portion of the sequence, and the circle a randomly selected target node.

1) Fire Truck Mutation Operators: We could not find examples in the literature of mutator operators applied directly to agent routes. Our primary goal for our mutation operators was to have sufficient richness to take any route and mutate it into any other route in some finite number of steps. Each operator directly alters the path of a single fire truck, either adding to the end of a route (Append), shortening a sub-sequence in the current route (Reduce), expanding a sub-sequence of the current route (Lengthen), or altering the end-point of the route (Redirect). These operations are illustrated in Figure 3. Work vector mutation consists of, with some probability, switching each entry.

2) Bulldozer Mutation Operators: We chose to use three standard mutation operators from the job-shop scheduling literature [10]: swapping task order in a single agent's schedule, swapping two tasks between two different agents' schedules, and reassigning a task from one agent to another. All mutation operations preserve the unique assignment of debris.

V. EXPERIMENTS AND RESULTS

In this section we describe our experimental setup and then present our comparisons of our time-extended approaches and several competing instantaneous allocation approaches.

A. Experimental Setup

Our simulation is designed to directly compare the different approaches. To make the comparison as accurate as possible we have agents in the different approaches operate on exactly the same domain instance: agents begin in the same locations, randomly generated debris are identically located, and the same set of fires must be addressed. This simulation method runs all agents within a single thread of computation.

We run all experiments in the seven-by-five road network shown in Figure 1, with a team consisting of three fire trucks and 12 bulldozers. 100 fires are distributed randomly, with initial reward values drawn from a normal distribution of $N(3000, 250)$ that linearly degrades at a rate drawn from $N(25, 5)$ units/cycle. Fires have zero duration, allowing trucks to extinguish fires immediately upon arrival; debris pile require ten cycles of bulldozer effort to clear. In our experiments we set four debris frequency levels of 0, 10, 200, 300; at the 300 debris level this translates to an average of over five debris per segment. In all cases debris are randomly scattered on the road network. For each debris frequency level we tested on ten randomly generated distinct domain instances.

B. Approach Specifications

We use two instantaneous allocation approaches that do not consider intra-path constraints during allocation to establish a baseline of performance. The first approach we call Independent Tasks. In this approach both fire trucks and bulldozers are assigned to move to the location of the nearest fire and nearest debris, respectively, taking the shortest path. We take care that assignments are unique. In this method there is no explicit coordination between agents. In our second baseline approach we use the "Allocate-then-coordinate" (ATC) approach as described in our previous work [6], where fires are assigned based on shortest path bids; after assignment trucks use the route sub-auction procedure described in Section III-A to find the fastest path to their assigned fire. The final two market-based approaches we compare are tiered auctions with instantaneous fire truck assignment (TA-IA) and tiered auctions with time-extended fire truck assignment (TA-TE). For both tiered auction approaches a value of 8 is set for $NumDebrisCons$. For the TA-TE approach we set $ClusterDistance = 2.5$, $ClusterDebrisMax = 5$, and $OppMaxDistance = 6.0$. In order to make sure that bulldozers are productively occupied even if not tasked by a fire truck we add an Idle behavior to all market-based approaches - if bulldozers are not tasked by fire trucks to clear particular debris they will be assigned to debris piles using the same method taken in the Independent Tasks approach. We test with three different parametrizations of genetic algorithms: 5000 population, 200 optimization generations; 10,000 population, 400 generations; and 20,000 population, 400 generations.

C. Results

We illustrate solution quality in Figures 4 and 5. Figure 4 shows approach reward averaged in proportion to an absolute upper bound on available reward. Performance differences are

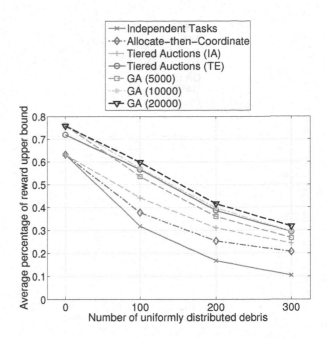

Fig. 4. Approach performance in a simulated emergency response domain. Values are shown as a proportion of an upper bound of available reward, where the bound is the summed reward of all issued tasks if they were addressed immediately.

clearer in Figure 5, where we plot how the approaches perform as an averaged proportion of the reward achieved by the best performer, GA-20000. TA-TE outperforms TA-IA by 24% averaged over the trials with debris, exceeding the performance of the baseline instantaneous allocation approaches by an even larger margin. Genetic algorithms with a population of 20,000 outperform TA-TE by 7% over the trials with debris. With a population of 10,000 the GA achieves roughly similar performance to TA-TE, and with a population of 5,000 and fewer generations TA-TE outperforms the GA approach by 7% averaged over the trials with debris.

Total time taken to produce coordination solutions is shown in Figure 6. Time-extended tiered auctions use substantially more computation time than do tiered auctions with instantaneous allocation except at the 300 debris level, where restrictive bounding makes the time-extended approach more efficient. All three settings of genetic algorithms use orders of magnitude more computation than any of the market-based approaches, underscoring that the heuristic techniques offers a good balance between performance and computation.

VI. CONCLUSION AND FUTURE WORK

We have shown two different methods for improving performance in a precedence-constrained domain by reasoning about time-extended allocation. Our time-extended tiered auction system uses two heuristics methods to allocate groups of tasks that share preconditions; this approach substantially outperforms a tiered auction approach that uses only instantaneous allocation though may require additional computation time. We also present a genetic algorithm approach to time-extended coordination, the first we know of that uses genetic algorithms

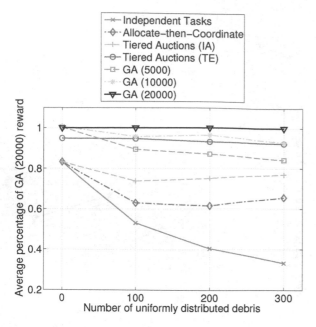

Fig. 5. Average proportion of reward achieved versus the performance of the GA-20000 approach on a per trial basis.

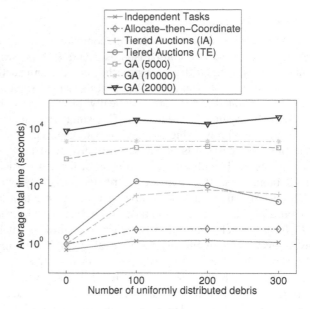

Fig. 6. Average total time taken to produce coordination solutions. Trials were all run in a single thread and allowed 100% of a CPU on a quad-core Intel Xeon 3.8 GHz CPU.

in a domain where allocation and path planning are coupled. The genetic algorithm approach with a sufficiently large population achieves better performance than time-extended tiered auctions at a cost of much greater computation time.

There are many avenues of future work that we would like to consider. First, we would like to extend our work to domains with higher uncertainty. One form of uncertainty we are particularly interested in addressing is dynamic task issue, where tasks at new locations are constantly being discovered. Time-extended coordination in such domains is especially difficult, as a balance must be struck between planning into the future and reacting to new information. We also are interested in extending this work to other forms of intra-path constraints, particularly those that involve simultaneity.

ACKNOWLEDGMENTS

This work was sponsored by the U.S. Army Research Laboratory, under contract "Robotics Collaborative Technology Alliance" (contract number DAAD19-01-2-0012). The views and conclusions contained in this document are those of the authors and should not be interpreted as representing the official policies or endorsements of the the U.S. Government.

REFERENCES

[1] M. Berhault, H. Huang, P. Keskinocak, S. Koenig, W. Elmaghraby, P. Griffin, and A. Kleywegt. Robot exploration with combinatorial auctions. In *Proceedings of the IEEE/RSJ International Conference on Intelligent Robots and Systems (IROS)*, 2003.

[2] K. Dahal, A. Hossain, B. Varghese, A. Abraham, F. Xhafa, and A. Daradoumis. Scheduling in multi-processor system using genetic algorithms. In *Proceedings of the Genetic Engineering and Evolutionary Computation Conference*, June 2005.

[3] M. B. Dias, R. Zlot, M. Zinck, J. P. Gonzalez, and A. Stentz. A versatile implementation of the *TraderBots* approach to multirobot coordination. In *Proceedings of the International Conference on Intelligent Autonomous Systems (IAS)*, 2004.

[4] B. P. Gerkey and M. J. Matarić. Sold!: Auction methods for multi-robot control. *IEEE Transactions on Robotics and Automation Special Issue on Multi-Robot Systems*, 18(5), 2002.

[5] E. G. Jones, M. Dias, and A. Stentz. Learning-enhanced market-based task allocation for oversubscribed domains. In *Proceedings of the IEEE/RSJ International Conference on Intelligent Robots and Systems (IROS)*, 2007.

[6] E. G. Jones, M. B. Dias, and A. Stentz. Tiered auctions for multi-agent coordination in domains with precedence constraints. In *Proceedings of the 26th Army Science Conference*, December 2008.

[7] M. Koes, I. Nourbakhsh, and K. Sycara. Heterogeneous multirobot coordination with spatial and temporal constraints. In *Proceedings of the National Conference on Artificial Intelligence (AAAI)*, 2005.

[8] T. Lemaire, R. Alami, and S. Lacroix. A distributed tasks allocation scheme in multi-UAV context. In *Proceedings of the IEEE International Conference on Robotics and Automation (ICRA)*, 2004.

[9] D. C. MacKenzie. Collaborative tasking of tightly constrained multi-robot missions. In *Multi-Robot Systems: From Swarms to Intelligent Automata: Proceedings of the 2003 International Workshop on Multi-Robot Systems*, volume 2. Kluwer Academic Publishers, 2003.

[10] K. Mesghouni, S. Hammadi, and P. Borne. Evolutionary algorithms for job-shop scheduling. *International Journal of Applied Math and Computer Science*, 14(1):91–103, 2004.

[11] M. A. Russell and G. B. Lamont. A genetic algorithm for unmanned aerial vehicle routing. In *Proceedings of the Genetic Engineering and Evolutionary Computation Conference (GECCO)*, June 2005.

[12] S. Sariel, T. Balch, and N. Erdogan. Incremental multi-robot task selection for resource constrained and interrelated tasks. In *Proceedings of the IEEE/RSJ International Conference on Intelligent Robots and Systems (IROS)*, 2007.

[13] T. Shima, S. Rasmussen, A. Sparks, and K. Passino. Multiple task assignments for cooperating uninhabited aerial vehicles using genetic algorithms. *Computers and Operations Research*, 33:3252–3269, 2006.

[14] F. Tang and L. Parker. A complete methodology for generating multi-robot task solutions using AsyMTRe-D and market-based task allocation. In *Proceedings of the IEEE International Conference on Robotics and Automation (ICRA)*, 2007.

[15] L. Vig and J. A. Adams. Multi-robot coalition formation. In *IEEE Transactions on Robotics*, volume 22 (4), 2006.

Bridging the Gap between Passivity and Transparency

Michel Franken[1], Stefano Stramigioli[2], Rob Reilink[2], Cristian Secchi[3] and Alessandro Macchelli[4]

[1] Institute of Technical Medicine, University of Twente, Enschede, The Netherlands
m.c.j.franken@utwente.nl
[2] IMPACT Institute, University of Twente, Enschede, The Netherlands
[3] DISMI, University of Modena and Reggio Emilia, Reggio Emilia, Italy
[4] DEIS, University of Bologna, Bologna, Italy

Abstract—In this paper a structure will be given which in a remarkably simple way offers a solution to the implementation of different telemanipulation schemes for discrete time varying delays by preserving passivity and allowing the highest transparency possible. This is achieved by splitting the communication channel in two separate ones, one for the energy balance which will ensure passivity and one for the haptic information between master and slave and which will address transparency. The authors believe that this structure is the most general up to date which preserves passivity under discrete time varying delays allowing different control schemes to address transparency.

I. INTRODUCTION

A telemanipulation chain is composed of a user, a master system, a communication channel, a slave system and an environment to act upon. Such a chain can be used to interact with materials in environments which are remote and/or dangerous or to augment the user's capabilities to facilitate the manipulation. In a bilateral telemanipulation chain the control algorithm usually tries to synchronize the motions of the slave device with the motions of the master device enforced by the user and to reflect the interaction forces between the slave device and the environment to the user by means of the master device. A proper haptic feedback is likely to increase the performance of the user with respect to effectiveness, accuracy and safety in many practical applications, e.g. robotic surgery [1].

Because the master and slave systems can be located at different sites, it is likely that time delays will be present in the communication channel. It is well known that due to these time delays it is not simply possible to exchange the power variables (forces and velocities) between the master and slave system. A direct exchange of these variables would create "fictitious" energy in the presence of time delays turning the communication channel into an active element. This energy could let the system become unstable.

An extensive overview of various approaches to the telemanipulation problem is given in [2]. Two criteria the "solution" to this problem should satisfy are transparency and stability, where transparency relates to how well the complete system is able to convey to the user the perception of directly interacting with the environment. A solution to the stability problem with respect to time delays can be found in the expression of the communication channel in scattering variables, which makes the system passive. It is shown that with these variables it is possible to "code" the power variables in such a way that no energy is virtually generated in the communication channel [3] by effectively resembling the physics of an analog transmission line. This approach however has often been criticized with respect to the attainable transparency [4] and much effort has and is being put in developing adaptations of the original formulations to improve the transparency [2]. In any case, guaranteeing passivity of the telemanipulation chain is an effective and elegant solution to the stability problem.

Clearly, as a time delay is present, there will be limits to the achievable transparency, but in current schemes this limit is the result of the fact that both the energy exchanged between master and slave and the information which should transmit the haptic behavior for transparency are transmitted/coded in the same variables. This results in algorithms based on scattering variables to be passive but also limited in the achievable transparency, while other proposed methods without passivity property achieve better transparency but have discussible stability margins [4].

In this article we will introduce a new layered algorithm which shows that it is possible, in a remarkably simple way, to combine the benefits of passivity with the desire to achieve transparency. A two-layer framework is defined, where the top layer tries to satisfy the goals of the telemanipulation chain and the bottom layer has to make sure that the commands originating from the top layer do not violate the passivity condition.

In the rest of this paper we assume an impedance causality for both the master and slave device (velocities as input and forces as output to the robotic devices). For these devices the energy exchanged with the outside world can precisely be determined by the measured change of position, which is at the heart of Section IV [5]. The paper is organized as follows: Section II discusses the basic concepts of passivity and transparency and related work. Section III introduces the framework of the proposed algorithm. Sections IV and V contain the theory and proposed structures of the various parts of the framework and in Section VI an example is discussed. The paper ends with conclusions and a discussion of future work in Sections VII and VIII.

II. BASIC CONCEPTS

Passivity and transparency are basic concepts in telemanipulation and for completeness will be explained shortly in this section. Afterwards, an overview of related work will be given to mark the differences with the proposed approach in Section III.

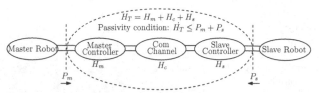

Fig. 1.   Energy balance of the telemanipulation chain

A. Passivity

A system is said to be passive if the energy which can be extracted from it is bounded from below by the injected and initial stored energy. Passivity of a system is a sufficient condition for stability and any proper combination of passive systems will again be passive. Physical energy exchange during operation is taking place between the user and the master system and between the slave system and the environment. Independent of anything else, including the goal of the telemanipulation chain, the only requirement therefore necessary to ensure passivity of the control system is:

$$\dot{H}_T(t) \leq P_m(t) + P_s(t) \tag{1}$$

where $H_T(t)$ represents the total energy present in the control system at instant t which is composed of all the energy present on the master controller $H_m(t)$, all the energy present on the slave controller $H_s(t)$ and the energy present in the communication channel $H_c(t)$, Fig. 1. $P_m(t)$ and $P_s(t)$ are respectively the power flowing from the master and slave robot into the master and slave controller.

B. Transparency

A system is said to be transparent if the user gets the experience that he is manipulating the environment directly. An interaction can be defined by the associated power variables, so a perfectly transparent system without time delays ensures:

$$\begin{aligned} \dot{q}_s(t) &= \dot{q}_m(t) \\ \tau_m(t) &= \tau_s(t) \end{aligned} \tag{2}$$

where $\dot{q}_m, \dot{q}_s$ are the velocities of the master and slave device and τ_m, τ_s are the forces associated with the interaction between the master device and the user and between the environment and the slave device. The transparency of such port-based systems can be compared as described in [6]. When increased time delays are introduced in the communication channel between the master and slave device, it is not enough to simply reflect the measured interaction forces towards the user, not even considering the possible problems with stability, as in the ideal situation this would lead to:

$$\begin{aligned} \dot{q}_s(t+T) &= \dot{q}_m(t) \\ \tau_m(t+T) &= \tau_s(t) \end{aligned} \tag{3}$$

and a mismatch in the display of the interaction occurs which becomes worse with increasing time delays. An example of this is presented in [7], which states that a wave variable travelling from the master to the slave system can be interpreted as a general *Move or Push* command and the returning wave describes the response of the slave to the received command. If the haptic feedback is introduced in order to safely interact with soft and delicate tissue then clearly this situation is not desirable as the tissue could be damaged before the imposed forces on the tissue are reflected to the user. Therefore a truly transparent system in the presence of arbitrary time delays should ensure:

$$\begin{aligned} \dot{q}_s(t+T) &= \dot{q}_m(t) \\ \tau_m(t) &= \tau_s(t+T) \end{aligned} \tag{4}$$

meaning that the behavior at both interaction ports is equal but delayed in time. As this requires an acausal system, true transparency cannot be achieved. The best achievable result requires a predictor at the side of the master device that predicts the future interaction forces between the slave device and the remote environment, which requires a local model of the remote environment.

C. Related Work

In this section, literature will be discussed that bears relevance to the approach proposed in Section III. In [8] a wave variable based algorithm is presented in which a Smith Predictor is used to reduce the mismatch in time from Eq. 3 by correcting the incoming wave variable with the predicted future value of that variable without loss of passivity. Several algorithms based on the haptic interaction with a local model of the remote environment in the time domain, generally referred to as *Impedance Reflection* algorithms, have already been proposed, e.g. in [9] and [10], but offer no passivity of the telemanipulation chain which is a desirable property as it guarantees stability. A first attempt to extend such a transparency strategy with a passivity property is described in [11], where a time domain Passivity Observer is used to adapt the locally computed feedback force based on the actual measured, but delayed, interaction force to make the system passive. Also interesting is the work in [12] where an online energy bounding algorithm is presented for a spring-damper controller at the slave side.

III. PROPOSED FRAMEWORK

In the opinion of the authors the most abstract view of an algorithm which combines the discussed transparency and passivity properties would be a two-layer structure as shown in Fig. 2. The *Transparency Layer* tries to satisfy the goals of the telemanipulation chain: movement synchronization on the slave side and force reflection on the master side, by computing a desired torque, τ_d based on the measured interaction data, $F_m, x_m, \dot{x}_m, ...$ and $F_s, x_s, \dot{x}_s, ...$. The *Passivity Layer* on the other hand has to make sure that the commands originating from the *Transparency Layer* do not violate the passivity condition. The benefit of this strict seperation in

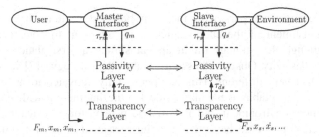

Fig. 2. Two layer algorithm for bilateral telemanipulation

layers is that the strategy used to ensure optimal transparency does not depend on the strategy used to ensure passivity and vice versa. Most schemes only incorporate a single, possibly mixed layer and as a result a single two-way communication connection between the master and slave system. In this algorithm, however, the passivity and transparency are dealt with in seperate layers and therefore we can easily define two two-way communication channels between the master and slave system. One channel is used to communicate energy exchange related information between the *Passivity Layers* and the other to communicate information related to the behavior between the *Transparency Layers*.

IV. PASSIVITY LAYER

This is the lowest-level control layer, whose only concern is to maintain passivity of the telemanipulation chain. Every movement the slave device makes will have an associated energetic cost and this energy will therefore have to be present at the slave side at the moment the movement is executed. In order to maintain passivity according to Eq. 1, the same energy will also have to be injected previously by the user at the master side. This clearly requires the transport of energy from the master to the slave system. To this end, the concept of a lossless energy tank is introduced in the *Passivity Layer* at both the master and the slave side, which can exchange energy. The level of these tanks can be interpreted as a tight energy budget from which actions can be fueled and which are continuously being replenished by the user at the master side.

If the energy level in the tanks is low, the movements the devices can make are restricted. An extreme situation occurs when the tank at the slave side is completely empty in which situation the slave device cannot make a controlled movement at all. In [12] also an energy tank is defined at the slave side to maintain passivity, but there the dissipated energy by the damping in the controller is accumulated, whereas in this framework the user directly makes energy available at the slave side regardless of the implemented controller for the desired behavior, this strict separation of energetic passivity and desired behavior is the key property of this two layered approach.

Suppose for the moment that no energy is exchanged between the user and the master system and no energy is exchanged between the slave system and the environment. In such a case an energy balance which would preserve passivity

would be obtained if:

$$H_m(t) + H_c(t) + H_s(t) = 0 \qquad (5)$$

where in general, instead of 0 any positive constant could be chosen and $H_m(t)$ and $H_s(t)$ represent the energy levels of the tanks at the master and slave side and $H_c(t)$ is the total energy in the communication channel at time instant t.

As far as energy exchange is concerned, we can consider the possibility to send energy quanta from master to slave when there is energy available in the tank at that side and vice versa in the form of packets containing the amount of energy sent. When these packages arrive at the other side they are stored in a receiving queue. Both master and slave can implement completely asynchronously at any instant of time the following operations which, if we indicate it for simplicity for the master, would be:

$$H_{m+}(t) = \sum_{i \in Q_m(t)} \bar{H}_m(i) \qquad (6)$$

where $Q_m(t)$ represents the set of all energy packets present in the queue of the master at time t, $\bar{H}_m(i)$ represents the packet (energy quantum) i present in that queue and $H_{m+}(t)$ represents the total amount of energy which is present in the receiving queue.

At time instant t the receiving queue is then emptied, meaning that the energy present in the receiving queue is added to the energy tank, and if necessary a certain amount of energy is transmitted to the other side. The energy level of the tank at the master side after these operations, $H_m(t_+)$ will therefore be:

$$\begin{cases} H_m(t_+) = H_m(t_-) - H_{m-}(t) + H_{m+}(t) \\ Q_m(t_+) = \emptyset \end{cases} \qquad (7)$$

where $H_m(t_-)$ is the energy level of the tank before the operations at time t, $\emptyset$ represents an empty set and $H_{m-}(t)$ is the amount of energy which is sent via the channel to the other side in the form of a packet and which should clearly be any value smaller or equal to $H_m(t_-) + H_{m+}$ not to lose passivity of the tank itself. An illustration of this procedure is given in Fig. 3. It can be easily seen that by construction, this algorithm satisfies Eq. 5 and allows energy to be sent from the master to the slave and vice versa without losing passivity.

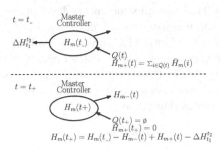

Fig. 3. Energy distribution at master side

A. Interaction with the Physical World

Unlike Eq. 7, Fig. 3 contains an energetic interaction port with the physical world. On the master and slave side, the controllers will have to control two robots which will interact with the user and the environment. A Zero Order Hold (ZOH) is used to connect the discrete time controller to the actuators and will hold the control signal constant until a new value is set.

Let us indicate with $q(t)$ the position vector representing the angles/positions of the actuators at time t, with $\dot{q}(t)$ their velocities and with $\bar{\tau}(t)$ the torques/forces they are exerting on the mechanism, set by the controllers and held constant by ZOH during the sample interval. Suppose at a certain instant of time t_1 to have measured $q(t_1)$ and to have set a certain value $\bar{\tau}(t_1)$ for the actuators. At a subsequent time instant t_2 it is possible to exactly calculate the physical energy exchanged with the robot, by measuring $q(t_2)$:

$$\Delta H_{t_1}^{t_2} = \int_{t_1}^{t_2} \bar{\tau}(t_1)\dot{q}(t)dt = \bar{\tau}(t_1)(q(t_2) - q(t_1)) \quad (8)$$

where $\bar{\tau}(t_1)$ indicates the constant value held by the ZOH between t_1 and t_2. This is following the line of the method proposed originally in [5].

Even if not strictly necessary, we can now synchronize the reading of $q(t_2)$ and the setting of a new value of $\bar{\tau}$ with the operations expressed in Eq. 7 extending them as in Fig. 3:

$$\begin{cases} H_m(t_{2+}) = H_m(t_{2-}) - H_{m-}(t_2) + H_{m+}(t_2) - \Delta H_{t_1}^{t_2} \\ Q(t_{2+}) = \emptyset \end{cases}$$
$$(9)$$

With this algorithm we are therefore able to ascertain the energy balance at each instant of time when sampling occurs.

B. Energy Tanks

We have shown that it is possible to exchange discrete energy quanta between the master and slave system in a passive manner with arbitrary time delays in the communication channel and also to quantify the energy exchange between the controllers and the external world (user/environment). Now a method is required to extract energy from the user to fill the energy tanks and to compute the energy quanta which needs to be transmitted to the other side.

In order to extract energy from the user a Tank Level Controller (TLC) is defined in the *Passivity Layer* of the master system. This TLC monitors the energy level of the local tank H_m and tries to maintain this at a desired level H_d. Whenever H_m is lower than H_d the TLC applies a small opposing force F_{TLC} to the user's movement to extract energy from the user into the energy tank. This force is superimposed on the force to obtain the desired behavior and could be implemented as a modulated viscous damper as:

$$\begin{aligned} F_{TLC}(k) &= -d(k)\dot{q}_m(k) \\ d(k) &= \begin{cases} \alpha(H_d - H_m(k)) & \text{if } H_m(k) < H_d \\ 0 & \text{otherwise} \end{cases} \end{aligned} \quad (10)$$

where α is a parameter that can be used to tune the replenishment of the energy tank. It is important to note that although this strategy might appear similar at first glance to the Passivity Observer/Passivity Controller strategy [13] it is in fact very different. This damper is only activated to make energy available at the slave side for the interaction with the remote environment and not to dissipate virtually generated energy. It should be noted that the presented strategy is but one way to extract energy from the user and that the framework can accommodate many alternatives.

Now that energy is being extracted from the user, this energy needs to be transported to the slave system. A simple way to accomplish this is for each tank to transmit a fixed fraction, β of its energy level (when energy is available) to the other system. If both systems run at the same sample frequency and there is no energy exchange between the slave tank and the physical system, the energy levels in the two tanks will converge to the same value, which in this case will be the set desired level for the master tank. This is illustrated in Fig. 4.

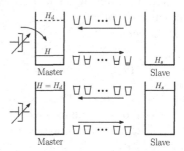

Fig. 4.   Illustration of level synchronization between energy tanks

This means that energy quanta are continuously being exchanged between the master and slave system. Assuming equal sample frequencies f_s at master and slave and constant time delays between master and slave, ΔT_{ms}, and vice versa, ΔT_{sm}, the total amount of energy present in the communication channel in steady state operation (no energy exchange with the physical systems) will be:

$$\begin{aligned} H_c &= H_{ms} + H_{sm} \\ H_{ms} &= \underline{\Delta T_{ms} f_s} \beta H_d \\ H_{sm} &= \underline{\Delta T_{sm} f_s} \beta H_d \end{aligned} \quad (11)$$

where $\underline{\ast}$ indicates a round down operation as there is a discrete number of packages present in the communication channel, H_{ms} represents the energy traveling from the master to the slave and H_{sm} the energy traveling in the other direction. The total energy in the communication channel H_c can therefore become quite large for large time delays. Future research will look into this issue.

C. Saturating the Controlled Effort

The value which will be set as controlled torque $\tau_{d\ast}$ (the subscript $\ast$ replaced with m refers to the master side and with s to the slave side) for the interval to come, will be specified by the hierarchically higher level of control, the *Transparency*

284

Layer, which will be treated in the next section. At both sides the *Passivity Layer* is used to limit the applied torque to the device with respect to what the *Transparency Layer* at that side requests in order to maintain passivity. Three limits to which τ_{d*} can be subjected are given in Eq. 12, 13, 14.

When no energy is available in the *Passivity Layer*, the applied torque is set to zero:

$$\tau_{r*} = \begin{cases} \tau_{d*} & \text{if } H_* > 0 \\ 0 & \text{otherwise} \end{cases} \quad (12)$$

Between two samples there will be no way to detect, act upon and therefore prevent a possible loss of passivity. If we know that the interval before a next sample will last T seconds, the available energy is H_* and the worst case motion velocity (highest in module) of the mechanism would be v_*, we can estimate an upper bound for the value of the applied torque τ_{r*} to which it should be constrained not to lose passivity. This tries to enforce:

$$T \bar{\tau}_{r*} v_* \leq H_* \quad (13)$$

An additional saturation method that can be useful is to define a mapping, f_m, from the current available energy in the tank to the maximum torque that can be applied. Meaning:

$$\tau_{r*(max)}(x) = f_m(H_*) \quad (14)$$

This mapping can be designed in such a way that a safe interaction in complex situations is guaranteed. Assume for instance that the slave is stationary but grasping an object in the environment and as a result no energy is exchanged between the slave and the environment. If at some point in time a loss of communication occurs, it could be desirable that the slave will smoothly release the object not to damage it by a continuous application of force. As each iteration the described algorithm for the energy tank sends an energy quantum to the other side, the energy level in the tank will drop even if no energy is exchanged with the environment. This means that the force exerted on the object in the environment will decrease over time and the storage function can be used to shape the manner in which it is released.

V. TRANSPARENCY LAYER

In the previous section we have described an algorithm to maintain passivity in the presence of arbitrary time delays and to make energy available at the slave system. On top of this layer we can implement *any* algorithm to ensure that the desired behavior is displayed and transparency is obtained.

A first proposal for the structure of the *Transparency Layer* is given in Fig. 5. This structure arises from the discussion in Section II-B where it was derived that a local model of the environment is required at the master side to obtain the best transparency in the presence of time delays. In the following sections the functionality of each block of the structure in Fig. 5 will be discussed.

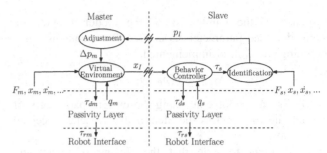

Fig. 5.   Structure of the Transparency Layer

A. Virtual Environment

This block contains a model which tries to mimic the remote environment and with which the user is interacting haptically. Based on the full state and interaction variables $F_m, x_m, \dot{x}_m, \ldots$ of the physical interaction the appropriate torque to apply to the master device is determined. The model is adaptable so that time varying environments can properly be displayed.

Many possibilities can be chosen for the structure of a model of the environment. A structure rich enough for many applications may be the one displayed in Fig. 6, where the right-end bar represents the interaction point between the slave robot and the environment.

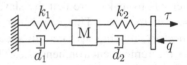

Fig. 6.   A virtual model structure for the remote environment

k_1, d_1 and M determine the dynamic behavior of the manipulated object in the remote environment and k_2 and d_2 the surface contact dynamics of the object itself. Simpler models can be implemented, e.g. based on a priori knowledge of the remote environment, to facilitate the parameter identification.

The model is running on a discrete medium with most likely a fixed sample frequency. This means that in *Impedance Reflection* algorithms using Euler discretized models without passivity properties, there exists an upper bound on the stiffness that can be reflected stably to the user [14]. Here the *Passivity Layer* will enforce passivity of the system and thus keep the system stable. However to minimize the interference of the *Passivity Layer* an elegant solution would be to implement a passive version of the model according to [5].

B. Adjustment

At the master side the parameters of the local model of the remote environment have to be adjusted to match the identified parameters from the slave side. The delay in the communication channel in combination with the movement by the user can result in the generation of impulsive forces when the received parameters are directly introduced into the local model. These impulsive forces can lead to undesired contact behaviors at the master side and therefore a gradual convergence of the parameters of the model towards the received identified parameters is desired. To accomplish this, a smoothing function is implemented:

$$\Delta p_m = f(p_m, p_I) \tag{15}$$

where p_m is the vector containing the current parameters of the model and p_I contains the identified parameters received from the slave.

It is important to note that the change of parameters like mass, stiffness and length of a spring has energetical consequences in the model. When a passive implementation of the model is used this change of internal energy has to be accounted for. In order to have information on this energetical consequences, we can proceed as follows. Suppose that the state of the model would be $x \in \Re^n$. The energy function representing the total energy would then be a scalar function of x: $H(x) \in \Re$. We can express the energy then as a function of p_m as well to get $H(x, p_m)$. By requiring a change in parameters Δp_m, the change in internal energy would be $\Delta H = H(x, p_m + \Delta p_m) - H(x, p_m)$. This energy can be drawn from the energy tank in the *Passivity Layer*.

C. Behavior Controller

This controller has to make sure that the slave exhibits the desired behavior by the user. This can either be following an imposed trajectory, moving at a certain speed, exerting a certain force on the remote environment, etc. This controller receives the full state information about the interaction between the user and the master device. Based on the difference between this received information and the measured state information of the slave an appropriate control action is computed so that the slave mimics the behavior dictated by the user.

The dynamics of the behavior displayed by the user can be time-varying and/or uncertain and the same can apply for the dynamics of the environment. To avoid oscillations and maintain a proper reference tracking the implemented control structure has to be able to cope with this variation/uncertainty. Suitable control structures to implement here will therefore be either robust or adaptive control structures, of which the adaptive controller might be more suitable as the magnitude of the variation/uncertainty can be difficult to determine a priori.

D. Identification

In order to accurately reflect the interaction forces to the user, the parameters of the local model have to be identified. A lot of research is being done in the field of impedance estimation of for instance soft tissue environments [15]. Any suitable identification algorithm can be implemented here as long as it is able to estimate the parameters of the model. This means that the complexity of the implemented model of the environment will largely determine the required complexity of the identification algorithm.

Several important characteristics of a suitable identification algorithm can be identified. The remote environment might contain delicate structures, e.g. soft tissues, so it is desirable that the identification algorithm does not require a strong excitation of the environment. The slave is probably used to manipulate objects in the remote environment so that hard and soft contacts and free space movements will be interchanging. During this process the stability of the estimation will have to be guaranteed.

VI. EXAMPLE

In order to illustrate the algorithm described above a simple simulation example will be presented here. The values of all the parameters are listed in Table I. The master and slave devices are modelled as two identical one-dimensional arms having a mass m_* and a small viscous friction c_*, where like before the subscript $*$ replaced with m refers to the master side and with s to the slave side.

$$\begin{aligned} \tau_h + \tau_m &= m_m \ddot{q}_m + c_m \dot{q} \\ \tau_e + \tau_s &= m_s \ddot{q}_s + c_s \dot{q} \end{aligned} \tag{16}$$

The user imposes a force τ_h on the master device to make it follow a sinusoidal motion. This is modelled as a PD-controller driven by a sinusoidal reference, q_d:

$$\begin{aligned} \tau_h &= k_h e_h + b_h \dot{e}_h \\ e_h &= q_d - q_m \\ q_d &= -Q_d \sin(\omega t) \end{aligned} \tag{17}$$

The environment contains a wall at position q_w with visco-elastic contact properties located on the trajectory the user tries to follow, q_p is the penetration of the slave device into this wall:

$$\begin{aligned} \tau_e &= k_e q_p + b_e |q_p| \dot{q}_p \\ q_p &= \begin{cases} 0 & \text{if } q_s > q_w \\ q_w - q_s & \text{otherwise} \end{cases} \end{aligned} \tag{18}$$

The proposed algorithm is implemented and runs at a sample frequency of 1000 Hz. The communication channels between the master and slave devices are lossless and have a fixed time delay of $1s$. All simulations have been carried out in the software program 20-sim [16].

A. Passivity Layer

The *Passivity Layer* is implemented as discussed in Section IV. At the slave side a mapping in the form of a simple linear spring with stiffness k_{ss} is associated with the energy level to saturate the control effort:

$$|\tau_{max}| = \sqrt{2 H_s k_{ss}} \tag{19}$$

B. Transparency Layer

For this simple initial example, some a priori knowledge about the environment is used. With respect to Fig. 6 the stiffness k_1 is known to be infinitely stiff and k_2 a linear spring and d_2 a non-linear damper where the damping coefficient depends linearly on the compression of spring k_2. The following implementations were chosen:

1) Virtual Environment:

The sample frequency of the control algorithm is more than sufficient to stably reflect the chosen stiffness of the environment. Therefore a simple discretized model of the environment with adjustable parameters x_0, k_m, d_m is implemented:

$$
\begin{aligned}
p_w(k) &= q_m(k) - x_0 \\
\dot{p}_w(k) &= \frac{p_w(k) - p_w(k-1)}{\Delta T} \\
\tau_{dm}(k) &= \begin{cases} -k_m p_w(k) - b_m \left| p_w \right| \dot{p}_w(k) & \text{if } p_w(k) < 0 \\ 0 & \text{otherwise} \end{cases}
\end{aligned}
\tag{20}
$$

2) Adjustment:

The model is not passively implemented. Therefore there are no energetic consequences connected with parameter changes which need to be handled. The smoothing function limits the change in parameters to a percentage of the difference between the currently used and identified parameters:

$$
\Delta p_m = \gamma(p_I - p_m)
\tag{21}
$$

3) Behavior Controller:

The behavior we want the slave device to mimic is trajectory tracking. The environment and user dynamics in this example are time invariant and therefore the controller parameters are fixed and tuned specifically for this example. Only the position of the master device q_m is transmitted and $q_{\hat{m}}(k)$ is the position of the master system which is received at time instant k by the slave system:

$$
\begin{aligned}
e_s(k) &= q_{\hat{m}}(k) - q_s(k) \\
\dot{e}_s(k) &= \frac{e_s(k) - e_s(k-1)}{\Delta T} \\
\tau_{ds}(k) &= -k_s e_s(k) - b_s \dot{e}_s(k)
\end{aligned}
\tag{22}
$$

4) Identification:

The identification algorithm implemented here is a linear regression algorithm based on [15]. The estimator tries to minimize the cost function:

$$
\begin{aligned}
V_N(k_e, d_e) &= \frac{1}{N} \sum_{k=1}^{N} \epsilon^2(k) \\
\epsilon(k) &= \tau_e(k) - (\hat{k}_e \hat{x}_s(k) + \hat{b}_e \left| \hat{x}_s(k) \right| \dot{\hat{x}}_s(k))
\end{aligned}
\tag{23}
$$

and is implemented by computing the following recursive equations each iteration:

$$
\begin{aligned}
\hat{\theta}(k) &= \hat{\theta}(k-1) + Q(k)[\phi(k) - \mu^T(k)\hat{\theta}(k-1)] \\
Q(k) &= R(k-1)\mu(k)[\beta_e + \mu^T(k)R(k-1)\mu(k)]^{-1} \\
R(k) &= \frac{1}{\beta_e}[\mathbb{I} - Q(k)\mu^T(k)]R(k-1)
\end{aligned}
\tag{24}
$$

where $\theta = \begin{vmatrix} \hat{k}_e \\ \hat{b}_e \end{vmatrix}$, $\mu = \begin{vmatrix} x_s \\ |\hat{x}_s|\dot{\hat{x}}_s \end{vmatrix}$, β_e is a forgetting factor to limit the estimation to more recent measurements, $\mathbb{I}$ is a 2×2 identity matrix and $\mathbf{Q}$ and $\mathbf{R}$ are 2×1 and 2×2 matrices. R is initialized as an identity matrix and no prior information about the parameters of the environment is assumed: $\hat{x}_0(0) = \hat{k}_e = \hat{d}_e = 0$, β_e is chosen as 0.99.

Parameter	Value	Parameter	Value
$m_{m/s}$	0.1 kg	$c_{m/s}$	0.1 Ns/m
k_h	5 N/rad	b_h	1 Ns/rad
Q_d	0.3 rad	ω	0.31 rad/s
k_e	25 N/rad	b_e	10 Ns/rad
q_w	-0.2 rad	H_d	0.01 J
α	100	β	0.01
k_{ss}	10 N^2/J	γ	0.01
k_s	100	b_s	10

TABLE I

PARAMETERS OF THE SIMULATION

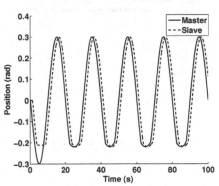

Fig. 7. Position tracking

C. Simulation Results

Fig. 7 shows the position tracking performance of the system. It is clearly visible that the user can initially complete the sinusoidal movement as the parameters of the local model are not yet set. The force exerted on the environment, Fig. 8, however, is limited due to the saturation strategy of Eq. 19. After this start-up phase, a very good correspondence between the master and slave trajectory is obtained. Fig. 8 shows the

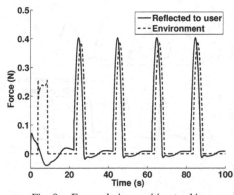

Fig. 8. Forces during position tracking

interaction forces. At the beginning the user is moving in free space, as the model parameters are initially not set, and the force applied is used to extract energy from the user to fill the tanks. After the initialization period there is also a good correspondence between the reflected force to the master and the remote interaction force. The reflected force is indeed leading in time with respect to the interaction force. The peak value of the reflected force is slightly higher and this is caused due to a deeper penetration in the virtual wall by the user than really occurs in the remote environment. Some spurious forces are present in the reflected force which accounts for the energy dissipated by the friction present in the slave system.

That the system is indeed passive can be seen in Fig. 9. The energy which flows into the remote environment is bounded by the energy injected by the user. The difference between the energy flows is stored in the energy tanks and the communication channel between the *Passivity Layers*. As mentioned, this energy is at the moment a quite large part of the energy injected by the user ($0.04J$).

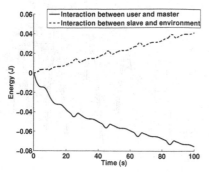

Fig. 9.   Energy exchange at interaction ports to the external world

A situation that demonstrates the benefit of the inclusion of the *Passivity Layers* is a loss of communication between the master and slave system. Fig. 10 shows the slave system response and at $t = 46$s all communication between master and slave system is eliminated. At this time instant the slave robot is still in contact with the remote environment but smoothly moves to a position where no force is applied. When the communication is restored at $t = 64$s the slave robot smoothly restarts following the master system.

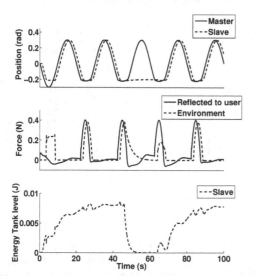

Fig. 10.   System response when communication loss occurs

VII. Conclusions

A structure is presented that allows the combination of passivity and transparency in a very intuitive manner. The layered approach can contain different strategies and allows the separate design of the passivity and transparency properties of the telemanipulation chain. Simulation results of a first implementation show that this approach is very promising and

deals very well with large time delays. It should be noted that the achievable performance in the proposed implementation relies heavily on the ability to properly characterize the remote environment, which can be problematic in practical implementation. The proposed framework, however, can accommodate many variations of the proposed implementation.

VIII. Future Work

The initial implementation as discussed in this paper will be implemented on a physical master-slave system to compare its performance with existing bilateral telemanipulation algorithms. As the structure can easily accomodate different implementations of the *Passivity* and *Transparency Layer*, further research will investigate different implementations based on this framework. One of the first additions to the *Passivity Layer* will be a friction compensation technique to extend this approach to manipulators with high internal friction.

References

[1] B. Bethea, A. Okamura, M. Kitagawa, T. Fitton, S. Cattaneo, V. Gott, W. Baumgartner, and D. Yuh, "Application of haptic feedback to robotic surgery," *Journal of Laparoendoscopic & Advanced Surgical Techniques*, vol. 14, no. 3, pp. 191–195, 2004.

[2] P. Hokayem and M. Spong, "Bilateral teleoperation: An historical survey," *Automatica*, vol. 42, pp. 2035–2057, 2006.

[3] S. Stramigioli, A. van der Schaft, B. Maschke, and C. Melchiorri, "Geometric scattering in robotic telemanipulation," *IEEE Transactions on Robotics and Automation*, vol. 18, no. 4, pp. 588–596, August 2002.

[4] P. Arcara and C. Melchiorri, "Control schemes for teleoperation with time delay: A comparative study," *Robotics and Autonous Systems*, vol. 38, no. 1, pp. 49–64, 2002.

[5] S. Stramigioli, C. Secchi, A. van der Schaft, and C. Fantuzzi, "Sampled data systems passivity and discrete port-hamiltonian systems," *IEEE Transactions on Robotics*, vol. 21, no. 4, pp. 574–587, August 2005.

[6] C. Secchi, S. Stramigioli, and C. Fantuzzi, "Transparency in port-hamiltonian-based telemanipulation," *IEEE Transactions on Robotics*, vol. 24, no. 4, pp. 903–910, August 2008.

[7] G. Niemeyer, "Using wave variables in time delayed force reflecting teleoperation," Ph.D. dissertation, Massachusetts Institute of Technology, September 1996.

[8] H. Ching and W. J. Book, "Internet-based bilateral teleoperation based on wave variable with adaptive predictor and direct drift control," *Journal of Dynamic Systems, Measurement, and Control*, vol. 128, no. 1, pp. 86–93, 2006.

[9] B. Hannaford, "A design framework for teleoperators with kinesthetic feedback," *IEEE Transactions on Robotics and Automation*, vol. 5, no. 4, pp. 426–434, 1989.

[10] C. Tzafestas, S. Velanas, and G. Fakiridis, "Adaptive impedance control in haptic teleoperation to improve transparancy under time-delay," *Proceedings of ICRA 2008*, pp. 212–219, 2008.

[11] K. Kawashima, K. Tadano, G. Sankaranarayanan, and B. Hannaford, "Model-based passivity control for bilateral teleoperation of a surgical robot with time delay," *Proceedings of IROS 2008*, pp. 1427–1432, 2008.

[12] D. Lee and K. Huang, "Passive position feedback over packet-switching communication network with varying-delay and packet-loss," *Symposium on haptic interfaces for virtual environment and teleoperator systems, 2008*, pp. 335–342, March 2008.

[13] J.-H. Ryu, D.-S. Kwon, and B. Hannaford, "Stable teleoperation with time-domain passivity control," *IEEE Transactions on Robotics*, vol. 20, no. 2, pp. 365–373, 2004.

[14] J. Colgate, P. Grafting, M. Stanley, and G. Schenkel, "Implementation of stiff virtual walls in force reflecting interfaces," *Virtual Reality Annual International Symposium*, pp. 202–208, 1993.

[15] N. Diolaiti, C. Melchiorri, and S. Stramigioli, "Contact impedance estimation for robotic systems," *IEEE Transactions on Robotics*, vol. 21, no. 5, pp. 925–935, 2005.

[16] Controllab Products B.V., "20-sim version 4.0," http://www.20sim.com/, 2008.

Feedback Control for Steering Needles Through 3D Deformable Tissue Using Helical Paths

Kris Hauser*, Ron Alterovitz†, Nuttapong Chentanez*, Allison Okamura‡, and Ken Goldberg*

*IEOR and EECS Departments, University of California, Berkeley, {hauser,nchentan,goldberg}@berkeley.edu
†Computer Science Department, University of North Carolina, Chapel Hill, ron@cs.unc.edu
‡Mechanical Engineering Department, Johns Hopkins University, aokamura@jhu.edu

Abstract—Bevel-tip steerable needles are a promising new technology for improving accuracy and accessibility in minimally invasive medical procedures. As yet, 3D needle steering has not been demonstrated in the presence of tissue deformation and uncertainty, despite the application of progressively more sophisticated planning algorithms. This paper presents a feedback controller that steers a needle along 3D helical paths, and varies the helix radius to correct for perturbations. It achieves high accuracy for targets sufficiently far from the needle insertion point; this is counterintuitive because the system is highly underactuated and not locally controllable. The controller uses a model predictive control framework that chooses a needle twist rate such that the predicted helical trajectory minimizes the distance to the target. Fast branch and bound techniques enable execution at kilohertz rates on a 2GHz PC. We evaluate the controller under a variety of simulated perturbations, including imaging noise, needle deflections, and curvature estimation errors. We also test the controller in a 3D finite element simulator that incorporates deformation in the tissue as well as the needle. In deformable tissue examples, the controller reduced targeting error by up to 88% compared to open-loop execution.

I. INTRODUCTION

Needles are used in medicine for a wide range of diagnostic and therapy delivery procedures. Many applications, such as biopsies and prostate brachytherapy, require the needle tip to be positioned accurately at a target in the tissue. Such procedures require a great deal of clinician skill and/or trial and error, and can be difficult even under image guidance. Errors are introduced when positioning the needle for insertion, and the needle can be deflected from its intended path by tissue inhomogeneities. Furthermore, the target may shift due to tissue deformations caused by the needle or patient motion.

Asymmetric-tip flexible needles are a new class of needles, developed in collaboration between Johns Hopkins and U.C. Berkeley, that can steer through soft tissue [24]. As it is inserted, the needle travels along a curved path due to the asymmetric cutting forces generated by either a bevel cutting tip or a pre-bent kink. The arc direction can be controlled by twisting the base of the needle, which allows the needle to travel in circular arcs [24] (by holding twist constant) and helical trajectories (by simultaneously twisting and inserting the base at constant velocity) [12]. In principle, these inputs can be used to correct for needle placement errors and deflections during insertion. They may also steer around obstacles such as bones or sensitive organs to reach targets that are inaccessible to standard rigid needles.

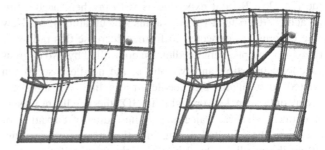

Fig. 1. Needle steering in a soft block of tissue. Left: the needle pierces the surface. Dotted curve is the predicted helical path of the needle. Right: feedback control corrects for errors due to deformation. More examples can be found at http://automation.berkeley.edu/projects/needlesteering/.

A variety of techniques have been used for 3D path planning of steerable needle trajectories [12, 13, 20, 26]. But these techniques assume an idealized rigid tissue model; in practice, the needle will deviate from the planned path due to deformation effects (of both tissue and needle), tissue inhomogeneities, and patient variability. Planners that address uncertainty and deformations are currently limited to 2D needle steering, and perform extensive precomputations [1, 2].

Our work breaks with the trend toward increasingly precise modeling and more sophisticated (and computationally expensive) planning techniques. We present a closed-loop, model predictive controller that steers the needle along helical paths to reach a desired target position. Perturbations are sensed using imaging feedback, and the controller reacts by adjusting the radius and heading of the helix to pass as close as possible to the target. The optimization step searches only the trajectory space of constant controls, and does not predict future changes in the control input. A fast branch and bound search enables trajectory corrections to be computed at kilohertz rates.

We evaluate the controller on a wide variety of simulation experiments in both rigid tissue and in a finite element deformable tissue simulator (Figure 1). In light of the fact that steerable needles are underactuated and non-controllable, our results are surprising. Assuming no perturbations, the controller achieves high accuracy (less than 1% of the needle's radius of curvature r) for all targets sufficiently far from the insertion point (approximately twice r). It attains reasonable accuracy (less than 5% of r) under a variety of simulated Markovian and non-Markovian perturbations, including imaging noise, needle deflection, and curvature estimation errors. In deformable tissue, the controller can compensate for deformation effects to achieve much higher accuracy than open-loop execution. In the example of Figure 1, the controller reduces targeting error by 88%.

II. RELATED WORK

Several mechanisms for needle steering have been proposed. Symmetric-tip flexible needles can be steered by translation and rotation at the needle base [10, 14]. Bevel-tip flexible needles achieve steering with a tip that produces asymmetric cutting forces, causing them to bend when inserted into soft tissue [24, 25]. Recent experiments with pre-bent needle tips have achieved up to a 6.1 cm radius of curvature [22]. New needle designs are expected to further decrease this radius.

Bevel-tip and pre-bent needles are controlled by two degrees of freedom: insertion and twisting at the needle base. When restricted to a plane, the needle moves like a Dubins car that can only turn left or right. In 3D, the needle moves in approximately helical trajectories. The radius of curvature of the path can be adjusted during insertion using duty cycling, enabling the needle to be driven straight forward [18].

Planning motions of 1D flexible objects has been studied in the context of redundant manipulators [6, 7, 8], ropes [23] and symmetric-tip flexible needles [10, 14]. Several researchers have developed motion planners for steerable needles in 3D tissues [12, 13, 20, 26]. These methods have all assumed that tissue is rigid. Duindam et al. derived a closed-form inverse kinematics solution for reaching a desired position orientation in 3D [13]. The solution uses up to three stop-and-turn motions in which the needle, between turns, follows a circular arc. Inverse kinematics solutions for helical paths have not been found. Planning with helices in 3D has been achieved using numerical optimization [12], an error diffusion technique [20], and a sampling-based planner [26].

Needle and tissue deformation will deflect the needle from an idealized trajectory. In such cases a controller may be used to keep a needle on its intended path. Kallem and Cowan developed controllers to stabilize a needle to a given plane [17]. For 2D, Alterovitz et al. developed an optimization-based planner that predicts tissue deformations using finite element simulation [2], although this assumes no uncertainty in the motion model or tissue properties. Alterovitz et al. developed a 2D motion planner for needle steering with obstacles and Markov motion uncertainty [1], but did not model large-scale tissue deformations or curvature estimation errors.

III. PROBLEM STATEMENT AND MODELING ASSUMPTIONS

A. Problem Statement

We wish to place the needle tip at a certain goal location $\mathbf{p}_g$ in the tissue. The desired needle orientation is not specified. We assume we have access to an imaging system that periodically provides an estimate of the world-space coordinate of $\mathbf{p}_g$ as it moves due to tissue deformations, as well as the position, heading, and bevel direction of the needle tip. We also assume a given workspace $\mathcal{W} \subseteq \mathbb{R}^3$ in which the needle

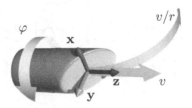

Fig. 2. The coordinate frame attached to the needle tip. Figure adapted with permission from [13].

is permitted to travel. In this work, $\mathcal{W}$ is taken to be either all of $\mathbb{R}^3$ or a bounding box.

We assume an insertion velocity is chosen for the duration of the procedure (on the order of 1 cm/s). The controller outputs a twist rate to be applied at the needle base. We impose a maximum twist rate φ_{max} (specified in rad/m) because fast spinning may damage the tissue. A robot or physician will simultaneously insert and twist for Δt seconds at the twist rate computed by the controller. The controller then repeats.

B. Rigid Tissue Motion Model

Though the controller is designed to operate in deformable tissue, it uses a purely kinematic rigid tissue motion model [24] for computational efficiency. This model assumes the dynamics of the needle are fully determined at the needle tip, such that the tip travels with constant curvature $1/r$ in the bevel direction. Thus, if inserted without twisting, the needle trajectory is a circle with radius of curvature r. Experiments suggest that this assumption approximately holds in homogeneous materials [24]. The model also assumes that a twist at the base of the needle is transmitted directly to the needle tip, and does not deflect the tip position. This assumption neglects torsional friction along the needle shaft, which causes a "lag" before twists are felt at the tip. Control techniques have been developed to counteract these effects [21].

We place a frame $\mathbf{x}, \mathbf{y}, \mathbf{z}$ at the needle tip such that $\mathbf{z}$ is forward and $\mathbf{y}$ is the bevel direction (Figure 2). Inserting the needle causes the tip to move along $\mathbf{z}$ with velocity $v(t)$ and rotate about $\mathbf{x}$ with angular velocity $v(t)/r$. A twist rotates the tip about $\mathbf{z}$ with angular velocity $\varphi(t)$. In screw notation, the instantaneous velocity of the frame is given by:

$$\hat{V}(t) = \begin{bmatrix} 0 & -\varphi(t) & 0 & 0 \\ \varphi(t) & 0 & -v(t)/r & 0 \\ 0 & v(t)/r & 0 & v(t) \\ 0 & 0 & 0 & 1 \end{bmatrix}$$

If $v(t) \neq 0$, then we can factor out $v(t)$ so that the needle travels at unit velocity, and the only variable to control is the twist rate $\varphi(t)$. If the twist rate is held constant, then after inserting the needle a distance d, the tip frame is transformed by the quantity $T(d) = \exp(d\hat{V})$ [12].

We model a trajectory as a sequence of screw motions with piecewise constant twist rates $\varphi(d)$ parameterized by insertion distance. Denote $\mathbf{t}_{\varphi(d)}(d)$ as the trajectory of the tip position under twists $\varphi(d)$. We drop the subscript if the twists are unimportant.

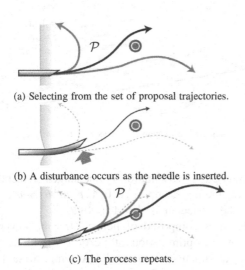

(a) Selecting from the set of proposal trajectories.

(b) A disturbance occurs as the needle is inserted.

(c) The process repeats.

Fig. 3. Overview of the control policy. Curves emanating from the needle depict proposal trajectories. Circle depicts the target.

IV. FEEDBACK CONTROLLER WITH HELICAL PATHS

A. Controller Framework

We assume a model predictive control (MPC) framework that iterates the following steps (Figure 3):

1) *Propose.* Generate a set $\mathcal{P}$ of proposal trajectories that start from the current estimated needle state.
2) *Select.* Find the trajectory in $\mathcal{P}$ with control $\varphi(d)$ that achieves the minimal distance to the target.
3) *Execute.* The needle is then inserted with twists according to $\varphi(d)$ and constant velocity for time Δt.

The process repeats until no trajectory can improve the distance to the target by more than ϵ (we set $\epsilon = 0.2\% r$).

The controller meets basic stability criteria: first, it will terminate, because the predicted distance to the target must decrease by at least ϵ every iteration; second, the final error will be no larger than the distance from the target to $\mathcal{P}$. Convergence is difficult to prove, and depends on the selection of $\mathcal{P}$. Empirical results in Section VI suggest that setting $\mathcal{P}$ to the set of constant-twist-rate helices is sufficient to reach most target locations with high accuracy, even under perturbations.

This is surprising and counterintuitive for two reasons. First, the system is not locally controllable, even if the needle were allowed to move backward [17]. Even though our controller cannot reduce errors asymptotically to zero, it often has low error upon termination. Second, traditional MPC techniques use expensive numerical optimizations at every time step to compute $\mathcal{P}$ (e.g. receding horizon optimal control). Our controller performs well even without considering future changes in twist rate, so each time step is computed efficiently.

In the remainder of this section, we derive a closed form expression for the helical trajectories followed by the needle tip when inserted with constant velocity and twist rate φ. We let the set of proposal trajectories $\mathcal{P}$ contain all such helices for $\varphi \in [-\varphi_{max}, \varphi_{max}]$. Using a visualization of $\mathcal{P}$, we argue that the controller converges for sufficiently distant targets.

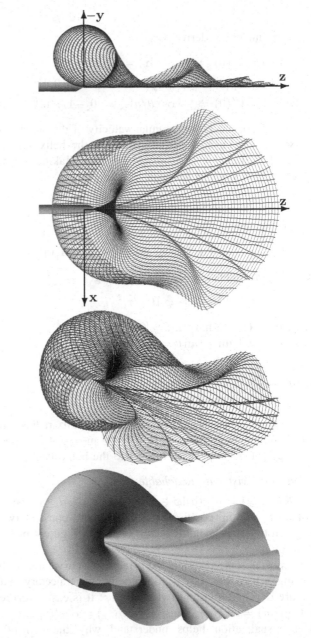

Fig. 4. Four views of the constant-twist-rate reachable set: $z, -y$ plane, z, x plane, perspective, and surface shaded. Twist rate in $(-\infty, \infty)$, insertion distance in $[0, 2\pi r]$.

B. Constant-Twist-Rate Helical Paths

Let $\mathbf{t}_\varphi(d)$ be the needle tip trajectory followed under a constant twist rate φ. We wish to find the helix that describes this trajectory relative to the initial frame of the tip. A helix with radius a, slope θ, and oriented in the $\mathbf{z}$ axis has the arc-length parameterization

$$\mathbf{h}(d) = \begin{bmatrix} a\cos((d/a)\cos\theta) \\ a\sin((d/a)\cos\theta) \\ d\sin\theta \end{bmatrix}. \quad (1)$$

Here we derive the helix parameters a, θ and a rigid transformation A such that $A\mathbf{h}(d) = \mathbf{t}_\varphi(d)$. Denote $\mathbf{h_x}, \mathbf{h_y}, \mathbf{h_z}$, as the coordinate axes of A, and denote $\mathbf{h_o}$ as its origin. We proceed by matching derivatives:

$$Ah(0) = t_\varphi(0) \Rightarrow a\mathbf{h_x} + \mathbf{h_o} = [0,0,0]^T \quad (2)$$

$$Ah'(0) = t'_\varphi(0) \Rightarrow \cos\theta\mathbf{h_y} + \sin\theta\mathbf{h_z} = [0,0,1]^T \quad (3)$$

$$Ah''(0) = t''_\varphi(0) \Rightarrow -\cos^2\theta/a\mathbf{h_x} = [0,-1/r,0]^T \quad (4)$$

Let $\omega = [1/r, 0, \varphi]^T$ be the angular velocity of the needle tip. The axis of rotation must be aligned with the helix axis, so $\mathbf{h_z} = \omega/\|\omega\|$. Let $q = \|\omega\| = \sqrt{1/r^2 + \varphi^2}$. Taking the dot product of both sides of (3) with $\mathbf{h_z}$, we get

$$\sin\theta = \varphi/q \quad \text{and} \quad \cos\theta = 1/(rq).$$

From (4) we see

$$a = r\cos^2\theta = 1/(rq^2) \quad \text{and} \quad \mathbf{h_x} = [0,1,0]^T.$$

Setting $\mathbf{h_y}$ orthogonal to $\mathbf{h_x}$ and $\mathbf{h_z}$, we get

$$\mathbf{h_y} = [-\varphi, 0, 1/r]^T/q.$$

Finally, from (2) we have $\mathbf{h_o} = [0, -a, 0]^T$.

In summary, letting $\tan\theta = r\varphi$, we have

$$t_\varphi(d) = \begin{bmatrix} -a\sin\theta\sin((d/r)\csc\theta) + d\sin\theta\cos\theta \\ a\cos((d/r)\csc\theta) - a \\ a\cos\theta\sin((d/r)\csc\theta) + d\sin^2\theta \end{bmatrix} \quad (5)$$

Where $a = r\cos^2\theta = r/(1/r^2 + \varphi^2)$ is the helix radius, and $[\cos\theta, 0, \sin\theta]^T$ is the helix axis. The tip moves at the speed $\sin\theta = \varphi/\sqrt{1/r^2 + \varphi^2}$ with respect to the helix axis.

C. Constant-Twist-Rate Reachable Set

Let $\mathcal{R}(\varphi_{max}) = \{t_\varphi(d) | \varphi \in [-\varphi_{max}, \varphi_{max}], d \geq 0\}$ be the set of all needle tip positions reachable under constant twist rate, given twist rate bounds $[-\varphi_{max}, \varphi_{max}]$. Figure 4 plots a portion of $\mathcal{R}(\infty)$.

$\mathcal{R}(\varphi_{max})$ is scale-invariant w.r.t. radius of curvature, in the following sense. Let $\tilde{t}_{\alpha\varphi}(d)$ denote the tip trajectory with curvature α/r and constant twist rate $\alpha\varphi$. It is easily verified that $\tilde{t}_{\alpha\varphi}(\alpha d) = \alpha t_\varphi(d)$. Thus, $\mathcal{R}(\alpha\varphi_{max}) = \alpha\mathcal{R}(\varphi_{max})$.

This visualization helps understand why the controller should converge. The projection of $\mathcal{R}(\infty)$ on the x, z plane covers the entire $z > 0$ halfplane, and all twists produce axes of rotation that are parallel to the x, z plane. Thus, if $\mathcal{R}(\infty)$ is rotated by a full turn about any axis, it will sweep out a large portion of $\mathbb{R}^3$. Most trajectories make a full turn quickly — in fact, constructing a trajectory that does not ever make a full turn is extremely difficult. For any target point in this swept volume, errors will converge to zero.

D. Boosting Accuracy with Alternating-Twist Maneuvers

For a finite maximum twist rate φ_{max}, $\mathcal{R}(\varphi_{max})$ has a gap of coverage along the $\mathbf{z}$ axis, shaped approximately like a wedge of angle $|\arctan r\varphi_{max}|$. Consider filling this gap with a maneuver that makes a full turn of the helix with twist rate φ_{max}, and another with twist rate $-\varphi_{max}$ (Figure 5).

Intuitively, the maneuver allows the needle to travel approximately straight forward. After each full turn, the needle tip is displaced by $(2\pi r\sin\theta\cos^2\theta, 0, 2\pi r\sin^2\theta\cos\theta)$ where

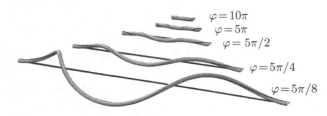

Fig. 5. Alternating-twist maneuvers for various twist rates. Units in rad/r.

$\theta = \arctan r\varphi_{max}$. Orientation is unchanged. Executing both turns places the needle tip at $(0, 0, 4\pi r\sin^2\theta\cos\theta)$. The path length of the maneuver is $d = 4\pi r\cos\theta$.

Because it is composed of helices, the maneuver can be included in the optimization of Section V with only minor changes. Experiments in Section VI-A demonstrate that accuracy is improved when $\mathcal{P}$ is augmented with this maneuver.

V. FAST TRAJECTORY SELECTION

The speed of computing the proposal and selection steps limit the speed at which the needle can be inserted. If more than Δt time is spent, the needle must be halted while the computation finishes. Thus, it is critical for the closest point optimization in the selection step to be fast. In this section we present an efficient branch-and-bound technique for computing the closest point. We also present a bounding volume method for incorporating workspace constraints.

A. Branch-and-Bound Closest Point Computation

As can be seen in Figure 4, the reachable set $\mathcal{R}(\varphi_{max})$ has many valleys and ridges, so local optimization techniques easily get stuck in local minima. Instead, we use branch-and-bound optimization to ensure that a global optimum is found.

We seek to minimize the function $f(\varphi, d) = \|t_\varphi(d) - \mathbf{p}\|$ over the space $\mathcal{S} = \{(\varphi, d) \mid |\varphi| \leq \varphi_{max}, d > 0\}$. Here $\mathbf{p}$ denotes the coordinates of $\mathbf{p}_g$ relative to the needle tip frame. We use an auxiliary function $f_L(R)$, described in Section V-C, that computes a lower bound on the function value attained in a region $R \subseteq \mathcal{S}$. A search tree is then built by recursively splitting the space into subregions which may contain the optimum. When a subregion is split, we also test the value of f at its midpoint. We maintain the helix $\varphi^\star$ and insertion distance $d^\star$ that give the minimum value of $\underline{f}$ seen thus far. If $f_L(R)$ is found to be larger than $\underline{f}$, then subregion R does not contain the optimum and can be safely pruned. The process continues until $\underline{f}$ achieves an ϵ tolerance.

In subsequent iterations of the control loop, the closest point does not change very much, so we can exploit this temporal coherence to help prune many intervals quickly. We keep the optimal twist $\varphi^\star$ and insertion distance $d^\star$ from the previous iteration, and initialize $\underline{f} = f(\varphi^\star, d^\star)$. In our experiments this reduces overall running time by about a factor of four.

The efficiency of branch and bound depends greatly on the tightness of the lower bound. Interval analysis could be used to automatically compute the bounds [16], but the bounds can be extremely poor, depending on how (5) is formulated. In our experiments, the running time of a standard interval analysis

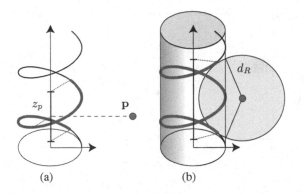

(a)   (b)

Fig. 6. Two initial intervals for finding the closest point from $\mathbf{p}$ to a helix. (a) The interval containing half a turn of the helix, centered about the $z = z_p$ plane. (b) The interval containing the z-range of the intersection between a sphere of radius d_R centered around $\mathbf{p}$, and a cylinder surrounding the helix.

implementation ranged from milliseconds to several seconds, and was sensitive to the location query point.

Our implementation is as follows. We partially discretize $\mathcal{S}$ by selecting $n = 100$ possible values for φ. For a more uniform coverage of the reachable set given a fixed n, we discretize evenly in the $\theta = \arctan r\varphi$ coordinate. To include other proposal trajectories in the search, we divide them into helix segments, and add the segments to $\mathcal{S}$. Then, for each helix, we narrow $\mathcal{S}$ further by computing a narrow initial interval of insertion distances that is guaranteed to contain the closest point. We then employ a good lower bound on the distance from a helix segment to a target point.

B. Initial Intervals for Helix Closest Point

We derive two intervals that must contain the closest point using geometrical reasoning; their intersection provides a tight initial interval for the search. Reparameterize (1) such that $u = (d/a)\cos\theta$, and let $b = a\tan\theta$. Write the helix and point $\mathbf{p}$ in cylindrical coordinates with respect to the helix origin. Denoting cylindrical coordinates with the superscript $\cdot^c$, we have $\mathbf{h}^c(u) = [a, u, ub]^T$. Let $\mathbf{p}^c = [r_p, \theta_p, z_p]^T$.

First, the helix attains z-coordinate z_p at $u = z_p/b$. The closest point to $\mathbf{p}$ must be within a half turn of this parameter; that is, $|u^\star - z_p/b| \leq \pi$ (Figure 6a). The proof is by contradiction: if $u^\star$ were actually outside this interval, then a full turn around the helix could move the z coordinate closer to z_p while keeping x and y fixed. Therefore $u^\star$ is not optimal.

Second, we examine the cylinder bounding the helix. If d_R is the distance from $\mathbf{p}$ to any point on the reachable set (say, $\mathbf{h}(z_p/b)$), then the closest point must lie within a ball of radius d_R around $\mathbf{p}$. Projecting the intersection of the bounding cylinder with the sphere onto the $\mathbf{z}$ axis, we see that the z-coordinate $z^\star$ of the closest point must satisfy $|z_p - z^\star| \leq \sqrt{d_R^2 - (r_p - a)^2}$ (Figure 6b). This bound performs well when d_R is close to the actual helix-point distance.

C. Lower Bounding the Helix-Point Distance

Now we describe a lower bound f_L for $u \in [\underline{u}, \bar{u}]$ and φ given. We compute the distance from the point to a patch (in

cylindrical coordinates) that contains the portion of the helix $[a, a] \times [\underline{u}, \bar{u}] \times [\underline{u}b, \bar{u}b]$. Let $C(x, [a, b])$ denote the clamping function $\max(a, \min(b, x))$, and implicitly assume operations in the θ coordinate are modulo 2π. Then the closest point to this patch is given by

$$\mathbf{q}^c = \begin{bmatrix} a \\ C(\theta_p, [\underline{u}, \bar{u}]) \\ C(z_p, [\underline{u}b, \bar{u}b]) \end{bmatrix} \quad (6)$$

The Cartesian distance $\|\mathbf{q} - \mathbf{p}\|$ is a lower bound.

D. Workspace Bounds

We consider workspace constraints using a simple extension to the selection step. Before computing the helix-point distance, we truncate the trajectory $\mathbf{t}(d)$ if it collides with the workspace boundary. In other words, we find the maximum D for which $\mathbf{t}(d)$ remains in $\mathcal{W}$. Then, the initial branch-and-bound search interval for the closest point (Section V-B) is restricted to lie in $[0, D]$.

We use an recursive bisection to find D up to tolerance ϵ. We start the recursion with interval $I = [0, \bar{d}]$, where $\bar{d}$ is the upper bound on the insertion distance from Section V-B. We then recurse as follows. If the size of I is less than ϵ we set D to its lower bound, as we cannot rule out the possibility of collision in I. Otherwise, we compute the axis-aligned bounding box containing the portion of $\mathbf{t}(d)$ for all $d \in I$ using interval computations. If the box lies within $\mathcal{W}$, the trajectory segment lies entirely in the workspace. If not, we bisect I, and recurse on the lower half. If no collision is reported in the lower half, then we recurse on the upper half.

E. Running Time

Since no interval will have size less than ϵ, the number of intervals examined by the branch and bound algorithm is no more than $\lceil 2L/\epsilon \rceil$, where L is the total length of all initial intervals. Each of the n initial intervals has length at most $2\pi r$, so the number of intervals is bounded by $\lceil 4\pi rn/\epsilon \rceil$. In practice, this bound is extremely loose; the number of examined intervals rarely exceeds a few hundred.

We performed timing experiments on a 2 GHz laptop PC. With an unbounded workspace, the average controller iteration was computed in 0.11 ms, with standard deviation of 1.3 ms. On rare occasions, particularly as the target nears the needle tip, it is much slower, at one point taking 63 ms. More computation is required for bounded workspaces with obstacles. In a box domain with 10 spherical obstacles, each iteration averaged 0.41 ms, with standard deviation 2.52 ms.

VI. SIMULATIONS IN RIGID TISSUE

This section performs a set of simulation experiments to evaluate the accuracy of the controller, as a function of the target position. We simulate the needle using the rigid tissue motion model, first under varying simulation parameters, and then under a variety of perturbations. We define accuracy as the final distance from the needle tip to the target when the controller terminates.

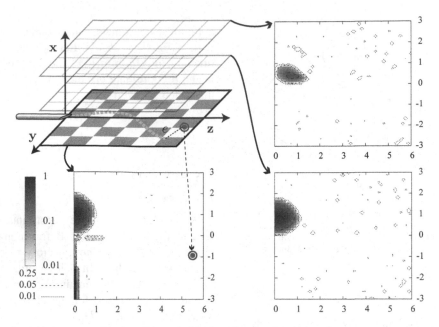

Fig. 7. Controller accuracy in rigid tissue as a function of target's initial relative location. For a given target, the reported accuracy is the final error e achieved by the controller when the needle is inserted at $(0, 0, 0)$. Targets are varied over a 50 x 50 grid in the region $[0r, 6r] \times [-3r, 3r]$ in the $z, -y$ plane, and in three slices along the x-axis at $x=0r$, $1r$, and $2r$. Darker shading indicates higher error. Axes are labeled in multiples of r.

A. Accuracy Without Perturbations

For comparison, we instantiate a *reference controller* that uses the following parameters: a refresh occurs every $2\%r$ of insertion distance, maximum twist rate $\varphi_{max} = 10\pi$ rad/r, and the workspace $\mathcal{W}$ is all of $\mathbb{R}^3$.

Our first experiment evaluates how accuracy of the reference controller varies over the space of targets, under the rigid-tissue motion model. Figure 7 plots accuracy as the target position varies in 2D slices of space. Grids are defined parallel to the $z, -y$ plane (see Figure 7). We then exhaustively enumerate each grid cell, set the target $\mathbf{p}_g$ to the cell center, and then run the controller from the initial configuration to completion. The final error from the needle tip to the target is recorded as the accuracy for that cell. The plots show convergence to sub-millimeter distances almost everywhere, except for a narrow strip along the $+y$ axis, and a circular spot directly above the needle tip in the $-y$ direction. As the x value deviates further from 0, the radius of this spot shrinks. Comparing to Figure 4, we see this region corresponds to the large "lobe" of the reachable set.

The next experiments evaluate accuracy when the controller still has an exact motion model, but with the following control parameters varied:

1) Maximum twist rate reduced to 2.5πrad/r.
2) Without alternating-twist maneuvers (see Section IV-D).
3) In a bounded workspace $[-3r, 3r] \times [-3r, 3r] \times [0r, 6r]$.

For each setting of parameters, we evaluate the accuracy distribution for targets in a region R in space. We chose R to be the rectangle $[0, 0] \times [-3r, 3r] \times [1.5r, 6r]$, which is the rightmost 3/4 of the $x = 0r$ plot of Figure 7. This region was chosen to exclude the circular spot where the reference controller does not converge. For each point in a 50×50 grid

TABLE I
ACCURACY UNDER VARYING CONTROLLER PARAMETERS.

Simulation	Parameters	Accuracy (std), $\% r$
Reference	—	0.2 (0.2)
Reduced twist rate	$\varphi_{max} = 2.5\pi$	2.9 (1.6)
No maneuver	—	1.3 (0.8)
Bounded workspace	—	1.0 (2.1)

over R, we set the target to the point, and ran the controller from start to finish. Table I reports the average accuracy over all grid points and its standard deviation.

B. Under Perturbations and Modeling Errors

The second set of experiments introduce a variety of perturbations and modeling errors into the simulation. The perturbations are discussed in detail below. Table II summarizes the results. Accuracy is measured exactly as in Table I.

All perturbations degrade accuracy to some extent, but rarely drive the error above $10\%r$. This suggests that the controller is stable. Noise in the target position and needle pose estimation can cause significant reductions in accuracy, but in practice a smoothing filter will help mitigate these effects. Twist lag causes high variance in accuracy, in some cases exceeding $20\%r$ error.

1) Noise introduced into the target position estimation: Tracking the target position may require the use of deformable image registration techniques (e.g. [3]), so target position estimates will contain errors due to imaging noise and deformation modeling errors. To simulate, we randomly perturb the target position estimate at each time step with isotropic Gaussian noise with standard deviation σ. We performed trials with $\sigma = 2\%r$ and with $\sigma = 4\%r$. The random number generator

TABLE II

Simulation	Parameters	Accuracy (std), %r
Reference	—	0.2 (0.2)
Target noise	$\sigma = 2\%r$	2.8 (1.5)
Target noise	$\sigma = 4\%r$	6.0 (4.6)
Needle pose noise	$\sigma_p = 4\%r$, $\sigma_o = 2°$	5.4 (4.3)
Needle deflections	$\sigma_p = 44\%$, $\sigma_o = 11°/r$	3.2 (3.1)
Curvature errors	$\tilde{r} = 80\%r$	0.9 (1.2)
Curvature errors	$\tilde{r} = 120\%r$	0.7 (3.8)
Twist lag	$\eta = 6°/r$	2.2 (12)

is seeded with the result of the C `time` procedure at the beginning of each trial.

2) Noise introduced into the needle pose estimation: Limited resolution of medical imaging (approximately 0.8 mm for modern ultrasound [11]) will cause errors in estimating the tip position and orientation. The position and direction of the tip may be estimated directly from the image, but more sophisticated state estimation models are needed to estimate the twist about the needle shaft because the bevel tip is too small to be seen [17]. To simulate these errors, we randomly perturb the needle position by isotropic Gaussian noise with standard deviation $\sigma_p = 2\%r$ and orientation by a rotation about a random axis and a random rotation angle with standard deviation $\sigma_o = 2°$.

3) Random deflections of the needle tip: We simulate random deflections of the needle tip that accumulate over time, and can be mathematically described as a Wiener process. The standard deviation of the position and orientation deflections are respectively $\sigma_p = 0.44r$ and $\sigma_o = 11.1°$ per each r distance the needle is inserted.

4) Curvature estimation errors: We set the actual radius of curvature to $\tilde{r} = 0.8r$ and $\tilde{r} = 1.2r$, while the estimated radius remains at r.

5) Twist lag: As a needle is twisted inside tissue, needle-tissue friction causes internal torsion to build up along the length of the needle. This causes twist lag, a phenomenon where twists applied at the base are not directly commanded to the tip [21]. We model twist lag by nullifying all base twists if $|\theta_b - \theta_t| < \eta L$, where θ_b and θ_t are the base and tip angle, L is the length of needle inside the tissue, and η is a constant (with units of angle over distance). In other words, we evolve θ_b and θ_t with the equations $\dot{\theta}_b = \varphi$, and $\dot{\theta}_t = 0$ if $|\theta_b - \theta_t| < \eta L$, or φ otherwise. We set $\eta = 6°/r$, which is consistent with the experimental data in [21] assuming r=5 cm.

VII. SIMULATIONS IN DEFORMABLE TISSUE

We evaluate the convergence of our controller in a finite element (FEM) simulation of a flexible needle in deformable tissue. The simulator is presented in detail in [5]. Past work has addressed simulation of rigid needles in 2D and 3D deformable FEM models [9, 15, 19], and symmetric tip flexible needles using a linear beam model [14]. Our simulator couples a discrete elastic rod flexible needle model with a FEM tissue simulation. The needle is modeled as a piecewise linear curve connecting a sequence of nodes. These nodes are coupled

to nodes of the tissue, which is modeled as a tetrahedral mesh. Viscous and sticking friction along the needle shaft is simulated, but torsional friction is not.

We performed experiments with an r=5 cm needle in a 10 cm cube of tissue constrained on the bottom face. The tissue mesh is discretized at 1.25 cm intervals, and contains 729 nodes and 3072 tetrahedra. Tissue constitutive properties were set to approximately those of prostate tissue (e.g., elastic modulus 60 kPa) [27]. The friction coefficient between the tissue and needle was set to 0.1 and the minimum cutting force was set to 1 N. We tuned needle stiffness to achieve a turning radius of approximately 5 cm ($\pm$0.3 cm) on a small number of experiments, where we varied the insertion point and angle. The tissue cube is used as the workspace $\mathcal{W}$.

A single trial is shown in Figure 8. The needle tip, initially at (0 cm,5 cm,5 cm) and pointing horizontally, reaches the target (8 cm,2 cm,8 cm) with 0.88 mm accuracy. Open-loop execution of a trajectory planned in rigid tissue achieved 5.9 mm error, so the controller's improvement is 85%. Figure 9.a plots accuracy for targets in the x, z slice at y=5 cm, showing good agreement with accuracy in rigid tissue (Figure 9.b).

We also tested our controller under exaggerated deformation. In Figure 1 we increased the cutting force, friction coefficient, and viscous damping by factor of 10. The tip entered the tissue at a downward angle of 30°. The final error achieved by open loop execution is 13.5 mm, while the controller achieved 1.59 mm, an 88% improvement.

Clinical imaging may impart excessive radiation when used continuously, so we tested controller performance when the imaging rate is reduced. Experiments show less than 2 mm error in both of the above examples even when up to 30 mm of needle travel occurs between imaging refreshes.

VIII. CONCLUSION

We presented a closed-loop control policy for steering bevel-tip flexible needles in deformable tissue. It rapidly iterates a single rule: execute the helical trajectory that minimizes the distance to the target. Fast branch and bound techniques that enable the controller to run in real-time on standard PC hardware. We evaluated the accuracy of the controller in rigid-tissue simulations under simulated disturbances as well as simulations of 3D finite element deformable tissue. The controller reaches relatively distant targets with high accuracy, which is surprising because steerable needles have nonholonomic and non-controllable dynamics.

A primary goal for future work is integration and evaluation on robotic hardware and tissue phantoms. We also plan to address obstacles in the workspace. Preliminary work suggests that the presented controller can already avoid simple obstacles by excluding them from the workspace. Complex workspaces might be addressed by decomposition into convex regions, or more sophisticated planners to generate obstacle-avoiding proposal trajectories. We will also seek to understand the effects of different proposal trajectories, which may enable better workspace coverage, or enable reaching targets with orientation specified. Finally, traditional asymptotic stability

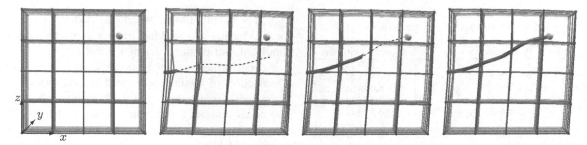

Fig. 8. A simulated trial of the controller in deformable tissue: initial setup, cutting through the surface, tracking the target, final needle configuration. Bottom plane is fixed. Grid used for visualization is at half the resolution of the simulation mesh. Dotted line depicts the trajectory predicted by the controller.

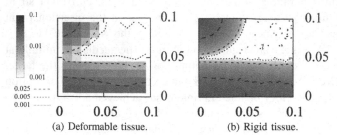

(a) Deformable tissue.   (b) Rigid tissue.

Fig. 9. Accuracy images for a deformable 10 cm cube compared to a rigid cube. Needle was inserted at (0 cm,5 cm). Slice taken down the middle of the cube. Because deformable tissue simulation is computationally expensive, a lower resolution grid was used. Units in meters.

criteria (e.g., [4]) do not apply to our system, so we are pursuing alternate methods to further understand why and when the controller converges.

ACKNOWLEDGMENT

This work is supported in part by NIH grant R01 EB 006435. We thank the reviewers for their helpful comments. We thank our collaborators and colleagues N. Cowan and G. Chirikjian of Johns Hopkins, J. O'Brien, J. Shewchuk, V. Duindam, and R. Jansen of U.C. Berkeley, and J. Pouliot, I-C. Hsu, and A. Cunha of UCSF for their valuable insights and suggestions.

REFERENCES

[1] R. Alterovitz, T. Simeon, and K. Goldberg, "The stochastic motion roadmap: A sampling framework for planning with markov motion uncertainty," in *Robotics: Science and Systems*, June 2007.

[2] R. Alterovitz, K. Goldberg, and A. M. Okamura, "Planning for steerable bevel-tip needle insertion through 2D soft tissue with obstacles," in *IEEE Intl. Conf. on Robot. and Autom.*, Apr. 2005, pp. 1652–1657.

[3] A. Bowden, R. Rabbitt, and J. Weiss, "Anatomical registration and segmentation by warping template finite element models," in *Proc. SPIE (Laser-Tissue Interaction IX)*, vol. 3254, 1998, pp. 469–476.

[4] H. Chen and F. Allgöwer, "Nonlinear model predictive control schemes with guaranteed stability," in *R. Berber and C. Kravaris, editors, Nonlinear Model Based Process Control*. Dodrecht: Kluwer Academic Publishers, 1998, pp. 465–494.

[5] N. Chentanez, R. Alterovitz, D. Ritchie, L. Cho, K. Hauser, K. Goldberg, J. Shewchuk, and J. O'Brien, "Interactive simulation of surgical needle insertion and steering," in *ACM SIGGRAPH*, Aug 2009.

[6] G. Chirikjian and J. Burdick, "A geometric approach to hyper-redundant manipulator obstacle avoidance," *ASME Journal of Mechanical Design*, vol. 114, pp. 580–585, December 1992.

[7] ——, "A modal approach to hyper-redundant manipulator kinematics," *IEEE Trans. Robot. and Autom.*, vol. 10, pp. 343–354, 1994.

[8] G. Chirikjian and I. Ebert-Uphoff, "Numerical convolution on the euclidean group with applications to workspace generation," *IEEE Trans. Robot. and Autom.*, vol. 14, no. 1, pp. 123–136, February 1998.

[9] S. DiMaio and S. Salcudean, "Needle insertion modelling and simulation," in *IEEE Intl. Conf. on Robot. and Autom.*, vol. 2, 2002, pp. 2098–2105.

[10] ——, "Needle steering and motion planning in soft tissues," *IEEE Trans. Biomedical Engineering*, vol. 52, no. 6, pp. 965–974, 2005.

[11] M. Ding and H. Cardinal, "Automatic needle segmentation in three-dimensional ultrasound image using two orthogonal two-dimensional image projections," *Medical Physics*, vol. 30, no. 2, pp. 222–234, 2003.

[12] V. Duindam, R. Alterovitz, S. Sastry, and K. Goldberg, "Screw-based motion planning for bevel-tip flexible needles in 3d environments with obstacles," in *IEEE Intl. Conf. on Robot. and Autom.*, May 2008, pp. 2483–2488.

[13] V. Duindam, J. Xu, R. Alterovitz, S. Sastry, and K. Goldberg, "3d motion planning algorithms for steerable needles using inverse kinematics," in *Workshop on the Algorithmic Foundations of Robotics*, 2008.

[14] D. Glozman and M. Shoham, "Flexible needle steering for percutaneous therapies," *Computer Aided Surgery*, vol. 11, no. 4, pp. 194–201, 2006.

[15] O. Goksel, S. E. Salcudean, S. P. DiMaio, R. Rohling, and J. Morris, "3d needle-tissue interaction simulation for prostate brachytherapy," in *Medical Image Computing and Computer-Assisted Intervention MIC-CAI 2005*, 2005, pp. 827–834.

[16] E. R. Hansen, *Global Optimization Using Interval Analysis*. New York: Marcel Dekker, 1992.

[17] V. Kallem and N. J. Cowan, "Image-guided control of flexible bevel-tip needles," in *IEEE Intl. Conf. on Robot. and Autom.*, 2007.

[18] D. Minhas, J. A. Engh, M. M. Fenske, and C. Riviere, "Modeling of needle steering via duty-cycled spinning," in *Proc. 29th Annual Intl. Conf. of the IEEE Eng. in Medicine and Biology Society*, August 2007, pp. 2756–2759.

[19] H.-W. Nienhuys and A. van der Stappen, "A computational technique for interactive needle insertions in 3d nonlinear material," in *IEEE Intl. Conf. on Robot. and Autom.*, vol. 2, May 2004, pp. 2061–2067.

[20] W. Park, Y. Liu, Y. Zhou, M. Moses, and G. S. Chirikjian, "Kinematic state estimation and motion planning for stochastic nonholonomic systems using the exponential map," *Robotica*, vol. 26, pp. 419–434, July 2008.

[21] K. B. Reed, "Compensating for torsion windup in steerable needles," in *Proc. 2nd Biennial IEEE/RAS-EMBS Intl. Conf. on Biomedical Rob. and Biomechatronics*, Scottsdale, Arizona, 2008.

[22] K. B. Reed, V. Kallem, R. Alterovitz, K. Goldberg, A. M. Okamura, and N. J. Cowan, "Integrated planning and image-guided control for planar needle steering," in *Proc. 2nd Biennial IEEE/RAS-EMBS Intl. Conf. on Biomedical Rob. and Biomechatronics*, Scottsdale, Arizona, 2008.

[23] M. Saha and P. Isto, "Manipulation planning for deformable linear objects," *IEEE Trans. Robot.*, vol. 23, no. 6, pp. 1141–1150, Dec. 2007.

[24] R. J. Webster III, N. J. Cowan, G. Chirikjian, and A. M. Okamura, "Nonholonomic modeling of needle steering," in *9th International Symposium on Experimental Robotics*, June 2004.

[25] R. J. Webster III, J. Memisevic, and A. M. Okamura, "Design considerations for robotic needle steering," in *IEEE Intl. Conf. on Robot. and Autom.*, April 2005, pp. 3588–3594.

[26] J. Xu, V. Duindam, R. Alterovitz, and K. Goldberg, "Motion planning for steerable needles in 3d environments with obstacles using rapidly-exploring random trees and backchaining," in *Proc. IEEE Conf. on Automation Science and Engineering*, August 2008, pp. 41–46.

[27] H. Yamada, *Strength of Biological Materials*. Baltimore: Williams & Wilkins., 1990.

Large Scale Graph-based SLAM
Using Aerial Images as Prior Information

Rainer Kümmerle   Bastian Steder   Christian Dornhege
Alexander Kleiner   Giorgio Grisetti   Wolfram Burgard
Department of Computer Science, University of Freiburg, Germany

Abstract—To effectively navigate in their environments and accurately reach their target locations, mobile robots require a globally consistent map of the environment. The problem of learning a map with a mobile robot has been intensively studied in the past and is usually referred to as the simultaneous localization and mapping (SLAM) problem. However, existing solutions to the SLAM problem typically rely on loop-closures to obtain global consistency and do not exploit prior information even if it is available. In this paper, we present a novel SLAM approach that achieves global consistency by utilizing publicly accessible aerial photographs as prior information. Our approach inserts correspondences found between three-dimensional laser range scans and the aerial image as constraints into a graph-based formulation of the SLAM problem. We evaluate our algorithm based on large real-world datasets acquired in a mixed in- and outdoor environment by comparing the global accuracy with state-of-the-art SLAM approaches and GPS. The experimental results demonstrate that the maps acquired with our method show increased global consistency.

I. INTRODUCTION

The ability to acquire accurate models of the environment is widely regarded as one of the fundamental preconditions for truly autonomous robots. In the context of mobile robots, these models typically are maps of the environment that support different tasks including localization and path planning. The problem of estimating a map with a mobile robot navigating through and perceiving its environment has been studied intensively and is usually referred to as the simultaneous localization and mapping (SLAM) problem.

In its original formulation, SLAM does not require any prior information about the environment and most SLAM approaches seek to determine the most likely map and robot trajectory given a sequence of observations without taking into account any prior information about the environment. However, there are certain scenarios, in which one wants a robot to autonomously arrive at a specific location described in global terms, for example, given by a GPS coordinate. Consider, for example, rescue or surveillance scenarios in which one requires specific areas to be covered with high probability to minimize the risk of potential casualties. Unfortunately, GPS typically suffers from outages or occlusions so that a robot only relying on GPS information might encounter substantial positioning errors. Even sophisticated SLAM algorithms cannot fully compensate for these errors as there still might be lacking constraints between some observations combined with large odometry errors that introduce a high uncertainty in the

current position of the vehicle. However, even in situations with substantial overlaps between consecutive observations, the matching processes might result in errors that linearly propagate over time and lead to substantial absolute errors. Consider, for example, a mobile robot mapping a linear structure (such as a corridor of a building or the passage between two parallel buildings). Typically, this corridor will be slightly curved in the resulting map. Whereas this is not critical in general as the computed maps are generally locally consistent [13], they often still contain errors with respect to the global coordinate system. Thus, when the robot has to arrive at a position defined in global coordinates, the maps built using a standard SLAM algorithm might be sub-optimal.

In this paper, we present an approach that overcomes these problems by utilizing aerial photographs for calculating global constraints within a graph-representation of the SLAM problem. In our approach, these constraints are obtained by matching features from 3D point clouds to aerial images.

Compared to traditional SLAM approaches, the use of a global prior enables our technique to provide more accurate solutions by limiting the error when visiting unknown regions. In contrast to approaches that seek to directly localize a robot in an outdoor environment, our approach is able to operate reliably even when the prior is not available, for example, because of the lack of appropriate matches. Therefore, it is suitable for mixed indoor/outdoor operation. Figure 1 shows a motivating example and compares the outcome of our approach with the ones obtained by applying a state-of-the-art SLAM algorithm and a pure localization method using aerial images.

The approach proposed in this paper relies on the so called graph formulation of the SLAM problem [18, 22]. Every node of the graph represents a robot pose and an observation taken at this pose. Edges in the graph represent relative transformations between nodes computed from overlapping observations. Additionally, our system computes its global position for every node employing a variant of Monte-Carlo localization (MCL) which uses 3D laser scans as observations and aerial images as reference maps. The use of 3D laser information allows our system to determine the portions of the image and of the 3D scan that can be reliably matched by detecting structures in the 3D scan which potentially correspond to intensity variations in the aerial image.

GPS is a popular device for obtaining accurate position

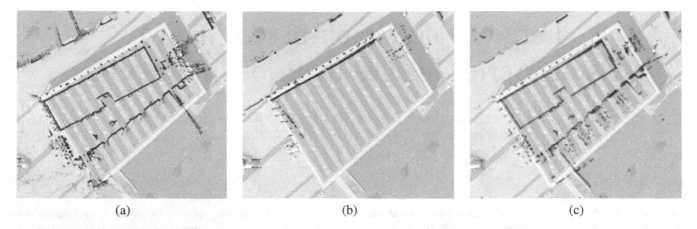

Fig. 1. Motivating example comparing standard SLAM (a), localization using aerial imagery as prior information (b), and our combined approach (c). Note the misalignment relative to the outer wall of the building in (a). Whereas the localization applied in (b), which relies on aerial images, yields proper alignments, it cannot provide accurate estimates inside the building. Combining the information of both algorithms yields the best result (c).

estimates. Whereas it has also been used to localize mobile vehicles operating outdoors, the accuracy of this estimate is usually not accurate enough to obtain a precise map. Generally, the position estimate provided by GPS substantially decreases when the robot moves close to buildings or in narrow streets. Our approach to deal with this problem is to use aerial images and to match measurements obtained by the robot to obtain an accurate estimate of the global position.

The approach proposed in this paper works as follows: we apply a variant of Monte Carlo localization [3] to localize a robot by matching 3D range scans to aerial images of the environment. To achieve this, our approach selects the portions of the scan and of the image which can be reliably matched. These correspondences are added as constraints in a graph-based formulation of the SLAM problem. Note that our system preserves the flexibility of traditional SLAM approaches and can also be used in absence of the prior information. However, when the prior is available our system provides highly accurate solutions also in pathological datasets (i.e., when no loop closures take place). We validate the results with a large-scale dataset acquired in a mixed in- and outdoor environment. We furthermore compare our method with state-of-the-art SLAM approaches and with GPS.

This paper is organized as follows. After discussing related work, we will give an overview over our system followed by a detailed description of the individual components in Section III. We then will present experiments designed to evaluate the quality of the resulting maps obtained with our algorithm in Section IV. In this section, we furthermore compare our approach with a state-of-the-art SLAM system that does not use any prior information.

II. RELATED WORK

SLAM techniques for mobile robots can be classified according to the underlying estimation technique. The most popular approaches are extended Kalman filters (EKFs) [16, 23], sparse extended information filters [7, 26], particle filters [19], and least square error minimization approaches [18, 9, 12].

The effectiveness of the EKF approaches comes from the fact that they estimate a fully correlated posterior about landmark maps and robot poses [16, 23]. Their weakness lies in the strong assumptions that have to be made upon both, the robot motion model and the sensor noise. If these assumptions are violated, the filter is likely to diverge [14, 27].

An alternative approach designed to find maximum likelihood maps is the application of least square error minimization. The idea is to compute a network of relations given the sequence of sensor readings. These relations represent the spatial constraints between the poses of the robot. In this paper, we also follow this approach. Lu and Milios [18] first applied this technique in robotics to address the SLAM problem by optimizing the whole network at once. Gutmann and Konolige [12] proposed an effective way for constructing such a network and for detecting loop closures while running an incremental estimation algorithm.

All the SLAM methods discussed above do not take into account any prior knowledge about the environment. On the other hand, several authors addressed the problem of utilizing prior knowledge to localize a robot outdoors. For example, Korah and Rasmussen [15] used image processing techniques to extract roads on aerial images. This information is then applied to improve the quality of GPS paths using a particle filter by calculating the particle weight according to its position relative to the streets. Leung *et al.* [17] presented a particle filter system performing localization on aerial photographs by matching images taken from the ground by a monocular vision system. Correspondences between aerial images and ground images have been detected by matching line features. These have been generated from aerial images by a Canny edge detector and Progressive Probabilistic Hough Transform (PPHT). A vanishing point analysis for estimating building wall orientations was used on the monocular vision. In contrast to laser-based approaches, their method maximally achieved an average positioning accuracy within several meters. Ding *et al.* [4] use vanishing point analysis to extract 2D corners from aerial images and inertial tracking data, and they also extract

298

2D corners from LiDAR generated depth maps. The extracted corners from LiDAR are matched with those from the aerial image in a multi-stage process. Corresponding matches are used to gain a fine estimation of the camera pose that is used to texture the LiDAR models with the aerial images. Chen and Wang [2] use an energy minimization technique to merge prior information from aerial images and mapping. Mapping is performed by constructing sub-maps consisting of 3D point clouds, that are constrained by relations. Using a Canny edge detector, they compute a vector field from the image that models force towards the detected edges. The sum of the forces applied to each point is used as an energy measure in the minimization process, when placing a sub-map into vector field of the image. Dogruer *et al.* [5] utilized soft computing techniques for segmenting aerial images into different regions, such as buildings, roads, and forests. They applied MCL on the segmented maps. However, compared to the approach presented in this paper, their technique strongly depends on the color distribution of the aerial images since different objects on these images might share similar color characteristics.

Früh and Zakhor [10] introduced the idea of generating edge images out of aerial photographs for 2D laser-based localization. As they stated in their paper, localization errors might occur if rooftops seen on the aerial image significantly differ from the building footprint observed by the 2D scanner. The method proposed in this paper computes a 2D structure from a 3D scan, which is more likely to match with the features extracted from the aerial image. This leads to an improved robustness in finding location correspondences. Additionally, our system is not limited to operate in areas where the prior is available. In this cases, our algorithm operates without relevant performance loss compared to SLAM approaches which do not utilize any prior. This allows our system to operate in mixed indoor/outdoor scenarios.

Sofman *et al.* [24] introduced an online learning system predicting terrain travel costs for unmanned ground vehicles (UGVs) on a large scale. They extracted features from locally observed 3D point clouds and generalized them on overhead data such as aerial photographs, allowing the UGVs to navigate on less obstructed paths. Montemerlo and Thrun [20] presented an approach similar to the one presented in this paper. The major difference to our technique is that they used GPS to obtain the prior. Due to the increased noise which affects the GPS measurements this prior can result in larger estimation errors.

III. GRAPH-SLAM WITH PRIOR INFORMATION FROM AERIAL IMAGES

Our system relies on a graph-based formulation of the SLAM problem. It operates on a sequence of 3D scans and odometry measurements. Every node of the graph represents a position of the robot at which a sensor measurement was acquired. Every edge stands for a constraint between the two poses of the robot. In addition to direct links between consecutive poses, it can integrate prior information (when

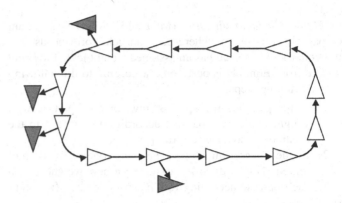

Fig. 2. The graph representation used by our approach. In contrast to the standard approach, we additionally integrate global constraints (shown in gray) given by the prior information.

available) which in our case is given in form of an aerial image.

This prior information is introduced to the graph-SLAM framework as global constraints on the nodes of the graph, as shown in Figure 2. These global constraints are absolute locations obtained by Monte Carlo localization [3] on a map computed from the aerial images. These images are captured from a viewpoint significantly different from the one of the robot. However, by using 3D-scans we can extract the 2D information which is more likely to be consistent with the one visible in the reference map. In this way, we can prevent the system from introducing inconsistent prior information. To initialize the particle filter, we draw the particle positions from a Gaussian distribution, where the mean was determined by GPS. We use 1,000 particles to approximate the posterior.

In the following we explain how we adapted MCL to operate on aerial images and how to select the points in the 3D scans to be considered in the observation model. Subsequently we describe our graph-SLAM framework.

A. Monte Carlo Localization

To estimate the pose $\mathbf{x}$ of the robot in its environment, we consider probabilistic localization, which follows the recursive Bayesian filtering scheme. The key idea of this approach is to maintain a probability density $p(\mathbf{x}_t \mid \mathbf{z}_{1:t}, \mathbf{u}_{0:t-1})$ of the location $\mathbf{x}_t$ of the robot at time t given all observations $\mathbf{z}_{1:t}$ and all control inputs $\mathbf{u}_{0:t-1}$. This posterior is updated as follows:

$$p(\mathbf{x}_t \mid \mathbf{z}_{1:t}, \mathbf{u}_{0:t-1}) = \qquad (1)$$
$$\alpha \cdot p(\mathbf{z}_t \mid \mathbf{x}_t) \cdot \int p(\mathbf{x}_t \mid \mathbf{u}_{t-1}, \mathbf{x}_{t-1}) \cdot p(\mathbf{x}_{t-1}) \, d\mathbf{x}_{t-1}.$$

Here, α is a normalization constant which ensures that $p(\mathbf{x}_t \mid \mathbf{z}_{1:t}, \mathbf{u}_{0:t-1})$ sums up to one over all $\mathbf{x}_t$. The terms to be described in Eqn. (1) are the *prediction model* $p(\mathbf{x}_t \mid \mathbf{u}_{t-1}, \mathbf{x}_{t-1})$ and the *sensor model* $p(\mathbf{z}_t \mid \mathbf{x}_t)$. One contribution of this paper is an appropriate computation of the sensor model in the case that a robot equipped with a 3D range sensor operates in a map generated from a birds-eye view.

For the implementation of the described filtering scheme, we use a sample-based approach which is commonly known as *Monte Carlo localization* (MCL) [3]. MCL is a variant of particle filtering [6] where each particle corresponds to a possible robot pose and has an assigned weight $w^{[i]}$. The *belief update* from Eqn. (1) is performed according to the following two alternating steps:

1) In the **prediction step**, we draw for each particle with weight $w^{[i]}$ a new particle according to $w^{[i]}$ and to the prediction model $p(\mathbf{x}_t \mid \mathbf{u}_{t-1}, \mathbf{x}_{t-1})$.

2) In the **correction step**, a new observation $\mathbf{z}_t$ is integrated. This is done by assigning a new weight $w^{[i]}$ to each particle according to the sensor model $p(\mathbf{z}_t \mid \mathbf{x}_t)$.

Furthermore, the particle set needs to be re-sampled according to the assigned weights to obtain a good approximation of the pose distribution with a finite number of particles. However, the re-sampling step can remove good samples from the filter which can lead to particle impoverishment. To decide when to perform the re-sampling step, we calculate the number N_{eff} of effective particles according to the formula proposed in [6]

$$N_{eff} = \frac{1}{\sum_{i=1}^{N} \left(\widetilde{w^{[i]}} \right)^2}, \quad (2)$$

where $\widetilde{w^{[i]}}$ refers to the normalized weight of sample i and we only re-sample if N_{eff} drops below the threshold of $\frac{N}{2}$ where N is the number of samples. In the past, this approach has already successfully been applied in the context of the simultaneous localization and mapping (SLAM) problem [11].

So far we described the general framework of MCL. In the next section, we will describe our sensor model for determining the likelihood $p(\mathbf{z} \mid \mathbf{x})$ of perceiving the 3D scan $\mathbf{z}$ from a given robot position $\mathbf{x}$ within an aerial image.

B. Sensor Model for 3D Range Scans in Aerial Images

The task of the sensor model is to determine the likelihood $p(\mathbf{z} \mid \mathbf{x})$ of a reading $\mathbf{z}$ given the robot is at pose $\mathbf{x}$. In our current system, we apply the so called endpoint model or likelihood fields [25]. Let z_k be the k-th measurement of a 3D scan $\mathbf{z}$. The endpoint model computes the likelihood of z_k based on the distance between the scan point z_k' corresponding to z_k re-projected onto the map according to the pose $\mathbf{x}$ of the robot and the point in the map d_k' which is closest to z_k' as:

$$p(\mathbf{z} \mid \mathbf{x}) = f(\|z_1' - d_1'\|, \ldots, \|z_k' - d_k'\|). \quad (3)$$

By assuming that the beams are independent and the sensor noise is normally distributed we can rewrite (3) as

$$f(\|z_1' - d_1'\|, \ldots, \|z_k' - d_k'\|) \propto \prod_j e^{\frac{(z_j' - d_j')^2}{\sigma^2}}. \quad (4)$$

Since the aerial image only contains 2D information about the scene, we need to select a set of beams from the 3D scan, which are likely to result in structures, that can be identified and matched in the image. In other words, we need to transform both, the scan and the image in a set of 2D points which can be compared via the function $f(\cdot)$.

To extract these points from the image we employ the standard Canny edge extraction procedure [1]. The idea behind this is, that if there is a height gap in the aerial image, there will often also be a visible change in intensity in the aerial image. This intensity change will be detected by the edge extraction procedure. In an urban environment, such edges typically correspond to borders of roofs, trees, fences, or other structures. Of course, the edge extraction procedure returns a lot of false positives that do not represent any actual 3D structure, like street markings, grass borders, shadows, and other flat markings. All these aspects have to be considered by the sensor model. Figure 3 shows an aerial image and the extracted canny image along with the likelihood-field.

To transform the 3D scan into a set of 2D points which can be compared to the canny image, we select a subset of points from the 3D scan and consider their 2D projection in the ground plane. This subset should contain all the points which may be visible in the reference map. To perform this operation we compute the z-buffer [8] of a scan from a bird's eye perspective. In this way we discard those points which are occluded in the bird's eye view from the 3D scan. By simulating this view, we handle situations like overhanging roofs, where the house wall is occluded and therefore is not visible in the aerial image in a more sophisticated way.

The regions of the z-buffer which are likely to be visible in the canny image are the ones which correspond to relevant depth changes. We construct a 2D scan by considering the 2D projection of the points in these regions. This procedure is illustrated by the sequence of images in Figure 4.

An implementation purely based on a 2D scanner (like the approach proposed by Früh and Zakhor [10]) would not account for occlusions due to overhanging objects. An additional situation where our approach is more robust is in the presence of trees. In this case a 2D view would only sense the trunk, whereas the whole crown is visible in the aerial image.

In our experiments, we considered variations in height of $0.5\,\mathrm{m}$ and above as possible positions of edges that could also be visible in the aerial image. The positions of these variations relative to the robot can then be matched against the Canny edges of the aerial image in a point-by-point fashion and in a similar way like matching of 2D-laser scans against an occupancy grid map.

This sensor model has some limitations. It is susceptible to visually cluttered areas, since it then can find random correspondences in the Canny edges. There is also the possibility of systematic errors, when a wrong line is used for the localization, e.g., in the case of shadows. Since we use position tracking, this is not critical, unless the robot moves through such areas for a long period. The main advantages of the end point model in this context are that it ignores possible correspondences outside of a certain range and implicitly deals with edge points that do not correspond to any 3D structure.

The method, of course, also depends on the quality of the aerial images. Perspective distortions in the images can easily introduce errors. However, in our experiments we did not find

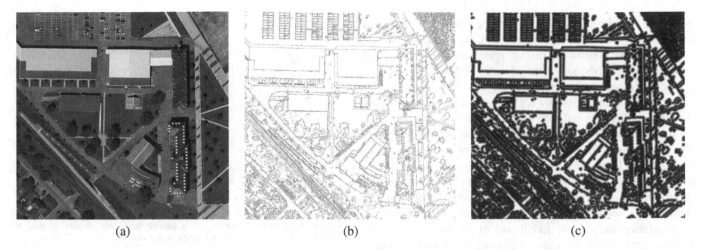

(a)   (b)   (c)

Fig. 3.   Google Earth image of the Freiburg campus (a), the corresponding Canny image (b), and the corresponding likelihood field computed from the Canny image (c). Note that the structure of the buildings and the vertical elements is clearly visible despite of the considerable clutter.

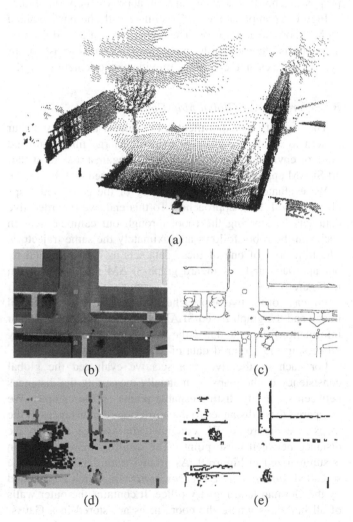

(a)

(b)   (c)

(d)   (e)

Fig. 4.   A 3D scan represented as a point cloud (**a**), the aerial image of the corresponding area (**b**), the Canny edges extracted from the aerial image (**c**), the 3D scene from (a) seen from the top (**d**) (gray values represent the maximal height per cell, the darker a pixel, the lower the height. Areas not visible in the 3D scan are marked in bright gray), and positions extracted from (d), where a high variation in height occurred (**e**).

evidence that this is a major complicating factor.

Note that we employ a heuristic to detect when the prior is not available, i.e., when the robot is inside of a building or under overhanging structures. This heuristic is based on the 3D perception. If there are range measurements whose endpoints are above the robot, no global constraints from the localization are integrated, since we assume that the area the robot is sensing is not visible in the aerial image. While a more profound solution regarding place recognition is clearly possible, this conservative heuristic turned out to yield sufficiently accurate results.

C. Graph-based Maximum Likelihood SLAM

This section describes the basic algorithm for obtaining the maximum likelihood trajectory of the robot. We use a graph-based SLAM technique to estimate the most-likely trajectory, i.e., we seek for the maximum-likelihood (ML) configuration like the majority of approaches to graph-based SLAM.

In our approach, each node $\mathbf{x}_i$ models a robot pose. The spatial constraints between two poses are computed by matching laser scans. An edge between two nodes i and j is represented by the tuple $\langle \delta_{ji}, \Omega_{ji} \rangle$, where δ_{ji} and Ω_{ji} are the mean and the information matrix of the measurement respectively. Let $e_{ji}(\mathbf{x}_i, \mathbf{x}_j)$ be the error introduced by the constraint $\langle j, i \rangle$. Assuming the independence of the constraints, a solution to the SLAM problem is given by

$$(\mathbf{x}_1, \ldots, \mathbf{x}_n)^* = \operatorname*{argmin}_{(\mathbf{x}_1, \ldots, \mathbf{x}_n)} \sum_{\langle j,i \rangle} e_{ji}(\mathbf{x}_i, \mathbf{x}_j)^T \Omega_{ji} e_{ji}(\mathbf{x}_i, \mathbf{x}_j). \quad (5)$$

To account for the residual error in each constraint, we can additionally consider the prior information by incorporating the position estimates of our localization approach. To this end, we extend Eqn. (5) as follows:

$$(\mathbf{x}_1, \ldots, \mathbf{x}_n)^* = \operatorname*{argmin}_{(\mathbf{x}_1, \ldots, \mathbf{x}_n)} \sum_{\langle j,i \rangle} e_{ji}(\mathbf{x}_i, \mathbf{x}_j)^T \Omega_{ji} e_{ji}(\mathbf{x}_i, \mathbf{x}_j)$$
$$+ \sum_{i \in \mathcal{G}} (\mathbf{x}_i - \hat{\mathbf{x}}_i)^T R_i (\mathbf{x}_i - \hat{\mathbf{x}}_i), \quad (6)$$

where $\hat{\mathbf{x}}_i$ denotes the position as it is estimated by the localization using the bird's eye image and R_i is the information matrix of this constraint. In our approach, we compute R_i based on the distribution of the samples in MCL. We use non-linear conjugate gradient to efficiently optimize Eqn. (6). The result of the optimization is a set of poses which maximizes the likelihood of all the individual observations. Furthermore, the optimization also accommodates the prior information about the environment to be mapped whenever such information is available. In particular, the objective function encodes the available pose estimates as given by our MCL algorithm described in the previous section. Intuitively the optimization deforms the solution obtained by the relative constraints path to maximize the overall likelihood of all the observations, including the priors. Note that including the prior information about the environment yields a globally consistent estimate of the trajectory.

If the likelihood function used for MCL is a good approximation of the global likelihood of the process, one can refine the solution of graph-based SLAM by an iterative procedure. During one iteration, new global constraints are computed based on the gradient of the likelihood function estimated at the current robot locations. The likelihood function computed by our approach is based on a satellite image from which we extract "potential" 3D structures which may be seen from the laser. Whereas the heuristics described in Section III-B reject most of the outliers, there may be situations where our likelihood function approximates the true likelihood only locally. We observed that applying the iterative procedure discussed above may worsen the initial estimate generated by our approach.

IV. EXPERIMENTS

The approach described above has been implemented and evaluated on real data acquired with a *MobileRobots Powerbot* with a *SICK LMS* laser range finder mounted on an *Amtec* wrist unit. The 3D data used for the localization algorithm has been acquired by tilting the laser up and down while the robot moves. The maximum translational velocity of the robot during data acquisition was 0.35 m/s. This relatively low speed allows our robot to obtain 3D data which is sufficiently dense to perform scan matching without the need to acquire the scans in a stop-and-go fashion. During each 3D scan the robot moved up to 2 m. We used the odometry to account for the distortion caused by the movement of the platform. Figure 5 depicts the setup of our robot. Although the robot is equipped with an array of sensors, in the experiments we only used the devices mentioned above.

A. Experiment 1 - Comparison to GPS

This experiment aims to show the effectiveness of the localization on aerial images compared with the one achievable with GPS. We manually steered our robot along a 890 m long trajectory through our campus, entering and leaving buildings. The robot captured 445 3D scans that were used for

Fig. 5. The robot used for carrying out the experiments is equipped with a laser range finder mounted on a pan/tilt unit. We obtain 3D data by continuously tilting the laser while the robot moves.

localization. We also recorded the GPS data for comparison purposes. The data acquisition took approximately one hour.

Figure 7 compares the GPS estimate with the one obtained by MCL on the aerial view. The higher error of the GPS-based approach is clearly visible. Note that GPS, in contrast to our approach, does not explicitly provide the orientation of the robot.

B. Experiment 2 - Global Map Consistency

The goal of this experiment is to evaluate the ability of our system to create a consistent map of a large mixed in- and outdoor environment and to compare it against a state-of-the-art SLAM approach similar to the one proposed by Olson [21].

We evaluate the global consistency of the generated maps obtained with both approaches. To this end, we recorded five data sets by steering the robot through our campus area. In each run the robot follows approximately the same trajectory. The trajectory of one of these data sets as it is estimated by our approach and a standard graph-SLAM method is shown in Figure 6.

For each of the two approaches (our method using the aerial image and the graph-based SLAM technique that uses no prior information) we calculated the maximum likelihood map by processing the acquired data of each run.

For each of the five data sets we evaluated the global consistency of the maps by manually measuring the distances between six easily distinguishable points on the campus. We compared these distances to the corresponding distances in the maps (see Figure 8). We computed the average error in the distance between these points. The result of this comparison is summarized in Figure 9. As ground-truth data we used the so-called *Automatisierte Liegenschaftskarte* which is provided by the German land registry office. It contains the outer walls of all buildings where the coordinates are stored in a Gauss-Krüger reference frame.

Compared to SLAM without prior information, our approach has a smaller error and it does not require frequent loop closures to limit the error of the estimate. Note that using our approach loop closures are not required to obtain a globally consistent map. Additionally, the standard deviation of the

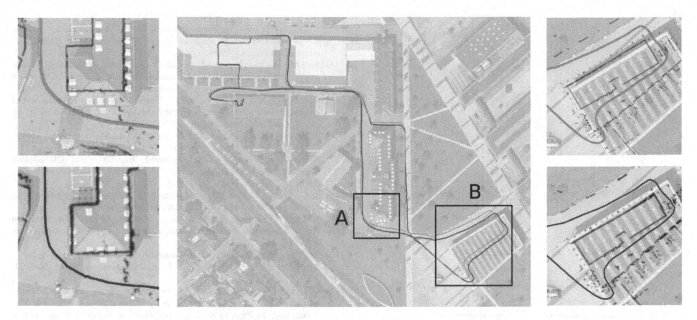

Fig. 6. Comparison of our system to a standard SLAM approach in a complex indoor/outdoor scenario. The center image shows the trajectory estimated by the SLAM approach (light gray) and the trajectory generated by our approach (dark gray) overlaid on the Google Earth image used as prior. On the left and right side, detailed views of the areas marked in the center image are shown, each including the trajectory and map. The upper images show the results of the standard SLAM approach; detail A on the left and B on the right. The lower images show the results of our system (A on the left side and B on the right). It is clearly visible, that, in contrast to the SLAM algorithm without prior information, the map generated by our approach is accurately aligned with the aerial image.

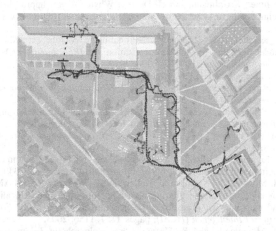

Fig. 7. Comparison between GPS measurements (crosses) and global poses from the localization in the aerial image (circles). Dashed lines indicate transitions through buildings, where GPS and aerial images are unavailable.

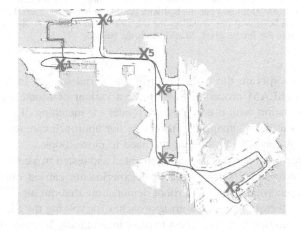

Fig. 8. The six points (corners on the buildings) we used for evaluation are marked as crosses on the map.

estimated distances is substantially smaller than the standard deviation obtained with a graph-SLAM approach that does not utilize prior information. Our approach is able to robustly estimate a globally consistent map on each data set.

C. Experiment 3 - Local Alignment Errors

Ideally, the result of a SLAM algorithm should perfectly correspond to the ground truth. For example, the straight wall of a building should lead to a straight structure in the resulting map. However, the residual errors in the scan matching process typically lead to a slightly bended wall. We investigated this in our data sets for both SLAM algorithms by analyzing an

approximately 70 m long building on our campus. To measure the accuracy, we approximated the first part of the wall by a line and extended this line to the other end of the building. In a perfectly estimated map, both corners of the building are located on this line. Figure 10 depicts a typical result. On average the distance between the horizontal line and the corner of the building for graph-SLAM is 0.5 m whereas it is 0.2 m for our approach.

V. CONCLUSION

In this paper, we presented an approach to solve the SLAM problem in mixed in- and outdoor environments based on 3D range information and using aerial images as prior information.

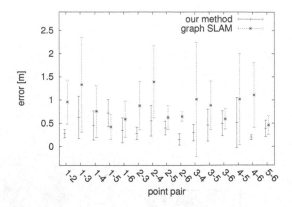

Fig. 9. Error bars ($\alpha = 0.05$) for the estimated distances between the six points used for evaluation of the map consistency.

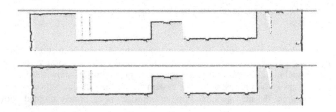

Fig. 10. Close up view of an outer wall of a building as it is estimated by graph-SLAM (top) and our method with prior information (bottom). In both images a horizontal line visualizes the true orientation of the wall. As can be seen from the image, graph-SLAM bends the building.

To incorporate the prior given by the aerial images into the graph-SLAM procedure, we utilize a variant of Monte-Carlo localization with a novel sensor model for matching 3D laser scans to aerial images. In this way, our approach can achieve global consistency without the need to close loops.

Our approach has been implemented and tested in a complex indoor/outdoor setting. Practical experiments carried out on data recorded with a real robot demonstrate that our algorithm outperforms state-of-the-art approaches for solving the SLAM problem that have no access to prior information. In situations, in which no global constraints are available, our approach is equivalent to standard graphical SLAM techniques. Thus, our method can be regarded as an extension to existing solutions of the SLAM problem.

ACKNOWLEDGMENTS

This paper has partly been supported by the DFG under contract number SFB/TR-8 and the European Commission under contract numbers FP6-2005-IST-6-RAWSEEDS and FP7-231888-EUROPA.

REFERENCES

[1] J. Canny. A computational approach to edge detection. *IEEE Transactions on Pattern Analysis and Machine Intelligence*, 8(6):679–698, 1986.

[2] C. Chen and H. Wang. Large-scale loop-closing by fusing range data and aerial image. *Int. Journal of Robotics and Automation*, 22(2):160–169, 2007.

[3] F. Dellaert, D. Fox, W. Burgard, and S. Thrun. Monte carlo localization for mobile robots. In *Proc. of the IEEE Int. Conf. on Robotics & Automation (ICRA)*, 1998.

[4] M. Ding, K. Lyngbaek, and A. Zakhor. Automatic registration of aerial imagery with untextured 3d lidar models. In *Proc. of the IEEE Conf. on Computer Vision and Pattern Recognition (CVPR)*, 2008.

[5] C. U. Dogruer, K. A. Bugra, and M. Dolen. Global urban localization of outdoor mobile robots using satellite images. In *Proc. of the Int. Conf. on Intelligent Robots and Systems (IROS)*, 2007.

[6] A. Doucet, N. de Freitas, and N. Gordan, editors. *Sequential Monte-Carlo Methods in Practice*. Springer Verlag, 2001.

[7] R. Eustice, H. Singh, and J.J. Leonard. Exactly sparse delayed-state filters. In *Proc. of the IEEE Int. Conf. on Robotics & Automation (ICRA)*, pages 2428–2435, 2005.

[8] J. D. Foley, A. Van Dam, K Feiner, J.F. Hughes, and Phillips R.L. *Introduction to Computer Graphics*. Addison-Wesley, 1993.

[9] U. Frese, P. Larsson, and T. Duckett. A multilevel relaxation algorithm for simultaneous localisation and mapping. *IEEE Transactions on Robotics*, 21(2):1–12, 2005.

[10] C. Früh and A. Zakhor. An automated method for large-scale, ground-based city model acquisition. *Int. Journal of Computer Vision*, 60:5–24, 2004.

[11] G. Grisetti, C. Stachniss, and W. Burgard. Improving grid-based SLAM with rao-blackwellized particle filters by adaptive proposals and selective resampling. In *Proc. of the IEEE Int. Conf. on Robotics & Automation (ICRA)*, pages 2443–2448, 2005.

[12] J.-S. Gutmann and K. Konolige. Incremental mapping of large cyclic environments. In *Proc. of the IEEE Int. Symposium on Computational Intelligence in Robotics and Automation (CIRA)*, 1999.

[13] A. Howard. Multi-robot mapping using manifold representations. In *Proc. of the IEEE Int. Conf. on Robotics & Automation (ICRA)*, pages 4198–4203, 2004.

[14] S. Julier, J. Uhlmann, and H. Durrant-Whyte. A new approach for filtering nonlinear systems. In *Proc. of the American Control Conference*, pages 1628–1632, 1995.

[15] T. Korah and C. Rasmussen. Probabilistic contour extraction with model-switching for vehicle localization. In *IEEE Intelligent Vehicles Symposium*, pages 710–715, 2004.

[16] J.J. Leonard and H.F. Durrant-Whyte. Mobile robot localization by tracking geometric beacons. *IEEE Transactions on Robotics and Automation*, 7(4):376–382, 1991.

[17] K. Y. K. Leung, C. M. Clark, and J. P. Huissoon. Localization in urban environments by matching ground level video images with an aerial image. In *Proc. of the IEEE Int. Conf. on Robotics & Automation (ICRA)*, 2008.

[18] F. Lu and E. Milios. Globally consistent range scan alignment for environment mapping. *Journal of Autonomous Robots*, 4:333–349, 1997.

[19] M. Montemerlo, S. Thrun D. Koller, and B. Wegbreit. FastSLAM 2.0: An improved particle filtering algorithm for simultaneous localization and mapping that provably converges. In *Proc. of the Int. Conf. on Artificial Intelligence (IJCAI)*, pages 1151–1156, 2003.

[20] M. Montemerlo and S. Thrun. Large-scale robotic 3-d mapping of urban structures. In *Proc. of the Int. Symposium on Experimental Robotic(ISER)*, pages 141–150, 2004.

[21] E. Olson. *Robust and Efficient Robotic Mapping*. PhD thesis, Massachusetts Institute of Technology, Cambridge, MA, USA, June 2008.

[22] E. Olson, J. Leonard, and S. Teller. Fast iterative optimization of pose graphs with poor initial estimates. In *Proc. of the IEEE Int. Conf. on Robotics & Automation (ICRA)*, pages 2262–2269, 2006.

[23] R. Smith, M. Self, and P. Cheeseman. Estimating uncertain spatial realtionships in robotics. In I. Cox and G. Wilfong, editors, *Autonomous Robot Vehicles*, pages 167–193. Springer Verlag, 1990.

[24] B. Sofman, E. L. Ratliff, J. A. Bagnell, N. Vandapel, and T. Stentz. Improving robot navigation through self-supervised online learning. In *Proc. of Robotics: Science and Systems (RSS)*, August 2006.

[25] S. Thrun, W. Burgard, and D. Fox. *Probabilistic Robotics*. MIT Press, 2005.

[26] S. Thrun, Y. Liu, D. Koller, A.Y. Ng, Z. Ghahramani, and H. Durrant-Whyte. Simultaneous localization and mapping with sparse extended information filters. *Int. Journal of Robotics Research*, 23(7/8):693–716, 2004.

[27] J. Uhlmann. *Dynamic Map Building and Localization: New Theoretical Foundations*. PhD thesis, University of Oxford, 1995.

Highly Scalable Appearance-only SLAM – FAB-MAP 2.0

Mark Cummins and Paul Newman
Oxford University Mobile Robotics Group

Abstract—We describe a new formulation of appearance-only SLAM suitable for very large scale navigation. The system navigates in the space of appearance, assigning each new observation to either a new or previously visited location, without reference to metric position. The system is demonstrated performing reliable online appearance mapping and loop closure detection over a 1,000 km trajectory, with mean filter update times of 14 ms. The 1,000 km experiment is more than an order of magnitude larger than any previously reported result. The scalability of the system is achieved by defining a sparse approximation to the FAB-MAP model suitable for implementation using an inverted index. Our formulation of the problem is fully probabilistic and naturally incorporates robustness against perceptual aliasing. The 1,000 km data set comprising almost a terabyte of omni-directional and stereo imagery is available for use, and we hope that it will serve as a benchmark for future systems.

I. INTRODUCTION

This paper is concerned with the problem of appearance-based place recognition at very large scale. We refer to the problem as "appearance-only SLAM" because our aim is not limited to localization. New observations can be determined to originate from previously unseen locations. Thus the system incrementally constructs a map, and so is effectively a SLAM technique. However, instead of estimating the positions of landmarks in metric coordinates, the system estimates their positions in an appearance space. Because distinctive places can be recognised even after unknown vehicle motion, appearance-only SLAM techniques provide a natural solution to the problems of loop-closure detection, multi-session mapping and kidnapped robot problems. The approach is thus complementary to metric SLAM methods that are typically challenged by these scenarios.

We have addressed these issues previously in [1], [2]. This paper advances the state of the art by describing a new formulation of the approach which extends the applicable scale of the techniques by at least two orders of magnitude. We validate the work on a 1,000 km data set; the largest experiment conducted with systems of this kind by a considerable margin. The data set, including omni-directional imagery, 20Hz stereo imagery and 5Hz GPS, is available for use by other researchers and is intended to serve as a benchmark for future systems.

II. RELATED WORK

The foundations of appearance-based navigation have a long history within robotics, however there have been a number of impressive technical advances quite recently. For example, in the largest appearance-based navigation experiment we are aware of [3], a set of biologically inspired approaches is employed. The system achieved successful loop closure detection

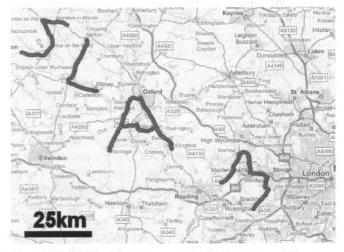

Figure 1: Segments of the 1,000km evaluation trajectory (showing ground truth positions).

and mapping in a collection of more than 12,000 images from a 66 km trajectory, with processing time of less than 100 ms per image. The appearance-recognition component of the system is based on direct template matching, so scales linearly with the size of the environment. Operating at a similar scale, Bosse and Zlot describe a place recognition system based on distinctive keypoints extracted from 2D lidar data [4], and demonstrate good precision-recall performance over an 18 km suburban data set. Related results, though based on a less scalable correlation-based submap matching method, were also described in [5]. Olson described an approach to increasing the robustness of general loop closure detection systems by using both appearance and relative metric information to select a single consistent set of loop closures from a larger number of candidates [6]. The method was evaluated over several kilometers of urban data and shown to recover high-precision loop closures even with the use of artificially poor image features.

Other appearance-based navigation methods we are aware of have generally been applied only at a more modest scale. Many systems have now been demonstrated operating at scales around a kilometer [7], [8], [9], [10]. Indeed, place recognition systems very similar in character to the one described here have become an important component even of single-camera SLAM systems designed for small-scale applications [11].

Considerable relevant work also exists on the more restricted problem of global localization. For example, Schindler et al. describe a city-scale location recognition system [12] based on the vocabulary tree approach of [13]. The system was demonstrated on a 30,000 image data set from 20 km of urban streets, with retrieval times below 200 ms. Also of direct relevance is the research on content-based image retrieval systems in the computer vision community, where systems have been described that deal with more than a million images [14], [15], [13]. However, the problem of retrieval from a fixed index is considerably easier than the full loop-closure problem, because it is possible to tune the system directly on the images to be recognised, and the difficult issue of new place detection does not arise. We believe the results presented in this paper represent the largest scale system that fully addresses these issues of incrementality and perceptual aliasing.

III. SYSTEM DESCRIPTION

A. Probabilistic Model

The probabilistic model employed in this paper builds directly on the scheme outlined in [1]. For completeness, we recap it briefly here.

The basic data representation used is the bag-of-words approach developed in the computer vision community [16]. Features are detected in raw sensory data, and these features are then quantized with respect to a *vocabulary*, yielding *visual words*. The vocabulary is learned by clustering all feature vectors from a set of training data. The Voronoi regions of the cluster centres then define the set of feature vectors that correspond to a particular visual word. The continuous space of feature vectors is thus mapped into the discrete space of visual words, which enables the use of efficient inference and retrieval techniques. In this paper, the raw sensor data of interest is imagery, processed with the SURF feature detector [17], though in principle the approach is applicable to any sensor or combination of sensors, and we have explored multi-sensory applications elsewhere [18].

FAB-MAP, our appearance-only SLAM system, defines a probabilistic model over the bag-of-words representation. An observation of local scene appearance captured at time k is denoted $Z_k = \{z_1, \ldots, z_{|v|}\}$, where $|v|$ is the number of words in the visual vocabulary. The binary variable z_i, which we refer to as an observation component, takes value 1 when the ith word of the vocabulary is present in the observation. $\mathcal{Z}^k$ is used to denote the set of all observations up to time k.

At time k, our map of the environment is a collection of n_k discrete and disjoint locations $\mathcal{L}^k = \{L_1, \ldots, L_{n_k}\}$. Each location has an associated appearance model, given by the set

$$\{p(e_1 = 1|L_i), \ldots, p(e_{|v|} = 1|L_i)\} \tag{1}$$

The variable e_i models feature existence, as distinct from z_i which models feature observation. A detector model relates existence e_i to detection z_i. The detector model captures the rate of false positive and false negative word detections, and is specified by

$$\mathcal{D}: \begin{cases} p(z_i = 1|e_i = 0), & \text{false positive probability.} \\ p(z_i = 0|e_i = 1), & \text{false negative probability.} \end{cases} \tag{2}$$

A further salient aspect of the data is that visual words do not occur independently – indeed, word occurrence tends to be highly correlated. For example, words associated with car wheels and car doors are likely to be observed simultaneously.

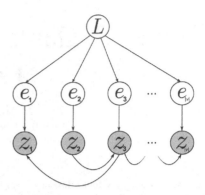

Figure 2: Graphical model of the system. Locations L independently generate existence variables e. Observed variables z_i are conditioned on existence variables e_i via the detector model, and on each other via the Chow Liu tree.

We capture these dependencies by learning a tree-structured Bayesian network using the Chow Liu algorithm [19], which yields the optimal approximation to the joint distribution over word occurrence within the space of tree-structured networks. Importantly, tree-structured networks also permit efficient learning and inference even for very large visual vocabulary sizes. The graphical model of the system is shown in Figure 2.

Given our probabilistic appearance model, localization and mapping can be cast as a recursive Bayes estimation problem, closely analogous to metric SLAM. A pdf over location given the set of observations up to time k is given by:

$$p(L_i|\mathcal{Z}^k) = \frac{p(Z_k|L_i, \mathcal{Z}^{k-1})p(L_i|\mathcal{Z}^{k-1})}{p(Z_k|\mathcal{Z}^{k-1})} \tag{3}$$

Here $p(L_i|\mathcal{Z}^{k-1})$ is our prior belief about our location, $p(Z_k|L_i, \mathcal{Z}^{k-1})$ is the observation likelihood, and $p(Z_k|\mathcal{Z}^{k-1})$ is a normalizing term. We briefly discuss the evaluation of each of these terms below. Full details can be found in [1].

Observation Likelihood: To evaluate the observation likelihood, we assume independence between the current and past observations conditioned on the location, and make use the Chow Liu model of the joint distribution, yielding:

$$p(Z_k|L_i) = p(z_r|L_i) \prod_{q=2}^{|v|} p(z_q|z_{p_q}, L_i) \tag{4}$$

where z_r is the root of the Chow Liu tree and z_{p_q} is the parent of z_q in the tree. Each term in the product can be further expanded as:

$$p(z_q|z_{p_q}, L_i) = \sum_{s_{e_q} \in \{0,1\}} p(z_q|e_q = s_{e_q}, z_{p_q})p(e_q = s_{e_q}|L_i) \tag{5}$$

which can be evaluated explicitly.

Location Prior: The location prior $p(L_i|\mathcal{Z}^{k-1})$ is obtained by transforming the previous position estimate via a simple motion model. The model assumes that if the vehicle is at location i at time $k-1$, it is likely to be at one of the topologically adjacent locations at time k.

Normalization: In contrast to a localization system, a SLAM system requires an explicit evaluation of the normalizing term $p(Z_k|\mathcal{Z}^{k-1})$, which incorporates the probability that the current observation comes from a previously unknown location. If we divide the world into the set of mapped locations M and the unmapped locations $\overline{M}$, then

$$p(Z_k|\mathcal{Z}^{k-1}) = \sum_{m \in M} p(Z_k|L_m)p(L_m|\mathcal{Z}^{k-1}) \quad (6)$$

$$+ \sum_{u \in \overline{M}} p(Z_k|L_u)p(L_u|\mathcal{Z}^{k-1}) \quad (7)$$

The second summation cannot be evaluated directly because it involves all possible unknown locations. However, if we have a large set of randomly collected location models L_u, (readily available from previous runs of the robot or other suitable data sources such as, for our application, Google Street View), we can approximate the summation by Monte Carlo sampling. Assuming a uniform prior over the samples, this yields:

$$p(Z_k|\mathcal{Z}^{k-1}) \approx \sum_{m \in M} p(Z_k|L_m)p(L_m|\mathcal{Z}^{k-1}) \quad (8)$$

$$+ p(L_{new}|\mathcal{Z}^{k-1}) \sum_{u=1}^{n_s} \frac{p(Z_k|L_u)}{n_s} \quad (9)$$

where n_s is the number of samples used, and $p(L_{new}|\mathcal{Z}^{k-1})$ is our prior probability of being at a new location.

Data Association: Once the pdf over locations is computed, a data association decision is made. The observation Z_k is used either to initialize a new location, or update the appearance model of an existing location. Each component of the appearance model is updated according to:

$$p(e_i = 1|L_j, \mathcal{Z}^k) = \frac{p(Z_k|e_i = 1, L_j)p(e_i = 1|L_j, \mathcal{Z}^{k-1})}{p(Z_k|L_j)}$$

$$(10)$$

In the case of new locations, the values $p(e_i = 1|L)$ are initialized to the marginal probability $p(e_i = 1)$ derived from the training data, and then the update is applied.

B. Efficient Large Scale Implementation - FAB-MAP 2.0

The probabilistic model defined above has been used previously in [20], [1], and an approximate inference procedure for it was described in [2]. A key contribution of this paper is to describe a modified version of the model which extends its applicability by more than two orders of magnitude in scale.

For a highly scalable system, we turn to an inverted index retrieval architecture. In computational terms, the inverted index approach essentially scales indefinitely [21]. However, FAB-MAP is not directly implementable using an inverted index structure, because the appearance likelihood $p(Z_k|L_i)$ requires evaluation of Equation 4, $\prod_{q=2}^{|v|} p(z_q|z_{p_q}, L_i)$. The computation pattern is illustrated in Figure 3. Every observation component contributes to the appearance likelihood, including *negative* observations – those where $z_q = 0$, words not detected in the current image. As such, it does not have the sparsity structure that enables inverted index approaches to

scale. Perhaps surprisingly, we have found that simply ignoring the negative observations has a detrimental impact on place recognition performance. Thus we seek a formulation that will enable efficient implementation, but preserve the information inherent in the negative observations.

To enable an inverted index implementation, we modify the probabilistic model in two ways. Firstly, we place some restrictions on the probabilities in the location models. Recalling Equation 1, locations models are parametrized as $\{p(e_1 = 1|L_j), \ldots, p(e_{|v|} = 1|L_j)\}$, that is, by a set of beliefs about the existence of features that give rise to observations of the words in the vocabulary. Let $p(e_i|L_j)|_{\{0\}}$ denote one of these beliefs, where the subscript $\{0\}$ indicates the history of observations that have been associated with the location. Thus $\{0\}$ denotes one associated observation with $z_i = 0$, and $\{0, 0, 1\}$ denotes three associated observations, with $z_i = 1$ in one of those observations. Further, let $p(e_i|L_j)|_0$ indicate that in all observations associated with the location, $z_i = 0$.

In the general probabilistic model described in Section III-A, $p(e_i|L_j)|_0$ can take on a range of values - for example, $p(e_i|L_j)|_{\{0\}} \neq p(e_i|L_j)|_{\{0,0\}}$, as the belief in the non-existence of the feature increases as more supporting observations become available. In FAB-MAP 2.0, the model is restricted so that $p(e_i|L_j)|_0$ must take the same value for all locations; it is clamped at the value $p(e_i|L_j)|_{\{0\}}$. This restriction enables an efficient likelihood calculation, illustrated in Figure 4.

Consider the calculation of one component of the observation likelihood, as per Equation 5, across all locations in the map. In the unrestricted model, this will involve a computation for each location, as illustrated in Figure 4(a). In the restricted model, Figure 4(b), the value of the observation likelihood in all locations where $p(e_i|L_j)|_0$ is the same. Working with log-likelihoods, and given that the distribution will later be normalized, the calculation can be reorganized so that it has a sparse structure, Figure 4(c), which allows for efficient implementation using an inverted index.

We emphasize the fact that this restriction placed on the model is rather slight, and most of the power of the original model is retained. During the exploration phase, when only one observation is associated with each location, the two schemes are identical.[1] The restricted terms $p(e_i|L_j)|_0$ can (and do) vary with i (word ID), and also with time. Treatment of correlations between words, of perceptual aliasing, and of the detector model remains unaffected.

The second change we make to the model concerns data association. Previously, data association was carried out via Equation 10, updating the beliefs $p(e|L)$. Effectively this amounts to capturing the average appearance of a location. For example, if a location has a multi-modal distribution over word occurrence, such as a door that may be either open or shut, then the location appearance model will approach the mean of this distribution. In FAB-MAP 1.0, when computation increased swiftly with the number of appearance models to be evaluated, this was a reasonable design choice. For FAB-MAP 2.0 we switch to representing locations in a sample-based

---

[1]Assuming the detector model does not change with time.

(a) FAB-MAP 1.0

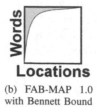
(b) FAB-MAP 1.0
with Bennett Bound

(c) FAB-MAP 2.0

Figure 3: Illustration of the amount of computation performed by the different models. The shaded region of each block represents the appearance likelihood terms $p(z_q|z_{p_q}, L_i)$ which must be evaluated. In (a), FAB-MAP 1.0 [1], the likelihood must be computed for all words in all locations in the map. Using the Bennett bound approximate inference procedure defined in [2], unlikely locations are discarded during the computation, yielding the evaluation pattern shown in (b). The restrictions imposed in FAB-MAP 2.0 allow a fully sparse evaluation, (c).

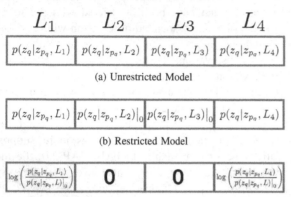

(a) Unrestricted Model

(b) Restricted Model

(c) Sparse likelihood update in the restricted model.

Figure 4: Illustration of calculation of one term of Equation 5, the observation likelihood, for a map with four locations. In (a), the model is unrestricted, and the observation likelihood can take a different value for each location. In (b), the restricted model, the likelihood in all locations where the currently considered word was not previously observed is constrained to take the same value. The calculation can now be organized so that it has a sparse structure, (c).

fashion, which better handles these multi-modal appearance effects. Locations now consist of a set of appearance models as defined in Equation 1, with each new observation associated with the location defining a new such model. This change is necessary because of the restrictions placed on word existence probabilities, but is also largely beneficial. While it means that inference time now increases with every observation collected, the system is sufficiently scalable that this is not of immediate relevance, and the greater ability to deal with variable location appearance is preferred.

Algorithm 1 gives pseudo-code for the calculation. The calculation is divided into two parts, A and B. Part A deals with the positive observations, those for which $z_i = 1$, while Part B deals with negative observations for which $z_i = 0$. The complexity of Part B is $O(\#vocab)$, whereas Part A depends only on the number of words present in a given observation, which is empirically a small constant largely independent of vocabulary size. The straightforward implementation (combining A and B) is in fact fast enough for use, however it can be improved by a caching scheme which eliminates Part B by pre-computing negative votes at the time when the location is added to the map. The negative votes are then adjusted based

Algorithm 1 Log-likelihood update using the inverted index.

Update, Part A (Positive Observations):
```
for z_i in Z, such that z_i = 1 do:
    //Get all locations where word i
    //was observed
    locations = inverted_index[z_i]
    for L_j in locations do:
        //Update the loglikelihood
        //of each of these locations
        loglikelihood[L_j] += log( p(z_i|z_{p_i},L_j) / p(z_i|z_{p_i},L)|_0 )
```
Update, Part B (Negative Observations):
```
for z_i in Z, such that z_i = 0 do:
    locations = inverted_index[z_i]
    for L_j in locations do:

        loglikelihood[L_j] += log( p(z_i|z_{p_i},L_j) / p(z_i|z_{p_i},L)|_0 )
```

on the current observation, but do not need to be recomputed entirely. This yields a scheme where the overall complexity is independent of the size of the vocabulary.

C. Geometric Verification

While a navigation system based entirely on the bag-of-words likelihood is possible, we have found, in common with others [14], that a post-verification stage which checks that the matched images satisfy geometric constraints considerably improves performance. The impact is particularly noticeable as data set size increases - it is helpful on the 70 km data set but almost essential on the 1,000 km set.

We apply the geometric verification to the 100 most likely locations (those which maximize $p(Z_k|L_i, \mathcal{Z}^{k-1})p(L_i|\mathcal{Z}^{k-1})$) and to the 100 most likely samples (the location models used to evaluate the normalizing term $p(Z_k|\mathcal{Z}^{k-1})$). For each of these locations we check geometric consistency with the current observation using RANSAC. Candidate interest point matches are derived from the bag-of-words assignment already computed. Because our aim is only to verify approximate geometric consistency rather than recover exact pose to pose transformations, we assume that the transformation between poses is a pure rotation about the vertical axis. A single point correspondence then defines a transformation, leading to a very rapid verification stage. We accommodate translation between poses by allowing large inlier regions for point correspondences. We have found this simplified model to be effective for increasing navigation performance, while keeping computation requirements to a minimum. Having recovered a set of inliers using RANSAC we recompute the location's likelihood by setting $z_i = 0$ for all those visual words not part of the inlier set. A likelihood of zero is assigned to all locations not subject to geometric verification. For the 1,000 km experiment, the maximum time taken to rerank all 200 locations was 145 ms, and mean time was 10 ms.

D. Visual Vocabulary Learning

1) Clustering: A number of challenges arise in learning visual vocabularies at large scale. The number of SURF features extracted from training images is typically very large; our relatively small training set of 1,921 images produces 2.5 million 128-dimensional SURF descriptors occupying 3.2 GB. Even the most scalable clustering algorithms such as k-means are too slow to be practical. Instead we apply the fast approximate k-means algorithm discussed in [14], where, at the beginning of each k-means iteration, a randomized forest of kd-trees [22], [23] is constructed over the cluster centres, which is then used for fast (approximate) distance calculations. This procedure has been shown to outperform alternatives such as hierarchical k-means [13] in terms of visual vocabulary retrieval performance.

As k-means clustering converges only to a local minima of its error metric, the quality of the visual vocabulary is sensitive to the initial cluster locations supplied to k-means. Nevertheless, random initial locations are commonly used. We have found that this leads to poor visual vocabularies, because there are very large density variations in the feature space. In these conditions, randomly chosen cluster centres tend to lie largely within the densest region of the feature space, and the final clustering over-segments the dense region, with poor clustering elsewhere. For example, in our vehicle-collected data, huge numbers of very similar features are generated by road markings, whereas rarer objects (more useful for place recognition) may only have a few instances in the training set. Randomly initialized k-means yields a visual vocabulary where a large fraction of the words correspond to road markings, with tiny variations between words.

To avoid these effects, we choose the initial cluster centres for k-means using a fixed-radius incremental pre-clustering, where the data points are inspected sequentially, and a new cluster centre is initialized for every data point that lies further than a fixed threshold from all existing clusters. This is similar to the furthest-first initialization technique [24], but more computationally tractable for large data sets. We also modify k-means by adding a cluster merging heuristic. After each k-means iteration, if any two cluster centres are closer than a fixed threshold, one of the two cluster centres is reinitialized to a random location.

2) Chow Liu Tree Learning: Chow Liu tree learning is also challenging at large scale. The standard algorithm for learning the Chow Liu tree involves computing a (temporary) mutual information graph of size $|v|^2$, so the computation time is quadratic in the vocabulary size. For the 100,000

word vocabulary discussed in Section V, to relevant graph would require 80 GB of storage. Happily, there is an efficient algorithm for learning Chow Liu trees when the data of interest is sparse [25]. Meilă's algorithm has complexity $O(s^2 \log s)$, where s is a sparsity measure, equal to the maximum number of visual words present in any training image. Visual word data is typically highly sparse, with only a small fraction of the vocabulary present in any given image. This allows efficient Chow Liu tree learning even for large vocabulary sizes. For example, the tree of the 100,000 word vocabulary used in Section V was learned in 31 minutes on a 3GHZ Pentium IV.

For both the clustering and Chow Liu learning, we use external memory techniques to deal with the large quantities of data involved [26].

IV. DATA SET

For a truly large scale evaluation of the system, the experiments in this paper make use of a 1,000 km data set. The data was collected by a car-mounted sensor array, and consists of omni-directional imagery from a Point Grey Ladybug2, 20Hz stereo imagery from a Point Grey Bumblebee,[2] and 5Hz GPS data. Omni-directional image capture was triggered every 4 meters on the basis of GPS.

The data set was collected over six days in December, with a total length of slightly less than 21 hours, and includes a mixture of urban, rural and motorway environments. The total set comprises 803 GB of imagery (including stereo) and 177 GB of extracted features. There are 103,256 omni-directional images, with a median distance of 8.7 m between image captures – this is larger than the targeted 4 m because the Ladybug2 could not provide the necessary frame rate during faster portions of the route. The median time between image captures is 0.48 seconds, which provides our benchmark for real-time image retrieval performance.

Two supplemental data sets were also collected. A set of 1,921 omni-directional images collected 30 m apart was used to train the visual vocabulary and Chow Liu tree, and also served as the sampling set for the Monte Carlo integration required in Equation 8. The area where this training set was collected did not overlap with the data sets used to test the system. A second smaller data set of 70 km was also collected in August, four months prior to the main 1,000 km dataset. This serves as a smaller-scale test and can also be used for testing cross-season matching performance. The data sets are summarized in Table I.

The 1,000 km data set, collected in mid-December, provides an extremely challenging benchmark for place recognition systems. Due to the time of year, the sun was low on the horizon, so that scenes typically have high dynamic range and quickly varying lighting conditions. We developed custom auto-exposure controllers for the cameras that largely ensured good image quality, however, there is unavoidable information loss in such conditions. Additionally, large sections of the route feature highly self-similar motorway environments, which provide a challenging test of the system's ability to deal with perceptual aliasing. The smaller data set collected during

---

[2]Not used in these results.

August features more benign imaging conditions and will demonstrate the performance that can be typically expected from the system.

Finally, collecting a data set of this magnitude highlights some practical challenges for any truly robust field robotics deployment. We encountered significant difficulty in keeping the camera lenses clean – in winter from accumulating moisture and particulate matter, in summer from fly impacts. For this experiment we periodically cleaned the cameras manually – a more robust solution seems a worthy research topic.

All data sets are available to researchers upon request.

V. RESULTS

The system was tested on the two datasets, respectively 70 km and 1,000 km. As input the system, we used 128D non-rotationally invariant SURF descriptors. The features were quantized to visual words using a randomized forest of eight kd-trees. The visual vocabulary was trained using the system described in Section III-D and the 1,921 image training set described in Section IV. In order to ensure an unbiased Chow Liu tree, the images in the training set were collected 30m apart, so that as far as possible they do not overlap in viewpoint, and thus approximate independent samples from the distribution over images.

We investigate two different visual vocabularies, of 10,000 and 100,000 words respectively. The detector model (Equation 2), the main user-configurable parameter of our system, was determined by a grid search to maximize average precision on a set of training loop closures. The detector model primarily captures the effects of variability in SURF interest point detection and feature quantization error. For the 10,000 word vocabulary we set $p(z = 1|e = 1) = 0.39$ and $p(z = 1|e = 0) = 0.005$. For the 100,000 word vocabulary, the values were $p(z = 1|e = 1) = 0.2$ and $p(z = 1|e = 0) = 0.005$. We also investigate the importance of learning the Chow Liu tree by comparing against a Naive Bayes formulation which neglects the correlations between words. We refer to these different system configurations as "100k, CL" and "100k, NB", and similarly for the 10k word vocabulary.

Performance of the system was measured against ground truth loop closures determined from the GPS data. GPS errors and dropouts were corrected manually. Any pair of matched images that were separated by less than 40 m on the basis of GPS was accepted as a correct correspondence. Note that while 40m may seem too distant for a correct correspondence, on divided highways the minimum distance between correct loop closing poses was sometimes as large as this. Almost all loop closures detected by the system are well below the 40 m limit – 89% were separated by less than 5m, and 98% by less than 10 m.

We report precision-recall metrics for the system. Precision is defined as the ratio of true positive loop closure detections to total detections. Recall is the ratio of true positive loop closure detections to the number of ground truth loop closures. Note that images for which no loop closure exists cannot contribute to the true positive rate, however they can generate false positives. Likewise true loop closures which are incorrectly assigned to a "new place" depress recall but do not impact

our precision metric. These metrics provide a good indication of how useful the system would be for loop closure detection as part of a metric SLAM system – recall at 100% precision indicates the percentage of loop closures that can be detected without any false positives that would cause filter divergence. Note that a typical loop closure consists of a sequence of several images, so even a recall rate of 20% or 30% is sufficient to detect most loop closure events, provided the detections have uniform spatial distribution.

Overall, we found the system to have excellent performance over the 70 km dataset, while the 1,000 km data set was more challenging. Precision recall curves for the two dataset are shown in Figures 5 and 6, and given numerically in Table II. Loop closing performance is also visualized in the maps shown in Figures 7 and 8. Loop closures are often detected even in the presence of large changes in appearance, a typical example is shown in Figure 10.

The performance contributions of the motion model and the geometric verification step are analysed in Figure 5. The geometric check in particular is useful in maintaining recall at higher levels of precision. The motion model is largely unnecessary on the 70km set, however it makes a more noticeable contribution on the 1,000 km set. The effect of vocabulary size and the Chow Liu tree is shown in Figure 6. Performance increases strongly with vocabulary size. The Chow Liu tree also boosts performance on all datasets and at all vocabulary sizes. The effect is weaker at the very highest levels of precision. The reason for this effect is that while the Chow Liu tree will on average improve the likelihood estimates assigned, some individual likelihoods may get worse. The recall at 100% precision is determined by the likelihood assigned to the very last false positive to be eliminated. While on average we expect the Chow Liu tree to improve this likelihood estimate, the opposite may be observed in some fraction of data sets. Below 100% precision the results are sensitive to the likelihood estimates for a larger number of false positives, and so the improvement due to the Chow Liu tree is more robustly observable.

The recall rate for the 70 km dataset is 48.4% at 100% precision. The spatial distribution of these loop closures is uniform over the trajectory – thus essentially every pose will be either detected as a loop closure, or a lie within a few meters of a loop closure. There are two short segments of the trajectory where this is not the case, one in a forest with poor lighting conditions, another in open fields with few visual landmarks. For practical purposes this dataset can be considered "solved". By contrast, the recall for the 1,000 km data set at 100% precision is only 3.1%. However, this figure requires careful interpretation – the data set contains hundreds of kilometers of highways, where the environment contains essentially no distinctive visual content. It is perhaps not reasonable to expect appearance-based loop closure detection in such conditions. To examine performance more closely, we segmented the data set into portions where the vehicle is travelling below 50 km/h (urban), and others (highways, etc). In urban areas (31% of the dataset) the recall is 6.5% at 100% precision, rising to 18.5% at 99% precision. The sharp drop in recall between 99% and 100% precision is

Table I: Data set summary.

Distance	No. of Images	Median distance between images	Size of Extracted Features	Environment
1,000 km	103,256	8.7 m	177 GB	Highways, Urban, Rural
70 km	9,575	6.7 m	16 GB	Urban, Rural

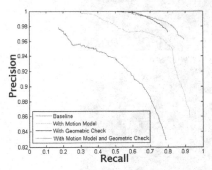

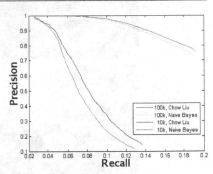

Figure 5: Precision-recall curves for the 70 km dataset, showing the effect of the different system components. Note the scaling on the axes. "Baseline" refers to the system without the geometric check and with a uniform position prior at each timestep. On this dataset the combination of motion model and geometric check is only slightly better than geometric check alone. The benefit is more significant for the 1,000 km dataset.

due to the fact that the highest confidence false positives are caused by particularly challenging cases of perceptual aliasing such as encountering distinctive brand logos multiple times in different locations. Given that the loop closures have an even distribution over the trajectory (Figure 7), even a recall rate of 6.5% is probably sufficient to support a good metric SLAM system.

Timing performance is presented in Figure 9. The time quoted is for inference and geometric verification, as measured on a single core of a 2.40 GHZ Intel Core 2 processor. SURF feature extraction and quantization adds an overhead of 484 ms on average. Recent GPU-based implementations can largely eliminate this overhead [27]. However, even including feature detection overhead, our real time requirement of 480 ms could be achieved by simply spreading the processing over two cores.

In comparison to prior work [1], the new system's inference times are on average 4,400 times faster, with comparable precision-recall performance. Equally important, the sparse representation means that location models now require only $O(1)$ memory, as opposed to $O(\#vocabulary)$. For the 100k vocabulary, a typical sparse location model requires 4 KB of memory as opposed to 400 KB previously. This enables the use of large vocabularies which improve performance, and is crucial for large scale operation because the size of the mappable area is effectively limited by available RAM.

VI. CONCLUSIONS

This paper has outlined a new, highly scalable architecture for appearance-only SLAM. The framework is fully probabilistic, and deals with challenging issues such as perceptual aliasing and new place detection. We have demonstrated the system on two extremely extensive data sets, of 70 km and 1,000 km. Both experiments are larger than any existing result we are aware of. Our approach shows very strong performance

Figure 6: Precision-recall curves for the 1,000 km data set, showing the effect of the vocabulary size and the Chow Liu tree on performance. Note the scaling on the axes. Performance increases strongly with vocabulary size. The Chow Liu tree also increases performance, for all vocabulary sizes. The 70 km set shows the same effects.

Dataset	70 km		1,000 km			1,000 km Urban	
Precision	*100%*	*99%*	*100%*	*99%*	*90%*	*100%*	*99%*
Recall							
100k CL	48.4	73.2	3.1	8.3	14.3	6.5	18.5
100k NB	49.1	70.0	3.7	7.9	13.5	7.5	17.9
10k CL	37.0	52.3	-	2.7	4.7	-	5.2
10k NB	30.1	51.5	-	-	4.4	-	-

Table II: Recall figures at specified precision for varying vocabulary size and with/without the Chow Liu tree. Recall improves with increasing vocabulary size at all levels of precision. The Chow Liu tree also improves recall in all cases with the exception of the 100k vocabulary at 100% precision. See text for discussion.

on the 70 km experiment. The 1,000 km experiment is more challenging, and we do not consider it solved, nevertheless system performance is already sufficient to provide a useful competency for an autonomous vehicle operating at this scale. Our data sets are available to the research community, and we hope that they will serve as a benchmark for future systems.

REFERENCES

[1] M. Cummins and P. Newman, "FAB-MAP: Probabilistic Localization and Mapping in the Space of Appearance," *Int. J. Robotics Res*, 2008.
[2] ——, "Accelerated appearance-only SLAM," in *Proc. IEEE Int. Conf. on Robotics and Automation*, Pasadena, April 2008.
[3] M. Milford and G. Wyeth, "Mapping a Suburb With a Single Camera Using a Biologically Inspired SLAM System," *Robotics, IEEE Transactions on*, vol. 24, no. 5, pp. 1038–1053, 2008.
[4] M. Bosse and R. Zlot, "Keypoint design and evaluation for global localization in 2d lidar maps," in *Robotics: Science and Systems Conference : Inside Data Association Workshop*, 2008.
[5] ——, "Map matching and data association for large-scale two-dimensional laser scan-based slam," *Int. J. Robotics Res*, 2008.
[6] E. Olson, "Robust and efficient robotic mapping," Ph.D. dissertation, Massachusetts Institute of Technology, June 2008.
[7] A. Angeli, D. Filliat, S. Doncieux, and J. Meyer, "A Fast and Incremental Method for Loop-Closure Detection Using Bags of Visual Words," *IEEE Transactions On Robotics, Special Issue on Visual SLAM*, 2008.
[8] T. Goedemé, T. Tuytelaars, and L. V. Gool, "Visual topological map building in self-similar environments," *Int. Conf. on Informatics in Control, Automation and Robotics*, 2006.
[9] K. L. Ho and P. Newman, "Detecting loop closure with scene sequences," *Int. J. Computer Vision*, vol. 74, pp. 261–286, 2007.
[10] F. Fraundorfer, C. Engels, and D. Nistér, "Topological mapping, localization and navigation using image collections," in *International Conference on Intelligent Robots and Systems*, 2007.

Figure 10: A correct loop closure from the 70 km data set. This is not an unusual match – the system typically finds correct matches in the presence of considerable scene change when the image content is distinctive.

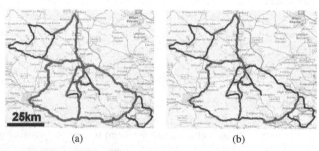

(a)   (b)

Figure 7: Loop closure maps for the 1,000 km data set. Sections of the trajectory where loop closures exists are coloured lighter. (a) The ground truth. (b) Loop closures detected by FAB-MAP (100k CL), showing 99.8% precision and 5.7% recall. There are 2,819 correct loop closures and six false positives. The long section on the right with no detected loop closures is a highway at dusk.

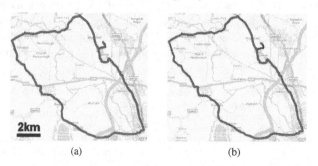

(a)   (b)

Figure 8: Loop closure maps for the 70 km dataset. Sections of the trajectory where loop closures exists are coloured lighter. (a) The ground truth. (b) Detected loop closures using FAB-MAP (100k CL), at 100% precision. The recall rate is 48.4%. A total of 2,300 loop closures are detected, with no false positives.

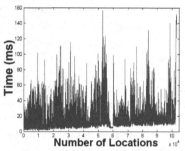

Figure 9: Filter update times on the 1,000 km dataset for the 100k vocabulary. Mean filter update time is 14 ms and maximum update time is 157 ms. The cost is dominated by the RANSAC geometric verification, which has $O(1)$ complexity. The core ranking stage excluding RANSAC exhibits linear complexity, but with a very small constant - taking 25 ms on average with 100,000 locations in the map.

[11] E. Eade and T. Drummond, "Unified loop closing and recovery for real time monocular slam," in *Proc. British Machine Vision Conf.*, 2008.

[12] G. Schindler, M. Brown, and R. Szeliski, "City-Scale Location Recognition," in *Conf. Computer Vision and Pattern Recognition*, 2007.

[13] D. Nister and H. Stewenius, "Scalable recognition with a vocabulary tree," in *Conf. Computer Vision and Pattern Recognition*, 2006.

[14] J. Philbin, O. Chum, M. Isard, J. Sivic, and A. Zisserman, "Object retrieval with large vocabularies and fast spatial matching," in *Proc. CVPR*, 2007.

[15] O. Chum, J. Philbin, J. Sivic, M. Isard, and A. Zisserman, "Total recall: Automatic query expansion with a generative feature model for object retreival," in *International Conference on Computer Vision*, 2007.

[16] J. Sivic and A. Zisserman, "Video Google: A text retrieval approach to object matching in videos," in *ICCV*, Nice, 2003.

[17] H. Bay, T. Tuytelaars, and L. Van Gool, "SURF: Speeded Up Robust Features," in *Proc European Conf on Computer Vision*, Graz, 2006.

[18] I. Posner, M. Cummins, and P. Newman, "Fast probabilistic labeling of city maps," in *Proc. Robotics: Science and Systems*, Zurich, June 2008.

[19] C. Chow and C. Liu, "Approximating discrete probability distributions with dependence trees," *IEEE Transactions on Information Theory*, vol. IT-14, no. 3, May 1968.

[20] M. Cummins and P. Newman, "Probabilistic appearance based navigation and loop closing," in *Proc. IEEE International Conference on Robotics and Automation (ICRA'07)*, Rome, April 2007.

[21] S. Brin and L. Page, "The anatomy of a large-scale hypertextual Web search engine," *Computer Networks and ISDN Systems*, vol. 30, 1998.

[22] C. Silpa-Anan and R. Hartley, "Optimised kd-trees for fast image descriptor matching," in *Computer Vision and Pattern Recognition*, 2008.

[23] M. Muja and D. Lowe, "Fast approximate nearest neighbors with automatic algorithm configuration," in *International Conference on Computer Vision Theory and Applications*, 2009.

[24] S. Dasgupta and P. Long, "Performance guarantees for hierarchical clustering," in *Conf. on Computational Learning Theory*, 2002.

[25] M. Meilǎ, "An accelerated Chow and Liu algorithm: Fitting tree distributions to high-dimensional sparse data," in *Int. Conf. on Machine Learning*, San Francisco, 1999.

[26] R. Dementiev, L. Kettner, and P. Sanders, "STXXL: Standard Template Library for XXL Data Sets," *Lecture Notes in Computer Science*, vol. 3669, p. 640, 2005.

[27] N. Cornelis and L. Van Gool, "Fast scale invariant feature detection and matching on programmable graphics hardware," in *CVPR 2008 Workshop CVGPU*, 2008.